London Mathematical Society Lecture Note Series: 495

Polygraphs: From Rewriting to Higher Categories

DIMITRI ARA
Aix-Marseille Université

ALBERT BURRONI
Université Paris Cité

YVES GUIRAUD
INRIA, Université Paris Cité

PHILIPPE MALBOS
Université Claude Bernard Lyon 1

FRANÇOIS MÉTAYER
Université Paris Cité

SAMUEL MIMRAM
École Polytechnique

CAMBRIDGE
UNIVERSITY PRESS

CAMBRIDGE
UNIVERSITY PRESS

Shaftesbury Road, Cambridge CB2 8EA, United Kingdom

One Liberty Plaza, 20th Floor, New York, NY 10006, USA

477 Williamstown Road, Port Melbourne, VIC 3207, Australia

314–321, 3rd Floor, Plot 3, Splendor Forum, Jasola District Centre, New Delhi – 110025, India

103 Penang Road, #05–06/07, Visioncrest Commercial, Singapore 238467

Cambridge University Press is part of Cambridge University Press & Assessment,
a department of the University of Cambridge.

We share the University's mission to contribute to society through the pursuit of
education, learning and research at the highest international levels of excellence.

www.cambridge.org
Information on this title: www.cambridge.org/9781009498982
DOI: 10.1017/9781009498968

When citing this work, please include a reference to the DOI 10.1017/9781009498968

First published 2025

Printed in the United Kingdom by TJ Books Limited, Padstow Cornwall

A catalogue record for this publication is available from the British Library.

A Cataloging-in-Publication data record for this book is available from the Library of Congress.

ISBN 978-1-009-49898-2 Paperback

LONDON MATHEMATICAL SOCIETY LECTURE NOTE SERIES

Managing Editor: Professor Endre Süli, Mathematical Institute, University of Oxford, Woodstock Road, Oxford OX2 6GG, United Kingdom

The titles below are available from booksellers, or from Cambridge University Press at www.cambridge.org/mathematics

385 Entropy of hidden Markov processes and connections to dynamical systems, B. MARCUS, K. PETERSEN & T. WEISSMAN (eds)
386 Independence-friendly logic, A.L. MANN, G. SANDU & M. SEVENSTER
387 Groups St Andrews 2009 in Bath I, C.M. CAMPBELL *et al* (eds)
388 Groups St Andrews 2009 in Bath II, C.M. CAMPBELL *et al* (eds)
389 Random fields on the sphere, D. MARINUCCI & G. PECCATI
390 Localization in periodic potentials, D.E. PELINOVSKY
391 Fusion systems in algebra and topology, M. ASCHBACHER, R. KESSAR & B. OLIVER
392 Surveys in combinatorics 2011, R. CHAPMAN (ed)
393 Non-abelian fundamental groups and Iwasawa theory, J. COATES *et al* (eds)
394 Variational problems in differential geometry, R. BIELAWSKI, K. HOUSTON & M. SPEIGHT (eds)
395 How groups grow, A. MANN
396 Arithmetic differential operators over the p-adic integers, C.C. RALPH & S.R. SIMANCA
397 Hyperbolic geometry and applications in quantum chaos and cosmology, J. BOLTE & F. STEINER (eds)
398 Mathematical models in contact mechanics, M. SOFONEA & A. MATEI
399 Circuit double cover of graphs, C.-Q. ZHANG
400 Dense sphere packings: a blueprint for formal proofs, T. HALES
401 A double Hall algebra approach to affine quantum Schur–Weyl theory, B. DENG, J. DU & Q. FU
402 Mathematical aspects of fluid mechanics, J.C. ROBINSON, J.L. RODRIGO & W. SADOWSKI (eds)
403 Foundations of computational mathematics, Budapest 2011, F. CUCKER, T. KRICK, A. PINKUS & A. SZANTO (eds)
404 Operator methods for boundary value problems, S. HASSI, H.S.V. DE SNOO & F.H. SZAFRANIEC (eds)
405 Torsors, étale homotopy and applications to rational points, A.N. SKOROBOGATOV (ed)
406 Appalachian set theory, J. CUMMINGS & E. SCHIMMERLING (eds)
407 The maximal subgroups of the low-dimensional finite classical groups, J.N. BRAY, D.F. HOLT & C.M. RONEY-DOUGAL
408 Complexity science: the Warwick master's course, R. BALL, V. KOLOKOLTSOV & R.S. MACKAY (eds)
409 Surveys in combinatorics 2013, S.R. BLACKBURN, S. GERKE & M. WILDON (eds)
410 Representation theory and harmonic analysis of wreath products of finite groups, T. CECCHERINI-SILBERSTEIN, F. SCARABOTTI & F. TOLLI
411 Moduli spaces, L. BRAMBILA-PAZ, O. GARCÍA-PRADA, P. NEWSTEAD & R.P. THOMAS (eds)
412 Automorphisms and equivalence relations in topological dynamics, D.B. ELLIS & R. ELLIS
413 Optimal transportation, Y. OLLIVIER, H. PAJOT & C. VILLANI (eds)
414 Automorphic forms and Galois representations I, F. DIAMOND, P.L. KASSAEI & M. KIM (eds)
415 Automorphic forms and Galois representations II, F. DIAMOND, P.L. KASSAEI & M. KIM (eds)
416 Reversibility in dynamics and group theory, A.G. O'FARRELL & I. SHORT
417 Recent advances in algebraic geometry, C.D. HACON, M. MUSTAŢĂ & M. POPA (eds)
418 The Bloch–Kato conjecture for the Riemann zeta function, J. COATES, A. RAGHURAM, A. SAIKIA & R. SUJATHA (eds)
419 The Cauchy problem for non-Lipschitz semi-linear parabolic partial differential equations, J.C. MEYER & D.J. NEEDHAM
420 Arithmetic and geometry, L. DIEULEFAIT *et al* (eds)
421 O-minimality and Diophantine geometry, G.O. JONES & A.J. WILKIE (eds)
422 Groups St Andrews 2013, C.M. CAMPBELL *et al* (eds)
423 Inequalities for graph eigenvalues, Z. STANIĆ
424 Surveys in combinatorics 2015, A. CZUMAJ *et al* (eds)
425 Geometry, topology and dynamics in negative curvature, C.S. ARAVINDA, F.T. FARRELL & J.-F. LAFONT (eds)
426 Lectures on the theory of water waves, T. BRIDGES, M. GROVES & D. NICHOLLS (eds)
427 Recent advances in Hodge theory, M. KERR & G. PEARLSTEIN (eds)
428 Geometry in a Fréchet context, C.T.J. DODSON, G. GALANIS & E. VASSILIOU
429 Sheaves and functions modulo p, L. TAELMAN
430 Recent progress in the theory of the Euler and Navier–Stokes equations, J.C. ROBINSON, J.L. RODRIGO, W. SADOWSKI & A. VIDAL-LÓPEZ (eds)
431 Harmonic and subharmonic function theory on the real hyperbolic ball, M. STOLL
432 Topics in graph automorphisms and reconstruction (2nd Edition), J. LAURI & R. SCAPELLATO
433 Regular and irregular holonomic D-modules, M. KASHIWARA & P. SCHAPIRA
434 Analytic semigroups and semilinear initial boundary value problems (2nd Edition), K. TAIRA
435 Graded rings and graded Grothendieck groups, R. HAZRAT
436 Groups, graphs and random walks, T. CECCHERINI-SILBERSTEIN, M. SALVATORI & E. SAVA-HUSS (eds)
437 Dynamics and analytic number theory, D. BADZIAHIN, A. GORODNIK & N. PEYERIMHOFF (eds)
438 Random walks and heat kernels on graphs, M.T. BARLOW
439 Evolution equations, K. AMMARI & S. GERBI (eds)

440 Surveys in combinatorics 2017, A. CLAESSON *et al* (eds)
441 Polynomials and the mod 2 Steenrod algebra I, G. WALKER & R.M.W. WOOD
442 Polynomials and the mod 2 Steenrod algebra II, G. WALKER & R.M.W. WOOD
443 Asymptotic analysis in general relativity, T. DAUDÉ, D. HÄFNER & J.-P. NICOLAS (eds)
444 Geometric and cohomological group theory, P.H. KROPHOLLER, I.J. LEARY, C. MARTÍNEZ-PÉREZ & B.E.A. NUCINKIS (eds)
445 Introduction to hidden semi-Markov models, J. VAN DER HOEK & R.J. ELLIOTT
446 Advances in two-dimensional homotopy and combinatorial group theory, W. METZLER & S. ROSEBROCK (eds)
447 New directions in locally compact groups, P.-E. CAPRACE & N. MONOD (eds)
448 Synthetic differential topology, M.C. BUNGE, F. GAGO & A.M. SAN LUIS
449 Permutation groups and cartesian decompositions, C.E. PRAEGER & C. SCHNEIDER
450 Partial differential equations arising from physics and geometry, M. BEN AYED *et al* (eds)
451 Topological methods in group theory, N. BROADDUS, M. DAVIS, J.-F. LAFONT & I. ORTIZ (eds)
452 Partial differential equations in fluid mechanics, C.L. FEFFERMAN, J.C. ROBINSON & J.L. RODRIGO (eds)
453 Stochastic stability of differential equations in abstract spaces, K. LIU
454 Beyond hyperbolicity, M. HAGEN, R. WEBB & H. WILTON (eds)
455 Groups St Andrews 2017 in Birmingham, C.M. CAMPBELL *et al* (eds)
456 Surveys in combinatorics 2019, A. LO, R. MYCROFT, G. PERARNAU & A. TREGLOWN (eds)
457 Shimura varieties, T. HAINES & M. HARRIS (eds)
458 Integrable systems and algebraic geometry I, R. DONAGI & T. SHASKA (eds)
459 Integrable systems and algebraic geometry II, R. DONAGI & T. SHASKA (eds)
460 Wigner-type theorems for Hilbert Grassmannians, M. PANKOV
461 Analysis and geometry on graphs and manifolds, M. KELLER, D. LENZ & R.K. WOJCIECHOWSKI
462 Zeta and L-functions of varieties and motives, B. KAHN
463 Differential geometry in the large, O. DEARRICOTT *et al* (eds)
464 Lectures on orthogonal polynomials and special functions, H.S. COHL & M.E.H. ISMAIL (eds)
465 Constrained Willmore surfaces, Á.C. QUINTINO
466 Invariance of modules under automorphisms of their envelopes and covers, A.K. SRIVASTAVA, A. TUGANBAEV & P.A.GUIL ASENSIO
467 The genesis of the Langlands program, J. MUELLER & F. SHAHIDI
468 (Co)end calculus, F. LOREGIAN
469 Computational cryptography, J.W. BOS & M. STAM (eds)
470 Surveys in combinatorics 2021, K.K. DABROWSKI *et al* (eds)
471 Matrix analysis and entrywise positivity preservers, A. KHARE
472 Facets of algebraic geometry I, P. ALUFFI *et al* (eds)
473 Facets of algebraic geometry II, P. ALUFFI *et al* (eds)
474 Equivariant topology and derived algebra, S. BALCHIN, D. BARNES, M. KĘDZIOREK & M. SZYMIK (eds)
475 Effective results and methods for Diophantine equations over finitely generated domains, J.-H. EVERTSE & K. GYŐRY
476 An indefinite excursion in operator theory, A. GHEONDEA
477 Elliptic regularity theory by approximation methods, E.A. PIMENTEL
478 Recent developments in algebraic geometry, H. ABBAN, G. BROWN, A. KASPRZYK & S. MORI (eds)
479 Bounded cohomology and simplicial volume, C. CAMPAGNOLO, F. FOURNIER-FACIO, N. HEUER & M. MORASCHINI (eds)
480 Stacks Project Expository Collection (SPEC), P. BELMANS, W. HO & A.J. DE JONG (eds)
481 Surveys in combinatorics 2022, A. NIXON & S. PRENDIVILLE (eds)
482 The logical approach to automatic sequences, J. SHALLIT
483 Rectifiability: a survey, P. MATTILA
484 Discrete quantum walks on graphs and digraphs, C. GODSIL & H. ZHAN
485 The Calabi problem for Fano threefolds, C. ARAUJO *et al*
486 Modern trends in algebra and representation theory, D. JORDAN, N. MAZZA & S. SCHROLL (eds)
487 Algebraic combinatorics and the Monster group, A.A. IVANOV (ed)
488 Maurer–Cartan methods in deformation theory, V. DOTSENKO, S. SHADRIN & B. VALLETTE
489 Higher dimensional algebraic geometry, C. HACON & C. XU (eds)
490 C^∞ 8-algebraic geometry with corners, K. FRANCIS-STAITE & D. JOYCE
491 Groups and graphs, designs and dynamics, R.A. BAILEY, P.J. CAMERON & Y. WU (eds)
492 Homotopy theory of enriched Mackey functors, N. JOHNSON & D. YAU
493 Surveys in combinatorics 2024, F. FISCHER & R. JOHNSON (eds)
494 K-theory and representation theory, R. PLYMEN & M.H. ŞENGÜN (eds)

Contents

Preface

Presentations of higher categories. A group is generally defined by characteristic properties of its elements, such as the group Σ_n of all permutations of the set $\{1, \ldots, n\}$, or the group of all isometries of a cube. However, to perform actual computations in a group, we usually pick a subset of *generators* among its elements and write down certain *relations* satisfied by these generators, in such a way that each element of the group is a product of some generators or their inverses, and each equality between two elements is derivable from the relations. This leads to a purely syntactic way of defining a group, called a *group presentation* by generators and relations, which describes the group as a free group over a set of generators quotiented by some relations. Presentations are not limited to groups, and they have been adapted to many other algebraic structures: monoids, categories, Lawvere theories (this is the subject of universal algebra), commutative (or not) algebras (where the relations are usually specified by an ideal), or higher algebras such as operads, product categories, and linear monoidal categories, to name a few.

The subject of this monograph is the concept of *polygraph*, which is the notion of presentation adapted to higher categories, and encompasses the previously mentioned settings as particular cases. Polygraphs were first introduced by Street [333] under the name of *computads* in their 2-dimensional version, in order to study 2-categorical limits, and then generalized in arbitrary dimension: this was first published in [304], but the generalization was already known and implicitly used in [334]. The terminology we adopt here comes from Burroni [71, 72], who independently developed the concept in order to study generalizations of the word problem and provide an equational presentation of cartesian categories. The name *polygraph* is meant to suggest a higher-dimensional analogue of the oriented graphs. It should be mentioned that these ideas were developed on both sides and informally circulated long before publication.

The word problem. Given a presentation of a group, the completeness of a set of relations means that for any two sequences $a_1, \ldots, a_n$ and $b_1, \ldots, b_m$ of generators, also called *words*, their respective products $a_1 \cdots a_n$ and $b_1 \cdots b_m$ are equal in the presented group if and only if one can transform the first into the second by using the relations. This very naturally led, early on, to the following algorithmic question, known as the *word problem*: given a finite (or recursive) presentation and two words as above, can we decide whether they are equal or not in the presented group? From a computational theory point of view: can we implement a computer program that automatically determines the equality of words in a presented group? However, this question predates by many years the invention of computers. It was first raised as an important one by Dehn in 1911 [105], who subsequently managed to provide an algorithm for a certain class of groups (the fundamental groups of closed orientable surfaces of genus greater than or equal to 2) [106]. For some time it was hoped that the problem could be solved for all groups, but it was in fact shown to be undecidable around 1955 by Novikov [293] and Boone [51].

The word problem has also been considered in other settings where presentations exist. For monoids, it was first studied by Thue in 1914 [344], leading to the emergence of the notion of *string rewriting system*, also called semi-Thue systems. It was much later, in 1947, that the word problem was shown to be undecidable by Post [301] and Markov [271]. In the case of universal algebra, the undecidability of the word problem follows from the undecidability of conversion in combinatory logic [102], which is closely related to the undecidability of β-conversion in λ-calculus [86].

Rewriting theory. In most cases where the general word problem is undecidable, we can nevertheless find many specific presentations with algorithmically decidable equality. For instance, if we ask a small child whether $2 + 3 + 4$ is equal to $2 + 2 + 5$, he will progressively compute both sums and observe that the results are the same:

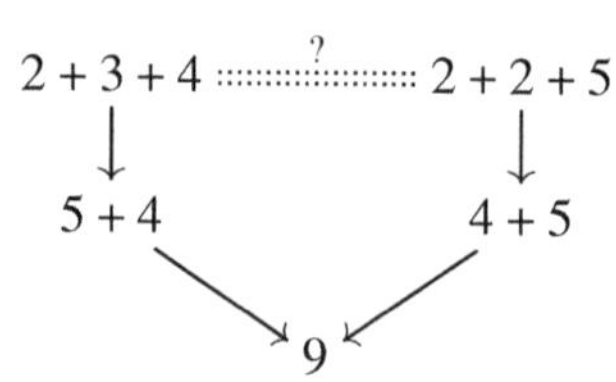

The starting point of rewriting theory is to provide each relation of a given presentation with an orientation taking one side of the equality to a simpler expression on the other side. The resulting structure is called a *rewriting system* and the oriented rules are called *rewriting rules*. Fundamental examples of rewriting systems are *string* rewriting systems for presentations of monoids [50] and *term* rewriting systems for presentations of Lawvere theories [342, 20].

Each rewriting system now comes with a notion of *normal form*, that is, an expression that cannot be further simplified by applying rewriting rules. This immediately suggests the *normal form algorithm*, consisting, given two expressions, in applying the two following steps:

1. simplify the two expressions in order to obtain normal forms, and
2. compare the normal forms.

The normal form algorithm only decides the word problem if the rewriting system satisfies the two following properties.

1. Each sequence of rewriting rules eventually reaches a normal form after a finite number of steps, in which case we call the system *terminating*.
2. If an expression can be reduced in two different ways, we can further reduce two expressions into a common one, in which case we call the system *confluent*.

A rewriting system is *convergent* when it is both terminating and confluent.

Proofs of termination often rely on embedding the reduction order into a partial well-founded order over the set of expressions. As for confluence, in cases where the system is already known to be terminating, it suffices to check a simpler condition called *local confluence*: this is the content of *Newman's lemma* [290]. This lemma holds in *abstract rewriting systems* and therefore does not depend on the particular formalism under consideration (rewriting on strings, terms, etc.). We should also mention that convergent rewriting systems have independently been discovered in the setting of presentations of (commutative) algebras in the 1960s by Shirshov [325] and Buchberger [65], where they are known as *Gröbner bases*. In this context, Newman's lemma is known as the *diamond lemma* [39].

Tietze transformations and completion. A presentation of a group – or any other type of algebraic structure – may be thought of as a particular implementation of it, thus allowing for computations. As two presentations of the same object may have very different computational properties, it is worth considering the family of *all* presentations of a given object. This has first been done in the case of groups by Tietze in 1908 [345] (and later on generalized to other settings): he introduced a family of operations on group presentations, now called *Tietze transformations*, such that any two presentations of the same group can be transformed into each other by means of those operations.

Precisely, Tietze transformations are sequences built from two elementary steps and their converse, namely:

1. to add a new generator which is equal to a product of preexisting generators, and
2. to add a relation which is derivable from the already present relations.

Tietze transformations will be used to turn a given rewriting system into a new one presenting the same structure, possibly with better computational properties. In particular, the idea behind *completion algorithms* is to add rules (or generators) to a rewriting system in order to turn it into a confluent one. These ideas were already present at the beginning of string rewriting systems [344] and gained much popularity when they were developed by Buchberger for Gröbner bases [65] (which eventually lead to very efficient algorithms [124]), by Knuth and Bendix for term rewriting systems [218], and by Nivat for string rewriting systems [292]. The new rules to be added to the rewriting system are determined by computing *critical branchings* (also called *S-polynomials* in the linear settings), which are minimal obstructions to confluence.

The universality problem. We have seen that, even though the word problem is undecidable in general, there are many cases where the presentations are convergent and the word problem can be decided with the normal form algorithm. This leads to the following question, first formulated by Jantzen in the context of string rewriting systems [202, 203], and sometimes referred to as the *universality problem* for convergent rewriting: given a finitely presented monoid with a decidable word problem, does it always admit a presentation by a finite convergent string rewriting system?

A first answer to this question was brought by Kapur and Narendran [213] by considering the Artin presentation of the positive braid group B_3^+ (with two generators a and b and one relation $aba = bab$), which has a decidable word problem: they showed that one cannot obtain a convergent presentation from it, by adding or removing relations. However, this does not settle the general question, because we are only using Tietze transformations of type (ii) here, and in fact, the same authors also observed that, by using a transformation of type (i), one can indeed obtain a convergent presentation.

Homotopy and homology. A complete solution to the universality problem was eventually found by using ideas coming from algebraic topology, which, in a nutshell, consists in assigning discrete invariants to continuous shapes. Such ideas already appear in the works of Euler, but their systematic development starts with Poincaré [299]. Among the first invariants to be considered are the *fundamental group* of a space, which consists of classes of loops up to continuous deformations (or homotopies) and the sequence of *homology groups*, providing information about the "holes" of various dimensions in a given space.

The fact that homological invariants are not just numbers but rather bear a structure of abelian group was first recognized by Emmy Noether (see [349, p. 478] or [181]), and independently by Vietoris, whose paper [352] contains the first definition of homology groups ever published. Homology groups are amenable to effective computations by using classical tools from linear algebra. On the other hand, one may consider a group or a monoid as a geometric object, by defining the corresponding *classifying space*. Thus, invariants of spaces may be applied to groups and monoids. These notions have been vastly generalized over the years and now apply to algebraic structures whose geometrical content is much less obvious, such as rings or algebras [186, 258].

Squier's homological and homotopical conditions. The universality problem was answered negatively by Squier in 1987 [326, 327]. Squier's argument is based on the homology of monoids and relies on two fundamental observations. First, any presentation of a monoid determines a sequence of homology groups, but these groups are independent of the particular presentation we use: they are invariants of the monoid itself. Second, a finite convergent presentation of a monoid M always yields a third homology group $H_3(M)$ of finite rank. Now, Squier was able to produce an explicit example of a finitely presented monoid M with a decidable word problem, whose third homology group is not of finite rank and therefore does not admit a finite convergent presentation.

In more precise terms, the homology of a monoid M is computed by building a *resolution* of the trivial $\mathbb{Z}M$-module $\mathbb{Z}$, that is, an exact sequence

$$\cdots \xrightarrow{d_4} C_3 \xrightarrow{d_3} C_2 \xrightarrow{d_2} C_1 \xrightarrow{d_1} C_0 \xrightarrow{\varepsilon} \mathbb{Z} \longrightarrow 0$$

of projective $\mathbb{Z}M$-modules. Tensoring the above sequence by $\mathbb{Z}$ over $\mathbb{Z}M$ gives a chain complex of abelian groups, which is not exact anymore in the general case; its homology groups only depend on M and not on the particular resolution we chose. Now, if we start with a convergent presentation of M, we obtain a partial resolution of the form

$$\cdots \xrightarrow{d_4} \mathbb{Z}M[P_3] \xrightarrow{d_3} \mathbb{Z}M[P_2] \xrightarrow{d_2} \mathbb{Z}M[P_1] \xrightarrow{d_1} \mathbb{Z}M[P_0] \xrightarrow{\varepsilon} \mathbb{Z} \longrightarrow 0,$$

where P_0 is the set with one element, P_1 is the set of generators of the presentation, P_2 the set of relations, and P_3 the set of critical branchings. For a finite rewriting system, the set P_3 of critical branchings is always finite. Hence the abelian group $\mathbb{Z}M[P_3] \otimes_{\mathbb{Z}M} \mathbb{Z}$ obtained by tensoring by $\mathbb{Z}$ over $\mathbb{Z}M$ has finite rank and so has the third homology group $H_3(M)$. Note that the finiteness condition for H_3 comes here from the strictly stronger statement that M is of type *left*-FP$_3$, which means that it admits a partial resolution of length 3 by finitely generated projective left $\mathbb{Z}M$-modules.

The above algebraic constructions can be interpreted in geometric terms [238], at least in the case where the monoid M is a group. The free resolution (C_i, d_i) comes from a cellular decomposition of a *contractible* space X on which M acts freely and transitively. Tensoring by $\mathbb{Z}$ over $\mathbb{Z}M$ amounts to quotient X by the action of M, thus obtaining a cellular decomposition of the classifying space of M itself. Starting with a presentation of M, the cellular complex we get is built dimensionwise: in dimension 0 there is a unique point, corresponding to the element of $\star \in P_0$. In dimension 1, each generator of P_1 gives a loop on $\star$. In dimension 2, each relation in P_2 gives a disk attached to the 1-dimensional paths determined by the products of the generators involved. For instance, with the Artin presentation of B_3^+ recalled above, we would obtain the space on the left, together with a disk attached between the paths corresponding to the words aba and bab, as pictured on the right:

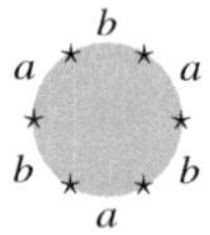

There may be different ways to fill the gap between two expressions representing the same element of M by using disks as above. Geometrically, this corresponds to 3-dimensional holes in the corresponding 2-dimensional complex. These holes have to be filled by appropriate 3-dimensional cells, and it turns out that, in the case of a convergent presentation, a set of 3-cells coming from the critical branchings is sufficient for that. Of course, the construction has to be pursued in higher dimensions in order to obtain the correct topology.

About the same time when Squier studied monoid resolutions, other authors developed similar ideas: Anick constructed a resolution of algebras in order to study Koszulness properties [8], Brown managed to use discrete Morse theory in order to "reduce" the standard resolution to a small one [59, 60], and Kobayashi extended Squier's partial resolution into a full one [219]. Finally, we should mention that Squier subsequently provided a "homotopical" variant of his condition, which was published after his death [328]; see also [233]. A finite presentation is said to be of *finite derivation type* when the *full congruence*, which identifies any two witnesses of equality between two given words, is finitely generated. It can be shown that this property is an invariant of the monoid and is implied by having a finite convergent presentation. Moreover, one can construct an explicit example of a monoid that has a finite presentation with decidable word problem but does not have finite derivation type.

Polygraphs as higher-dimensional presentations. A monoid is a category with only one object, and it is thus natural to extend the notion of presentation, and associated theorems and techniques, from monoids to small categories. Precisely, a small category C will be presented by a set $P_0 = C_0$ of objects, a set P_1 of 1-generators, that is, a subset $P_1 \subseteq C_1$ of morphisms generating all morphisms in C by composition, together with a set P_2 of relations between certain pairs of composites of 1-generators. This pattern generalizes to higher categories, resulting in the notion of *n-polygraph* [72, 333], which consists of the following data:

- for each $i \in \{0, \ldots, n\}$, a set P_i, freely generating a set P_i^* of i-dimensional cells; and
- for each $i \in \{1, \ldots, n\}$, a pair of maps associating to each i-generator its source and target in P_{i-1}^*.

Given an $(n+1)$-polygraph P, the source and target maps defined on P_{n+1} generate an equivalence relation on P_n^*, whose quotient set C_n is the set of n-morphisms of the n-category C *presented* by P.

Right from the beginning, a polygraph was thought of as a "higher-dimensional rewriting system" [72, 120, 337, 338] and the associated theory of rewriting was subsequently developed, in particular by some of the authors of this book [158, 161, 164, 165, 235, 283]. It turns out that many of the classical theorems go through but, starting at dimension $n = 3$, there is a major difference with the classical setting: a finite rewriting system might give rise to an infinite number of critical pairs, thus preventing easy generalizations of Squier-type theorems.

The fact that polygraphs are higher-dimensional rewriting systems should be taken in a very strong sense here: they are about rewriting rewriting paths. Namely, an $(n + 1)$-polygraph consists of an n-polygraph together with $(n + 1)$-dimensional rewriting rules, which specify how to rewrite rewriting paths in the underlying n-polygraph. For instance, consider the abstract rewriting system on the left, which is a 1-polygraph:

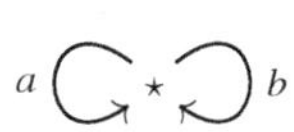

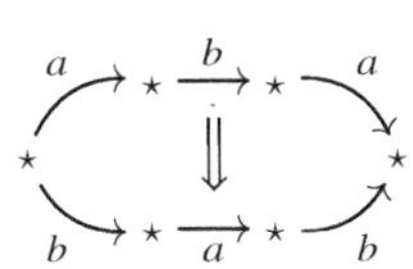

Here, we have one object $\star$ and two rewriting rules a and b rewriting $\star$ to itself. We can obtain a 2-polygraph by adjoining a 2-dimensional rewriting rule which rewrites the path aba into the path bab, as pictured on the right, thus providing a polygraphic presentation of the monoid B_3^+: in this way, we can see that string rewriting is secretly rewriting rewriting paths in abstract rewriting systems! Similarly, in the next dimension, we can see that term rewriting, and more generally rewriting of diagrams, is an instance of rewriting rewriting paths for string rewriting systems.

Several particular generalizations of the notion of polygraph have also been investigated (for (n, p)-categories, linear higher-categories [160], cartesian higher-categories, etc.). Most of them are particular instances of the notion of *T-polygraph* introduced by Batanin [28] in order to define polygraphs adapted to weak higher categories: he defines a notion of polygraph parametrized by a globular monad T, whose various instantiations allow recovering the previously mentioned variants of polygraphs.

Coherence. Given a presentation of an n-category C by an $(n{+}1)$-polygraph P, the elements of P^*_{n+1} witness the equalities between n-morphisms of C. Now, different $(n{+}1)$-cells may witness the same equality: this defines a congruence on P^*_{n+1}. If we extend P by a set P_{n+2} generating this congruence, we get a *coherent presentation* of M by an $(n{+}2)$-polygraph. For instance, from a convergent presentation of a monoid, we build a coherent presentation by a 3-polygraph P in which the generators in P_3 correspond to the critical branchings.

Coherent presentations provide the "right" generalization (in a homotopical sense detailed below) of a structure in higher dimensions. For instance, the theory of monoids can be described by a 3-polygraph. If we extend this polygraph into a coherent one, we obtain a theory corresponding to pseudomonoids, which is the expected notion of monoid in a monoidal 2-category.

Resolutions with polygraphs. The construction of a coherent presentation of an n-category may be infinitely pursued in higher dimensions by introducing, for each $m \geqslant n$, a set of $(m{+}1)$-cells generating all desired congruences between m-cells. More generally, starting with any ω-category C, we may build a polygraph P together with an ω-functor $p\colon P^* \to C$ satisfying the following properties.

1. For each dimension $n \geqslant 0$, P_n generates C_n, in the sense that $p_n\colon P^*_n \to C_n$ is surjective.
2. For any two parallel cells x, y in P^*_n such that $p_n x = p_n y = u \in C_n$, there is an $(n{+}1)$-cell $z \in P^*_{n+1}$ with source x and target y such that $p_{n+1} z = 1_u$.

We call such a map $p\colon P^* \to C$ a *polygraphic resolution* of C by the polygraph P [278]. It turns out that two polygraphic resolutions of the same ω-category are equivalent up to a suitable notion of homotopy. Polygraphic resolutions are cofibrant replacements in the "canonical" model structure on the category of small ω-categories, as introduced in [237]. As a consequence, there is a well-defined notion of homology for ω-categories: to each polygraphic resolution of C by P corresponds a chain complex $(\mathbb{Z}P_n, \partial_n)_{n\geqslant 0}$ of abelian groups, whose homology only depends on C. This illustrates the general principle according to which many constructions are better behaved when performed on *free* objects. In the particular case where C is a monoid M seen as an ω-category, this homology coincides with the one computed via free resolutions of $\mathbb{Z}$ by $\mathbb{Z}M$-modules [236, 155]. These constructions transfer to strict ω-groupoids, or more generally (ω, n)-categories [16], where all cells of dimension strictly above n are invertible, yielding appropriate notions of polygraphic resolutions.

Structure of the book(s). This book informally splits into two books, which, however strongly connected, can be read separately, according to one's taste and objectives.

Low-dimensional book. The first book explores the theory of polygraphs in low dimensions and its applications. It is meant to be very progressive, with little requirements on the background of the reader, apart from basic category theory, and is illustrated with algorithmic computations on algebraic structures. We namely study polygraphs in dimension 1 (Chapter 1), in dimension 2 (Chapters 2 to 9), and in dimension 3 (Chapters 10 to 13).

In all the cases, we introduce the notion of polygraph as well as the associated notions of generated and presented categories (Sections 1.1, 1.2, and 10.1 and Chapter 2), develop the theory of rewriting (Sections 1.3 and 10.2 and Chapter 4), Tietze transformations and completion procedures (Section 1.2 and Chapter 5), termination techniques (Sections 1.3 and 4.4 and Chapter 11) and coherent presentations (Chapters 7 and 12), with the last notion requiring the introduction of higher-dimensional notions of polygraphs. We also present the homotopical and homological invariants (Chapters 8 and 9) they allow to compute and introduce variants of the notion of polygraph, namely linear (Chapter 6) and cartesian polygraphs (Chapter 13).

Higher-dimensional book. The second book goes at a faster pace and supposes that the reader is familiar with category theory. Moreover, acquaintance with strict higher categories, as well as the notion of model category, can be helpful, even though these notions are recalled. The beginning of this book introduces and studies the general notion of n-polygraph (Chapters 14 to 18). The remainder of the book deals with the homotopy theory of these polygraphs. We construct the "folk" model structure on the category of ω-categories (Chapters 19 to 21), in which polygraphs are precisely the cofibrant objects. This model structure is used to define a homology theory for ω-categories as a derived functor (Chapter 22). Finally, we study the variant of $(\omega, 1)$-polygraphs (Chapter 23), which allows to formulate a higher-dimensional generalization of the coherence results developed in the "low-dimensional book".

Appendix. The book is followed by a number of chapters containing additional material. Some of them perform a review of classical – or not – examples of polygraphs, in order to illustrate their diversity and applications: 2-polygraphs (Appendix A), coherent 2-polygraphs (Appendix B), and 3-polygraphs (Appendix C). Some other chapters recall elements of classical topics used throughout the second part of the book: free n-categories (Appendix D), homology (Appendix E), locally presentable categories (Appendix G), and model categories (Appendix H).

Acknowledgments. The writing of this book started in 2015, under the code name of *polybook*, and the present exposition has benefited much from interactions with other researchers and friends. We would like to warmly thank the following people who, voluntarily or not, contributed to its elaboration: Mathieu Anel, Thibaut Benjamin, Pierre Cagne, Cameron Calk, Cyrille Chenavier, Pierre-Louis Curien, Benjamin Dupont, Eric Finster, Simon Forest, Jonas Frey, Andrea Gagna, Léonard Guetta, Amar Hadzihasanovic, Simon Henry, Hugo Herbelin, Eric Hoffbeck, Cédric Ho Thanh, Jan Willem Klop, Yves Lafont, Chaitanya Leena-Subramaniam, Maxime Lucas, Georges Maltsiniotis, Paul-André Melliès, Jovana Obradovic, Vincent van Oostrom, Viktorya Ozornova, Jacques Penon, and all participants of the Higher Category Seminar at IRIF.

String diagrams were first typeset using catex [156], and then satex [285].

The early stages of the book also benefited from funding provided by the projects Focal IDEX and CATHRE ANR.

Our working sessions took place over the years in various locations, and we especially acknowledge the hospitality of the IHP in Paris, the CIRM in Marseille, and the Bois-Maréchal in the middle of the Mâconnais vineyard.

PART I

FUNDAMENTALS OF REWRITING

1

Abstract Rewriting and 1-Dimensional Polygraphs

We begin by discussing 1-polygraphs, which are simply directed graphs, thought of here as abstract rewriting systems: they consist of vertices, which represent the objects of interest, and arrows, which indicate that we can rewrite one object into another. After formally introducing those in Section 1.1, we will see in Section 1.2 that they provide a notion of *presentation* for sets, by generators and relations. Of course, presentations of sets are of little interest in themselves, but they are merely used here as a gentle introduction to some of the main concepts discussed in this work: in particular, we introduce the notion of Tietze transformations which generate the equivalence between two presentations of the same set. In this context, an important question consists in deciding when two objects are equivalent, i.e., represent the same element of the presented set. In order to address it, we develop the theory of abstract rewriting systems in Section 1.3. Most notably, we show that when the rewriting system satisfies the two properties of *termination* and *confluence*, equivalence classes of objects admit a unique canonical representative, the *normal form*, and equivalence of objects can thus be decided by comparing the associated normal forms. Finally, in Section 1.4, we detail the more advanced method of decreasing diagrams, which can be used to show confluence in the absence of termination.

1.1 The Category of 1-Polygraphs

A 0-*polygraph* is simply another name for a set. Since there is not much to do with those, we move on to 1-polygraphs.

1.1.1 Definition. A 1-*polygraph* P consists of a 0-polygraph P_0, whose elements are called 0-*generators*, together with a set P_1 of 1-*generators* and two functions $s_0^P, t_0^P : P_1 \to P_0$ respectively associating to each 1-generator its *source* and *target* 0-cell. We often write $\langle P_0 \mid P_1 \rangle$ for such a polygraph and $a\colon x \to y$ for a 1-generator a in P_1 such that $s_0^P(a) = x$ and $t_0^P(a) = y$. A 1-polygraph P is *finite* when both P_0 and P_1 are.

The notion of 1-polygraph is simply another name for the notion of *graph*, by which we always mean a directed multigraph, which we sometimes also call a 1-*graph*. Indeed, a polygraph P as above is a graph with P_0 as set of vertices P_1 as set of edges, an edge $a \in P_1$ having $s_0^P(a)$ as source and $t_0^P(a)$ as target. Thus, any terminology pertaining to oriented graphs, such as the notion of *path*, immediately applies to 1-polygraphs.

1.1.2 Example. The directed graph

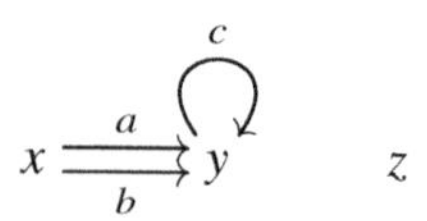

(1.1)

can be encoded as the 1-polygraph P with $P_0 = \{x, y, z\}$, $P_1 = \{a, b, c\}$ and

$$s_0(a) = s_0(b) = x, \qquad t_0(a) = t_0(b) = y, \qquad s_0(c) = t_0(c) = y,$$

which can be more concisely denoted as

$$P = \langle\, x, y, z \mid a\colon x \to y, b\colon x \to y, c\colon y \to y \,\rangle\,.$$

1.1.3 The category of 1-polygraphs. A *morphism* $f\colon P \to Q$ between 1-polygraphs P and Q consists of a pair of functions $f_0\colon P_0 \to Q_0$ and $f_1\colon P_1 \to Q_1$ respectively sending the 0- and 1-cells of P to those of Q and preserving sources and targets:

$$s_0^Q \circ f_1 = f_0 \circ s_0^P, \qquad t_0^Q \circ f_1 = f_0 \circ t_0^P.$$

We write $\mathbf{Pol}_1$ for the category of 1-polygraphs and their morphisms. Again, this is simply another name for the usual category of directed graphs and their morphisms.

1.2 Presenting Sets

A 1-polygraph P can be seen as a *presentation* of a set X, in the following sense. Each element x of P_0 denotes an element $\overline{x}$ of X, in such a way that each element of X has at least one “name” in P_0, and each element $a\colon x \to y$ in P_1 represents the renaming of $\overline{x}$ by $\overline{y}$. The elements of P_0 and P_1 are often respectively called *generators* and *relations*.

1.2.1 *P*-congruence. The *P-congruence* $\approx^P$ associated with a 1-polygraph P is the smallest equivalence relation on P_0 such that $x \approx^P y$ for every 1-generator $a\colon x \to y$ in P_1.

1.2.2 The presented set. The *set $\overline{P}$ presented* by a 1-polygraph P is the set $P_0/{\approx^P}$ obtained by quotienting P_0 by the P-congruence $\approx^P$, what we usually simply write P_0/P_1. More generally, a set X is *presented* by a 1-polygraph P when X is isomorphic to $\overline{P}$, and in this case P is called a *presentation* of X. Geometrically speaking, X amounts to the set of connected components of the graph P.

1.2.3 Example. In Example 1.1.2, the relation $\approx^P$ identifies x and y, and the presented set is the set with two elements, corresponding to the equivalence classes $\{x, y\}$ and $\{z\}$.

More abstractly, the set presented by a 1-polygraph P can be characterized by the following universal property:

1.2.4 Lemma. *For any set X and function $f\colon P_0 \to X$ such that $f(x) = f(y)$ for every 1-generator $a\colon x \to y$ in P_1, there exists a unique function $\overline{f}\colon \overline{P} \to X$ such that $\overline{f} \circ q = f$*

$$\begin{array}{ccc} P_0 & \xrightarrow{f} & X \\ {\scriptstyle q}\downarrow & \nearrow_{\overline{f}} & \\ \overline{P} & & \end{array},$$

where $q\colon P_0 \to \overline{P}$ is the function sending an element to its equivalence class.

1.2.5 Tietze transformations. At this point, a natural question to ask is: when do two polygraphs present the same set? For instance, the set with two elements can also be presented by the polygraph

$$x \xrightarrow{d} x' \xrightarrow{e} y \qquad z \tag{1.2}$$

which looks quite different from (1.1), and it is not obvious how the two are related. This question was first studied by Tietze for presentations of groups [345], as we shall see in Chapter 5, but similar results already hold for plain sets as we now explain.

We call *elementary Tietze transformations* the following operations transforming a 1-polygraph P into a 1-polygraph Q:

(T1) *adding a definable generator*: given $x \in P_0$, $y \notin P_0$, and $a \notin P_1$, we define

$$Q = \langle\, P_0, y \mid P_1, a\colon x \to y \,\rangle,$$

(T2) *adding a derivable relation*: given $x, y \in P_0$ and $a \notin P_1$ such that $x \approx^P y$, we define

$$Q = \langle\, P_0 \mid P_1, a\colon x \to y \,\rangle.$$

A *Tietze transformation* from P to Q is a zigzag of elementary Tietze transformations, i.e., a finite sequence of polygraphs $(P_i)_{0 \leqslant i \leqslant n}$ with $P_0 = P$ and $P_n = Q$, together with, for each index $0 \leqslant i < n$, an elementary Tietze transformation either from P_i to P_{i+1} or from P_{i+1} to P_i. The *Tietze equivalence* is the smallest equivalence relation on 1-polygraphs, identifying any two polygraphs related by an elementary Tietze transformation and closed by isomorphism; otherwise said, two polygraphs are Tietze equivalent when there exists a Tietze transformation between them, up to isomorphism.

1.2.6 Lemma. *Two Tietze equivalent* 1*-polygraphs present isomorphic sets.*

Proof. By induction on the length of Tietze transformations, it is enough to show that two polygraphs P and Q related by an elementary Tietze transformation present the same set. Using the same notations as above, in the case of the transformation (T1), we have

$$\overline{Q} = (P_0 \sqcup \{y\})/{\approx^Q} = ((P_0 \sqcup \{y\})/(x \approx y))/{\approx^P} = P_0/{\approx^P} = \overline{P},$$

where $x \approx y$ denotes the smallest equivalence relation identifying x and y. In the case of the transformation (T2), the relations generated by P_1 and Q_1 are the same and we have

$$\overline{Q} = Q_0/{\approx^Q} = P_0/{\approx^P} = \overline{P}.$$ □

We will see in Theorem 1.2.12 that the converse also holds: these operations exactly axiomatize when two finite 1-polygraphs are presenting the same set.

1.2.7 Example. Using the above lemma, one can deduce that the two polygraphs (1.1) and (1.2) present the same set, by building a series of Tietze transformations relating them:

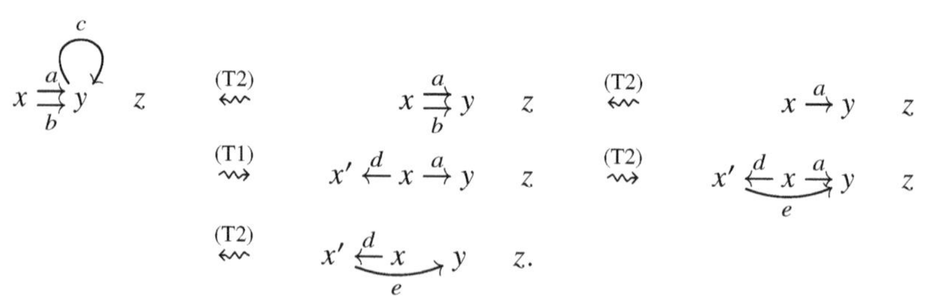

In the first step, $y \approx y$ can be shown without resorting to the relation $c\colon y \to y$ (this is because, by definition, $\approx$ is an equivalence relation), and therefore the relation h can be removed using the Tietze transformation (T2) backward. Other steps can be justified similarly. Of course, in this case, it is very easy to compute the sets presented by the two polygraphs (1.1) and (1.2) and to see that they are isomorphic (both have two elements), but it will no longer be the case when generalizing to higher dimensions.

1.2.8 Backward Tietze transformations. A Tietze transformation is a zigzag of elementary Tietze transformations. It can alternatively be seen as a sequence of elementary Tietze transformations or the following transformations, that we call *backward elementary Tietze transformations*, corresponding to using an elementary Tietze transformation in the "backward direction":

$\overline{\text{(T1)}}$ *removing a definable generator*: given a polygraph P of the form

$$P = \langle\, P'_0, x \,|\, P'_1, a\colon x \to y \,\rangle,$$

where x does not occur in any relation of P'_1, we define

$$Q = \langle\, P'_0 \,|\, P'_1 \,\rangle,$$

$\overline{\text{(T2)}}$ *removing a derivable relation*: given a polygraph P of the form

$$P = \langle\, P_0 \,|\, P'_1, a\colon x \to y \,\rangle,$$

we define

$$Q = \langle\, P_0 \,|\, P'_1 \,\rangle$$

whenever $x \approx^Q y$.

1.2.9 Remark. Given an elementary Tietze transformation from P to Q, there is an obvious inclusion of P into Q that induces a morphism of 1-polygraphs $P \to Q$. However, for a backward elementary Tietze transformation from P to Q there is no canonical morphism $P \to Q$. For instance, consider the transformation

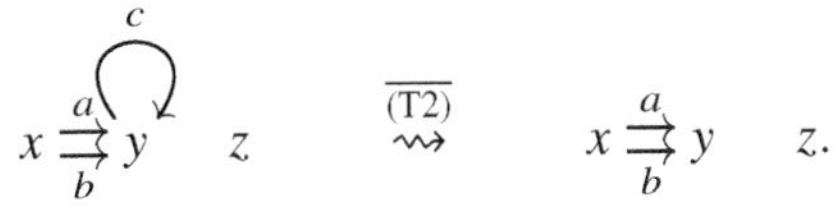

The only reasonable choice would be to send the 1-generator $c\colon y \to y$ to an identity on y, which is not possible with a morphism of 1-polygraph (those send 1-generators to 1-generators). This is one of the reasons why we take the elementary Tietze transformations (as opposed to the backward ones) as more primitive.

1.2.10 Minimal presentations. It can be noted that Tietze transformations consisting only of elementary transformations (T1) and (T2) make the presentations larger (in terms of number of generators and relations), whereas those consisting only of $\overline{\text{(T1)}}$ and $\overline{\text{(T2)}}$ make them smaller. We thus sometimes respectively call *Tietze expansions* and *Tietze reductions* these two families of Tietze transformations and say that a polygraph P *Tietze expands* (resp. *Tietze reduces*) to a polygraph Q if Q can be obtained from P by applying a series of Tietze expansions (resp. Tietze reductions). One may wonder if, by applying only the second kind of transformations, we eventually always reach a *minimal* presentation with respect to both generators and relations, and whether two such minimal presentations are necessarily isomorphic. We will see that it is indeed the case for finite polygraphs. First, note that a 1-polygraph P without relations (i.e., $P_1 = \emptyset$) is always minimal.

1.2.11 Lemma. *Any finite* 1*-polygraph P Tietze reduces to a polygraph isomorphic to* $\langle\, \overline{P} \mid \,\rangle$.

Proof. By induction on the cardinal of P_1, we show that we can remove a 1-generator using Tietze transformations, unless P_1 is empty. Suppose that P contains a non-directed cycle, i.e., a non-empty non-directed path from a 0-generator x to itself. We can assume that this path does not use the same edge twice; otherwise, we can choose a smaller cycle. Given a 1-generator $a\colon x \to y$ occurring in this cycle, there exists a non-directed path from x to y that is not using a. Therefore, we can apply a Tietze transformation $\overline{\text{(T2)}}$ to remove a. Otherwise, there is no cycle, and consider a maximal non-directed path in P. Since P is finite and acyclic, this path will end by a 1-generator $a\colon x \to y$ such that either x or y is incident to no other edge. Therefore, we can use a Tietze transformation $\overline{\text{(T1)}}$ to remove x or y, along with a. □

In the case of *finite* 1-polygraphs, the above lemma implies the converse of Lemma 1.2.6:

1.2.12 Theorem. *Two finite* 1*-polygraphs present isomorphic sets if and only if they are Tietze equivalent.*

Proof. Suppose given two polygraphs P and Q such that $\overline{P} \simeq \overline{Q}$. By the previous lemma, P is Tietze equivalent to $\langle\, \overline{P} \mid \,\rangle$, and similarly Q is Tietze equivalent to $\langle\, \overline{P} \mid \,\rangle$. Finally, the presentations $\langle\, \overline{P} \mid \,\rangle$ and $\langle\, \overline{Q} \mid \,\rangle$ are easily seen to be Tietze equivalent because $\overline{P}$ and $\overline{Q}$ are isomorphic. □

1.2.13 Remark. Note that, given the above definition of Tietze transformations, the previous theorem does not generalize to infinite presentations. For instance, the 1-polygraphs $\langle\, x \mid \,\rangle$ and $\langle\, x_i \mid a_i\colon x_i \to x_0 \,\rangle_{i\in\mathbb{N}}$ both present the set with one element but are not Tietze equivalent since we can only add or remove a finite number of relations using Tietze equivalences (the notation on the right means that i ranges over $\mathbb{N}$ both in generators x_i and relations a_i). In order to overcome this counter-example, one might be naively tempted to allow infinite sequences of Tietze transformations between 1-polygraphs, but this does not preserve presented sets. For instance, consider the 1-polygraph

$$\langle\, x_i, y \mid a_i : x_{i+1} \to x_i, b_i : x_i \to y \,\rangle_{i\in\mathbb{N}},$$

i.e., the graph

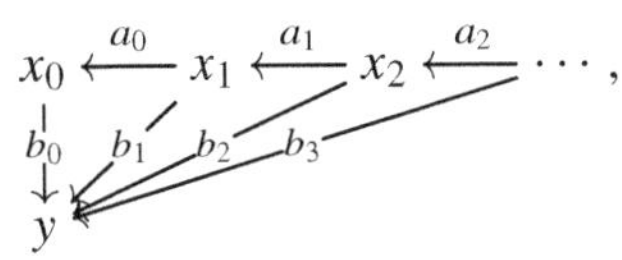

presenting the set with one element. Using Tietze transformations, any finite number of relations b_i can be removed from the polygraph, since they are derivable. However, if we remove all of them the resulting polygraph presents the set with two elements.

In order to account for infinite presentations, the notion of Tietze equivalence has to be generalized as follows. Firstly, we say that a 1-polygraph P Tietze expands to Q if there is a transfinite sequence of elementary Tietze expansions from P to Q; secondly, we define Tietze equivalence as the smallest equivalence relation containing Tietze expansions. Two (not necessarily finite) 1-polygraphs are Tietze equivalent in this sense if and only if they present isomorphic sets. We do not dwell further on infinite polygraphs, because we are mostly interested in finite polygraphs in this book; details can be found in [178].

We will see that Lemma 1.2.11 does not generalize in dimensions higher than 1, where arbitrary finite sequences of Tietze transformations, interleaving Tietze reductions and expansions, might be required in order to show that two polygraphs present the same object. However, an analogous of Theorem 1.2.12 will still hold, but its proof has to be carried over differently, as explained in Chapter 5.

1.3 Abstract Rewriting Systems

The orientations of the relations do not really matter in a 1-polygraph, with respect to the presented set: if we reverse an edge, the presented set is the same. This is easily shown using the following series of Tietze transformations:

$$\langle\, P_0 \,|\, P_1', x \to y \,\rangle \overset{\text{(T2)}}{\rightsquigarrow} \langle\, P_0 \,|\, P_1', x \to y, y \to x \,\rangle \overset{\overline{\text{(T2)}}}{\rightsquigarrow} \langle\, P_0 \,|\, P_1', y \to x \,\rangle,$$

which are based on the fact that $\approx$ is an equivalence relation and thus symmetric.

However, the orientations can still be useful to *decide equality* between generators, i.e., answer the following question:

> Given two generators, do they represent the same element of the presented set? Or, equivalently, are they related by $\approx$?

We will see that in good cases, one can come up with canonical representatives of equivalence classes under $\approx$, in such a way that the representative of an arbitrary generator can easily be computed. In those situations, the equivalence of two generators can be tested by checking whether their representatives are equal or not. In order to come up with representatives, we use the orientation of the 1-generators. Given two 0-generators x and y such that there is a 1-generator $a\colon x \to y$, we have $x \approx y$, and the orientation of the 1-generator will be interpreted as indicating that y is a "more canonical" representative than x in the equivalence class under $\approx$. With respect to this, the "most canonical" elements, which are called normal forms, are good candidates for being representatives of equivalence classes with good properties: under reasonable assumptions, it can be shown that every class admits exactly one such representative. This point of view is the starting point of *rewriting theory* [20, 342].

1.3.1 Terminology and notations. We have seen that a 1-polygraph P is simply another name for a graph. Since people in rewriting theory like to think about it from a different point of view, they give it yet another name and call it an *abstract rewriting system*. In this context, the elements of P_0 are called *objects* and those of P_1 are called *rewriting rules* (or *rewriting steps*). A *rewriting path* is simply a path, i.e., a sequence

$$x_0 \xrightarrow{a_1} x_1 \xrightarrow{a_2} x_2 \xrightarrow{a_3} \cdots \xrightarrow{a_n} x_n$$

of composable rewriting steps. The 0-cells x_0 and x_n are respectively called the *source* and *target* of the path, and we write $f\colon x \overset{*}{\to} y$ for a path f from x to y. One also writes $x \to y$ (resp. $x \overset{*}{\to} y$) when there exists a rewriting step (resp. a rewriting path) from x to y, and the notation $x \overset{*}{\leftrightarrow} y$ is often used instead of $x \approx y$.

1.3.2 Normal forms. A 0-cell $x \in P_0$ is a *normal form* when there is no rule $a\colon x \to y$ in P_1 with x as source.

We can distinguish the following situations concerning normal forms in equivalence classes under $\approx$ of 0-cells in a polygraph P: we say that P has

- the *existing normal form property* when every equivalence class contains at least one normal form; i.e., for every $x \in P_0$ there exists a normal form $y \in P_0$ such that $x \overset{*}{\leftrightarrow} y$;
- the *unique normal form property* when every equivalence class contains at most one normal form; i.e., for every normal form $x, y \in P_0$, $x \overset{*}{\leftrightarrow} y$ implies $x = y$; and
- the *canonical form property* when every equivalence class contains exactly one normal form, called the *canonical representative* of the class; i.e., it satisfies both the existing and the unique normal form property.

1.3.3 Example. Consider the following 1-polygraphs:

(1) (2) (3)

(1) and (3) have the unique normal form property, (2) and (3) have the existing normal form property, and (3) has the canonical form property.

We are interested here in providing practical conditions on P that ensure that the canonical form property holds, as well as that we are able to efficiently compute the canonical form associated to the class of a 0-cell. We will see that termination of a 1-polygraph implies the existing normal form, that confluence implies the unique normal form property, and moreover that confluence can be checked locally for terminating 1-polygraphs.

1.3.4 Normalizability. A polygraph is *normalizing* when every 0-cell x rewrites to a normal form. We sometimes write $\widehat{x}$ for an arbitrary choice of such a normal form. From the definition, we deduce the following result.

1.3.5 Lemma. *A normalizing* 1*-polygraph has the existing normal form property.*

The converse does not hold, as illustrated in Example 1.3.20.

1.3.6 Termination. In practice, in order to show that a 1-polygraph is normalizing, one often uses the following property. A 1-polygraph P is *terminating* (or *well-founded* or *noetherian* or *strongly normalizing*) when there is no infinite sequence of rewriting steps

$$x_0 \xrightarrow{a_1} x_1 \xrightarrow{a_2} x_2 \xrightarrow{a_3} \cdots .$$

For instance, in Example 1.3.3, (2) and (3) are terminating but not (1).

Starting from a 0-cell x in a terminating 1-polygraph, we can define a sequence of 0-cells by induction by $x_0 = x$, and x_{i+1} is the target of an arbitrary rewriting rule $x_i \to x_{i+1}$ with x_i as source; we stop if there is no such rewriting rule. Termination ensures that this process will end after a finite number of steps, and the last 0-cell x_n is necessarily a normal form. We have just shown the following.

1.3.7 Lemma. *A terminating* 1*-polygraph is normalizing.*

The converse does not hold, as illustrated in Example 1.3.20.

In practice, the termination of a 1-polygraph P can be shown using the following lemma. We recall that a poset $(N, \leqslant)$ is *well-founded* when every decreasing sequence $n_1 \geqslant n_2 \geqslant \ldots$ is eventually stationary: there exists $k \in \mathbb{N}$ such that for every $i, j \in \mathbb{N}$ with $i \geqslant j \geqslant k$ one has $n_i = n_j$. Equivalently, the poset is well-founded when there exists no infinite strictly decreasing sequence $n_1 > n_2 > \ldots$ of elements of N. The typical example of such an order is $(\mathbb{N}, \leqslant)$, or any ordinal.

1.3.8 Lemma. *Given a rewriting system P the following statements are equivalent.*

1. *The rewriting system P is terminating.*
2. *There exists a well-founded order on P_0 such that $x > y$ for every* 1*-generator $a\colon x \to y$ in P_1.*
3. *There exists a function $f\colon P_0 \to N$, where N is a well-founded poset, such that $f(x) > f(y)$ for every* 1*-generator $a\colon x \to y$ in P_1.*

Proof. Suppose that P is terminating. Then the preorder relation on P_0 defined by $x \geqslant y$ whenever $x \overset{*}{\to} y$ is a well-founded partial order that shows that 1 implies 2, and taking $f\colon P_0 \to P_0$ to be the identity shows that 2 implies 3. Finally, 3 implies 1, for if there was an infinite reduction sequence in P, the image of the objects under f would be an infinite strictly decreasing sequence of elements of N. □

1.3.9 Well-founded induction. Suppose given a predicate $\mathcal{P}$ on the 0-cells of a terminating polygraph P. In order to show that $\mathcal{P}$ holds all the elements of P_0, it is often useful to use the following *well-founded induction principle*: if

$$\forall x \in P_0, \qquad \big((\forall y \in P_0,\ x \to y \text{ implies } \mathcal{P}(y)) \quad \text{implies} \quad \mathcal{P}(x)\big) \tag{1.3}$$

then $\forall x \in P_0,\ \mathcal{P}(x)$ holds.

1.3.10 Proposition. *If P is a terminating 1-polygraph then the well-founded induction principle holds.*

Proof. By contradiction, suppose that the well-founded induction principle does not hold: there is a predicate $\mathcal{P}$, such that the hypothesis (1.3) holds but not the conclusion, i.e., $\mathcal{P}(x_0)$ does not hold for some $x_0 \in P_0$. By repeated use of (1.3), we can construct a family $(x_i)_{i\in\mathbb{N}}$ of elements of P_0 such that $\mathcal{P}(x_i)$ does not hold for any $i \in \mathbb{N}$, and $x_0 \to x_1 \to \cdots$. This contradicts the fact that P is terminating. □

1.3.11 Quasi-termination. Following [111], we introduce the following variant of the termination condition. We say that a 1-polygraph P is *quasi-terminating* if every sequence $(x_i)_{i\in\mathbb{N}}$ of 0-cells, with $x_i \to x_{i+1}$ for every index $i \in \mathbb{N}$, contains an infinite number of occurrences of the same 0-cell: there exists a 0-cell x such that for every $i \in \mathbb{N}$, there exists $j > i$ such that $x_j = x$.

Let P be a 1-polygraph. A 0-cell x is called a *quasinormal form* if for any rewriting step $x \to y$, there exists a rewriting path from y to x. If P is quasi-terminating, any 0-cell x rewrites to a quasi-normal form. Note that this quasi-normal form is neither irreducible nor unique in general. We say that P is *quasi-convergent* if it is confluent and it quasi-terminates.

1.3.12 Example. The following 1-polygraph

$$x \longrightarrow y \rightleftarrows z$$

is quasi-terminating and quasi-convergent. Both y and z are quasi-normal forms.

The above termination and normalizability conditions ensure the existing normal form property. We now investigate conditions implying the unique normal form property.

1.3.13 Joinability. Two 0-cells $x, y \in P_0$ of a polygraph P are *joinable* when there exists 0-cell z such that there are rewriting paths $f : x \overset{*}{\to} z$ and $g : y \overset{*}{\to} z$:

$$\begin{array}{ccc} x & & y. \\ & {\scriptstyle *}\searrow \quad \swarrow{\scriptstyle *} & \\ & z & \end{array}$$

1.3.14 The Church–Rosser property. A 1-polygraph P has the *Church–Rosser property* when any two 0-cells $x, y \in P_0$ which are equivalent are joinable:

$$\begin{array}{ccc} x & \overset{*}{\longleftrightarrow} & y. \\ & {\scriptstyle *}\searrow \quad \swarrow{\scriptstyle *} & \\ & z & \end{array}$$

1.3.15 Proposition. *A 1-polygraph with the Church–Rosser property has the unique normal form property.*

Proof. Suppose given two normal forms x and y such that $x \approx y$. By the Church–Rosser property, there exists a 0-cell z and rewriting paths $x \overset{*}{\to} z$ and $y \overset{*}{\to} z$. Since x and y are normal forms, these two paths are necessarily empty, and thus $x = y$. □

The converse property is not true, as illustrated by the 1-polygraph

$$x \longleftarrow y \longrightarrow z \circlearrowleft$$

where x and z are equivalent but cannot be rewritten to a common 0-cell, even though there is a unique normal form x.

In the following, we present more "local" properties which imply the Church–Rosser property, and thus the unique normal form property.

1.3.16 Branchings. In a 1-polygraph P, a pair (a, a') of coinitial 1-generators $a \colon x \to y$ and $a' \colon x \to y'$ in P is called a *local branching*; a pair (f, f') of coinitial rewriting paths $f \colon x \overset{*}{\to} y$ and $f' \colon x \overset{*}{\to} y'$ is called a *branching*. The 0-cell x is called the *source* of the branching.

1.3.17 Confluence. A branching (f, f') as above is *confluent* when y and y' are joinable:

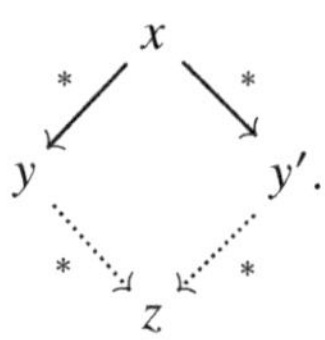

In this situation, we say that the branching is *confluent.* A 1-polygraph is *confluent* (resp. *locally confluent*) when every branching (resp. local branching) is confluent. Note that a confluent 1-polygraph is necessarily locally confluent.

The above confluence conditions can be summarized graphically as follows:

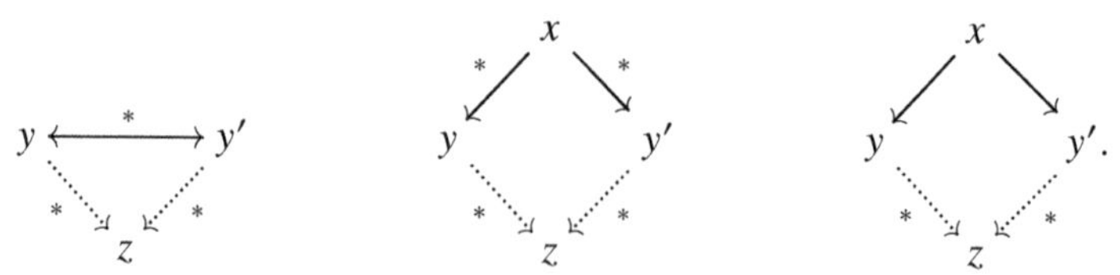

Church–Rosser confluence local confluence

1.3.18 Proposition. *A 1-polygraph has the Church–Rosser property if and only if it is confluent.*

Proof. The left-to-right direction is immediate. For the right-to-left direction, suppose that x and y are two equivalent 0-cells: this means that there exists rewriting paths $f_i : y_i \overset{*}{\to} x_i$ and $g_i : y_i \overset{*}{\to} x_{i+1}$ in P, with $0 \leqslant i < n$, where $x_0 = x$ and $x_n = y$, forming a diagram as below (ignoring the dotted arrows, z and z'):

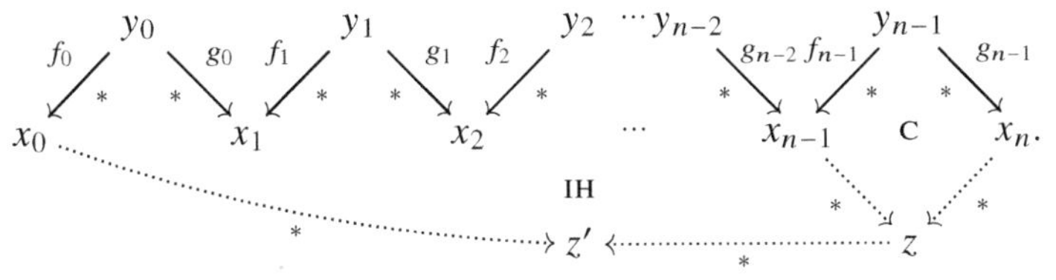

By induction on $n \in \mathbb{N}$, we show that x_0 and x_n can be joined. The result is immediate when $n = 0$, and otherwise the diagram can be completed as above using the confluence hypothesis for C and the induction hypothesis for IH. □

As a direct corollary, we deduce:

1.3.19 Lemma. *A confluent* 1*-polygraph has the unique normal form property.*

Confluence is difficult to show in practice, whereas local confluence is much more tractable. Clearly confluence of a rewriting system implies its local confluence, and one could hope that both properties are equivalent. This is however not the case: local confluence does not imply confluence in general illustrated by the following example attributed by Hindley to Kleene, see [188, Figure 6b] and [342, Section 1.2].

1.3.20 Example. Consider the following 1-polygraph:

$$x' \longleftarrow x \rightleftarrows y \longrightarrow y'.$$

It is locally confluent (it is easy to check all the possible cases), but not confluent: we have $x \overset{*}{\to} x'$ and $x \overset{*}{\to} y'$, but there is no 0-cell to which both x' and y' rewrite.

In the previous example, it can be noted that the rewriting system is not terminating since there is a directed cycle between the vertices x and y. It was shown in a famous lemma by Newman [290], also known as the *diamond lemma*, that local confluence and confluence are equivalent when restricting to terminating rewriting systems, thus providing us with simple ways of checking for their confluence.

1.3.21 Lemma. *A terminating* 1*-polygraph is confluent if and only if it is locally confluent.*

Proof. We show the right-to-left direction, the other one being immediate. We say that a 1-polygraph is *confluent* (resp. *locally confluent*) *at a* 0*-cell* x when every branching (resp. local branching) with x as source is joinable. By well-founded induction, whose use is justified by Proposition 1.3.10 based on the hypothesis that the 1-polygraph is terminating, we show that the local confluence property at a vertex x implies the confluence property at x. The base cases are immediate. Otherwise, we have a diagram of the form

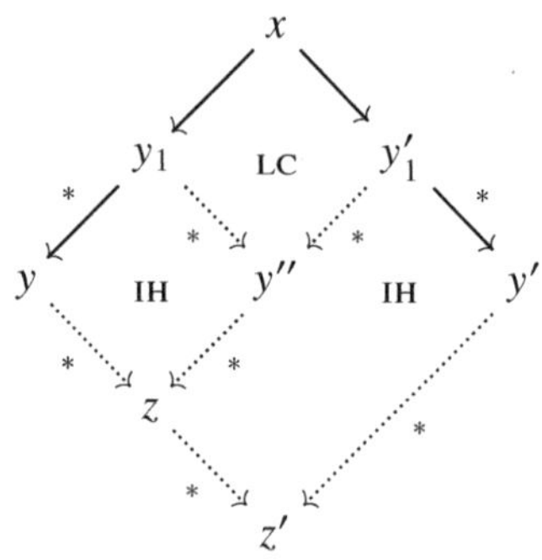

which can be closed using the local confluence hypothesis for LC and the induction hypothesis for IH (which provides confluence at y_1 and y'_1, respectively). □

1.3.22 Remark. Showing termination and local confluence is the most usual way of proving that an abstract rewriting system is confluent, but it is not the only one. We refer to standard rewriting textbooks for other properties which imply confluence [20, 342]. For instance, an abstract rewriting system has the *diamond property* when for every pair of coinitial rewriting steps $a\colon x \to y$ and $b\colon x \to y'$ there exists a pair of cofinal rewriting steps (i.e., rewriting paths of length one) $a'\colon y \to x$ and $b'\colon y' \to x$. Graphically,

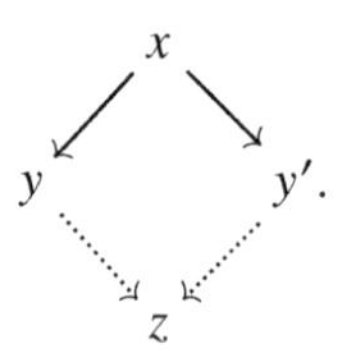

In this case, the abstract rewriting system is always confluent (this can be shown using a variant of the proof of Lemma 1.3.21) even if it is not terminating.

1.3.23 Convergence. A 1-polygraph is *convergent* when it is both terminating and confluent.

1.3.24 Proposition. *A convergent* 1*-polygraph has the canonical form property.*

Proof. Suppose given a convergent 1-polygraph. Since it is terminating, it is normalizing by Lemma 1.3.7 and thus has the existing normal form property by Lemma 1.3.5. Since it is confluent, Lemma 1.3.19 ensures that it also has the unique normal form property. □

1.3.25 Remark. A polygraph can have the canonical form property without being convergent:

$$x \rightleftarrows y \longrightarrow z.$$

Here, all the 0-cells are equivalent and z is the only normal form, which shows the canonical form property. The polygraph is not terminating (there is a cycle between x and y) and thus not convergent.

1.3.26 Deciding equality. Give a finite 1-polygraph P, the *equality decision problem*, or the *word problem*, for P consists in answering the following question:

Given two 0-cells $x, y \in P_0$, do we have $x \approx y$?

Since we only consider only finite 1-polygraphs, this problem is *decidable*, meaning that there is a program which takes P, x and y as input and outputs whether $x \approx y$ holds or not. Namely, we can implement a program which will construct all acyclic paths starting from x, which are in finite number, and check whether one of those paths ends at y. We will see that if we assume additional properties on P, this can be performed much more efficiently.

When the 1-polygraph P has the canonical form property, the equivalence class of x (resp. y) contains a unique normal form denoted $\widehat{x}$ (resp. $\widehat{y}$), and we have $x \approx y$ if and only if we have $\widehat{x} = \widehat{y}$. In this case, the equality decision problem can be decided by comparing normal forms. In particular, in the case where the 1-polygraph is convergent, we have seen in Proposition 1.3.24 that it has the canonical form property, and moreover the normal form $\widehat{x}$ associated to a 0-cell x can be computed easily. A maximal path starting from x

$$x = x_0 \xrightarrow{a_0} x_1 \xrightarrow{a_1} x_2 \xrightarrow{a_2} \cdots \xrightarrow{a_{n-1}} x_n$$

exists because P is terminating, and the fact that it is maximal means that its target is a normal form, i.e., $x_n = \widehat{x}$. In order to decide whether x and y are equivalent, we can thus use the *normal form algorithm* which consists in

1. rewriting x as much as possible in order to obtain a normal form $\widehat{x}$, and similarly compute a normal form $\widehat{y}$ for y; and
2. checking whether $\widehat{x} = \widehat{y}$ holds or not.

Formally, this is justified as follows:

1.3.27 Proposition. *In a convergent* 1*-polygraph, two* 0*-cells* x *and* y *are equivalent if and only if they have the same normal form:* $x \approx y$ *if and only if* $\widehat{x} = \widehat{y}$.

Proof. Since the polygraph is terminating, it is normalizing by Lemma 1.3.7: x rewrites to a normal form $\widehat{x}$, and similarly y rewrites to a normal form $\widehat{y}$. If $\widehat{x} = \widehat{y}$, then clearly x and y are equivalent:

$$x \xrightarrow{*} \widehat{x} = \widehat{y} \xleftarrow{*} y.$$

Conversely, suppose that x and y are equivalent, and thus that $\widehat{x}$ and $\widehat{y}$ are also equivalent:

$$\widehat{x} \xleftarrow{*} x \xleftrightarrow{*} y \xrightarrow{*} \widehat{y}.$$

The confluence of the polygraph implies that it has the Church–Rosser property by Proposition 1.3.18, and thus the unique normal form property by Proposition 1.3.15. Since $\widehat{x}$ and $\widehat{y}$ are equivalent normal forms, we deduce that they are equal. □

1.3.28 Deciding confluence. As a direct corollary of the above proposition, we also have a practical method for checking whether a terminating 1-polygraph is confluent (and thus convergent):

1.3.29 Proposition. *A terminating* 1*-polygraph is confluent if and only if for every local branching* $x \to y$ *and* $x \to z$*, we have* $\hat{y} = \hat{z}$.

1.4 Decreasing Diagrams

The main method we have seen so far in order to show the confluence of a 1-polygraph is provided by Newman's lemma (Lemma 1.3.21), which requires supposing termination of the polygraph. As a more advanced topic, we explain here the method of *decreasing diagrams*, introduced by van Oostrom [350], see also [342, Section 14.2], which can be used in order to show the confluence of a 1-polygraph which is non-terminating. Stronger versions of this method have been introduced more recently [126, 351].

1.4.1 Multisets. Given a set A, a *multiset* on A is a function $\mu\colon A \to \mathbb{N}$ which is null almost everywhere, i.e., the set $\{a \in A \mid \mu(a) \neq 0\}$ is finite. The set A is called the *domain* of the multiset. Given an element $a \in A$, the natural number $\mu(a)$ is called its *multiplicity* in the multiset: μ should be thought of as a collection of elements of A where each element a occurs $\mu(a)$ times. We denote by $A^\sharp$ the set of all multisets on A.

We write $\emptyset$ for the *empty multiset* on A, i.e., the constant function $\emptyset\colon A \to \mathbb{N}$ equal to 0. Given two multisets μ and ν on A, their *union* or *sum* is the multiset $\mu \sqcup \nu$ on A such that $(\mu \sqcup \nu)(a) = \mu(a) + \nu(a)$ for every element $a \in A$. The operation $\sqcup$ equips $A^\sharp$ with a structure of commutative monoid, with $\emptyset$ as neutral element, which characterizes multisets over A. Given an element $a \in A$, we often write $\{a\}$ for the multiset with a as only element. Given two multisets μ and ν, we say that μ is *included* in ν, what we write $\mu \sqsubseteq \nu$ when $\mu(a) \leqslant \nu(a)$ for every $a \in A$. This is the case precisely when there is a multiset μ' such that $\mu \sqcup \mu' = \nu$. This relation makes $A^\sharp$ into a poset which is well-founded.

A partial order $\leqslant$ on a set A induces an order $\leqslant^\sharp$ on $A^\sharp$, called its *multiset extension*, defined by $\mu \leqslant^\sharp \nu$ if and only if

$$\forall b \in A, \qquad \mu(b) > \nu(b) \quad \text{implies} \quad \exists a \in A, \quad a > b \text{ and } \mu(a) < \nu(a).$$

Let us spell it out: for μ to be smaller than ν, it is fine to have more b's as long as ν has more of something greater than b. The following result is due to Dershowitz and Manna [112]:

1.4.2 Proposition. *Given a well-founded poset $(A, \leqslant)$, its multiset extension $(A^\sharp, \leqslant^\sharp)$ is also well-founded.*

1.4.3 Labeled 1-polygraphs. A *labeled* 1-*polygraph* $(P, \mathcal{L}, \leqslant, \ell)$ consists of

— a 1-polygraph P,
— a set $\mathcal{L}$ of *labels* equipped with a well-founded ordering $\leqslant$, and
— a function $\ell\colon P_1 \to \mathcal{L}$ associating a label to each rewriting step.

1.4.4 Lexicographic maximum measure. Let $(P, \mathcal{L}, \leqslant, \ell)$ be a fixed labeled 1-polygraph. We write $\mathcal{L}^*$ for the sets of words over $\mathcal{L}$, i.e., finite sequences of elements of $\mathcal{L}$. The empty word is noted 1, and the concatenation of two words w and w is noted ww': these operations equip the sets of words with a structure of monoid. Following [350, Definition 3.1], we define the *lexicographic maximum measure* $\|w\|$ of a word $w \in \mathcal{L}^*$ as the multiset defined inductively by

$$\|1\| = \emptyset, \qquad\qquad \|lw\| = \{l\} \sqcup \|w^{\not< l}\|.$$

Above, $w^{\not< l}$ is the subword of w whose letters are not strictly below l, which is formally defined by induction by

$$1^{\not< l} = 1, \qquad\qquad (aw)^{\not< l} = \begin{cases} w^{\not< l} & \text{if } a < l, \\ aw^{\not< l} & \text{otherwise.} \end{cases}$$

Informally, the multiset $\|w\|$ thus consists of the letters of w which are not dominated by some letter on their left.

The measure $\|\cdot\|$ is extended to the set of finite rewriting paths of P by setting, for every rewriting path $a_1 \dots a_n$,

$$\|a_1 \dots a_n\| = \|\ell(a_1) \dots \ell(a_n)\|,$$

where $\ell(a_1) \dots \ell(a_n)$ is the product in the monoid $\mathcal{L}^*$. Finally, the measure $\|\cdot\|$ is extended to the set of finite branchings (a, b) of P, by setting

$$\|(a, b)\| = \|a\| \sqcup \|b\|.$$

1.4.5 Decreasing diagrams. A diagram of rewriting paths of the form

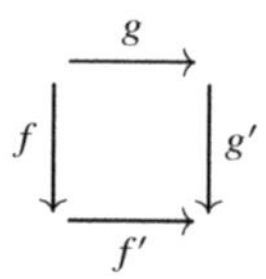

is *decreasing* if

$$\|ff'\| \leqslant^\# \|f\| \sqcup \|g\| \qquad \text{and} \qquad \|gg'\| \leqslant^\# \|f\| \sqcup \|g\|.$$

In the case where $f = a$ and $g = b$ are both 1-generators, it can be shown that the diagram is decreasing if and only if it is of the form

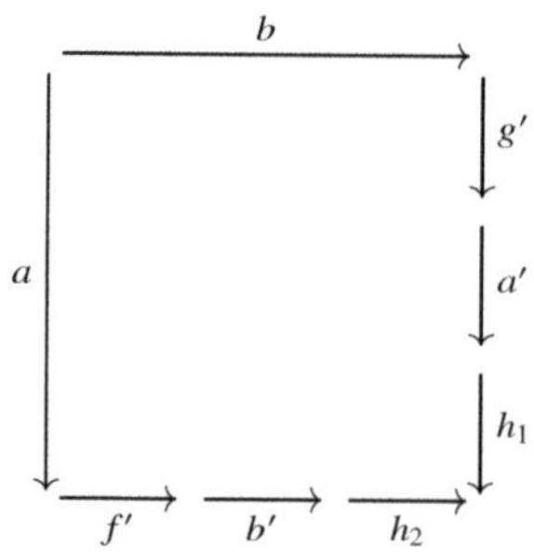

(1.4)

where

— $l < \ell(a)$ for every label l of a rewriting step in f',
— $l < \ell(b)$ for every label l of a rewriting step in g',
— a' is either an identity or a rewriting step labeled by $\ell(a)$,
— b' is either an identity or a rewriting step labeled by $\ell(b)$, and
— $l < \ell(a)$ or $l < \ell(b)$ for every label l of a transition in h_1 (resp. in h_2).

A labeled 1-polygraph is *locally decreasing* when every local branching (a, b) can be completed as a locally decreasing diagram (1.4). We can now recall van Oostrom's theorem [350, Theorem 3.7], whose proof follows the one of Newman's Lemma 1.3.21:

1.4.6 Theorem. *A locally decreasing* 1*-polygraph is confluent.*

This method is complete, in the sense that given a 1-polygraph with countably many 0-cells which is confluent, there is always a way to choose a well-founded poset $\mathcal{L}$ of labels so that the polygraph is locally decreasing [342, Theorem 14.2.32]. Moreover, we can always choose the set $\mathcal{L} = \{0, 1\}$ with $0 < 1$ as a set of labels, see [122].

2

Two-Dimensional Polygraphs

This chapter is dedicated to the definition of 2-polygraphs, which are a 2-dimensional generalization of the 1-polygraphs presented in the previous chapter. Before introducing this notion of 2-polygraph, we first give in Section 2.1 a refined viewpoint over 1-polygraphs. Instead of merely focusing on the set presented by a 1-polygraph P as a set of equivalence classes of P_0 modulo the relations in P_1, we now consider the free category generated by P, whose set of objects is P_0 and whose morphisms are all the rewriting paths obtained by composing the elements of P_1. A variant of this construction is the notion of free groupoid generated by a 1-polygraph P, where all 1-generators are supposed to be invertible.

The notion of 2-polygraph, introduced in Section 2.2, naturally appears as soon as arbitrary, non-necessarily free, small categories are considered. In order to present such a category C, one starts as above with a polygraph P such that the elements of P_1 generate the morphisms of C, but now we must take account of the relations induced by C among the morphisms of the free category generated by $\langle\, P_0 \mid P_1 \,\rangle$. These relations will be generated by a set P_2 of 2-generators, consisting in certain pairs of morphisms we want to equalize in C, as explained in Section 2.3.

Following the same pattern, we finally explain in Section 2.4 that a 2-polygraph can also be seen as a system of generators for a free 2-category, thus preparing the study of 3-polygraphs. We also examine, in Section 2.5, the variant where we freely generate a $(2, 1)$-category, that is, a 2-category in which every 2-cell is invertible. This framework allows to invoke explicit witnesses for the confluence of rewriting paths, which are used in the proofs of coherence results on abstract rewriting systems.

2.1 Generating Categories and Groupoids

We now define the free category and the free groupoid generated by a 1-polygraph.

2.1.1 Underlying polygraph of a category. Any small category C has an *underlying* 1*-polygraph* P with P_0 being the set of objects of C, P_1 being the set of morphisms of C, and the source and target of a morphism $f: x \to y$ of C being, respectively, x and y. This construction extends in the expected way into a functor $V: \mathbf{Cat} \to \mathbf{Pol}_1$ which to every category C associates its underlying polygraph VC.

2.1.2 Freely generated category. A 1-polygraph P induces a category P^*, called the category *freely generated* by P, or the *free category on* P, defined as follows:

- its objects are the 0-cells of P,
- its morphisms from x to y are composable sequences of 1-generators

$$x = x_0 \xrightarrow{a_1} x_1 \xrightarrow{a_2} x_2 \xrightarrow{a_3} \dots \xrightarrow{a_n} x_n = y$$

  which, more precisely, consist of a sequence $(x_i)_{0 \leqslant i \leqslant n}$ of elements of P_0, with $x_0 = x$ and $x_n = y$, together with a sequence $(a_i)_{0 < i \leqslant n}$ of elements of P_1, for some $n \geqslant 0$, such that $s_0(a_{i+1}) = x_i$ and $t_0(a_{i+1}) = x_{i+1}$ for $0 \leqslant i < n$,
- identities are morphisms as above with $n = 0$, and
- the composition of two morphisms

$$x = x_0 \xrightarrow{a_1} x_1 \xrightarrow{a_2} x_2 \xrightarrow{a_3} \dots \xrightarrow{a_n} x_n = y$$

  and

$$y = y_0 \xrightarrow{b_1} y_1 \xrightarrow{b_2} y_2 \xrightarrow{b_3} \dots \xrightarrow{b_m} y_m = z$$

  is

$$x = x_0 \xrightarrow{a_1} x_1 \xrightarrow{a_2} \dots \xrightarrow{a_n} x_n = y_0 \xrightarrow{b_1} y_1 \dots \xrightarrow{b_m} y_m = z.$$

In the terminology of directed graphs, the morphisms of the category P^* are the *directed paths* in P, with the natural number n for a path as above being its *length*, identities are empty paths, and composition is given by *concatenation* of paths. Following the terminology of rewriting systems, we will also call a morphism in P^* a *rewriting path*.

The category P^* can be characterized, up to isomorphism, as the category satisfying the following universal property:

2.1.3 Lemma. *For any category C and morphism of 1-polygraphs $f\colon P \to VC$, there exists a unique functor $f^*\colon P^* \to C$ such that the following diagram commutes*

$$\begin{array}{ccc} P & \xrightarrow{\ f\ } & VC \\ {\scriptstyle i}\big\downarrow & \nearrow_{Vf^*} & \\ VP^* & & \end{array}$$

where $i\colon P \to VP^*$ is the morphism of 1-polygraphs sending a 0-generator to itself and a 1-generator $a\colon x \to y$ to the corresponding path of length 1.

By classical theorems [261, Section IV.1], this is equivalent to the fact that the operation which to every polygraph P associates the freely generated category P^* extends to a functor $-^*\colon \mathbf{Pol}_1 \to \mathbf{Cat}$ which is left adjoint to the functor V.

2.1.4 Freely generated groupoid. The previous construction can be modified in order to describe the groupoid freely generated by a polygraph. We recall that a *groupoid* C is a category in which every morphism is invertible, i.e., for every morphism $f\colon x \to y$ of C there exists a morphism $f^-\colon y \to x$ such that $f^- \circ f = 1_x$ and $f \circ f^- = 1_y$. We write **Gpd** for the category of groupoids and functors between them.

There is an obvious forgetful functor $\mathbf{Gpd} \to \mathbf{Cat}$, which admits a left adjoint [52, Proposition 5.2.2]. Given a 1-polygraph P, we write $P^\top$ for the *free groupoid* on the category P^*. It can be described as the category whose objects are the elements of P_0, and morphisms from x to y are composable sequences

$$x = x_0 \xleftrightarrow{f_1} x_1 \xleftrightarrow{f_2} x_2 \xleftrightarrow{f_3} \ldots \xleftrightarrow{f_n} x_n = y,$$

where f_{i+1} is a 1-generator which is either of the form $a_{i+1}\colon x_i \to x_{i+1}$ or $a_{i+1}\colon x_{i+1} \to x_i$, quotiented by the equivalence relation identifying

$$\cdots \xleftrightarrow{f_{i-1}} x \xrightarrow{\ a\ } y \xleftarrow{\ a\ } x \xleftrightarrow{f_{i+1}} \cdots \qquad \text{with} \qquad \cdots \xleftrightarrow{f_{i-1}} x \xleftrightarrow{f_{i+1}} \cdots$$

and

$$\cdots \xleftrightarrow{f_{i-1}} y \xleftarrow{\ a\ } x \xrightarrow{\ a\ } y \xleftrightarrow{f_{i+1}} \cdots \qquad \text{with} \qquad \cdots \xleftrightarrow{f_{i-1}} y \xleftrightarrow{f_{i+1}} \cdots$$

which means that we can remove two adjacent occurrences of a 1-generator a in different directions. Identities are empty sequences and composition behaves as in §2.1.2. This explicit construction will be presented in more detail in §3.2.1. Note that, in the terminology of graphs, a morphism as above is called a *non-directed path* in P.

The interest of this construction lies in the fact that the morphisms of $P^\top$ are "witnesses" for the P-congruence of 0-cells, as defined in §1.2.1:

2.1.5 Lemma. *Two 0-cells $x, y \in P_0$ are P-congruent if and only if there exists a morphism $f : x \to y$ in $P^\top$.*

2.1.6 Three functors. To sum up, we have defined the following sequence of functors:

$$\mathbf{Pol}_1 \to \mathbf{Cat} \to \mathbf{Gpd} \to \mathbf{Set},$$

where

- **Pol**$_1$ $\to$ **Cat** associates to a 1-polygraph the category it freely generates,
- **Cat** $\to$ **Gpd** associates to a category the groupoid it freely generates, and
- **Gpd** $\to$ **Set** associates to a groupoid C the corresponding quotient set (obtained from the set of objects of C by identifying any two objects between which there is a morphism).

The composite functor **Pol**$_1$ $\to$ **Set** was explicitly described in Section 1.2.

2.2 The Category of 2-Polygraphs

We are now ready to introduce 2-polygraphs, which consist in a 1-polygraph together with a set of globular 2-generators between *parallel* 1-cells, that is, 1-cells having same source and target. Those were introduced by Street [333]. They are sometimes also called *linear sketches* [27, Section 4.6] when seen as a particular class of sketches; see Section G.1.

2.2.1 Notations on 1-cells. Given a 1-polygraph P, we write P_1^* for the set of 1-cells of the category P^*, i.e., the set of paths in P. The composite of two 1-cells $u : x \to y$ and $v : y \to z$ is written $uv : x \to z$, or $u *_0 v : x \to z$, and the empty path on a 0-generator x is written 1_x. We also write $i_1 : P_1 \to P_1^*$ for the canonical inclusion sending a 1-generator to the corresponding path of length 1. Given a 1-generator $a \in P_1$, we generally simply write a instead of $i_1(a)$. The source and target functions $s_0, t_0 : P_1 \to P_0$ canonically extend to functions $s_0^*, t_0^* : P_1^* \to P_0$ such that

$$s_0^* \circ i_1 = s_0 \qquad \text{and} \qquad t_0^* \circ i_1 = t_0 \tag{2.1}$$

respectively sending a path $f: x \to y$ to its source x and its target y, what we often picture as a "commuting" diagram of form

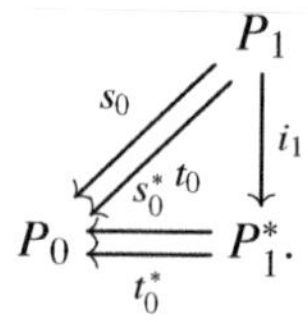

2.2.2 Definition. A 2-*polygraph* consists of

– a 1-polygraph P, i.e.,

$$P_0 \underset{t_0}{\overset{s_0}{\leftleftarrows}} P_1, \text{ and}$$

– a set P_2 of 2-*generators* together with two functions

$$s_1, t_1 \colon P_2 \to P_1^*,$$

associating to each relation its *source* and *target*, which is such that

$$s_0^* \circ s_1 = s_0^* \circ t_1 \qquad \text{and} \qquad t_0^* \circ s_1 = t_0^* \circ t_1. \tag{2.2}$$

A 2-polygraph thus consists of a diagram of sets and functions

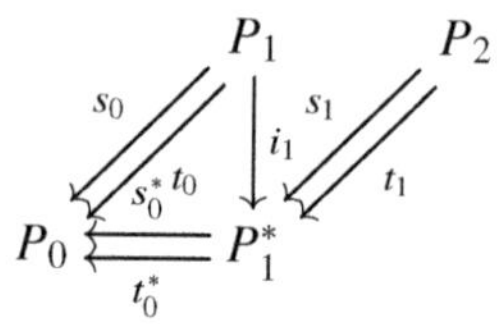

together with compositions and identities on P_1^*, which "commutes" in the sense that the relations (2.1) and (2.2) hold. We often write $\alpha\colon u \Rightarrow v$ for a 2-generator α in P_2 such that $s_1(\alpha) = u$ and $t_1(\alpha) = v$, and picture it as a 2-cell

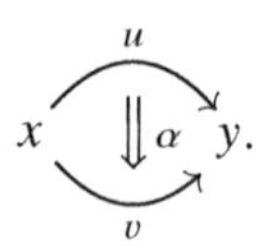

Moreover, we sometimes write $\langle\, P_0 \mid P_1 \mid P_2 \,\rangle$ to indicate the generators of a 2-polygraph P. A 2-polygraph is *finite* when the sets P_0, P_1, and P_2 are. The *underlying* 1-*polygraph* of a 2-polygraph P is denoted $P_{\leqslant 1}$.

2.2.3 The category of 2-polygraphs. A *morphism* $f\colon P \to Q$ between two 2-polygraphs P and Q consists of a morphism $f\colon P_{\leqslant 1} \to Q_{\leqslant 1}$ between the underlying 1-polygraphs together with a function $f_2\colon P_2 \to Q_2$ such that $s_1^Q \circ f_2 = f_1 \circ s_1^P$ and $t_1^Q \circ f_2 = f_1 \circ t_1^P$. These compose in the expected way, and we write $\mathbf{Pol}_2$ for the category of 2-polygraphs and their morphisms.

2.3 Presenting Categories

2.3.1 Quotient categories. Given a category C, a *congruence* $\approx$ on C is an equivalence relation on the morphisms of C such that

- given $u\colon x \to y$ and $u'\colon x' \to y'$, $u \approx u'$ implies $x = x'$ and $y = y'$, and
- given morphisms $u\colon x' \to x$, $v, v'\colon x \to y$ and $w\colon y \to y'$, if $v \approx v'$ then $uvw \approx uv'w$:

$$x \underset{v'}{\overset{v}{\rightrightarrows}} y \ (\wr\wr) \qquad \text{implies} \qquad x' \xrightarrow{u} x \underset{v'}{\overset{v}{\rightrightarrows}} y \xrightarrow{w} y'.$$

In such a situation, one defines the *quotient category* $C/{\approx}$ as the category whose objects are those of C, whose morphisms are the equivalence classes of morphisms of C, and where composition and identities are induced by those of C.

2.3.2 P-congruence. Given a 2-polygraph P, the *P-congruence* $\approx^P$, also sometimes noted $\overset{*}{\Leftrightarrow}$, is the smallest congruence on P^* such that $u \approx^P v$ for every 2-generator $\alpha\colon u \Rightarrow v$ in P_2.

2.3.3 Presented category. The category $\overline{P}$ *presented* by a 2-polygraph P is the category $\overline{P} = P^*_{\leqslant 1}/{\approx^P}$ obtained by quotienting the category freely generated by the underlying polygraph by the P-congruence, what we usually write $P^*_{\leqslant 1}/P_2$. It can be characterized by the following universal property:

2.3.4 Lemma. *Given any category C and functor $f\colon P^*_{\leqslant 1} \to C$ such that $f(u) = f(v)$ for any 2-generator $\alpha\colon u \Rightarrow v$ in P_2, there exists a unique functor $\overline{f}\colon \overline{P} \to C$ such that $\overline{f} \circ q = f$*

$$\begin{array}{ccc} P^*_{\leqslant 1} & \xrightarrow{f} & C \\ {\scriptstyle q}\downarrow & \nearrow_{\overline{f}} & \\ \overline{P}, & & \end{array}$$

where q is the quotient functor *which is the identity on objects and sends a morphism to its equivalence class under $\approx^P$.*

Note that, with the notations of the above lemma, two morphisms u and v in P_1^* are such that $u \approx^P v$ if and only if $q(u) = q(v)$.

We say that a category C is *presented* by a 2-polygraph P, or that P is a *presentation* of C, when C is isomorphic to $\overline{P}$.

2.3.5 Presenting monoids. Any monoid M can canonically be seen as a category with only one object $\star$, the morphisms of the category being the elements of M, and compositions and identities being given by multiplication and unit of the monoid. This construction extends as a functor **Mon** $\to$ **Cat** from the category of monoids to the category of small categories, which is full and faithful. In the following, by a *presentation of a monoid*, we will always implicitly mean a presentation of the associated category. Those provide a most abundant source of examples of presentations; see Appendix A.

2.3.6 Example. There are exactly two monoids with two elements. They are presented by the following two 2-polygraphs:

$$P = \langle \star \mid a\colon \star \to \star \mid \alpha\colon aa \Rightarrow 1 \rangle,$$
$$Q = \langle \star \mid a\colon \star \to \star \mid \alpha\colon aa \Rightarrow a \rangle.$$

The monoid presented by P is $\mathbb{N}/2\mathbb{N}$; see Example 4.5.2 for details.

2.3.7 Example. The *symmetric group* S_n is the monoid whose elements are bijections on a set with n elements, sometimes also called *permutations*, with composition as multiplication and identities as neutral elements. It admits a presentation by a 2-polygraph P with $P_0 = \{\star\}$, $P_1 = \{a_0, \ldots, a_{n-1}\}$, with each 1-generator having $\star$ as source and target, and the relations in P_2 are

- $a_i a_i \Rightarrow 1$, for $0 \leqslant i < n$,
- $a_i a_{i+1} a_i \Rightarrow a_{i+1} a_i a_{i+1}$, for $0 \leqslant i < n-1$, and
- $a_i a_j \Rightarrow a_j a_i$, for $0 \leqslant i < j < n$ with $i+1 < j$.

A generator a_i corresponds here to the transposition exchanging the ith and the $(i+1)$-th element of the set with n elements, and the reader is encouraged to check for himself that these relations make sense; see §§A.1.19 and C.1.3 for details.

2.3.8 A characterization of presentations. The following lemma characterizes when a 2-polygraph is a presentation of a category C. In practice, it is quite cumbersome to use and more practical tools will be introduced in Chapter 4, based on rewriting. We recall that the underlying 1-polygraph VC of a category C was defined in §2.1.1.

2.3.9 Lemma. *A* 2*-polygraph P is a presentation of a category C if and only if there is a morphism of* 1*-polygraphs* $f\colon P_{\leqslant 1} \to VC$ *such that the following three conditions hold.*

1. *The map* $f_0\colon P_0 \to C_0$ *is a bijection between* 0*-generators and objects of* C*.*
2. *For any generator* $\alpha\colon u \Rightarrow v$ *in* P_2*,* $f(u) = f(v)$*.*
3. *The function* $f_1^*\colon P_1^* \to C_1$ *induces a bijection between* P_1^*/P_2 *and* C_1*.*

Proof. By postcomposition with the counit $(VC)^* \to C$ of the adjunction described in §2.1.2, the functor $f^*\colon P_{\leqslant 1}^* \to (VC)^*$ induces a functor $P_{\leqslant 1}^* \to C$ that we still write f^*. The second condition amounts to require that it induces, by Lemma 2.3.4, a well-defined quotient functor $\overline{f^*}\colon \overline{P} \to C$. The first condition amounts to require that this functor is bijective on objects, and the third that it is full and faithful. □

2.3.10 Models. Given categories C and S, the *category of models* of C in S is the category

$$\mathbf{Mod}_S(C) = \mathbf{Cat}(C, S)$$

of functors from C to S and natural transformations between those. We simply write $\mathbf{Mod}(C)$ in the case $S = \mathbf{Set}$.

2.3.11 Lemma. *If P is a* 2*-polygraph and C category, the category* $\mathbf{Mod}_C(\overline{P})$ *is isomorphic to the category whose objects consist of*

- *a family* $(f_x)_{x \in P_0}$ *of objects of C indexed by* 0*-generators in* P_0*, and*
- *a family* $(f_a\colon f_x \to f_y)_{a:x\to y \in P_1}$ *of morphisms in C indexed by* 1*-generators in* P_1*,*

such that for every 2*-generator* $\alpha\colon a_1 \ldots a_m \Rightarrow b_1 \ldots b_n$ *in* P_2*, we have*

$$f_{a_1} \cdots f_{a_m} = f_{b_1} \cdots f_{b_n}$$

and a morphism ϕ between two objects f and g consists of a family

$$(\phi_x\colon f_x \to g_x)_{x \in P_0}$$

of morphisms of C indexed by the 0*-generators of P such that, for every* 1*-generator* $a\colon x \to y \in P_1$*, the following diagram commutes*

$$\begin{array}{ccc} f_x & \xrightarrow{\phi_x} & g_x \\ {\scriptstyle f_a}\downarrow & & \downarrow{\scriptstyle g_a} \\ f_y & \xrightarrow[\phi_y]{} & g_y. \end{array}$$

2.3.12 Example. Consider the category C with two objects X and Y, and a single morphism in each hom-set:

$$1_X \circlearrowleft X \underset{G}{\overset{F}{\rightleftarrows}} Y \circlearrowright 1_G.$$

This category admits a presentation by the 2-polygraph

$$P = \langle\, x, y \mid a\colon x \to y, b\colon y \to x \mid \alpha\colon ab \Rightarrow 1_x, \beta\colon ba \Rightarrow 1_y \,\rangle,$$

which can be shown by using Lemma 2.3.9. We define a morphism of 1-polygraphs $f\colon P_{\leqslant 1} \to VC$ by

$$f(x) = X, \qquad f(y) = Y, \qquad f(a) = F, \qquad f(b) = G$$

and check the conditions of Lemma 2.3.9.

1. The map $f_0\colon \{x, y\} \to \{X, Y\}$ is a bijection.
2. The map f^* preserves the 2-generator $\alpha\colon ab \Rightarrow 1_x$:

   $$f^*(ab) = f^*(a)f^*(b) = FG = 1_X = 1_{f^*(x)} = f^*(1_x).$$

   And similarly for β,
3. The morphisms of P_1^* are of the form

   $$(ab)^n\colon x \to x, \quad (ab)^n a\colon x \to y, \quad (ba)^n b\colon y \to x, \quad (ba)^n\colon y \to y,$$

   for some $n \in \mathbb{N}$, and those are respectively equivalent to

   $$1_x\colon x \to x, \qquad a\colon x \to y, \qquad b\colon y \to x, \qquad 1_y\colon y \to y,$$

   by induction on $n \in \mathbb{N}$, because the presence of α and β respectively imply that we have $ab \approx 1_x$ and $ba \approx 1_y$. These are distinct (they have different types) and are in bijection with the morphisms of C (there is one in each hom-set), in a way compatible with source and target.

By Lemma 2.3.11, a model of C in a category S consists of

- two objects f_x and f_y of S,
- two morphisms $f_a\colon f_x \to f_y$ and $f_b\colon f_y \to f_x$ of S,
- $f_b \circ f_a = 1_{f_x}$ and $f_a \circ f_b = 1_{f_y}$.

The algebras of C in S are thus precisely the isomorphisms in S. Otherwise said, the category C represents the functor **Cat** $\to$ **Set** sending a category to its set of isomorphisms. For this reason, C is sometimes called the *walking isomorphism*.

2.3.13 Free categories. A category C is *free* when it admits a presentation by a 2-polygraph P which has no relations (i.e., $P_2 = \emptyset$): in this case, the category C can be obtained as the category freely generated by the underlying 1-polygraph of P. Contrary to the case of sets, see Lemma 1.2.11, not every category is free: the relations are really needed in order to have a presentation for every category.

Consider $\mathbb{Z}$ as the category with one object $\star$, with the morphisms being the integers with composition given by addition and identity by zero. Suppose given a presentation without relations of this category. This presentation necessarily has exactly one 0-generator $\star$, and at least two generators a, b: if there was zero (resp. one) generator, the presented category would be the terminal one (resp. the one corresponding to the additive monoid $\mathbb{N}$). Writing P for the underlying polygraph of the presentation, there is an isomorphism $f\colon P^* \to C$. Since $\mathbb{Z}$ is abelian, we have $f(a)f(b) = f(b)f(a)$ and therefore $ab = ba$ in P^*, which does not actually hold in P^*. By contradiction, every presentation of $\mathbb{Z}$ has at least one relation, i.e., $\mathbb{Z}$ is not free. An actual presentation of $\mathbb{Z}$ is given in §A.1.14. For similar reasons, the category corresponding to the monoid $\mathbb{N} \times \mathbb{N}$ (or in fact any abelian monoid, excepting $\mathbb{N}$ and the trivial monoid) is not free.

2.3.14 Canonical and standard presentations. Any category admits a presentation, in the following way. Suppose given a category C and write $P = VC$ for its underlying 1-polygraph, whose 0-generators are the objects of C and 1-generators are the morphisms of C; see §2.1.1. The identity morphism $1_P\colon P \to VC$ extends, by Lemma 2.1.3, as a functor $1_P^*\colon P^* \to C$, which sends a 1-cell in P^*, i.e., a formal composite of morphisms in C, to the result of its composition. The 2-polygraph with P as underlying 1-polygraph and whose set of 2-generators is

$$P_2 = \left\{(u, v) \in P_1^* \times P_1^* \mid 1_P^*(u) = 1_P^*(v)\right\},$$

with $s_1(u, v) = u$ and $t_1(u, v) = v$, is a presentation of C called its *canonical presentation*.

If, in the previous presentation, we restrict the set P_2 to the 2-generators which are either of the form (ab, c) or of the form $(1_x, a)$, with $a, b, c \in P_1$, we obtain another presentation of C, which is smaller, called its *standard presentation*, detailed in §4.5.5.

2.4 Generating 2-Categories

In the same way a 1-polygraph generates a category which is a "graph with compositions", a 2-polygraph also generates a 2-category which is a "2-graph with compositions". Here, a 2-graph consists of a graph together with "2-cells" which have edges as source and target.

2.4.1 2-graphs. A 2-*graph* C, or 2-*globular set*,

$$C_0 \underset{t_0}{\overset{s_0}{\leftleftarrows}} C_1 \underset{t_1}{\overset{s_1}{\leftleftarrows}} C_2$$

consists in sets

- C_0 of 0-*cells*,
- C_1 of 1-*cells* together with functions $s_0, t_0 \colon C_1 \to C_0$ respectively associating to each 1-cell its source and target 0-cell, and
- C_2 of 2-*cells* together with functions $s_1, t_1 \colon C_2 \to C_1$ respectively associating to each 2-cell its source and target 1-cell,

such that

$$s_0 \circ s_1 = s_0 \circ t_1, \qquad\qquad t_0 \circ s_1 = t_0 \circ t_1.$$

We often write $a \colon x \to y$ for a 1-cell a with $s_0(a) = x$ and $t_0(a) = y$, and

$$\alpha \colon a \Rightarrow b \colon x \to y,$$

for a 2-cell α with

$$s_1(\alpha) = a, \quad t_1(\alpha) = b, \quad s_0(a) = s_0(b) = x, \quad \text{and} \quad t_0(a) = t_0(b) = y.$$

Any 2-graph has an *underlying* 1-*graph* with C_0 as vertices and C_1 as edges.

2.4.2 Example. The 2-graph C with

$$C_0 = \{x, y, z\},$$
$$C_1 = \{a \colon x \to y, b_1 \colon y \to z, b_2 \colon y \to z, b_3 \colon y \to z\}, \text{ and}$$
$$C_2 = \{\alpha \colon b_1 \Rightarrow b_2, \beta \colon b_2 \Rightarrow b_3\},$$

can be depicted as

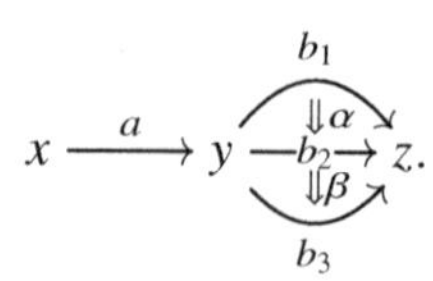

2.4.3 2-categories. A 2-*category* consists in a 2-graph C together with

- for each 0-cell x an *identity* 1-cell

$$1_x : x \to x,$$

- for each 1-cells $f : x \to y$ and $g : y \to z$, a *horizontal composite* 1-cell

$$f *_0 g : x \to z,$$

- for each 2-cells $\alpha : f \Rightarrow f' : x \to y$ and $\beta : g \Rightarrow g' : y \to z$, a *horizontal composite* 2-cell

$$\alpha *_0 \beta : (f *_0 g) \Rightarrow (f' *_0 g') : x \to z,$$

- for each 1-cell $f : x \to y$, an *identity* 2-cell

$$1_f : f \Rightarrow f,$$

- for each 2-cells $\alpha : f \Rightarrow g : x \to y$ and $\beta : g \Rightarrow h : x \to y$, a *vertical composite* 2-cell

$$\alpha *_1 \beta : f \Rightarrow h : x \to y,$$

such that

- the compositions $*_0$ and $*_1$ are associative and admit identities as neutral elements,
- the *exchange law* holds: given 1-cells $f : x \to y$ and $g : y \to z$, one has

$$1_f *_0 1_g = 1_{f *_0 g}, \tag{2.3}$$

  and given 2-cells

$$\alpha : f \Rightarrow f' : x \to y, \qquad \alpha' : f' \Rightarrow f'' : x \to y,$$
$$\beta : g \Rightarrow g' : y \to z, \qquad \beta' : g' \Rightarrow g'' : y \to z,$$

  we have

$$(\alpha *_1 \alpha') *_0 (\beta *_1 \beta') = (\alpha *_0 \beta) *_1 (\alpha' *_1 \beta'). \tag{2.4}$$

  Graphically, the equations (2.3) and (2.4) can be pictured as the "commutation" of the diagrams

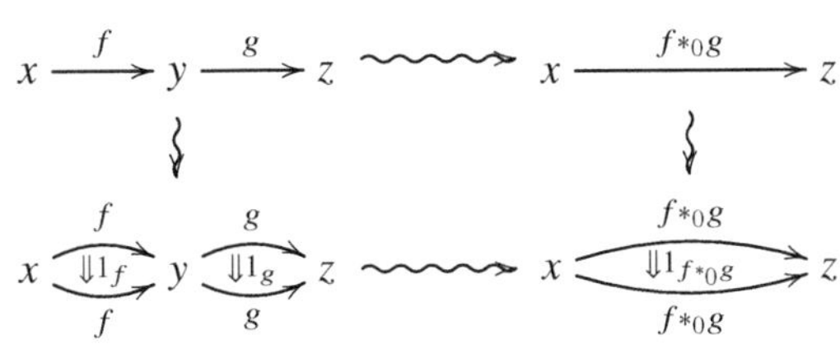

and

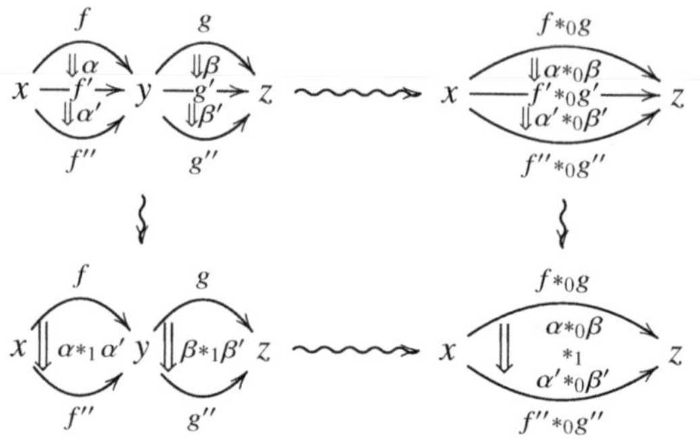

Note that every 2-category C has an underlying category, sometimes denoted by $C_{\leqslant 1}$, with C_0 as set of objects and C_1 as set of morphisms.

2.4.4 Notation. In the following, when considering morphisms in 2-categories, we often omit writing identities and horizontal compositions $*_0$, and write $*$ for vertical composition $*_1$: for instance, we write

$$u\alpha w * \beta \qquad \text{instead of} \qquad (1_u *_0 \alpha *_0 1_w) *_1 \beta.$$

2.4.5 2-functors. A *2-functor* $f\colon C \to D$ between 2-categories C and D consists of functions

$$f_0\colon C_0 \to D_0, \qquad f_1\colon C_1 \to D_1, \qquad f_2\colon C_2 \to D_2,$$

which are

- compatible with sources and targets:

$$s_0 \circ f_1 = f_0 \circ s_0, \qquad t_0 \circ f_1 = f_0 \circ t_0,$$
$$s_1 \circ f_2 = f_1 \circ s_1, \qquad t_1 \circ f_2 = f_1 \circ t_1,$$

- compatible with compositions: for every composable pair of 1-cells (u, v) (resp. 2-cells (α, β)),

$$f_1(u *_0 v) = f_1(u) *_0 f_1(v), \qquad f_2(\alpha *_0 \beta) = f_2(\alpha) *_0 f_2(\beta),$$
$$f_2(\alpha *_1 \beta) = f_2(\alpha) *_1 f_2(\beta),$$

- compatible with identities: for every 0-cell x and 1-cell u,

$$f_1(1_x) = 1_{f_0(x)}, \qquad f_2(1_u) = 1_{f_1(u)}.$$

In the following, we generally omit writing the subscript i from the components f_i of a 2-functor f.

2.4.6 Freely generated 2-categories. The *free 2-category* P^* generated by a 2-polygraph P is the 2-category C with $P^*_{\leqslant 1}$ as underlying category, i.e.,

- $C_0 = P_0$,
- $C_1 = P_1^*$ with source and target given by $s_0^C = s_0^*$ and $t_0^C = t_0^*$,

whose 2-cells are freely generated by P_2, i.e., C_2 is the smallest set containing P_2, together with source and target given by the functions s_1 and t_1 of P, such that

- for every $\alpha, \beta \in C_2$ such that $t_0 \circ t_1(\alpha) = s_0 \circ s_1(\beta)$, there is an element $\alpha *_0 \beta \in C_2$,
- for every $\alpha, \beta \in C_2$ such that $t_1(\alpha) = s_1(\beta)$, there is an element $\alpha *_1 \beta \in C_2$,

quotiented by the axioms required to form a 2-category; see §2.4.3. We write P_2^* for the set C_2 of 2-cells of P^*.

Another construction for this 2-category will be presented in §4.1.8. It can be characterized by the following universal property:

2.4.7 Lemma. *Suppose given a 2-polygraph P, a 2-category C, and functions*

$$f_0 \colon P_0 \to C_0, \qquad f_1 \colon P_1 \to C_1, \qquad f_2 \colon P_2 \to C_2,$$

such that

- *for every 1-generator $a \colon x \to y$, we have $f_1(a) \colon f_0(x) \to f_0(u)$,*
- *for every 2-generator $\alpha \colon u \Rightarrow v$, we have $f_2(\alpha) \colon f_1^*(u) \Rightarrow f_1^*(v)$,*

where $f_1^*(a_1 \dots a_n) = f_1(a_1) \dots f_1(a_n)$. Then there exists a 2-functor

$$f^* : P^* \to C,$$

which is unique up to isomorphism, such that

- *for every 0-generator $x \in P_0$, $f^*(x) = f_0(x)$,*
- *for every 1-generator $a \in P_1$, $f^*(a) = f_1(a)$,*
- *for every 2-generator $\alpha \in P_2$, $f^*(\alpha) = f_2(\alpha)$.*

2.4.8 String diagrams. A convenient and intuitive notation for morphisms in free 2-categories is provided by *string diagrams*. Those were originally introduced by Feynman [127] and Penrose [297] in physics, and formally studied by Joyal and Street [208]; see [24] for a detailed historical account and [324] for a panorama of the possible variations.

Suppose fixed a 2-polygraph P. A 2-generator

$$\alpha \colon a_1 \dots a_m \Rightarrow b_1 \dots b_n,$$

where the $a_i : x_{i-1} \to x_i$ and $b_i : y_{i-1} \to y_i$ are 1-generators, can be thought of as some kind of device with m inputs, with a_i as types, and n outputs, with b_i as types. This suggests that, instead of using the usual depiction

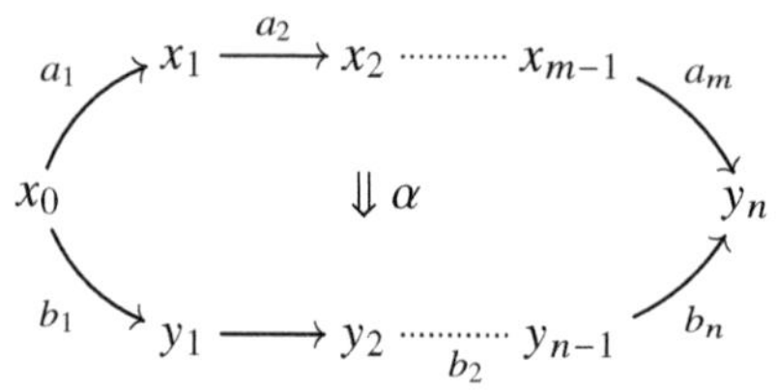

we can use the alternative graphical notation

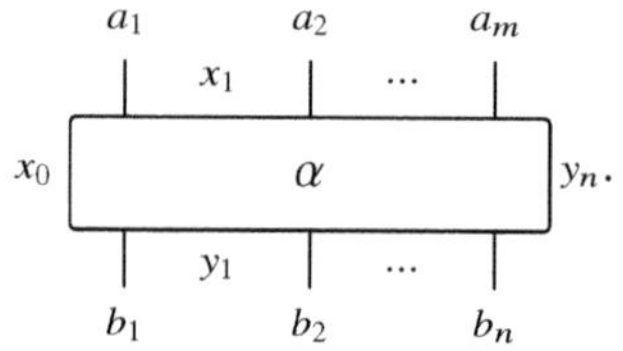

This notation, called a *string diagram* because the 1-cells become some kind of strings, is inspired of electric circuits where electronic components are linked with conductive wires. Namely, a 2-cell will more generally be depicted as various gates linked together, for instance:

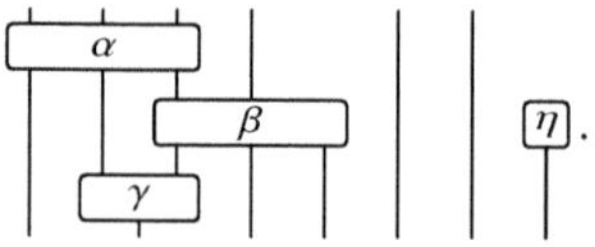

This notation is "dual" (in the sense of Poincaré duality) to the traditional one: 0-cells are pictured as 2-dimensional region of the plane, 1-cells are pictured as wires, and 2-cells are pictured as points (the above rectangles, which are here in order to be able to put labels inside them, should be pictured as points if we were drawing string diagrams by the book).

String diagrams can be composed in the various ways expected in a 2-category, as we now describe. Horizontal composition of 2-cells of P^*

$$\phi : a_1 \ldots a_m \Rightarrow a'_1 \ldots a'_{m'} \qquad \text{and} \qquad \psi : b_1 \ldots b_n \Rightarrow b'_1 \ldots b'_{n'}$$

amounts to horizontal juxtaposition of the corresponding diagrams, as shown on the left

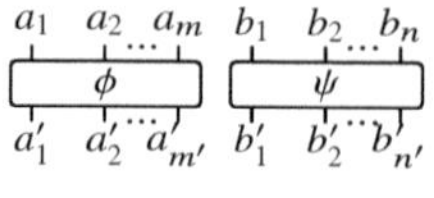

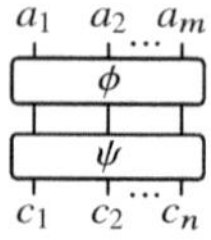

and vertical composition of

$$\phi\colon a_1 \dots a_m \Rightarrow b_1 \dots b_k \qquad \text{and} \qquad \psi\colon b_1 \dots b_k \Rightarrow c_1 \dots c_n$$

is obtained by vertically juxtaposing the corresponding diagrams and linking the wires, as shown on the right above, identities simply being wires. These diagrams are to be considered up to "planar isotopy", by which we mean continuous deformations, fixing boundaries, preserving direction, and forbidding wires to cross. For instance, the three diagrams below are considered to be equal:

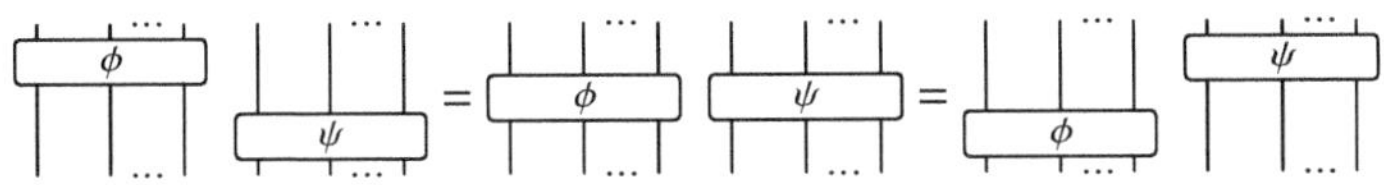

This is detailed in [208], where it is proved that the axioms of 2-categories are satisfied and the following universal property is satisfied:

2.4.9 Theorem. *String diagrams (up to planar isotopy) are the* 2*-cells of a* 2*-category which is the free* 2*-category P^* on the* 2*-polygraph P.*

Given a 0-cell x, the string diagram corresponding to the 2-cell 1_{1_x} (the identity on the identity on x) is the *empty* one. Since this can be confusing, we generally use the diagram

in order to represent it.

2.4.10 Strict monoidal categories. We introduce here useful categorical structures, which can be seen as particular cases of 2-categories.

A *strict monoidal category* $(C, \otimes, i)$ consists of a category C together with a *tensor product* functor $\otimes\colon C \times C \to C$ and a *unit* object $i \in C$, such that the product is associative and admits i as unit, which means that for any objects $x, y, z \in C$, we should have

$$(x \otimes y) \otimes z = x \otimes (y \otimes z) \qquad \text{and} \qquad i \otimes x = x = x \otimes i.$$

Any 2-category with only one 0-cell induces a strict monoidal category with C_1 as set of objects, C_2 as set of morphisms, with composition being given by $*_1$ and tensor product by $*_0$, and this extends to an isomorphism between 2-categories with one fixed 0-cell and strict monoidal categories: in this way strict monoidal categories can be seen as particular cases of 2-categories (and we will often implicitly make use of this identification in the following).

Among strict monoidal categories, the following class is a source of particularly useful examples (see Chapter 10 and Appendix C). A *PRO* is a strict monoidal category whose set of objects is $\mathbb{N}$ and such that the tensor product of two objects m and n is given by addition $m \otimes n = m + n$. The terminology is an abbreviation of "PROduct category" [262].

2.4.11 Bicategories. The axioms for 2-categories are sometimes too strong for situations encountered in practice: it happens in many situations that composition of 1-cells is not strictly associative, but only associative up to a coherent invertible 2-cell, and similarly identity 1-cells are not strictly neutral elements for composition. This motivates the introduction of the following generalization of the notion of 2-category.

A *bicategory* C consists in a 2-graph C together with identity 1-cells and 2-cells, horizontal composition of 1- and 2-cells, vertical composition of 2-cells, as in §2.4.3, and moreover natural families of invertible 2-cells

$$\alpha_{f,g,h}\colon (f *_0 g) *_0 h \Rightarrow f *_0 (g *_0 h), \qquad \lambda_f\colon 1_x *_0 f \Rightarrow f,$$
$$\rho_g\colon f *_0 1_y \Rightarrow f,$$

indexed by 1-cells $f\colon x \to y$, $g\colon y \to z$, and $h\colon z \to w$, respectively called *associator* and *left* and *right unitors*, such that

- vertical composition $*_1$ is associative on 2-cells and admits identities as neutral elements,
- the exchange law between horizontal and vertical composition holds, and
- the two following coherence axioms hold for every composable 1-cell f, g, h, and i:

$$\begin{array}{ccccc} & & (f *_0 g) *_0 (h *_0 i) & & \\ & \overset{\alpha_{f *_0 g,h,i}}{\nearrow} & & \overset{\alpha_{f,g,h *_0 i}}{\searrow} & \\ ((f *_0 g) *_0 h) *_0 i & & & & f *_0 (g *_0 (h *_0 i)) \\ & \underset{\alpha_{f,g,h} *_0 i}{\searrow} & & \underset{f *_0 \alpha_{g,h,i}}{\nearrow} & \\ & (f *_0 (g *_0 h)) *_0 i & \xRightarrow[\alpha_{f,g *_0 h,i}]{} & f *_0 ((g *_0 h) *_0 i) & \end{array}$$

$$\begin{array}{ccc} (f *_0 1_y) *_0 g & \xRightarrow{\alpha_{f,1_y,g}} & f *_0 (1_y *_0 g). \\ & \underset{\rho_f *_0 g}{\searrow} \quad \underset{f *_0 \lambda_g}{\swarrow} & \\ & f *_0 g & \end{array}$$

Obviously, any 2-category can be seen as a bicategory where $\alpha_{f,g,h}$, λ_f and ρ_f are all identity 2-cells. A typical non-trivial example of bicategory is the bicategory $\mathbf{Span}(C)$ of spans in a category C with pullbacks; see §3.3.11.

An important particular case of the previous construction is the following one. We have seen in §2.4.10 that a 2-category with one 0-cell corresponds to a strict monoidal category. Similarly, a bicategory with one 2-cell corresponds to the notion of *monoidal category*, which generalizes the one of strict monoidal category. The definition is recalled in §12.4.1, and more details can be found in [261, Chapter VII].

A fundamental theorem for bicategories is the coherence theorem which ensures that it can always be "replaced" by a 2-category:

2.4.12 Theorem. *Every bicategory is biequivalent to a 2-category.*

This result is shown for instance in [303] and a proof of coherence for monoidal categories, which are equivalent to bicategories with only one 0-cell, is given in Section 12.4.

2.5 Coherent Confluence of 1-Polygraphs

In §2.1.4, we have seen that the morphisms in the free groupoid generated by a 1-polygraph could be seen as representatives of the congruence associated to the polygraph. A similar construction can be performed for 2-polygraphs as follows, from which we will be able to define a notion of coherent presentation. Here, rather than presenting a category (as in Section 2.3) or generating a 2-category (as in Section 2.4), we think of a 2-polygraph as a presentation of a set by the underlying 1-polygraph (see Section 1.2) together with additional 2-dimensional coherence data provided by the 2-generators.

2.5.1 Freely generated (2,1)-category. A *(2,1)-category* C is a 2-category in which every 2-cell is invertible, that is, for every 2-cell $\alpha : u \Rightarrow v$ in C there exists a 2-cell $\alpha^- : v \Rightarrow u$ in C such that $\alpha^- *_1 \alpha = 1_u$, and $\alpha *_1 \alpha^- = 1_v$.

Given a 2-polygraph P, with underlying 1-polygraph $P_{\leqslant 1}$, one can generate the *free* (2,1)*-category* $P^\top$ with $P^*_{\leqslant 1}$ as underlying category and containing the 2-generators in P_2 as 2-cells. This construction can be performed in a similar way as in §2.4.6.

- The set of 0-cells is the set P_0.
- The set of 1-cells is P_1^*.
- The set of 2-cells $P_2^\top$ is the set of formal horizontal and vertical composites of elements of P_2 and identities of elements of P_1^*, and their inverses, quotiented by the axioms of (2,1)-categories.

The following lemma motivates why one can think of 2-cells in this 2-category as witnesses of equivalence of rewriting paths:

2.5.2 Lemma. *Given a 2-polygraph P, two 1-cells $f, g\colon x \to y$ are P-congruent if and only if there exists a 2-cell $\phi\colon f \Rightarrow g$ in $P^\top$.*

For any 2-polygraph P, there is thus a canonical 2-functor $q_P\colon P^\top \to \overline{P}$ from the generated $(2,1)$-category to the presented category, which is the identity on 0-cells, sends every 1-cell to its equivalence class under the P-congruence, and sends every 2-cell to the identity.

Any two presentations of a given category generate equivalent $(2,1)$-categories in the following sense:

2.5.3 Lemma. *Suppose given two 2-polygraphs P and Q both presenting a same category C. There exist two 2-functors*

$$f\colon P^\top \to Q^\top, \qquad\qquad g\colon Q^\top \to P^\top,$$

between the free $(2,1)$-categories generated by P and Q and, for every 1-cell u of $P^\top$ and v of $Q^\top$, there exist 2-cells

$$\phi_u\colon gf(u) \Rightarrow u, \qquad\qquad \psi_v\colon fg(v) \Rightarrow v,$$

in $P^\top$ and $Q^\top$, such that the two following conditions are satisfied.

1. *The 2-functors f and g induce the identity through the canonical projections q_P and q_Q onto C:*

$$\begin{array}{ccc} P^\top & \xrightarrow{f} & Q^\top \\ {\scriptstyle q_P}\downarrow & & \downarrow{\scriptstyle q_Q} \\ C & \xrightarrow[1_C]{} & C, \end{array} \qquad\qquad \begin{array}{ccc} Q^\top & \xrightarrow{g} & P^\top \\ {\scriptstyle q_Q}\downarrow & & \downarrow{\scriptstyle q_P} \\ C & \xrightarrow[1_C]{} & C. \end{array}$$

2. *The 2-cells ϕ_u and ψ_v are functorial in u and v, that is,*

$$\phi_{uu'} = \phi_u *_0 \phi_{u'}, \qquad\qquad \phi_{1_x} = 1_{1_x},$$

for any 0-composable 1-cells u and u' and 0-cell x and

$$\psi_{vv'} = \psi_v *_0 \psi_{v'}, \qquad\qquad \psi_{1_y} = 1_{1_y},$$

for any 0-composable 1-cells v and v' and 0-cell y.

Proof. We construct the 2-functor f, the case of the 2-functor g being similar. For a 0-cell x, we set $f(x) = q_Q^{-1}q_P(x)$. If $a\colon x \to y$ is a 1-generator of P, we choose, in an arbitrary way, a 1-cell $f(a)\colon f(x) \to f(y)$ in $Q^\top$ such that $q_Q f(a) = q_P(a)$. Then, we extend f to every 1-cell of $P^\top$ by functoriality. Let $\alpha\colon u \Rightarrow u'$ be a 2-generator of P. Since P is a presentation of C, we have $q_P(u) = q_P(u')$ so that $q_Q f(u) = q_Q f(u')$ holds. Using the fact that Q is a presentation of C, we arbitrarily choose a 2-cell $f(\alpha)\colon f(u) \Rightarrow f(u')$ in $Q^\top$. Then, we extend f to every 2-cell of $P^\top$ by functoriality.

Now, let us define the 2-cells ϕ_u, the case of 2-cells ψ_v being similar. Let a be a 1-generator of P. By construction of the 2-functors f and g, we have:

$$q_P g f(a) = q_Q f(a) = q_P(a).$$

Since P is a presentation of C, there exists a 2-cell $\phi_a\colon gf(a) \Rightarrow a$ in $P^\top$. We extend ϕ to every 1-cell u of $P^\top$ by functoriality. □

2.5.4 Remark. Condition 2 of the lemma exactly expresses that ϕ and ψ are *icons* (which is the acronym for "identity component oplax natural transformations", i.e., oplax natural transformations with whose 1-cell components are identities) in the sense of [228].

2.5.5 Freely generated 2-groupoid. A 2-*groupoid*, also called $(2,0)$-*category*, is a 2-category where both 1-cells and 2-cells are invertible. As a variant of the previous construction, a 2-polygraph P freely generates a 2-groupoid, whose set of 0-cells is P_0, whose set of 1-cells is $P_1^\top$ (the morphisms of the free 1-groupoid generated by the underlying 1-polygraph, as described in §2.1.4), and whose set of 2-cells is the set of formal horizontal and vertical composites of elements of P_2 and identities of elements of $P_1^\top$, and their inverses, quotiented by the axioms of 2-groupoids.

2.5.6 Coherent 2-polygraphs. Given a 2-polygraph P, the 1-cells $P_1^\top$ of the freely generated 2-groupoid may be thought of as witnesses for equivalence of 0-cells (see §2.1.4) and the 2-cells $P_2^\top$ as witnesses of "equivalences between equivalences" (as a variant of Lemma 2.5.2). A 2-polygraph is said to be coherent when equivalences do not bring essential information, in the sense that there is at most one equivalence between two given 0-cells, up to equivalence between equivalences. Formally, a 2-polygraph P is *coherent* when for every pair of 0-cells $x, y \in P_0$ and pair of 1-cells $f, g\colon x \to y$ in $P_1^\top$ there is a 2-cell $\alpha\colon f \Rightarrow g$ in $P_2^\top$ in the free 2-groupoid generated by P.

The notion of coherence can be generalized in the expected way to variants of 2-polygraphs where the source and target of 2-generators are cells in $P_2^\top$ (as opposed to P_2^*): those are called $(2,0)$-polygraphs and formally defined in Section 15.3. In this context, following the terminology introduced in Section 7.1, we can also say that P_2 is a *cellular extension* of the category $P_1^\top$: the 2-polygraph P is then coherent precisely when this extension is *acyclic*.

2.5.7 Coherent confluence. Our aim here is to provide techniques to show the coherence of a 2-polygraph by adapting the rewriting concepts presented in Chapter 1 in order to take 2-dimensional information into account, as an explicit witness of the commutation of diagrams. This approach first appeared under the terminology of *commuting diagrams* [190, Section 4.3].

A 2-polygraph P is *coherently confluent* when for every pair of coinitial 1-cells $f: x \to y$ and $f': x \to y'$ in P_1^* there is a pair of cofinal 1-cells $g: y \to z$ and $g': y' \to z$ in P_1^* and a 2-cell $\alpha: fg \Rightarrow f'g'$ in $P_2^\top$ (in the free $(2,1)$-category it generates):

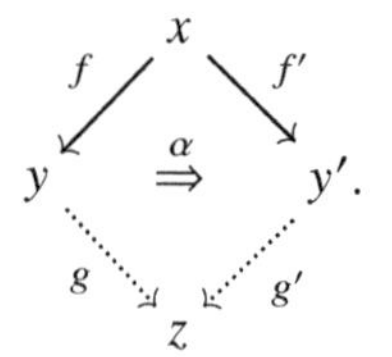

(2.5)

In such a situation, we also say that the branching (f, f') is *coherently joinable*. Similarly, P is *locally coherently confluent* when for every pair of coinitial 1-generators $a: x \to y$ and $b: x \to y'$ in P_1 there is a pair of cofinal 1-cells $g: y \to z$ and $g': y' \to z$ in P_1^* and a 2-cell $\alpha: ag \Rightarrow bg'$ in $P_2^\top$:

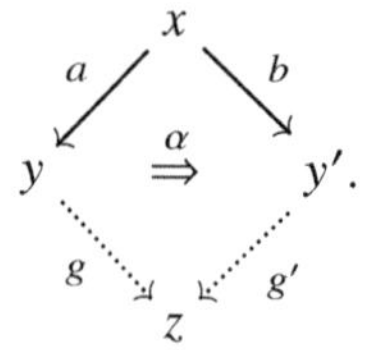

We say that a 2-polygraph is *terminating* when the underlying 1-polygraph is, in the sense of §1.3.6. We can now adapt Newman's lemma (see Lemma 1.3.21) to this extended notion of confluence as follows:

2.5.8 Lemma. *A terminating 2-polygraph is coherently confluent if and only if it is locally coherently confluent.*

Proof. The left-to-right implication is immediate; let us show the right-to-left implication. Supposing that the terminating polygraph P is locally coherently confluent, we show that it is coherently confluent at x (i.e., that every branching at x is coherently joinable) by well-founded induction on x. Suppose given two coinitial 1-cells $f: x \to y$ and $f': x \to y$. If one of them is the identity, say f', we can close the diagram as in (2.5) with $g = 1_y$, $g' = f$, and $\alpha = 1_f$. Otherwise, we have $f = af_1$ and $f' = a'f_1'$ for some 1-generators a and a' and 1-cells f and f':

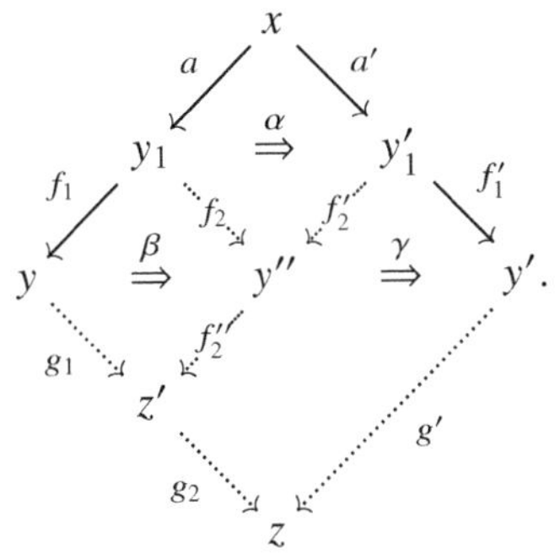

By local confluence we deduce the existence of 1-cells f_2 and f_2 and a 2-cell $\alpha\colon af_2 \Rightarrow af_2'$ as above. By induction hypothesis, we obtain the existence of 1-cells g_1 and f_2'' together with a 2-cell $\beta\colon f_1g_1 \Rightarrow f_2f_2''$. By induction hypothesis again, we deduce the existence of 1-cells g_2 and g' and a 2-cell $\gamma\colon f_2'f_2''g_2 \Rightarrow f_1'g'$. We have shown the existence of cofinal 1-cells $g_1g_2\colon y \to z$ and $g'\colon y' \to z$, and of a 2-cell

$$a\beta g_2 * \alpha f_2'' g_2 * a'\gamma\colon af_1g_1g_2 \Rightarrow a'f_1'g'$$

and the branching $(af_1, a'f_1')$ is thus coherently joinable. □

2.5.9 Proposition. *A terminating and coherently confluent 2-polygraph P is coherent.*

Proof. Since the polygraph is terminating, for every 0-cell x there is a normal form $\widehat{x}$ and a rewriting path $n_x\colon x \to \widehat{x}$ in P_1^*. Given a 1-cell $f\colon x \to y$ in P_1^*, since the polygraph is convergent we have $\widehat{x} = \widehat{y}$ and, since the branching (n_x, fn_y) is coherently joinable there is a 2-cell $\alpha_f\colon fn_y \Rightarrow n_x$, as on the left:

In the free 2-groupoid generated by P, we also define the 2-cell $\alpha_{f^-} = f^- *_0 \alpha_f^-$, which can be pictured as on the right above. Now consider a 1-cell $f\colon x \to y$ in $P_1^\top$. It factors as

$$f = f_1^- g_1 f_2^- g_2 \ldots f_k^- g_k,$$

for some suitably composable morphisms f_i and g_i in P_1^*. We then define $\alpha_f\colon fn_y \Rightarrow n_x$ as the composite

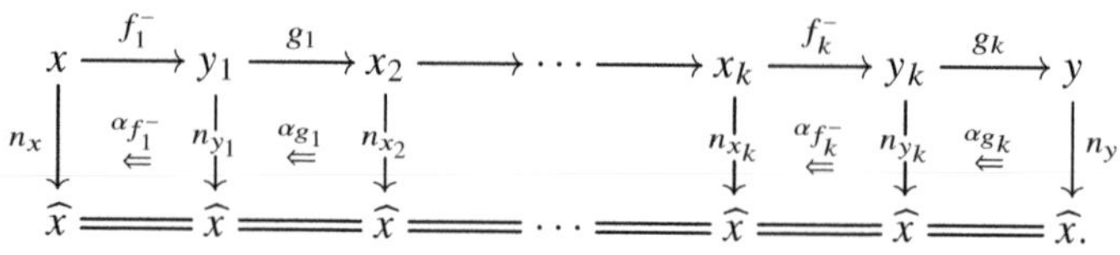

Finally, for any pair of parallel 1-cells $f, g\colon x \to y$ in $P_1^\top$, the composite 2-cell

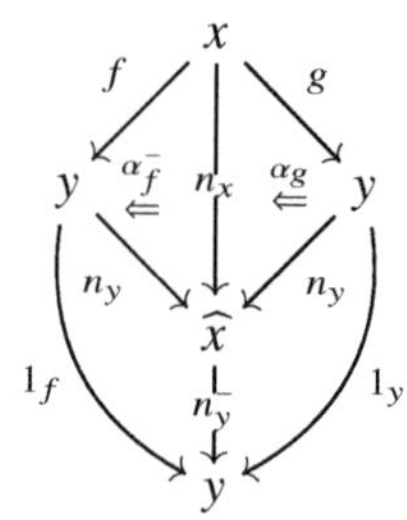

has type $f \Rightarrow g$. We thus conclude that the 2-polygraph P is coherent. □

Composing Lemma 2.5.8 and Proposition 2.5.9, we obtain the *coherent Squier theorem* for 1-polygraphs [328, Theorem 5.2]:

2.5.10 Theorem. *Let P be a terminating 2-polygraph. If, for every pair of coinitial 1-generators $a\colon x \to y_1$ and $b\colon x \to y_2$ in P_1, there is a pair of cofinal 2-cells $f\colon y_1 \to z$ and $g\colon y_2 \to z$ in P_1^* and a 2-cell $\phi\colon af \Rightarrow bg$ in $P_2^\top$,*

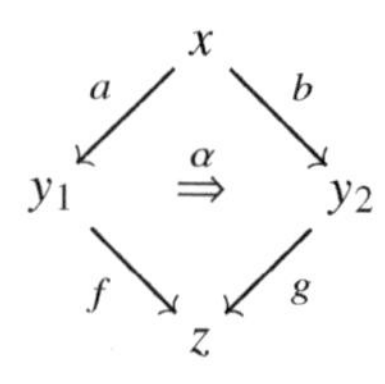

then P is coherent.

This theorem is extended to 2-polygraphs in Chapter 7 and to higher-dimensional polygraphs in Chapter 23: we will see that it provides us with a canonical way of extending a convergent presentation into a coherent one using the homotopical completion procedure; see Section 7.5.

3

Operations on Presentations

The usefulness and richness of 2-polygraphs is confirmed by the large number and variety of categories they present. Some examples of presentations were given in Section 2.3 and many more are described in Appendix A. In order to show that a given polygraph is a presentation of a given category, one can either tackle the issue directly, by using the rewriting tools of Chapter 4, or take a modular approach, by combining already known presentations: this is the route taken in the present chapter.

Three significant applications are given. We first address, in Section 3.1, the presentation of limits and colimits by means of given presentations of the base categories and precisely show how to systematically build presentations of products, coproducts, and pushouts. Next, in Section 3.2, we show how to add formal inverses to some morphisms of a category at the level of presentations. Finally, Section 3.3 is about distributive laws, in relation to factorization systems on categories. We introduce a notion of composition along a distributive law between two small categories sharing the same set of objects and show how to derive a presentation of this composite from presentations of the components.

3.1 Limits and Colimits of Presented Categories

3.1.1 Initial category. The initial category (with no object nor morphism) admits the presentation $\langle\ |\ |\ \rangle$.

3.1.2 Coproducts of categories. Given two categories C and D respectively presented by 2-polygraphs P and Q, their coproduct $C \sqcup D$ is presented by the 2-polygraph with $P_0 \sqcup Q_0$ as 0-generators, $P_1 \sqcup Q_1$ as 1-generators and $P_2 \sqcup Q_2$ as 2-generators, with expected source and target maps (this is the coproduct of P and Q in the category of 2-polygraphs).

3.1.3 Coproducts with fixed objects. Given a fixed set O, consider the subcategory $\mathbf{Cat}_O$ of $\mathbf{Cat}$ whose objects are categories with O as objects, and whose morphisms are functors which are identities on objects. Given two categories C and D in $\mathbf{Cat}_O$, respectively presented by 2-polygraphs P and Q, their coproduct in $\mathbf{Cat}_O$ is presented by the 2-polygraph with $P_0 = Q_0 = O$ as 0-generators, $P_1 \sqcup Q_1$ as 1-generators, and $P_2 \sqcup Q_2$ as 2-generators, with expected source and target maps. In particular, when $O = \{\star\}$, the category $\mathbf{Cat}_O$ is isomorphic to the category of monoids and their coproduct is the *free product*.

3.1.4 Coequalizers. Suppose given two categories C and D respectively presented by 2-polygraphs P and Q and two functors

$$f, g \colon C \to D.$$

Reformulating the results of [37], the category obtained as their coequalizer is presented by the following 2-polygraph R. We write $\sim$ for the smallest equivalence relation on Q_0 such that $f(x) \sim g(x)$ for every 0-generator $x \in P_0$, and $[y]$ for the equivalence class of a 1-cell $y \in Q_0$. The coequalizer of f and g is the category presented by the polygraph R such that

- the 0-generators are the equivalence classes $[x]$ of 0-generators in Q_0,
- the 1-generators are of the form $a\colon [x] \to [y]$ for 1-generators $a\colon x \to y$ in Q_1, and
- the 2-generators are either of the form
  - $\alpha\colon u \Rightarrow v$ for 2-generators $\alpha\colon u \Rightarrow v$ in Q_2, or
  - $\alpha_a\colon u \Rightarrow v$ for 1-generators $a\colon x \to y$ in P_1 such that $f(a) = \overline{u}$ and $g(a) = \overline{v}$.

This construction allows to quotient a presented category both on objects and morphisms. Note that the relations in P_2 are not used in this construction; we only need a generating graph for C.

3.1.5 Pushouts. All finite colimits of presented categories can be constructed from coproducts and coequalizers [261, Section V.2]. For instance, the pushout of the diagram of categories on the left

$$C \xleftarrow{\;f\;} B \xrightarrow{\;g\;} D, \qquad\qquad B \underset{i_D \circ g}{\overset{i_C \circ f}{\rightrightarrows}} C + D,$$

can be computed as the coequalizer of the diagram on the right, where the morphisms $i_C\colon C \to C + D$ and $i_D\colon D \to C + D$ are the canonical inclusions. In particular, when both f and g are inclusions of polygraphs, the pushout is simply given by (non-disjoint) union on sets of n-generators for $n = 0, 1, 2$.

3.1.6 Terminal category. We now turn to limits. The most simple example is the terminal category which admits the presentation $\langle \star \mid \mid \rangle$.

3.1.7 Products. Suppose given two categories C and D along with respective presentations by 2-polygraphs P and Q. A presentation R of the product category $C \times D$ is given by the 2-polygraph R with

- $R_0 = P_0 \times Q_0$ as set of 0-generators,
- $R_1 = P_1 \times Q_0 \sqcup P_0 \times Q_1$ as set of 1-generators with

$$(a, y) \colon (x, y) \to (x', y), \qquad\qquad (x, b) \colon (x, y) \to (x, y'),$$

  with $a \colon x \to x'$ in P_1 and y in Q_0 (resp. x in P_0 and $b \colon y \to y'$ in Q_1),
- $R_2 = P_2 \times Q_0 + P_1 \times Q_1 + P_0 \times Q_2$ as set of 2-generators: a 2-generator is either

$$(\alpha, y) \colon (u, y) \Rightarrow (u', y) \colon (x, y) \to (x', y),$$

  with $\alpha \colon u \Rightarrow u' \colon x \to x'$ in P_2 and $y \in Q_0$, or

$$(a, b) \colon (x, b)(a, y') \Rightarrow (a, y)(x', b) \colon (x, y) \to (x', y'),$$

  with $a \colon x \to x'$ in P_1 and $b \colon y \to y'$ in Q_1, or

$$(x, \beta) \colon (x, v) \Rightarrow (x, v') \colon (x, y) \to (x, y'),$$

  with $x \in P_0$ and $\beta \colon v \Rightarrow v' \colon y \to y'$ in Q_2.

Above, given $u = a_1 a_2 \ldots a_n$ in P_1^* (with the a_i being generators in P_1) and $y \in Q_0$, the 1-cell (u, y) is a notation for $(u, y) = (a_1, y)(a_2, y) \ldots (a_n, y)$ and the notation (x, v), for $x \in P_0$ and $v \in P_1^*$, is similar.

In particular, when C and D are both monoids (or groups), their product in the above sense is often called their *direct product.*

3.2 Localizations of Presented Categories

3.2.1 Free groupoid. The forgetful functor **Gpd** $\to$ **Cat** witnessing for the fact that a groupoid is a particular category (with invertible morphisms) admits a left adjoint, constructing the *free groupoid* $C^\top$ (also called the *enveloping groupoid*) over a category C. Given a 2-polygraph P presenting C, the groupoid $C^\top$ admits a presentation (as a category) by the 2-polygraph Q with

- $Q_0 = P_0$ as set of 0-generators,

- $Q_1 = P_1 \sqcup P_1^-$ as set of 1-generators, with

$$P_1^- = \{a^- : y \to x \mid a : x \to y \in W\},$$

- $Q_2 = P_2 \sqcup I_2$ as set of 2-generators with

$$I_2 = \left\{aa^- \Rightarrow 1_x, a^-a \Rightarrow 1_y \,\middle|\, a : x \to y \in W\right\},$$

where $W = P_1$.

A morphism in Q_1^* is *reduced* when it is not of the form uaa^-v or ua^-av for some $u, v \in Q_1^*$ and $a \in P_1$. The equivalence classes of elements of Q_1^* modulo the congruence generated by I_2 contain exactly one reduced morphism, which is often convenient to choose as a canonical representative; this is detailed in Example 4.3.13.

3.2.2 Localization. As a generalization of the previous construction, given a category C and a class W of morphisms of C, we can consider the category $C[W^{-1}]$, called the *localization* of C by W, obtained by formally inverting the morphisms of W; see §H.2.1 for a proper definition. Given a category C presented by a 2-polygraph P and a set $W \subseteq P_1$ of 1-generators, the localization $C[\overline{W}^{-1}]$ of C by equivalence classes of elements of W is presented by the polygraph Q defined exactly as in the previous section. In particular, we recover the free groupoid on C as $C^\top = C[P_1^{-1}]$.

3.3 Distributive Laws

In this section, we present a very useful tool in order to build presentations in a modular fashion. The typical situation we want to address here is when the category E we want to present is "built" from two subcategories C and D, in the sense that every morphism of E factors a composite of morphisms in the two subcategories: in this case, we can expect to be able to construct a presentation of E from presentations of C and D. The way the category E can be obtained as a composite of C and D can be encoded in a distributive law. This notion was introduced by Beck [36], related to categories and strict factorization systems by Rosebrugh and Wood [316], and applied to presentations of categories by Lack [230]. We begin by recalling this setting and then presenting the generalizations necessary to handle situations arising in practice.

3.3.1 Strict factorization system. A *strict factorization system* on a category E consists of two subcategories C and D of E, with the same objects as E, such that every morphism h of E factorizes uniquely as $h = g \circ f$ with f in C and g in D:

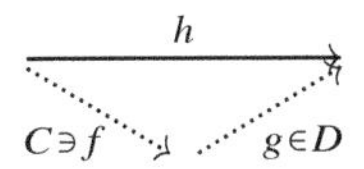

This structure can equivalently be encoded through operations which help expressing every morphism E as one of C composed with one of D, as we now explain.

3.3.2 Distributive law. A *distributive law* ℓ between two categories C and D having the same objects is a function, often noted

$$\ell \colon D \otimes C \to C \otimes D$$

(the notation as a tensor will be formally justified in §3.3.13), which to every "composable" pair of morphisms

$$g \colon x \to y \in D, \qquad f \colon y \to z \in C,$$

associates an object ${}^{g}y^{f}$ of C and D, and morphisms

$$f^{g} \colon x \to {}^{g}y^{f} \in C, \qquad {}^{f}g \colon {}^{g}y^{f} \to z \in D,$$

which can be pictured as

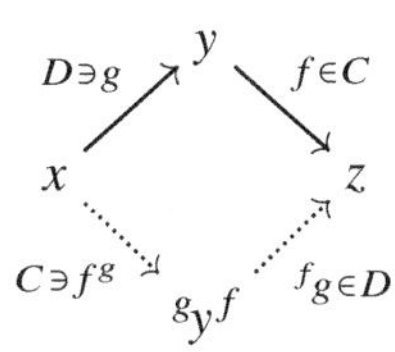

in a way compatible with compositions

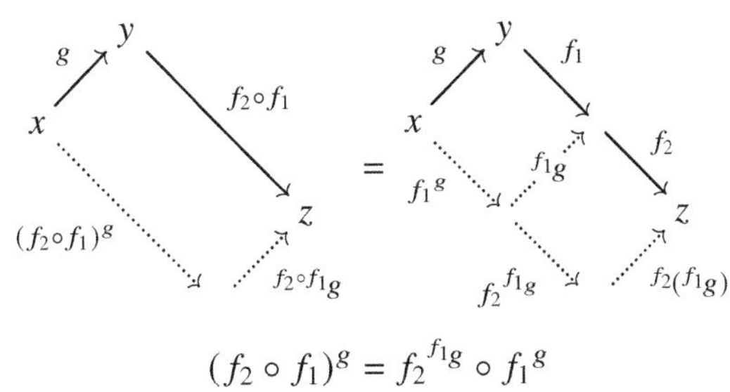

$$(f_2 \circ f_1)^{g} = f_2{}^{{}^{f_1}g} \circ f_1{}^{g}$$
$${}^{f_2 \circ f_1}g = {}^{f_2}({}^{f_1}g)$$

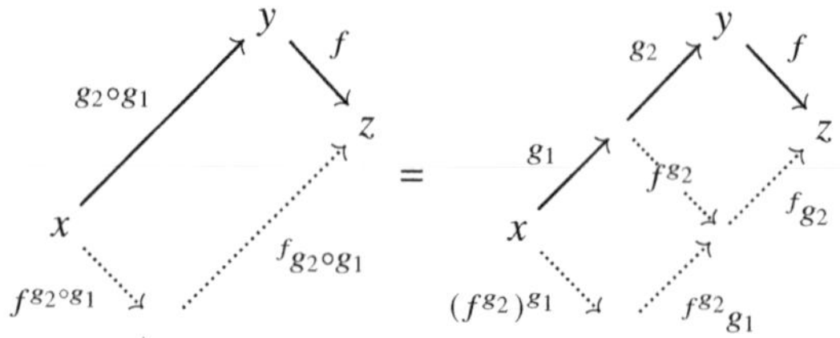

$$f^{g_2 \circ g_1} = (f^{g_2})^{g_1}$$

$${}^{f}(g_2 \circ g_1) = {}^{f}g_2 \circ {}^{f^{g_2}}g_1$$

and identities

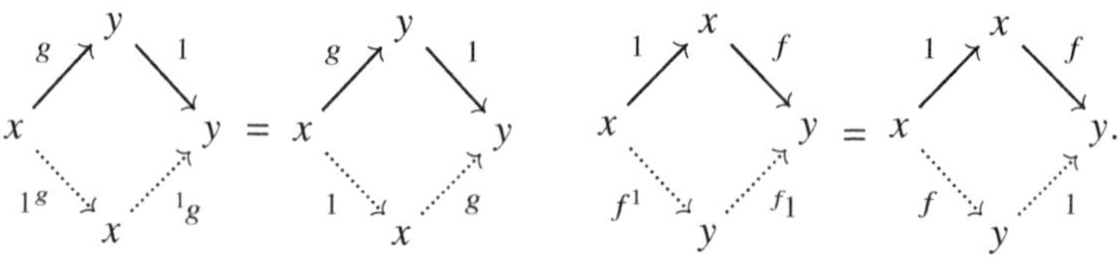

$$1^g = 1 \qquad\qquad f^1 = f$$

$${}^{1}g = g \qquad\qquad {}^{f}1 = 1$$

3.3.3 Composite category. Given a distributive law $\ell\colon D \otimes C \to C \otimes D$, we can *compose* the categories C and D along ℓ and obtain a new category, noted $C \otimes_\ell D$: it has the same objects as C and D, a morphism from $x \to z$ is a pair of morphisms (f, g) with $f\colon x \to y$ in C and $g\colon y \to z$ in D for some object y, identities are pairs of identities, and compositions are induced in the expected way by the distributive law:

$$(f', g') \circ (f, g) = (f'^{g} \circ f, g' \circ {}^{f'}g)$$

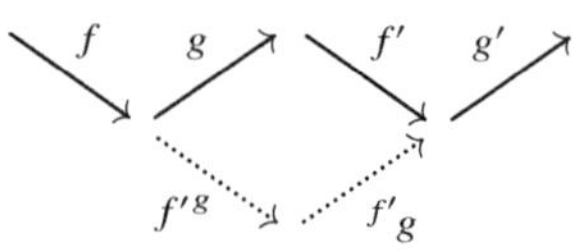

The fact that the axioms of categories are satisfied follows from the axioms of distributive laws.

3.3.4 Proposition. *Given categories C, D, E with the same objects, the following statements are equivalent.*

1. *The categories C and D form a strict factorization system on E.*
2. *There is a distributive law $\ell\colon D \otimes C \to C \otimes D$ such that $C \otimes_\ell D = E$.*

Proof. In the case where C and D form a strict factorization system on E, we define the distributive law ℓ which maps a composable pair of morphisms $(g, f) \in D \times C$ to the pair of morphisms obtained by factorizing $f \circ g \in E$: the axioms of distributive laws follows from the unique factorization of morphisms in E and the functor $C \otimes_\ell D \to E$ which is the identity on objects and sends a composable pair (f, g) to $g \circ f$ is easily seen to be an isomorphism. Conversely, given the distributive law $\ell \colon D \otimes C \to C \otimes D$, the functor $C \to C \otimes_\ell D$ which is the identity on objects and sends a morphism $f \colon x \to y$ to $(f, 1_y)$ is faithful: the category C can be seen as a subcategory of $C \otimes_\ell D$, and similarly for D. Moreover, C and D form a strict factorization system for $C \otimes_\ell D$: for every morphism (f, g) of $C \otimes_\ell D$, we have $(f, g) = (1, g) \circ (f, 1)$, and this is the unique such factorization. □

3.3.5 Presenting composite categories. Because of compatibility with composition, we expect that a distributive law is uniquely determined by the image of pairs of generators for morphisms of the two subcategories. In the case of presented categories, the composite category can thus be presented as follows.

3.3.6 Theorem. *Suppose given two 2-polygraphs P, Q and a distributive law $\ell \colon \overline{Q} \otimes \overline{P} \to \overline{P} \otimes \overline{Q}$ between the presented categories. Then the category $\overline{P} \otimes_\ell \overline{Q}$ is presented by the polygraph R with*

$$R_0 = P_0 = Q_0, \qquad R_1 = P_1 \sqcup Q_1, \qquad R_2 = P_2 \sqcup Q_2 \sqcup R_2^\ell,$$

where R_2^ℓ contains a 2-generator

$$\alpha_{u',v'} \colon v'u' \Rightarrow uv, \tag{3.1}$$

for every pair of composable 1-cells $v' \in Q_1^*$ and $u' \in P_1^*$ such that we have $\ell(\overline{v}', \overline{u}') = (\overline{u}, \overline{v})$, for some $u \in P_1^*$ and $v \in Q_1^*$.

Moreover, if the rewriting relation on 1-cells induced by R_2^ℓ is terminating then one can restrict R_2^ℓ to 2-generators of the form

$$\alpha_{a,b} \colon ba \Rightarrow uv, \tag{3.2}$$

indexed by pairs of 1-generators $a \in P_1$ and $b \in Q_1$.

Proof. We have a functor $f \colon \overline{R} \to \overline{P} \otimes_\ell \overline{Q}$ which is the identity on objects and sends the class of a 1-generator $a \in P_1$ (resp. $b \in Q_1$) to the morphism $(\overline{a}, 1)$ (resp. $(1, \overline{b})$). This functor is full since every morphism of $\overline{P} \otimes_\ell \overline{Q}$ is of the form $(\overline{u}, \overline{v})$, with $u \in P_1^*$ and $v \in Q_1^*$, which is the image of $\overline{uv}$. Moreover, by the rules $\alpha_{u',v'}$ every morphism $w \in R_1^*$ is equivalent to one of the form uv with $u \in P_1^*$ and $v \in Q_1^*$, from which the faithfulness of the functor follows easily.

When the rewriting relation generated by R_2' is terminating, a normal form of a morphism $v'u'$, with $v' \in Q_1^*$ and $u' \in P_1^*$ is necessarily of the form uv

with $u \in P_1^*$ and $v \in Q_1^*$ and therefore a relation of the form (3.1) is derivable for every $u' \in P_1^*$ and $v' \in Q_1^*$, and we conclude as above. □

Note that the first part of the theorem usually gives rise to infinite presentations (because R_2^ℓ is infinite), whereas the reduction provided by the second part produces finite presentations from finite presentations (provided that the termination condition is satisfied).

3.3.7 Example. The additive monoids $C = \mathbb{N}/2\mathbb{N}$ and $D = \mathbb{N}/3\mathbb{N}$ respectively admit the following presentations; see also §A.1.5:

$$\langle \star \mid a \mid aa = 1 \rangle, \qquad\qquad \langle \star \mid b \mid bbb = 1 \rangle.$$

The product monoid $C \times D$ contains C and D as submonoids: an element $n \in C$ can be seen as $(m, 0) \in C \times D$, and similarly for D. Moreover, every element $(m, n) \in C \times D$ can be seen, in a unique way, as a product of an element of C and one of D, namely $(m, n) = (m, 0) + (0, n)$, and therefore C and D form a factorization system for $C \times D$. We deduce that $C \times D$ admits the presentation

$$\langle \star \mid a, b \mid aa = 1, bbb = 1, ba = ab \rangle.$$

The presented monoid is $\mathbb{N}/6\mathbb{N}$ (the generators a and b respectively get interpreted as 3 and 2) and we have embeddings

$$\begin{aligned} \mathbb{N}/2\mathbb{N} &\to \mathbb{N}/6\mathbb{N} \\ p &\mapsto 3p \end{aligned} \qquad\qquad \begin{aligned} \mathbb{N}/3\mathbb{N} &\to \mathbb{N}/6\mathbb{N} \\ q &\mapsto 2q \end{aligned}$$

which induce the strict factorization system corresponding to the distributive law: one readily verifies that every element $n \in \mathbb{N}/6\mathbb{N}$ can be written in a unique way as $n = 3p + 2q$ with $p \in \mathbb{N}/2\mathbb{N}$ and $q \in \mathbb{N}/3\mathbb{N}$.

We can more generally recover in this way the presentation for products of monoids given in §A.1.12. Note that the distributive law induced between C and D is not the only possible one. For instance, the presentation

$$\langle \star \mid a, b \mid aa = 1, bbb = 1, ba = abb \rangle$$

induces another one, which is not isomorphic (an argument for this is that it is not commutative since $ab \neq ba$ can easily be shown, based on the fact that the presentation is convergent; see Section 4.2).

3.3.8 Example. Starting from two 2-polygraphs P and Q, we can take their union and add relations of the form (3.2) and hope that the resulting 2-polygraph R will present a composite category. This is not the case in general. For instance, consider the situation with

$$P = \langle \star \mid a \mid \ \rangle, \quad Q = \langle \star \mid b \mid bb = 1 \rangle, \quad R = \langle \star \mid a, b \mid bb = 1, ba = 1 \rangle.$$

The 2-polygraphs P and Q respectively present the monoids $\mathbb{N}$ and $\mathbb{N}/2\mathbb{N}$. In the 2-polygraph R, the relation $a = bba = b$ is derivable and thus R presents $\mathbb{N}/2\mathbb{N}$: the functor $\overline{P} \to \overline{R}$ is not faithful and $\overline{P}$ and $\overline{Q}$ thus do not form a strict factorization system for $\overline{R}$.

3.3.9 Example. The following (counter-)example illustrates the need for the termination hypothesis in the second part of Theorem 3.3.6. Consider first the polygraph R whose underlying graph is shown on the left, together with the relations on the right:

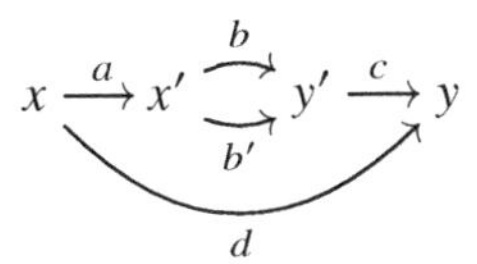

$$ab \Rightarrow ab', \qquad abc \Rightarrow d,$$
$$b'c \Rightarrow bc, \qquad ab'c \Rightarrow d.$$

We then write P and Q for the polygraphs, with no relations, whose respective underlying graphs are

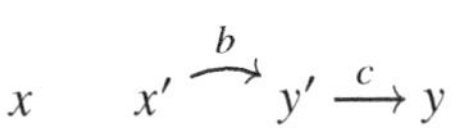

and

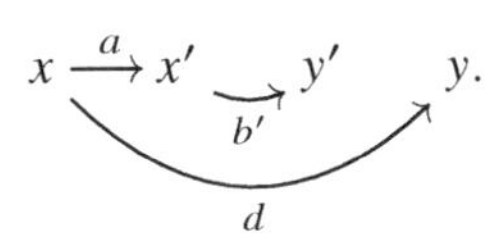

One easily checks that the canonical inclusions $\overline{P} \to \overline{R}$ and $\overline{Q} \to \overline{R}$ are faithful and that every morphism of $\overline{R}$ factorizes uniquely as one from $\overline{P}$ followed by one from $\overline{Q}$. However, if one restricts to relations of the form (3.2), the only relations left are $ab \Rightarrow ab'$ and $b'c \Rightarrow bc$, from which the two relations $abc \Rightarrow d$ and $ab'c \Rightarrow d$ are not derivable. Here, the termination hypothesis of Theorem 3.3.6 is clearly not satisfied since we have the infinite sequence of reductions

$$abc \Rightarrow ab'c \Rightarrow abc \Rightarrow ab'c \Rightarrow \dots$$

3.3.10 Example. There is a strict factorization on the augmented simplicial category Δ_+, presented in detail in §4.5.6: every morphism factorizes as an epimorphism followed by a monomorphism. Writing Δ_μ (resp. Δ_η) for the subcategory of Δ_+, with the same objects, whose morphisms are surjective (resp. injective) functions, we thus have $\Delta_+ = \Delta_\mu \otimes_\ell \Delta_\eta$ for some distributive law ℓ. The categories Δ_μ and Δ_η respectively admit the presentations

$$\left\langle \star \,\middle|\, s_i^n : n+1 \to n \,\middle|\, s_i^{n+1} s_j^n = s_{j+1}^{n+1} s_i^n \right\rangle_{n\in\mathbb{N},0\leqslant i\leqslant j<n}$$
$$\left\langle \star \,\middle|\, d_i^n : n \to n+1 \,\middle|\, d_j^n d_i^{n+1} = d_i^n d_{j+1}^{n+1} \right\rangle_{n\in\mathbb{N},0\leqslant i\leqslant j\leqslant n}$$

and by applying Theorem 3.3.6, we recover the presentation of Δ_+ given in §4.5.6; see also §C.2 for a 2-dimensional analysis of the situation.

3.3.11 Spans. We now briefly recall the construction of span bicategories, which will turn out to be useful in order to explain the axioms for distributive laws, as well as provide a rich source of examples for distributive laws between a category and its opposite. We refer the reader to [316, 230] for details.

Suppose given a category C with pullbacks. A *span* (f, g) from x to y is a pair of coinitial morphisms

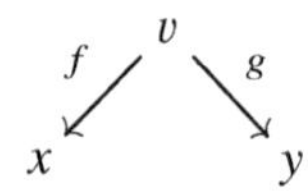

in C. Given a span (f, g) from x to y and (h, i) from y to z, one can define a composite span from x to z by taking the pullback of the two arrows in the middle

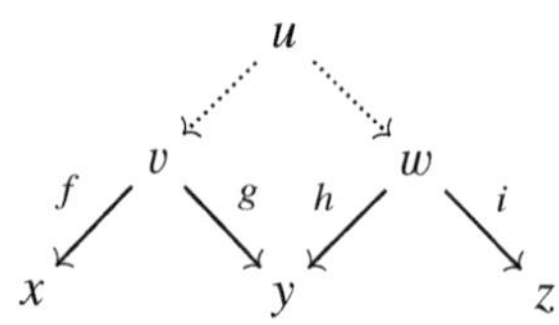

and given an object x one defines the identity span on x as

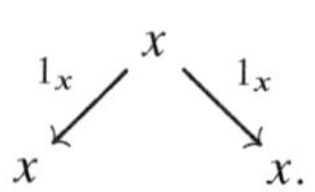

A morphism h between two spans (f, g) and (f', g') from x to y is a morphism of C making the following diagram commute:

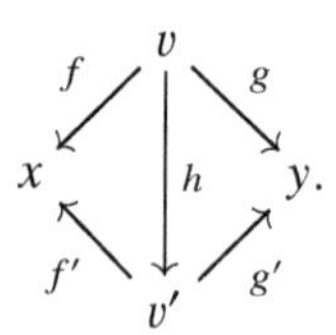

Because of the way composition was defined, it is generally not strictly associative, but rather associative up to isomorphism: we can form a bicategory **Span**(C) whose 0-cells are the objects of C, 1-cells are spans, and 2-cells are morphisms of spans.

Of course, when the category C has pushouts, one can dually define a bicategory $\mathbf{Cospan}(C)$ of *cospans* in C, i.e., diagrams of the form

$$\begin{array}{ccccc} & & v & & \\ & {\scriptstyle f}\nearrow & & \nwarrow{\scriptstyle g} & \\ x & & & & y \end{array}$$

in the category C.

3.3.12 Distributive laws between monads. Given a bicategory $\mathcal{B}$, one can consider a *monad* (t, μ, η) in $\mathcal{B}$ (also called a *monoid* in $\mathcal{B}$), which consists of an endomorphism $t \colon x \to x$ together two 2-cells $\mu \colon tt \Rightarrow t$ and $\eta \colon 1_x \Rightarrow t$, respectively called multiplication and unit, which are associative and unital:

$$\begin{array}{ccccc} (tt)t & \xRightarrow{\sim} & t(tt) & \xRightarrow{t\mu} & tt \\ {\scriptstyle \mu t}\Downarrow & & & & \Downarrow{\scriptstyle \mu} \\ tt & & \xRightarrow[\mu]{} & & t \end{array} \qquad \begin{array}{ccccc} 1t & \xRightarrow{\eta t} & tt & \xLeftarrow{t\eta} & t1. \\ & {\scriptstyle\sim}\searrow & \Downarrow{\scriptstyle \mu} & \swarrow{\scriptstyle\sim} & \\ & & t & & \end{array}$$

In particular, a monad in the 2-category **Cat** is a monad in the usual sense. We write $\mathbf{Mon}(\mathcal{B})$ for the category of monads in $\mathcal{B}$, with the expected notion of morphism.

Given two monads $s \colon x \to x$ and $t \colon x \to x$ the composite tu is not in general a monad. The missing piece of data in order to properly compose those was introduced by Beck [36]: a *distributive law* between two monads t and u is a 2-cell $\lambda \colon ut \Rightarrow tu$ making the diagrams

$$\begin{array}{ccccccccccc} u(tt) & \xRightarrow{\sim} & (ut)t & \xRightarrow{\lambda t} & (tu)t & \xRightarrow{\sim} & t(ut) & \xRightarrow{t\lambda} & t(tu) & \xRightarrow{\sim} & (tt)u \\ {\scriptstyle u\mu}\Downarrow & & & & & & & & & & \Downarrow{\scriptstyle \mu u} \\ ut & & & & & \xRightarrow[\lambda]{} & & & & & tu \end{array}$$

$$\begin{array}{ccccccccccc} (uu)t & \xRightarrow{\sim} & u(ut) & \xRightarrow{u\lambda} & u(tu) & \xRightarrow{\sim} & (ut)u & \xRightarrow{\lambda u} & (tu)u & \xRightarrow{\sim} & t(uu) \\ {\scriptstyle \mu t}\Downarrow & & & & & & & & & & \Downarrow{\scriptstyle t\mu} \\ ut & & & & & \xRightarrow[\lambda]{} & & & & & tu \end{array}$$

$$\begin{array}{ccccc} u & \xRightarrow{\sim} & u1 & \xRightarrow{u\eta} & ut \\ {\scriptstyle \sim}\Downarrow & & & & \Downarrow{\scriptstyle \lambda} \\ 1u & & \xRightarrow[\eta u]{} & & tu \end{array} \qquad \begin{array}{ccccc} t & \Longrightarrow & 1t & \xRightarrow{\eta t} & ut \\ {\scriptstyle \sim}\Downarrow & & & & \Downarrow{\scriptstyle \lambda} \\ t1 & & \xRightarrow[t\eta]{} & & tu \end{array}$$

commute. When equipped with a distributive law, one can define a monad structure on tu, called the *composite monad* of t and u, with multiplication and unit being respectively

$$tutu \xRightarrow{t\lambda u} ttuu \xRightarrow{\mu\mu} tu \qquad \text{and} \qquad 1 \xRightarrow{\eta\eta} tu$$

(omitting coherence isomorphisms). A more detailed description is given in Section C.3.

3.3.13 Monads in spans. Interestingly, a monad in **Span**(**Set**) precisely corresponds to a small category: x is the set of objects of the category, the endomorphism $t\colon x \to x$ is a span of the form

$$x \xleftarrow{f} u \xrightarrow{g} x$$

providing the underlying graph of the category (where x and u are respectively the sets of objects and morphisms of the category and f and g are respectively the source and target functions), and μ and η respectively describe compositions and identities of the category.

A distributive law between two categories seen as spans in this way corresponds precisely to the notion of distributive law defined in §3.3.2, and the notation $C \otimes D$ corresponds to the composite of the underlying 1-cells (i.e., graphs) of the monads corresponding to categories C and D: concretely, $C \otimes D$ is the graph with the objects of C (or equivalently D) as vertices, and pairs (f, g) with $f\colon x \to y$ in C and $g\colon y \to z$ in D as edges $x \to z$.

3.3.14 Categories of spans. Given a category C with pullbacks, a category Span(C) can be defined from the bicategory **Span**(C) by quotienting 1-cells under isomorphisms and discarding 2-cells. This provides a rich source of examples of distributive laws between a category and its opposite, as we now illustrate. A category Cospan(C) of cospans can be defined similarly and, of course, satisfies dual results.

There are canonical functors

$$C^{\mathrm{op}} \to \mathrm{Span}(C) \qquad \text{and} \qquad C \to \mathrm{Span}(C)$$

respectively sending a morphism $f\colon x \to y$ to the class of the span $(f, 1_x)$ and $(1_x, f)$, which are both faithful: the categories C^{op} and C can be considered as subcategories of Span(C). Moreover, in the category of spans, we have $(f, g) = (1, g) \circ (f, 1)$ so that every morphism is the composite of a morphism in C^{op} followed by one in C. When this factorization is unique, we have a strict factorization system; in particular, this is always the case when the category C has no non-trivial isomorphism, because in this case the quotient constructing Span(C) from **Span**(C) will be trivial. In such a situation, when we have a presentation for C, we clearly also have one for C^{op}, and thus also for $\mathrm{Span}(C)$ by Theorem 3.3.6. The general case corresponds to a generalized notion of distributive law, presented in §3.3.17.

3.3.15 Example. As a simple example, consider the monoid $\mathbb{N}$. The pullback of two morphisms m and n

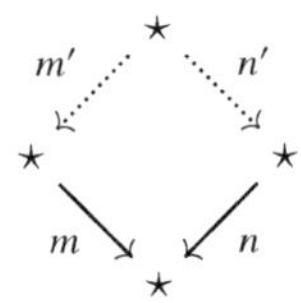

is given by $m' = \max(m, n) - m$ and $n' = \max(m, n) - n$, and the only isomorphism is the identity 0. From the presentation of $\mathbb{N}$ as the free monoid on one generator – see §A.1.4 – we deduce that a presentation of $\mathrm{Span}(\mathbb{N})$ is $\langle\, a, b \mid ba = 1 \,\rangle$, i.e., this is the bicyclic monoid – see §A.1.13 – where the relation is deduced from the fact that the pullback of 1 with 1 is given by 0 and 0.

Other examples are presented in Section 4.6, in the slightly different language of residuals (which provide techniques in order to show on the presentation that the presented category actually has pushouts) and in Section 10.5 using 3-polygraphs.

3.3.16 Iterated distributive laws. Composing more than two monads in a bicategory can be achieved if one assumes that there are distributive laws between any pair of monads and every triple of distributive laws is compatible in the following sense; see [82] for details.

Suppose given three monads $t, u, v \colon x \to x$ in a bicategory and distributive laws

$$\ell_{tu} \colon ut \Rightarrow tu, \qquad \ell_{tv} \colon vt \Rightarrow tv, \qquad \ell_{uv} \colon vu \Rightarrow uv,$$

which are *compatible* in the sense that the following diagram commutes:

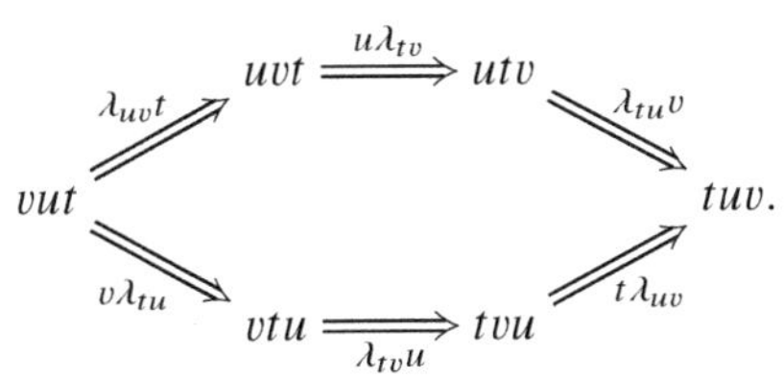

In this situation, there are distributive laws

$$vtu \xRightarrow{\lambda_{tv}u} tvu \xRightarrow{t\lambda_{uv}} tuv \qquad \text{and} \qquad uvt \xRightarrow{u\lambda_{tv}} utv \xRightarrow{\lambda_{tu}v} tuv$$

respectively between tu and v, and t and uv, which both induce the same structure of monad on $(tu)v = t(uv)$.

3.3.17 More general compositions. The notion of strict factorization system (or equivalently of distributive law) is sometimes too restrictive: in many situations, the desirable factorization is not strictly unique, but only unique up to an isomorphism (or even up to some subclass of morphisms). For instance, consider the category E which is the full subcategory of sets, with finite sets $\{0, \ldots, n-1\}$ as objects for $n \in \mathbb{N}$ (this category will be denoted $\mathbf{F}$ in Section C.2). Generalizing the situation of Example 3.3.10, consider the categories C and D which are the subcategories of E whose morphisms are respectively surjective and injective functions: the categories C and D "almost" form a factorization system for E. Namely, every function h factorizes as $h = g \circ f$ where f is surjective and g is injective, and this factorization is "almost" unique in the sense that for every other factorization $h = g' \circ f'$ there exists an isomorphism w making the following diagram commute:

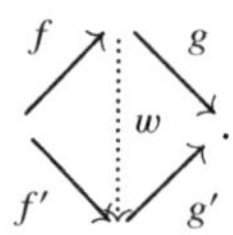

Writing W for subcategory W of isomorphisms of E, we notice that both the categories C and D contain W as subcategory. Thus, if we know presentations for both C and D, we can expect to deduce a presentation for E by taking the union of the presentations of C and D and adding distributivity relations as before (Theorem 3.3.6), but we should moreover identify the presentation of bijections (i.e., the subcategories W) in C and in D. In the above situation, note that the category W "acts" on the left (resp. on the right) on C: for any morphism $w\colon x' \to x$ (resp. $w\colon y \to y'$) of W and $f\colon x \to y$ of C one can obtain a new morphism wf (resp. fw) of C and the situation is the same for D. The distributive law corresponding to the above situation should now have the form

$$\ell\colon D \otimes_W C \to C \otimes_W D$$

where $C \otimes_W D$ is defined as the quotient of $C \otimes D$ above where the right action of W on C is identified with the left action of W on D.

Other typical situations where we would like to identify subcategories W of C and D is when those are symmetric monoidal categories (in which W is the actions of symmetric groups) or Lawvere theories (in which case $W = \mathbf{F}^{\mathrm{op}}$; see Chapter 13). Proper generalizations of distributive laws techniques in order to encompass such situations were given by Lack [230] and detailed and generalized by Cheng [84]; we briefly present those below.

3.3.18 A (non-strict) *factorization system* on a category E consists of subcategories W, C, and D with the same objects as E such that

- W is a subcategory of both C and D,
- every morphism of E factorizes as $g \circ f$ with $f \in C$ and $g \in D$, and
- any two factorizations $g \circ f$ and $g' \circ f'$ of a given morphism in E are W-equivalent: there is a morphism $w \in W$ making the diagram

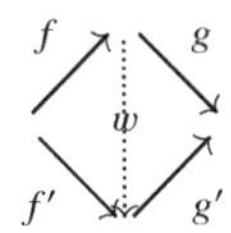

commute.

Above, the *W-equivalence* is the smallest equivalence relation on composable pairs of morphisms in $C \otimes D$ such that for every morphisms $f: x \to y$ in C, $w: y \to y'$ in W and $g: y' \to z$ in D the pairs (fw, g) and (f, wg) are W-equivalent.

3.3.19 Given a bicategory $\mathcal{B}$ with coequalizers, we write $\mathbf{Mod}(\mathcal{B})$ for the bicategory of *bimodules* in $\mathcal{B}$ where

- a 0-cell is a monad in $\mathcal{B}$,
- given monads $t: x \to x$ and $u: y \to y$, a 1-cell $f: t \to u$ is a *bimodule* in $\mathcal{B}$, i.e., a 1-cell $f: x \to y$ of $\mathcal{B}$ together with two 2-cells of $\mathcal{B}$

  $$\lambda: tf \Rightarrow f \qquad \text{and} \qquad \rho: fu \Rightarrow u$$

  respectively called *left* and *right action* making the following diagrams commute:

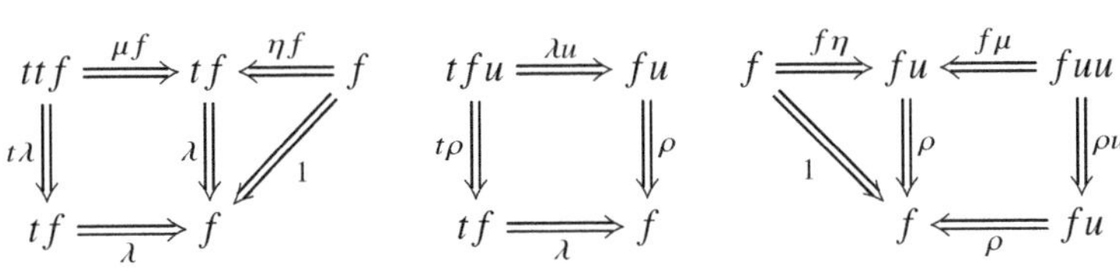

- a 2-cell $\phi: f \Rightarrow g: t \to u$ is a 2-cell $\phi: f \Rightarrow g$ in $\mathcal{B}$ making the following diagram commute:

$$\begin{array}{ccccc} tf & \overset{\lambda}{\Longrightarrow} & f & \overset{\rho}{\Longleftarrow} & fu \\ {\scriptstyle t\phi}\Downarrow & & \Downarrow{\scriptstyle \phi} & & \Downarrow{\scriptstyle \phi u} \\ tg & \underset{\rho}{\Longrightarrow} & g & \underset{\rho}{\Longleftarrow} & gu \end{array}$$

– the composite $f \otimes_u g$ of 1-cells $f\colon t \to u$ and $g\colon u \to v$ is given by the following coequalizer in $\mathcal{B}$:

$$fug \overset{\rho g}{\underset{f\lambda}{\rightrightarrows}} fg \dashrightarrow f \otimes_u g.$$

Other compositions and identities are the expected ones.

In particular, $\mathbf{Mod}(\mathbf{Set})$ is biequivalent to the usual bicategory of profunctors, where a 0-cell is a category and a 1-cell $f\colon C \to D$ is a functor $f\colon C^{\mathrm{op}} \times D \to \mathbf{Set}$ (called a *profunctor* from C and D). More interestingly for our matters, consider the bicategory $\mathbf{Mod}(\mathbf{Span}(\mathbf{Set}))$. By the correspondence given in §3.3.13, the 0-cells are categories $(V, W, \dots)$, and a 1-cell $C\colon V \to W$ is a span

$$V_0 \xleftarrow{s} C \xrightarrow{t} W_0$$

that can be seen as a set C of "arrows" with source (resp. target) being an object of V (resp. W) on which V (resp. W) acts by precomposition (resp. postcomposition). The horizontal composite $C \otimes_W D$ thus consists of the set of pairs (f, g) of composable arrows in $C \times D$ quotiented by the equivalence relation identifying (fw, g) and (f, wg) for composable $f \in C$, $w \in W$ and $g \in D$.

3.3.20 Fix a category W. A *category under* W consists of a category C, with the same objects as W, together with a functor $f\colon W \to C$ which is the identity on objects. We can think of C as having W as distinguished subcategory, at least when the functor f is faithful.

Given a monad $t\colon x \to x$ in a bicategory $\mathcal{B}$ with colimits, there is always an equivalence of categories

$$\mathbf{Mon}(\mathbf{Mod}(\mathcal{B})(t,t)) \quad \cong \quad t/\mathbf{Mon}(\mathcal{B}(x,x)).$$

Instantiated to the case where $\mathcal{B} = \mathbf{Span}(\mathbf{Set})$ and t is a category W, this says that we have a correspondence between monads in W-bimodules and categories under W.

Given two categories C and D under W, we can now define a *distributive law*

$$\ell\colon D \otimes_W C \to C \otimes_W D \tag{3.3}$$

as being a distributive law between C and D, seen as monads in W-bimodules. Explicitly, it consists of a function which maps a W-equivalence class of a composable pair (g, f) of morphisms $g \in D$ and $f \in C$ to a W-equivalence class of a composable pair $(f^g, {}^f g)$ with $f^g \in C$ and ${}^f g \in D$ in a way compatible with compositions and identities, in a similar fashion as for distributive laws; see §§3.3.12 and C.3. As before, we write $C \otimes_\ell D$ for the resulting composite category.

Every factorization system (in the sense of §3.3.18) induces a distributive law in the above sense, and conversely a distributive law induces a factorization system when the functors $W \to C$ and $W \to D$ are faithful.

3.3.21 A generalization of Theorem 3.3.6 can also be given as follows. Suppose given two 2-polygraph P and Q with the same 0-cells, and write $W = P \cap Q$ for the 2-polygraph such that

$$W_0 = P_0 = Q_0, \qquad W_1 = P_1 \cap Q_1, \qquad W_2 = P_2 \cap Q_2.$$

Above, we suppose that source and target maps agree in P and Q for elements of W_1 and of W_2, and that they induce those of W. The inclusion $W \to P$ induces a functor $\overline{W} \to \overline{P}$ making $\overline{P}$ a category under $\overline{W}$, and similarly for $\overline{Q}$. Suppose given a distributive law

$$\ell\colon \overline{Q} \otimes_{\overline{W}} \overline{P} \to \overline{P} \otimes_{\overline{W}} \overline{Q}$$

between the presented categories. Then the category $\overline{P} \otimes_\ell \overline{Q}$ is presented by the polygraph R with

$$R_0 = P_0 = Q_0, \qquad R_1 = P_1 \cup Q_1, \qquad R_2 = (P_2 \cup Q_2) \sqcup R_2^\ell,$$

(note that some unions are not disjoint) where R_2^ℓ contains a 2-generator

$$\alpha_{u',v'}\colon v'u' \Rightarrow uv,$$

for every composable 1-cells $v' \in Q_1^*$ and $u' \in P_1^*$ such that $\ell(\overline{b}, \overline{a}) = (\overline{u}, \overline{v})$, for some $u \in P_1^*$ and $v \in Q_1^*$. When the rewriting relation induced by R_2^ℓ is moreover terminating, this set can be further reduced as in Theorem 3.3.6.

3.3.22 Example. Suppose given a category C and write W for the subcategory of C, with the same objects and the isomorphisms of C as morphisms. When C has pullbacks, there is a distributive law

$$\ell\colon C \otimes_W C^{\mathrm{op}} \to C^{\mathrm{op}} \otimes_W C,$$

which to a pair of morphism (g', f'^{op}) associates the pullback (f^{op}, g):

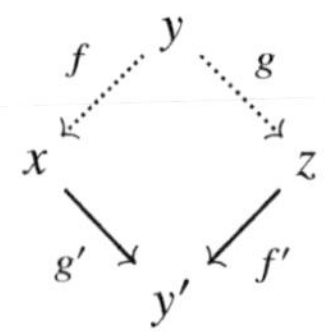

and the composite category is $C^{\mathrm{op}} \otimes_{\ell} C = \mathrm{Span}(C)$, with the category of isomorphism classes spans described in §3.3.14; see [230, 316, 359]. Dually, when C has pushouts, the category $\mathrm{Cospan}(C)$ can be obtained as $C \otimes_{\ell} C^{\mathrm{op}}$ where ℓ is given by pushout.

Other examples and applications are given in Section 10.5.

4
String Rewriting and 2-Polygraphs

We recast the notion of *string rewriting system* into the language of polygraphs. This notion, which consists of a set of pairs of words called *relations* or *rewriting rules* over a fixed alphabet, can be traced back to Thue. In his 1914 paper [344], he introduces the notion of *word problem*: this is the question of deciding whenever two words are equivalent with respect to the congruence generated by the relations. He also shows that the word problem is decidable when the associated rewriting system is terminating and confluent, and even introduces a completion algorithm in order to make a system confluent (an accessible presentation of the paper, along with an English translation can be found in [302]). For this reason, string rewriting systems are also sometimes called *semi-Thue systems* (the "semi" here is to distinguish with *Thue systems* which are defined in the same way, but where the relations are not oriented). Unexpectedly at the time, the word problem was shown to be undecidable for those systems in 1947 by Post [301] and Markov [271]. Of course, this does not preclude us from deciding the word problem for subclasses of monoids, and this is precisely what rewriting is about. The notion of string rewriting system is a variant of the notion of presentation for groups, which is adapted to monoids and where the relations are oriented. Group presentations have been introduced by Dehn [105] in 1911 along with the corresponding word problem for finitely presented groups and Dehn's algorithm for solving the word problem in favorable cases. However, the general word problem for groups has been shown undecidable by Novikov [293] and Boone [51]. We do not intend to give a complete presentation of those early works, nor of the recent developments, and we refer the reader to the standard textbooks [20, 50, 204, 342] for an in-depth treatment. We rather explain here how string rewriting systems can be seen as a particular case of 2-polygraphs, and how the polygraphic rewriting techniques generalize traditional ones.

The notion of string rewriting system – and the more general variant adapted to categories – is introduced in Section 4.1, where we show that the rewriting paths form the morphisms of a sesquicategory, in which we can instantiate the concepts for abstract rewriting systems developed in Chapter 1. In Section 4.2, we introduce the word problem and show that it can be efficiently solved for convergent, i.e., confluent and terminating rewriting systems. In practice, confluence can be checked by inspecting the *critical branchings* of the rewriting system, presented in §4.3.6, and termination by introducing a suitable *reduction order*, as defined in Section 4.4. The convergence of a rewriting system is also useful to show that it forms a presentation of a given category, as illustrated in Section 4.5. Finally, in Section 4.6, we introduce *residuation* techniques which allow proving useful properties of categories (such as the existence of pushouts) by performing computations on their presentations.

4.1 String Rewriting Systems

We have seen in Section 2.3 that a 2-polygraph P can be considered as a notion of presentation for the category $\overline{P}$, obtained from the category freely generated by the underlying 1-polygraph $P_{\leqslant 1}$, by quotienting the 1-cells under the congruence $\approx$ generated by P_2. By Lemma 2.5.2, this congruence is the smallest equivalence relation identifying two 1-cells u and v whenever there is a 2-cell $\phi\colon u \Rightarrow v$ in P_2^*. In such a situation, u and v are thus two representatives of the same 1-cell in $\overline{P}$, and if we adopt the point of view developed in Chapter 1, we can think of $\phi\colon u \Rightarrow v$ as indicating that v is a "more canonical representative" of the equivalence class than u. All this suggests that a 2-polygraph can be considered as a form of rewriting system, where the objects of interest are the 1-cells in P_1^*, and where the generators $\alpha\colon u \Rightarrow v$ in P_2 are rewriting rules indicating that u can be rewritten to v.

If we consider the particular case of a 2-polygraph P with only one 0-generator, say $P_0 = \{\star\}$, the presented category $\overline{P}$ has only 0-cell and can thus be considered as a monoid, as explained in §2.3.5. We will see that, if we restrict to such polygraphs, the associated notion of rewriting system corresponds precisely to string rewriting systems, thus establishing 2-polygraphs as a mild generalization of those, in which letters are "typed" and only well-typed words are considered: in practice, this extra generality does not bring any major complication and we develop here the traditional theory of rewriting in full generality.

4.1.1 Terminology. A 2-polygraph P, when considered as a rewriting system, is sometimes called a *categorical string rewriting system*, or a 1-*dimensional* *rewriting system*. The terminology *string rewriting system* is reserved to the particular case where $P_0 = \{\star\}$. The underlying 1-polygraph $P_{\leqslant 1}$ is called the *signature* and is composed of *sorts* (the elements of P_0) and *letters* (the elements of P_1). The 1-cells in P_1^* freely generated by this signature are called *words* or *strings*, and an identity is sometimes referred to as an *empty word*. The 2-generators are the *rewriting rules* of the rewriting system.

4.1.2 Rewriting step. Suppose fixed a 2-polygraph P. A *rewriting step of P*

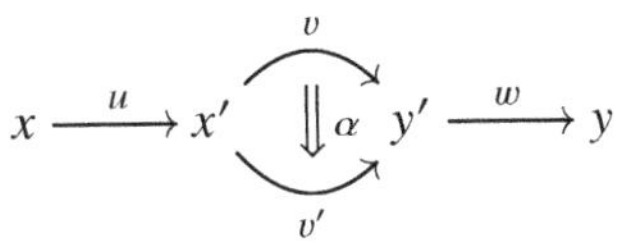

consists in a 2-generator $\alpha\colon v \Rightarrow v'\colon x' \to y'$ in P_2, together with two 1-cells $u\colon x \to x'$ and $w\colon y' \to y$ in P_1^*. Such a rewriting step will be denoted

$$u\alpha w\colon uvw \Rightarrow uv'w\colon x \to y \tag{4.1}$$

and pictured as

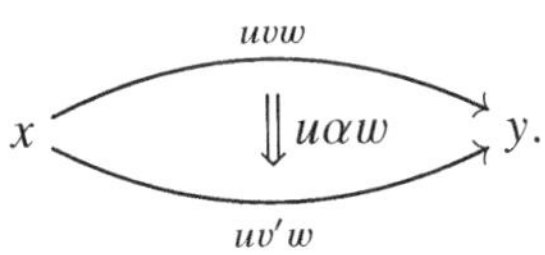

(4.2)

The 1-cell uvw (resp. $uv'w$) in P_1^* is called its *source* (resp. *target*). In this situation, we say that uvw is *reducible* by α. The pair (u, w) of 1-cells in P_1^* is sometimes called the *context* or *whisker* in which the rule α applies to the 1-cell uvw. We sometimes write $u \Rightarrow v$ to indicate that there exists a rewriting step of P from u to v.

4.1.3 Rewriting path. A *rewriting path of P* is a sequence ϕ

$$u_1\alpha_1 w_1, u_2\alpha_2 w_2, \ldots, u_n\alpha_n w_n \tag{4.3}$$

of rewriting steps of P

$$u_i\alpha_i w_i\colon u_i v_i w_i \Rightarrow u_i v_i' w_i\colon x \to y$$

which is composable, in the sense that $u_i v_i' w_i = u_{i+1} v_{i+1} w_i$ for $1 \leqslant i < n$. The natural number n is called the *length* of the rewriting path ϕ and is denoted by $|\phi|$. The 1-cells $u_1 v_1 w_1$ (resp. $u_n v_n' w_n$) are called the *source* (resp. *target*) of the rewriting path, what we write

$$\phi\colon u_1 v_1 w_1 \Rightarrow u_n v_n' w_n\colon x \to y.$$

By convention, an empty path has a determined source (which is the same as its target). We sometimes write $u \overset{*}{\Rightarrow} v$ when there exists a rewriting path from u to v, in which case we say that u *rewrites to* v. Given two rewriting paths $\phi: u \Rightarrow v$ and $\psi: v \Rightarrow w$, we write $\phi * \psi$ for their concatenation. The rewriting path (4.3) can therefore be written as a composition of rewriting steps:

$$\phi = (u_1\alpha_1 w_1) * (u_2\alpha_2 w_2) * \ldots * (u_n\alpha_n w_n). \tag{4.4}$$

Given two 1-cells $u: x' \to x$ and $w: y \to y'$ in P_1^*, we extend the notation (4.1) and write $u\phi w$ for the rewriting path

$$u\phi w = ((uu_1)\alpha_1(w_1 w)) * ((uu_2)\alpha_2(w_2 w)) * \ldots * ((uu_n)\alpha_n(w_n w)). \tag{4.5}$$

These operations equip the 0-cells, 1-cells and rewriting paths in a polygraph with the structure of a sesquicategory, see §4.1.5.

4.1.4 Support. Any 2-cell ϕ in P_2^* can be written as a 1-composite of finitely many rewriting steps, of the form (4.4). We define the *support* of the 2-cell ϕ as the multiset, denoted by $\mathrm{supp}_2^\sharp(f)$, consisting of the 2-cells α_i occurring in this decomposition. The support is well-defined because any two decompositions of ϕ in P_2^* into a 1-composite of rewriting steps involve the same rewriting steps. We have seen in §1.4.1 that multiset inclusion is a well-founded order on supports, allowing us to prove some properties by induction on the support of 2-cells.

4.1.5 Sesquicategory. A *sesquicategory* C consists of

- a 2-graph C (see §2.4.1),
- a structure of category C' on the underlying 1-graph of C,
- a functor $C(-,-): C'^{\mathrm{op}} \times C' \to \mathbf{Cat}$,

such that the composite of the functor $C(-,-)$ with the forgetful functor $\mathbf{Cat} \to \mathbf{Set}$, which to a category associates its set of objects, coincides with the hom functor $C'(-,-): C'^{\mathrm{op}} \times C' \to \mathbf{Set}$.

The notion of sesquicategory was introduced by Street [338]. Let us detail the operations available in such a structure. Given 0-*cells* x, y (i.e., objects of C'), we have a category $C(x, y)$ whose objects are the morphisms $f: x \to y$ of C', called 1-*cells*, morphisms $\alpha: f \Rightarrow g: x \to y$ are called the 2-*cells*, and composition is denoted $*$ and called (vertical) composition. Given a 2-cell $\alpha: g \Rightarrow g': x \to y$ and 1-cells $f: x' \to x$ and $h: y \to y'$, we have a 2-cell $C(f, h)(\alpha)$ that will be denoted

$$f\alpha h: fgh \Rightarrow fg'h: x' \to y'$$

and pictured as

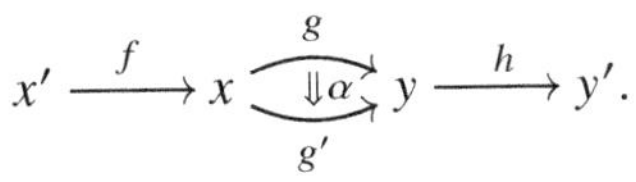

The functoriality of $C(-,-)$ ensures that this is a proper left and right action of 1-cells on 2-cells: in a situation such as

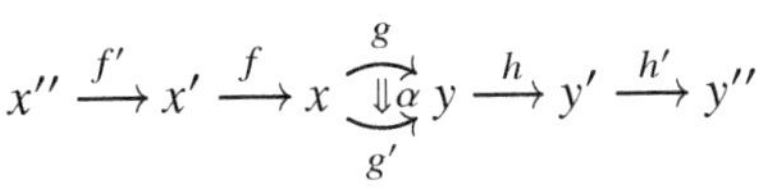

or

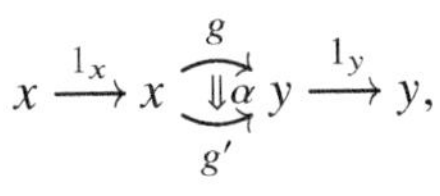

we have

$$f'(f\alpha h)h' = (f'f)\alpha(hh') \qquad \text{and} \qquad 1_x\alpha 1_y = \alpha.$$

The following observation is a reformulation in the language of polygraphs of observations originating in [332, 338]:

4.1.6 Lemma. *Any 2-polygraph P, induces a sesquicategory with $P^*_{\leqslant 1}$ as underlying category, rewriting paths of P as 2-cells and left and right actions defined as in* (4.5).

Any 2-category C induces a sesquicategory in the expected way, where C' is the category underlying C (with C_0 as objects and C_1 as morphisms) and for every $x, y \in C_0$, $C(x,y)$ is the hom-category whose objects are 1-cells $f\colon x \to y$ in C_1 and morphisms are 2-cells $\alpha\colon f \Rightarrow g\colon x \to y$ in C_2, vertical composition $*$ is $*_1$, and the action of 1-cells on 2-cells is given by $f\alpha g = 1_f *_0 \alpha *_0 g$. Moreover, the horizontal composition $*_0$ of the original 2-category can be recovered from the vertical composition and the action since, given $\alpha\colon f \Rightarrow f'\colon x \to y$ and $\beta\colon g \Rightarrow g'\colon y \to z$, we have

$$(1_x\alpha g) * (f'\beta 1_z) \quad = \quad \alpha *_0 \beta \quad = \quad (f\beta 1_z) * (1_x\alpha g')$$

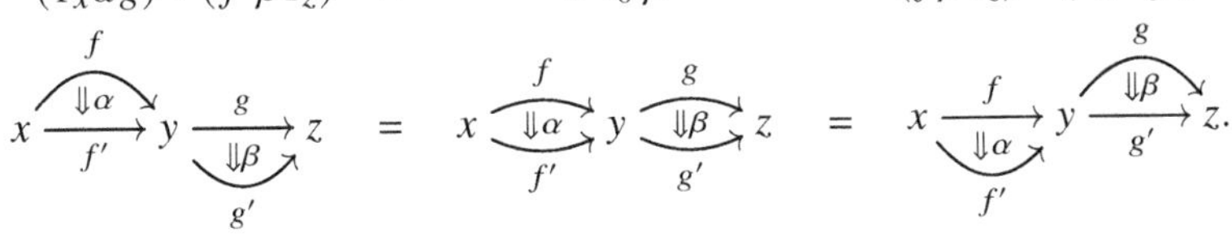

In a general sesquicategory, the left and right members of the above equality are not necessarily equal, and sesquicategories in which this is always the case are precisely 2-categories:

4.1.7 Proposition. *A 2-category is a sesquicategory such that for every 2-cells $\alpha\colon f \Rightarrow f'\colon x \to y$ and $\beta\colon g \Rightarrow g'\colon y \to y'$ we have*

$$(\alpha g) * (f'\beta) = (f\beta) * (\alpha g'). \tag{4.6}$$

This explains the name *sesqui*category, meaning a 1½-category: a sesquicategory is almost a 2-category excepting that the exchange law is not required to hold.

4.1.8 Freely generated 2-category. From the alternative description of a 2-category provided by Proposition 4.1.7, one can come up with an alternative construction of the 2-category P^* generated by a 2-polygraph P (see §2.4.6): P^* is the 2-category, with $P^*_{\leqslant 1}$ as underlying category, whose 2-cells are rewriting paths of P considered up to the congruence generated by (4.6). This means that we do not take in account the order of rewriting steps operating at disjoint positions and consider them up to the congruence identifying two rewriting paths of length two of the form

$$x \xrightarrow{u_1} x' \underset{v'}{\overset{v}{\Downarrow\alpha}} y' \xrightarrow{u_2} y \underset{w'}{\overset{w}{\Downarrow\beta}} z \xrightarrow{u_3} z' \;=\; x \xrightarrow{u_1} x' \underset{v'}{\overset{v}{\Downarrow\alpha}} y' \xrightarrow{u_2} y \underset{w'}{\overset{w}{\Downarrow\beta}} z \xrightarrow{u_3} z'. \tag{4.7}$$

It can be shown that the sesquicategory constructed in Lemma 4.1.6 is free on the polygraph in the expected sense, akin to §2.4.6. One of the main advantage of considering sesquicategories instead of 2-categories here is that the 2-cells are much easier to represent by data structures, thus making those amenable to mechanized computations: the presence of the quotient (4.7) makes everything more difficult.

4.1.9 Rewriting properties of 2-polygraphs. Given a 2-polygraph P, we write here P^{rs} for its set of rewriting steps and $s_1, t_1 : P^{\mathrm{rs}} \to P_1^*$ for the functions which to a rewriting step respectively associates its source and target. Any 2-polygraph P thus induces an abstract rewriting system

$$P_1^* \underset{t_1}{\overset{s_1}{\leftleftarrows}} P^{\mathrm{rs}},$$

with 1-cells as vertices and rewriting steps

$$u\alpha w \colon uvw \Rightarrow uv'w,$$

as in (4.1), as edges from uvw to $uv'w$. We always use double arrows to denote the edges of this rewriting system. Note that, with this point of view, the two rewriting paths shown in (4.7) are not considered to be equivalent.

This construction allows us to extend the properties of Section 1.3 to 2-polygraphs. In particular, a 2-polygraph is

terminating / quasi-terminating / Church–Rosser /

confluent / locally confluent / decreasing
convergent / quasi-convergent

when the associated abstract rewriting system is. Moreover, the properties of Section 1.3 immediately extend to our case. We list below such constructions and properties, reformulated in the framework of 2-polygraphs.

4.1.10 Branching. A *branching* in a 2-polygraph P is a pair (ϕ_1, ϕ_2) of coinitial rewriting paths $\phi_1 \colon u \Rightarrow v_1$ and $\phi_2 \colon u \Rightarrow v_2$ in P_2^*, which we sometimes write $(\phi_1, \phi_2) \colon u \Rightarrow (v_1, v_2)$.

Such a branching is *local* when both ϕ_1 and ϕ_2 are rewriting steps. It is *confluent* when there exist cofinal rewriting paths $\psi_1 \colon v_1 \Rightarrow w$ and $\psi_2 \colon v_2 \Rightarrow w$ which "close" the diagram:

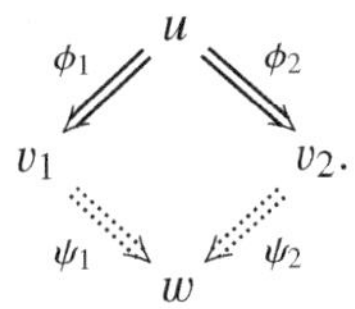

(4.8)

We sometimes write $(\psi_1, \psi_2) \colon (v_1, v_2) \Rightarrow w$ for such a pair.

The goal of this chapter is to provide conditions which are sufficient to ensure that a 2-polygraph is locally confluent (Section 4.3) and terminating (Section 4.4). When both properties are satisfied, we can apply Newman's Lemma 1.3.21 to conclude that it is confluent.

4.2 Deciding Equality

One of the main applications of showing that a 2-polygraph P is convergent is to show that the *equality decision problem*, or *word problem*, for P is decidable for those. We have already handled this situation in the case of 1-polygraphs in §1.3.26.

4.2.1 The word problem. In the context of 2-polygraphs, the equality decision problem for a 2-polygraph P is often called the *word problem* for P, and consists in answering the following question:

Given two 1-cells $u, v \in P_1^*$, do we have $u \approx v$?

Above, we recall that $\approx$ denotes the congruence generated by P_2, see §2.3.2. The problem was originally introduced by Thue [344], as well as Dehn [104] in the closely related context of group presentations.

A 2-polygraph has *decidable word problem* if there is an algorithm answering the above question: this algorithm consists in a procedure, taking the 1-cells u and v as arguments, and answering `true` or `false` depending on whether $u \approx v$ holds or not. We should insist on the fact that we require that the procedure terminates on every input, i.e., provides an answer after some finite amount of time. When there is no such procedure, the problem is said to be *undecidable*.

4.2.2 Undecidability. We always restrict to finite polygraphs P (since the algorithm needs to use this polygraph, the latter must be encoded in a finite way, although we could consider the more general case of recursively enumerable presentations). Contrarily to the case of dimension 1, the set P_1^* of 1-cells is generally infinite, even though the polygraph P is supposed to be finite. Therefore, the argument used in §1.3.26 for showing that the problem is decidable cannot be used anymore: the naive procedure, consisting in computing the equivalence class of u and checking whether v belongs to it or not, is not guaranteed to terminate since the class might not be finite. In fact, the word problem was shown by Post [301] and Markov [271] to be undecidable in general: there exists a finite 2-polygraph P for which there is no algorithm deciding the word problem. A concrete example of such a polygraph is given in §A.1.32.

Having a decidable word problem is however a property of the monoid, not of a particular presentation:

4.2.3 Proposition. *Let P and Q be two finite Tietze equivalent 2-polygraphs. Then the word problem for P is decidable if and only if the word problem for Q is decidable.*

Proof. Suppose given two finite Tietze equivalent 2-polygraphs P and Q such that the word problem is decidable for Q. Given two parallel 1-cells u and v in P_1^*, the Tietze equivalence allows the effective construction for every 1-generator $a \in P_1$ of a 1-cell $[a] \in Q_1^*$ such that $\overline{a} = \overline{[a]}$. Extending the operation $[-]$ as a functor $P_1^* \to Q_1^*$, we have that $u \approx v$ in P if and only if $[u] \approx [v]$ in Q, thus allowing us to conclude. □

4.2.4 The normal form algorithm. When the 2-polygraph P of interest is finite and convergent, the word problem can be decided as in the case of dimension 1 presented in §1.3.26. Namely, the normal form $\widehat{u}$ of a 1-cell u can be computed by maximally rewriting u, and two 1-cells u and v are equivalent if and only if their normal forms $\widehat{u}$ and $\widehat{v}$ are equal.

We shall now present this algorithm in practice. A 1-cell $u \in P_1^*$ can be encoded as being either

– a non-empty sequence of elements of P_1, or

– an element x of P_0, which we write $\mathtt{id}(x)$, representing the identity over x.

In the following, we will not insist on the handling of identities and assimilate those to empty lists in order to simplify the writing of algorithms. We denote by $\mathtt{len}(u)$ the length of u, with $\mathtt{len}(\mathtt{id}(x))$ being 0 by convention. A natural number $i \in \mathbb{N}$ is a *position* in u when $0 \leqslant i < \mathtt{len}(u)$, and in this case we write $u[i]$ for the i-th letter of u. Given $i, k \in \mathbb{N}$ such that i and $i + k - 1$ are positions in u, we write $\mathtt{sub}(u, i, k)$ for the subword of u of length k starting at position i, i.e., the word

$$\mathtt{sub}(u, i, k) = u[i]u[i + 1] \dots u[i + k - 1].$$

We say that a word v *matches* u at position i whenever v is a subword of u starting at position i. This can be tested with the following first procedure:

```
def matches(u,i,v) =
  return (sub(u,i,len(v)) = v)
```

Given 1-cells $u, v \in P_1^*$ such that $\mathtt{tgt}(u)$ and $\mathtt{src}(v)$ are the same, we write $\mathtt{concat}(u,v)$ for their composition, which is simply the concatenation of the two sequences. More generally, we allow ourselves to consider the composition $\mathtt{concat}(u_1, \dots, u_k)$ of k composable 1-cells $u_1, \dots, u_k$.

Given a rule $\alpha : u \Rightarrow v$ in P_2, we write $\mathtt{src}(\alpha)$ for its source u and $\mathtt{tgt}(\alpha)$ for its target v. The normal form of a word u can be computed with the following recursive procedure, expressed in a language which should look familiar to anyone accustomed to imperative programming languages:

```
def rec normalize(P,u) =
  for α ∈ P₂ do
    v = src(α)
    for i = 0 to len(u)−len(v) do
      if matches(u,i,v) then
        w₁ = sub(u,0,i)
        w₂ = tgt(α)
        w₃ = sub(u,i+len(v),len(u)−len(v)−i)
        return normalize(P,concat(w₁,w₂,w₃))
  return u
```

Finally, equality can be decided by the *normal form algorithm* which can be implemented as

```
def equal(P,u,v) =
  return (normalize(P,u) = normalize(P,v))
```

4.2.5 Complexity of the word problem. Let us mention the following result obtained by Avenhaus and Madlener [18, 19] for presentations of groups, but the proof can be applied to presentation of monoids.

4.2.6 Theorem. *Let P and Q be two Tietze equivalent finite 2-polygraphs. If the word problem can be decided for P in time $O(f(n))$, then the word problem for Q can be solved in time $O(f(cn))$ for some constant natural number $c > 0$.*

For a finite convergent 2-polygraph P, consider a function $f : \mathbb{N} \to \mathbb{N}$ such that for any 1-cell u of length $n = |u|$ in P_1^*, the leftmost reduction sequence from u to its normal form contains at most $f(n)$ many steps (here, the leftmost reduction means that we always reduce words with a reduction which rewrites a subword as much on the left as possible). In [49], Book proves that for a finite convergent and reduced 2-polygraph P, the normal form for u in P_1^* can be computed in time $O(n + f(n))$. We say that a polygraph P is *length-reducing* when for every rewriting rule $\alpha : u \Rightarrow v$ in P_2 we have $|u| > |v|$. As a consequence of previous result, if a 2-polygraph P is length-reducing and confluent, then its word problem is decidable in linear time.

4.2.7 Other undecidable problems. The word problem is far from being the only difficult one for 2-polygraphs [50, 281]. It is undecidable, given a finite 2-polygraph P, to determine whether

- it is terminating, locally confluent or confluent,
- there is a finite convergent polygraph presenting the same category,
- it is presenting the terminal category, a finite category, a free category, a cancellative category, or a commutative monoid.

4.3 Critical Branchings

In order to show that a 2-polygraph is confluent using the standard techniques developed in §1.3.17 (by using Newman's lemma, as stated in Lemma 1.3.21), one has to check that all its local branchings are confluent. Contrarily to the case of 1-polygraphs, a finite 2-polygraph usually has an infinite number of local branchings. We however show here that it is enough to check for the confluence of a finite subset of those, the critical branchings.

4.3.1 Example. Consider the 2-polygraph

$$P = \langle \star \mid a \mid \alpha : aa \Rightarrow a \rangle.$$

In order to verify that it is confluent, one can check that all the local branchings

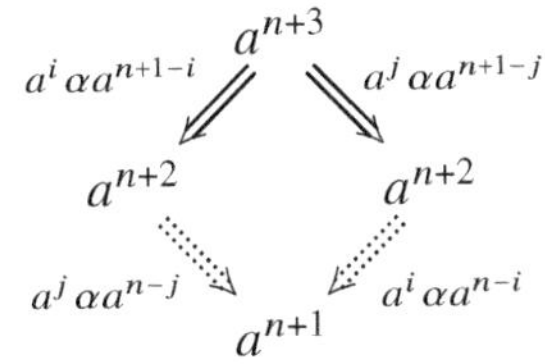

can be closed, for $n \in \mathbb{N}$ and $0 \leqslant i < j \leqslant n+1$, where a^n denotes the composite $aa\ldots a$ of n instances of a.

In order to ease those checks, people are often interested in *critical branchings*, which are minimal possible obstructions of confluence, which will index a finite subset of the above confluence diagrams, enough to ensure the confluence of the 2-polygraph. In the next sections, we introduce a classification of local branchings in order to define the critical ones. Such a classification in view of the critical branching lemma first appeared in [218] for terms rewriting systems and [292] for string rewriting systems.

4.3.2 Trivial branchings. A branching $(\phi_1, \phi_2)\colon u \Rightarrow (v_1, v_2)$ is *trivial* when $\phi_1 = \phi_2$. Such a branching is always confluent since we can take $w = v_1 = v_2$ and $\phi_1' = \phi_2' = 1_w$ to close the diagram:

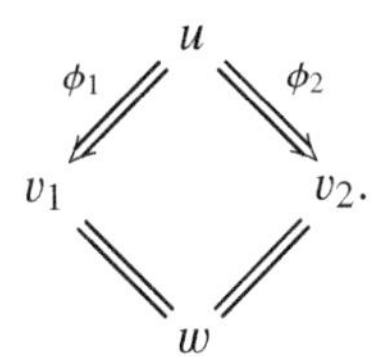

4.3.3 Orthogonal branchings. A local branching $(\phi_1, \phi_2)\colon u \Rightarrow (v_1, v_2)$ is *orthogonal* when it is of the form

$$u = u_1vu_2wu_3, \qquad \phi_1 = u_1\alpha u_2wu_3, \qquad \phi_2 = u_1vu_2\beta u_3,$$

for some words u_1, v, u_2, w, u_3 and rules $\alpha\colon v \Rightarrow v'$ and $\beta\colon w \Rightarrow w'$ (or of the symmetric form, obtained by exchanging the roles of ϕ_1 and ϕ_2). Such a branching is always confluent:

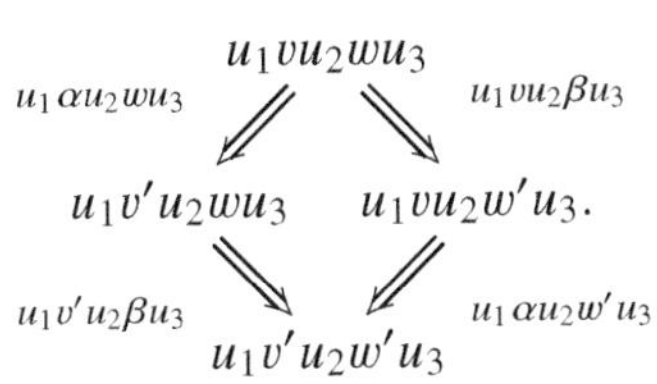

Informally, it corresponds to rewriting two independent parts of the word u. Note that the above diagram corresponds precisely to the equality (4.7). Orthogonal branchings are also sometimes called *Peiffer* branchings in reference to the corresponding notions for spherical diagrams in Cayley complexes associated to presentations of groups [257].

4.3.4 Overlapping branchings. A local branching is *overlapping* when it is not trivial nor orthogonal.

4.3.5 Minimal branchings. We define a partial order on branchings by setting $(\phi_1, \phi_2) \sqsubseteq (\phi'_1, \phi'_2)$ whenever the second branching can be obtained by putting the first one in context. Formally, writing $v\colon x \to y$ for the source of the branching (ϕ_1, ϕ_2), we have $(\phi_1, \phi_2) \sqsubseteq (\phi'_1, \phi'_2)$ whenever there are words $u\colon x' \to x$ and $w\colon y \to y'$ such that $\phi'_1 = u\phi_1 w$ and $\phi'_2 = u\phi_2 w$. In such a situation, the confluence of the first branching $(\phi_1, \phi_2)\colon v \Rightarrow (v_1, v_2)$, say by $(\psi_1, \psi_2)\colon (v_1, v_2) \Rightarrow v'$, implies the confluence of (ϕ'_1, ϕ'_2), since we have

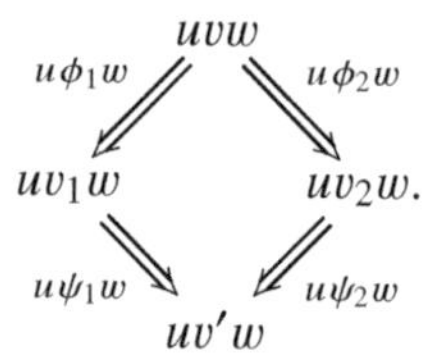

A branching is *minimal* when it is minimal with respect to this order.

4.3.6 Critical branchings. A local branching is *critical* when it is overlapping and minimal. The following lemma is sometimes called the *critical branching lemma*:

4.3.7 Lemma. *A 2-polygraph is locally confluent if and only if all its critical branchings are confluent.*

Proof. Suppose given a local branching (ϕ_1, ϕ_2). If this branching is critical, then it is confluent by hypothesis. Otherwise, it is either trivial, or orthogonal, or non-minimal. In the first two cases, we have seen that the branching is always confluent (§§4.3.2 and 4.3.3). When the branching is non-minimal, it is greater (with respect to $\sqsubseteq$) than a critical branching (since those are the minimal ones), which is confluent by hypothesis, and we have seen that this implies the confluence of the branching (§4.3.5). $\square$

4.3.8 Remark. Note that the confluence of a branching (ϕ_1, ϕ_2) also implies the confluence of the branching (ϕ_2, ϕ_1). We could thus further reduce the number of critical branchings by considering them up to symmetry. We will refrain from doing so in the following, because it obfuscates the formulation of the algorithms without bringing significant improvements (it “only” divides by two the number of critical branchings).

4.3.9 Classification of critical branchings. Suppose given two rewriting steps

$$u_1 \alpha u_2 \colon u_1 u u_2 \Rightarrow u_1 u' u_2, \qquad\qquad v_1 \beta v_2 \colon v_1 v v_2 \Rightarrow v_1 v' v_2,$$

with $\alpha \colon u \Rightarrow u'$ and $\beta \colon v \Rightarrow v'$, forming a local branching, i.e., such that $u_1 u u_2 = v_1 v v_2$. We now study when such a branching is critical.

If the 1-cells u_1 and v_1 are both non-identities, they are necessarily of the form $u_1 = wu'_1$ and $v_2 = wu'_2$ for some non-identity 1-cell w and the branching is thus not minimal. We deduce that either u_1 or v_1 must be an identity, and similarly either u_2 or v_2 must be an identity. Since moreover, the branching should be overlapping, the situation must be of one of the four forms given in Figure 4.1, for some 1-cells w_1, w_2, and w_3. In the two first cases, we suppose that w_2 is not an identity (otherwise the branching is orthogonal). We also suppose that the branching is not trivial, i.e., that we are not in a situation where w_1 and w_3 are identities and $\alpha = \beta$. The last two cases are called *inclusion* branchings because of the relative positions of the rewriting rules as shown in the above figures.

From the above classification, it should be clear that there is a simple algorithm, which is detailed below, to compute critical branchings: for any two rules $\alpha \colon u \Rightarrow u'$ and $\beta \colon v \Rightarrow v'$, we try to overlap u and v at various offsets which are small enough to deduce the possible w_1, w_2, and w_3 for the decompositions of the above form, and remove those which are trivial. In particular, we have as a consequence:

4.3.10 Lemma. *Given a 2-polygraph P with a finite set P_2 of rewriting rules, the number of critical branchings is finite.*

4.3.11 Example. Consider the 2-polygraph

$$P = \langle \star \mid a, b, c \mid \alpha \colon abc \Rightarrow a, \beta \colon ca \Rightarrow a \rangle.$$

In order to compute the critical branchings, we consider pairs of rules and examine how they can overlap. Suppose that we choose α and β. The relative positions of the source abc of α and the source ca of β can be

$$\begin{array}{l} abc \\ ca \end{array} \qquad \begin{array}{l} abc \\ ca \end{array} \qquad \begin{array}{l} abc \\ ca \end{array} \qquad \begin{array}{l} abc \\ ca \end{array} \qquad \begin{array}{l} abc \\ ca \end{array}$$

u_1	u	u_2	v_1	v	v_2	Diagram
1	w_1w_2	w_3	w_1	w_2w_3	1	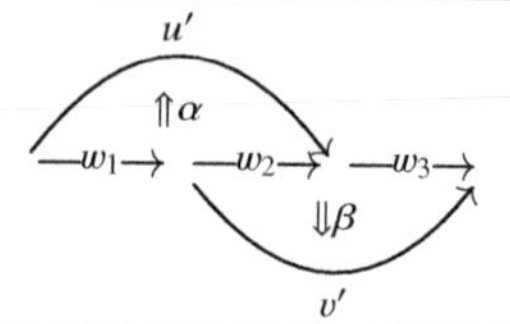
w_1	w_2w_3	1	1	w_1w_2	w_3	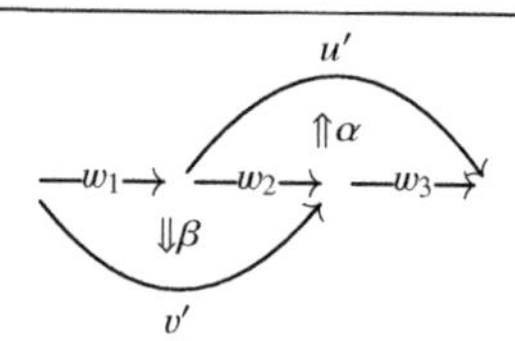
1	$w_1w_2w_3$	1	w_1	w_2	w_3	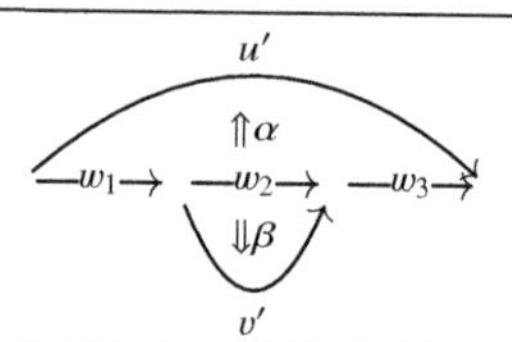
w_1	w_2	w_3	1	$w_1w_2w_3$	1	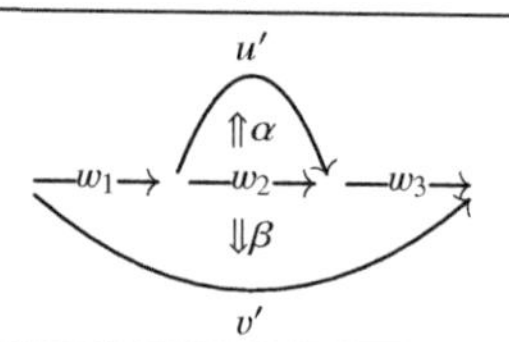

Figure 4.1 Classification of critical branchings.

Among those, only the first and the last one are valid overlappings, which means that the vertically aligned letters are the same, giving rise to the two following critical branchings:

$$(c\alpha, \beta bc), \qquad\qquad (\alpha a, ab\beta),$$

which can also be pictured as

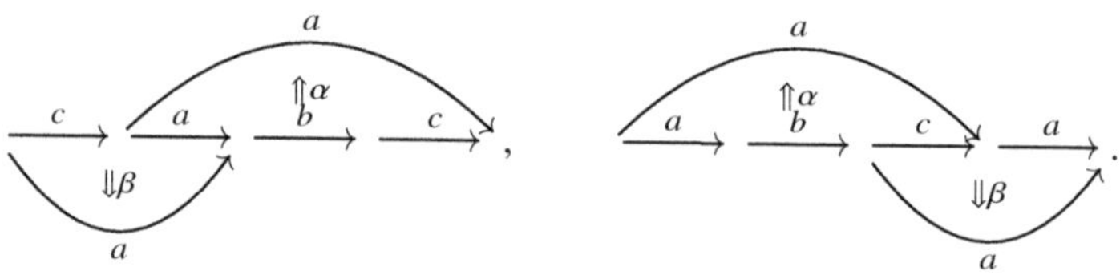

It can be checked that these are the only critical branchings (there is no critical branching involving α with α or β with β). The first branching is confluent, but not the second one:

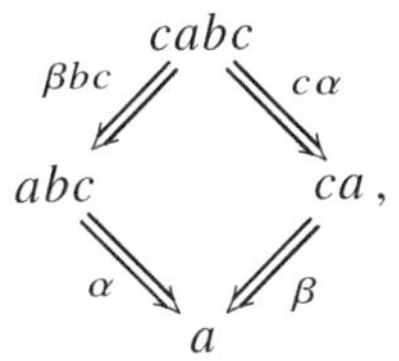

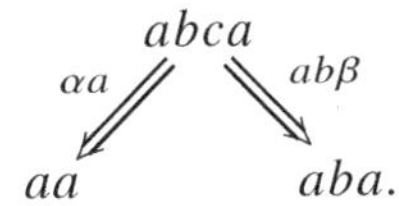

4.3.12 Example. Consider the 2-polygraph of Example 4.3.1 again:

$$P = \langle \star \mid a \mid \alpha\colon aa \Rightarrow a \rangle.$$

The only critical branching

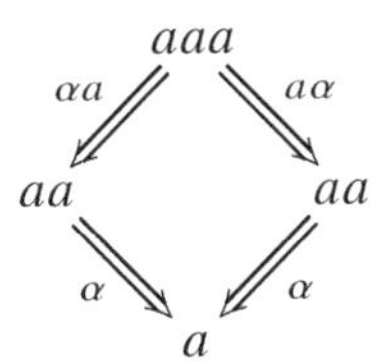

is confluent, therefore the 2-polygraph is locally confluent (by Lemma 4.3.7). Each rewriting step $u\alpha v\colon uaav \Rightarrow uav$ has a source whose length is one less than the length of the source and therefore the system is terminating. Finally, we deduce that the 2-polygraph is convergent (by Lemma 1.3.21 and §4.1.9).

4.3.13 Example. Given a set X, the free group on this set can be presented, as a monoid, by the polygraph

$$P = \langle \star \mid a, \overline{a} \mid \lambda\colon \overline{a}a \Rightarrow 1, \rho\colon a\overline{a} \Rightarrow 1 \rangle_{a \in X}.$$

This polygraph is always locally confluent since the two critical branchings

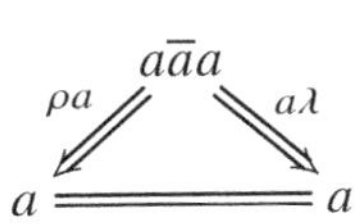

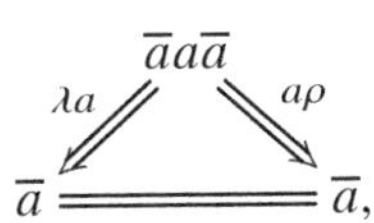

indexed by $a \in X$ are confluent. The rules decrease length and the polygraph is also terminating. Normal forms, also sometimes called *reduced words*, are words which do not contain factors of the form $\overline{a}a$ or $a\overline{a}$ for some $a \in X$. This construction easily extends to present the free groupoid on a graph.

4.3.14 Algorithm. The critical branchings of a 2-polygraph P can be computed thanks to the following algorithm, which tries to unify the sources of all pairs of rules (α, β) in P_2. For simplicity, we suppose here that P is a presentation of a monoid, i.e., there is only one possible identity 1-cell denoted `empty`. The output is a set of pairs $(u_1, \alpha, u_2), (v_1, \beta, v_2)$, with $\alpha\colon u' \Rightarrow u''$ and $\beta\colon v' \Rightarrow v''$ forming a critical branching

$$u_1 u'' u_2 \xLeftarrow{u_1 \alpha u_2} u_1 u' u_2 = v_1 v' v_2 \xRightarrow{v_1 \beta v_2} v_1 v'' v_2.$$

The procedure in peudo-code is

```
def critical_branchings(P) =
  cp = ∅
  for (α,β) ∈ P₂ × P₂ do
    u = src(α)
    v = src(β)
    for i = 1 − len(v) to len(u) − 1 do
      if α ≠ β or i > 0 then
        j = len(v) + i − len(u)
        i′ = max(0,i)
        j′ = max(0,j)
        l = len(v) + i − i′ − j′
        u′ = sub(u,i′,l)
        v′ = sub(v,i′−i,l)
        if u′ = v′ then
          u₁ = if i ⩾ 0 then empty else sub(v,0,−i)
          v₁ = if i ⩾ 0 then sub(u,0,i) else empty
          u₂ = if j ⩾ 0 then sub(v,len(v)−j,j) else empty
          v₂ = if j ⩾ 0 then empty else sub(u,len(u)+j,−j)
          cp = cp ∪ {((u₁,α,u₂),(v₁,β,v₂))}
  return cp
```

4.4 Reduction Orders

In order to show that a 2-polygraph P is terminating, one has to consider a well-founded order on 1-cells which is compatible with composition.

4.4.1 Definition. A *reduction order* on a category C is a partial order $\preccurlyeq$ relating pairs of parallel morphisms in C which is

- well-founded: every weakly decreasing sequence of morphisms is eventually stationary,
- compatible with composition: for every morphisms $u\colon x' \to x$, $v, v' \colon x \to y$ and $w\colon y \to y'$, we have that $v \succ v'$ implies $uvw \succ uv'w$.

Given a 2-polygraph P, a reduction order $\preccurlyeq$ on P_1^* is said to be *compatible* with the rules of P when $u \succ v$ for every rule $\alpha\colon u \Rightarrow v$ in P_2. In this case, the order $\preccurlyeq$ is called a *termination order* for P.

4.4.2 Proposition. *A 2-polygraph P is terminating if and only if it admits a termination order.*

Proof. Suppose that P is terminating. Then the following relation $\preccurlyeq$ is a reduction order compatible with P, where given 1-cells $u, v \in P_1^*$ we have $u \succcurlyeq v$ if and only if u rewrites to v. Conversely, in a 2-polygraph equipped with a reduction order compatible with P, every rewriting step is of the form $uvw \Rightarrow uv'w$ for some rule $\alpha\colon v \Rightarrow v'$. In such a situation, we have $v \succ v'$ because the order is compatible with the rules, and thus $uvw \succ uv'w$ because the order is compatible with composition. Therefore, the existence of an infinite sequence of reductions in P would imply the existence of an infinite decreasing sequence in the order, contradicting its well-foundedness. □

4.4.3 Remark. In a terminating 2-polygraph, an identity is necessarily a normal form. Namely, suppose that we have $1_x \Rightarrow u$ for some $x \in P_0$ and $u \neq 1_x$ in P_1. Then we would have the infinite sequence of rewriting steps

$$1_x \Rightarrow u = 1_x u \Rightarrow uu = 1_x uu \Rightarrow \dots$$

4.4.4 Constructing reduction orders. There is no general rule to construct a reduction order witnessing that a 2-polygraph is terminating: in fact, deciding termination is an undecidable problem [50, Section 2.5]. Fortunately, there is however a "standard toolbox", which we now introduce, from which one is able to construct orders in many useful cases.

4.4.5 Reduction function. The most usual method to show the termination of a 2-polygraph P is to provide a function $f\colon P_1^* \to N$, called a *reduction function*, where $(N, \leqslant)$ is a well-founded poset, which is compatible with composition: for every 1-cells $u\colon x' \to x$, $v, v'\colon x \to y$ and $w\colon y \to y'$, we have that $f(v) > f(v')$ implies $f(uvw) > f(uv'w)$. Such a reduction function induces a reduction order $\preccurlyeq$ on P_1^* defined by $u \preccurlyeq v$ if and only if $f(u) \leqslant f(v)$.

A reduction function is *monotone* when $f(u) > f(v)$ for every rule $\alpha\colon u \Rightarrow v$. With such a reduction function, the associated reduction order is a termination order and thus, by Proposition 4.4.2:

4.4.6 Lemma. *A polygraph equipped with a monotone reduction function is terminating.*

4.4.7 Example. The function which to every word u associates its *length* $|u|$ in $\mathbb{N}$, also called its *degree* in this context, is a reduction function. If a polygraph is *length-decreasing*, in the sense that for every rule $\alpha\colon u \Rightarrow v$ we have $|u| > |v|$, we can thus conclude that it is terminating by Lemma 4.4.6. This is for instance the argument we have been using to show termination in Example 4.3.12.

4.4.8 Example. Functions other than length can also be useful. For instance, consider the presentation

$$P = \langle\, \star \mid a, b \mid \alpha\colon ba \Rightarrow aaa \,\rangle.$$

The 2-polygraph intuitively terminates because each application of a rules decreases the number of occurrences of b in a word. In order to formalize this, consider the function $f\colon P_1^* \to \mathbb{N}$ defined by $f(a) = 0$, $f(b) = 1$, and extended as a morphism of monoids, i.e., $f(1) = 0$ and $f(uv) = f(u) + f(v)$. We have $f(ba) = 1 > 0 = f(a)$ and the function is compatible with composition since it is a morphism of monoids. Therefore the 2-polygraph is terminating. Note that there are no critical branchings, therefore all of them are trivially confluent and the 2-polygraph is convergent.

4.4.9 Remark. Instead of reduction functions, one could more generally consider the notion of a *reduction 2-functor* which is a 2-functor $f\colon P^* \to N$ where

- N is a 2-category such that for each pair of objects $x, y \in N$ the category $N(x, y)$ is a well-founded poset: there is at most one morphisms between two objects and every decreasing sequence is eventually stationary,
- f is injective on parallel 1-cells: for every 1-cells $u, v\colon x \to y$ in P_1^*, $f(u) = f(v)$ implies $u = v$.

4.4.10 Lexicographic product. Given two posets $(M, \leqslant_M)$ and $(N, \leqslant_N)$, one can equip their product $M \times N$ with an order $\leqslant_{M\times N}$, called the *lexicographic product* of the two orders, such that $(m, n) \leqslant_{M\times N} (m', n')$ whenever

- $m <_M m'$, or
- $m = m'$ and $n \leqslant_N n'$.

When the two original orders are well-founded, their lexicographic product is always well-founded. Namely, from every decreasing sequence

$$(m_0, n_0) \geqslant_{M\times N} (m_1, n_1) \geqslant_{M\times N} (m_2, n_2) \geqslant_{M\times N} \cdots$$

the sequence of $(m_i)_{i\in\mathbb{N}}$ is decreasing with respect to $\leqslant_M$ and thus eventually stationary, and similarly for the sequence $(n_i)_{i\in\mathbb{N}}$, and thus the sequence of $(m_i, n_i)_{i\in\mathbb{N}}$ is also eventually stationary. As a consequence, the lexicographic product of two reduction orders is a reduction order.

Given a well-founded poset $(N, \leqslant)$ and $n \in \mathbb{N}$, a partial order $\leqslant_n$ can thus be defined on N^n (the product of n copies of N) by induction on n:

- N^0 is equipped with the trivial order,
- $N^1 = N$ is equipped with the order $\leqslant$, and
- $N^{n+1} = N \times N^n$ is equipped with the lexicographic product of $\leqslant$ and $\leqslant_n$.

Finally, one can define a well-founded partial order $\leqslant_{\text{lex}}$ on $\coprod_{n\in\mathbb{N}} N^n$, called the *lexicographic order* induced by $\leqslant$, by $u \leqslant_{\text{lex}} v$ whenever both u and v belong to N^n for some $n \in \mathbb{N}$ and $u \leqslant_n v$. For instance, given $N = \{a, b\}$ with $a \leqslant b$, one has $abb \leqslant_{\text{lex}} baa$ and $bbba \leqslant_{\text{lex}} bbaa$. This is easily adapted to the setting of polygraphs: given a 2-polygraph P and a well-founded partial order $\leqslant$ on P_1, one can define its lexicographic extension as above, where P_1^n is now the set of composable sequences of length n of elements of P_1. The resulting order $\leqslant_{\text{lex}}$ on P_1^* is always a reduction order.

The variant of the lexicographic order where letters are compared from right to left (instead of from left to right) is also useful and called the *colexicographic order*.

Note that, in the above definition of the lexicographic order, two words of different lengths are always incomparable. One can define a variant of the lexicographic order, sometimes called the *dictionary order*, which is such that $u \leqslant v$ whenever u is a prefix of v, or u and v admit respective prefixes u' and v', of the same length, such that $u' <_{\text{lex}} v'$. This order is not in general well-founded, even if the order on the letters is. For instance, with $a < b$, one has

$$b > ab > aab > aaab > \dots$$

Another variant of the lexicographic order, not suffering from this problem, is presented in next section.

4.4.11 Deglex order. Suppose given a polygraph P equipped with a well-founded partial order $\leqslant_1$ on P_1. We have seen in Example 4.4.7 that the length on words is a reduction function and thus induces a reduction order on P_1^*, as explained in §4.4.5. Moreover, the lexicographic order induced by $\leqslant_1$ is also a reduction order on P_1^*. By taking the lexicographic product of these two reduction orders, we obtain a new reduction order $\preccurlyeq$ on P_1^* called the *deglex order* associated to $\leqslant_1$. Explicitly, given two words $u = a_1 \dots a_m$ and $v = b_1 \dots b_n$, we have $u \prec v$ whenever

- $m < n$, or
- $m = n$, and there exists i with $1 \leqslant i \leqslant n$ such that $a_i \lessdot_1 b_i$ and $a_j = b_j$ for every $j < i$.

4.4.12 Example. Consider the 2-polygraph

$$P = \langle \star \mid a, b \mid \alpha \colon ab \Rightarrow ba \rangle.$$

If we order the letters by $a > b$, the induced deglex order is a reduction order such that $ab > ba$. We can therefore apply Proposition 4.4.2 and deduce that the 2-polygraph is terminating. Since there is no critical branching, the 2-polygraph is convergent.

4.4.13 Derivation. In order to construct a reduction function or a reduction order, one sometimes needs to propagate information from the left or the right of the word. This idea is nicely captured by the classical notion of derivation. For instance, consider the 2-polygraph

$$P = \langle \star \mid a, b, c \mid \alpha \colon cba \Rightarrow aabc \rangle. \tag{4.9}$$

Most of the simple techniques above (considering the length, the number of letters, or a deglex order) do not apply here to show the termination of the rewriting system: for instance, with any deglex order we have $cba < aabc$ because the second word is longer than the first one. However, one can justify the termination of the rewriting system by noticing that a rewriting step always decreases the number of "occurrences of c on the left of an occurrence of b". The purpose of derivation is precisely to formulate such definitions obtained by propagating information (here, the number of occurrences of c) and summing over each letter a quantity obtained from the propagated information (here, the number of occurrence of c on the left for each b and 0 for each a or c). For simplicity, we consider only the case of presentations of monoids here, but it extends seamlessly to presentations of categories.

Given a monoid $(M, \cdot, 1)$, an *M-bimodule N* consists of a commutative monoid $(N, +, 0)$ together with a function $M \times N \times M \to N$, called an *action* of M on N, the image of a triple (u, n, v) being written $u \cdot n \cdot v$, which is

- linear: for every $u, v \in M$ and $n, n' \in N$,

$$u \cdot (n + n') \cdot v = u \cdot n \cdot v + u \cdot n' \cdot v, \qquad u \cdot 0 \cdot v = 0,$$

- associative: for every $u', u, v, v' \in M$ and $n \in N$,

$$u' \cdot (u \cdot n \cdot v) \cdot v' = (u' \cdot u) \cdot n \cdot (v \cdot v'), \qquad 1 \cdot n \cdot 1 = n.$$

A *derivation* of M with values in N is a function $d \colon M \to N$ such that, for $u, v \in M$, one has

$$d(u \cdot v) = 1 \cdot d(u) \cdot v + u \cdot d(v) \cdot 1 \qquad \text{and} \qquad d(1) = 1.$$

When N is equipped with a partial order, the derivation is *monotone* when

- the addition is monotone: $n > n'$ implies $m_1 + n + m_2 > m_1 + n' + m_2$ for every $m_1, n, n', m_2 \in N$,
- the action is monotone: $n > n'$ implies $u \cdot n \cdot v > u \cdot n' \cdot v$ for every $u, v \in M$ and $n, n' \in N$.

In the following, we will be mostly interested in derivations in the case where M is the monoid P_1^* for some fixed 2-polygraph P with one 0-generator. In such a situation, we say that a derivation d is *adapted* to P when $d(u) > d(v)$ for every 2-generator $\alpha\colon u \Rightarrow v$ in P_2.

4.4.14 Lemma. *Suppose given a monoid N equipped with a well-founded partial order, together with a structure of P_1^*-bimodule. A derivation $d\colon P_1^* \to N$ of P_1^* with values in N which is monotone and adapted to P, is a reduction function.*

Proof. Suppose given words u, v, v', w such that $d(v) > d(v')$, we have

$$
\begin{aligned}
d(uvw) &= 1 \cdot d(u) \cdot vw + u \cdot d(v) \cdot w + uv \cdot d(w) \cdot 1 \\
&> 1 \cdot d(u) \cdot vw + u \cdot d(v') \cdot w + uv \cdot d(w) \cdot 1 = d(uv'w).
\end{aligned}
$$ □

Since P_1^* is free, an action onto a monoid N is specified by its effect on generators. Namely, any P_1^*-bimodule N induces, by restriction of the action, two functions

$$
\begin{aligned}
l\colon P_1 \times N &\to N \\
(a, n) &\mapsto a \cdot n \cdot 1
\end{aligned}
\qquad\qquad
\begin{aligned}
r\colon N \times P_1 &\to N \\
(n, b) &\mapsto 1 \cdot n \cdot b
\end{aligned}
$$

which satisfy for $a, b \in P_1$ and $n \in N$,

$$
r(l(a, n), b) = l(a, r(n, b)), \tag{4.10}
$$

both members of the equality being equal to $a \cdot n \cdot b$. Conversely, any such pair of functions $l\colon P_1 \times N \to N$ and $r\colon N \times P_1 \to N$ satisfying the above equality extend uniquely as an action. Similarly, a derivation $d\colon P_1^* \to N$ is uniquely determined by the function $d\colon P_1 \to N$ obtained as its restriction, and any such function extends uniquely as a derivation.

The purpose of the action is intuitively to specify which information is propagated sideways and the derivation determines how the propagated information is used.

4.4.15 Example. The termination of the rewriting system (4.9) can be shown as follows. Consider the monoid $(\mathbb{N} \times \mathbb{N}, +, (0, 0))$ equipped with the componentwise addition, i.e., $(m, n) + (m', n') = (m + m', n + n')$, and the partial order is the lexicographic of the standard order on $\mathbb{N}$ by itself. The first component of $d(u)$ will count the number of b in a word u and the second component the number of c before a b. The action is given, for $(m,n) \in \mathbb{N} \times \mathbb{N}$, by

$$
\begin{aligned}
a \cdot (m,n) \cdot 1 &= (m,n), & 1 \cdot (m,n) \cdot a &= (m,n),\\
b \cdot (m,n) \cdot 1 &= (m,n), & 1 \cdot (m,n) \cdot b &= (m,n),\\
c \cdot (m,n) \cdot 1 &= (m,n+m), & 1 \cdot (m,n) \cdot c &= (m,n).
\end{aligned}
$$

The first column specifies the function l and the second one specifies r, and those two functions are easily checked to be compatible in the sense that (4.10) holds. The left equation on the last line can be read as: given a word u with m letters b and n occurrences of c before a b, the word au has m letters b and $n+m$ occurrences of c before a b; other equations are similar. The derivation $d\colon P_1^* \to \mathbb{N} \times \mathbb{N}$ is defined on generators by

$$
d(a) = (0,0), \qquad d(b) = (1,0), \qquad d(c) = (0,0).
$$

The rule α is decreasing with respect to the derivation as above: we have

$$
\begin{aligned}
d(cba) &= 1 \cdot d(c) \cdot ba + c \cdot d(b) \cdot a + cb \cdot d(a) \cdot 1\\
&= (0,0) + (1,1) + (0,0) = (1,1)
\end{aligned}
$$

and

$$
\begin{aligned}
d(aabc) &= 1 \cdot d(a) \cdot abc + a \cdot d(a) \cdot bc + aa \cdot d(b) \cdot c + aab \cdot d(c) \cdot 1\\
&= (0,0) + (0,0) + (0,0) + (0,0) = (0,0).
\end{aligned}
$$

Therefore, the rewriting system is terminating (and convergent since it has no critical branching).

4.5 Constructing Presentations of Categories

In order to show that a given category C is presented by a given 2-polygraph P, one must show that the 1-cells of C are in bijection with equivalence classes of 1-cells in P_1^* under the congruence generated by relations in P_2, see Lemma 2.3.9 for a formal statement. Without further hypothesis on the polygraph this is usually difficult because one has little control over the equivalence classes. However, in the case where the polygraph P is convergent, equivalence classes of 1-cells have normal forms as canonical representatives, which greatly simplifies the situation. We explore this here by providing a method in order to show that a given convergent polygraph is a presentation of a given category, based on the observation that, in this case, Lemma 2.3.9 can be reformulated as follows. We recall that VC denotes the underlying 1-polygraph of a category C, as defined in §2.1.1.

4.5.1 Lemma. *A convergent* 2*-polygraph* P *is a presentation of a category* C *if and only if there is a morphism of* 1*-polygraphs* $f\colon P_{\leqslant 1} \to VC$ *such that*

1. $f_0\colon P_0 \to C_0$ *is a bijection between the* 0*-generators and the objects of* C,
2. *for any* 2*-generator* $\alpha\colon u \Rightarrow v$ *in* P_2, $f(u) = f(v)$,
3. *the function* $f_1^*\colon P_1^* \to C_1$ *restricts to a bijection between normal forms in* P_1^* *and* C_1.

In the following, we sometimes write $[\![x]\!]$ instead of $f(x)$ for the image of a generator in P_0 or P_1 and call it the *interpretation* of x in C.

4.5.2 Example. Let us show that the 2-polygraph

$$P = \langle \star \mid a \mid \alpha\colon aa \Rightarrow 1 \rangle$$

of Example 2.3.6 is presenting the monoid $\mathbb{N}/2\mathbb{N}$. The polygraph is terminating since the only rule decreases the length of the words and confluent since the only critical branching is confluent:

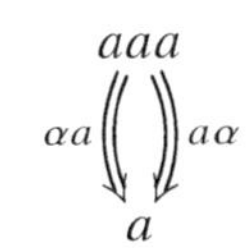

We can thus apply Lemma 4.5.1. There is an obvious bijection between $P_0 = \{\star\}$ and the only object of the monoid (recall that a monoid is considered as a category with only one object), and we define a morphism $f\colon P_{\leqslant 1} \to \mathbb{N}/2\mathbb{N}$ by interpreting the generator a as $f(a) = 1$. This morphism is compatible with the relation, since $f(aa) = 1 + 1 = 0 = f(1)$. Finally, the words of P_1^* in normal form are 1 and a, and those are in bijection with the elements of $\mathbb{N}/2\mathbb{N}$, allowing us to conclude.

4.5.3 Example. Let us use Lemma 4.5.1 in order to give a simpler construction of the presentation of the category of Example 2.3.12. We want to show that the posetal category

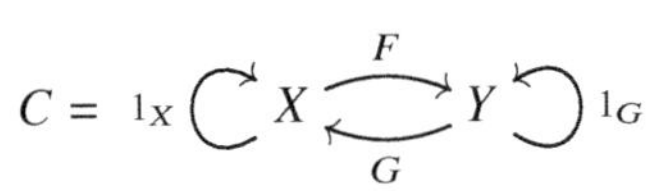

is presented by the 2-polygraph

$$P = \big\langle\, x, y \,\big|\, a\colon x \to y, b\colon y \to x \,\big|\, \alpha\colon ab \Rightarrow 1_x, \beta\colon ba \Rightarrow 1_y \,\big\rangle.$$

The polygraph is convergent since the rules decrease the length and the two critical branchings are confluent:

We define a morphism of 1-polygraphs $P' \to VC$ by

$$[\![x]\!] = X, \qquad [\![y]\!] = Y, \qquad [\![a]\!] = F, \qquad [\![b]\!] = G,$$

which obviously induces a bijection between $P_0 = \{x, y\}$ and $C_0 = \{X, Y\}$. Finally, the 1-cells in P_1^* which are in normal form are the words over the alphabet $\{a, b\}$ which do not contain aa nor bb (because a cannot be composed with a, and similarly for b) nor ab nor ba (because the word would not be in normal form since the rule α or β would apply) as a factor. Thus, there are four normal forms in P_1^*: 1_x, 1_y, a, and b. They are respectively sent by f to 1_X, 1_Y, F, and G, and thus we have a bijection between normal forms and 1-cells of C. We conclude that P is a presentation of C, i.e., $C \simeq \overline{P}$.

4.5.4 Remark. Note that the method given by Lemma 4.5.1 would work with any notion of "canonical form" for the elements of P_1^* modulo $\approx$, not necessarily corresponding to the normal forms for a rewriting system. Namely, for a 2-polygraph P and a category C, suppose given a morphism of 1-polygraphs $f\colon P_{\leqslant 1} \to VC$ satisfying the two first conditions of Lemma 4.5.1 and a set $N \subseteq P_1^*$, whose elements are called *canonical forms*, such that

- every element of P_1^* is equivalent to an element of N,
- f induces a bijection between N and C_1,

then P is a presentation of C. Note that the second condition ensures that every element of P_1^* is equivalent to a unique canonical form.

4.5.5 The standard presentation. Any category C admits a convergent presentation P, called the *standard presentation*, introduced in §2.3.14, which is defined by

- P_0 is the set of 0-cells of C,
- P_1 contains a 1-generator $\widehat{f}\colon x \to y$ for every 1-cell $f\colon x \to y$ in C,
- P_2 contains 2-cells of the form

$$\eta\colon \widehat{1_x} \Rightarrow 1_x : x \to x \qquad \text{and} \qquad \mu_{f,g} : \widehat{f}\widehat{g} \Rightarrow \widehat{fg} : x \to z,$$

which can be represented as

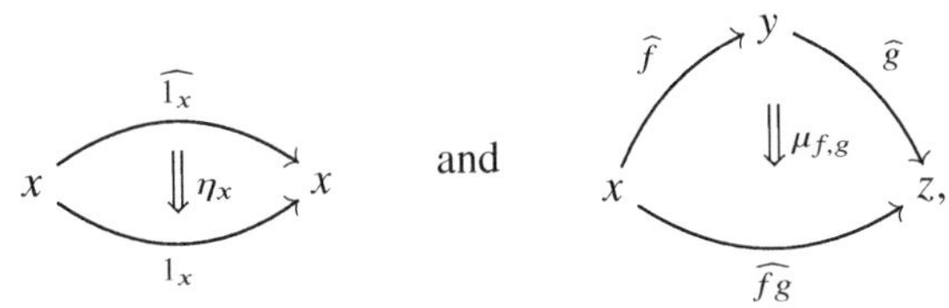

for every 0-cell x of C and pair of composable 1-cells $f : x \to y$ and $g : y \to z$ in C.

Above, note the subtle distinction between $\widehat{f}\widehat{g}$ and $\widehat{fg}$: the source of the 2-cell $\mu_{f,g}$ is a path of length 2 (consisting of the edges $\widehat{f}$ and $\widehat{g}$), whereas its target is a path of length 1 (consisting of the edge $\widehat{fg}$, the generator associated to the composite 1-cell fg). Similarly, a 2-cell η_x has a path of length 1 (resp. 0) as source (resp. target).

The proof that P presents C can be performed using the above method. The 2-polygraph P is terminating because the rules decrease the length of the 1-cells. It is also convergent: its critical branchings are of the form

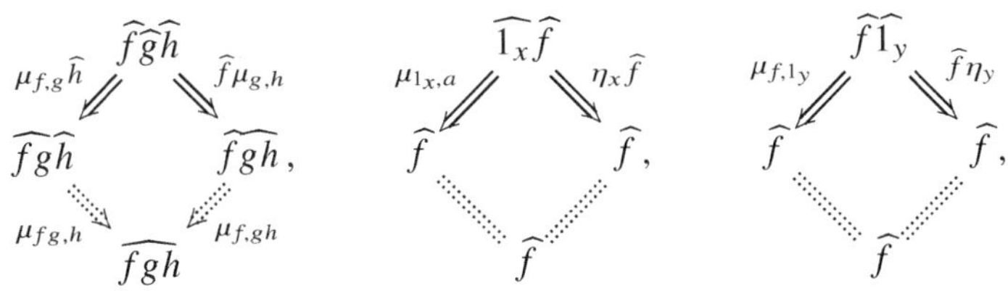

for some composable 1-cells $f : x \to y$, $g : y \to z$ and $h : z \to t$ of C, and are thus confluent. The normal forms are either empty paths (identities) or paths of length 1 consisting of a 1-cell a which is not an identity. Finally, we define a morphism of 1-polygraphs $P_{\leqslant 1} \to VC$ such that the function on objects $P_0 \to C_0$ is the identity and the function on morphisms $P_1 \to C_1$ is the identity, which is obviously compatible with the relations in P_2. This functor clearly induces a bijection between normal forms and 1-cells of C.

A variant where the orientation of the 2-generators η_x is reversed, i.e., $\eta_x : 1_x \Rightarrow \widehat{1_x}$, is more commonly found in the literature. It is Tietze equivalent to the above one, and thus also a presentation of the category C, although not a convergent one since identities are not normal forms (see Remark 4.4.3).

4.5.6 The simplicial category. As a concrete, non-trivial, and useful example, we recall here the well-known presentation of the *augmented simplicial category* Δ_+, as given in [261, Proposition VII.5.2]. Its objects are natural numbers $n \in \mathbb{N}$ and a morphism $f : m \to n$ is a weakly increasing function

$f\colon [m] \to [n]$, where $[n]$ denotes the finite ordinal $\{0, \ldots, n-1\}$. We claim that this category admits a presentation by the 2-polygraph with

- 0-generators: natural numbers $n \in \mathbb{N}$,
- 1-generators: for $n \in \mathbb{N}$,

$$s_i^n : n+1 \to n,$$

  with $0 \leqslant i < n$, and

$$d_i^n : n \to n+1,$$

  with $0 \leqslant i \leqslant n$,
- 2-generators:

$$\begin{array}{lrll} \sigma: & s_i^{n+1} s_j^n & \Rightarrow s_{j+1}^{n+1} s_i^n & \text{for } 0 \leqslant i \leqslant j < n, \\ \delta: & d_j^n d_i^{n+1} & \Rightarrow d_i^n d_{j+1}^{n+1} & \text{for } 0 \leqslant i \leqslant j \leqslant n, \\ \gamma: & d_i^{n+1} s_j^{n+1} & \Rightarrow s_{j-1}^n d_i^n & \text{for } 0 \leqslant i < j \leqslant n, \\ & & \Rightarrow 1_n & \text{for } i = j \text{ or } i = j+1, \\ & & \Rightarrow s_j^n d_{i-1}^n & \text{for } 0 \leqslant j+1 < i \leqslant n+1. \end{array}$$

We consider the order on generators such that, for $i, j, m, n \in \mathbb{N}$, we have

- $s_i^n \geqslant s_j^n$ for $i < j$,
- $d_j^n \geqslant d_i^n$ for $j \geqslant i$,
- $d_i^n \geqslant s_j^m$.

This order is easily shown to be well-founded, and all the rules are strictly decreasing according to the associated deglex order. By Proposition 4.4.2, the polygraph is thus terminating. For simplicity, from now on, we omit the superscripts from generators.

The critical branchings of the rewriting system are

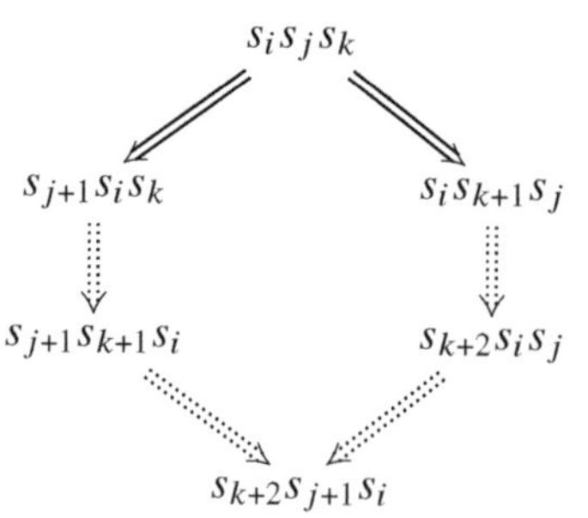

for $i \leqslant j \leqslant k$

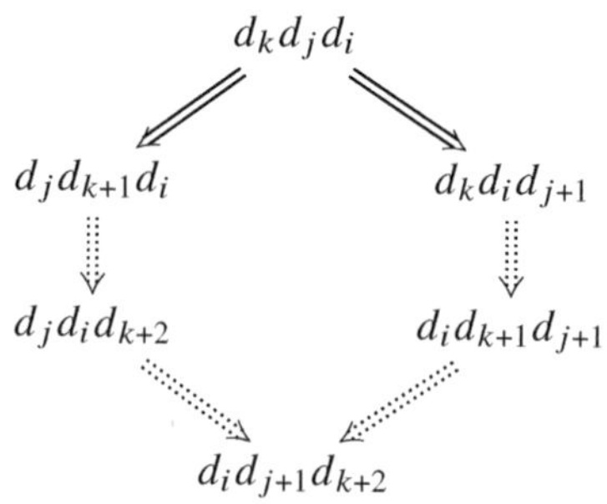

for $i \leqslant j \leqslant k$

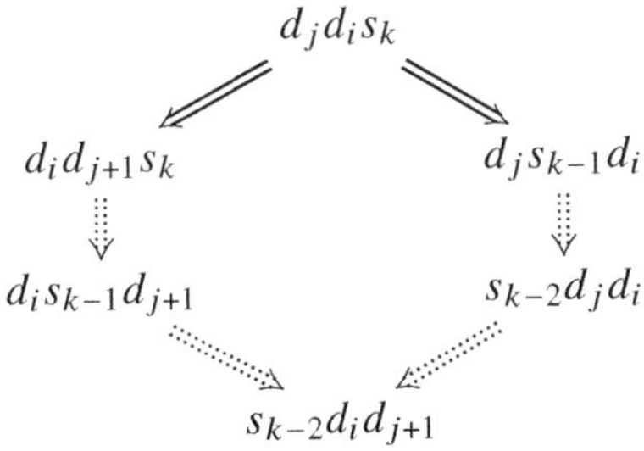

for $i \leqslant j < k-1$

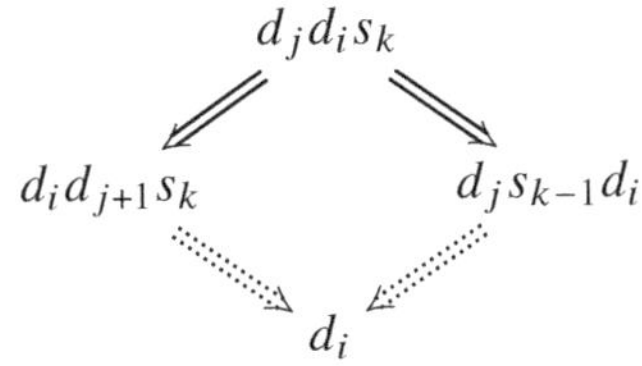

for $i < k$, and $j = k-1$ or $j = k$

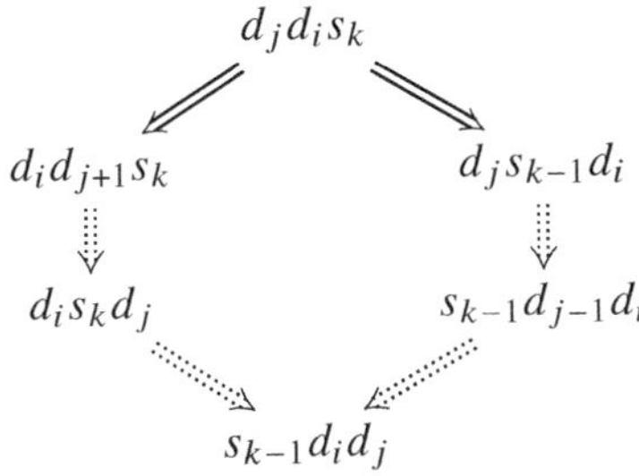

for $i < k < j$

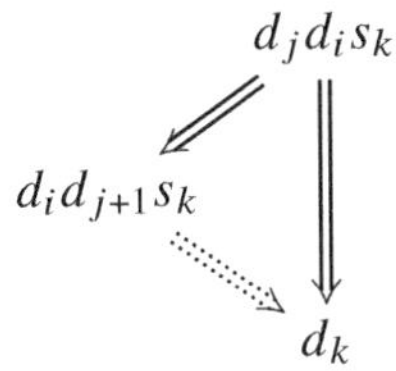

for $i = j = k$

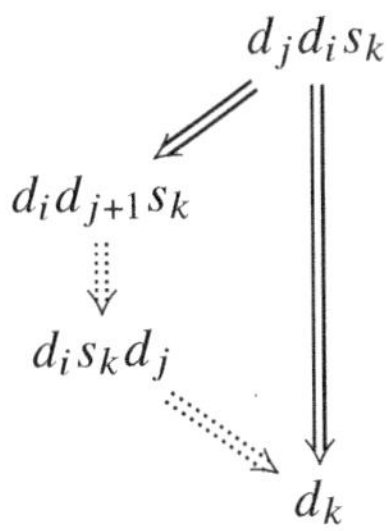

for $i = k$ or $i = k+1$, and $k < j$

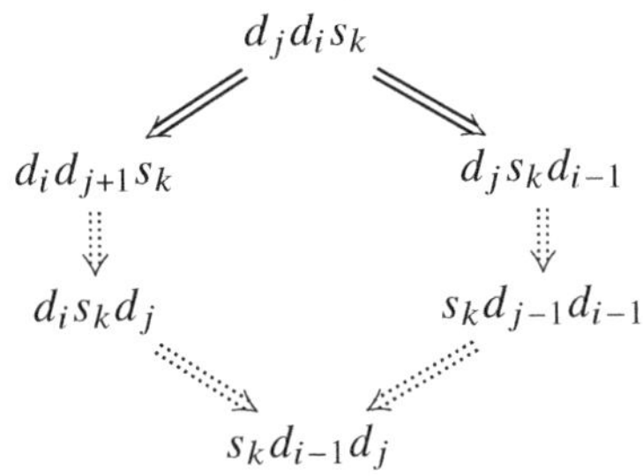

for $k+1 < i \leqslant j$

The 1-generators

$$s_i^n : n+1 \to n \quad \text{and} \quad d_i^n : n \to n+1$$

of P_1 are respectively interpreted as the morphisms

$$[\![s_i^n]\!] : n+1 \to n \quad \text{and} \quad [\![d_i^n]\!] : n \to n+1$$

of Δ_+, which are the functions defined by

$$[\![s_i^n]\!](k) = \begin{cases} k & \text{if } 0 \leqslant k \leqslant i, \\ k-1 & \text{if } i < k \leqslant n, \end{cases} \quad \text{and} \quad [\![d_i^n]\!](k) = \begin{cases} k & \text{if } 0 \leqslant k < i, \\ k+1 & \text{if } i \leqslant k < n. \end{cases}$$

For instance, the graphs of $[\![s_2^4]\!]$ and $[\![d_2^3]\!]$ are respectively

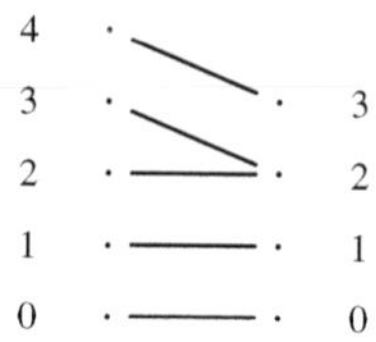

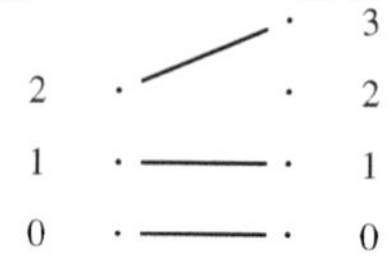

It can be checked that the interpretation is compatible with the relations in P_2. For instance, for the rule $d_j d_i \Rightarrow d_i d_{j+1}$, with $0 \leqslant i \leqslant j$, we have

$$
\begin{aligned}
[\![d_j d_i]\!]\,(k) = [\![d_i]\!] \circ [\![d_j]\!]\,(k) &= \begin{cases} [\![d_i]\!]\,(k) & \text{if } k < j, \\ [\![d_i]\!]\,(k+1) & \text{if } j \leqslant k, \end{cases} \\
&= \begin{cases} k & \text{if } k < i \leqslant j, \\ k+1 & \text{if } i \leqslant k < j, \\ k+2 & \text{if } i \leqslant j \leqslant k, \end{cases} = \begin{cases} [\![d_{j+1}]\!]\,(k) & \text{if } k < i, \\ [\![d_{j+1}]\!]\,(k+1) & \text{if } i \leqslant k, \end{cases} \\
&= [\![d_{j+1}]\!] \circ [\![d_i]\!]\,(k) = [\![d_i d_{j+1}]\!]\,(k).
\end{aligned}
$$

The normal forms are of the form

$$
s_{i_0}^{n+p} s_{i_1}^{n+p-1} \dots s_{i_p}^{n} d_{j_0}^{n} d_{j_2}^{n+1} \dots d_{j_q}^{n+q}, \tag{4.11}
$$

for $n, p, q \in \mathbb{N}$, and

$$
n + p > i_0 > i_1 > \ldots > i_p \geqslant 0, \qquad 0 \leqslant j_0 < j_1 < \ldots < j_q \leqslant n + q.
$$

Namely, the rule γ imposes that there is no d_j before a s_i, σ (resp. δ) imposes that the indices of successive s_i (resp. d_i) are increasing (resp. decreasing).

Every morphism $f\colon m \to m'$ of Δ_+ is the interpretation of exactly one such a normal form. We can namely observe that, f being a weakly increasing function, it is uniquely determined by

- the set of "merged" elements, i.e., the set

$$
\{i_0, i_1, \ldots, i_p\} \quad \subseteq \quad [m]
$$

  of elements such that $f(i_k) = f(i_k + 1)$, and
- its image, or equivalently its complement, i.e., the set

$$
\{j_0, j_1, \ldots, j_q\} \quad \subseteq \quad [m']
$$

  of elements j_k which are not in the image of f.

Finally, with the above notations, and writing $n = m - p = m' - q$, it is easily checked that f is precisely the interpretation of the normal form (4.11).

4.6 Residuation

The notion of residual, which intuitively specifies what "remains" of a morphism after another one, provides a powerful tool in order to derive properties of a presented category, from combinatorial properties of its presentation. Namely, by studying the properties of residuals, through rewriting systems, one is often able to show interesting properties of the presented category such as the existence of pushouts, the fact that morphisms are mono, or that it embeds into its enveloping groupoid. The exposition provided here is adapted from classical techniques in rewriting theory originating in Lévy's thesis [249, 191], see [342, Section 8.7] in the context of term rewriting systems, [276] for a modern presentation, [109, Section II.4] in the context of presentations of groups, and [87] of which the current presentation is inspired.

4.6.1 Residuation structure. A *residuation structure* on a category is a function which to every pair of coinitial morphisms $f: x \to y_1$ and $g: x \to y_2$ associates a morphism

$$f/g: y_2 \to z,$$

called the *residual* of f after g, satisfying the three following conditions.

1. The morphisms f/g and g/f are cofinal and satisfy

   $$f(g/f) = g(f/g),$$

   i.e.,

   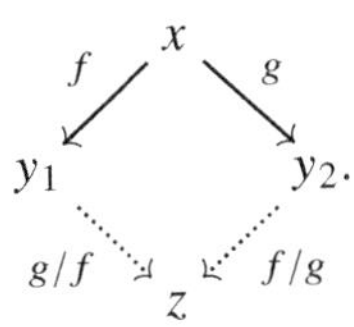

2. Residuation is compatible with composition: given a morphism $f: x \to y$ and morphisms $g: x \to z$ and $h: z \to z'$,

   $$f/1_x = f, \qquad f/(gh) = (f/g)/h,$$
   $$1_x/f = 1_y, \qquad (gh)/f = (g/f)(h/(f/g)),$$

   i.e.,

   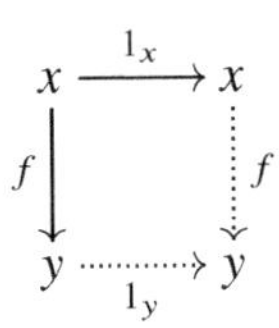

   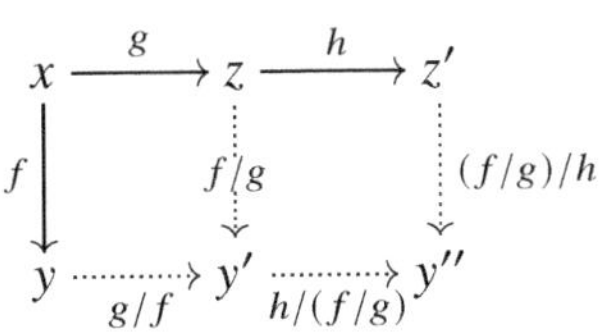

3. Self-residuation is trivial: for every morphism $f\colon x \to y$,

$$f/f = 1_y$$

i.e.,

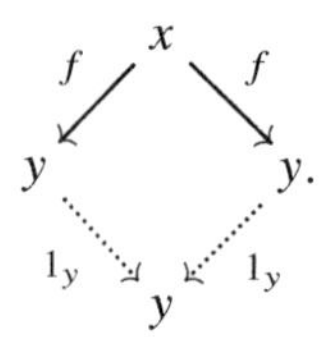

A residuation structure thus provides a witness of confluence for branchings, which is compatible with the categorical structure.

4.6.2 Remark. A residuation structure is precisely a distributive law

$$\ell\colon C^{\mathrm{op}} \otimes C \to C \otimes C^{\mathrm{op}}$$

as developed §3.3.2, such that for every morphism $f\colon x \to y$ in C we have $\ell(f^{\mathrm{op}}, f) = (1_y, 1_y)$, see also §3.3.14.

4.6.3 Proposition. *In a category equipped with a residuation structure, every morphism is epi.*

Proof. We show that a morphism $f\colon x \to y$ is necessarily epi. Suppose given morphisms $g, h\colon y \to z$ such that $fg = fh$. We have

$$(fg)/f = (f/f)(g/(f/f)) = 1_y(g/1_y) = g.$$

Thus,

$$g = (fg)/f = (fh)/f = h$$

and the morphism f is epi. □

4.6.4 Proposition. *In a category equipped with a residuation structure, every pair of coinitial morphisms admits a pushout.*

Proof. Given coinitial morphisms $f\colon x \to y_1$ and $g\colon x \to y_2$, we claim that the morphisms $g/f\colon y_1 \to z$ and $f/g\colon y_2 \to z$ form a pushout cocone. Suppose given morphisms $f'\colon y_1 \to z'$ and $g'\colon y_2 \to z'$ such that $ff' = gg'$:

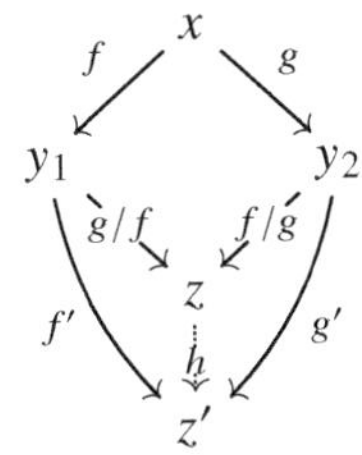

The morphism $h = f'/(g/f)$ makes the two triangles commute. Namely, we have

$$(g/f)/f' = g/(ff') = g/(gg') = 1_{y_2}/g' = 1_{z'}$$

from which follows the commutation of the left triangle:

$$(g/f)h = (g/f)(f'/(g/f)) = f'((g/f)/f') = f'1_{z'} = f'.$$

Moreover, we have

$$h = f'/(g/f) = (ff')/(f(g/f)) = (gg')/(g(f/g)) = g'/(f/g)$$

from which we deduce that the right triangle commutes as above, by exchanging the roles of f and g:

$$(f/g)h = (f/g)(g'/(f/g)) = g'((f/g)/g') = g'.$$

Conversely, given a morphism $h\colon z \to z'$ such that $(g/f)h = f'$ and $(f/g)h = g'$, we necessarily have

$$h = ((g/f)h)/(g/f) = f'/(g/f).$$ □

4.6.5 Proposition. *Suppose given a category C such that both C and C^{op} are equipped with a residuation structure. Then the canonical functor $C \to C^{\top}$, from C to its enveloping groupoid, is faithful.*

Proof. Because C admits a residuation structure, the collection W of all morphisms of C forms a *calculus of left fractions* in the sense of [143]:

- this collection contains identities and is closed under composition,
- for any pair of coinitial morphisms f and g there exists morphisms f' and g' such that $ff' = gg'$ (namely, we can take $f' = g/f$ and $g' = f/g$),
- for any morphisms $h\colon x \to y$ and $f, g\colon y \to z$ such that $hf = hg$, there exists a morphism $h'\colon z \to z'$ such that $fh' = gh'$:

$$x \xrightarrow{h} y \overset{f}{\underset{g}{\rightrightarrows}} z \overset{h'}{\dashrightarrow} z'$$

(namely, by Proposition 4.6.3 every morphism of C is epi and we can take $h' = 1_z$).

The enveloping groupoid $C^\top$ can thus be described as the category of left fractions $C[W^{-1}]$. Since C^{op} admits a residuation structure, by Proposition 4.6.3, every morphism of C (and thus of W) is mono, and in this case, the canonical functor $C \to C[W^{-1}]$ is easily shown to be faithful. □

4.6.6 Residuated presentation. In practice, it is difficult to directly exhibit a residuation structure on a category C and show that it satisfies the required axioms. We provide here a general methodology in order to show that C admits a residuation structure in the case where it is equipped with a presentation by a 2-polygraph. Namely, in this case, we can specify the residuation structure on generators and extend it to other morphisms by functoriality.

A *residuated presentation* P is a 2-polygraph together with, for every pair of coinitial generators $a\colon x \to y_1$ and $b\colon x \to y_2$ in P_1, a morphism

$$a/b\colon y_1 \to z$$

in P_1^*, in such a way that

- a/b and b/a have the same target,
- the morphisms $a(b/a)$ and $b(a/b)$ are P-congruent:

$$\begin{array}{ccc} & x & \\ {}^{a}\swarrow & & \searrow^{b} \\ y_1 & \overset{*}{\Leftrightarrow} & y_2 \\ {}_{b/a}\searrow & & \swarrow_{a/b} \\ & z & \end{array} \tag{4.12}$$

- for every 1-generator $a \in P_1$, we have $a/a = 1$,
- for every generators $a\colon x \to x'$ and $\alpha\colon u \Rightarrow v\colon x \to y$, respectively in P_1 and P_2, we have

$$a/u = a/v$$

and there is a 2-generator

$$\alpha/a\colon u/a \Rightarrow v/a\colon x' \to y'$$

in P_2:

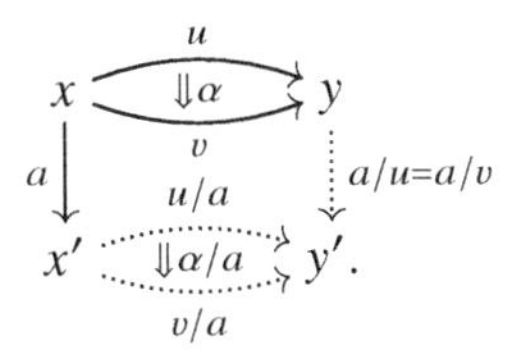

(4.13)

4.6.7 Residuation of morphisms. Suppose fixed a residuated presentation P. We can extend the residuation operation in order to define the residual $u/v \in P_1^*$ of a morphism $u \in P_1^*$ after another morphism $v \in P_1^*$. The definition is performed by induction on u and v by

$$u/1 = u, \qquad u/(vv') = (u/v)/v', \tag{4.14}$$
$$1/u = 1, \qquad (uu')/v = (u/v)(u'/(v/u)).$$

We will eventually see in Theorem 4.6.15 that, under suitable hypothesis, this induces a residuation structure on the presented category $\overline{P}$.

4.6.8 Example. Consider the presentation

$$\langle \star \mid a, b \mid ab = baa \rangle.$$

The relation can be pictured as

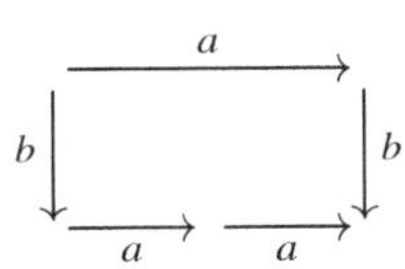

and the only possible residuation structure is defined by

$$a/b = aa \qquad \text{and} \qquad b/a = b.$$

For instance, we have

$$\begin{aligned} ab/bb &= (ab/b)/b = ((a/b)(b/(b/a)))/b = (aa(b/b))/b = aa1/b \\ &= aa/b = (a/b)(a/(b/a)) = aa(a/b) \\ &= aaaa. \end{aligned}$$

Graphically,

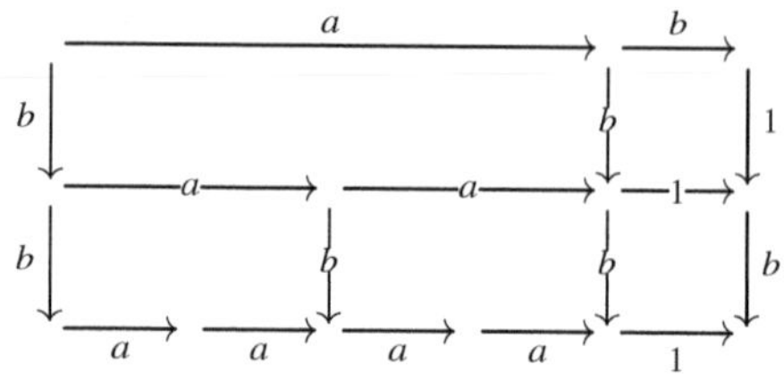

and similarly, we have $bb/ab = b$.

It remains to check that the above definition is sound in the sense that we can always compute a value for the residual using the relations of §4.6.7, and that residual is uniquely defined, i.e., the computed value for u/v does not depend on the way we bracket u and v or the order in which we use the equalities (4.14). Compatibility with bracketing is easily handled:

4.6.9 Lemma. *Residuation of morphisms is compatible with the axioms of categories.*

Proof. Residuation is compatible with associativity since

$$\begin{aligned}((uu')u'')/v &= ((uu')/v)(u''/(v/(uu')))\\ &= (u/v)(u'/(v/u))(u''/((v/u)/u'))\\ &= (u/v)((u'u'')/(v/u))\\ &= (u(u'u''))/v\end{aligned}$$

and

$$u/((vv')v'') = (u/(vv'))/v'' = ((u/v)/v')/v'' = (u/v)/(v'v'') = u/(v(v'v'')).$$

Similarly, it is compatible with left and right-unitality since

$$(1u)/v = (1/v)(u/(v/1)) = u/v = (u/v)(1/(v/u)) = (u1)/v$$

and

$$u/(1v) = (u/1)/v = u/v = (u/v)/1 = u/(v1).$$

□

However, the definition given in §4.6.7 is not sound in general, because it can be vacuous, as illustrated by the following example.

4.6.10 Example. Consider the presentation

$$\langle \star \mid a, b, c, d \mid ba = ab, ca = ac, da = abd, cb = bac, db = bd, dc = cd \rangle,$$

whose relations can be pictured as

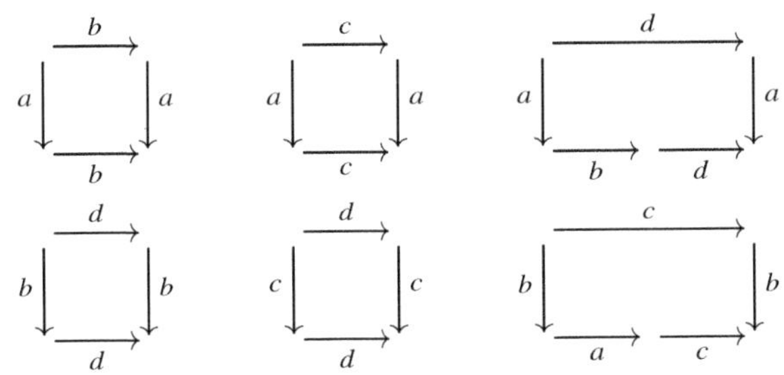

and consider the residuation structure defined by

$$
\begin{array}{llllll}
a/b = a, & a/c = a, & a/d = a, & b/a = b, & b/c = b, & b/d = b, \\
c/a = c, & c/b = ac, & c/d = c, & d/a = bd, & d/b = d, & d/c = d.
\end{array}
$$

The process of computing the residual ac/bd by applying, from left to right, the relations of §4.6.7 defining residuation of morphisms does not terminate. Namely, the first two steps of this computation are

$$ac/bd = aac/d = ac/bd,$$

which clearly leads to a loop since the left and the right member are the same. This can be illustrated as follows:

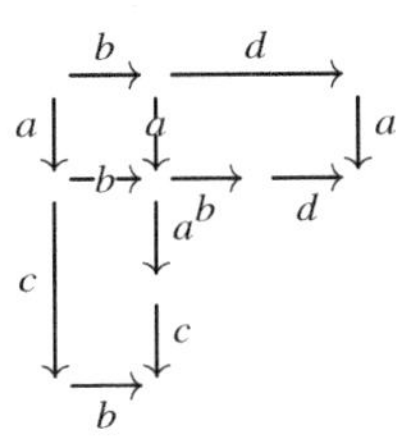

4.6.11 Termination of residuation. Let P be a fixed residuated presentation. In order to ensure that the process of computing the residual is well-defined, we follow the technique of considering "reversed words" introduced by Dehornoy [108], and consider the following polygraph Q defined from P by

$$
\begin{aligned}
Q_0 &= P_0, \\
Q_1 &= \{a\colon x \to y, a^- : y \to x \mid a\colon x \to y \in P_1\}, \\
Q_2 &= \{a^- b \Rightarrow vu^- \mid a, b \in P_1, u = a/b, v = b/a\},
\end{aligned}
$$

where $(a_1 \dots a_n)^-$ is a notation for $a_n^- \dots a_1^-$, and a^- should be thought of as a formal inverse for the generator a. A morphism in Q_1^* is a composite of generators in P_1, some of which might be formally inverted, and rewriting step corresponds to taking residuals in P:

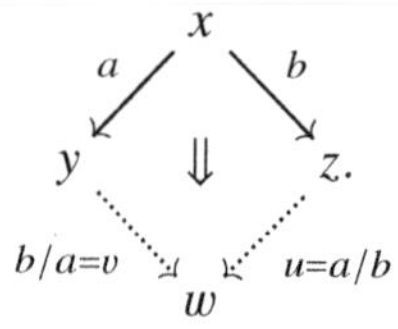

We say that a residuated presentation P is *terminating* when the associated 2-polygraph Q is terminating in the usual sense.

4.6.12 Lemma. *Given a terminating residuated presentation P, the associated 2-polygraph Q is convergent and the residuation operation is well-defined on morphisms of P_1^*. Moreover, for morphisms $u\colon x \to y$ and $v\colon x \to y'$, the morphisms $u(v/u)$ and $v(u/v)$ are P-congruent:*

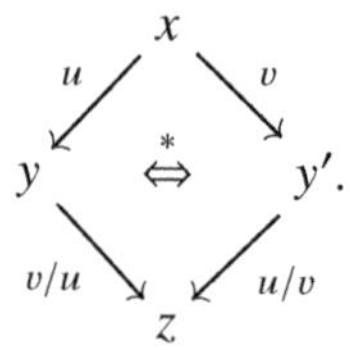

Proof. The polygraph Q has no critical pair, it is thus locally confluent by Lemma 4.3.7 and confluent by Lemma 1.3.21 since it is assumed to be terminating. By well-founded induction, we can show that the normal form of a word $u^- v$ is a word of the form $v' u'^-$ with $v' = v/u$ and $u' = u/v$, and that any word of this form is a normal form. The last part of the lemma follows by induction from the assumption (4.12). □

In practice, various practical conditions are sufficient to ensure the termination of the polygraph Q, see [109, 87]. For instance,

4.6.13 Lemma. *Suppose given a function $\omega\colon P_1 \to \mathbb{N}$, which we extend as a function $\omega\colon P_1^* \to \mathbb{N}$ by $\omega(1) = 0$ and $\omega(uv) = \omega(u)+\omega(v)$. Suppose moreover that we have $\omega(a/b) < \omega(a)$ for every pair of generators $a, b \in P_1$. Then the 2-polygraph Q is terminating.*

Proof. We define a function $\omega'\colon Q_1^* \to \mathbb{N}$ by $\omega'(a) = \omega(a)$ and $\omega(a^-) = 0$ for $a \in P_1$, $\omega'(uv) = \omega'(u) + \omega'(v)$, $\omega(1) = 0$. This function is a reduction order on the 2-polygraph Q and we conclude by Lemma 4.4.6. □

Finally, we can check that residuation is well-defined on morphisms of the presented category $\overline{P}$, i.e., that it is compatible with P-congruence on morphisms.

4.6.14 Lemma. *Given a terminating residuated presentation P, for every morphisms $u, v\colon x \to y$ and $w\colon x \to x'$ in P_1^*, we have that*

$$u \overset{*}{\Leftrightarrow} v \qquad \text{implies} \qquad u/w \overset{*}{\Leftrightarrow} v/w \quad \text{and} \quad w/u = w/v.$$

Graphically,

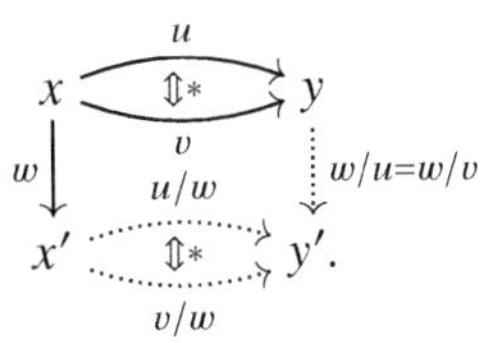

Proof. The assumption that u and v are P-congruent means that there exists a sequence of 2-cells of the form

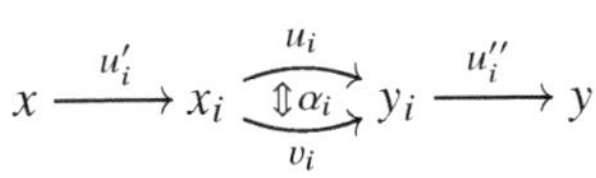

with $1 \leqslant i \leqslant n$, $u_i, u'_i, u''_i, v_i \in P_1^*$ and $\alpha_i \in P_2$, with either $\alpha_i\colon u_i \Rightarrow v_i$ or $\alpha_i\colon v_i \Rightarrow u_i$, such that $u'_1 u_1 u''_1 = u$, $u'_{i+1} u_{i+1} u''_{i+1} = u'_i v_i u''_i$ and $u'_n v_n u''_n = v$. We have

$$w/(u'_i u_i u''_i) = ((w/u'_i)/u_i)/u''_i = ((w/u'_i)/v_i)/u''_i = w/(u'_i v_i u''_i)$$

where the equality $(w/u'_i)/u_i = (w/u'_i)/v_i$ can be shown by recurrence on the length of w/u'_i using axioms (4.13). Also, by recurrence on w/u'_i and using axioms (4.13), we have the existence of a 2-generator between $u_i/(w/u'_i)$ and $v_i/(w/u'_i)$:

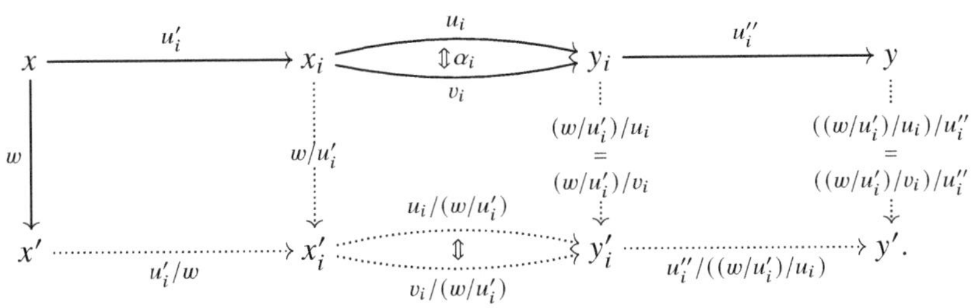

We conclude, by performing a recurrence on n. □

4.6.15 Theorem. *Given a terminating residuated presentation P, the presented category $\overline{P}$ admits a residuation structure.*

Proof. The residuation operation is well-defined on morphisms in P_1^* modulo P-congruence by previous lemmas and immediately satisfies the axioms of a residuation structure. □

The axiomatization presented in this section has the advantage of being relatively simple to state and prove, but more advanced generalizations are often required in practice. For instance, in many situations, not every pair of coinitial morphisms f and g admit a residual, but only those which are *bounded*, i.e., for which there exists f' and g' with $ff' \overset{*}{\Leftrightarrow} gg'$. Also, it is useful to weaken axiom (4.13) and require that for every 1-generator $a\colon x \to x'$ and 2-generator $\alpha\colon u \Rightarrow v\colon x \to y$, we have an 2-cell

$$a/\alpha\colon a/u \Rightarrow a/v\colon y \to y'$$

and a 2-cell

$$\alpha/a\colon u/a \Rightarrow v/a\colon x' \to y'$$

in P_2^* (or even in $P_2^\top$):

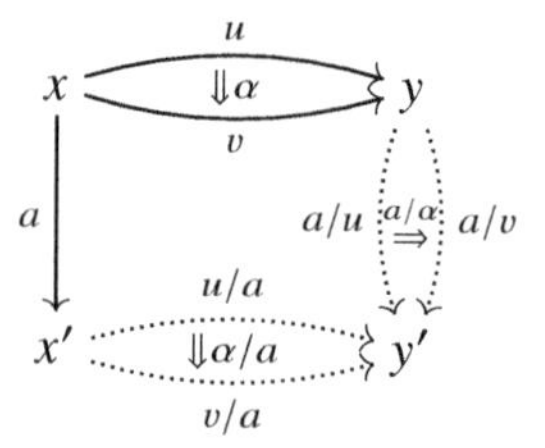

(4.15)

axiom (4.13) being the particular case where we further impose that a/α is an identity and α/a is a whiskered 2-generator. In this case, in order for Lemma 4.6.14 to hold, one has to impose further termination conditions. Given two cofinal morphisms f and g, we write $g|f$ whenever there exists h with $hg = f$, and in this case we say that g *divides* f on the right. A category is *right noetherian* when every infinite sequence (f_i) of cofinal morphisms $f_{i+1}|f_i$ is eventually stationary. In particular, a category presented by a 2-polygraph whose relations are homogeneous (i.e., preserve the length of words) necessarily has this property. The following theorem is due to Dehornoy: see [109, Section II.4] for detailed statement and proof.

4.6.16 Theorem. *Given a residuated presentation with generalized axiom* (4.15), *whose presented category is right Noetherian, every bounded pair of morphisms u and v in P_1^* admits a residual.*

4.6.17 Example. Consider the positive braid monoid B_4^+, see §A.1.21, which admits a presentation by a 2-polygraph with three generators a_0, a_1, a_2 and three relations

$$\alpha_{01}\colon a_0a_1a_0 \Rightarrow a_1a_0a_1, \quad \alpha_{12}\colon a_1a_2a_1 \Rightarrow a_2a_1a_2, \quad \alpha_{02}\colon a_0a_2 \Rightarrow a_2a_0,$$

which can respectively be pictured as

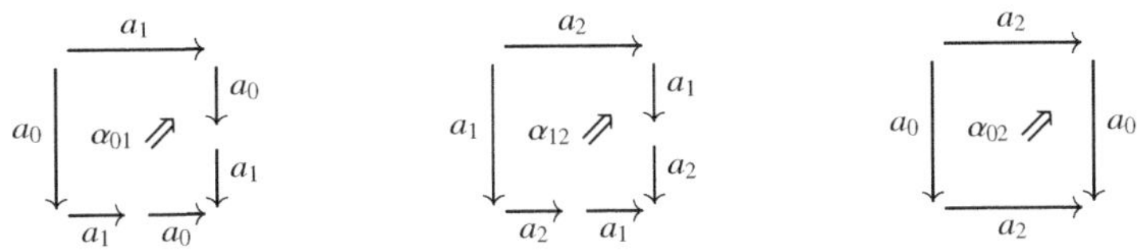

We define residuation on generators by

$$a_0/a_0 = 1, \qquad a_1/a_0 = a_1a_0, \qquad a_2/a_0 = a_2,$$
$$a_0/a_1 = a_0a_1, \qquad a_1/a_1 = 1, \qquad a_2/a_1 = a_2a_1,$$
$$a_0/a_2 = a_0, \qquad a_1/a_2 = a_1a_2, \qquad a_2/a_2 = 1.$$

We can check axiom (4.13), i.e., that residuation of 1-generators is compatible with 2-generators. For instance, for the relation α_{01}, the residuals of a_0 (a_1 is similar) and a_2 after the source and the target are respectively

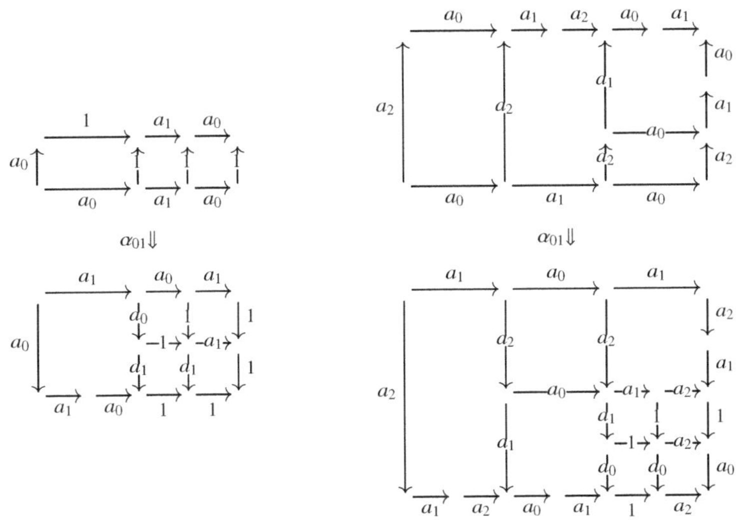

and we can thus take $a_0/\alpha_{01} = 1_1$, $a_2/\alpha_{01} = 1_{a_2a_1a_0}$, $\alpha_{01}/a_0 = 1_{a_1a_0}$ and α_{01}/a_2 to be the 2-cell

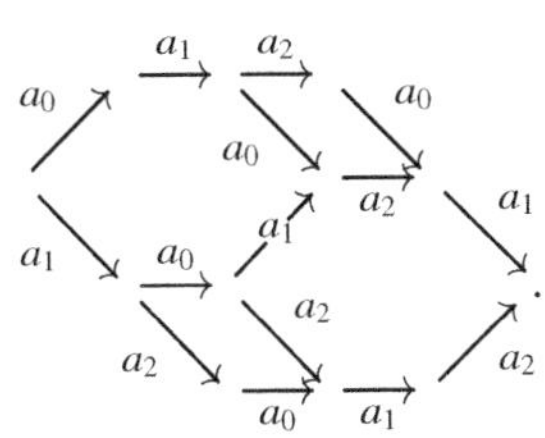

Other cases are left to the reader. Being homogeneous, this presentation is right Noetherian and residuation is always terminating [107]. The category B_4^+ is thus residuated. The argument generalize to all positive braid monoids B_n^+.

4.6.18 Deciding equality. As a last remark, note that for residuated presentations P the word problem can be solved in the following way. Given two morphisms $u, v \colon x \to y$ in P_1^*, we have $u \overset{*}{\Leftrightarrow} v$ if and only if

$$u/v \overset{*}{\Leftrightarrow} 1_y \qquad \text{and} \qquad v/u \overset{*}{\Leftrightarrow} 1_y.$$

This follows easily from the fact that residuals corresponds to pushouts cocones by Proposition 4.6.4. In particular, when P has no 2-generator with an identity as source or as target, we have $u \overset{*}{\Leftrightarrow} v$ if and only if

$$u/v = 1_y \qquad \text{and} \qquad v/u = 1_y.$$

5

Tietze Transformations and Completion

In this chapter, we introduce a notion of *Tietze transformation* for 2-polygraphs, generalizing the one introduced in Section 1.2 for 1-polygraphs. The Tietze transformations are elementary operations on 2-polygraphs, which preserve the presented category, and such that any two finite 2-polygraphs presenting the same category can be transformed into one another by applying a series of such transformations. Our notion, introduced in Section 5.1, is very close to the one first introduced by Tietze for presentations of groups [345]. We refer to [257, 263] for more details on the notion of Tietze transformation in combinatorial group theory; see also [80] for a historical account. The notion of Tietze transformation was developed in the polygraphic language in [145].

By using Tietze transformations, one seeks to turn a given presentation of a category into another one, possessing better computational properties. In particular, the *Knuth–Bendix completion* procedure described in Section 5.2 applies those transformations to turn a presentation into a confluent one.

We have seen in §1.3.26 how convergent presentations lead to a solution of the word problem: for those, the equivalence between two words is immediately decided by comparing their normal forms. In order to tackle the word problem for an arbitrary presentation, a good strategy thus consists in trying to transform it into a convergent one by using Tietze transformations. From this point of view, we naturally ask ourselves whether a finite presentation of a category with decidable word problem can always be turned into a convergent one by applying Tietze transformations. This problem, called *universality of convergent presentations*, is introduced in Section 5.3. We will see in Chapters 8 and 9 that the answer to this question is negative.

5.1 Tietze Transformations

5.1.1 Definition. The *elementary Tietze transformations* are the following transformations of a 2-polygraph P into a 2-polygraph Q:

(T1) *adding a definable* 1*-generator*: given $a \notin P_1$, $u\colon x \to y \in P_1^*$, and $\alpha \notin P_2$ we define

$$Q = \langle\, P_0 \mid P_1, a\colon x \to y \mid P_2, \alpha\colon a \Rightarrow u \,\rangle\,,$$

(T2) *adding a derivable relation*: given $u, v \in P_1^*$ such that $u \approx v$ and $\alpha \notin P_2$, we define

$$Q = \langle\, P_0 \mid P_1 \mid P_2, \alpha\colon u \Rightarrow v \,\rangle\,.$$

The *Tietze equivalence* is the smallest equivalence relation on 2-polygraphs which is stable under isomorphisms and Tietze transformations. We respectively write $\overline{\text{(T1)}}$ and $\overline{\text{(T2)}}$ for operations (T1) and (T2) performed backward.

These local transformations completely axiomatize the property of presenting the same categories. This was first shown by Tietze [345] for presentations of groups, and the proof extends to the case of 2-polygraphs.

5.1.2 Theorem. *Two finite 2-polygraphs P and Q present the same category, i.e., $\overline{P} \simeq \overline{Q}$, if and only if they are Tietze equivalent.*

Proof. Let P be a 2-polygraph. If the 2-polygraph Q is either isomorphic to P or obtained by performing transformations (T1) or (T2) on P, then $\overline{Q}$ is isomorphic to $\overline{P}$. Therefore, any two Tietze equivalent 2-polygraphs present the same category.

Conversely, suppose that P and Q present the same category C. Up to isomorphism, that is, renaming of generators, we may suppose that $P_0 = Q_0$, $P_1 \cap Q_1 = \emptyset$, and $P_2 \cap Q_2 = \emptyset$. We write $q^P\colon P_1^* \to (P_1^*/{\approx_P}) = C$ for the quotient functor (see Section 2.3): this functor is full and such that $u \approx^P v$ precisely when $q^P(u) = q^P(v)$. Similarly, we also consider the quotient functor $q^Q\colon Q_1^* \to C$. Starting from the presentation P, we apply the following series of Tietze equivalences.

1. Given a 1-generator $a \in Q_1$, its image $q^Q(a)$ is a morphism of C and therefore has a representative in P_1^*: since q^P is full, there exists $u_a \in P_1^*$ satisfying $q^P(u_a) = q^Q(b)$. By a transformation (T1), we add to P the 1-generator a and the relation $u_a \Rightarrow a$. Performing this for every generator $a \in Q_1$, we obtain the 2-polygraph

$$P' = \langle\, C_0 \mid P_1 \cup Q_1 \mid P_2 \cup \{u_a \Rightarrow a \mid a \in Q_1\} \,\rangle.$$

2. Given a 1-cell $u = a_1 \dots a_n \in Q_1^*$, by construction of P', we have that $q^{P'}(u_a) = q^Q(a)$, which implies

$$\begin{aligned} q^{P'}(u) &= q^{P'}(a_1)\dots q^{P'}(a_n) \\ &= q^{P'}(u_{a_1})\dots q^{P'}(u_{a_n}) \\ &= q^Q(a_1)\dots q^Q(a_n) \\ &= q^Q(u). \end{aligned}$$

For each relation $\alpha : u \Rightarrow v$ in Q_2, we have $q^Q(u) = q^Q(v)$, which implies $q^{P'}(u) = q^{P'}(v)$ by the above, and therefore $u \approx^{P'} v$. By a transformation (T2), we can thus add to the previous 2-polygraph the derivable relation $\alpha : u \Rightarrow v$. Performing this for every relation $\alpha \in Q_2$, we obtain the 2-polygraph

$$P'' = \langle\, C_0 \mid P_1 \cup Q_1 \mid P_2 \cup Q_2 \cup \{u_a \Rightarrow a \mid a \in Q_1\}\,\rangle.$$

3. Suppose given a 1-generator $a \in P_1$. For similar reasons as in first step, there exists $v_a \in Q_1^*$ such that $q^Q(v_a) = q^P(a)$, which implies $q^{P''}(v_a) = q^{P''}(a)$, i.e., $v_a \approx^{P''} a$. By a transformation (T2), we can therefore add the derivable relation $v_a \Rightarrow a$. Performing this for every generator $a \in P_1$, we obtain the 2-polygraph P''' which is

$$\langle\, C_0 \mid P_1 \cup Q_1 \mid P_2 \cup Q_2 \cup \{v_a \Rightarrow a \mid a \in P_1\} \cup \{u_a \Rightarrow a \mid a \in Q_1\}\,\rangle.$$

By exchanging the roles of P and Q, one shows that Q is also Tietze equivalent to the same polygraph P'''. Therefore, the 2-polygraphs P and Q are Tietze equivalent. □

5.1.3 Remark. Similarly to the case of 1-polygraphs (Remark 1.2.13), Tietze transformations can be extended to account for infinite 2-polygraphs. The notion of Tietze transformation has to be refined in the following way: we say that a polygraph P *Tietze expands* to a polygraph Q when there is a transfinite sequence of Tietze transformations from P to Q, and we define *Tietze equivalence* as the smallest equivalence relation containing Tietze expansion. The above proof can be adapted in order to show that two polygraphs (of arbitrary cardinality) present the isomorphic categories if and only if they are Tietze equivalent.

5.1.4 Example. Consider the symmetric group S_3 on 3 elements. We consider it here as a monoid (in which elements happen to have inverses). It can be described as the category with only one object, whose morphisms are bijections $f : [3] \to [3]$, where $[3]$ denotes the set $\{0, 1, 2\}$ with three elements, equipped with usual composition and identity. This group is generated by the two transpositions s and t whose graphs are respectively

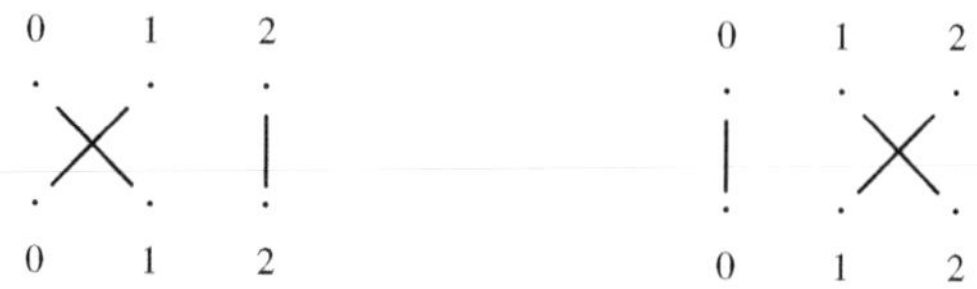

and working out the relations which are satisfied by those generators, one can come up with the following presentation of S_3:

$$P = \langle \star \mid s, t \mid ss = 1, tt = 1, sts = tst \rangle$$

See §§5.2.7 and A.1.19 for details.

The group S_3 can also be considered as the group of symmetries of an equilateral triangle

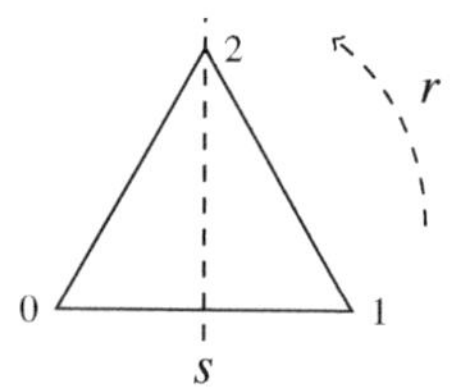

Namely, any bijection between the set of vertices determines a unique symmetry. As such, it can be generated by a symmetry s about a vertical axis and a rotation r of angle $2\pi/3$: those respectively correspond to the bijections between vertices whose graphs are

Note that the interpretation of s is the same as previously, and that r can be expressed in terms of the previous generators as $r = ts$. Working out the relations satisfied by those generators, one obtains the following presentation:

$$Q = \langle \star \mid s, r \mid rrr = 1, ss = 1, rsrs = 1 \rangle,$$

which is the usual presentation of the dihedral group D_3; see §A.1.24.

Since the two above 2-polygraphs P and Q present the same group, Theorem 5.1.2 asserts that they are Tietze equivalent. For instance, the presentation P

can be transformed into Q by the following series of Tietze transformations. Starting from the 2-polygraph P,

(T1) add the definable generator $r = ts$:

$$\langle \star \mid r, s, t \mid ss = 1, tt = 1, sts = tst, r = ts \rangle,$$

(T2) add the relation $rrr = 1$ (derivable since $rrr = tststs = ttstts = ss = 1$):

$$\langle \star \mid r, s, t \mid rrr = 1, ss = 1, tt = 1, sts = tst, r = ts \rangle,$$

(T2) add the relation $rsrs = 1$ (derivable since $rsrs = tsstss = tt = 1$):

$$\langle \star \mid r, s, t \mid rrr = 1, ss = 1, tt = 1, rsrs = 1, sts = tst, r = ts \rangle,$$

(T2) add the relation $t = rs$ (derivable since $t = tss = rs$):

$$\langle \star \mid r, s, t \mid rrr = 1, ss = 1, tt = 1, rsrs = 1, sts = tst, r = ts, t = rs \rangle,$$

$\overline{\text{(T2)}}$ remove the relation $r = ts$ (derivable since $r = rss = ts$):

$$\langle \star \mid r, s, t \mid rrr = 1, ss = 1, tt = 1, rsrs = 1, sts = tst, t = rs \rangle,$$

$\overline{\text{(T2)}}$ remove the relation $tt = 1$ (derivable since $tt = rsrs = 1$):

$$\langle \star \mid r, s, t \mid rrr = 1, ss = 1, rsrs = 1, sts = tst, t = rs \rangle,$$

$\overline{\text{(T2)}}$ remove the relation $sts = tst$, which is derivable since

$$sts = srss = sr = rrrsrss = rrs = rssrs = tst$$

to obtain

$$\langle \star \mid r, s, t \mid rrr = 1, ss = 1, rsrs = 1, t = rs \rangle.$$

$\overline{\text{(T1)}}$ finally, remove the definable generator t (which does not occur in any relation other than $t = rs$) to obtain Q.

5.1.5 Example. The monoid $(\mathbb{N}/3\mathbb{N}) \times (\mathbb{N}/2\mathbb{N})$ admits the presentation

$$\langle \star \mid s, t \mid s^3 = 1, t^2 = 1, ts = st \rangle$$

but it also admits the presentation

$$\langle \star \mid r \mid r^6 = 1 \rangle$$

(hint: define r by $r = ts$). This shows that we have an isomorphism

$$(\mathbb{N}/3\mathbb{N}) \times (\mathbb{N}/2\mathbb{N}) \simeq \mathbb{N}/6\mathbb{N}$$

as already noted in Example 3.3.7. More generally, one can show that the monoids

$$(\mathbb{N}/p\mathbb{N}) \times (\mathbb{N}/q\mathbb{N}) \simeq \mathbb{N}/pq\mathbb{N}$$

are isomorphic when p and q are relatively prime natural numbers.

5.1.6 Reduced 2-polygraphs. Tietze equivalences allow one to simplify presentations without changing the presented category. In particular, one can, without loss of generality, restrict to the following class of 2-polygraphs, which are often easier to handle than general 2-polygraphs. Those were studied by Metivier [280] for term rewriting systems, and Squier [326, Theorem 2.4] for string rewriting systems.

A 2-polygraph P is

- *left reduced* when, for every rule $\alpha\colon u \Rightarrow v$ in P_2, u is not reducible by any rule other than α,
- *right reduced* when, for every rule $\alpha\colon u \Rightarrow v$ in P_2, v is not reducible by any rule,
- *reduced* when it is both left and right reduced.

Note that a left reduced 2-polygraph never has inclusion critical branchings, as defined in §4.3.9, which often simplifies the study of branchings.

5.1.7 Theorem ([326, Theorem 2.4]). *Every convergent 2-polygraph P is Tietze equivalent to a reduced convergent 2-polygraph.*

Proof. Starting from the 2-polygraph P, we successively apply the following Tietze transformations.

1. Replace every 2-cell $\alpha\colon u \Rightarrow v$ by $\alpha\colon u \Rightarrow \widehat{u}$, where $\widehat{u}$ is the normal form of u:

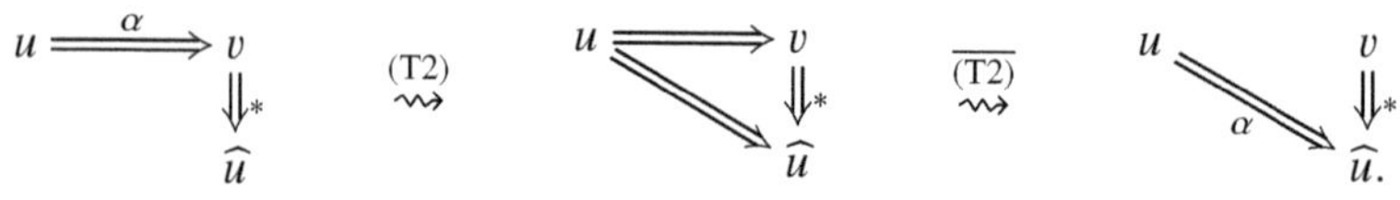

2. If the resulting 2-polygraph contains parallel 2-cells, remove all but one:

3. Finally, remove, in the resulting 2-polygraph, every 2-cell whose source is reducible by another 2-cell:

$$
\begin{array}{ccc}
uvw & \overset{\alpha}{\Longrightarrow} & \widehat{uvw} \\
& {\scriptstyle u\beta w}\searrow\!\!\searrow & \Uparrow \\
& & u\widehat{v}w
\end{array}
\quad \overset{\text{(T2)}}{\rightsquigarrow} \quad
\begin{array}{ccc}
uvw & & \widehat{uvw} \\
& {\scriptstyle u\beta w}\searrow\!\!\searrow & \Uparrow \\
& & u\widehat{v}w.
\end{array}
$$

These steps all correspond to Tietze transformations of type (T2) and the resulting polygraph is clearly reduced. □

5.1.8 Example. Consider the following presentation of the symmetric group S_3:

$$
\left\langle \star \;\middle|\; r, s, t \;\middle|\; \begin{array}{l} \sigma\colon\ ss \Rightarrow 1 \quad \gamma\colon\ sts \Rightarrow tst, \quad \rho\colon ts \Rightarrow r \\ \tau\colon\ \ tt \Rightarrow 1,\ \gamma'\colon\ sts \Rightarrow sstst \end{array} \right\rangle.
$$

It is not reduced because the target of γ' is not reduced (its normal form is tst) and the sources of γ and γ' are reducible by ρ. Applying the procedure described in the proof of Theorem 5.1.7, we obtain the following reduced, Tietze equivalent, presentation:

$$
\langle \star \mid r, s, t \mid \sigma\colon ss \Rightarrow 1, \tau\colon tt \Rightarrow 1, \gamma\colon sr \Rightarrow rt, \rho\colon ts \Rightarrow r \rangle.
$$

5.1.9 The reduced standard presentation. In §4.5.5, we have seen that every category C admits a canonical presentation, the standard presentation. One can actually achieve a smaller presentation by not adding identities as 1-generators. The *reduced standard polygraphic presentation* of C is the 2-polygraph R where

- R_0 is the set of objects of C,
- R_1 is the set of morphisms of C which are not identities,
- R_2 contains 2-cells of the form

$$
\mu_{a,b}\colon ab \Rightarrow (b \circ a)\colon x \to z,
$$

for every object x of C and pair of composable morphisms $a\colon x \to y$ and $b\colon y \to z$ in C such that $b \circ a$ is not an identity, and 2-cells of the form

$$
\mu'_{a,b}\colon ab \Rightarrow x\colon x \to x,
$$

for every object x of C and pair of composable morphisms $a\colon x \to y$ and $b\colon y \to x$ in C such that $b \circ a = 1_x$.

The proof that this is indeed a presentation of C can be performed by adapting the rewriting argument provided in §4.5.5. Another way to show this, since we know that the standard presentation P of C is a presentation of C, is to show that R is Tietze equivalent to C. Starting from P, this can be done by using the following series of Tietze transformations.

– For each 2-cell $\mu_{a,b}\colon ab \Rightarrow (b \circ a)\colon x \to x$ such that $b \circ a = 1_x$ is an identity one can add the derivable 2-cell $\mu'_{a,b}\colon ab \Rightarrow x$ and remove the 2-cell $\mu_{a,b}$ by using transformations (T2):

Note the subtle difference between $\mu_{f,g}$ and $\mu'_{f,g}$: in the first case the target is the path of length one consisting of the 1-generator 1_x, whereas in the second case it is the path of length zero at x.

– For each $x \in P_0$, remove the 1-generator 1_x along with the 2-cell η_x by using transformation (T1): this can be done because this 1-generator does not occur in the source or target of any 2-cell other than η_x.

The resulting 2-polygraph is the reduced standard presentation. In fact, this presentation is precisely the one that one would obtain by applying the procedure described in the proof of Theorem 5.1.7.

5.1.10 Tietze reductions. There are two kinds of Tietze transformations: (T1) adding a definable generator and (T2) adding a derivable relation. During a Tietze equivalence, those can also be performed backward: $\overline{\text{(T1)}}$ removing a definable generator and $\overline{\text{(T2)}}$ removing a derivable relation. A Tietze equivalence using only the two backward transformations is called a *Tietze reduction* and consists in making the presentation smaller by suitable removing generators and relations. It would be nice if two 2-polygraphs P and Q were Tietze equivalent if and only if they reduce to a common 2-polygraph: this would mean that we do not have to come up with new generators or relations in order to study Tietze equivalence. We have seen in §1.2.10 that this holds in the case of 1-polygraphs, but we show here that this is not the case for 2-polygraphs. This explains why the proof of Theorem 5.1.2 proceeds by transforming two polygraphs into a bigger one, which contains both, and not a smaller one.

Consider the presentation

$$P = \langle \star \mid a, b \mid \alpha\colon aa \Rightarrow a, \beta\colon bb \Rightarrow b, \gamma\colon aa \Rightarrow bb \rangle.$$

One can apply to it the following Tietze transformations:

(T2) add the derivable relation $a = b$:

$$\langle \star \mid a, b \mid aa = a, bb = b, aa = bb, a = b \rangle$$

(the relation is derivable by $a = aa = bb = b$),

$\overline{\text{(T2)}}$ remove the derivable relation $bb = b$:

$$\langle\, \star \mid a, b \mid aa = a, aa = bb, a = b \,\rangle$$

(the relation is derivable by $bb = aa = a = b$),

$\overline{\text{(T2)}}$ remove the derivable relation $aa = bb$:

$$\langle\, \star \mid a, b \mid aa = a, a = b \,\rangle$$

(respectively derivable since $a = b$),

$\overline{\text{(T1)}}$ remove the definable generator b:

$$P' = \langle\, \star \mid a \mid aa = a \,\rangle.$$

The polygraphs P and P' are thus Tietze equivalent and P presents the free monoid with an idempotent element: this monoid has two elements 1 and a, with multiplication given by $1a = a1 = aa = a$.

The polygraph P is *Tietze minimal*, in the sense that no non-trivial Tietze reduction can be applied to it; otherwise said, in order to prove a non-trivial Tietze equivalence, one has to begin by adding definable generators or derivable relations. Namely, the 1-generator a cannot be removed along the relation α because a occurs in the source of γ, and similarly for β. Finally, we can show that no relation is derivable by contradiction as follows.

- Suppose that α is derivable. This means that the 2-polygraph P is Tietze equivalent to the 2-polygraph

$$Q = \langle\, \star \mid a, b \mid \beta\colon bb \Rightarrow b, \gamma\colon aa \Rightarrow bb \,\rangle$$

and we have $\overline{P} \simeq \overline{Q}$. The 2-polygraph Q is not convergent, but it is Tietze equivalent to the convergent 2-polygraph

$$Q' = \langle\, \star \mid a, b \mid \beta\colon bb \Rightarrow b, \gamma'\colon aa \Rightarrow b, \delta\colon ba \Rightarrow ab \,\rangle$$

(the relation δ is derivable by $ba = bba = aaa = ab$; see also Example 5.2.3). The termination of Q' can be shown using the deglex order generated by $b > a$, and the critical branchings are confluent:

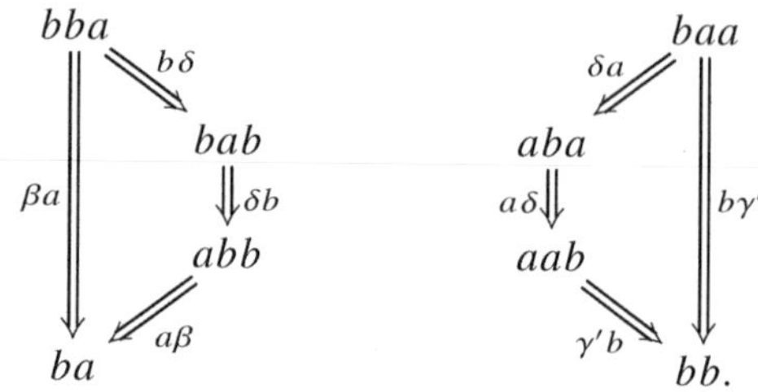

The normal forms are 1, a, b and ab; i.e., there are four morphisms in $\overline{Q}$ whereas there are only two in $\overline{P}$, contradicting the isomorphism $\overline{P} \simeq \overline{Q}$.

- By exchanging the role of a and b in previous case, and reversing the orientation of γ (which does not change the presented category), the relation β is not derivable either.
- Suppose that γ is derivable. This means that P is Tietze equivalent to the 2-polygraph

$$Q = \langle \star \mid a, b \mid \alpha \colon aa \Rightarrow a, \beta \colon bb \Rightarrow b \rangle.$$

Again, Q is convergent (rules decrease the length of morphisms and there is no critical pair) and every word of the form $ababab\ldots$ as normal form, whereas $\overline{P}$ has only two elements.

Since the presentation P' is also (obviously) minimal, we see that there is no way to show that P and P' are Tietze equivalent by Tietze reducing both to a common 2-polygraph.

5.1.11 Tietze transformations up to equivalence. We have seen in Theorem 5.1.2 that Tietze transformations generate the following equivalence relation on polygraphs: two polygraphs are equivalent when they present isomorphic categories. We consider here the following variant of the notion of equivalence: two polygraphs are *equivalent* when they present equivalent categories. A corresponding notion of Tietze transformation can be obtained as a variant of those presented in §5.1.1, by adding the following kind of transformation:

(T0) *adding an isomorphic* 0*-generator*: given $x \in P_0$, $y \notin P_0$, $a, b \notin P_1$, $\alpha, \beta \notin P_2$, we define

$$Q = \left\langle\, P_1, y \,\middle|\, P_1, a \colon x \to y, b \colon y \to x \,\middle|\, P_2, \alpha \colon ab \Rightarrow 1_x, \beta \colon ba \Rightarrow 1_y \,\right\rangle.$$

5.2 The Knuth–Bendix Completion Procedure

We have seen that convergent 2-polygraphs are very convenient to work with. When given a polygraph which does not have this property, one can in many cases use Tietze transformations to turn it into one which does, preserving the presented category. We present here a procedure due to Knuth and Bendix [218] (in the setting of term rewriting systems) whose purpose is to perform this transformation in an automated way: starting from a 2-polygraph with a reduction order, it adds definable 2-generators until possibly reaching a convergent 2-polygraph, which is Tietze equivalent to the original one. We use the terminology of "procedure" and not an "algorithm", because there is no guarantee that it will eventually stop, although it very often does in practice.

This procedure is based on two observations. The first one is that, in a terminating 2-polygraph, the completion of any confluent critical branching can always be chosen to be convergent toward a normal form. In fact, suppose that a critical pair (ϕ_1, ϕ_2) is closed by (ϕ'_1, ϕ'_2) as shown on the left diagram below:

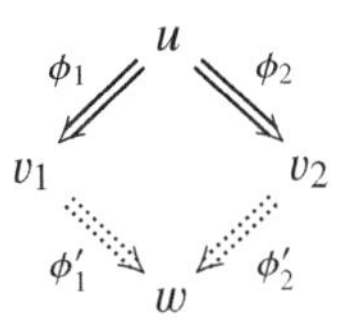

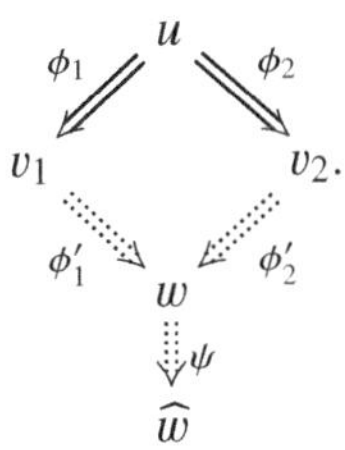

The termination property yields a normalization path ψ from w to a normal form $\widehat{w}$ of w, so we may close the diagram by the new pair $(\psi \circ \phi'_1, \psi \circ \phi'_2)$ as shown above on the right.

The second observation is that, given a non-confluent critical pair as on the left below,

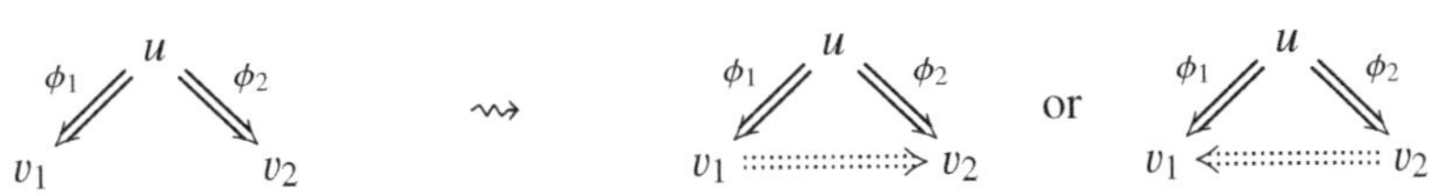

we have $v_1 \approx v_2$ and it is therefore possible to add the definable relation $v_1 \Rightarrow v_2$ or $v_2 \Rightarrow v_1$ to the polygraph without changing the presented category since this is a Tietze transformation of type (T2). The new presentation is "more confluent" in the sense that the above critical pair is now confluent. We are thus tempted to add new rules in this way for every critical branching. However, newly added rules can create new non-confluent branchings and we want therefore to add as few of them as possible. For instance, in the above situation, suppose that we have added a rule $v_1 \Rightarrow v_2$ and that there was already another reduction $v_1 \Rightarrow v_1'$, making a non-confluent branching, as shown on the left below:

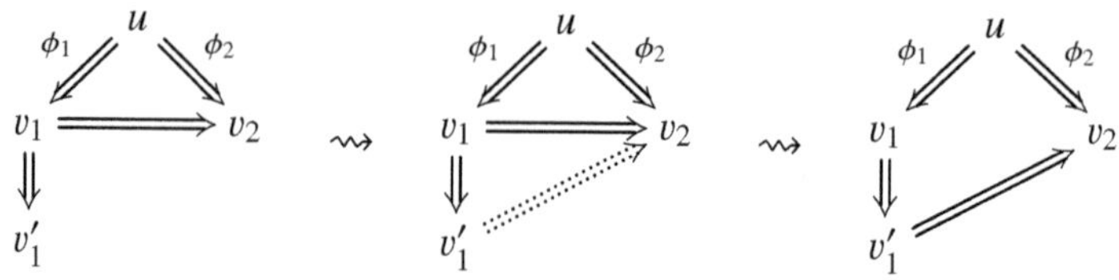

In order to make the polygraph confluent, we are now forced to add a new rule between v_1' and v_2, say, $v_1' \Rightarrow v_2$, making the former rule $v_1 \Rightarrow v_2$ useless: it would have been preferable to directly add the rule $v_1' \Rightarrow v_2$ instead of $v_1 \Rightarrow v_2$. For this reason, given a critical branching as above, we only add new rules $\widehat{v_1} \Rightarrow \widehat{v_2}$ (or $\widehat{v_2} \Rightarrow \widehat{v_1}$), between normal forms $\widehat{v_1}$ (resp. $\widehat{v_2}$) of v_1 (resp. v_2). Finally the newly added rules must be oriented without breaking the termination of the original polygraph. This is usually done by orienting rules according to a reduction order.

5.2.1 The completion procedure. Suppose given a finite 2-polygraph P, equipped with a total reduction order $\preccurlyeq$ which is compatible with P, i.e., $u \succ v$ for every 2-generator $\alpha : u \Rightarrow v$ in P_2^*. By Proposition 4.4.2, the polygraph P is necessarily terminating.

The *Knuth–Bendix completion procedure* starts with the 2-polygraph P and iteratively transforms it by adding definable relations, as follows.

1. For every critical branching

$$v \overset{\phi}{\Longleftarrow} u \overset{\psi}{\Longrightarrow} w$$

we compute reduction paths $\phi' : v \overset{*}{\Rightarrow} \widehat{v}$ and $\psi : w \overset{*}{\Rightarrow} \widehat{w}$ to some normal forms $\widehat{v}$ and $\widehat{w}$ of v and w respectively, until finding one with $\widehat{v} \neq \widehat{w}$. If there is none the procedure halts and returns the computed polygraph.
2. With the normal forms computed in previous step, we either have $\widehat{v} \succcurlyeq \widehat{w}$, in which case we add a 2-generator $\alpha : \widehat{v} \Rightarrow \widehat{w}$ to P, or $\widehat{v} \preccurlyeq \widehat{w}$, in which case we add a 2-generator $\alpha : \widehat{w} \Rightarrow \widehat{v}$ to P:

3. Go back to step 1.

If the procedure stops, it returns a 2-polygraph, which we denote as $\mathrm{KB}(P)$ and call a *Knuth–Bendix completion* of P. In the case where the procedure does not terminate, it constructs an infinite sequence $P = P^0, P^1, P^2, \ldots$ of 2-polygraphs, where P^{i+1} is obtained from P^i by adding a derivable relation. This sequence is thus increasing, in the sense that we have $P^i \subseteq P^j$ for $i \leqslant j$, and thus admits an inductive limit $\bigcup_i P^i$, which we still denote as $\mathrm{KB}(P)$.

5.2.2 Theorem ([218, 189]). *The Knuth–Bendix completion* $\mathrm{KB}(P)$ *of a 2-polygraph* P *is a convergent presentation of the category* $\overline{P}$.

Proof. Since all the rules respect the termination order by construction, the reduction order $\preccurlyeq$ is a termination order, and the polygraph $\mathrm{KB}(P)$ is thus terminating by Proposition 4.4.2. Moreover, step 1 ensures that all the critical branchings are confluent, and the polygraph $\mathrm{KB}(P)$ is thus locally confluent by Lemma 4.3.7 and confluent by Lemma 1.3.21. Finally, the procedure proceeds by adding derivable transformations at step 2, i.e., by performing Tietze transformations of type (T2). By Theorem 5.1.2, the polygraph $\mathrm{KB}(P)$ thus presents the same category as P. □

Note that the above theorem applies in both the cases where the procedure terminates and where it does not. It can moreover be noted that the 2-polygraph $\mathrm{KB}(P)$ is finite if and only if the 2-polygraph P is finite and the Knuth–Bendix completion procedure halts. For implementation purposes, we are thus mostly interested in the cases where the procedure computes a result after a finite amount of time, but for theoretical purposes it is still useful when it runs indefinitely. It is also interesting to remark that if the starting 2-polygraph P is already convergent, we immediately have $\mathrm{KB}(P) = P$.

5.2.3 Example. Consider the 2-polygraph

$$P = \langle \star \mid a, b \mid bb \Rightarrow b, aa \Rightarrow bb \rangle$$

already encountered in §5.1.10, equipped with the deglex order generated by $b > a$, which is compatible with P. The two critical branchings are

and the dotted arrows are chosen normalization 1-cells. In the first case, the two normal forms are equal, but not in the second one. Since $ba > ab$, the Knuth–Bendix procedure adds a rule $ba \Rightarrow ab$, thus obtaining the 2-polygraph

$$P = \langle \star \mid a, b \mid bb \Rightarrow b, aa \Rightarrow bb, ba \Rightarrow ab \rangle.$$

Once this new rule added, all the critical pairs are confluent (see §5.1.10), so that the procedure halts on the above convergent 2-polygraph.

5.2.4 Example. Consider the following 2-polygraph from [238]

$$P = \langle \star \mid a, b, c, d \mid \alpha_0 \colon ab \Rightarrow a, \beta \colon da \Rightarrow ac \rangle,$$

equipped with the deglex order associated to the reverse alphabetic order. The Knuth–Bendix completion does not terminate and gives rise to the infinite convergent presentation

$$P = \langle \star \mid a, b, c, d \mid \alpha_n \colon ac^n b \Rightarrow ac^n, \beta \colon da \Rightarrow ac \rangle_{n \in \mathbb{N}}.$$

Namely, at the n-th step of the procedure the rule α_{n+1} is added by closing the critical branching

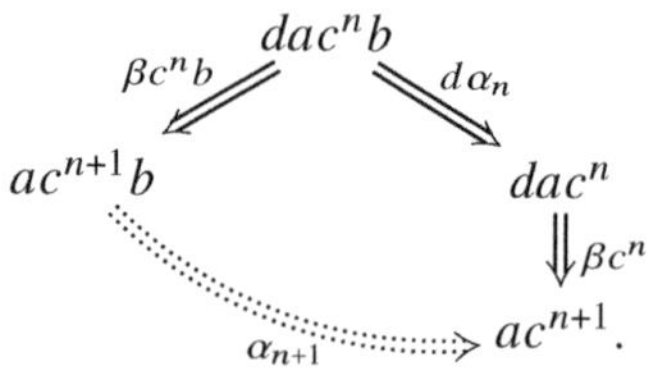

It can be remarked that if we take the converse orientation for rule β

$$P = \langle \star \mid a, b, c, d \mid \alpha_0 \colon ab \Rightarrow a, \beta \colon ac \Rightarrow da \rangle$$

and equip the polygraph with the deglex order associated to the reverse alphabetic order, the procedure halts immediately since there is no critical branching.

As illustrated in the above example, the procedure depends on many parameters, each of which can have a strong influence on the output of the procedure, i.e., how small the completed polygraph will be, or even the termination of the procedure: the termination order, the order in which critical pairs are studied in step 1, the normal forms $\widehat{v}$ and $\widehat{w}$ chosen for each critical pair in step 1.

5.2.5 Detailed description of the procedure. The procedure can be improved so that it produces reduced polygraphs, by combining it with the procedure presented in §5.1.6. It can also be more efficiently implemented by observing that if a pair is confluent at some stage, then it is still confluent if new rules are added; therefore one can restrict step 1 to consider only critical branchings formed by newly added rules. In a more operational way, close to the presentation given by Huet [189], the resulting improved procedure can be described as follows.

We generalize here slightly the situation of the previous section and suppose given a 2-polygraph P together with a reduction order which is not necessarily compatible with P (the procedure will reorient the rules anyway) and may be partial. The procedure will modify the following variables:

- a set E of equations, i.e., pairs $u = v$ with $u, v \in P_1^*$, whose initial value is

$$E = \{u = v \mid \alpha \colon u \Rightarrow v \in P_2\}$$

- a polygraph Q, which is initially the polygraph P where the set of rules has been replaced by the empty set:

$$Q_0 = P_0 \qquad\qquad Q_1 = P_1 \qquad\qquad Q_2 = \emptyset.$$

The procedure repeats the following steps until we have $E = \emptyset$:

1. pick an equation $u = v$ in E and remove it from E,
2. compute normal forms $\widehat{u}$ and $\widehat{v}$ of u and v,
3. if $\widehat{u} = \widehat{v}$ then go back to step 1,
4. if neither $\widehat{u} < \widehat{v}$ or $\widehat{v} < \widehat{u}$ then fail,
5. if $\widehat{u} < \widehat{v}$ then exchange u and v (and $\widehat{u}$ and $\widehat{v}$) so that $\widehat{u} > \widehat{v}$,
6. for each rule $\alpha_i \colon u_i \Rightarrow v_i$ in Q_2 such that u_i rewrites to u_i' by the rule $u \Rightarrow v$

   a.b remove α_i from Q_2,
   b.b add $u_i' = v_i$ to E,

7. add $\alpha \colon u \Rightarrow v$ to Q_2,
8. replace each rule $\alpha_i \colon u_i \Rightarrow v_i$ of Q_2 by $\alpha_i \colon u_i \Rightarrow \widehat{v_i}$, where $\widehat{v_i}$ is a normal form of v_i with respect to Q as computed in previous step,
9. in the polygraph Q, for each critical branching

$$u' \overset{\phi}{\Leftarrow} w \overset{\psi}{\Rightarrow} v'$$

   where ϕ consists of the rule α in context, add $u' = v'$ to E.

In the end, i.e., when $E = \emptyset$ is reached after a finite number of steps, the procedure returns the polygraph Q. This polygraph is reduced and Tietze equivalent to the original polygraph P. It is also possible to reasonably define a notion of outcome of the procedure when it does not terminate; see [189] for details.

5.2.6 Example. Consider the presentation

$$\langle \star \mid a, b, c \mid aba \Rightarrow bab, ba \Rightarrow c \rangle$$

obtained from the usual presentation of B_3^+, see §A.1.21, by adding a generator c along with its definition $ba = c$. We consider the deglex order induced by $a > b > c$, which is compatible with the rules. The procedure will

- replace $aba \Rightarrow bab$ by $ac \Rightarrow cb$,
- add the rule $bcb \Rightarrow cc$ coming from the non-confluent critical branching

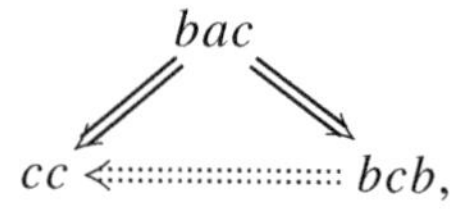

- add the rule $bcc \Rightarrow cca$ coming from the non-confluent critical branching

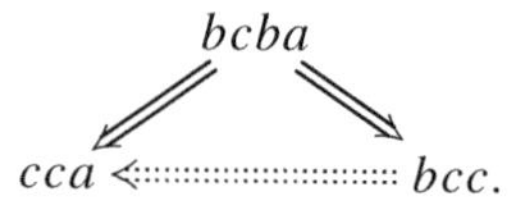

We finally obtain the convergent Tietze equivalent presentation

$$\langle \star \mid a, b, c \mid ba \Rightarrow c, ac \Rightarrow cb, bcb \Rightarrow cc, bcc \Rightarrow cca \rangle.$$

5.2.7 The symmetric group. Let us work out a fundamental and non-trivial example of a presentation of a monoid. Given $n \in \mathbb{N}$, we consider the symmetric group S_{n+1} of bijections on a set with $n + 1$ elements. We claim that it admits a presentation by the 2-polygraph

$$P = \left\langle \star \,\middle|\, a_0, \dots, a_{n-1} \,\middle|\, \alpha_i, \beta_i, \gamma_{i,j} \right\rangle$$

where the 2-generators are

$$\begin{array}{rrl} \alpha_i : & a_i a_i \Rightarrow 1 & \text{for } 0 \leqslant i < n, \\ \beta_i : & a_{i+1} a_i a_{i+1} \Rightarrow a_i a_{i+1} a_i & \text{for } 0 \leqslant i < n-1, \\ \gamma_{i,j} : & a_j a_i \Rightarrow a_i a_j & \text{for } 0 \leqslant i < i+1 < j < n. \end{array}$$

See §A.1.19 for details. Our strategy to show this result is based on the following two steps.

1. We use the Knuth–Bendix completion procedure to compute a convergent 2-polygraph Q presenting the same category as P.
2. We use the techniques presented in Section 4.5 to show that Q is a presentation of S_{n+1}, by showing that elements of Q_1^* in normal form are in bijection with the elements of S_{n+1}.

In order to apply the Knuth–Bendix procedure, we equip the polygraph P with the deglex reduction order $\preccurlyeq$ induced by $a_j > a_i$ whenever $j > i$, which is compatible with the rules. After a finite amount of steps, the Knuth–Bendix procedure terminates, producing the convergent 2-polygraph Q with the same 0- and 1-generators, and with rules

$$\begin{array}{rlrl} \alpha_i : & a_i a_i \Rightarrow 1 & & \text{for } 0 \leqslant i < n, \\ \beta_i^k : & a_{i+k+1} \ldots a_i a_{i+k+1} \Rightarrow a_{i+k} a_{i+k+1} \ldots a_i & & \text{for } 0 \leqslant k < n, \\ & & & \text{and } 0 \leqslant i < n-k-1, \\ \gamma_{i,j} : & a_j a_i \Rightarrow a_i a_j & & \text{for } 0 \leqslant i < i+1 < j < n, \end{array}$$

where $a_{i+k+1} \ldots a_i$ denotes the sequence of a_j, with indices j decreasing one by one between $i+k+1$ and i. The reader is advised to compute this by himself or refer to [247] for details.

In an element of Q_1^* in normal form, because of the rules α_i and $\gamma_{i,j}$, if we have a factor $a_i a_j$, then we have $i < j$ or $i = j + 1$. Taking the rules $\beta_{i,j}$ in account too, we see that the normal forms are the words of the form

$$w_0 w_1 w_2 \ldots w_{n-1} \qquad \text{with} \qquad w_i = a_i a_{i-1} a_{i-2} \ldots a_{i-k_i}.$$

Now, let us show that those normal forms are in bijective correspondence with the elements of S_{n+1}, i.e., bijections $f\colon [n+1] \to [n+1]$. First, we interpret the generator a_i as the bijection $[\![a_i]\!] : [n+1] \to [n+1]$ which exchanges i and $i+1$, and can be depicted as

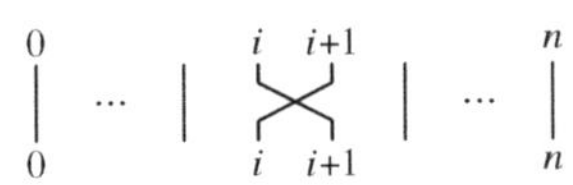

To any bijection $f\colon [n+1] \to [n+1]$, we associate a 1-cell $u_f \in P_1^*$ defined by induction on n. We set $u_f = 1$ whenever $n = 0$. Otherwise, we write $f'\colon [n] \to [n]$ for the function obtained from f by “removing” n from the source of f and $f(n)$ from its image, i.e.,

$$f'(i) = \begin{cases} f(i) & \text{if } f(i) < f(n) \\ f(i) - 1 & \text{if } f(i) > f(n) \end{cases}$$

and define

$$u_f = u_{f'} a_{n-1} a_{n-2} \ldots a_{f(n)}.$$

For instance, consider the bijection $f\colon [6] \to [6]$ such that the images of 0, 1, 2, 3, 4, and 5 are respectively 4, 1, 0, 5, 2, and 3. Its associated word u_f is $a_0 a_1 a_0 a_3 a_2 a_4 a_3$, which can be pictured as

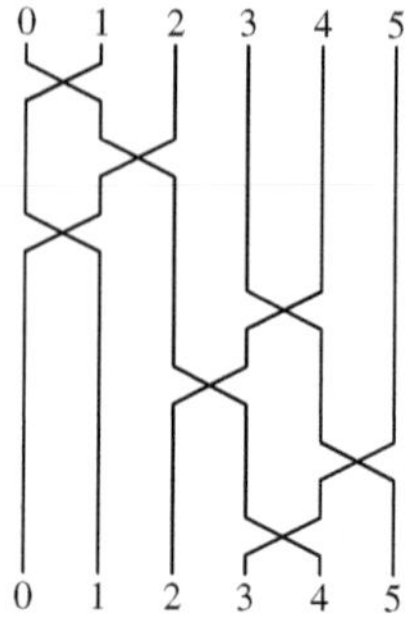

Finally, using the above description of normal forms, it can be shown that u_f is a normal form for any bijection f, and that this provides a bijection between normal forms and elements of S_{n+1}. Other examples of such completions for finite groups can be found in [145, 166, 247].

5.2.8 Generated subcategories. As an application of the previously developed techniques, consider the following situation. We suppose given a category C presented by a 2-polygraph P and a set G of morphisms of C, whose elements are called *generators*. The *category generated* by G, denoted $\langle G\rangle$, is the smallest subcategory of C which contains the elements of G as morphisms (and is closed under identities and composition, source and target of morphisms in C). Our goal here is to compute a presentation of it: we will provide a method to perform this in the case where C admits a suitable convergent presentation. Before addressing the general case, we look at the following example: let C be the monoid $\mathbb{N}/6\mathbb{N}$, presented by

$$\langle\, \star \mid a \mid a^6 \Rightarrow 1 \,\rangle$$

and let us compute the category generated by $G = \{a^2\}$.

First, note that we can always suppose that each morphism $f \in G$ admits a 1-generator $b \in P_1$ as a representative. Otherwise, given a representative $u \in P_1^*$ of f, we can apply to P the Tietze transformation which consists in adding a new generator b together with the rule $u \Rightarrow b$. In our example, this amounts to considering the presentation

$$\langle\, \star \mid a, b \mid a^6 \Rightarrow 1, a^2 \Rightarrow b \,\rangle.$$

For this reason, we will suppose in the following that the set G of generating morphisms is a subset of the 1-generators, i.e., $G \subseteq P_1$ (in the above example, we have $G = \{b\}$). Moreover, we can suppose that the presentation is convergent and reduced: if it is not the case, we can apply the Knuth–Bendix procedure, and hope that it succeeds. For instance, with the previous presentation, consider the deglex order with $a < b$. The critical branchings

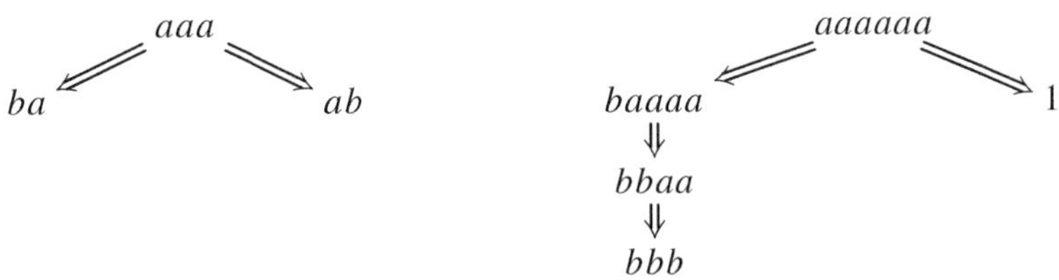

are not confluent, and it can be checked that adding the induced relations $ba \Rightarrow ab$ and $b^3 \Rightarrow 1$ makes the presentation convergent and the rule $a^6 \Rightarrow 1$ superfluous. We thus consider the alternative, convergent, presentation

$$\langle \star \mid a, b \mid a^2 \Rightarrow b, ba \Rightarrow ab, b^3 \Rightarrow 1 \rangle$$

of $\mathbb{N}/6\mathbb{N}$. Using Proposition 5.2.9 below, we can finally deduce that the category $\langle a^2 \rangle$ admits the presentation

$$\langle \star \mid b \mid b^3 \Rightarrow 1 \rangle.$$

It is thus the monoid $\mathbb{N}/3\mathbb{N}$, as expected.

Below, given $G \subseteq P_1$, we write $G^* \subseteq P_1^*$ for the set of morphisms in P_1^* which can be expressed as composites of generators in G.

5.2.9 Proposition. *Suppose given a convergent 2-polygraph P together with a set $G \subseteq P_1$ of generators, such that for every rule $u \Rightarrow v$ in P_2 with $u \in G^*$ we have $v \in G^*$. Then the category generated by G admits a presentation by the polygraph Q where*

- *$Q_0 \subseteq P_0$ consists of the sources and targets of elements of G,*
- *$Q_1 = G \subseteq P_1$,*
- *$Q_2 \subseteq P_2$ consists of the rules $u \Rightarrow v$ in P_2 such that $u \in G^*$ and $v \in G^*$.*

Proof. Since $\langle G \rangle$ has to be closed under taking the source and target of morphisms in G, it contains at least Q_0 as objects, and conversely, any composite of morphisms in G will have elements of Q_0 as source and target; Q_0 is thus precisely the set of 0-cells of $\langle G \rangle$. The morphisms in $\langle G \rangle$ contain the equivalence classes $\overline{a}$ of 1-generators $a \in G$, and since it is closed under composition and identities, its morphisms are precisely the equivalence classes of morphisms in $G^* \subseteq P_1^*$. Finally, given $u, v \in G^*$ such that $\overline{u} = \overline{v}$, since P is convergent both u and v rewrite to a common element $w \in P_1^*$, i.e., $u \overset{*}{\Rightarrow} w$ and $v \overset{*}{\Rightarrow} w$. By induction, the two rewriting paths contain only rules in P_1 and $w \in G^*$; thus Q_2 is sufficient to generate the required equivalence on elements of G^*. □

As a bonus, note that the 2-polygraph Q in the previous proposition is necessarily convergent, because P is supposed to be so.

5.2.10 Remark. Suppose that we start with a 2-polygraph P together with a set $G \subseteq P_1$, such that the following property is satisfied: for every rule $u \Rightarrow v$ in P_2, $u \in G^*$ implies $v \in G^*$. In order to be able to apply previous proposition, we need to ensure that P is convergent and, if this is not the case, we can apply the Knuth–Bendix completion procedure in order to obtain a convergent polygraph. However, in general, the completed polygraph will not satisfy the property anymore. In order to improve this, the Knuth–Bendix procedure can be modified in order not to produce "bad rules", i.e., rules of the form $u \Rightarrow v$ with $u \in G^*$ and $v \in P_1^* \setminus G^*$, which prevent the resulting polygraph from satisfying the required property. Namely, the completion procedure adds new rules of the form $u \Rightarrow v$ where both u and v are normal forms. In the case such a rule is "bad", it can be useful to add instead a rule $u' \Rightarrow v$ where $u' \overset{*}{\Rightarrow} u$ and $u' \in P_1^* \setminus G^*$.

5.2.11 Exercise. A presentation for the symmetric groups S_n was constructed in §5.2.7. Deduce from it a presentation for the alternating groups A_3 and A_4; see §A.1.20.

5.3 Universality of Finite Convergent Rewriting

We have seen in Section 4.2 that a finite convergent rewriting system always has decidable word problem. The question of *universality of convergent rewriting* is the converse question, first asked by Jantzen [202]; see also [32, 113, 203, 204]:

> Given a category C admitting a finite presentation with decidable word problem, does it always admit a finite convergent presentation?

The answer to this question is negative, but showing this requires more tools than we have at our disposal for now and will be handled in Chapters 8 and 9. We however study here restricted forms of the question.

5.3.1 Universality of Knuth–Bendix completion. A more restricted variant of the above question consists in wondering whether it is always possible to add or remove relations to a 2-polygraph so that it becomes convergent. Kapur and Narendran [213] have shown that this is not the case, by considering the usual presentation of the braid monoid B_3^+, detailed in §A.1.21:

$$P = \langle \star \mid a, b \mid aba \Rightarrow bab \rangle. \tag{5.1}$$

They show that there is no finite convergent presentation of this monoid on the same generators; see Proposition 5.3.3 below. As a consequence, for such a presentation, the Knuth–Bendix procedure will never end whichever reduction order or strategy for considering rules is adopted.

5.3.2 Lemma. *The polygraph P has decidable word problem.*

Proof. Since the only relation preserves the length of 1-cells, equivalence classes contain 1-cells of the same length and are therefore finite. □

5.3.3 Proposition. *There is no finite convergent rewriting system which is Tietze equivalent to the polygraph P by a sequence of Tietze transformation consisting only in adding or removing derivable relations.*

Proof. First notice that $abbab \overset{*}{\Leftrightarrow} babba$ is derivable in P since we have

$$abbab \Leftarrow ababa \Rightarrow abbab.$$

More generally, by induction, it can be shown that

$$a^{i+1}b^{j+2}ab \overset{*}{\Leftrightarrow} bab^{i+2}a^{j+1} \tag{5.2}$$

for every $i, j \in \mathbb{N}$. Namely, the base case where $i = j = 0$ is handled above, and if we suppose that (5.2) holds for some i and j, we have

$$a^{i+2}b^{j+2}ab \overset{*}{\Leftrightarrow} abab^{i+2}a^{j+1} \Rightarrow bab^{i+3}a^{j+1}$$

and

$$a^{i+1}b^{j+2}ab \Leftarrow a^{i+1}b^{j+1}aba \overset{*}{\Leftrightarrow} bab^{i+2}a^{j+2},$$

which constitute the induction step on i and j, respectively. Another easy remark is that, for $n \in \mathbb{N}$, any word u such that the relation $u \overset{*}{\Leftrightarrow} b^nab$ (resp. $u \overset{*}{\Leftrightarrow} bab^n$) is derivable is of the form $u = b^{n-i}aba^i$ (resp. $u = a^ibab^{n-i}$) for some i with $0 \leqslant i \leqslant n$. Writing $\overline{u}$ for the equivalence class of a word u under $\overset{*}{\Leftrightarrow}$, we thus have

$$\overline{b^nab} = \left\{b^{n-i}aba^i \mid 0 \leqslant i \leqslant n\right\} \qquad \overline{bab^n} = \left\{a^ibab^{n-i} \mid 0 \leqslant i \leqslant n\right\}.$$

We now proceed by contradiction. Suppose given a finite 2-polygraph Q convergent and Tietze equivalent to P by a sequence of Tietze transformations consisting only in adding or removing derivable relations. By Theorem 5.1.7, we can suppose that Q is reduced. Since, $\{aba, bab\}$ is an equivalence class, Q should contain either $aba \Rightarrow bab$ or $bab \Rightarrow aba$. We suppose that we are in the former case, with the other one being similar. Writing l for the length of the longest left-hand side of a rule in Q, and

$$u = a^{l+1}b^{l+2}ab \qquad\qquad v = bab^{l+2}a^{l+1}$$

we have $u \overset{*}{\Leftrightarrow} v$ and therefore both u and v should reduce to a common word. The only factors in those words whose equivalence class is not a singleton are of the form $b^n ab$ or bab^n. Therefore, we must have rules of the form $b^n ab \Rightarrow w$ or $bab^n \Rightarrow w$. By the preceding remark, the word w has to be of the form $b^{n-i}aba^i$ (resp. $a^i bab^{n-i}$) with $0 < i \leqslant n$ and therefore is reducible by the rule $aba \Rightarrow bab$, which contradicts the assumption that the rewriting system is reduced. □

As an alternative example, it is shown in [203] that the monoid (in fact, group) presented by

$$\langle\, \star \mid a, b \mid abba = 1 \,\rangle$$

admits no finite convergent presentation on the same generators.

5.3.4 Other Tietze transformations. The result of Proposition 5.3.3 can be restated as follows: there is a finite 2-polygraph P which cannot be transformed into a finite convergent one by using Tietze transformations (T2) only. However, this does not bring a definitive answer to the original question of universality of rewriting raised at the beginning of this section, since it does not rule out the possibility of turning a presentation into a convergent one by using both transformations (T1) and (T2). In fact, this is the case for the presentation (5.1) of B_3^+. Namely, if we use a transformation (T1) to introduce a generator c and a relation $ba = c$, the resulting presentation can be completed into a convergent one; this was already detailed in Example 5.2.6: adding a superfluous generator allows the Knuth–Bendix procedure to produce a convergent presentation of B_3^+. Finding a counter-example to the problem of universality in full generality is much more difficult and will be addressed in Chapters 8 and 9.

The situation encountered for B_3^+, where the introduction of a definable generator improves the properties of the presentation is not an "isolated case". For instance for every natural number $n > 3$, the plactic monoid P_n of type A does not have a finite presentation on the usual generators [226]. However, if we add the *column generators*, we get a finite presentation [46, 73, 172]; see Section B.2 for details on convergent presentations of plactic monoids. Modified Knuth–Bendix completion procedures have been proposed in order to exploit this and allow for adding generators to handle such situations [166].

5.3.5 Conditions for convergence. Since not every monoid admits a presentation by a finite convergent rewriting system, a natural question is whether there are natural conditions on monoids which ensure that this is the case.

Diekert [113] has addressed this question in the case of abelian groups: he derived a whole class of finite string rewriting systems presenting abelian groups with decidable word problem, which are not Tietze equivalent to a finite convergent string rewriting system on the same alphabet. Moreover, he constructed necessary and sufficient conditions for the existence of a convergent presentation for finitely generated abelian groups. However, the question for general monoids was still open at this time and new methods had to be introduced to solve this problem, which concerns intrinsic properties of the presented monoid. In this direction, Squier introduced in [326, 328] homotopical and homological approaches to formulate necessary conditions for a finitely presented monoid to have a finite convergent presentation. The homotopical construction is presented in Chapter 8 and the homological one in Chapter 9.

6

Linear Rewriting

This chapter presents rewriting techniques for associative algebras. We look here for algorithms turning a given presentation by generators and relations into a rewriting system by orienting the latter, thereby producing linear bases of the presented algebra. In particular, this approach applies to various fundamental decision problems, such as the word problem and ideal membership, or to compute quadratic bases, e.g., Poincaré–Birkhoff–Witt bases (PBW-bases), Hilbert series, syzygies of presentations, homology groups, and Poincaré series. However, if we require rewriting rules to be compatible with the linear structure, we immediately face the following problem: for any rule $u \to v$, we also have $-u \to -v$ and thus

$$v = -u + (u + v) \to -v + (u + v) = u.$$

Therefore, $u \to v$ implies $v \to u$ and thus no rewriting system can be terminating. In order to fix this problem, one can either restrict rewriting to be decreasing with respect to a monomial order, as in the non-commutative Gröbner basis approach [39, 44, 289], or consider the structure of linear polygraph introduced in [160] with an appropriate notion of reduction. It is the latter notion that we present in this chapter.

We first introduce linear polygraphs as a framework for linear rewriting in Section 6.1. We then study the confluence properties of linear polygraphs in Section 6.2. Finally, in Section 6.3, we express Gröbner bases and PBW bases in the setting of linear polygraphs. The polygraphic approach presented in this chapter subsumes many linear rewriting models developed throughout the 20th century. We present a brief historical overview of these works in Section 6.4.

The way to define rewriting in associative algebras depends on the definition considered for the associative algebra structure, either as an internal monoid in the category of vector spaces or a linear category with a single object [287]. In this chapter, we consider the first point of view, as introduced in [160].

6.1 Linear Rewriting

In this section, we introduce the notion of rewriting in associative algebras.

6.1.1 Associative algebras. Suppose fixed a ground field $\mathbb{k}$. An *(associative) algebra* (A, m, e) consists of a $\mathbb{k}$-vector space A together with an operation $m\colon A \otimes A \to A$ and an element $e \in A$ such that the operation m is associative and admits e as neutral element. Otherwise said, an algebra is a monoid object (see Example 10.1.5) in the category **Vect** of vector spaces and linear maps. A *morphism of algebras* $\phi\colon (A, m, e) \to (B, m', e')$ is a linear map $\phi\colon A \to B$ which is compatible with operations m and m' and the neutral elements:

$$\phi(m(x, y)) = m'(\phi(x), \phi(y)), \qquad\qquad \phi(e) = e',$$

for all x, y in A. We denote by **Alg** the category of algebras and their morphisms.

6.1.2 Free algebras. Given a set P_0, we will denote by P_0^ℓ the free algebra over P_0. A *monomial of* P_0^ℓ is an element of the free monoid P_0^* over P_0. The monomials of P_0^ℓ form a linear basis of the algebra P_0^ℓ; thus every 0-cell p of P_0^ℓ can be uniquely written as a linear combination

$$p = \sum_{i=1}^{k} \lambda_i u_i$$

of pairwise distinct monomials $u_1, \ldots, u_k$ of P_0^ℓ, with $\lambda_1, \ldots, \lambda_p$ non-zero scalars, called *the canonical decomposition of* p. We define the *support of* p as the set $\operatorname{supp}(p) = \{u_1, \ldots, u_k\}$.

6.1.3 Linear 1-polygraphs. A *linear* 1-*polygraph* consists of a set P_0, together with a set P_1 equipped with two functions $s_0, t_0\colon P_1 \to P_0^\ell$. Such a polygraph is thus characterized by a diagram of sets and functions

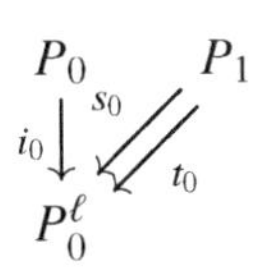

(6.1)

where P_0^ℓ is the free algebra over a set P_0 and $i_0\colon P_0 \to P_0^\ell$ is the canonical inclusion. We often write

$$\langle\, x_i \mid \alpha_i\colon u_i \to v_i \,\rangle$$

for a 1-polygraph with the x_i as elements of P_0 and the α_i as elements of P_1 with $s_0(\alpha_i) = u_i$ and $t_0(\alpha_i) = v_i$.

6.1.4 One-dimensional algebras. A 1-*algebra* is a category internal to **Alg**. It thus consists of a diagram

$$A_0 \underset{t_0}{\overset{s_0}{\Longleftarrow}} A_1$$

comprising two algebras A_0 and A_1, whose elements are respectively called 0- and 1-*cells*, with two algebra morphisms $s_0, t_0 : A_1 \to A_0$ respectively providing the *source* and *target* of a 1-cell, together with an algebra morphism $i : A_0 \to A_1$ which to every 0-cell p associates the *identity* $i(p)$ on p, and an algebra morphism $m : A_1 \times_{A_0} A_1 \to A_1$ which to every pair of composable 1-cells associates their composite, in such a way that composition is associative and admits identities as neutral elements. According to our notations for categories, we set $m(\phi, \phi') = \phi *_0 \phi'$ for any pair ϕ, ϕ' of composable 1-cells.

6.1.5 Lemma. *Let A be* 1*-algebra, then*

– *for all composable* 1*-cells ϕ and ϕ' in A,*

$$\phi *_0 \phi' = \phi - t_0(\phi) + \phi', \tag{6.2}$$

– *every* 1*-cell ϕ in A is invertible with inverse*

$$\phi^- = s_0(\phi) - \phi + t_0(\phi),$$

– *the product of two* 1*-cells ϕ, ϕ' in A decomposes into*

$$\begin{aligned} \phi\phi' &= \phi s_0(\phi') + t_0(\phi)\phi' - t_0(\phi)s_0(\phi') \\ &= s_0(\phi)\phi' + \phi t_0(\phi') - s_0(\phi)t_0(\phi'). \end{aligned} \tag{6.3}$$

Proof. For any composable 1-cells ϕ and ϕ' in A, we have

$$\phi *_0 \phi' = (\phi - s_0(\phi') + s_0(\phi')) *_0 (t_0(\phi) - t_0(\phi) + \phi').$$

By linearity of the 0-composition, this implies

$$\phi *_0 \phi' = \phi *_0 t_0(\phi) - s_0(\phi') *_0 t_0(\phi) + s_0(\phi') *_0 \phi'$$

and by neutrality of identities we get (6.2).

The second condition is deduced from the first one. Let ϕ be a 1-cell in A, we set $\phi^- = s_0(\phi) - \phi + t_0(\phi)$. We have $s_0(\phi^-) = t_0(\phi)$ and $t_0(\phi^-) = s_0(\phi)$. Moreover, from (6.2), we have $\phi *_0 \phi^- = s_0(\phi)$ and $\phi^- *_0 \phi = t_0(\phi)$. We have thus proved that ϕ^- is 0-inverse of ϕ.

Let us prove the third condition. Let ϕ, ϕ' be 1-cells in A. The product of these two 1-cells in A_1 decomposes into

$$\phi\phi' = (\phi *_0 t_0(\phi))(s_0(\phi') *_0 \phi').$$

With the 0-composition being an algebra morphism, we deduce that

$$\phi\phi' = \phi s_0(\phi') *_0 t_0(\phi)\phi'.$$

From (6.2), we deduce the first equality in (6.3). The second equality is proved symmetrically. □

6.1.6 Free 1-algebras. A linear 1-polygraph P generates a *free 1-algebra*, denoted by P^ℓ, with P_0^ℓ as algebra of 0-cells and an algebra P_1^ℓ of 1-cells that we now describe. A *1-monomial* of P^ℓ is a triple

$$u\alpha v \tag{6.4}$$

with $u, v \in P_0^*$ monomials and $\alpha \in P_1$. We respectively define the source and target of such a monomial by

$$s_0^\ell(u\alpha v) = us_0(\alpha)v, \qquad t_0^\ell(u\alpha v) = ut_0(\alpha)v.$$

We consider $(P_0^\ell \otimes \mathbb{k}P_1 \otimes P_0^\ell) \oplus P_0^\ell$ the free P_0^ℓ-bimodule on 1-monomials, and we form the P_0^ℓ-bimodule

$$P_1^\ell = (P_0^\ell \otimes \mathbb{k}P_1 \otimes P_0^\ell) \oplus P_0^\ell/\sim,$$

whose elements are linear combinations of the form

$$\phi = \sum_i \lambda_i\phi_i + 1_p, \tag{6.5}$$

where the ϕ_i are distinct monomials and 1_p is a formal identity on a 0-cell $p \in P_0^\ell$, quotiented by the relation $\sim$ generated by the relations

$$\phi s_0^\ell(\psi) + t_0^\ell(\phi)\psi - t_0^\ell(\phi)s_0^\ell(\psi) = s_0^\ell(\phi)\psi + \phi t_0^\ell(\psi) - s_0^\ell(\phi)t_0^\ell(\psi), \tag{6.6}$$

where ϕ and ψ range over 1-monomials. The relation (6.6) encodes a linear version of the exchange law. The multiplication of the algebra structure in P_1^ℓ precisely associates to two 1-cells ϕ and ψ the cell defined by either member of (6.6). The source and target maps are the above functions s_0^ℓ and t_0^ℓ on monomials, extended by linearity. Given a 1-cell ϕ in P^ℓ, its *size* is the minimum number of 1-monomials ϕ_i occurring in a decomposition of the form (6.5) of ϕ. In particular, a monomial is of size 1.

If we write $i_1 : P_1 \to P_1^\ell$ for the canonical inclusion, sending $\alpha \in P_1$ to the monomial (6.4) where u and v are the empty words, we obtain a diagram

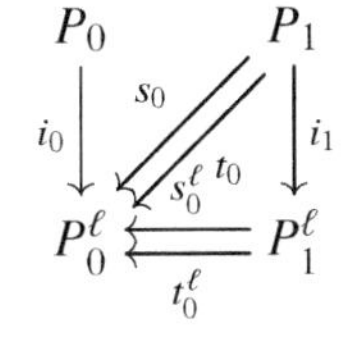

which "commutes" in the sense that we have $s_0^\ell \circ i_1 = s_0$ and $t_0^\ell \circ i_1 = t_0$. In the following, we often simply write respectively $s(\phi)$ and $t(\phi)$ instead of $s_0^\ell(\phi)$ and $t_0^\ell(\phi)$ for the source and target of a 1-cell ϕ.

The free 1-algebra P^ℓ is characterized by the following universal property:

6.1.7 Lemma. *Suppose given a linear 1-polygraph P, a 1-algebra A with P_0^ℓ as underlying algebra of 0-cells and a function $f\colon P_1 \to A_1$ such that for every 1-generator $\alpha\colon u \to v$ in P_1, we have $f(\alpha)\colon u \to v$. Then there exists a unique morphism of 1-algebras*

$$f^*\colon P^\ell \to A$$

such that $f^*(\alpha) = f(\alpha)$ for every α in P_1, seen as a 1-cell in P^ℓ.

In the sequel, we will use the following decomposition result.

6.1.8 Lemma. *Let P be a linear 1-polygraph. Then, every non-identity 1-cell ϕ of P^ℓ admits a decomposition $\phi = \phi_1 \star_0 \cdots \star_0 \phi_k$, for some $k \in \mathbb{N}$, where the ϕ_i are 1-cells of size 1 in P^ℓ.*

Proof. The 1-cell ϕ decomposes into $\phi = \lambda_1\psi_1 + \ldots + \lambda_k\psi_k + 1_p$. When $k = 1$, the 1-cell ϕ is of size 1. Otherwise, for any $i \in \{1, \ldots, k\}$, we set

$$\alpha_i = \lambda_1 t(\psi_1) + \ldots + \lambda_i t(\psi_i), \qquad \beta_i = \lambda_1 s(\psi_1) + \ldots + \lambda_k s(\psi_k)$$

and $\alpha_0 = \beta_{p+1} = 0$. For each $i \in \{1, \ldots, k\}$, we define the 1-cell of size 1

$$\phi_i = \lambda_i\psi_i + 1_p + 1_{\alpha_{i-1}} + 1_{\beta_{i+1}}.$$

We have $s(\phi_i) = p + \alpha_{i-1} + \beta_i$ and $t(\phi_i) = p + \alpha_i + \beta_{i+1}$ so that $\phi_1 \star_0 \cdots \star_0 \phi_k$ is a well-defined 1-cell of P^ℓ. Following relation (6.2), we deduce

$$\phi_1 \star_0 \cdots \star_0 \phi_p = \sum_{i=1}^{k} \lambda_i\psi_i + \sum_{i=1}^{k}(1_p + 1_{\alpha_{i-1}} + 1_{\beta_{i+1}}) - \sum_{i=1}^{k-1}(\lambda_i 1_{t(\psi_i)} + 1_p + 1_{\alpha_{i-1}} + 1_{\beta_{i+1}}).$$

We conclude thanks to $\alpha_{k-1} = \lambda_1 t(\psi_1) + \ldots + \lambda_{k-1} t(\psi_{k-1})$, and $\beta_{k+1} = 0$. □

6.1.9 Presentations and ideals of linear polygraphs. Let P be a linear 1-polygraph. The *algebra presented by* P is the quotient algebra $\overline{P} = P_0^\ell / P_1$ of the algebra P_0^ℓ by the congruence generated by the 1-generators in P_1. We will denote by $\overline{p}$ the image of a 0-cell p of P_0^ℓ through the canonical projection. We say that an algebra A is *presented by* P, or that P is a *presentation of* A, if A is isomorphic to $\overline{P}$. Two linear 1-polygraphs P and Q are *Tietze equivalent* when they present isomorphic algebras: $\overline{P} \simeq \overline{Q}$.

We define the *boundary* of a 1-cell ϕ in the free 1-algebra P^ℓ, as the 0-cell

$$d(\phi) = t_0(\phi) - s_0(\phi).$$

We denote by $I(P)$ the ideal of the algebra P_0^ℓ generated by the boundaries of the 1-cells in P_1. Since the algebra P_0^ℓ is free, the ideal $I(P)$ consists of all the linear combinations

$$\sum_{i=1}^{k} \lambda_i u_i d(\alpha_i) v_i,$$

where the $u_i \alpha_i v_i$ are pairwise distinct 1-monomials of P^ℓ, and the λ_i are non-zero scalars so that the algebra $\overline{P}$ is isomorphic to the quotient of P_0^ℓ by $I(P)$.

6.1.10 Example. The *Weyl algebra* of dimension n over a field $\Bbbk$ of characteristic zero is the algebra presented by the linear 1-polygraph whose 0-cells are

$$x_1, \dots, x_n, \partial_1, \dots, \partial_n$$

and with the following 1-cells:

$$x_i x_j \to x_j x_i \qquad \partial_i \partial_j \to \partial_j \partial_i \qquad \partial_i x_j \to x_j \partial_i \qquad \text{for any } 1 \leqslant i < j \leqslant n,$$
$$\partial_i x_i \to x_i \partial_i + 1 \qquad \text{for any } 1 \leqslant i \leqslant n.$$

6.1.11 Lemma. *Let P be a linear 1-polygraph. For all 0-cells p and q of P_0^ℓ, the following two conditions are equivalent:*

1. *The 0-cell $q - p$ belongs to the ideal $I(P)$.*
2. *There exists a 1-cell $\phi\colon p \to q$ in the free 1-algebra P^ℓ.*

As a consequence, $I(P)$ exactly contains the 0-cells p of P^ℓ such that $\overline{p} = 0$ holds in $\overline{P}$.

Proof. Suppose that $q - p \in I(P)$, that is,

$$q - p = \sum_{1 \leqslant i \leqslant k} \lambda_i u_i d(\alpha_i) v_i.$$

Then the following 1-cell ϕ of P^ℓ has source p and target q:

$$\phi = \sum_{i=1}^{k} \lambda_i u_i \alpha_i v_i + \left(p - \sum_{i=1}^{k} \lambda_i u_i s(\alpha_i) v_i \right).$$

Conversely, let $\phi\colon p \to q$ be a 1-cell of P^ℓ. Using Lemma 6.1.8, we decompose ϕ into 1-cells of size 1:

$$\phi = \phi_1 *_0 \cdots *_0 \phi_k \qquad \text{with} \qquad \phi_i = \lambda_i u_i \alpha_i v_i + h_i.$$

Since $t(\phi_i) = s(\phi_{i+1})$, we have $q - p = d(\phi_1) + \cdots + d(\phi_p)$. Moreover, since $d(\phi_i) = \lambda_i u_i d(\alpha_i) v_i$ we have that each $d(\phi_i)$ belongs to $I(P)$, and thus so does $q - p$.

Finally, if one applies the equivalence to the case $p = 0$, since $\overline{0} = 0$ holds in $\overline{P}$, we get that q is in $I(P)$ if and only if we have $\overline{q} = 0$ in $\overline{P}$. □

6.1.12 Left-monomiality. A linear 1-polygraph P is *left-monomial* if, for every 1-generator α of P_1, the source of α is a monomial of P_0^ℓ that does not belong to $\operatorname{supp}(t(\alpha))$. Note that, from any linear 1-polygraph P, one obtains a Tietze equivalent left-monomial linear 1-polygraph as follows. For every 1-generator α in P_1, if the boundary $d(\alpha)$ is 0, discard α; otherwise, replace α with

$$\alpha' : u \to u - \frac{1}{\lambda} d(\alpha),$$

where u is any chosen monomial in $\operatorname{supp}(d(\alpha))$ and λ is the coefficient of u in $d(\alpha)$.

6.2 Rewriting Properties of Linear Polygraphs

In the linear setting, the definition of a rewriting step is more difficult than in the set-theoretic case, which can be explained as follows. In the set-theoretic case developed in previous chapters, a 1-polygraph P generates two different objects: a free 1-category P^* and a free 1-groupoid $P^\top$. In this situation, we define a rewriting step as a size-one 1-cell of P^*, and their compositions generate all the 1-cells of P^*. But, in the case of associative algebras, there is no difference between the free 1-category and the free 1-groupoid (see also Theorem 18.3.3), which is the cause of the problem mentioned in the introduction of the present chapter. For this reason, we need to adopt a different point of view to define rewriting steps and positive 1-cells. Here, we identify, among the 1-cells of P^ℓ, a set of positive 1-cells that will play the same role as the 1-cells of P^* with respect to $P^\top$ in the case of set-theoretic rewriting. When defining this set, we need to ensure that two conditions are satisfied. Firstly, the set of positive 1-cells should be big enough for every 1-cell of P^ℓ to factor into a composite of positive 1-cells and opposites of positive 1-cells, as given by Lemma 6.1.8 and Lemma 6.2.2. Secondly, the set of positive 1-cells should be small enough for preventing a non-trivial 1-cell and its inverse to be positive at the same time so that the polygraph has a chance to be terminating.

In this section, P denotes a left-monomial linear 1-polygraph.

6.2.1 Rewriting steps and normal forms. A *rewriting step of P* is a 1-cell $\lambda\phi + 1_p$ of size 1 of the free 1-algebra P^ℓ that satisfies the condition

$$\operatorname{supp}(\lambda s(\phi) + p) = \{s(\phi)\} \sqcup \operatorname{supp}(a),$$

that is, such that $\lambda \neq 0$ and $s(\phi) \notin \operatorname{supp}(p)$. A 1-cell of the free 1-algebra P^{ℓ} is called *positive* if it is a (possibly empty) 0-composite $\phi_1 *_0 \cdots *_0 \phi_k$ of rewriting steps of P.

6.2.2 Lemma. *Let P be a left-monomial linear 1-polygraph. Every 1-cell ϕ of size 1 of P^{ℓ} can be decomposed into $\phi = \psi *_0 \chi^{-}$, where each of ψ and χ is either an identity or a rewriting step of P.*

Proof. Write $\phi = \lambda\phi' + 1_q$, where $\phi' : u \to p$ is a 1-monomial of P^{ℓ}. Let μ be the coefficient of u in q, possibly zero, so that $q = \mu u + r$ with r such that $\operatorname{supp}(r)$ does not contain u. Put

$$\psi = (\lambda + \mu)\phi' + 1_r \qquad \text{and} \qquad \chi = \lambda 1_p + \mu\phi' + 1_r.$$

The linearity of the 0-composition of P^{ℓ} gives $\phi = \psi *_0 \chi^{-}$. Moreover, by hypothesis, u does not belong to any of $\operatorname{supp}(p)$ or $\operatorname{supp}(r)$. As a consequence, each of the 1-cells ψ and χ is either an identity (if $\lambda + \mu = 0$ for ψ, if $\mu = 0$ for χ) or a rewriting step. □

6.2.3 Reduced cells and normal forms. A 0-cell p of P_0^{ℓ} is called *reduced* if there is no rewriting step of P of source p. The reduced 0-cells of P_0^{ℓ} form a linear subspace of the free algebra P_0^{ℓ} which we denote by $\operatorname{Red}(P)$. Because P is left-monomial, the set of reduced monomials of P_0^{ℓ}, denoted by $\operatorname{Red}_m(P)$, forms a basis of $\operatorname{Red}(P)$.

If p is a 0-cell of P_0^{ℓ}, a *normal form of p* is a reduced 0-cell q of P_0^{ℓ} such that there exists a positive 1-cell of source p and target q in the free 1-algebra P^{ℓ}.

6.2.4 Binary relations on free algebras. Assume that $\vdash$ is a binary relation on the free monoid P_0^* generated by the set P_0. We say that $\vdash$ is *stable by context* if $u \vdash u'$ implies $vuw \vdash vu'w$ for all u, u', v, and w in P_0^*. We say that $\vdash$ is *compatible with P_1* if $u \vdash v$ holds for every 1-cell $\alpha : u \to p$ in P_1 and every monomial v in $\operatorname{supp}(p)$.

The relation $\vdash$ is extended to the 0-cells of the free algebra P_0^{ℓ} by setting $p \vdash q$ when the following two conditions hold:

1. $\operatorname{supp}(p) \setminus \operatorname{supp}(q) \neq \emptyset$,
2. for every v in $\operatorname{supp}(q) \setminus \operatorname{supp}(p)$, there exists u in $\operatorname{supp}(p) \setminus \operatorname{supp}(q)$, such that $u \vdash v$.

As a consequence, if u is a monomial and p is a 0-cell of P_0^{ℓ}, then $u \vdash p$ holds if and only if $u \vdash v$ holds for every v in $\operatorname{supp}(p)$. Hence, we use the same notation for the relation on P_0^* and for its extension to the 0-cells of P_0^{ℓ}.

The relation $\vdash$ on the 0-cells of P_0^ℓ corresponds to the restriction to finite subsets of P_0^* of the so-called *multiset relation* generated by $\vdash$. We refer to [20, Section 2.5] for the general definition and the main properties of multiset relations, and, in particular, the fact that $\vdash$ is well-founded on the 0-cells if and only if it is well-founded on the monomials; see also §1.4.1.

6.2.5 The termination order. Define $>_P$ as the smallest transitive binary relation on P_0^* that is stable by context and compatible with P_1. We say that the polygraph P *terminates* if the relation $>_P$ is well-founded. In that case, the reflexive closure $\geqslant_P$ of the relation $>_P$ is a well-founded order, called the *termination order of* P (this relation is also sometimes written $\overset{*}{\to}$). This notion of termination order on linear polygraphs corresponds to that defined for 2-polygraphs in Section 4.4.

Assume that the polygraph P terminates. Then the minimal 0-cells for the termination order of P are the reduced ones. Moreover, for every non-identity positive 1-cell p of P_1^ℓ, we have $s(p) >_P t(p)$. This implies that the 1-algebra P^ℓ contains no infinite sequence of 0-composable rewriting steps

$$p_0 \xrightarrow{\phi_1} p_1 \xrightarrow{\phi_2} \cdots \xrightarrow{\phi_{n-1}} a_{n-1} \xrightarrow{\phi_n} a_n \xrightarrow{\phi_{n+1}} \cdots$$

As a consequence, every 0-cell of P_0^ℓ admits at least one normal form. If P terminates, induction on the well-founded order $>_P$ is called *noetherian induction*.

6.2.6 Monomial orders. A well-founded total order $\leqslant$ on the free monoid P_0^* such that the relation $<$ is stable by context is called a *monomial order*. A classic example of a monomial order is given, for any well-founded total order relation $>$ on P_0, by the *deglex order generated by* $>$, as already introduced in §4.4.11, which is defined by

1. $u >_{\text{deglex}} v$ for all monomials u and v of P_0^* such that u has greater length than v, and
2. $uxv >_{\text{deglex}} uyw$ for all $x > y$ of P_0, and monomials u, v, and w of P_0^ℓ such that v and w have the same length.

Given a monomial order $\leqslant$ on P_0^ℓ. If p is a non-zero 1-cell of P_0^ℓ, the *leading monomial of* p is the maximum element of $\text{supp}(p)$ with respect to $\leqslant$ (or 0 if $\text{supp}(p)$ is empty), it is denoted by $\text{lm}_\leqslant(p)$. The *leading coefficient of* p is the coefficient $\text{lc}_\leqslant(p)$ of $\text{lm}_\leqslant(p)$ in p, and the *leading term of* p is the element $\text{lt}_\leqslant(p) = \text{lc}_\leqslant(p)\text{lm}_\leqslant(p)$ of P_0^ℓ. Observe that, for p and q in P_0^ℓ, we have $p < q$ if and only if either $\text{lm}_\leqslant(p) < \text{lm}_\leqslant(q)$ or ($\text{lt}_\leqslant(p) = \text{lt}_\leqslant(q)$ and $p - \text{lt}_\leqslant(p) < q - \text{lt}_\leqslant(q)$).

If there exists a monomial order $>$ on P_0^ℓ that is compatible with P_1, then the polygraph P terminates: the order $>$ is well-founded, and $p >_P q$ implies $p > q$ for all 0-cells p and q. However, the converse implication does not hold, as illustrated by the following example.

6.2.7 Example. The following linear 1-polygraph terminates:

$$P = \langle\, x, y, z \mid \gamma\colon xyz \to x^3 + y^3 + z^3 \,\rangle.$$

Indeed, for every monomial u of P_1^ℓ, denote by $A(u)$ the number of factors xyz that occur in u and by $B(u)$ the number of y that u contains, and we consider the function $C(u) = 3A(u) + B(u)$. It is sufficient to check that $C(uxyzv)$ is strictly greater than each of $C(ux^3v)$, $C(uy^3v)$, and $C(uz^3v)$, for all monomials u and v of P_1^ℓ; see [160, Example 3.2.4] for details. However, no monomial order on P_0^ℓ is compatible with P_1, because, for such an order $>$, one of the monomials x^3, y^3, z^3 is always greater than xyz.

6.2.8 Lemma. *If P is a terminating left-monomial linear 1-polygraph, then, as a vector space, P_0^ℓ admits the decomposition*

$$P_0^\ell = \operatorname{Red}(P) + I(P).$$

Proof. Since the polygraph P terminates, every 0-cell p of P_0^ℓ admits at least a normal form q. Let us write $p = q + (p - q)$, and note that q belongs to $\operatorname{Red}(P)$, by hypothesis, and that $p - q$ is in $I(P)$, by Lemma 6.1.11. □

6.2.9 Branchings. A *branching* of the polygraph P is a pair (ϕ, ψ) of positive 1-cells of the free 1-algebra P^ℓ with the same source, called the *source of* (ϕ, ψ). We do not distinguish the branchings (ϕ, ψ) and (ψ, ϕ). A branching (ϕ, ψ) of P is called *local* if both ϕ and ψ are rewriting steps of P^ℓ. For a branching (ϕ, ψ) of P of source p, define the branching

$$\lambda u(\phi, \psi)v + q = (\lambda u\phi v + q, \lambda u\psi v + q)$$

of P of source $\lambda upv + q$, for all scalar λ, monomials u and v and 0-cell q of P^ℓ. Note that if (ϕ, ψ) is local and $\lambda \neq 0$, then $\lambda u(\phi, \psi) + q$ is also local.

6.2.10 Classification of local branchings. Consider a local branching

$$(\lambda u_1 \alpha u_2 + p, \mu v_1 \beta v_2 + q)$$

of P. We have two main possibilities, depending on whether

$$u_1 s(\alpha) u_2 = v_1 s(\beta) v_2$$

holds or not. Moreover, in the case of equality, there are three different situations, depending on the respective positions of $s(\alpha)$ and $s(\beta)$ in this common monomial. This analysis leads to a partition of the local branchings of P into the following four families.

1. *Trivial* branchings: $\lambda(\phi, \phi)+q$, for all 1-monomial $\phi\colon u \to p$ of P^ℓ, non-zero scalar λ, and 0-cell q of P^ℓ, with $u \notin \mathrm{supp}(q)$.
2. *Additive* branchings: $(\lambda\phi+\mu v+r,\ \lambda u+\mu\psi+r)$, for all 1-monomials $\phi\colon u \to p$ and $\psi\colon v \to q$ of P^ℓ, non-zero scalars λ and μ, and 0-cell r of P^ℓ, with $u \neq v$ and $u, v \notin \mathrm{supp}(r)$.
3. *Multiplicative* branchings: $\lambda(\phi v, u\psi)+r$, for all 1-monomials $\phi\colon u \to p$ and $\psi\colon v \to q$ of P^ℓ, non-zero scalar λ, and 0-cell r of P^ℓ, with $u, v \notin \mathrm{supp}(r)$.
4. *Overlapping* branchings: $\lambda(\phi, \psi) + r$, for all 1-monomials $\phi\colon u \to p$ and $\psi\colon u \to q$ of P^ℓ such that (ϕ, ψ) is neither trivial nor multiplicative, every non-zero scalar λ, and every 0-cell r of P^ℓ, with $u \notin \mathrm{supp}(r)$.

The *critical branchings of P* are the overlapping branchings of P such that $\lambda = 1$ and $r = 0$, and that cannot be factored $(\phi, \psi) = u(\phi', \psi')v$ in a non-trivial way. Note that an overlapping branching has a unique decomposition $\lambda u(\phi, \psi)v + r$, with (ϕ, ψ) critical.

6.2.11 Confluence. Assume that P is a left-monomial linear 1-polygraph. A branching (ϕ, ψ) of P is called *confluent* if there exist positive 1-cells ϕ' and ψ' of P_1^ℓ as in

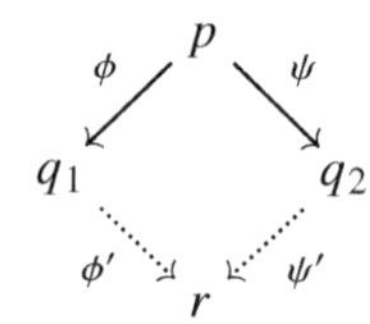

If p is a 0-cell of P_0^ℓ, we say that P is *confluent at p* (resp. *locally confluent at p*, resp. *critically confluent*) if every branching (resp. local branching, resp. critical branching) of P of source p is confluent. We say that P is *confluent* (resp. *locally confluent*, resp. *critically confluent*) if it is so at every 0-cell of P^ℓ. We say that P is *convergent* when it is both terminating and confluent.

When the polygraph P is confluent, then every 0-cell of P_0^ℓ admits at most one normal form, and when it is convergent then every 0-cell p of P_0^ℓ has a unique normal form, denoted by $\widehat{p}$, such that $\overline{p} = \overline{q}$ holds in $\overline{P}$ if and only if $\widehat{p} = \widehat{q}$ holds in P_0^ℓ. As a consequence, if P is a convergent presentation of an algebra A, the assignment of each element p of A to the normal form of any representative of p in P^ℓ, written $\widehat{p}$ by extension, defines a section $A \to P^\ell$ of the canonical projection, where A is seen as a 1-algebra with identity 1-cells only. Note that the section is linear, that is $\widehat{\lambda p + \mu q} = \lambda\widehat{p} + \mu\widehat{q}$, and it preserves the unit, that is, $\widehat{1} = 1$. However, in general, the equality $\widehat{pq} = \widehat{p}\widehat{q}$ does not hold.

6.2.12 Proposition. *Let P be a terminating left-monomial linear 1-polygraph. The following assertions are equivalent:*

1. *The polygraph P is confluent.*
2. *Every 0-cell of $I(P)$ admits 0 as a normal form.*
3. *As a vector space, P_0^ℓ admits the direct decomposition $P_0^\ell = \mathrm{Red}(P) \oplus I(P)$.*

Proof. $1 \Rightarrow 2$. By Lemma 6.1.11, if p is in $I(P)$, then there exists a 1-cell $\phi\colon p \to 0$ in P_1^ℓ. Since P is confluent, this implies that p and 0 have the same normal form, if any. And, since 0 is reduced, this implies that 0 is a normal form of p.

$2 \Rightarrow 3$. By Lemma 6.2.8, it is sufficient to prove that $\mathrm{Red}(P) \cap I(P)$ is reduced to 0. On the one hand, if p is in $\mathrm{Red}(P)$, then p is reduced and, thus, admits itself as only normal form. On the other hand, if p is in $I(P)$, then p admits 0 as a normal form by hypothesis.

$3 \Rightarrow 1$. Consider a branching (ϕ, ψ) of P, with $\phi\colon p \to q$ and $\psi\colon p \to r$. Since P terminates, each of q and r admits at least one normal form, say q' and r' respectively. Hence, there exist positive 1-cells $\phi'\colon q \to q'$ and $\psi'\colon r \to r'$ in P^ℓ. Note that the difference $q' - r'$ is also reduced. Moreover, the 1-cell $(\phi *_0 \phi')^- *_0 (\psi *_0 \psi')$ has q' as source and r' as target. This implies, by Lemma 6.1.11, that $q' - r'$ also belongs to $I(P)$. The hypothesis gives $q' - r' = 0$ so that (ϕ, ψ) is confluent. □

6.2.13 Theorem. *Let A be an algebra and P a convergent presentation of A. Then the set $\mathrm{Red}_m(P)$ of reduced monomials of P^ℓ is a linear basis of A. As a consequence, the vector space $\mathrm{Red}(P)$, equipped with the product defined by $p \cdot q = \widehat{pq}$, is an algebra that is isomorphic to A.*

Proof. If P is convergent, Proposition 6.2.12 implies that the following sequence of vector spaces is exact:

$$0 \longrightarrow I(P) \rightarrowtail P_0^\ell \twoheadrightarrow \mathrm{Red}(P) \longrightarrow 0.$$

Thus, since the algebra $P_0^\ell / I(P)$ is isomorphic to $\overline{P}$, convergence implies that the set $\mathrm{Red}_m(P)$ is a linear basis of $\overline{P}$. We deduce that $\mathrm{Red}(P)$ and $\overline{P}$ are isomorphic as vector spaces. There remains to transport the product of $\overline{P}$ to $\mathrm{Red}(P)$ to get the result. □

6.2.14 Proving confluence. The techniques developed in previous chapters for proving confluence in practical cases can be adapted to the setting of linear polygraphs. Namely, by a direct adaptation of the proof in the set-theoretic case, see Lemma 2.5.8, one can show an analogous of Newman's lemma: a terminating left-monomial linear 1-polygraph P which is locally confluent is confluent. The critical branching lemma, see Lemma 4.3.7, also generalizes to our setting. However, compared to the set-theoretic case, one has to add an extra termination assumption in order to accommodate with the linearity of contexts. The reason is explained in Remark 6.2.18 below, and we defer the proof to next chapter, where it will be proved in the more general setting of coherent presentations; see Lemma 7.6.5.

6.2.15 Lemma. *Suppose given a terminating left-monomial linear* 1*-polygraph* P*. If* P *is critically confluent, then* P *is locally confluent.*

A terminating left-monomial linear 1-polygraph P in which all critical branchings are confluent is thus necessarily confluent.

6.2.16 Example. Let A be the algebra presented by the linear 1-polygraph

$$P = \langle\, x, y \,|\, \alpha\colon xy \to x^2 \,\rangle .$$

This polygraph terminates, because $xy > x^2$ holds for the deglex order generated by $y > x$. This presentation is also confluent, because it has no critical branching; see Lemma 6.2.15. Hence, the set

$$\mathrm{Red}_m(P) = \{y^i x^j \mid i, j \in \mathbb{N}\}$$

is a linear basis of the algebra A. Moreover, the product defined by

$$y^i x^j \cdot y^k x^l = \begin{cases} y^i x^{j+k+l} & \text{if } j \geqslant k, \\ y^{i-j+k} x^{2j+l} & \text{if } j \leqslant k, \end{cases}$$

turns $\mathrm{Red}(P)$ into an algebra that is isomorphic to A.

Now, consider the presentation $Q = \langle\, x, y \,|\, \beta\colon x^2 \to xy \,\rangle$ of A. Termination of Q follows from the deglex order generated by $x > y$, but Q is not confluent, since it has a non-confluent critical branching:

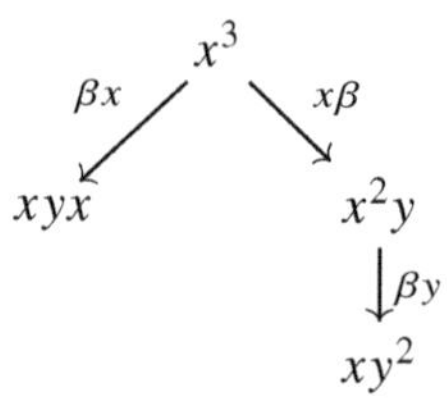

Thus the 0-cell $xyx - xy^2$ is both in $\mathrm{Red}(Q)$ and $I(Q)$, proving that the sum $\mathrm{Red}(Q) + I(Q)$ is not direct. As a consequence, $\mathrm{Red}_m(Q)$ is not a linear basis of A.

6.2.17 Example. The polygraph of Example 6.1.10 that presents the Weyl algebra of dimension n is convergent with the following six families of confluent critical branchings:

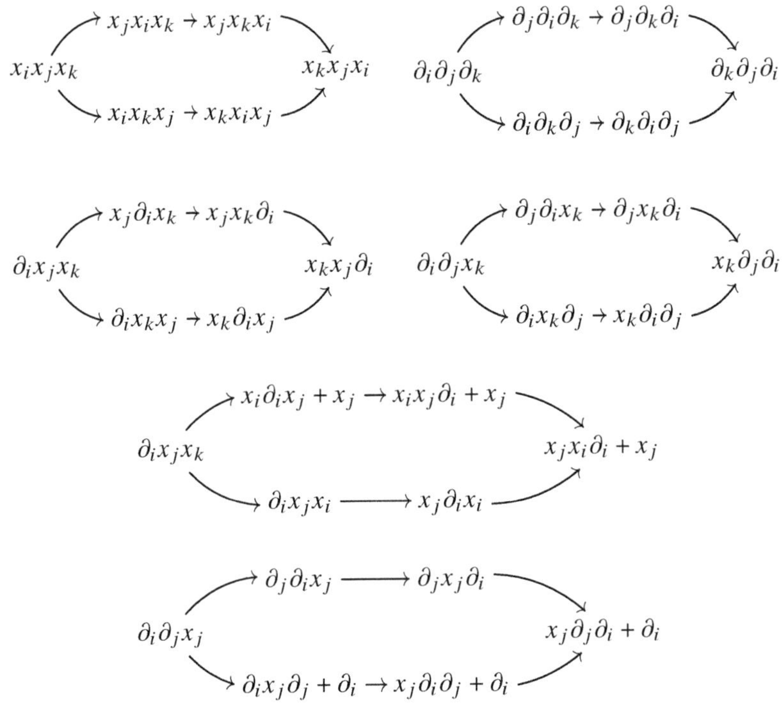

where $1 \leqslant i < j \leqslant n$.

6.2.18 Remark. The critical branching lemma for linear 1-polygraphs given in Lemma 6.2.15 differs from its set-theoretic counterpart because it requires the polygraph to be terminating, as noted in [160, Section 4.2]. Indeed, in the set-theoretic case, the termination hypothesis is not required, and non-overlapping branchings are always confluent, independently of critical confluence. The following two counterexamples show that the linear case is different. The termination assumption comes from the fact that the rewriting steps are modulo the vector space structure. We refer the reader to [81] for an explanation of the linear critical pair lemma in terms of modulo rewriting.

6.2.19 Example. On the one hand, some local branchings can be non-confluent without termination, even if critical confluence holds. Indeed, the linear 1-polygraph

$$\langle\, x, y, z, t \mid \alpha\colon xy \to xz, \beta\colon zt \to 2yt \,\rangle$$

has no critical branching, but it has a non-confluent additive branching:

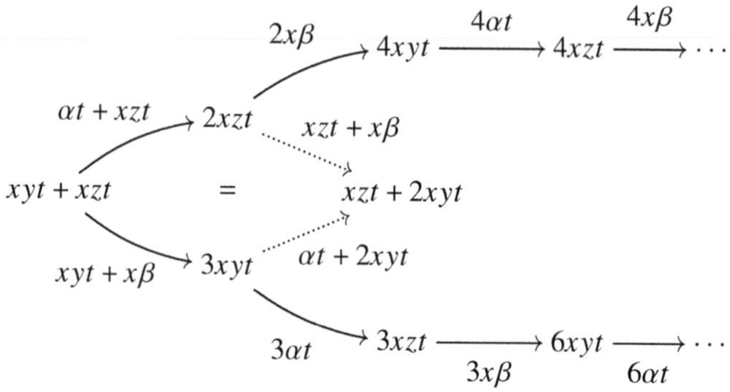

The only positive 1-cells of source $2xzt$ are alternating 0-compositions of $2^k x\beta$ and $2^{k+1}\alpha t$, whose targets are all the 0-cells $2^k xzt$ and $2^{k+1}xyt$, for $k \geqslant 1$. Similarly, the only positive 1-cells of source $3xyt$ have the 0-cells $3.2^k xyt$ and $3.2^k xzt$ as targets, for $k \geqslant 0$. The other possible 1-cells of source $2xzt$ and $3xyt$ are not positive, like the dotted ones. Here, it is the termination hypothesis that fails, as testified by the infinite sequences of rewriting steps in the previous diagram.

6.2.20 Example. On the other hand, the lack of critical confluence may imply that some non-overlapping local branchings are not confluent, even under the hypothesis of termination. For example, the linear 1-polygraph

$$\langle\, x, y, z \mid \alpha\colon xy \to 2x, \beta\colon yz \to z \,\rangle$$

terminates, but it has a non-confluent orthogonal branching:

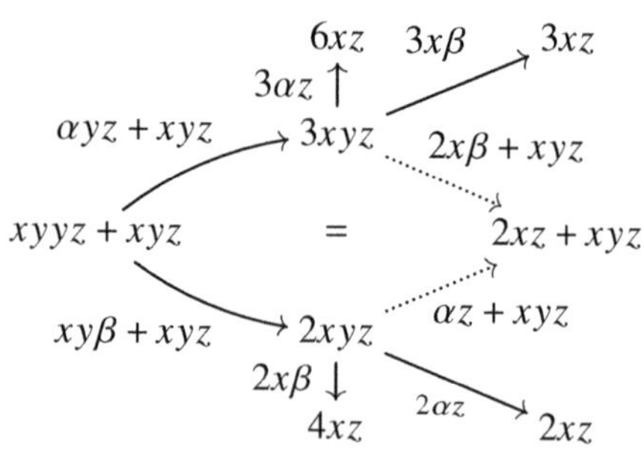

Here, it is the hypothesis on confluence of critical branchings that is not satisfied, since the critical branching $(\alpha z, x\beta)$ of source xyz is not confluent. As a consequence, the only 1-cells that would close the confluence diagram of the Peiffer branching are the dotted ones, which are not positive.

6.2.21 Reduced convergent presentations. As with 2-polygraphs in §5.1.6, without loss of generality, we can restrict to the class of reduced linear polygraphs. We say that a left-monomial linear polygraph P is *left-reduced* if, for every 1-cell α of P, the only rewriting step of P of source $s(\alpha)$ is α itself. We say that P is *right-reduced* if, for every 1-cell α of P, the 0-cell $t(\alpha)$ is reduced. We say that P is *reduced* if it is both left-reduced and right-reduced.

Using the same proof as for 2-polygraphs, Theorem 5.1.7, we prove that every convergent left-monomial linear 1-polygraph is Tietze equivalent to a reduced convergent one.

6.2.22 Completion of presentations. The completion procedure, developed by Buchberger for commutative algebras [65] and by Knuth and Bendix for term rewriting systems [218], see Section 5.2, adapts to terminating left-monomial linear 1-polygraphs as follows, to transform them into convergent ones.

Fix a left-monomial linear 1-polygraph P, and a well-founded strict order that is stable by context and compatible with P_1. For each non-confluent critical branching (ϕ, ψ) of P, consider $p = r - s$, where r and s are arbitrary normal forms of $t(\phi)$ and $t(\psi)$, respectively. If $\operatorname{supp}(p)$ contains a maximal element u, add the 1-cell $u \to q$ to P, where q is defined by $p = \lambda u + q$ and $u \notin \operatorname{supp}(q)$; otherwise, the procedure fails. After the exploration of all the critical branchings of P, the procedure, if it has not failed, yields a terminating left-monomial linear 1-polygraph Q such that $\overline{P} \simeq \overline{Q}$. If Q is not confluent, restart with Q. The procedure either stops when it reaches a convergent left-monomial linear 1-polygraph or runs forever.

6.3 Linear Bases Induced by Monomial Orders

In this section, we consider linear rewriting systems whose rewriting rules are oriented with respect to a fixed monomial order. Suppose fixed a monomial order on polynomials (a well-founded order suitably compatible with multiplication) and an ideal I. A polynomial $p = \sum_i \lambda_i u_i$ in the ideal I can of course be interpreted as a relation

$$\sum_i \lambda_i u_i = 0$$

However, supposing that u_0 is the monomial which is the greatest with respect to the fixed order, called the *leading monomial*, it can also be interpreted as a relation

$$u_0 = \frac{1}{\lambda_0} \sum_{i \neq 0} u_i$$

which, in turn, can be seen as a rewriting rule transforming the left member u_0 into the right member. A Gröbner basis is then a generating set for the ideal such that the associated rewriting rules form a confluent rewriting system, and Buchberger's algorithm to compute a basis can be seen as a form of Knuth–Bendix completion [67].

6.3.1 Gröbner bases. Let P_0 be a set, and let I be an ideal of the free algebra P_0^ℓ. A *Gröbner basis* for I with respect to a monomial order $\preccurlyeq$ on P_0^* is a subset $\mathcal{G}$ of I such that the ideals of P_0^ℓ generated by $\mathrm{lm}_{\preccurlyeq}(I)$ and by $\mathrm{lm}_{\preccurlyeq}(\mathcal{G})$ coincide.

6.3.2 Proposition. *If P is a convergent left-monomial linear 1-polygraph, and $\preccurlyeq$ is a monomial order on P_0^* that is compatible with P_1, then the set*

$$d(P_1) = \{d(\alpha) \mid \alpha \in P_1\}$$

of boundaries of 1-generators of P is a Gröbner basis for $(I(P), \preccurlyeq)$.

Conversely, let P_0 be a set, let $\preccurlyeq$ be a monomial order on P_0^ℓ, and let I be an ideal of P_0^ℓ and $\mathcal{G}$ be a subset of I. Define $P(\mathcal{G})$ the linear 1-polygraph whose set of 0-cells is P_0 and with one 1-cell

$$\alpha_p : \mathrm{lm}(p) \to \mathrm{lm}(p) - \frac{1}{\mathrm{lc}(p)} p,$$

for each p in $\mathcal{G}$. If $\mathcal{G}$ is a Gröbner basis for $(I, \preccurlyeq)$, then $P(\mathcal{G})$ is a convergent left-monomial presentation of the algebra P^ℓ/I, such that $I(P(\mathcal{G})) = I$, and $\preccurlyeq$ is compatible with $P(\mathcal{G})_1$.

Proof. If the polygraph P is convergent, then $d(\alpha)$ is in $I(P)$ for every 1-cell α of P. Since $\preccurlyeq$ is compatible with P_1, we have $\mathrm{lm}(d(\alpha)) = s(\alpha)$ for every 1-cell α of P. Now, if p is in $I(P)$, it is a linear combination

$$p = \sum_i \lambda_i u_i d(\alpha_i) v_i$$

of 1-cells $u_i d(\alpha_i) v_i$, where α_i is a 1-cell of P, and u_i and v_i are monomials of P_0^ℓ. This implies that

$$\mathrm{lm}(p) = u_i s(\alpha_i) v_i = u_i \mathrm{lm}(d(\alpha_i)) v_i$$

hold for some i. Thus $d(P_1)$ is a Gröbner basis for $(I(P), \preccurlyeq)$.

Conversely, assume that $\mathcal{G}$ is a Gröbner basis for $(I, \preccurlyeq)$. By definition, $\preccurlyeq$ is compatible with $P(\mathcal{G})_1$, so $P(\mathcal{G})$ terminates, and $I(P(\mathcal{G})) = I$ holds, so that the algebra presented by $P(\mathcal{G})$ is indeed isomorphic to P^ℓ/I. Moreover, the reduced monomials of $P(\mathcal{G})^\ell$ are the monomials of P^ℓ that cannot be decomposed as $u\mathrm{lm}(a)v$ with a in $\mathcal{G}$, and u and v monomials of P^ℓ. Thus, if a reduced 0-cell p of $P(\mathcal{G})^\ell$ is in I, its leading monomial must be 0, because $\mathcal{G}$ is a Gröbner basis of $(I, \preccurlyeq)$. As a consequence of Proposition 6.2.12, we get that $P(\mathcal{G})$ is confluent. □

By previous proposition and the critical branching lemma (Lemma 6.2.15), the notion of Gröbner basis can be related to confluence as follows. This is sometimes called *Buchberger's criterion* for determining whether a set of polynomials forms a Gröbner basis with respect to a fixed monomial order.

6.3.3 Proposition. *Let P_0 be a set, $\preccurlyeq$ be a monomial order on P_0^*, and I be an ideal of the free algebra P_0^ℓ. A subset $\mathcal{G}$ of I is a Gröbner basis for $(I, \preccurlyeq)$ if and only if the linear 1-polygraph $\mathrm{lm}(\mathcal{G})$ of Proposition 6.3.2 is critically confluent.*

6.3.4 Polygraphs for graded associative algebras. Let denote by **gVect** the category of (non-negatively) graded vector spaces over $\Bbbk$ and graded linear maps of degree 0. Recall that a graded vector space V admits a decomposition $V = \bigoplus_{i\in\mathbb{N}} V^{(i)}$, and the elements of $V^{(i)}$ are said to be *homogeneous of degree i*. A *graded associative algebra* is an internal monoid in the category **gVect**. Following [160, Section 2.2] we can define a notion of polygraph, called *graded linear polygraphs*, for presentation of graded algebras. Let us expand this notion in low dimensions.

A *graded linear* 1*-polygraph* is a data (P_0, P_1) made of

- a graded linear 0-polygraph P_0, that is, a graded set $P_0 = \coprod_{i\in\mathbb{N}} P_0^{(i)}$,
- a graded cellular extension P_1 of the free graded algebra P_0^ℓ generated by P_0, meaning that $P_1 = \coprod_{i\in\mathbb{N}} P_1^{(i)}$ and that the source and target of each 1-generator in $P_1^{(i)}$ are homogeneous of degree i.

If $N \geqslant 2$, a 1-polygraph P is called *N-homogeneous* if P_0 is concentrated in degree 1 and P_1 is concentrated in degree N. We say *quadratic* and *cubical* instead of 2-homogeneous and 3-homogeneous, respectively.

An algebra A is called *N-homogeneous* if it admits a presentation by an N-homogeneous graded linear 1-polygraph.

6.3.5 Poincaré–Birkhoff–Witt bases. Let A be an N-homogeneous algebra, for $N \geqslant 2$, let P_0 be a generating set of A, concentrated in degree 1, and let $\preccurlyeq$ be a monomial order on P_0^*. A *Poincaré–Birkhoff–Witt (PBW) basis for* $(A, P_0, \preccurlyeq)$ is a subset $\mathcal{B}$ of P_0^* satisfying the following conditions:

1. $\mathcal{B}$ is a linear basis of A, with $[u]_{\mathcal{B}}$ denoting the decomposition of an element u of P_0^* in the basis $\mathcal{B}$.

2. For all u and v in $\mathcal{B}$, we have $uv \succcurlyeq [uv]_{\mathcal{B}}$.
3. An element u of P_0^* belongs to $\mathcal{B}$ if and only if for every decomposition $u = vu'w$ of u in P_0^* such that u' has degree N, then u' is in $\mathcal{B}$.

6.3.6 Proposition. *If P is a convergent left-monomial N-homogeneous presentation of an algebra A, and $\preccurlyeq$ is a monomial order on P_0^ℓ that is compatible with P_1, then the set $\mathrm{Red}_m(P)$ of reduced monomials of P_0^ℓ is a PBW basis for $(A, P_0, \preccurlyeq)$.*

Conversely, let A be an N-homogeneous algebra, let P_0 be a generating set of A that is concentrated in degree 1, let $\preccurlyeq$ a monomial order on P_0^ℓ, and let $\mathcal{B}$ be a PBW basis of $(A, P_0, \preccurlyeq)$. Define $P(\mathcal{B})$ as the linear 1-polygraph with 0-cells P_0 and with one 1-cell

$$\alpha_{u,v} \colon uv \to [uv]_{\mathcal{B}}$$

for all u and v in $\mathcal{B}$ such that uv has degree N and $uv \neq [uv]_{\mathcal{B}}$. Then $P(\mathcal{B})$ is a convergent left-monomial N-homogeneous presentation of A, such that $\mathrm{Red}_m(P(\mathcal{B})) = \mathcal{B}$, and $\preccurlyeq$ is compatible with $P(\mathcal{B})_1$.

Proof. If P is a convergent left-monomial presentation of A, Theorem 6.2.13 implies that the set $\mathrm{Red}_m(P)$ of reduced monomials of P_0^* is a linear basis of A. The fact that $\preccurlyeq$ is compatible with P_1 implies Axiom 2 of a PBW basis, and Axiom 3 comes from the definition of a reduced monomial for an N-homogeneous left-monomial linear 1-polygraph.

Conversely, assume that $\mathcal{B}$ is a PBW basis for $(A, P_0, \preccurlyeq)$. By definition, $P(\mathcal{B})$ is N-homogeneous and left-monomial, and Axiom 2 of a PBW basis implies $\mathrm{Red}_m(P(\mathcal{B})) = \mathcal{B}$. Termination of $P(\mathcal{B})$ is given by Axiom 2 of a PBW basis, because $\preccurlyeq$ is well-founded. By Proposition 6.2.12, it is sufficient to prove that $\mathrm{Red}(P(\mathcal{B})) \cap I(P(\mathcal{B})) = 0$ to get confluence: on the one hand, a reduced 0-cell a of $\mathrm{Red}(P(\mathcal{B}))$ is a linear combination of 0-cells of $\mathcal{B}$ so that a is its only normal form; and, on the other hand, if a belongs to $I(P(\mathcal{B}))$, then a admits 0 as a normal form by Lemma 6.1.11. Finally, the algebra presented by $P(\mathcal{B})$ is isomorphic to $\mathrm{Red}(P(\mathcal{B}))$, that is, to $\Bbbk\mathcal{B}$, hence to A, by Theorem 6.2.13 and because $\mathcal{B}$ is a linear basis of A. □

6.4 Historical Account of Linear Rewriting

Gröbner basis theory for ideals in commutative polynomial rings was introduced by Buchberger in [65]. He defined the notion of *S-polynomial* to describe the obstructions to local confluence and gave an algorithm for computation of Gröbner bases, [65, 66, 68]; see also [67] for an historical account. In the commutative setting, any ideal of a polynomial ring has a finite Gröbner basis. Indeed, the Buchberger algorithm on a finite family of generators of an ideal always terminates and returns a Gröbner basis of the ideal. More recently, refined efficient algorithms have been proposed to achieve this task, such as Faugère's F_4 and F_5 algorithms [124, 125].

Shirshov introduced in [325] an algorithm to compute a linear basis of a Lie algebra defined by generators and relations. He used the notion of *composition* of elements in a free Lie algebra that corresponds to the notion of S-polynomial in the work of Buchberger. He gave an algorithm to compute bases in free algebras having the computational properties of the Gröbner bases. He proved that irreducible elements for such a basis form a linear basis of the Lie algebra. This result is called now the *Composition Lemma* for Lie algebras [45].

The Gröbner basis theory has been developed for other types of algebras, such as associative algebras by Bokut in [44] and by Bergman in [39]. They prove Newman's Lemma for rewriting systems in free associative algebras compatible with a monomial order, stating that local confluence and confluence are equivalent properties. This result was called *Composition Lemma* by Bokut and *Diamond Lemma* for ring theory by Bergman; see also [289, 347]. In general, the Buchberger algorithm does not terminate for ideals in a non-commutative multivariate polynomial ring. Indeed, its termination would give a decision procedure of the undecidable word problem. Even if the ideal is finitely generated it may not have a finite Gröbner basis. However, an infinite Gröbner basis can be computed over a ground field [289, 348]. The Buchberger algorithm is the analogue of the Knuth–Bendix completion procedure in a linear setting. Several frameworks unify Buchberger and Knuth–Bendix algorithms; in particular, a Gröbner basis corresponds to a confluent and terminating presentation of an algebra, see [67].

Finally, note that ideas in the style of Gröbner's basis approach appear in many independent works throughout the 20th century. Günter has defined a similar notion in 1913 [311]. Janet [199–201] and Thomas [343] developed the notion of involutive bases that are particular cases of Gröbner bases in the context of partial differential algebra. We refer to [197, 198] for an historical account on involutive bases and their applications to algebraic analysis of linear partial differential systems. Hironaka in [183] and Grauert in [150] compute bases of ideals in rings of power series having analogous properties to Gröbner bases but without a constructive method for computing such bases. In [93], Cohn gave a method to decide the word problem by a normal form algorithm based on a confluence property. Much more recently, Gröbner basis theory was developed in various non-commutative contexts such as Weyl algebras; see [318], or operads [115, 269].

PART II

COHERENT PRESENTATIONS

7

Coherence by Convergence

To any presentation of a category C by a 2-polygraph P corresponds a free $(2,1)$-category $P_2^\top$, as defined in §2.5.1. An *extended presentation* then consists in a choice of a family P_3 of 3-generators between some pairs of parallel 2-cells in $P_2^\top$. Since the category C is already entirely determined by the presentation $\langle P_0 \mid P_1 \mid P_2 \rangle$, we are mostly interested in the case where the congruence generated by P_3 is the *full* relation among pairs of parallel 2-cells in $P_2^\top$, that is, when each 2-sphere is filled with a 3-cell generated by P_3. An extended presentation satisfying this property is said to be *coherent.*

Any given presentation P of a category can be extended into a coherent one by taking all parallel pairs of 2-cells as 3-generators, but we are mainly interested in "small" coherent presentations, which are amenable to computations. The key result in building small coherent presentations is Theorem 7.3.5, a refined version of Newman's lemma, called here *Squier's homotopical theorem.* It states that a *convergent* presentation P can be extended to a coherent one by taking for P_3 a family of confluence diagrams of critical branchings. As a consequence, if P is *finite* convergent, then P_3 can be chosen finite.

We then introduce a notion of Tietze transformation preserving the coherence property and the presented category. This, combined with Squier's homotopical theorem, suggests the following general procedure to build a coherent extension of a given – not necessarily convergent – presentation P:

1. Use the Knuth–Bendix completion procedure to compute a convergent 2-polygraph Q presenting the same category as P.
2. Use Squier's homotopical theorem to extend Q into a coherent presentation $\tilde{Q}$.
3. Use Tietze transformations to reduce $\tilde{Q}$ into a smaller coherent presentation $\tilde{P}$, which extends P.

The first two steps can be performed at once, using what we call a *coherent completion procedure*, and a *transfer theorem* (Theorem 7.1.6), providing an immediate description of the coherent extension of P from the one of Q.

Coherent presentations will prove essential for computing, in low dimensions, homotopical and homological invariants of presented categories introduced in Chapters 8 and 9. Moreover, we shall see in Chapter 23 how they extend in any dimension to *polygraphic resolutions* (Chapter 19) of the presented category. In the language of homotopy theory, these resolutions are cofibrant replacements of a category by a free $(\omega, 1)$-category. Note also that the rewriting method for calculating coherent presentations can be applied in many algebraic contexts, as illustrated in Appendix B.

This chapter is organized as follows. In Section 7.1, we introduce the notion of *acyclic extension* of a 2-category, which consists of the additional data of 3-generators "filling all the spheres". This leads in Section 7.2 to the notion of *coherent presentation* of a category C, that is, a 2-polygraph P presenting C together with an acyclic extension of the free $(2,1)$-category on P. Coherent presentations are then constructed from convergent ones in Section 7.3. The appropriate notion of Tietze transformation between coherent presentations is studied in Section 7.4: this allows us in Section 7.5 to formulate a coherent variant of the Knuth–Bendix completion procedure, but also a reduction procedure, which can be used to obtain smaller coherent presentations. Finally, in Section 7.6, we study coherent presentations of algebras, thereby defining the proper notion of coherent extension for the linear polygraphs of Chapter 6.

7.1 Acyclic Extensions

7.1.1 Cellular extension of a 2-category. A *2-sphere* in a 2-category C is a pair (α, β) of parallel 2-cells in C, i.e., satisfying $s_1(\alpha) = s_1(\beta)$ and $t_1(\alpha) = t_1(\beta)$. A *cellular extension of* C is a set X equipped with two maps $s_2, t_2 \colon X \to C_2$ such that, for every A in X, the pair $(s_2(A), t_2(A))$ is a 2-sphere of C. More generally, we also call any such element A in X a *2-sphere* of C.

Every 2-category C has two canonical cellular extensions:

- the empty extension,
- the full one that contains all the 2-spheres of C, denoted by $\mathrm{Sph}(C)$.

7.1.2 Quotient 2-category. A *congruence* on a 2-category C is an equivalence relation $\approx$ on the 2-cells of C such that

- given $\phi \colon u \Rightarrow v$ and $\phi' \colon u' \Rightarrow v'$ in C_2, $\phi \approx \phi'$ implies $u = u'$ and $v = v'$,

– given 1-cells and 2-cells of C as in the following diagram

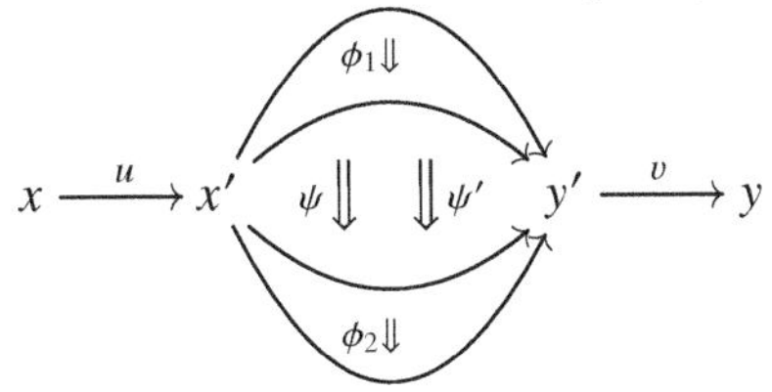

if $\psi \approx \psi'$, then

$$u *_0 (\phi_1 *_1 \psi *_1 \phi_2) *_0 v \approx u *_0 (\phi_1 *_1 \psi' *_1 \phi_2) *_0 v.$$

We define the *quotient 2-category* of a 2-category C by a congruence $\approx$ on C as the 2-category, denoted by $C/{\approx}$, whose 0-cells and 1-cells are those of C, and whose 2-cells are the equivalence classes of 2-cells of C modulo the congruence $\approx$, composition and identities being induced by those of C.

Given a cellular extension X of C, the *congruence generated by* X, denoted by $\approx^X$, is defined as the smallest congruence on C such that $\phi \approx^X \psi$, for every 2-sphere (ϕ, ψ) in X. That is, $\approx^X$ is the smallest equivalence relation on the parallel 2-cells compatible with all the compositions of C and relating ϕ and ψ, for every (ϕ, ψ) in X.

7.1.3 Acyclic extension. We say that a cellular extension X of a 2-category C is *acyclic*, or equivalently that C is X-*acyclic*, if $\phi \approx^X \psi$ holds for every 2-sphere (ϕ, ψ) of C. This is equivalent to say that the equality $\overline{\phi} = \overline{\psi}$ holds in the quotient 2-category $C/{\approx^X}$, where $\overline{\phi}$ and $\overline{\psi}$ denote the images of the 2-cells under the canonical projection $C \to C/{\approx^X}$. For instance, any 2-category C is $\mathrm{Sph}(C)$-acyclic.

7.1.4 Remark. A congruence on a free $(2, 1)$-category is called a *homotopy relation* by Squier in [328]. He noticed that these relations are not really the same as usual homotopies in the sense of algebraic topology and justified the terminology by saying that, for a homotopy relation generated by a set of 2-spheres, two "homotopic" paths can be transformed into one another by a finite sequence of elementary transformation steps. In [328], the relation $\approx^X$ is called a homotopy relation *generated* by X. The terminology *homotopy basis* for an acyclic cellular extension was introduced in [220, 147] and since then has been widely used by various authors. Note also that Squier did not formulate his results on the properties of homotopy relations in the categorical language we use in the present chapter. Instead of $(2, 1)$-categories, he considered 2-dimensional cellular complexes defined by directed graphs with inverses and whose 2-cells correspond to the exchange relation between compositions with respect to 0- and 1-composition. The categorical formulation of

Squier's constructions presented here was introduced in [161, 165]. Another formulation, using the structure of monoidal category, is given in [233].

7.1.5 Transfer theorem for acyclic extensions. Given two presentations P and Q of a 1-category C, by Lemma 2.5.3, there exist two 2-functors

$$f\colon P^{\top} \to Q^{\top} \qquad \text{and} \qquad g\colon Q^{\top} \to P^{\top}$$

and, for every 1-cell v in $Q^{\top}$, there exists a 2-cell $\psi_v\colon fg(v) \Rightarrow v$ in $Q^{\top}$ that satisfies the conditions given in Lemma 2.5.3. Let us define a cellular extension X_Q of the (2, 1)-category $Q^{\top}$ that contains one 3-generator

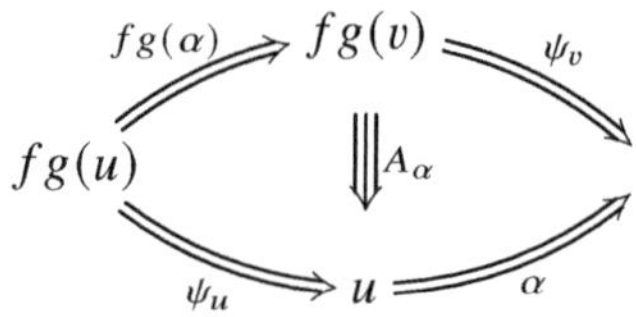

(7.1)

for every 2-generator $\alpha : u \Rightarrow v$ of Q. Furthermore, given a cellular extension X of the (2, 1)-category $P^{\top}$, we will denote by $f(X)$ the cellular extension of $Q^{\top}$ that contains one 3-generator

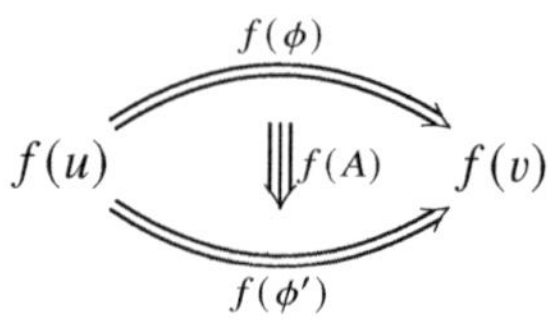

for every 3-generator $A : \phi \Rrightarrow \phi'$ of X. Using these notations, we can formulate the following transfer result among acyclic extensions of presentations of a given category.

7.1.6 Theorem. *Let P and Q be two presentations of the same category. If X is an acyclic cellular extension of the (2, 1)-category $P^{\top}$, then the cellular extension $f(X) \sqcup X_Q$ is an acyclic cellular extension of the (2, 1)-category $Q^{\top}$.*

The proof consists in extending the notation on 3-generators A_α of (7.1), where α is a 2-generator of Q, in a functorial way, to define a 3-cell of the shape

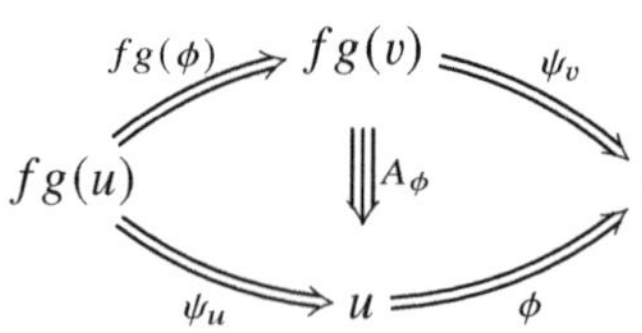

for any 2-cell ϕ in $Q^{\top}$. Then, given two parallel 2-cells $\phi, \phi' : u \Rightarrow v$ of $Q^{\top}$, one proves that $\phi \approx_{f(X)\sqcup X_Q} \phi'$ by constructing a 3-cell with source ϕ and target ϕ'

obtained by compositions along 0-cells, 1-cells and 2-cells of the 3-cells A_α. This construction is based on the notion of free $(3,1)$-category generated by a cellular extension, which is the aim of the following section. The full proof of Theorem 7.1.6 will be given in §7.2.5.

7.2 Coherent Presentations

7.2.1 (3, 1)-polygraphs. A $(3,1)$-*polygraph* is a pair (P, P_3) consisting of a 2-polygraph P and a cellular extension P_3 of the free $(2,1)$-category $P^\top$. It thus consists of a diagram of sets and functions

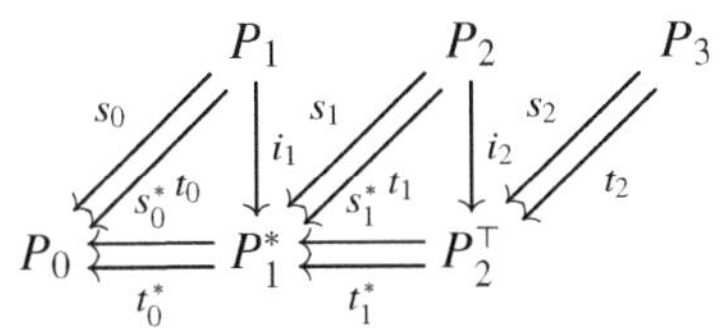

together with the compositions and identities of the underlying $(2,1)$-category

$$P_0 \underset{t_0^*}{\overset{s_0^*}{\leftleftarrows}} P_1^* \underset{t_1^*}{\overset{s_1^*}{\leftleftarrows}} P_2^\top$$

whose *source* and *target maps* s_i and t_i satisfy the globular relations

$$s_i^* \circ s_{i+1} = s_i^* \circ t_{i+1} \qquad \text{and} \qquad t_i^* \circ s_{i+1} = t_i^* \circ t_{i+1}$$

for every $i \in \{0,1\}$. The elements of the cellular extension P_3 are called the 3-*generators* of the $(3,1)$-polygraph (P, P_3). We write $A\colon \phi \Rrightarrow \psi$ for a 3-generator A in P_3 such that $s_2(A) = \phi$ and $t_2(A) = \psi$, often pictured as

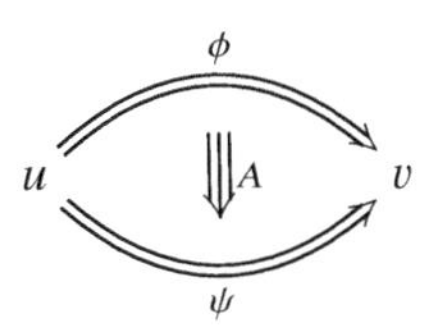

and, more generally, we will call A a 3-*generator* of the cellular extension. A $(3,1)$-polygraph P will be also denoted by

$$\langle\, P_0 \mid P_1 \mid P_2 \mid P_3 \,\rangle$$

and we will write $P_{\leqslant k}$ for its underlying k-polygraph for $0 \leqslant k \leqslant 2$. Morphisms of $(3,1)$-polygraphs are defined as for 3-polygraphs (§10.1.8) and we denote by $\mathbf{Pol}_{3,1}$ the resulting category.

7.2.2 Example. As a simple example, consider the (3, 1)-polygraph with only one generator in each dimension:

$$\langle \star \mid a \mid \alpha\colon aa \Rightarrow a \mid A\colon a\alpha * \alpha \Rrightarrow \alpha a * \alpha \rangle.$$

The 3-generator A can be represented by the diagram

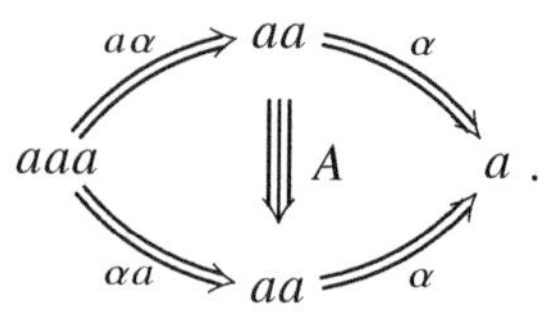

In Section 7.3, Example 7.3.6, we use a rewriting argument to show that the 3-generator A forms an acyclic extension of the free (2, 1)-category generated by $\langle \star \mid a \mid \alpha \rangle$.

7.2.3 Free (3, 1)-category. The definition of 3-*category* is adapted from the one of 2-category by replacing the hom-categories and the composition functors by hom-2-categories and composition 2-functors. We refer the reader to Chapter 14 for the complete definition of strict n-categories for all $n \geqslant 0$. In a 3-category, the 3-cells can be composed in three different ways:

– by $*_0$, along their 0-dimensional boundary:

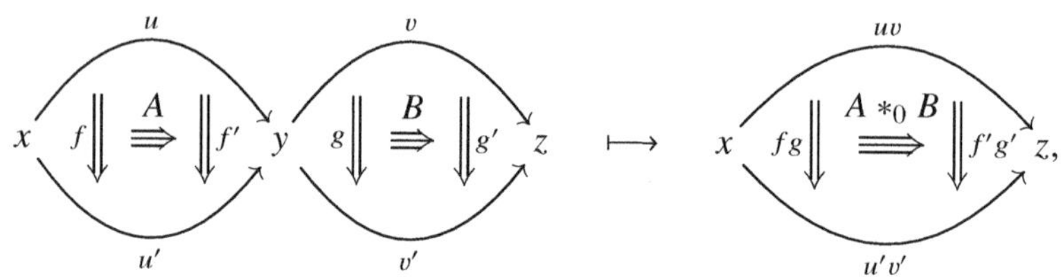

– by $*_1$, along their 1-dimensional boundary:

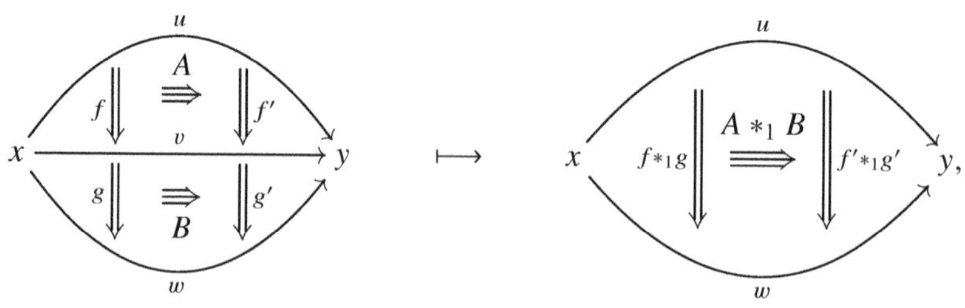

– by $*_2$, along their 2-dimensional boundary:

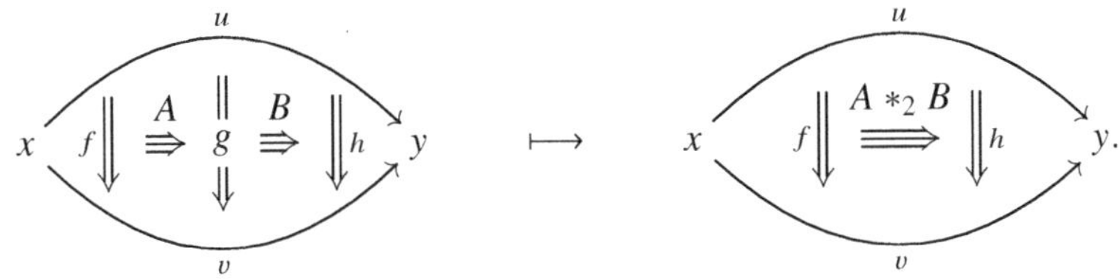

A $(3,1)$-*category* is a 3-category whose 2-cells are invertible with respect to the composition $*_1$ and whose 3-cells are invertible with respect to the composition $*_2$ (which implies their invertibility with respect to $*_1$).

The *free* $(3,1)$-*category* over a $(3,1)$-polygraph P is the $(3,1)$-category, denoted by $P^\top$, or $P^\top_{\leqslant 2}(P_3)$ in some contexts, whose

- underlying 2-category is the free $(2,1)$-category $P^\top_{\leqslant 2}$,
- set $P^\top_3$ of 3-cells consists of all formal compositions with respect to $*_0$, $*_1$ and $*_2$ of 3-generators of P, of their inverses, and of identities of 2-cells, considered up to associativity, identity, exchange, and inverse relations.

This construction will be detailed in arbitrary dimension $n \geqslant 1$ in Section 15.3.

We denote by A^- the inverse with respect to $*_2$ of a 3-cell A: it satisfies $A *_2 A^- = 1_{s_2(A)}$ and $A^- *_2 A = 1_{t_2(A)}$. Note that if a 3-cell A is invertible with respect to the composition $*_2$, and its 2-source and 2-target are invertible, then it is invertible with respect the composition $*_1$, with inverse given by

$$t_2(A)^- *_1 A^- *_1 s_2(A)^-.$$

Every 3-cell A of the $(3,1)$-category $P^\top$ of size $k \geqslant 1$ has a decomposition

$$A = C_1[A_1^{\epsilon_1}] *_2 \dots *_2 C_k[A_k^{\epsilon_k}],$$

with $\epsilon_1, \dots, \epsilon_k \in \{-,+\}$, and $A_1, \dots, A_k$ are 3-generators of P, where for every $1 \leqslant i \leqslant k$, $C_i[A_i^{\epsilon_i}]$ denotes a composition of the form

$$f_2 *_2 (f_1 *_0 A_i^{\epsilon_i} *_0 g_1) *_2 g_2,$$

where f_j, g_j are j-cells for $j = 1, 2$, and where A_i^+ is equal to A_i

7.2.4 Coherent presentations. A $(3,1)$-polygraph P is *coherent* when P_3 is an acyclic extension of the free $(2,1)$-category $P^\top_{\leqslant 2}$. This amounts to requiring that for every pair of parallel 2-cells ϕ and ψ in $P^\top_2$, there is a 3-cell $F\colon \phi \Rrightarrow \psi$ in $P^\top_3$.

An *extended presentation* of a 1-category C is a $(3,1)$-polygraph P whose underlying 2-polygraph $P_{\leqslant 2}$ is a presentation of C. A *coherent presentation* of C is an extended presentation P of C which is coherent.

7.2.5 Proof of Theorem 7.1.6. Let us denote by Y the cellular extension $f(X) \sqcup X_Q$. We construct, for every 2-cell $\phi\colon u \Rightarrow v$ of $Q^\top$, a 3-cell A_ϕ of the free (3, 1)-category $Q^\top(Y)$ with the following shape:

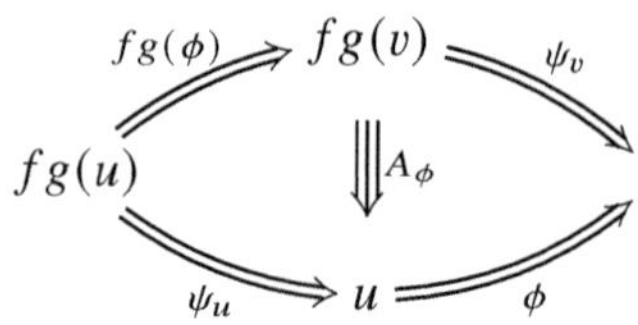

by extending the notation A_α, where α is a 2-generator of Q, in a functorial way, according to the following formulas:

$$A_{1_u} = 1_{\psi_u}, \qquad A_{\phi *_0 \phi'} = A_\phi *_0 A_{\phi'}, \qquad A_{\phi^-} = fg(\phi)^- *_1 A_\phi^- *_1 \phi^-,$$

$$A_{\phi *_1 \phi'} = \big(fg(\phi) *_1 A_{\phi'}\big) *_2 \big(A_\phi *_1 \phi'\big).$$

We prove that the 3-cells A_ϕ are well-defined, i.e., their definition is compatible with the relations on 2-cells, such as the exchange relation. Indeed, whenever the composition of the 2-cells ϕ_1, ϕ_2, ϕ_1', and ϕ_2' are defined in $Q^\top$, we have

$$\begin{aligned} A_{(\phi_1 *_0 \phi_2) *_1 (\phi_1' *_0 \phi_2')} &= ((fg(\phi_1) *_0 fg(\phi_2)) *_1 (A_{\phi_1'} *_0 A_{\phi_2'}) *_2 \\ &\qquad ((A_{\phi_1} *_0 A_{\phi_2}) *_1 (\phi_1' *_0 \phi_2')) \\ &= ((fg(\phi_1) *_1 A_{\phi_1'}) *_0 (fg(\phi_2) *_1 A_{\phi_2'})) *_2 \\ &\qquad ((A_{\phi_1} *_1 \phi_1') *_0 (A_{\phi_2} *_1 \phi_2')) \\ &= ((fg(\phi_1) *_1 A_{\phi_1'}) *_2 (A_{\phi_1} *_1 \phi_1')) *_0 \\ &\qquad ((fg(\phi_2) *_1 A_{\phi_2'}) *_2 (A_{\phi_2} *_1 \phi_2')) \\ &= A_{(\phi_1 *_1 \phi_1') *_0 (\phi_2 *_1 \phi_2')}. \end{aligned}$$

Now, let us consider two parallel 2-cells $\phi, \phi' : u \Rightarrow v$ of $Q^\top$. The 2-cells $g(\phi)$ and $g(\phi')$ are parallel in $P^\top$ so that, by X-acyclicity of $P^\top$, there exists a 3-cell

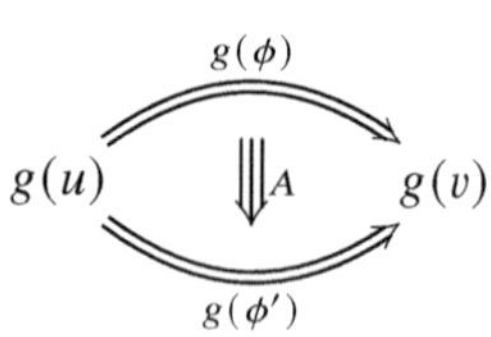

in the (3, 1)-category $P^\top(X)$. By definition of Y and functoriality of f, there exists a 3-cell

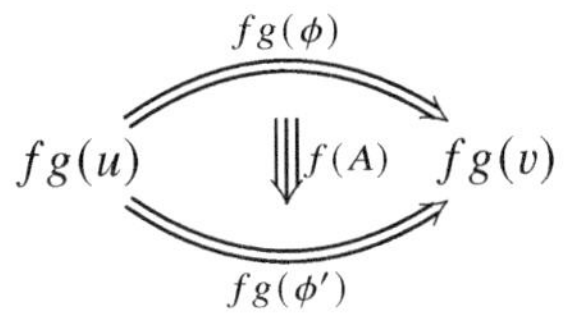

in the free (3, 1)-category $P^{\top}(Y)$. Using the 3-cells $f(A)$, A_ϕ and $A_{\phi'}$, we get the following 3-cell from ϕ to ϕ' in $P^{\top}(Y)$:

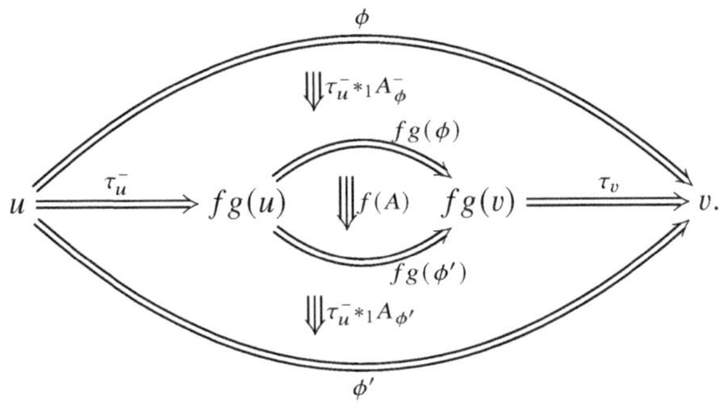

This concludes the proof that the (2, 1)-category $Q^{\top}$ is Y-acyclic.

7.2.6 Cofibrant replacements and coherent presentation. The notion of coherent presentation of a category corresponds to the notion of cofibrant replacement for the model structure for 2-categories introduced by Lack in [229, 231]. This will be detailed and generalized in Chapter 21, and we only give here a brief overview. In this model structure a 2-category is *cofibrant* if its underlying 1-category is free, and a 2-functor $F\colon C \to D$ is a *weak equivalence* if it satisfies the following two conditions.

1. Every 0-cell y of D is *equivalent* to a 0-cell $F(x)$ for x in C, i.e., there exist 1-cells $u\colon F(x) \to y$ and $v\colon y \to F(x)$ and invertible 2-cells $f\colon u *_1 v \Rightarrow 1_{F(x)}$ and $g\colon v *_1 u \Rightarrow 1_y$ in D.
2. For every 0-cells x and x' in C, the induced functor

$$F(x,x')\colon C(x,x') \to D(F(x),F(x'))$$

is an equivalence of categories.

In particular, an equivalence of 2-categories is a weak equivalence. A *cofibrant replacement* of a 2-category C is a cofibrant 2-category $\widetilde{C}$ that is weakly equivalent to C. The following theorem is proved in [145, Theorem 1.3.1]:

7.2.7 Theorem. *Let P be an extended presentation of a category C. Then the $(3,1)$-polygraph P is a coherent presentation of C if and only if the $(2,1)$-category $\overline{P}$ is a cofibrant replacement of C.*

Note that a given category C may admit other cofibrant replacements than the 2-categories presented by coherent presentations of C. For instance, consider the terminal category $\mathbf{1}^{\mathrm{cat}}$: it contains one 0-cell and the corresponding identity 1-cell only. This category $\mathbf{1}^{\mathrm{cat}}$ is cofibrant and, as a consequence, is a cofibrant replacement of itself: this cofibrant replacement corresponds to the coherent presentation of the terminal category given by the $(3,1)$-polygraph with one 0-generator and no higher-dimensional generators. But the terminal category also admits, as a cofibrant replacement, the 2-category with two 0-cells x, y, two 1-cells u, v as follows

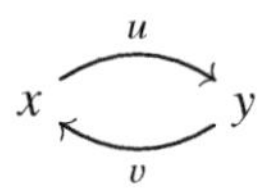

and two invertible 2-cells $f\colon uv \Rightarrow 1_x$ and $g\colon vu \Rightarrow 1_y$. However, this 2-category is not presented by a coherent presentation of the terminal category, since it has two 0-cells.

7.3 Coherent Confluence

In this section, we extend to $(3,1)$-polygraphs the results on coherent confluence given in Section 2.5 for $(2,0)$-polygraphs.

7.3.1 Coherent confluence. Let P be a $(3,1)$-polygraph. A branching (ϕ, ψ) of P is *coherently confluent* if there exist 2-cells ϕ' and ψ' in P_2^* and a 3-cell F in $P_3^{\top}$ of the form

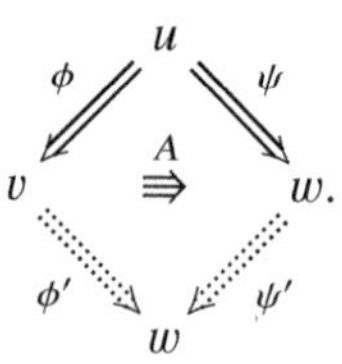

We say that P is *coherently confluent* (resp. *locally coherently confluent*, resp. *critically coherently confluent*) when every branching (resp. local branching, resp. critical branching) of P is coherently confluent. We say that P is *coherently convergent* if it terminates and is coherently confluent. Note that, given a 2-polygraph P, by taking $P_3 = \mathrm{Sph}(P^*)$ to be the set of all 2-spheres of P^* (see

§7.1.1), the notions of coherent confluence and coherent convergence in the $(3,1)$-polygraph (P, P_3) boil down to the ones of confluence and convergence of the 2-polygraph defined in §4.1.9.

The following result essentially amounts to a coherent version of Newman's lemma for 2-polygraphs (Lemma 1.3.21). Its proof is essentially the same as in the case of 1-polygraphs (Lemma 2.5.8). It is interesting to note that the geometric intuition behind it was already present in Newman's original article [290, Section 6].

7.3.2 Proposition. *Let P be a terminating $(3,1)$-polygraph. If P is locally coherently confluent, then P is coherently confluent.*

As above, we recover the Newman's lemma for 2-polygraphs by taking P_3 to be the set of all 2-spheres. Similarly, the following result is a coherent version of the critical branching lemma (Lemma 4.3.7).

7.3.3 Lemma. *Let P be a $(3,1)$-polygraph. If P is critically coherently confluent, then P is locally coherently confluent.*

Proof. We proceed by case analysis on the type of the local branchings of P. First, non-overlapping (i.e., trivial and orthogonal) branchings are always coherently confluent. Indeed, if $\phi\colon u \Rightarrow v$ is a rewriting step of P, then the trivial branching (ϕ, ϕ) is coherently confluent because of

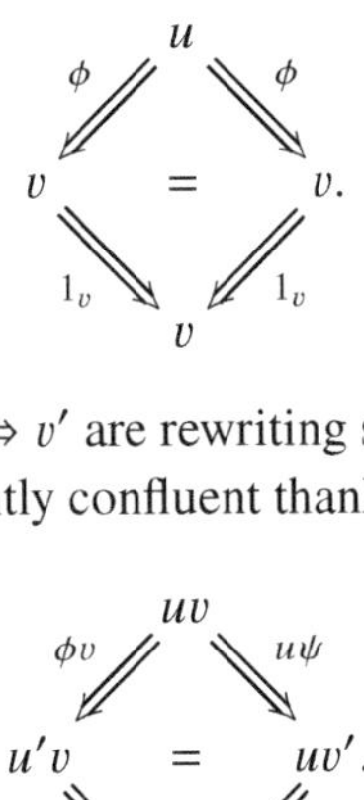

And, if $\phi\colon u \Rightarrow u'$ and $\psi\colon v \Rightarrow v'$ are rewriting steps of P, then the orthogonal branching $(\phi v, u\psi)$ is coherently confluent thanks to the following equality

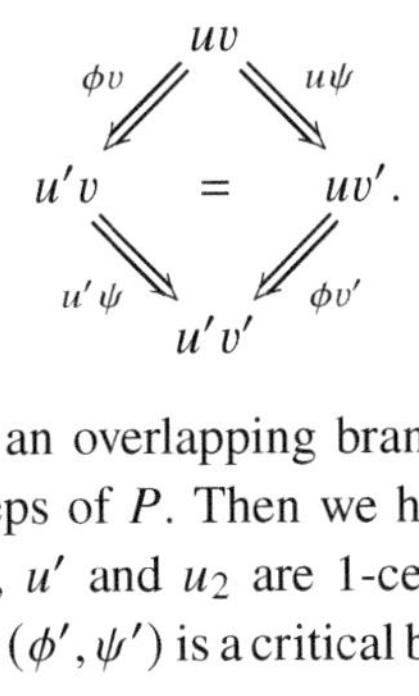

Now, assume that (ϕ, ψ) is an overlapping branching, where $\phi\colon u \Rightarrow v$ and $\psi\colon u \Rightarrow w$ are rewriting steps of P. Then we have $u = u_1 u' u_2$, $\phi = u_1 \phi' u_2$ and $\psi = u_1 \psi' u_2$, where u_1, u' and u_2 are 1-cells of P_1^*, and ϕ' and ψ' are rewriting steps of P such that (ϕ', ψ') is a critical branching of P. By hypothesis,

(ϕ', ψ') is critically coherently confluent, from which we deduce the existence of 2-cells ϕ'' and ψ'' of P_2^*, and of a 3-cell F of $P_3^\top$

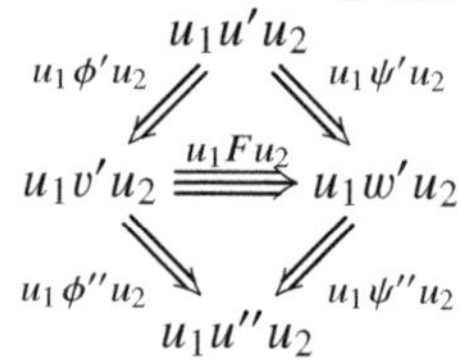

proving that (ϕ, ψ) is coherently confluent. □

Again, the critical branching lemma (Lemma 4.3.7) can be recovered by taking P_3 to be the set of 2-spheres in a 2-polygraph.

7.3.4 Proposition. *Let P be a $(3, 1)$-polygraph. If P is coherently convergent then P is coherent.*

As for 1-polygraphs we prove the following coherence result, called *Squier's homotopical theorem* [328, Theorem 5.2] (see also [161, 233]).

7.3.5 Theorem. *Let P be a convergent 2-polygraph, and P_3 be a cellular extension of the free $(2, 1)$-category $P^\top$. If P_3 contains, for every critical branching (ϕ, ψ) of P, one 3-generator of the form*

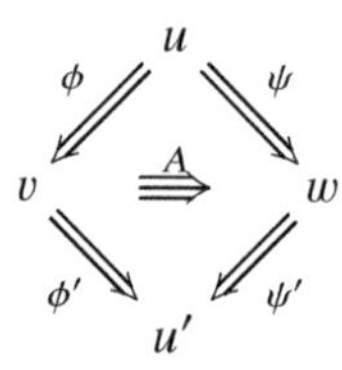

(7.2)

where ϕ' and ψ' are 2-cells in P_2^*, then the $(3, 1)$-polygraph (P, P_3) is coherent.

A 3-generator of the form (7.2), indexed by a critical branching of P, is called a *generating confluence* of the polygraph P. Theorem 7.3.5 states that the set of generating confluences of a convergent 2-polygraph P, indexed by all its critical branchings, forms an acyclic extension of the $(2, 1)$-category $P^\top$.

7.3.6 Example. Consider the monoid M with the convergent presentation

$$\langle \star \mid a \mid \alpha\colon aa \Rightarrow a \rangle.$$

This 2-polygraph has exactly one critical branching, whose corresponding generating confluence has the form:

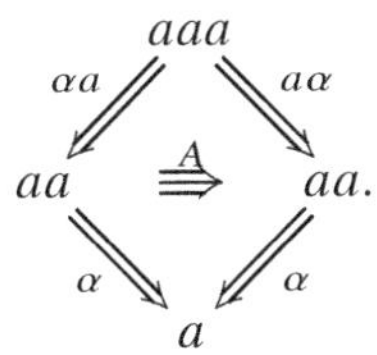

By Theorem 7.3.5, the (3, 1)-polygraph $\langle\, \star \mid a \mid \alpha \mid A \,\rangle$ defined in Example 7.2.2 is thus a coherent presentation of the monoid M.

7.3.7 The standard coherent presentation. Recall from §§2.3.14 and 4.5.5 that the *standard presentation* of a category C is the 2-polygraph $\mathrm{Std}_2(C)$ such that

- the 0-generators are the 0-cells of C,
- there is a 1-generator $\widehat{f}\colon x \to y$ for every 1-cell $f\colon x \to y$ of C,
- there is a 2-generator $\mu_{f,g}\colon \widehat{f}\widehat{g} \Rightarrow \widehat{fg}$ for all composable 1-cells f and g of C,
- there is a 2-generator $\eta_x\colon 1_x \Rightarrow \widehat{1}_x$ for every 0-cell x of C.

The *standard coherent presentation* $\mathrm{Std}_3(C)$ of C is the presentation $\mathrm{Std}_2(C)$ extended with the following 3-generators

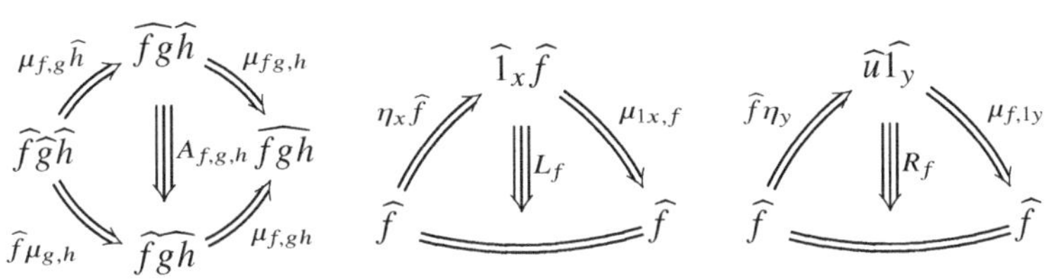

for all 1-cells $f\colon x \to y$, $g\colon y \to z$ and $h\colon z \to t$ of C. Those 3-generators can be shown to form an acyclic cellular extension of the free (2, 1)-category $\mathrm{Std}_2(C)^\top$ (by first reversing the orientation of the generators η_x, as explained in §4.5.5, and then applying Theorem 7.3.5).

7.4 Tietze Transformations of (3, 1)-Polygraphs

We extend here the notion of Tietze transformation presented in §1.2.5 for 1-polygraphs and in Section 5.1 for 2-polygraphs to (3, 1)-polygraphs.

7.4.1 Tietze transformations. If P is a $(3,1)$-polygraph, an *elementary Tietze transformation* on P is one of the following operations transforming P into a $(3,1)$-polygraph Q:

(T1) *adding a definable 1-cell*: Q is obtained from P by adding a 1-generator $a\colon x \to y$ together with a 2-generator $\alpha\colon u \Rightarrow a$ for some 1-cell $u\colon x \to y \in P_1^*$:

(T2) *adding a derivable 2-cell*: Q is obtained from P by adding a 2-generator $\alpha\colon u \Rightarrow v$ together with a 3-generator $A\colon \phi \Rrightarrow \alpha$ for some 3-cell $\phi\colon u \Rightarrow v \in P_2^\top$:

(T3) *adding a derivable 3-cell*: Q is obtained from P by adding a 3-generator $A\colon \phi \Rrightarrow \psi$ for some 3-cell $F\colon \phi \Rrightarrow \psi \in P_3^\top$:

A *Tietze transformation* between $(3,1)$-polygraphs P and Q is a finite sequence of polygraphs $P = P_1, P_2, \ldots, P_n = Q$ such that, for $1 \leqslant i < n$, either P_{i+1} is obtained from P_i by an elementary Tietze transformation, or P_i is obtained from P_{i+1} by an elementary Tietze transformation. Two $(3,1)$-polygraphs are *Tietze equivalent* when there is a Tietze transformation between them. As in §1.2.5 and Section 5.1, the notion of Tietze equivalence is supposed to be closed by isomorphism.

7.4.2 Functors induced by Tietze transformations. For any of the above elementary Tietze transformations from a $(3,1)$-polygraph P to a $(3,1)$-polygraph Q, there is a canonical morphism of polygraphs $P \to Q$, witnessing the inclusion of P into Q, which induces a 3-functor $F\colon P^\top \to Q^\top$ between the freely generated $(3,1)$-categories. This functor always admits a retraction, i.e., a 3-functor $G\colon Q^\top \to P^\top$ such that $G \circ F = 1_{P^\top}$. For instance, in the case of (T1), with the same notations as above, the functor G is such that $Ga = u$, $G\alpha = 1_u$, and G leaves the other generators of Q unchanged.

We recall the following result from [145, Theorem 2.1.3]:

7.4.3 Theorem. *Two finite* $(3, 1)$*-polygraphs* P *and* Q *are Tietze equivalent if and only if there is an equivalence between the presented* 2*-categories* $\overline{P}$ *and* $\overline{Q}$ *which induces a bijection between the respective sets of* 0*-cells.*

As a consequence, if a $(3, 1)$-polygraph P is a coherent presentation of a category C and if there exists a Tietze transformation from P to a $(3, 1)$-polygraph Q, then Q is also a coherent presentation of C.

7.4.4 Higher Nielsen transformations. As a particular subset of Tietze transformation, we identify the following family of transformations, which will prove useful in the following. The *elementary Nielsen transformations* on a $(3, 1)$-polygraph P are the following transformations:

(N1) the replacement of a 2-cell by a formal inverse (including in the source and target of every 3-cell),
(N2) the replacement of a 3-cell by a formal inverse,
(N3) the replacement of a 3-cell $F\colon \psi \Rrightarrow \psi'$ by a 3-cell

$$\tilde{F}\colon \phi \star_1 \psi \star_1 \chi \Rrightarrow \phi \star_1 \psi' \star_1 \chi$$

where ϕ and χ are 2-cells of $P^{\top}$.

The *Nielsen equivalence* on $(3, 1)$-polygraphs is the smallest equivalence relation identifying any two polygraphs between which there is an elementary Nielsen transformation. The following is shown in [145, Section 2.1.4]:

7.4.5 Lemma. *The elementary Nielsen transformations are Tietze transformations.*

7.4.6 Collapsible generators. Given a $(3, 1)$-polygraph P, we identify the following families of redundant generators in the polygraph. Following the terminology introduced by Brown [60], we say that a 2-generator α of P is *collapsible* if

- the target of α is a 1-generator $a \in P_1$, and
- the source of α is a 1-cell in which a does not occur.

Similarly, a 3-generator A of P is *collapsible* if

- the target of A is a 2-generator $\alpha \in P_2$, and
- the source of A is a 2-cell in which α does not occur.

A 3-*sphere* Φ is a pair (F, G) of 3-cells in $P_3^{\top}$ with the same source and with the same target, where F and G are respectively the source and target of the

3-sphere. Note that a 3-sphere can be seen as a 4-generator in a $(4, 1)$-polygraph, which consists of a $(3, 1)$-polygraph P equipped with a cellular extension P_4 of the freely generated $(3, 1)$-category $P^{\top}$, see Section 15.3. We thus denote by $\Phi\colon F \Rrightarrow G$ a 3-sphere from F to G. A 3-sphere $\Phi\colon F \Rrightarrow A$ whose target A is a 3-generator is said to be *collapsible*.

Given a collapsible 2-generator $\alpha\colon u \Rightarrow a$, we write P/α for the $(3, 1)$-polygraph with

- P_0 as 0-generators,
- $P_1 \setminus \{a\}$ as 1-generators,
- $P_2 \setminus \{\alpha\}$ as 2-generators, where every occurrence of a in the source or target of a 2-generator has been replaced by u,
- P_3 as 3-generators, where every occurrence of α in the source or target of a 3-generator has been replaced by 1_u.

Similarly, given a collapsible 3-generator $A\colon \phi \Rrightarrow \alpha$, we write P/A for the $(3, 1)$-polygraph with

- P_0 as 0-generators,
- P_1 as 1-generators,
- $P_2 \setminus \{\alpha\}$ as 2-generators,
- $P_3 \setminus \{A\}$ as 3-generators, where every occurrence of α in the source or target of a 3-generator has been replaced by ϕ.

Similarly, given a collapsible 3-sphere $\Phi\colon F \Rrightarrow A$, we write P/Φ for the $(3, 1)$-polygraph P_0, P_1, P_2, and $P_3 \setminus \{A\}$ as sets of 0-, 1-, 2-, and 3-generators, respectively.

In the above situation, the target generator of the collapsible cell is said to be *redundant*, and the polygraph P/α (resp. P/A, resp. P/Φ) is said to be obtained from P by *collapsing* α (resp. A, resp. Φ).

7.4.7 Example. In the polygraph

$$P = \langle\, \star \mid a, b, c \mid \alpha\colon ab \Rightarrow c, \beta\colon ac \Rightarrow bc, \gamma\colon ab \Rightarrow c \mid A\colon \alpha \Rrightarrow \gamma \,\rangle$$

the 2-generator α is collapsible and the polygraph resulting from its collapse is

$$P/\alpha = \langle\, \star \mid a, b \mid \beta\colon aab \Rightarrow bab, \gamma\colon ab \Rightarrow ab \mid A\colon 1_{ab} \Rrightarrow \gamma \,\rangle$$

The following is shown in [145, Section 2.3]:

7.4.8 Proposition. *Let P be a $(3, 1)$-polygraph. Given a collapsible 2-generator α (resp. 3-generator A, resp. 3-sphere Φ) of P, the $(3, 1)$-polygraphs P and P/α (resp. P/A, resp. P/Φ) are Tietze equivalent.*

7.4.9 Remark. The class of collapsible generators can be made larger and thus lead to more collapses by working "up to Nielsen equivalence". By this, we mean that one can consider that a generator is collapsible in a $(3, 1)$-polygraph P, when P is Nielsen equivalent to a polygraph Q in which the corresponding generator is collapsible. Namely, one can generalize the above notion of collapse to those generators.

7.5 Coherent Completion and Reduction

Given a convergent 2-polygraph P, Squier's homotopical theorem (Theorem 7.3.5) provides a way to extend it into a coherent presentation of the category $\overline{P}$. When the 2-polygraph P is not convergent, we can use the Knuth–Bendix completion procedure (Section 5.2) in order to obtain a convergent presentation of the category $\overline{P}$ and then apply Squier's theorem on it in order to finally obtain a coherent presentation of $\overline{P}$. We present the *coherent completion procedure* from [166, 145] which combines the two steps at once: it adds both 2- and 3-generators to the polygraph in order to obtain a coherent convergent presentation.

Often, the resulting coherent presentation of $\overline{P}$ is not minimal, in the sense that some of its generators are collapsible. In such a situation, it is desirable to remove those superfluous generators in order to obtain a smaller presentation. We also present here techniques to perform this, which, when combined with the procedure described above, give rise to a *coherent completion-reduction procedure*.

7.5.1 Family of generating confluences. Given a 2-polygraph P, a cellular extension X of the free $(2, 1)$-category $P^{\top}$ containing a 3-generator

$$\begin{array}{ccc} & u & \\ {}^{\phi}\swarrow & & \searrow^{\psi} \\ v & \overset{A_{\phi,\psi}}{\Rrightarrow} & w \\ {}_{\phi'}\searrow & & \swarrow_{\psi'} \\ & u' & \end{array} \tag{7.3}$$

for every critical branching (ϕ, ψ) of P is called a *family of generating confluences* for P.

A *Squier completion* of the polygraph P is a $(3, 1)$-polygraph, denoted by $\mathrm{Sq}(P)$, obtained from P by adding a 3-generator of the form (7.3) for every critical branching (ϕ, ψ). By Theorem 7.3.5, such a $(3, 1)$-polygraph is a coherent presentation of the category $\overline{P}$. Note that the notation is slightly abusive since a polygraph $\mathrm{Sq}(P)$ is not entirely determined by P. In particular, it depends on a choice of confluences for the critical branchings and the orientation of the 3-generator $A_{\phi,\psi}$. We will see in Chapter 23 that this choice can be encoded as an *ι-contraction* and can be extended in all higher dimensions.

7.5.2 Coherent completion of terminating 2-polygraphs. By extending the Knuth–Bendix completion procedure, see Section 5.2, we define a procedure that computes a coherent presentation of a category C starting with a terminating, but not necessarily confluent, presentation of C, by suitably adding 2- and 3-generators obtained from computing critical branchings. The procedure is defined as follows.

Given a terminating 2-polygraph P, equipped with a total termination order, the *coherent completion* of P is the $(3,1)$-polygraph obtained from P by successive applications of Knuth–Bendix and Squier completion steps, as follows. In this procedure, one considers each critical branching (ϕ, ψ) of P and performs the following operations:

- if the branching is confluent, the procedure adds a 3-generator

  $$A\colon \phi *_1 \phi' \Rrightarrow \psi *_1 \psi'$$

  to the polygraph (if such a generator is not already present):

  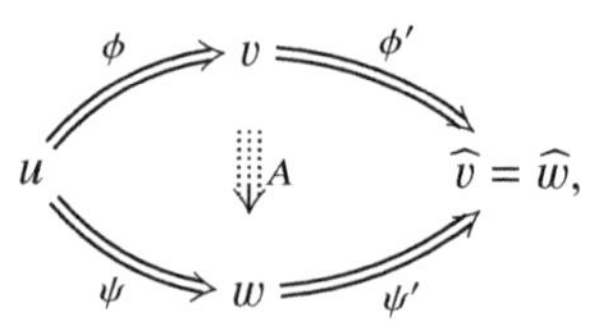

- if the branching is not confluent, the procedure coherently adds a 2-generator

  $$\alpha\colon \widehat{v} \Rightarrow \widehat{w} \quad \text{if } \widehat{v} > \widehat{w} \qquad \text{or} \qquad \alpha\colon \widehat{w} \Rightarrow \widehat{v} \quad \text{if } \widehat{w} > \widehat{v}$$

  and a 3-generator

  $$A\colon \phi *_1 \phi' \Rrightarrow \psi *_1 \psi'$$

  to the polygraph:

  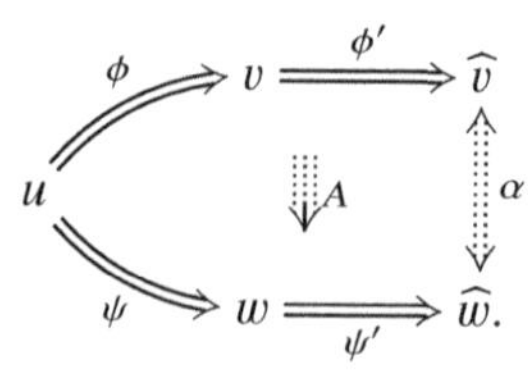

In the second case, the procedure adds a new 2-generator α, which can in turn create new critical branchings, which have to be inspected by the procedure. For this reason, like in the usual Knuth–Bendix procedure, the process is not guaranteed to terminate. In this situation, this defines an increasing sequence of $(3,1)$-polygraphs whose inductive limit is a potentially infinite $(3,1)$-polygraph.

As a consequence of Theorem 7.3.5, the $(3,1)$-polygraph constructed using this procedure satisfies the following property [145, Theorem 2.2.5]:

7.5.3 Theorem. *Let P be a terminating 2-polygraph. Any coherent completion of P is a coherent convergent presentation of the category $\overline{P}$.*

7.5.4 Generic homotopical reduction. In order to reduce the size of the $(3,1)$-polygraph obtained by a coherent completion of a terminating 2-polygraph, one can identify generators which can be collapsed and thus be removed without changing the presented category nor the coherence of the category (see Proposition 7.4.8). We formalize here the process of collapsing multiple such generators at once.

A *collapsible part* of a $(3,1)$-polygraph P is a family X of its generators that we can collapse together, in the sense introduced in §7.4.6. Explicitly, it consists in a triple $X = (X_2, X_3, X_4)$ made of a family X_2 of 2-generators of P, a family X_3 of 3-generators of P and a family X_4 of 3-spheres of the free $(2,1)$-category $P^\top$, such that the following conditions are satisfied:

- the elements of X_2, X_3, and X_4 are collapsible, potentially up to a Nielsen transformation (see Remark 7.4.9),
- no 2-generator of X_2 is the target of a 3-generator in X_3,
- no 3-generator of X_3 is the target of a 3-sphere in X_4,
- the following relations are well-founded:
  - the relation $<_1$ on P_1 such that $b <_1 a$ when there exists a 1-generator $\alpha\colon u \Rightarrow a$ in X_2 such that b occurs in u,
  - the relation $<_2$ on P_2 such that $\beta <_2 \alpha$ when there exists a 2-generator $A\colon \phi \Rrightarrow \alpha$ in X_2 such that β occurs in ϕ,
  - the relation $<_3$ on P_3 such that $B <_3 A$ when there exists a 3-sphere $\Phi\colon F \equiv\!\!> A$ in X_3 such that B occurs in F.

Given such a collapsible part X, one can define a $(3,1)$-polygraph P/X, obtained by successively collapsing all the elements of X, which is called the *homotopical reduction of P with respect to X*. By construction, the polygraph P/X is Tietze equivalent to P.

7.5.5 Generating triple confluences. The coherent elimination of 3-generators of a $(3,1)$-polygraph P by homotopical reduction requires a collapsible set of 3-spheres of $P^{\top}$. When P is convergent and coherent, its triple critical branchings provide a convenient way to build such a set.

A *local triple branching* is a triple (ϕ, χ, ψ) of 2-cells which are rewriting steps with a common source. Similarly to local branchings, local triple branchings are classified into three families:

- *trivial* triple branchings have two of the 2-cells equal,
- *orthogonal* triple branchings have at least one of their 2-cells that form an orthogonal branching with the other two,
- *overlapping* triple branchings are the remaining local triple branchings.

Local triple branchings are ordered by inclusion of their sources, similarly to branchings. A *critical* triple branching is an overlapping triple branching that is minimal for this inclusion. For a reduced 2-polygraph, such a triple branching can have two different shapes, where ϕ, ψ, and χ are 2-generators:

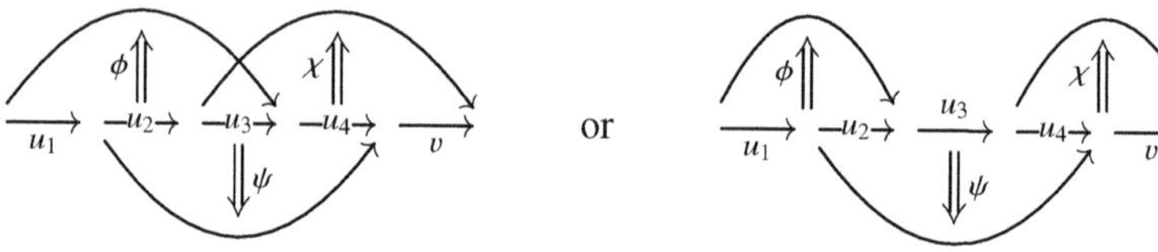

When the polygraph is not reduced, the other possible type of critical branchings, with an inclusion of one source into the other one, generates several other possibilities.

If P is a coherent and convergent $(3,1)$-polygraph, a *generating triple confluence of* P is a 3-sphere

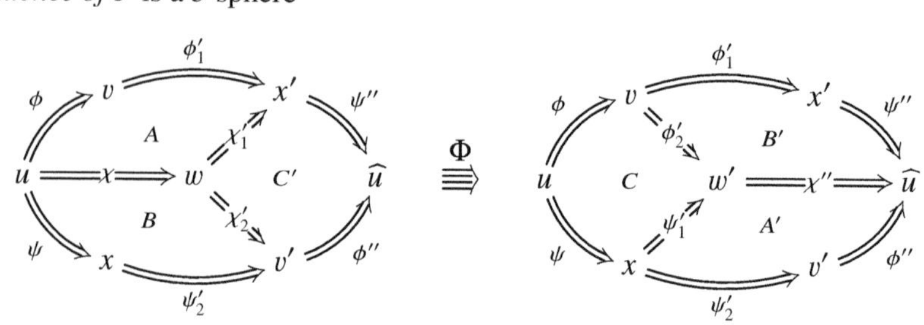

where (ϕ, χ, ψ) is a triple critical branching of P and the 3-cells are generated by the generating confluence induced by the critical branchings.

7.5.6 Coherent completion-reduction. In practice, we apply homotopical reduction to a coherent completion Q of a terminating 2-polygraph P. In such a situation, one can define a collapsible part X of Q whose elements are

- some of the generating triple confluences of Q,
- the 3-generators coherently adjoined with a 2-generator by coherent completion to reach confluence,
- some collapsible 2-generators or 3-generators already present in the initial presentation P.

In practice, the collapsible triple confluences are chosen among those in which some 3-generator A occurs in the source or the target without 1-dimensional whiskers and occurs exactly once. Similarly, the collapsible 3-generators are chosen among those where a 2-generator α occurs in the source or the target without whiskers and occurs exactly once. Finally, the collapsible 2-generators are chosen among those of the form $\alpha\colon u \Rightarrow a$ or $\alpha\colon a \Rightarrow u$ where a is a generator which does not occur in u. Moreover, one should check that the conditions of §7.5.5 are satisfied. In particular, one should be careful not to select too many such generators in order for the well-foundedness conditions to be satisfied. An illustration is given in Example 7.5.8 below.

If P is a terminating 2-polygraph, the *coherent completion-reduction of* P with respect to a collapsible part X of its completion Q is the $(3,1)$-polygraph the homotopical reduction Q/X of Q with respect to X.

7.5.7 Theorem. *Let* P *be a terminating 2-polygraph. A coherent completion-reduction of* P *is a coherent presentation of the category* $\overline{P}$.

We refer to Appendix B for examples of coherent completion-reduction calculations in the algebraic situations of Artin, plactic, and Chinese monoids. We end this section with a simple example to illustrate the method.

7.5.8 Example. Consider the following presentation of the braid monoid B_3^+, already encountered in Example 5.2.6 and §5.3.1, see also §A.1.21:

$$P = \langle \star \mid a, b, c \mid aba \Rightarrow bab, ba \Rightarrow c \rangle$$

and equipped with the deglex order induced by $a > b > c$. The 2-polygraph P is terminating and its coherent completion is the $(3,1)$-polygraph:

$$Q = \langle \star \mid a, b, c \mid \alpha, \beta, \gamma, \delta \mid A, B, C, D \rangle,$$

where $\alpha\colon ac \Rightarrow cb$, $\beta\colon ba \Rightarrow c$, $\gamma\colon bcb \Rightarrow cc$, $\delta\colon bcc \Rightarrow cca$ and A, B, C, and D are the following 3-generators, induced by completion of the critical branchings:

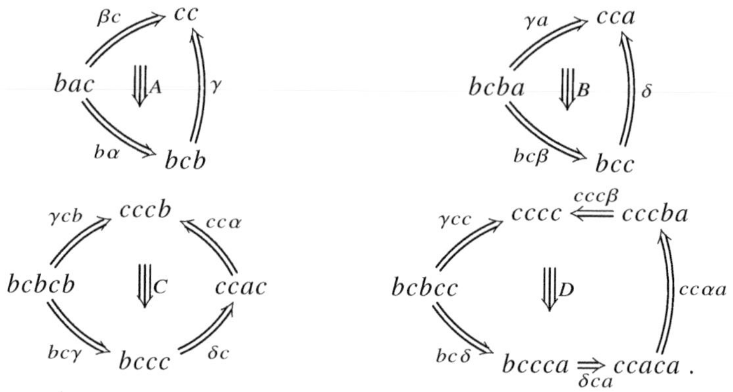

The coherent presentation Q of B_3^+ can be reduced using the collapsible part consisting of the following two generating triple confluences

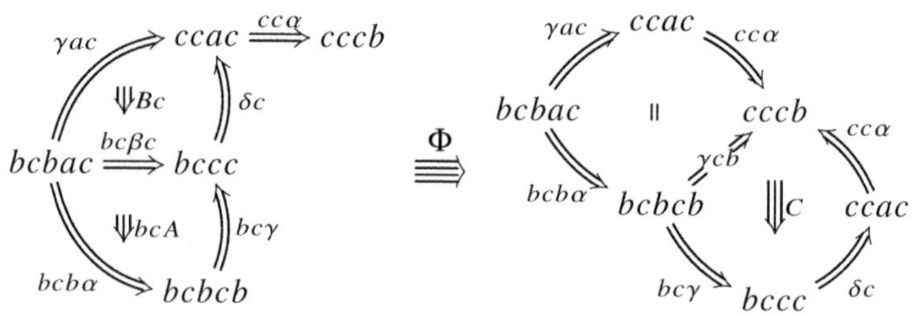

and

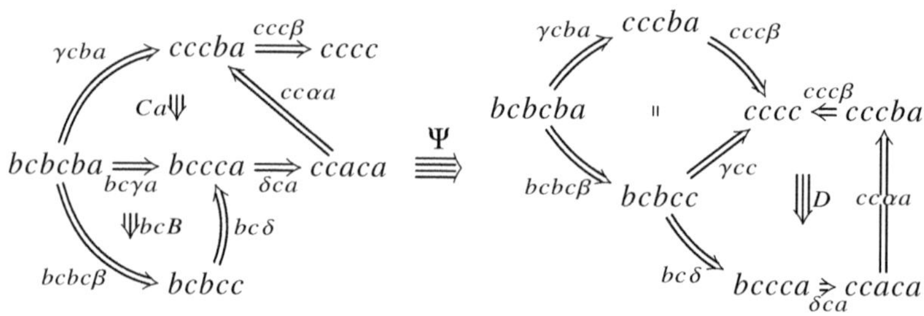

together with the 3-generators A and B coherently adjoined with the 2-generators γ and δ during coherent completion and the 2-generator $\beta\colon\ ba \Rightarrow c$ that defines the redundant generator c. The generators β, A, B, Φ, and Ψ are collapsible up to a Nielsen transformation, with respective redundant generators c, γ, δ, C, and D. We conclude that X is collapsible since the relations $<_1$, $<_2$, and $<_3$ are respectively included in the following well-founded total orders, and are thus well-founded:

$$c > b > a \qquad \delta > \gamma > \beta > \alpha \qquad D > C > B > A.$$

It follows that the homotopical reduction of the coherent presentation Q with respect to this collapsible part is the following coherent $(3, 1)$-polygraph:

$$R = \langle \star \mid b, a \mid aba \Rightarrow bab \mid \ \rangle.$$

By Theorem 7.5.7, we recover that the monoid B_3^+ admits a coherent presentation made of Artin's presentation and no 3-generator. This example is generalized in [145] where coherent presentations of Artin monoids are constructed, see also Appendix B.

7.6 Coherent Presentations of Associative Algebras

In this section, we define the notion of coherent presentation of an associative algebra by extending the notion of presentation of an algebra introduced in Chapter 6.

7.6.1 Extended presentations. A *cellular extension* of a 1-algebra A, with the notations of §6.1.4, is a set X equipped with functions $s_1, t_1 : X \to A_1$ such that $s_0 \circ s_1 = s_0 \circ t_1$. A *linear 2-polygraph* (P, P_2) consists of a linear 1-polygraph together with a cellular extension P_2 of the free 1-algebra P^ℓ generated by P. An *extended presentation* of an algebra A is a linear 2-polygraph whose underlying linear 1-polygraph presents A. A linear 2-polygraph is *left-monomial* when the underlying linear 1-polygraph is, in the sense of §6.1.12.

A 2-*algebra* is an internal 2-category in the category **Alg** of algebras. Note that contrarily to the set-theoretic case, we will not bother about distinguishing whether we take cells to be invertible or not: it can be shown that the notion of 2-algebra coincides with the notion of internal 2-groupoid in the category of algebras, see [160]. Any linear 2-polygraph P freely generates a linear 2-algebra that we denote as P^ℓ, and whose algebra of 2-cells is in particular written P_2^ℓ.

7.6.2 Coherent confluence and convergence. Let P be a left-monomial linear 2-polygraph. A branching (ϕ, ψ) of P is *coherently confluent* if there exist positive 1-cells ϕ' and ψ' in P_1^ℓ and a 2-cell F in P_2^ℓ as in

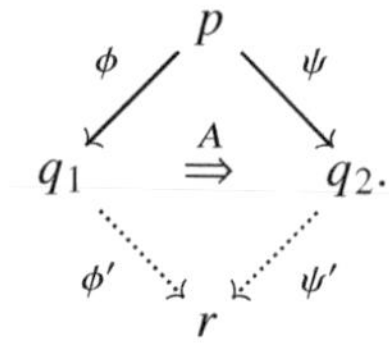

If p is a 0-cell of P_0^ℓ, say that P is *coherently confluent* (resp. *locally coherently confluent*, resp. *critically coherently confluent*) at p if every branching (resp. local branching, resp. critical branching) of P of source p is coherently confluent. Say that P is *coherently confluent* (resp. *locally coherently confluent*, resp. *critically coherently confluent*) if it is so at every 0-cell of P_0^ℓ, and that P is *coherently convergent* if it is terminating and coherently confluent.

7.6.3 Lemma. *Let P be a left-monomial linear 2-polygraph with a fixed 0-cell p, and suppose that P is coherently confluent at every 0-cell q such that $p \overset{*}{\to} q$. Let ϕ be a 1-cell of P_1^ℓ which admits a decomposition*

$$p_0 \overset{\phi}{\longrightarrow} p_k \quad = \quad p_0 \overset{\phi_1}{\longrightarrow} p_1 \overset{\phi_2}{\longrightarrow} \cdots \overset{\phi_k}{\longrightarrow} p_k$$

into 1-cells ϕ_i of size 1. If $p \overset{*}{\to} p_i$ holds for every $0 \leqslant i < k$, then there exist positive 1-cells ϕ' and ψ in P_1^ℓ and a 2-cell F in P_2^ℓ as in

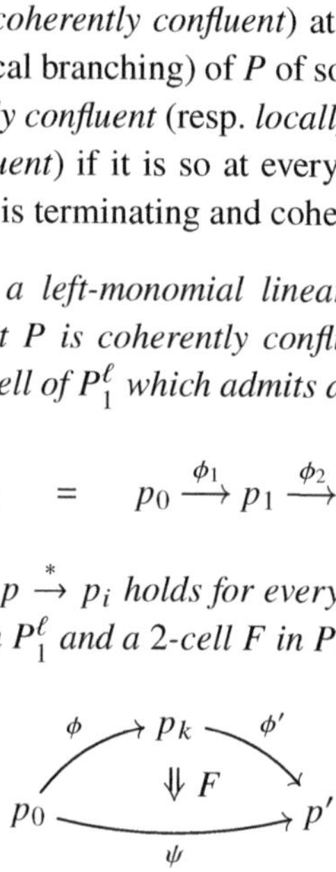

Proof. Proceed by induction on k. If $k = 0$, then ϕ is an identity, so taking $\phi' = \psi = 1_{p_0}$ and $F = 1_\phi$ proves the result. Otherwise, we construct

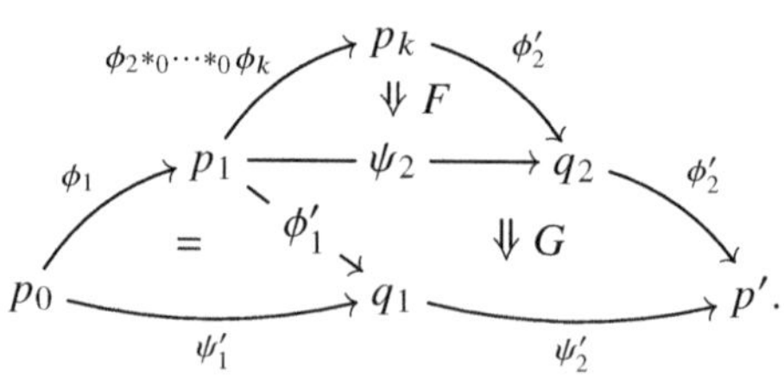

Apply Lemma 6.2.2 to the 1-cell ϕ_1 of size 1 to get the positive 1-cells ϕ_1' and ψ_1 such that $\phi_1 = \phi_1' *_0 \psi_1^-$. We have $p \overset{*}{\to} p_i$ for every $1 \leqslant i < k$, so the induction hypothesis applies to $\phi_2 *_0 \cdots *_0 \phi_k$, providing the positive 1-cells ϕ_2' and ψ_2, and the 2-cell F. Then, consider the branching (ϕ_1', ψ_2), whose source p_1 satisfies $p \overset{*}{\to} p_1$: by hypothesis, this branching is coherently confluent, giving the positive 1-cells ϕ_2' and ψ_2', and the 2-cell G. □

The following result is a formulation of coherent Newman's lemma for linear polygraphs. The proof is the same as in the set-theoretical case given by Proposition 7.3.2.

7.6.4 Proposition. *Let P be a terminating left-monomial linear 2-polygraph. If P is locally coherently confluent then it is coherently confluent.*

The following result is a formulation of the coherent critical branchings lemma, Lemma 7.3.3, for linear polygraphs. Due to the linearity of contexts, the termination is necessary and the proof differs from the set-theoretical case, as already explained in Remark 6.2.18.

7.6.5 Lemma. *Suppose given a terminating left-monomial linear 2-polygraph P. If P is critically coherently confluent, then P is locally coherently confluent.*

Proof. We proceed by Noetherian induction on the sources of the local branchings to prove that P is locally coherently confluent at every 0-cell of P_0^ℓ. We note that a reduced 0-cell cannot be the source of a local branching, so P is locally coherently confluent at reduced 0-cells. Now, fix a non-reduced 0-cell p of P_0^ℓ, and assume that P is locally coherently confluent at every 0-cell q with $p \overset{*}{\to} q$. With a termination-based argument similar to that of Proposition 7.3.2, we deduce that P is coherently confluent at every q. Then we proceed by case analysis on the type of the local branchings, noting that an aspherical branching $\lambda(\phi, \phi) + b$ is always coherently confluent.

For an additive branching, we construct

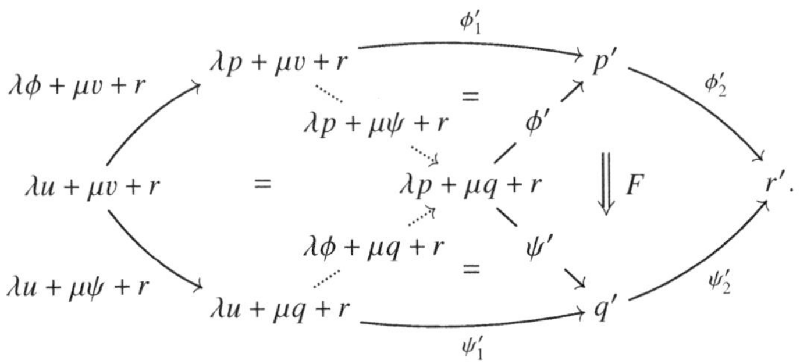

By linearity of the 0-composition, we have

$$(\lambda\phi + \mu v + r) *_0 (\lambda p + \mu\psi + r) = \lambda\phi + \mu\psi + r = (\lambda u + \mu\psi + r) *_0 (\lambda\phi + \mu q + r).$$

Note that the dotted 1-cells $\lambda p + \mu\psi + r$ and $\lambda\phi + \mu q + r$ are not positive in general since u can be in $\mathrm{supp}(q)$ or v in $\mathrm{supp}(p)$. However, those 1-cells are of size 1, and Lemma 6.2.2 applies to both of them to give positive 1-cells ϕ'_1, ψ'_1, ϕ', and ψ' that satisfy

$$\phi'_1 = (\lambda p + \mu\psi + r) \star_0 \phi' \qquad \psi'_1 = (\lambda\phi + \mu q + r) \star_0 \psi'.$$

Now, $u \overset{*}{\to} p$, $v \overset{*}{\to} q$, $\lambda \neq 0$ and $\mu \neq 0$ imply $\lambda u + \mu v + r \overset{*}{\to} \lambda p + \mu q + r$. Thus, the branching (ϕ', ψ') is coherently confluent by hypothesis, yielding the positive 1-cells ϕ'_2 and ψ'_2 and the 2-cell F.

Next, in the case of an orthogonal branching, we construct

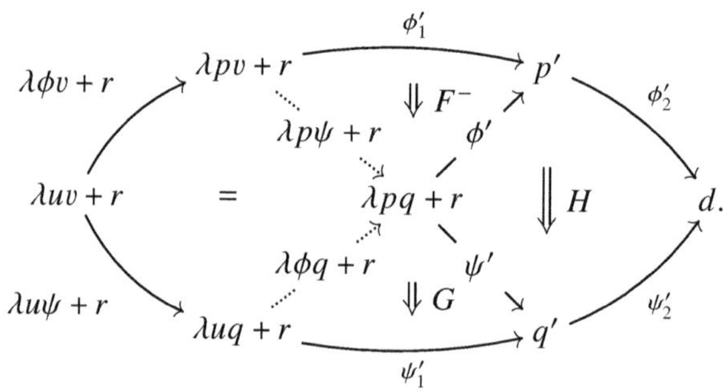

Use the linearity of the 0-composition to obtain

$$(\lambda\phi v + r) \star_0 (\lambda p\psi + r) = \lambda\phi\psi + r = (\lambda u\psi + r) \star_0 (\lambda\phi q + r).$$

Again, the dotted 1-cells $\lambda\phi q + r$ and $\lambda p\psi + r$ are not positive in general: this is the case, for example, if either $\mathrm{supp}(uq) \cap \mathrm{supp}(r)$ or $\mathrm{supp}(pv) \cap \mathrm{supp}(r)$ is not empty. Let $p = \sum_{i=1}^{k} \mu_i u_i$ be the canonical decomposition of p. By linearity of the 0-composition, the 1-cell $\lambda p\psi + r$ admits the following decomposition in 1-cells of size 1:

$$\lambda p\psi + r = \psi_1 \star_0 \cdots \star_0 \psi_k$$

with

$$\psi_j = \sum_{1 \leqslant i < j} \lambda\mu_i u_i q + \lambda\mu_j u_j \psi + \sum_{j < i \leqslant k} \lambda\mu_i u_i v + r.$$

We have $u \overset{*}{\to} u_i$ for every i, and $v \overset{*}{\to} b$, giving $\lambda uv + v \overset{*}{\to} t(\psi_j)$ for every j. Hence $\lambda p\psi + r$ is eligible for Lemma 7.6.3, yielding ϕ'_1, ϕ', and F. The cells ψ'_1, ψ', and G are obtained similarly from $\lambda\phi q + r$. Finally, $\lambda uv + r \overset{*}{\to} \lambda pq + r$ implies, by induction hypothesis, that (ϕ', ψ') is coherently confluent, giving ϕ'_2, ψ'_2, and H.

Finally, for an overlapping branching $(\lambda\phi + r, \lambda\psi + r)$, we construct

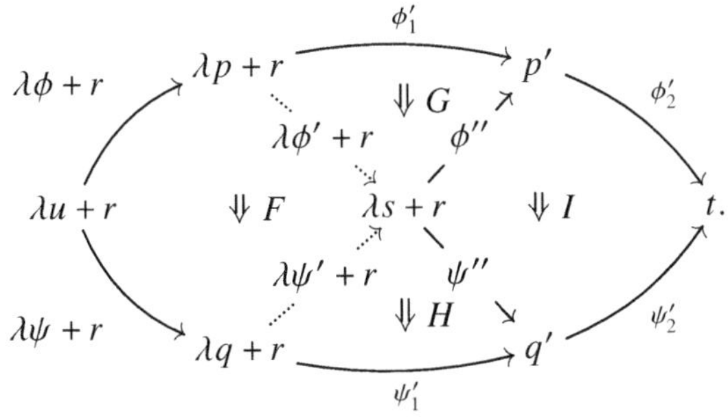

Consider the unique decomposition $(\phi, \psi) = v(\phi_0, \psi_0)w$, with (ϕ_0, ψ_0) critical. Since (ϕ_0, ψ_0) is coherently confluent by hypothesis, one obtains

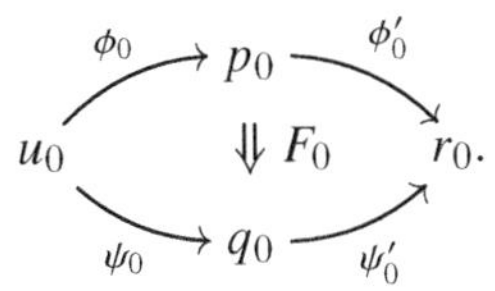

Define the positive 1-cells $\phi' = v\phi_0' w$ and $\psi' = v\psi_0' w$, and the 2-cell $F = vF_0 w$. As previously, the dotted 1-cells are not positive in general, if $\mathrm{supp}(c)$ intersects $\mathrm{supp}(p)$ or $\mathrm{supp}(q)$, for example. However, the 1-cell ϕ' is positive, so that it is a 0-composite $\phi' = \chi_1 *_0 \cdots *_0 \chi_k$ of rewriting steps. As a consequence, we have the chain of reductions

$$u \overset{*}{\to} p = s(\chi_1) \overset{*}{\to} \cdots \overset{*}{\to} s(\chi_k) \overset{*}{\to} s.$$

Since we have $\lambda \neq 0$ and $u \notin \mathrm{supp}(r)$ by hypothesis, the inequality

$$\lambda u + r \overset{*}{\to} \lambda s(\chi_i) + r$$

holds for every i, so that the following decomposition of the 1-cell $\lambda\phi' + r$ satisfies the hypotheses of Lemma 7.6.3:

$$\lambda\phi' + r = (\lambda\chi_1 + r) *_1 \cdots *_1 (\lambda\chi_k + r).$$

This gives ϕ_1', ϕ'', and G. Proceed similarly with the 1-cell $\lambda\psi' + r$ to obtain ψ_1', ψ'', and H. Finally, apply the induction hypothesis on (ϕ'', ψ''), since $\lambda u + r \overset{*}{\to} \lambda s + r$, to get ϕ_2', ψ_2', and I. □

Given a terminating left-monomial linear 1-polygraph P, taking P_2 to be the set of all 2-spheres, critical coherent confluence (resp. local coherent confluence) in the linear 2-polygraph (P, P_2) is the same as critical confluence (resp. local confluence) in P. We thus deduce the critical branching lemma for linear 1-polygraphs, already announced in Lemma 6.2.15, as a particular case.

With a proof similar to the one in the set-theoretical case, see Theorem 7.3.5, we have the coherent Squier theorem for linear polygraphs:

7.6.6 Theorem. *Let P be a convergent left-monomial linear 1-polygraph and P_2 be a cellular extension of P_1^ℓ that contains a 2-cell*

$$\begin{array}{ccccc} & & p & & \\ & {\scriptstyle\phi}\swarrow & & \searrow{\scriptstyle\psi} & \\ q & & \overset{A}{\Rightarrow} & & r \\ & {\scriptstyle\phi'}\searrow & & \swarrow{\scriptstyle\psi'} & \\ & & p' & & \end{array}$$

for every critical branching (ϕ, ψ) of P, with ϕ' and ψ' positive 1-cells of P_1^ℓ. Then the 2-polygraph (P, P_2) is coherent.

7.6.7 Example. We consider the quadratic algebra A presented by

$$\langle\, x, y, z \,|\, x^2 + yz = 0, x^2 + \lambda zy = 0 \,\rangle$$

where λ is a fixed scalar different from 0 and 1, from [300, Section 4.3]. Put $\mu = \lambda^{-1}$. The algebra A admits the presentation

$$P = \langle\, x, y, z \,\big|\, \alpha\colon yz \to -x^2, \beta\colon zy \to -\mu x^2 \,\rangle.$$

The deglex order generated by $z > y > x$ satisfies $yz > x^2$ and $zy > x^2$, proving that P terminates. However, P is not confluent. Indeed, it has two critical branchings:

$$\begin{array}{ccc} & \overset{\alpha y}{\nearrow} & -x^2y \\ yzy & & \\ & \underset{y\beta}{\searrow} & -\mu yx^2 \end{array} \qquad \text{and} \qquad \begin{array}{ccc} & \overset{\beta z}{\nearrow} & -\mu x^2 z \\ zyz & & \\ & \underset{z\alpha}{\searrow} & -zx^2 \end{array}$$

and neither of them is confluent because the monomials x^2y, yx^2, x^2z, and zx^2 are reduced. The adjunction of the 1-cells

$$\gamma\colon yx^2 \to \lambda x^2 y \qquad \text{and} \qquad \delta\colon zx^2 \to \mu x^2 z$$

gives a left-monomial linear 1-polygraph

$$Q = \left\langle\, x, y, z \,\middle|\, \begin{array}{ll} \alpha\colon\ yz \to -x^2, & \gamma\colon\ yx^2 \to \lambda x^2 y, \\ \beta\colon\ zy \to -\mu x^2, & \delta\colon\ zx^2 \to \mu x^2 z \end{array} \right\rangle$$

that also presents A, since γ and δ induce relations that already hold in $\overline{P}$, and that also terminates, because of $yx^2 > x^2y$ and $zx^2 > x^2z$. Moreover, each one of the four critical branchings of Q is confluent:

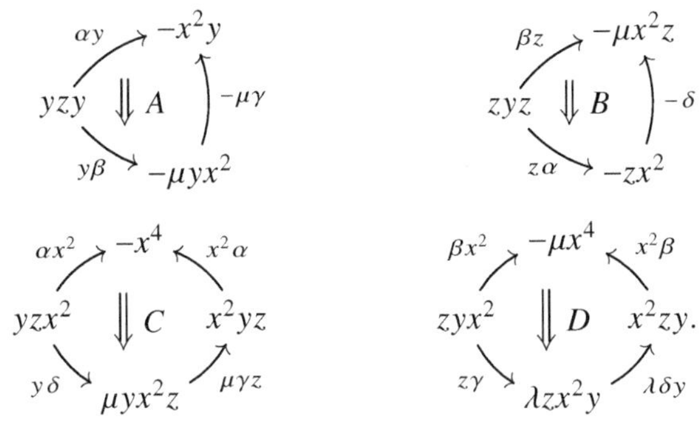

Theorem 7.6.6 implies that the 2-polygraph

$$\langle\, x, y, z \mid \alpha, \beta, \gamma, \delta \mid A, B, C, D \,\rangle$$

is a coherent presentation of A.

This coherent presentation can be reduced to a smaller one by a collapsing mechanism, similar to the one developed in §7.5.6 in the set-theoretic case, and hinted at in this example. First, some 2-cells may be removed without breaking acyclicity because their boundary can also be filled by a composite of other 2-cells. Here, the "critical 3-branchings", where three rewriting steps overlap, reveal two relations between 2-cells:

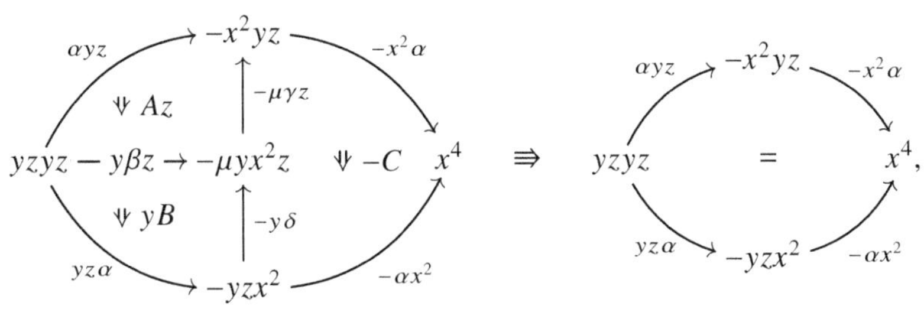

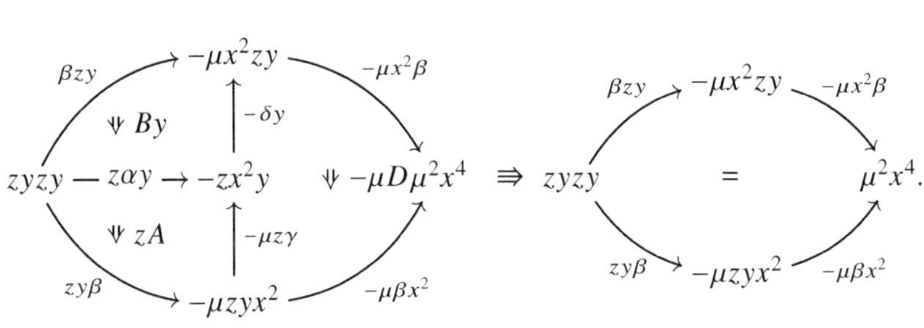

Since the boundaries of C and D can also be filled using A and B only, the 2-polygraph $\langle\, x, y, z \mid \alpha, \beta, \gamma, \delta \mid A, B \,\rangle$ is also a coherent presentation of A. Next, the 1-cells γ and δ are redundant because the corresponding relations can be derived from α and β, as testified by the 2-cells A and B: removing γ with A, and δ with B, proves that P_1^{ℓ} admits an empty acyclic cellular extension, so that $\langle\, x, y, z \mid \alpha, \beta \mid \,\rangle$ is actually a coherent presentation of A.

7.6.8 Example (The standard coherent presentation). Assume that $A = \Bbbk \oplus A_+$ is an augmented algebra, and fix a linear basis $\mathcal{B}$ of A_+. For u and v in $\mathcal{B}$, write $u \otimes v$ for the product of u and v in the free algebra over $\mathcal{B}$, and uv for their product in A. Consider the linear 1-polygraph $\mathrm{Std}(\mathcal{B})_1$ whose 0-cells are the elements of $\mathcal{B}$, and with a 1-cell

$$u \otimes v \xrightarrow{u|v} uv,$$

for all u and v in $\mathcal{B}$. Note that uv belongs to the free algebra over $\mathcal{B}$ because A is augmented. By definition, $\mathrm{Std}(\mathcal{B})_1$ is a presentation of A. Moreover, $\mathrm{Std}(\mathcal{B})_1$ terminates by a length argument: for all u and v in $\mathcal{B}$, the monomial $u \otimes v$ is a word of length 2 in the free monoid over $\mathcal{B}$, while uv is a word of length 1. Finally, $\mathrm{Std}(\mathcal{B})_1$ has one critical branching $(u|v \otimes w, u \otimes v|w)$ for each triple (u, v, w) of elements of $\mathcal{B}$, and this critical branching is confluent. Thus, extending $\mathrm{Std}(\mathcal{B})_1$ with a 2-cell

$$\begin{array}{ccccc} & \overset{u|v\otimes w}{\nearrow} & uv \otimes w & \overset{uv|w}{\searrow} & \\ u \otimes v \otimes w & & \Downarrow u|v|w & & uvw \\ & \underset{u\otimes v|w}{\searrow} & u \otimes vw & \underset{u|vw}{\nearrow} & \end{array}$$

for each triple (u, v, w) of elements of $\mathcal{B}$ produces, by Theorem 7.6.6, a coherent presentation of A, denoted by $\mathrm{Std}(\mathcal{B})_2$. Note that the free 2-algebra over $\mathrm{Std}(\mathcal{B})_2$ does not depend (up to isomorphism) on the choice of the basis $\mathcal{B}$.

This coherent presentation of A is extended in every dimension in §23.3.8 to obtain a polygraphic version of the standard resolution of an algebra. As in the previous example, the next dimension contains the 3-cells generated by the "critical 3-branchings" of $\mathrm{Std}_1(\mathcal{B})$: there is one such 3-cell $u|v|w|x$ for each quadruple (u, v, w, x) of elements of $\mathcal{B}$, with source

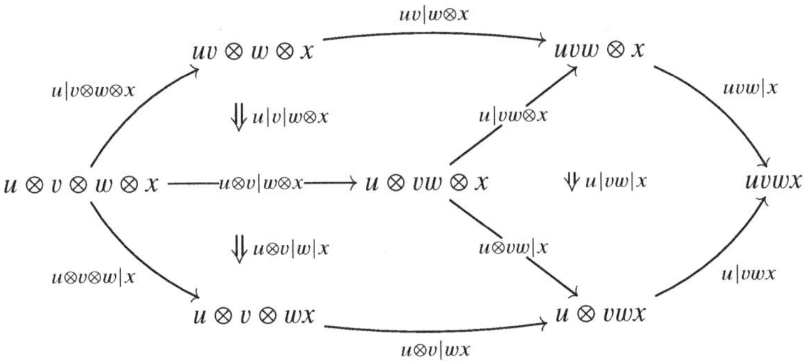
uv ⊗ w ⊗ x
uvw ⊗ x
u ⊗ v ⊗ w ⊗ x
u ⊗ vw ⊗ x
uvwx
u ⊗ v ⊗ wx
u ⊗ vwx
uv|w⊗x
u|v⊗w⊗x
u|v|w⊗x
u|vw⊗x
uvw|x
u⊗v|w⊗x
u|vw|x
u⊗v|w|x
u⊗vw|x
u⊗v⊗w|x
u|vwx
u⊗v|wx

and target

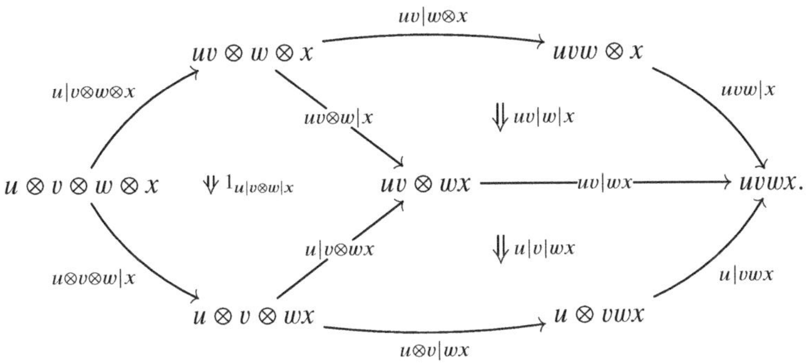
uv ⊗ w ⊗ x
uvw ⊗ x
u ⊗ v ⊗ w ⊗ x
uv ⊗ wx
uvwx.
u ⊗ v ⊗ wx
u ⊗ vwx
uv|w⊗x
u|v⊗w⊗x
uv⊗w|x
uv|w|x
uvw|x
1u|v⊗w|x
uv|wx
u|v⊗wx
u|v|wx
u⊗v⊗w|x
u|vwx
u⊗v|wx

8

Categories of Finite Derivation Type

In Chapter 7, we have seen a canonical and efficient way to extend a convergent presentation of a category C by a 2-polygraph P into a coherent one. Precisely, the 3-cells used in this extension procedure are in one-to-one correspondence with the confluence diagrams of critical branchings in P (Theorem 7.3.5). Now if P is finite, so is the set of its critical branchings and therefore the set of 3-cells generating coherence can be taken to be finite. In such a situation, we say that the polygraph P has *finite derivation type*, or *FDT*. The relevance of this concept lies in the following invariance property: if a category C admits a finite presentation P having finite derivation type, then *all* finite presentations of C also have FDT (Theorem 8.1.2). This invariance will prove essential to show that some finitely presented categories do *not* admit convergent presentations.

This finiteness condition, introduced by Squier [328], is of homotopical nature, and is, in some sense, a refinement of the homological condition introduced earlier in [326]. The latter will be discussed in the next chapter. Using these conditions, Squier managed to produce an explicit example of a finitely presented monoid, with decidable word problem, but having no finite convergent presentation. This provides a negative answer to the question of *universality of finite convergent rewriting* we raised in Section 5.3. Let us finally emphasize the power of the FDT invariant: by performing computations on *one* presentation of a monoid, we are able to deduce properties of *any* finite presentation of it!

The finiteness condition is introduced in Section 8.1 and studied in the case of convergent 2-polygraphs in Section 8.2. In Section 8.3, we define the notion of identities among relations for 2-polygraphs, generalizing those already known for presentations of groups: such identities are described by 2-spheres of the free (2, 1)-category on the 2-polygraph.

8.1 Finite Derivation Type

A 2-polygraph P has *finite derivation type*, or *FDT* for short, if it is finite and if the $(2,1)$-category $P^{\top}$ admits a finite acyclic cellular extension. Otherwise said, there is a finite coherent $(3,1)$-polygraph Q of which P is the underlying 2-polygraph. A category C has *finite derivation type* if it admits a finite coherent presentation.

8.1.1 Tietze invariance of the FDT property. Recall from Section 5.1 that two 2-polygraphs are *Tietze equivalent* when they present isomorphic categories. We say that a property $\mathcal{P}$ on 2-polygraphs is *Tietze invariant* when for every Tietze equivalent 2-polygraphs P and Q, the polygraph P satisfies the property $\mathcal{P}$ if and only if the polygraph Q does.

Given two Tietze equivalent 2-polygraphs P and Q whose sets P_2 and Q_2 of 2-generators are finite, consider a finite acyclic cellular extension X of the free $(2,1)$-category $P^{\top}$. By Theorem 7.1.6, the cellular extension X transfers to a finite cellular extension of the free $(2,1)$-category $Q^{\top}$. We may therefore state the following invariance result, first proved by Squier for monoids [328, Theorem 4.3] and revisited in [165, Theorem 4.2.3] in polygraphic terms.

8.1.2 Theorem. *Let P and Q be two Tietze equivalent 2-polygraphs such that P_2 and Q_2 are finite. Then P has finite derivation type if and only if Q has finite derivation type.*

This result shows that the property for a category C of having finite derivation type does not depend on the presentation, provided that it is finite.

The following result will help prove that a presentation admits no finite acyclic cellular extension, i.e., that the presented category does not have finite derivation type.

8.1.3 Proposition. *Let P be a 2-polygraph and let X be an acyclic extension of the free $(2,1)$-category $P^{\top}$. If $P^{\top}$ admits a finite acyclic cellular extension, then there exists a finite subset of X that is an acyclic cellular extension of $P^{\top}$.*

Proof. Suppose that $P^{\top}$ admits a finite acyclic cellular extension Y and let A be a 3-generator of Y. Since X is an acyclic extension of $P^{\top}$, there exists a 3-cell $F_A \colon s_2(A) \Rrightarrow t_2(A)$ in the free $(3,1)$-category $P^{\top}(X)$. This induces a 3-functor between free $(3,1)$-categories $f \colon P^{\top}(Y) \to P^{\top}(X)$, which is the identity on P and such that $f(A) = F_A$ for every 3-generator A of Y. Let X_Y be the subset of X containing all the 3-generators occurring in some 3-cell F_A, for A in Y. Since Y is finite and each 3-cell F_A can be written as a composition of finitely many 3-generators of X, we deduce that X_Y is finite.

Finally, consider a 2-sphere (ϕ, ψ) of $P^{\top}$. By hypothesis, there exists a 3-cell $A\colon \phi \Rrightarrow \psi$ in $P^{\top}(Y)$. By application of f, one obtains a 3-cell $f(A)\colon \phi \Rrightarrow \psi$ in $P^{\top}(X)$. Moreover, the 3-cell $f(A)$ is a composite of cells F_A, and the 3-cell $f(A)$ is thus in $X_Y^{\top}$. As a consequence, one has $\phi \approx^{X_Y} \psi$, so that X_Y is a finite acyclic cellular extension of $P^{\top}$. □

8.2 Convergence and Finite Derivation Type

Theorem 7.3.5 states that any family of generating confluences of a convergent 2-polygraph P forms an acyclic extension of the free $(2, 1)$-category $P^{\top}$. The set of critical branchings of a finite 2-polygraph being finite, we deduce that a finite convergent 2-polygraph has finite derivation type. Moreover, from Theorem 8.1.2, the property of having finite derivation type is Tietze invariant for finite 2-polygraphs. We thus obtain a finiteness condition for finitely presented categories to have a presentation by a finite convergent 2-polygraph.

8.2.1 Theorem. *If a category admits a finite convergent presentation, then it has finite derivation type.*

This result was first proved by Squier for finitely presented monoids [328, Theorem 5.3]. Several other proofs can be found in the literature: we refer to [233] for a reformulation of Squier's arguments and to [165] for a proof in the polygraphic language presented in this book.

Now suppose we want to show that some category does not admit a finite convergent presentation: by Theorem 8.2.1 it is sufficient to prove that it has no finite derivation type. The first example based on this argument, due to Squier [328], is presented in §8.2.3. Before that, we turn to a simplified version introduced by Lafont and Prouté in [233, 238].

8.2.2 Example. Consider the monoid M presented by the following 2-polygraph:

$$P = \langle\, \star \mid a, b, c, d, d' \mid \alpha_0\colon ab \Rightarrow a, \beta\colon da \Rightarrow ac, \beta'\colon d'a \Rightarrow ac \,\rangle.$$

This is a variant of the monoid already encountered in Example 5.2.4. It admits a finite presentation and has a decidable word problem, yet it does not have finite derivation type and, as a consequence, it does not admit a finite convergent presentation. To prove these facts, the 2-polygraph P is completed, by Knuth–Bendix procedure (see Section 5.2), into the following infinite convergent 2-polygraph

$$\tilde{P} = \langle\, \star \mid a, b, c, d, d' \mid \alpha_n, \beta, \beta' \,\rangle_{n \in \mathbb{N}},$$

with

$$\alpha_n : ac^n b \Rightarrow ac^n, \qquad \beta : da \Rightarrow ac, \qquad \beta' : d'a \Rightarrow ac.$$

Even though the polygraph $\tilde{P}$ is infinite, we can implement an algorithm to normalize 1-cells in $\tilde{P}$ by iteratively rewriting those, and therefore decide the word problem in $\tilde{P}$, and thus in P, by comparing normal forms. The 2-polygraph $\tilde{P}$ has two infinite families of critical branchings from which we deduce two infinite families of 3-generators:

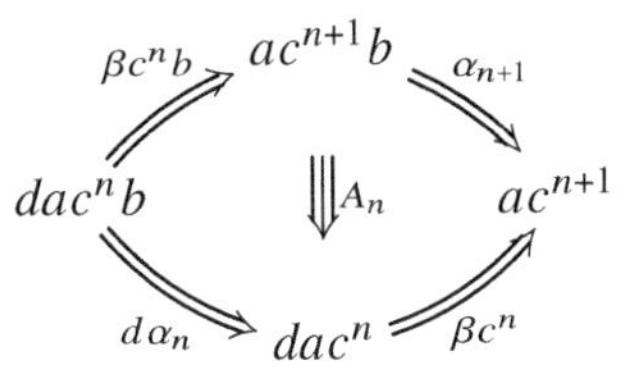

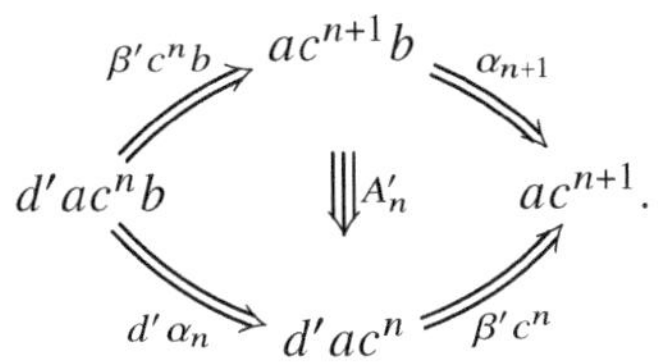

The 3-generators A'_n induce a projection functor $f : \tilde{P}^\top \to P^\top$ which is the identity on 0- and 1-generators, sends the 2-generators α_0, β and β' to themselves and the image of α_n, for $n > 0$, is defined by induction by

$$f(\alpha_{n+1}) = \beta'^- c^n b *_1 d' f(\alpha_n) *_1 \beta' c^n.$$

This functor is a retract of the canonical inclusion functor $g : P^\top \to \tilde{P}^\top$. By the transfer theorem (Theorem 7.1.6), the family

$$X = \{ f(A_n) \mid n \in \mathbb{N} \}$$

is thus an infinite acyclic cellular extension of the (2, 1)-category $P^\top$: in this case, the generators of the form (7.1) are superfluous because $f \circ g$ is the identity on $P^\top$.

By Proposition 8.1.3, in order to conclude that the polygraph P does not have FDT, it is enough to show that no finite subset of X forms an acyclic cellular extension of $P^\top$. A fully explicit direct proof of this fact is rather tedious. A complete proof is given in [233, Section 5] using an abelianized form of the category $P^\top$ in terms of monoidal groupoids. Note also that this can be shown indirectly by a homological argument outlined in the next chapter. Indeed, in Example 9.3.11 we show that the third integral homology group of the monoid M is not of finite type. This shows by Theorem 9.3.4 that the monoid M does not have FDT, and so in particular, we cannot extract a finite acyclic cellular extension of $P^\top$ from X.

8.2.3 Squier's monoids. We now recall Squier's original example of a finitely presented monoid that does not admit a finite convergent presentation and studied in [328] and [326] using homotopical and homological arguments respectively. Consider, for $k \geqslant 1$, the monoid S_k defined in [328, Example 4.5] and presented by

$$\langle \star \mid a, b, c, d_i, e_i \mid \alpha_n, \beta_i, \gamma_i, \delta_i, \varepsilon_i \rangle_{n\in\mathbb{N}, 1\leqslant i\leqslant k},$$

where the rules are $\alpha_n : ac^n b \Rightarrow 1$ for $n \in \mathbb{N}$, and for $1 \leqslant i \leqslant k$,

$$\beta_i : d_i a \Rightarrow acd_i,\ \gamma_i : d_i c \Rightarrow cd_i,\ \varepsilon_i : d_i e_i \Rightarrow 1,\ \delta_i : d_i b \Rightarrow bd_i.$$

Squier proves the following properties for the monoid S_1 in [328, Theorem 6.7, Corollary 6.8]. The proof is reworked in [233, Section 6] and in [165, Section 6] using polygraphs.

8.2.4 Theorem. *The monoid S_1 is a finitely presented monoid that has the following properties.*

1. *It has a decidable word problem.*
2. *It does not have finite derivation type.*
3. *It does not have a finite convergent presentation.*

Proof. The monoid S_1 has the following infinite presentation:

$$P = \langle \star \mid a, b, c, d, e \mid \alpha_n, \beta, \gamma, \delta, \varepsilon \rangle_{n\in\mathbb{N}}$$

with

$$\alpha_n : ac^n b \Rightarrow 1,\ \ \beta : da \Rightarrow acd,\ \ \gamma : dc \Rightarrow cd,\ \ \varepsilon : de \Rightarrow 1,\ \ \delta : db \Rightarrow bd.$$

This presentation is infinite, so the normal-form algorithm of §4.2.4 cannot be applied to decide the word problem in the monoid S_1. However, the sources of the 2-generators α_n are the elements of the regular language ac^*b. This implies that the sources of the 2-generators of the polygraph P form a regular language over the finite set $\{a, b, c, d, e\}$. Following [294, Proposition 3.6], this implies that the word problem for S_1 is decidable, which proves Condition 1.

Condition 3 is a consequence of Condition 2 and Theorem 8.2.1.

We sketch the main arguments of the proof of Condition 2, and we refer to [328] for the original proof and to [165, Section 6] for the proof presented here. We denote by $\gamma_n : dc^n \Rightarrow c^n d$ the 2-cell of P_2^* defined by induction on n as follows:

$$\gamma_0 = 1_x \qquad \text{and} \qquad \gamma_{n+1} = \gamma c^n *_1 c\gamma_n.$$

For every n, we write $\phi_n : dac^n b \Rightarrow ac^{n+1}bd$ the following composite in P_1^*

$$dac^n b \xRightarrow{\beta c^n b} acdc^n b \xRightarrow{ac\gamma_n b} ac^{n+1}db \xRightarrow{ac^{n+1}\delta} ac^{n+1}bd.$$

Considering for every natural number $n \geqslant 0$, the following 2-sphere of $P_2^{\top}$:

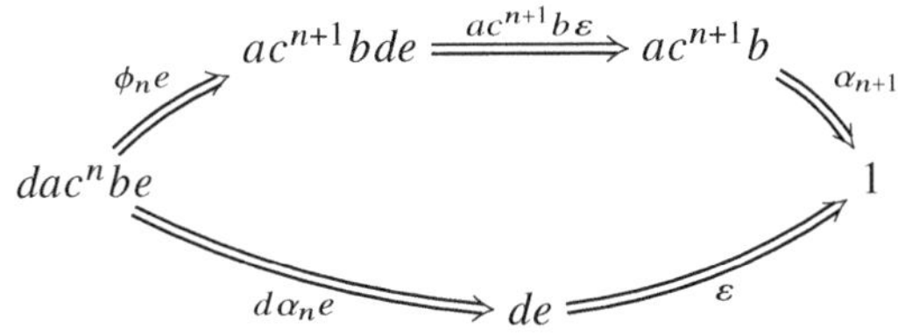

(8.1)

we prove that the monoid S_1 admits the following finite presentation

$$Q = \langle \star \mid a, b, c, d, e \mid \alpha_0, \beta, \gamma, \delta, \varepsilon \rangle.$$

We also prove that the 2-polygraph P is convergent and the Squier completion of P contains a 3-generator A_n with shape

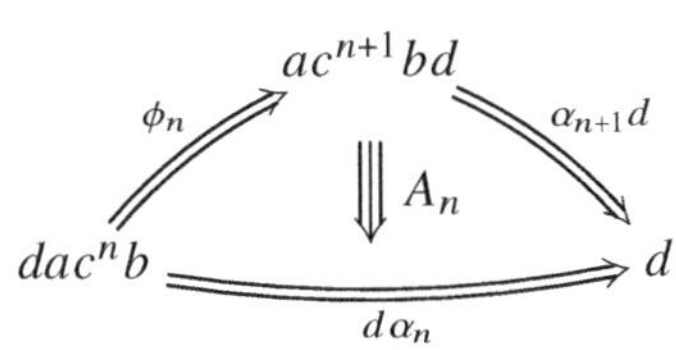

for every natural number n. In order to show that the monoid S_1 does not have FDT, by Theorem 8.1.2, it is sufficient to check that the polygraph Q admits no finite acyclic cellular extension. We denote by $g\colon P^{\top} \to Q^{\top}$ the projection that sends the 2-cells β, γ, δ, and ε to themselves and whose value on α_n is given by induction on n, thanks to (8.1), i.e.,

$$g(\alpha_0) = \alpha \qquad \text{and} \qquad g(\alpha_{n+1}) = (\phi_n e \star_1 ac^{n+1}b\varepsilon)^- \star_1 df(\alpha_n)e \star_1 \varepsilon.$$

By application of Theorem 7.1.6 to the canonical inclusion $f\colon Q^{\top} \to P^{\top}$ and g defined above, we deduce that the monoid S_1 admits the coherent presentation

$$\widetilde{Q} = \big\langle \star \,\big|\, a, b, c, d, e \,\big|\, \alpha_0, \beta, \gamma, \delta, \varepsilon \,\big|\, \tilde{A}_n \big\rangle_{n \in \mathbb{N}}$$

where $\tilde{A}_n$ is the 3-generator

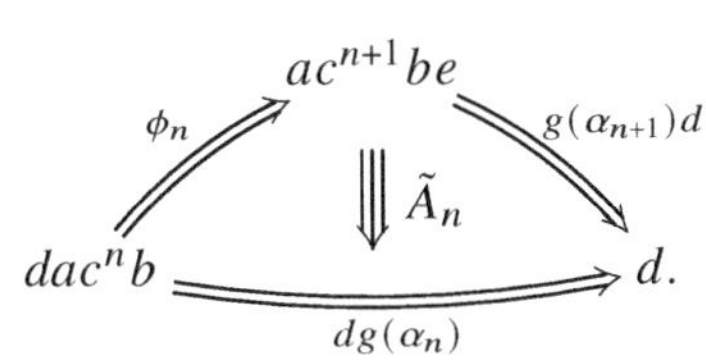

By studying the relations among 3-cells in the 3-category generated by the coherent presentation $\widetilde{Q}$ in terms of generating critical, see §7.5.5, we show that the polygraph Q is not Tietze equivalent to a polygraph having FDT.

Following Theorem 8.1.2, this shows that the monoid S_1 does not have FDT. We refer to [165] for more details on this proof. □

8.2.5 Higher-dimensional finite derivation type. With the aim of characterizing the class of finite presented decidable monoids admitting a finite convergent presentation, some refinements of the FDT condition were introduced, such as the 2-dimensional FDT property [274] and the infinite-dimensional FDT property [163], see also §23.4.1. Note that the characterization of this class by finiteness conditions is still an open problem.

8.3 Identities Among Relations

The 3-generators in a coherent presentation are closely related to the notion of *identity among relations*, which originates in the work of Peiffer and Reidemeister in combinatorial group theory [295, 310]. This notion is based on the one of crossed module, introduced by Whitehead, in algebraic topology, for the classification of homotopy 2-types [356, 357]. There exist several formulations of identities for presentations of groups: as homological 2-syzygies [64], as homotopical 2-syzygies [252], or as Igusa's pictures [212, 252]. One can also interpret identities as the critical pairs of a presentation of a group by a convergent string rewriting system [99]. The latter approach yields an algorithm based on Knuth–Bendix's completion procedure that computes a family of generators of the module of identities among relations [179].

In this section, we introduce the notion of identity among relations for a 2-polygraph. We relate the property for a polygraph of having a finite generating set of identities among relations to the property of having abelian finite derivation type. First, let us recall the notion of natural system used in this section.

8.3.1 Natural system. The *category of factorizations* of a small category C is the category, denoted by FC, whose 0-cells are the 1-cells of C and whose 1-cells from w to w' are pairs (u, v) of 1-cells of C such that the following diagram commutes in C:

$$\begin{array}{ccc} \cdot & \xleftarrow{\;u\;} & \cdot \\ {\scriptstyle w}\big\downarrow & & \big\downarrow{\scriptstyle w'} \\ \cdot & \xrightarrow[\;v\;]{} & \cdot \end{array}$$

The triple (u, w, v) is called a *factorization of* w'. A *natural system* on C is a functor $D\colon \mathrm{F}C \to \mathbf{Ab}$ with values in the category $\mathbf{Ab}$ of abelian groups.

We will denote by D_w the abelian group which is the image of w by D. The category of natural systems is denoted $\mathbf{Nat}(C, \mathbf{Ab})$. We refer to §F.2.2 for more details on this notion.

8.3.2 Identities among relations. Let P be a 2-polygraph. We define the natural system $\Pi(P)$ on the presented category $\overline{P}$ of *identities among relations* of P as follows.

- If u is a 1-cell of $\overline{P}$, the abelian group $\Pi(P)_u$ is generated by one element $\lfloor\phi\rfloor$, for each 2-cell $\phi\colon v \Rightarrow v$ of the $(2,1)$-category $P^\top$ such that $\overline{v} = u$, and subject to the relation
$$\lfloor \phi \ast_1 \psi \rfloor = \lfloor \phi \rfloor + \lfloor \psi \rfloor, \tag{8.2}$$
for every 2-cells $\phi\colon v \Rightarrow v$ and $\psi : v \Rightarrow v$ of $P^\top$, with $\overline{v} = u$, and
$$\lfloor \phi \ast_1 \psi \rfloor = \lfloor \psi \ast_1 \phi \rfloor, \tag{8.3}$$
for every 2-cells $\phi\colon v \Rightarrow w$ and $\psi : w \Rightarrow v$ of $P^\top$, with $\overline{v} = \overline{w} = u$.
- If $w' = uwv$ is a factorization in $\overline{P}$, then the homomorphism of groups $\Pi(P)_{(u,v)} : \Pi(P)_w \to \Pi(P)_{w'}$ is defined by
$$\Pi(P)_{(u,v)}(\lfloor\phi\rfloor) = \lfloor \widehat{u}\phi\widehat{v} \rfloor,$$
where $\widehat{u}$ and $\widehat{v}$ are any representative 1-cells of u and v in P_1^*, respectively.

Note that the value of $\Pi(P)_{(u,v)}$ does not depend on the choice of the representative 1-cells $\widehat{u}$ and $\widehat{v}$. This proves that $\Pi(P)$ is a natural system on $\overline{P}$. We will often write $\lfloor u\phi v\rfloor$ instead of $\lfloor \widehat{u}\phi\widehat{v} \rfloor$.

As a consequence of the defining relations of each group $\Pi(P)_u$, the relations
$$\lfloor 1_u \rfloor = 0, \qquad \lfloor \phi^- \rfloor = -\lfloor\phi\rfloor, \qquad \text{and} \qquad \lfloor \psi \ast_1 \phi \ast_1 \psi^- \rfloor = \lfloor\phi\rfloor$$
hold for every 1-cell u and every 2-cells $\phi\colon u \Rightarrow u$ and $\psi : v \Rightarrow u$ of the free $(2,1)$-category $P^\top$.

8.3.3 Loops and cellular extensions. In some situations, it is helpful to consider cellular extensions by means of *2-loops* in a 2-category C, i.e., 2-cells ϕ such that $s_1(\phi) = t_1(\phi)$. The following result will be useful in the sequel.

8.3.4 Lemma. *Let C be a $(2,1)$-category and let Y be a family of 2-loops in C. The following assertions are equivalent.*

1. *The cellular extension $\widetilde{Y} := \{\widetilde{\beta}\colon \beta \Rrightarrow 1_{s_1(\beta)},\ \beta \in Y\}$ of C is acyclic.*
2. *Every 2-loop ϕ in C has a decomposition*
$$\phi = \left(\psi_1 \ast_1 u_1\beta_1^{\epsilon_1}v_1 \ast_1 \psi_1^-\right) \ast_1 \cdots \ast_1 \left(\psi_p \ast_1 u_p\beta_p^{\epsilon_p}v_p \ast_1 \psi_p^-\right) \tag{8.4}$$

with, for every $1 \leqslant i \leqslant p$, β_i in Y, ϵ_i in $\{-,+\}$, u_i, v_i 1-cells of C and ψ_i a 2-cell of C.

Proof. Suppose that C is $\widetilde{Y}$-acyclic. Given a closed 2-cell $\phi\colon w \Rightarrow w$ in C, by hypothesis there exists a 3-cell $A\colon \phi \Rrightarrow 1_w$ in $C(\widetilde{Y})$. In the $(3,1)$-category $C(\widetilde{Y})$ the 3-cell A can be decomposed into

$$A = A_1 *_2 \cdots *_2 A_k,$$

where each A_i is a 3-cell of $C(\widetilde{Y})$ that contains exactly one generating 3-cell of Y. Thus each 3-cell A_i has the shape

$$\psi_i *_1 u_i \widetilde{\beta}_i^{\epsilon_i} v_i *_1 \psi_i'$$

with $\beta_i \in Y$, $\epsilon_i \in \{-,+\}$, u_i, v_i 1-cells of C and ψ_i, ψ_i', 2-cells of C. By hypothesis on A, we have $\phi = s_2(A)$, hence $\phi = \psi_1 *_1 u_1 s_2(\beta_1^{\epsilon_1}) v_1 *_1 \psi_1'$. For $\epsilon_1 = +$, we have:

$$\begin{aligned}
\phi &= \psi_1 *_1 u_1 \beta_1 v_1 *_1 \psi_1' \\
&= \left(\psi_1 *_1 u_1 \beta_1 v_1 *_1 \psi_1^-\right) *_1 \left(\psi_1 *_1 \psi_1'\right) \\
&= \left(\psi_1 *_1 u_1 \beta_1 v_1 *_1 \psi_1^-\right) *_1 s_2(A_2).
\end{aligned}$$

And, for $\epsilon_1 = -$, we have:

$$\begin{aligned}
\phi &= \psi_1 *_1 \psi_1' \\
&= \left(\psi_1 *_1 u_1 \beta_1^- v_1 *_1 \psi_1^-\right) *_1 \left(\psi_1 *_1 u_1 \beta_1 v_1 *_1 \psi_1'\right) \\
&= \left(\psi_1 *_1 u_1 \beta_1^- v_1 *_1 \psi_1^-\right) *_1 s_2(A_2).
\end{aligned}$$

We proceed by induction on k to prove that ϕ has a decomposition as in (8.4).

Conversely, we assume that every closed 2-cell ϕ in C has a decomposition as in (8.4). Then we have $\phi \approx^{\widetilde{Y}} 1_{s_1(\phi)}$ for every closed 2-cell ϕ in C. Let us consider two parallel 2-cells ϕ and ψ in C. Then $\phi *_1 \psi^-$ is a closed 2-cell, yielding $\phi *_1 \psi^- \approx^{\widetilde{Y}} 1_{s(\phi)}$. We compose both members by ψ on the right hand to get $\phi \approx^{\widetilde{Y}} \psi$. Thus $\widetilde{Y}$ is a homotopy basis of C. □

8.3.5 Abelian finite derivation type. A $(2,1)$-category C is called *abelian* if, for every 1-cell u of C, the group Aut_u^C of 2-loops of C with source u is abelian. For C an $(2,1)$-category, its *abelianization* C_{ab} is the quotient of C by the cellular extension that contains one 2-sphere $\phi *_1 \psi \Rrightarrow \psi *_1 \phi$ for all 2-loops ϕ and ψ of C with the same source.

One says that a 2-polygraph P has *abelian finite derivation type*, or $\mathrm{FDT}_{\mathrm{ab}}$ for short, when the abelian $(2,1)$-category $P^{\top}_{\mathrm{ab}}$ admits a finite acyclic extension.

8.3.6 Proposition. *Given a 2-polygraph P, there exists an isomorphism of natural systems on the free category P_1^*:*

$$\Pi(P)_{\pi\circ(-)} \xrightarrow{\simeq} \mathrm{Aut}_{(-)}^{P_{\mathrm{ab}}^{\top}}. \tag{8.5}$$

Proof. For a 1-cell u of $P_{\mathrm{ab}}^{\top}$, we define the morphism of groups

$$\Phi_u : \Pi(P)_{\overline{u}} \to \mathrm{Aut}_u^{P_{\mathrm{ab}}^{\top}}$$

given on generators by $\Phi_u(\lfloor\phi\rfloor) = \phi^{\psi}$, where ϕ is a 2-loop of $P_{\mathrm{ab}}^{\top}$ on a 1-cell v such that $\overline{v} = \overline{u}$ and $\psi : v \Rightarrow u$ is any 2-cell of $P_{\mathrm{ab}}^{\top}$. The morphism Φ_u is well-defined. Indeed, it is independent of the choice of ψ, and its definition is compatible with the relations (8.2) and (8.3) defining $\Pi(P)_{\overline{u}}$.

For the relation (8.2), let ϕ_1 and ϕ_2 be 2-loops of $P_{\mathrm{ab}}^{\top}$ on a 1-cell v such that $\overline{v} = \overline{u}$ and let $\psi : v \Rightarrow u$ be a 2-cell of $P_{\mathrm{ab}}^{\top}$. Then,

$$\begin{aligned}
\Phi_u(\lfloor\phi_1 \star_1 \phi_2\rfloor) &= (\phi_1 \star_1 \phi_2)^{\psi} \\
&= \phi_1^{\psi} \star_1 \phi_2^{\psi} \\
&= \Phi_u(\lfloor\phi_1\rfloor) \star_1 \Phi_u(\lfloor\phi_2\rfloor) \\
&= \Phi_u(\lfloor\phi_1\rfloor + \lfloor\phi_2\rfloor).
\end{aligned}$$

For the relation (8.3), we fix 2-cells $\phi_1 : v_1 \Rightarrow v_2$, $\phi_2 : v_2 \Rightarrow v_1$ and $\psi : v_1 \Rightarrow u$, with $\overline{v_1} = \overline{v_2} = \overline{u}$. Then,

$$\begin{aligned}
\Phi_u(\lfloor\phi_1 \star_1 \phi_2\rfloor) &= (\phi_1 \star_1 \phi_2)^{\psi} \\
&= (\psi^- \star_1 \phi_1) \star_1 (\phi_2 \star_1 \phi_1) \star_1 (\phi_1^- \star_1 \psi) \\
&= (\phi_2 \star_1 \phi_1)^{\psi^- \star_1 \phi_1} \\
&= \Phi_u(\lfloor\phi_2 \star_1 \phi_1\rfloor).
\end{aligned}$$

Thus Φ_u is a morphism of groups from $\Pi(P)_{\overline{u}}$ to $\mathrm{Aut}_u^{P_{\mathrm{ab}}^{\top}}$. Moreover, it admits $\phi \mapsto \lfloor\phi\rfloor$ as inverse and, as a consequence, is an isomorphism.

Let us prove that Φ_u is natural in u. Let K be a context of P_1^* such that $v = K[u]$, and prove the equality of the two morphisms $\Phi_v \circ \Pi(P)_{\overline{K}}$ and $\mathrm{Aut}_K^P \circ \Phi_u$. Let ϕ be a 2-loop of $P_{\mathrm{ab}}^{\top}$ with source u' such that $\overline{u}' = \overline{u}$. We fix a 2-cell $\psi : u' \to u$ in $P_{\mathrm{ab}}^{\top}$ and consider the 2-cell $K[\psi] : K[u'] \to v$ of $P_{\mathrm{ab}}^{\top}$. Then, we have

$$\begin{aligned}\Phi_v \circ \Pi(P)_{\overline{K}}(\lfloor \phi \rfloor) &= (K[\phi])^{K[\psi]} \\ &= K[\psi^-] *_1 K[\phi] *_1 K[\psi] \\ &= K\left[\psi^- *_1 \phi *_1 \psi\right] \\ &= K[\phi^{\psi}] \\ &= \operatorname{Aut}_K^{P^\top_{\mathrm{ab}}} \circ \Phi_u(\lfloor \phi \rfloor). \end{aligned}$$ □

Proposition 8.3.6 characterizes the natural system $\Pi(P)$ on the category $\overline{P}$ up to isomorphism. Using this characterization, we deduce the following result.

8.3.7 Proposition. *A 2-polygraph P has* $\mathrm{FDT_{ab}}$ *if and only if the natural system $\Pi(P)$ is finitely generated.*

Proof. Suppose that the 2-polygraph P has $\mathrm{FDT_{ab}}$. Then the abelian $(2,1)$-category $P^\top_{\mathrm{ab}}$ admits a finite acyclic extension X. Given a 3-generator $A : \phi \Rrightarrow \psi$ in X, we write $\partial A = \phi *_2 \psi^-$ and $\partial X = \{\partial A \mid A \in X\}$ for the set of 2-loops of $P^\top_{\mathrm{ab}}$ associated to 3-generators in X.

By Lemma 8.3.4, any 2-loop ϕ can be written in $P^\top_{\mathrm{ab}}$ as

$$\phi = \left(\psi_1 *_1 u_1 \partial A_1^{\epsilon_1} v_1 *_1 \psi_1^-\right) *_1 \ldots *_1 \left(\psi_p *_1 u_p \partial A_p^{\epsilon_p} v_p *_1 \psi_p^-\right),$$

with, for every $1 \leqslant i \leqslant p$, A_i in X, ϵ_i in $\{-1,+1\}$, u_i, v_i 1-cells of P_1^* and ψ_i a 2-cell of the free $(2,1)$-category $P^\top$. As a consequence, for any $\lfloor \phi \rfloor$ in $\Pi(P)$, we have the following decomposition:

$$\lfloor \phi \rfloor = \sum_{i=1}^{k} (-1)^{\epsilon_i} \left\lfloor \psi_i *_1 u_i \partial A_i v_i *_1 \psi_i^- \right\rfloor = \sum_{i=1}^{k} (-1)^{\epsilon_i} \overline{u}_i \left\lfloor \partial A_i \right\rfloor \overline{v}_i.$$

Thus, the elements of $\lfloor \partial X \rfloor$ form a finite generating set for the natural system of abelian groups $\Pi(P)$.

Conversely, suppose that the natural system $\Pi(P)$ is finitely generated. There exists a finite set X of 2-loops of the abelian $(2,1)$-category $P^\top_{\mathrm{ab}}$ such that, for every 1-cell $\overline{u}$ of $\overline{P}$ and every 2-loop ϕ with source w of $P^\top_{\mathrm{ab}}$ such that $\overline{w} = \overline{u}$, one can write

$$\lfloor \phi \rfloor = \sum_{i=1}^{p} \epsilon_i \overline{u}_i \left\lfloor \alpha_i \right\rfloor \overline{v}_i,$$

with, for every $1 \leqslant i \leqslant p$, α_i in X, ϵ_i an integer and u_i, v_i 1-cells of $\overline{P}$ such that, for every representative $\widehat{u}_i$ of $\overline{u}_i$ and $\widehat{v}_i$ of $\overline{v}_i$ in $P^\top_{\mathrm{ab}}$, $\widehat{u}_i \alpha_i \widehat{v}_i$ is a 2-loop of $P^\top_{\mathrm{ab}}$ whose source w_i that satisfies $\overline{w}_i = \overline{w}$. We fix, for every i, a 2-cell $\psi_i : w \Rightarrow w_i$ in $P^\top$. Then, the properties of $\Pi(P)$ imply:

$$
\begin{aligned}
\lfloor \phi \rfloor &= \sum_{i=1}^{p} \left\lfloor \psi_i *_1 \widehat{u}_i \alpha_i^{\epsilon_i} \widehat{v}_i *_1 \psi_i^- \right\rfloor \\
&= \left\lfloor \left(\psi_1 *_1 \widehat{u}_1 \alpha_1^{\epsilon_1} \widehat{v}_1 *_1 \psi_1^- \right) *_1 \ldots *_1 \left(\psi_p *_1 \widehat{u}_p \alpha_p^{\epsilon_p} \widehat{v}_p *_1 \psi_p^- \right) \right\rfloor .
\end{aligned}
$$

We use the isomorphism (8.5) and Lemma 8.3.4 to deduce that the cellular extension $\{A_\alpha \colon \alpha \Rightarrow 1_{s(\alpha)} \mid \alpha \in X\}$ of $P^\top_{\mathrm{ab}}$ is acyclic, proving that the 2-polygraph P has $\mathrm{FDT}_{\mathrm{ab}}$. □

The following result states that the property of being finitely generated for $\Pi(P)$ is Tietze invariant for polygraphs P having a finite set of 2-generators [164, Proposition 2.3.5].

8.3.8 Proposition. *Let P and Q be two Tietze equivalent 2-polygraphs such that P_2 and Q_2 are finite. Then the natural system $\Pi(P)$ is finitely generated if and only if the natural system $\Pi(Q)$ is finitely generated.*

From this result and Proposition 8.3.7, we deduce that the property $\mathrm{FDT}_{\mathrm{ab}}$ is Tietze invariant for finite polygraphs. As a consequence, we can define a category $\mathrm{FDT}_{\mathrm{ab}}$ if it admits a presentation by a finite 2-polygraph having $\mathrm{FDT}_{\mathrm{ab}}$.

We conclude this chapter with a remarkable property of the natural system of identities among relations from [164, Proposition 2.4.2], which is a consequence of Squier's homotopical theorem. By Theorem 7.3.5, the set of generating confluences of a convergent 2-polygraph P forms an acyclic extension of the $(2,1)$-category $P^\top$. Following the proof of Proposition 8.3.7, we transform this extension into a generating set for the natural system $\Pi(P)$, proving the following result.

8.3.9 Theorem. *Let P be a convergent 2-polygraph. The natural system $\Pi(P)$ is generated by the generating confluences of P.*

9

Homological Syzygies and Confluence

The main purpose of algebraic topology is the classification of topological spaces and continuous maps by means of discrete algebraic invariants preserving homotopy equivalence. Among those invariants, a particularly important one is homology, which assigns to each space a sequence of abelian groups. Starting from very geometric insights, homology has developed into a whole body of concepts and methods known as homological algebra and has been applied to the study of various algebraic structures, including groups and monoids [260]. For instance, the homology of a monoid is defined by first building a *resolution* of it, that is, an exact sequence of left-modules over the ring generated by the monoid, ending at the trivial module. Of course the soundness of this definition is based on the fact that the homology does not depend on the choice of the resolution.

Squier showed in his 1987 article [326] that a convergent presentation P of a monoid M yields a partial resolution generated by the set P_1 of generators in dimension 1, by the set P_2 of rules in dimension 2 and by the critical branchings in dimension 3. If moreover the presentation P is finite, the Squier resolution is finitely generated up to dimension 3. In this case, we say that the monoid M is of *homological type left*-FP_3. This property readily implies that the third integral homology group $H_3(M, \mathbb{Z})$ of the monoid is finitely generated. Therefore, a monoid whose third homology group is not finitely generated does not admit a finite convergent presentation. By explicitly exhibiting an example of this type, Squier first provided a negative answer to the question of universality of convergent rewriting.

The homological finiteness condition is of course linked to the homotopical one discussed in the previous chapter. Indeed, we will prove that for a monoid, the property of having FDT implies the property of having left-FP_3 (Theorem 9.3.4). In this sense, the homological finiteness condition is weaker than its homotopical counterpart. It however has the advantage of being simpler to compute.

We begin by introducing the finite homological type for monoids in Section 9.1. We study the case $n = 2$ in Section 9.2 and show that finitely presented monoids are left-FP_2 (Proposition 9.2.4). Then, we study the case $n = 3$ in Section 9.3 and show that monoids with finite convergent presentations have the left-FP_3 property (Theorem 9.3.5). We illustrate these results with several examples. The constructions of this chapter will be generalized in any homological dimension for 1-categories in Section 23.5. The case of monoids treated in this chapter corresponds to 1-categories with a single object.

The homological notions used in this chapter are recalled in Appendix E. In particular, homology of monoids is recalled in Section E.4. In this chapter, we study the homological type left-FP_3 relative to left modules, but the homological type right-FP_3 relative to right modules is treated in the same way. We refer to Sections F.3 and 9.3.14 for relationships between homology types according to the module categories considered.

9.1 Monoids of Finite Homological Type

9.1.1 Monoid ring. Let M be a monoid. The *ring generated by* M is the free abelian group over M, denoted by $\mathbb{Z}M$. Its elements are formal sums $\sum_{u\in M} n_u u$ of elements u of M with coefficients $n_u \in \mathbb{Z}$, finitely many of which are non-zero, and it is equipped with the canonical extension of the product of M:

$$\left(\sum_{u\in M} n_u u\right)\left(\sum_{v\in M} n_v v\right) = \sum_{u,v\in M} n_u n_v uv = \sum_{w\in M}\left(\sum_{uv=w} n_u n_v\right) w.$$

This construction coincides with the one of the free $\mathbb{Z}$-module, thus the notation.

9.1.2 Free modules. Given a monoid M and a set X, we write $\mathbb{Z}M[X]$ for the free left $\mathbb{Z}M$-module generated by X: its elements are formal sums of the form

$$\sum_{u\in M, x\in X} n_{u,x} u[x]$$

with $n_{u,x} \in \mathbb{Z}$, finitely many of which are non-zero, and other operations are defined in the expected way. Any function $f\colon X \to C$, where C is a $\mathbb{Z}M$-module extends uniquely as a morphism of $\mathbb{Z}M$-modules $f\colon \mathbb{Z}M[X] \to C$. Note that any $\mathbb{Z}M$-module is also canonically a $\mathbb{Z}$-module.

9.1.3 Resolutions. If M is a monoid, the *trivial* $\mathbb{Z}M$*-module* is the abelian group $\mathbb{Z}$ equipped with the trivial action $un = n$, for every u in M and n in $\mathbb{Z}$. A *partial resolution* of length n of this trivial module consists of a chain complex

$$C_n \xrightarrow{d_n} C_{n-1} \xrightarrow{d_{n-1}} \cdots \xrightarrow{d_2} C_1 \xrightarrow{d_1} C_0 \xrightarrow{d_0} \mathbb{Z} \longrightarrow 0 \tag{9.1}$$

where, for $0 \leqslant k \leqslant n$, the C_k are left $\mathbb{Z}M$-modules, and the d_k are $\mathbb{Z}M$-linear maps making the complex exact. By convention, $C_{-1} = \mathbb{Z}$ and $d_{-1}: \mathbb{Z} \to 0$ is the terminal map. The main properties on resolutions that we use in this chapter are recalled in Section E.3.

A contracting homotopy of a chain complex of the form (9.1) is a sequence

$$C_n \xleftarrow{i_n} C_{n-1} \xleftarrow{i_{n-1}} \cdots \xleftarrow{i_2} C_1 \xleftarrow{i_1} C_0 \xleftarrow{i_0} \mathbb{Z}$$

where the i_k are $\mathbb{Z}$-linear maps for $0 \leqslant k \leqslant n$, and such that

$$d_k \circ i_k + i_{k-1} \circ d_{k-1} = 1_{C_{k-1}}$$

holds for $0 \leqslant k \leqslant n$, see §E.2.6 for details. By convention, $i_{-1}: 0 \to \mathbb{Z}$ is the initial map. Any chain complex of the form (9.1) equipped with a contracting homotopy is necessarily a partial resolution, see Proposition E.2.7.

9.1.4 Homological type left-FP$_n$. A monoid M has *homological type left-*FP$_n$ (where FP$_n$ stands for "finitely n-presented"), for a natural number n, if there exists a partial resolution of length n of the trivial $\mathbb{Z}M$-module $\mathbb{Z}$ of the form (9.1), where the C_i are projective modules which are finitely generated. A monoid M has *homological type left-*FP$_\infty$ if it has homological type left-FP$_n$ for all $n \geqslant 0$.

We will use the following characterization given by Proposition F.3.6: a monoid M has homological type left-FP$_n$ if and only if there exists a free, finitely generated partial resolution of the trivial $\mathbb{Z}M$-module $\mathbb{Z}$ of length n:

$$F_n \longrightarrow F_{n-1} \longrightarrow \cdots \longrightarrow F_0 \longrightarrow \mathbb{Z}. \tag{9.2}$$

9.1.5 Finiteness homological type and homology. For a monoid M, having homological type left-FP$_n$ implies a finiteness property on its homology modules. First, we recall the definition of homology of a monoid and we refer to Section E.4 for more details. Given a free resolution

$$\cdots \longrightarrow F_{n+1} \xrightarrow{d_{n+1}} F_n \longrightarrow \cdots \longrightarrow F_1 \xrightarrow{d_1} F_0 \xrightarrow{\varepsilon} \mathbb{Z}$$

of the trivial $\mathbb{Z}M$-module $\mathbb{Z}$ by left $\mathbb{Z}M$-modules, the operation of tensoring by the trivial right $\mathbb{Z}M$-module $\mathbb{Z}$ gives the following complex of $\mathbb{Z}$-modules:

$$\cdots \longrightarrow \mathbb{Z} \otimes_{\mathbb{Z}M} F_{n+1} \xrightarrow{\widetilde{d}_{n+1}} \mathbb{Z} \otimes_{\mathbb{Z}M} F_n \longrightarrow \cdots \longrightarrow \mathbb{Z} \otimes_{\mathbb{Z}M} F_1 \xrightarrow{\widetilde{d}_1} \mathbb{Z} \otimes_{\mathbb{Z}M} F_0$$

where $\widetilde{d}_k$ denotes the map $1_{\mathbb{Z}} \otimes_{\mathbb{Z}M} d_k$, for all $k \geqslant 1$. The *n-th homology group* of M with integral coefficient $\mathbb{Z}$ is defined as the following $\mathbb{Z}$-module:

$$\mathrm{H}_n(M, \mathbb{Z}) = \ker \widetilde{d}_n / \mathrm{im}\, \widetilde{d}_{n+1},$$

with the convention that $\widetilde{d}_0 = 0$. By definition, for any monoid M, we have $\mathrm{H}_0(M, \mathbb{Z}) \simeq \mathbb{Z}$.

Now, suppose that the monoid M has homological type left-FP_n and consider a resolution of M of the form (9.2). Then the $\mathbb{Z}$-modules $\mathbb{Z} \otimes_{\mathbb{Z}M} F_i$ are finitely generated for $0 \leqslant i \leqslant n$. This proves the following result.

9.1.6 Proposition. *If a monoid M has homological type left-*FP_n *for some $n \in \mathbb{N}$, then the groups $H_k(M, \mathbb{Z})$ are finitely generated for* $0 \leqslant k \leqslant n$.

9.1.7 Homological type left-FP_0**.** Let M be a monoid. We write $P_0 = \{\star\}$ for a set with one element. We have that $\mathbb{Z}M[P_0] \simeq \mathbb{Z}M$ and sometimes implicitly identify the elements of these two modules. The *augmentation map* of $\mathbb{Z}M$ is the morphism of $\mathbb{Z}M$-modules

$$\varepsilon\colon \mathbb{Z}M[P_0] \to \mathbb{Z}$$

defined by $\varepsilon(u) = 1$ for any u in $\mathbb{Z}M$. The augmentation map is clearly surjective and thus the sequence

$$\mathbb{Z}M[P_0] \xrightarrow{\ \varepsilon\ } \mathbb{Z} \longrightarrow 0$$

is exact. It follows that every monoid has homological type left-FP_0.

9.1.8 Homological type left-FP_1**.** Let P be a presentation of a monoid M. We define a free partial resolution of length 1 of the trivial $\mathbb{Z}M$-module $\mathbb{Z}$ by $\mathbb{Z}M$-modules

$$\mathbb{Z}M[P_1] \xrightarrow{\ d_1\ } \mathbb{Z}M[P_0] \xrightarrow{\ \varepsilon\ } \mathbb{Z} \longrightarrow 0$$

where the morphism ε is the augmentation map and the morphism d_1 is defined, on any generator $[a]$, by

$$d_1([a]) = \overline{a} - 1.$$

A *section* of the canonical projection $\pi\colon P_1^* \to M$ is a map $M \to P_1^*$ sending every u in M to a 1-cell $\widehat{u}$ of P_1^* such that $\pi(\widehat{u}) = u$. In general, we do not assume that the chosen section is functorial, i.e., that $\widehat{u}\widehat{v} = \widehat{uv}$ holds in P_1^*. However, we assume that $\widehat{1} = 1$. For a 1-cell u of P_1^*, we simply write $\widehat{u}$ for $\widehat{\overline{u}}$.

9.1.9 Proposition. *If a monoid M is finitely generated, then it has homological type left-*FP_1*, and thus the group $H_1(M,\mathbb{Z})$ is finitely generated.*

Proof. We first note that the sequence is a chain complex. Indeed, exactness at $\mathbb{Z}$ was already observed in §9.1.7. Moreover, we have

$$\varepsilon d_1[a] = \varepsilon(\overline{a}) - \varepsilon(1) = 1 - 1 = 0$$

for every 1-generator a of P. In order to prove exactness at $\mathbb{Z}M[P_0]$, as explained in §9.1.3, we construct contracting homotopies

$$i_0 \colon \mathbb{Z} \to \mathbb{Z}M[P_0] \qquad \text{and} \qquad i_1 \colon \mathbb{Z}M[P_0] \to \mathbb{Z}M[P_1]$$

as follows. The morphism i_0 is simply defined by $i_0(1) = 1$ and extended by linearity. As for i_1 we first need to extend the bracket map $[-] \colon P_1 \to \mathbb{Z}M[P_1]$ to a map $[-] \colon P_1^* \to \mathbb{Z}M[P_1]$. This is done by induction on the length of the words in P_1^* by setting

$$[1] = 0 \qquad \text{and} \qquad [aw] = [a] + \overline{a}[w],$$

for $a \in P_1$ and $w \in P_1^*$ (technically, we extend the map as a derivation, see §4.4.13). It follows that the equation

$$d_1([w]) = \overline{w} - 1 \tag{9.3}$$

holds for all elements w of P_1^*, not just for generators. We reason by induction on the length of the words in P_1^*. One first has $d_1([1]) = 0 = \overline{1} - 1$. Let now $a \in P_1$ and $w \in P_1^*$ such that the equation (9.3) holds for w. Then

$$d_1([aw]) = d_1([a]) + \overline{a}d_1([w]) = \overline{a} - 1 + \overline{a}(\overline{w} - 1) = \overline{aw} - 1.$$

Now, we choose a section and define the morphism i_1 by setting

$$i_1(u) = [\widehat{u}]$$

and extending it by linearity. Finally, for any $u \in M$, we have $i_0\varepsilon(u) = 1$ and

$$d_1 i_1(u) = d_1[\widehat{u}] = \overline{\widehat{u}} - 1 = u - 1.$$

Thus, $d_1 i_1 + i_0\varepsilon = 1_{\mathbb{Z}M}$ and i_0, i_1 are a contracting homotopies. □

9.2 Monoids Having Homological Type Left-FP$_2$

9.2.1 Presentations and partial resolutions of length 2. Let P be a presentation of a monoid M. We define a partial resolution of length 2 of the trivial $\mathbb{Z}M$-module $\mathbb{Z}$ by free $\mathbb{Z}M$-modules

$$\mathbb{Z}M[P_2] \xrightarrow{d_2} \mathbb{Z}M[P_1] \xrightarrow{d_1} \mathbb{Z}M[P_0] \xrightarrow{\varepsilon} \mathbb{Z} \longrightarrow 0.$$

The morphisms ε and d_1 are those defined in the previous section. The morphism d_2 is defined, on generators of $\mathbb{Z}M[P_2]$, by

$$d_2([\alpha]) = [s(\alpha)] - [t(\alpha)],$$

for every α in P_2, and called the *Reidemeister-Fox Jacobian* of the presentation P.

9.2.2 Normalization strategies. Let P be a 2-polygraph with a given section. A *normalization strategy* σ for P is a map

$$\sigma\colon P_1^* \to P_2^\top$$

that sends every 1-cell w of P_1^* to a 2-cell

$$\sigma(w)\colon w \Rightarrow \widehat{w}$$

in $P_2^\top$, such that $\sigma(\widehat{w}) = 1_{\widehat{w}}$ holds for every 1-cell w of P_1^*. A normalization strategy σ is a *left* (resp. *right*) one if it also satisfies

$$\sigma(wv) = \sigma(w)v *_1 \sigma(\widehat{w}v) \qquad (\text{resp.}\quad \sigma(wv) = w\sigma(v) *_1 \sigma(w\widehat{v})\ )$$

that is

$\sigma(wv) =$ 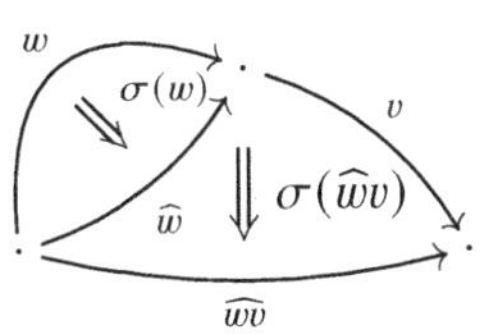

(resp. $\sigma(wv) =$ 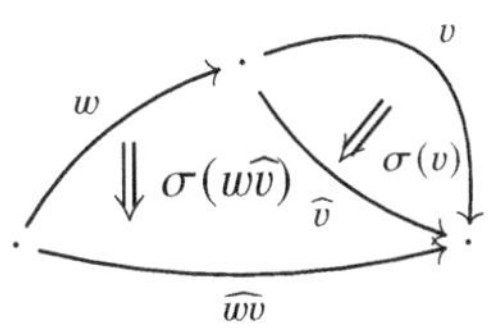

).

A 2-polygraph P always admits left and right normalization strategies. Let us prove this in the left case, the right case being treated in the same way. Let us arbitrarily choose a 2-cell $\sigma(wa)\colon wa \Rightarrow \widehat{wa}$ in $P_2^\top$, for every 1-cell w of P_1^* and every 1-generator a of P, such that $\widehat{w} = w$ and $\widehat{wa} \neq wa$. Then we extend σ into a left normalization strategy by setting $\sigma(w) = 1_w$ if $\widehat{w} = w$ (which implies $\sigma(1) = 1$), and

$$\sigma(w) = \sigma(v)a *_1 \sigma(\widehat{v}a)$$

if $\widehat{w} \neq w$ and $w = va$ with v in P_1^* and a in P_1.

9.2.3 Proposition. *Let M be a monoid and let P be a presentation of M. The sequence of $\mathbb{Z}M$-modules*

$$\mathbb{Z}M[P_2] \xrightarrow{d_2} \mathbb{Z}M[P_1] \xrightarrow{d_1} \mathbb{Z}M[P_0] \xrightarrow{\varepsilon} \mathbb{Z} \longrightarrow 0$$

is a partial free resolution of length 2 of $\mathbb{Z}$.

Proof. In Proposition 9.1.9, we have proved the exactness at $\mathbb{Z}$ and $\mathbb{Z}M[P_0]$, and exactness at $\mathbb{Z}M[P_1]$ remains to be shown. The equation $d_1 d_2 = 0$ is a consequence of (9.3). Indeed, we have

$$d_1 d_2[\alpha] = d_1[s(\alpha)] - d_1[t(\alpha)] = \overline{s(\alpha)} - \overline{t(\alpha)} = 0,$$

for every 2-generator α of P, where the last equality comes from the equality $\overline{s(\alpha)} = \overline{t(\alpha)}$, which holds because P is a presentation of M.

In order to prove the exactness at $\mathbb{Z}M[P_1]$, we construct a contracting homotopy of the complex. The morphisms of $\mathbb{Z}$-modules i_0 and i_1 are defined in the proof of Proposition 9.1.9, and the morphism of $\mathbb{Z}$-modules

$$i_2 \colon \mathbb{Z}M[P_1] \to \mathbb{Z}M[P_2]$$

is defined by fixing a left normalization strategy σ for the 2-polygraph P. Namely, we define the morphism of $\mathbb{Z}$-modules i_2 by its value on generic elements

$$i_2(u[a]) = [\sigma(\widehat{u}a)],$$

where the bracket $[-]$ is extended to every 2-cell of the free $(2, 1)$-category $P^{\top}$ by the following relations

$$[1_u] = 0, \qquad [u\phi v] = \overline{u}[\phi], \qquad [\phi \ast_1 \psi] = [\phi] + [\psi],$$

for all 1-cells u and v and 2-cells ϕ and ψ of $P^{\top}$ such that the composite $\phi \ast_1 \psi$ are defined.

We have, on the one hand,

$$i_1 d_1(u[a]) = i_1(u\overline{a} - u) = [\widehat{u\overline{a}}] - [\widehat{u}]$$

and, on the other hand,

$$d_2 i_2(u[a]) = d_2[\sigma(\widehat{u}a)] = [\widehat{u}a] - [\widehat{u\overline{a}}] = u[a] + [\widehat{u}] - [\widehat{u\overline{a}}].$$

For the equality in the middle, one proves that $d_2[\phi] = [s(\phi)] - [t(\phi)]$ holds for every 2-cell ϕ of $P^{\top}$ by induction on the size of ϕ. Hence we have

$$d_2 i_2 + i_1 d_1 = 1_{\mathbb{Z}M[P_1]},$$

thus concluding the proof. □

The previous proposition allows us to deduce:

9.2.4 Proposition. *If a monoid M admits a finite presentation, then it has homological type left-*FP_2*, and thus the group $H_2(M, \mathbb{Z})$ is finitely generated.*

9.2.5 Homological 2-syzygies. The kernel of the morphism d_2 defined in §9.2.1 is called the $\mathbb{Z}M$-module of *homological* 2-*syzygies* of the 2-polygraph P. Using natural systems as modules, we will establish in Section 23.5 an isomorphism between the homological 2-syzygies and the identities among relations for a 1-category presented by a 2-polygraph.

9.3 Homological Type Left-FP_3 and Confluence

9.3.1 Coherent presentations and partial resolutions of length 3**.** Let P be a coherent presentation of a monoid M. Let us extend the partial resolution of Proposition 9.2.3 into the resolution of length 3

$$\mathbb{Z}M[P_3] \xrightarrow{d_3} \mathbb{Z}M[P_2] \xrightarrow{d_2} \mathbb{Z}M[P_1] \xrightarrow{d_1} \mathbb{Z}M[P_0] \xrightarrow{\varepsilon} \mathbb{Z} \longrightarrow 0.$$

The boundary map d_3 is defined, for every 3-cell A of P, by

$$d_3[A] = [s_2(A)] - [t_2(A)].$$

The bracket notation $[-]$ is extended to 3-cells of $P^\top$ by setting

$$[uFv] = \overline{u}[F] \qquad [F *_1 G] = [F] + [G] \qquad [F *_2 G] = [F] + [G]$$

for all 1-cells u and v and 3-cells F and G of $P^\top$ such that the composites are defined. In particular, the latter relation implies $[1_\phi] = 0$ for every 2-cell ϕ of $P^\top$. We check, by induction on the size, that $d_3[F] = [s_2(F)] - [t_2(F)]$ holds for every 3-cell F of $P^\top$.

9.3.2 Proposition. *Let P be a coherent presentation of a monoid M. The sequence of $\mathbb{Z}M$-modules*

$$\mathbb{Z}M[P_3] \xrightarrow{d_3} \mathbb{Z}M[P_2] \xrightarrow{d_2} \mathbb{Z}M[P_1] \xrightarrow{d_1} \mathbb{Z}M[P_0] \xrightarrow{\varepsilon} \mathbb{Z} \longrightarrow 0$$

is a partial free resolution of length 3 *of $\mathbb{Z}$.*

Proof. We proceed with the same notations as in the proof of Proposition 9.2.3, with the extra hypothesis that σ is a left normalization strategy for P. This implies that $i_2(u[v]) = [\sigma(\widehat{u}v)]$ holds for all u in M and v in P_1^*, by induction on the length of v. We have $d_2d_3 = 0$ because $s_1s_2 = s_1t_2$ and $t_1s_2 = t_1t_2$. Then, we define the following morphism of $\mathbb{Z}$-modules $i_3 \colon \mathbb{Z}M[P_2] \to \mathbb{Z}M[P_3]$ by setting, for $u \in M$ and $\alpha \in P_3$,

$$i_3(u[\alpha]) = [\sigma(\widehat{u}\alpha)]$$

where $\sigma(\widehat{u}\alpha)$ is a 3-cell of $P^\top$ with the following shape, with $v = s(\alpha)$ and $w = t(\alpha)$:

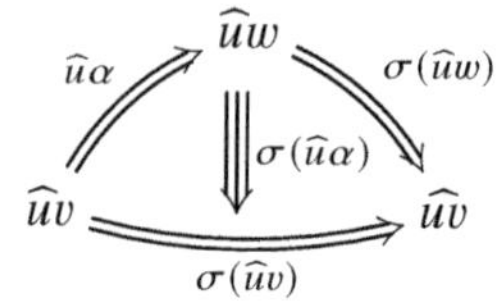

Let us note that such a 3-cell necessarily exists in $P^\top$ because P_3 is an acyclic cellular extension of $P^\top$. Then we have, on the one hand,

$$i_2 d_2(u[\alpha]) = i_2(u[v] - u[w]) = [\sigma(\widehat{u}v)] - [\sigma(\widehat{u}w)]$$

and, on the other hand,

$$\begin{aligned} d_3 i_3(u[\alpha]) &= [\widehat{u}\alpha *_1 \sigma(\widehat{u}w)] - [\sigma(\widehat{u}v)] \\ &= u[\alpha] + [\sigma(\widehat{u}w)] - [\sigma(\widehat{u}v)]. \end{aligned}$$

Hence $d_3 i_3 + i_2 d_2 = 1_{\mathbb{Z}M[P_2]}$, concluding the proof. □

9.3.3 Remark. The proof of Proposition 9.3.2 uses the fact that P_3 is an acyclic cellular extension to produce, for every 2-cell α of P_2 and every u in M, a 3-cell $\sigma(\widehat{u}\alpha)$ with the required shape. The hypothesis on P_3 could thus be modified to only require the existence of such a 3-cell in $P^\top$: however, it is proved in [163] that this implies that P_3 is an acyclic cellular extension.

The previous proposition has the following consequence, already noted in [306], [98, Theorem 3.2], and [233, Theorem 3]:

9.3.4 Theorem. *Let M be a finitely presented monoid. If M has finite derivation type, then it has homological type left-*FP_3*, and thus the group $H_3(M, \mathbb{Z})$ is finitely generated.*

By Theorem 8.2.1 and Proposition 9.1.6, this implies the following homological finiteness condition for finite convergence [326, Theorem 4.1]:

9.3.5 Theorem. *If a monoid M admits a finite convergent presentation, then it has homological type left-*FP_3*, and thus the group $H_3(M, \mathbb{Z})$ is finitely generated.*

The construction of this chapter will be generalized in Chapter 23 to produce a free resolution of infinite length, involving n-fold critical branchings for every natural number n (Theorem 23.3.3).

9.3.6 Example. Consider the monoid M with the convergent presentation

$$P = \langle \star \mid a \mid \mu\colon aa \Rightarrow a \rangle.$$

Writing $\widehat{w}$ for the normal form of a word w, we have $\widehat{w} = a$ for every non-identity 1-cell $w \in P_1^*$. With the leftmost normalization strategy σ, we get, writing the 2-cell μ as a string diagram $\triangledown$:

$\sigma(a) = 1_a$ $\qquad$ $\sigma(aa) =$ 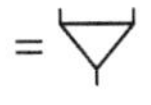 $\qquad$ $\sigma(aaa) = \mu a *_1 \mu =$ 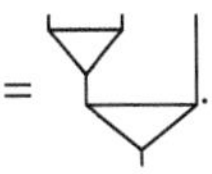.

The presentation has exactly one critical branching, whose corresponding generating confluence can be written in two equivalent ways

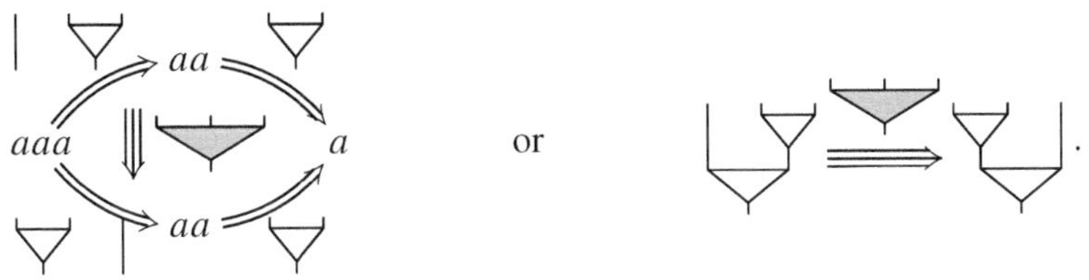

The $\mathbb{Z}M$-module $\ker d_2$ is generated by

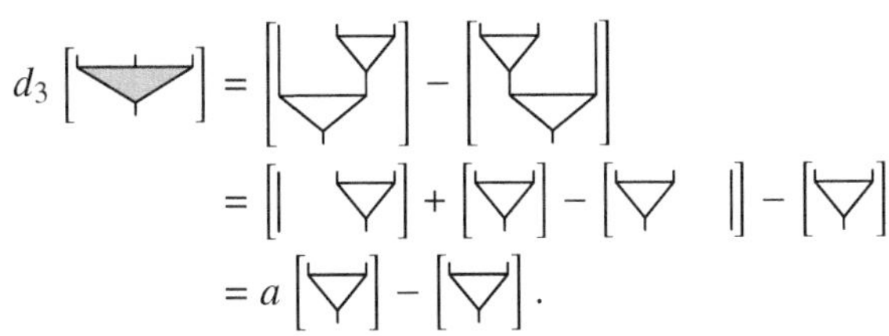

We will see in Chapter 23 that the construction of the partial resolution in Proposition 9.3.2 can be extended to arbitrary length. We only provide here a small generalization [326, Theorem 3.2], which is enough to imply a negative answer to the universality of finite convergent rewriting, see Example 9.3.11.

9.3.7 A short exact sequence. Theorem 7.3.5 states that any set P_3 of generating confluences of a convergent 2-polygraph P, indexed by all its critical branchings, forms an acyclic extension of the $(2,1)$-category $P^\top$. Following Proposition 9.3.2, this induces a partial free resolution of length 3 of $\mathbb{Z}$ by $\mathbb{Z}M$-modules

$$\mathbb{Z}M[P_3] \xrightarrow{d_3} \mathbb{Z}M[P_2] \xrightarrow{d_2} \mathbb{Z}M[P_1] \xrightarrow{d_1} \mathbb{Z}M[P_0] \xrightarrow{\varepsilon} \mathbb{Z} \longrightarrow 0.$$

We have also shown in §7.5.5 that the triple generating confluences of P generate the relations among the 3-cells of the free $(3,1)$-category $(P, P_3)^{\top}$. We will show in Sections 23.2 and 23.3 that this allows us to extend the previous resolution with a boundary map $d_4 \colon \mathbb{Z}M[P_4] \to \mathbb{Z}M[P_3]$ defined on the free module generated by a set of 4-chains P_4 indexed by generating triple confluences. In particular, when there are no critical triples, we recover the following result shown by Squier in [326, Theorem 3.2], see also Corollary 23.3.6.

9.3.8 Proposition. *Suppose given a convergent 2-polygraph P without critical 3-branching, and write P_3 for a set of 2-spheres containing a confluence diagram for every critical branching of P. Then the sequence of $\mathbb{Z}M$-modules*

$$0 \longrightarrow \mathbb{Z}M[P_3] \xrightarrow{d_3} \mathbb{Z}M[P_2] \xrightarrow{d_2} \mathbb{Z}M[P_1] \xrightarrow{d_1} \mathbb{Z}M[P_0] \xrightarrow{\varepsilon} \mathbb{Z} \longrightarrow 0$$

is a partial resolution of length 4 of $\mathbb{Z}$.

In the rest of this section, we show that this result turns out to be very useful for constructing examples of finitely presented monoids having an infinite third integral homology group while having a decidable word problem.

9.3.9 Example. Consider the monoid M presented by the 2-polygraph

$$P = \langle\, \star \mid a, b, c \mid \alpha_n \colon ac^n b \Rightarrow 1 \,\rangle_{n \in \mathbb{N}}.$$

The polygraph P is convergent without critical branchings. Hence, by Squier's Theorem 7.3.5 it can be extended into a coherent presentation with an empty set of 3-generators. Following Proposition 9.3.2, we have a partial free resolution of length 3 of $\mathbb{Z}$ by free $\mathbb{Z}M$-modules:

$$0 \longrightarrow \mathbb{Z}M[P_2] \xrightarrow{d_2} \mathbb{Z}M[P_1] \xrightarrow{d_1} \mathbb{Z}M[P_0] \xrightarrow{\varepsilon} \mathbb{Z} \longrightarrow 0.$$

We have

$$\widetilde{d_1}(a) = \widetilde{d_1}(b) = \widetilde{d_1}(c) = 0$$

and

$$\widetilde{d_2}([\alpha_n]) = [a] + n[c] + [b]$$

for all $n \geqslant 0$. As a consequence $H_1(M,\mathbb{Z}) = \mathbb{Z}$ and $\mathrm{H}_2(M,\mathbb{Z}) = \ker \widetilde{d_2}$ is the free $\mathbb{Z}$-module generated by

$$[\alpha_n] - n[\alpha_1] + (n-1)[\alpha_0]$$

for $n \geqslant 2$. Since $H_2(M)$ is not finitely generated, this shows that the finitely generated monoid M cannot be finitely presented by Propositions 9.1.6 and 9.2.4.

9.3.10 Example. Consider the monoid M presented by the following 2-polygraph considered in [238]:

$$P = \langle \star \mid a, b, c, d \mid \alpha_0 \colon ab \Rightarrow a, \beta \colon da \Rightarrow ac \rangle.$$

We have seen in Example 5.2.4 that using the Knuth–Bendix completion procedure, this polygraph can be completed into the following convergent polygraph with infinitely many 2-generators:

$$\widetilde{P} = \langle a, b, c, d \mid \alpha_n \colon ac^n b \Rightarrow ac^n, \beta \colon da \Rightarrow ac \rangle_{n\in\mathbb{N}}.$$

There are infinitely many critical branchings indexed by $n \in \mathbb{N}$:

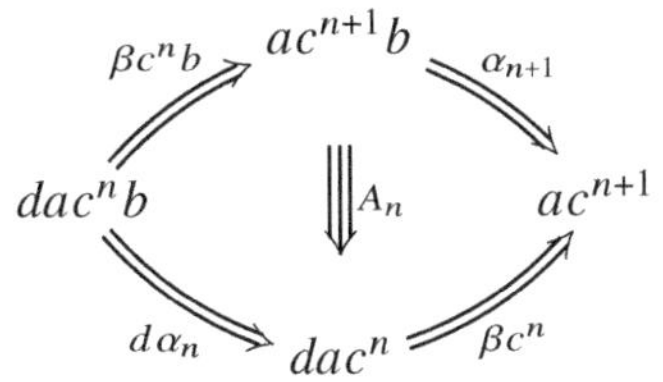

Denoting by $\widetilde{P}_3$ the set of 3-generators $\{A_n \mid n \in \mathbb{N}\}$, by Theorem 7.3.5, $\widetilde{P}_3$ extends $\widetilde{P}$ into a coherent presentation. This system has no critical 3-branching, thus by Proposition 9.3.8, we have an exact sequence

$$0 \longrightarrow \mathbb{Z}M[\widetilde{P}_3] \xrightarrow{d_3} \mathbb{Z}M[\widetilde{P}_2] \xrightarrow{d_2} \mathbb{Z}M[\widetilde{P}_1] \xrightarrow{d_1} \mathbb{Z}M[\widetilde{P}_0] \xrightarrow{\varepsilon} \mathbb{Z} \longrightarrow 0.$$

To calculate homology groups of M, we consider the maps $\widetilde{d}_k := 1_{\mathbb{Z}} \otimes_{\mathbb{Z}M} d_k$ defined on $\mathbb{Z}[\widetilde{P}_k]$ and with values in $\mathbb{Z}[\widetilde{P}_{k-1}]$. We have

$$\begin{aligned}
\widetilde{d}_1(a) = \widetilde{d}_1(b) = \widetilde{d}_1(c) = \widetilde{d}_1(d) &= 0,\\
\widetilde{d}_2([\alpha_n]) &= [a] + n[c] + [b] - ([a] - n[c]) = [b],\\
\widetilde{d}_2([\beta]) &= [d] + [a] - ([a] + [c]) = [d] - [c],\\
\widetilde{d}_3([A_n]) &= [\beta] + [\alpha_{n+1}] - ([\alpha_n] + [\beta]) = [\alpha_{n+1}] - [\alpha_n].
\end{aligned}$$

Thus

$$H_0(M,\mathbb{Z}) = \mathbb{Z}, \qquad H_1(M,\mathbb{Z}) = \mathbb{Z}^2, \qquad H_i(M,\mathbb{Z}) = 0, \qquad \text{for } i = 2, 3.$$

At this stage, therefore, we cannot use the finiteness condition of Theorem 9.3.5 to conclude the existence of a convergent presentation for the monoid M. As noted in [238, Section 3.5], we can nevertheless construct a finite convergent presentation of M with another orientation of the rule β. Indeed, the following polygraph presents the monoid M and has no critical branching:

$$\langle \star \mid a, b, c, d \mid \alpha_0 \colon ab \Rightarrow a, \gamma \colon ac \Rightarrow da \rangle.$$

It is therefore trivially convergent.

9.3.11 Example. Consider the monoid M of Example 8.2.2 presented by the following 2-polygraph:

$$P = \langle \star \mid a, b, c, d, d' \mid \alpha_0 \colon ab \Rightarrow a, \beta \colon da \Rightarrow ac, \beta' : d'a \Rightarrow ac \rangle.$$

We have seen that, by using the Knuth–Bendix completion procedure it can be completed into an infinite convergent polygraph, from which we deduce the following coherent presentation

$$\widetilde{P} = \left\langle \star \,\middle|\, a, b, c, d, d' \,\middle|\, \alpha_n, \beta, \beta' \,\middle|\, A_n, A'_n \right\rangle_{n\in\mathbb{N}},$$

with

$$\alpha_n \colon ac^n b \Rightarrow ac^n, \qquad \beta \colon da \Rightarrow ac, \qquad \beta' : d'a \Rightarrow ac,$$

and

$$A_n \colon \beta c^n b *_1 \alpha_{n+1} \Rrightarrow d\alpha_n *_1 \beta c^n, \quad A'_n \colon \beta' c^n b *_1 \alpha_{n+1} \Rrightarrow d'\alpha_n *_1 \beta' c^n.$$

There are no critical 3-branching and thus by Proposition 9.3.8 we have a partial resolution of length 4

$$0 \longrightarrow \mathbb{Z}M[\widetilde{P}_3] \xrightarrow{d_3} \mathbb{Z}M[\widetilde{P}_2] \xrightarrow{d_2} \mathbb{Z}M[\widetilde{P}_1] \xrightarrow{d_1} \mathbb{Z}M[\widetilde{P}_0] \xrightarrow{\varepsilon} \mathbb{Z} \longrightarrow 0.$$

The computations are similar to those of Example 9.3.10. The map $\widetilde{d}_1$ is zero on the $\mathbb{Z}$-module $\mathbb{Z}\widetilde{P}_1$, and we have

$$\widetilde{d}_2([\alpha_n]) = [b], \qquad \widetilde{d}_2([\beta]) = [d] - [c], \qquad \widetilde{d}_2([\beta']) = [d'] - [c],$$

$$\widetilde{d}_3([A_n]) = [\alpha_{n+1}] - [\alpha_n], \qquad \widetilde{d}_3([A'_n]) = [\alpha_{n+1}] - [\alpha_n].$$

We deduce that

$$H_0(M, \mathbb{Z}) = \mathbb{Z}, \qquad H_1(M, \mathbb{Z}) = \mathbb{Z}^2, \qquad H_2(M, \mathbb{Z}) = 0,$$

and the $\mathbb{Z}$-module $H_3(M, \mathbb{Z})$ is freely generated by the infinite family

$$([A_n] - [A'_n])_{n\geqslant 0}.$$

Following Theorem 9.3.5, we deduce that the monoid M does not have a finite convergent presentation. This example thus exhibits a finitely presented monoid with a decidable word problem, which does not admit a finite convergent presentation. It therefore illustrates the fact that string rewriting theory is not universal for deciding the word problem in monoids, see Section 5.3.

9.3.12 Example. Consider the coherent presentation of the monoid B_3^+ given in Example 7.5.8. By Proposition 9.3.2, it induces a resolution of the trivial $\mathbb{Z}B_3^+$-module $\mathbb{Z}$, from which we can compute the following homology groups:

$$H_0(B_3^+, \mathbb{Z}) = H_1(B_3^+, \mathbb{Z}) = H_2(B_3^+, \mathbb{Z}) = \mathbb{Z}, \qquad H_3(B_3^+, \mathbb{Z}) = 0.$$

9.3.13 Remark. Note that in combinatorial group theory several examples of finitely presented groups with a decidable word problem that do not have homological type FP$_3$ were discovered before Squier's work on homology of monoids. In particular, Stallings constructed in [329] a finitely presented group whose 3-dimensional homology group with integer coefficients is not finitely generated and thus it does not have homological type left-FP$_3$. The group is presented by

$$\langle\, \star \mid a, b, c, d, e \mid [d, a], [e, a], [d, b], [e, b], [a^- d, c], [a^- e, c], [b^- a, c]\,\rangle$$

where $[u, v]$ denotes the relation $uv = vu$. Bieri proved that this group has a decidable word problem [40]. It was not yet known that its word problem cannot be solved by the normal form algorithm.

9.3.14 Remarks on other homological finiteness conditions. In the definition of homological type left-FP$_n$ for a monoid M (§9.1.4), changing left modules to right modules, bimodules, or natural systems gives the definitions of the homological types *right*-FP$_n$, *bi*-FP$_n$ and FP$_n$. We refer the reader to [163, Section 5.2] for the relations between these different finiteness conditions, see also §F.3.3. In particular, for $n = 3$, all of these homotopical conditions are consequences of the finite derivation type property defined in Section 8.1. Moreover, all these homological finiteness properties are necessary conditions for finite convergence. The proof is similar to the one for the left-FP$_3$ property given in Section 9.3. In particular, for the right-FP$_3$ property, we consider right modules and, to get the contracting homotopy, we construct a right normalization strategy σ by defining a 3-cell $\sigma(\alpha\widehat{u})$ with shape

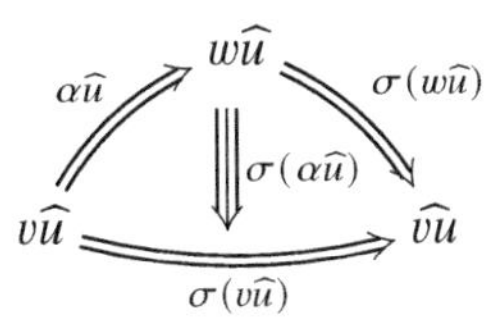

for any 2-generator $\alpha\colon v \Rightarrow w$ and element u in the monoid M.

To conclude this chapter, we summarize in the following theorem the properties of the family of monoids S_k, for $k \geqslant 0$, defined in §8.2.3, which is Squier's original example [326, 328]. This family illustrates the homological and homotopical finiteness conditions for the convergence of string rewriting systems studied in this and the previous chapter, see Theorems 8.2.1 and 9.3.5.

9.3.15 Theorem. *For $k \geqslant 1$, the monoid S_k is a finitely presented monoid that has the following properties.*

1. *It has a decidable word problem [326, Example 4.5].*
2. *For $k = 1$, it does not have finite derivation type [328, Theorem 6.7].*
3. *For $k = 1$, it has homological type left-*FP_∞ *[326, Example 4.5].*
4. *For $k \geqslant 2$, it does not have homological type left-*FP_3 *[326, Example 4.5].*
5. *It does not have a finite convergent presentation.*

Conditions 1, 2, and 5 are seen in Theorem 8.2.4 for $k = 1$. Theorem 9.3.4 proves that finite derivation type implies homological type left-FP_3. Conditions 2 and 3 on monoid S_1 prove that the converse implication is false in general. Note, however, that in the special case of groups, the property of having finite derivation type is equivalent to the homological finiteness condition left-FP_3 [99]. The latter result is based on the Brown-Huebschmann isomorphism between identities among relations and homological syzygies [64], see also Theorem 23.5.3.

PART III

DIAGRAM REWRITING

10

Three-Dimensional Polygraphs

We have seen in Chapter 1 that 1-polygraphs provide a notion of presentation for sets and in Chapter 2 that 2-polygraphs provide a notion of presentation for categories. We go on climbing the dimensional ladder and establish 3-polygraphs as a notion of presentation for 2-categories, see Section 10.1. As expected, those consist in generators for 0-, 1-, and 2-dimensional cells, together with relations between freely generated 2-cells, which are represented by generating 3-cells. As particular cases, let us mention the notions of presentation of monoidal category (when there is only one 0-generator) and of PRO (when there is only one 0-generator and one 1-generator). This includes 2-categories encoding theories for fundamental algebraic structures such as monoids, groups, etc. Note that in the point of view on $(3,1)$-polygraphs we adopt here, the 3-cells encode relations, as opposed to Chapter 7 where they encode coherences between relations.

Any 3-polygraph induces an abstract rewriting system, so that all general rewriting concepts still make sense in this setting: confluence, termination, etc. However, more specific tools have to be adapted to this context: the notion of *critical branching* is defined for 3-polygraphs in Section 10.2, along with the proof that confluence of critical branchings implies the local confluence of the polygraph (Lemma 10.2.8). In the case where the polygraph is terminating (techniques to show this will be presented in Chapter 11), local confluence implies confluence, and we thus have a systematic method to show the convergence of a 3-polygraph. When this is the case, normal forms give canonical representatives for 2-cells modulo the congruence generated by 3-cells, and we explain how to exploit this to show that a given 3-polygraph is a presentation of a given 2-category in Section 10.3. There is however a major difference with the case of 2-dimensional polygraphs: a finite convergent polygraph might give rise to an infinite number of critical branchings (Section 10.4). This prevents us from making direct generalizations of homotopical or homological finiteness conditions (Chapters 8 and 9) from 2- to 3-polygraphs. Finally, in Section 10.5, we provide some techniques for combining presentations of 2-categories and for building presentations of 2-categories in a modular way.

10.1 Three-Dimensional Polygraphs

10.1.1 Definition. A 3-*polygraph* (P, P_3) consists of a 2-polygraph P together with a cellular extension P_3 of the 2-category P^* freely generated by P, the elements of P_3 being referred to as 3-*generators*. Explicitly, a 3-polygraph consists of a diagram

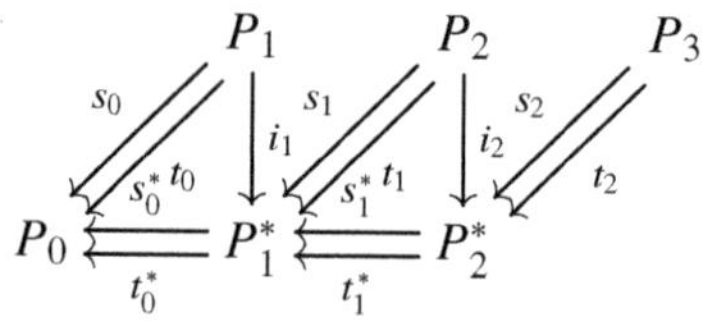

in **Set**, together with a structure of 2-category on the 2-graph

$$P_0 \underset{t_0^*}{\overset{s_0^*}{\leftleftarrows}} P_1^* \underset{t_1^*}{\overset{s_1^*}{\leftleftarrows}} P_2^*$$

such that, for $n \in \{0, 1, 2\}$,

- P_n^* is the set of n-cells of the n-category freely generated by the underlying n-polygraph,
- $i_n : P_n \to P_n^*$ is the canonical inclusion,
- s_n^* and t_n^* are the respective canonical extensions of s_n and t_n, satisfying

$$s_n^* \circ i_n = s_n \qquad \text{and} \qquad t_n^* \circ i_n = t_n,$$

- the globular identities are satisfied:

$$s_n^* \circ s_{n+1} = s_n^* \circ t_{n+1} \qquad \text{and} \qquad t_n^* \circ s_{n+1} = t_n^* \circ t_{n+1}.$$

We write $A : \phi \Rrightarrow \psi$ for a 3-generator $A \in P_3$ with $s_2(A) = \phi$ and $t_2(A) = \psi$. A 3-polygraph P is often concisely denoted

$$\langle\, P_0 \mid P_1 \mid P_2 \mid P_3 \,\rangle.$$

We write $P_{\leqslant 2}$ for the *underlying 2-polygraph* of a 3-polygraph P. Contrarily to previous chapters, we always respectively denote by $*_0$ and $*_1$ the horizontal and vertical compositions of a 2-category.

10.1.2 Example. The 3-polygraph Mon is

$$\mathrm{Mon} = \langle\, \star \mid a \mid \mu\colon a *_0 a \Rightarrow a,\ \eta\colon 1_\star \Rightarrow a \mid A, L, R \,\rangle$$

where the sources and targets of the 3-generators are given by

$$A\colon (\mu *_0 a) *_1 \mu \Rrightarrow (a *_0 \mu) *_1 \mu \qquad L\colon (\eta *_0 a) *_1 \mu \Rrightarrow a$$
$$R\colon (a *_0 \eta) *_1 \mu \Rrightarrow a$$

Using string diagrams (see §2.4.8), the 2-generators of Mon are pictured as

and its 3-generators A, L, and R respectively as

Many other examples of 3-polygraphs are given in Appendix C.

10.1.3 Presented 2-category. Let P be a 3-polygraph. The *2-category presented by P* is the 2-category, denoted by $\overline{P}$, obtained by quotienting the free 2-category over $P_{\leqslant 2}$ by the congruence $\approx^P$ generated by P_3 on 2-cells, as described in §7.1.2:

$$\overline{P} = P^*_{\leqslant 2}/P_3$$

If C is a 2-category, we say that P *presents* C if C is isomorphic to $\overline{P}$.

In particular, when the set P_0 is reduced to one element, the category presented by P has one 0-cell and is thus a strict monoidal category (see §2.4.10). Moreover, when both P_0 and P_1 are reduced to one element, the set P_1^* of 1-cells is the free monoid on one generator, i.e., $\mathbb{N}$, and the presented category is a PRO (see §2.4.10). This is for instance the case in Example 10.1.2.

Two 3-polygraphs P and Q are said *Tietze equivalent* when the presented 2-categories are isomorphic: $\overline{P} \simeq \overline{Q}$.

10.1.4 Models. Given a 2-category C, the category of *models* (or *algebras*) of C in a 2-category S is the category $\mathbf{Cat}_2(C, S)$ of 2-functors $C \to S$ and oplax 2-natural transformations between those (see §20.2.13 for a general definition). More explicitly, given a 3-polygraph P, a *model* of $\overline{P}$ in a 2-category C consists of

– a family

$$(f_x)_{x \in P_0}$$

of 0-cells of C indexed by the 0-generators of P,

– a family

$$(f_a\colon f_x \to f_y)_{a\colon x\to y\in P_1}$$

of 1-cells of C indexed by the 1-generators of P; if $u = a_1 \dots a_n$ is a 1-cell of P^*, we write f_u for the 1-cell $f_{a_1} \dots f_{a_n}$,

– a family

$$(f_\alpha\colon f_u \Rightarrow f_v)_{\alpha\colon u\Rightarrow v\in P_2}$$

of 2-cells of C indexed by the 2-generators of P; the notation f_ϕ is extended to any 2-cell of P^* by $f_{\phi *_0 \psi} = f_\phi *_0 f_\psi$, $f_{\phi *_1 \psi} = f_\phi *_1 f_\psi$ and $f_{1_u} = 1_{f_u}$,

such that, for every 3-generator $A\colon \phi \Rrightarrow \psi$ of P, we have

$$f_\phi = f_\psi.$$

10.1.5 Example. Let C be a monoidal category. The models of the 3-polygraph Mon of Example 10.1.2 in C (considered as a 2-category with only one 0-cell) are precisely monoids in C in the following sense. A *monoid* in a monoidal category C consists an object x of C and two morphisms

$$m\colon x \otimes x \to x \qquad\qquad e\colon i \to x$$

such that the diagrams

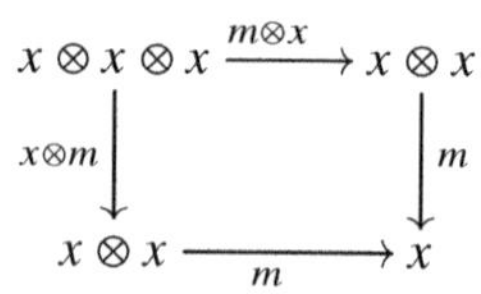

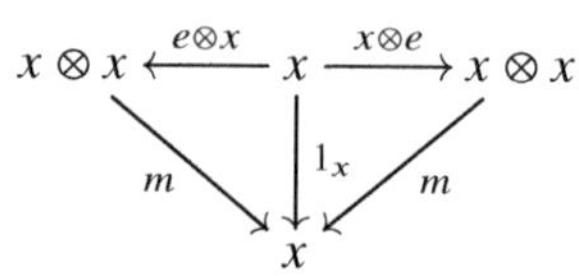

commute. A morphism $f\colon (x, m, e) \to (x', m', e')$ between monoids is a morphism $f\colon x \to x$ of C such that the diagrams

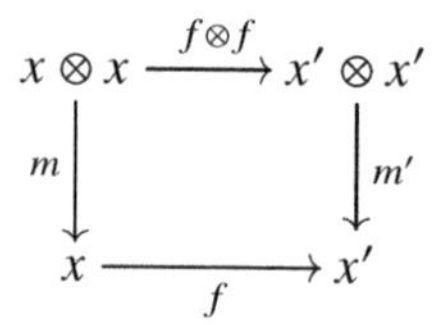

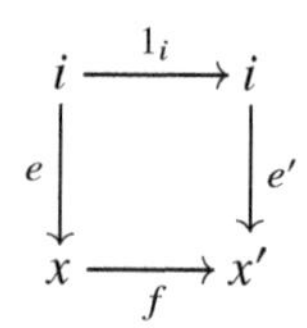

commute. In particular, a monoid in **Set** equipped with cartesian product as tensor product and terminal set as unit is precisely a monoid in the usual sense (by Mac Lane's coherence theorem, Theorem 12.4.4, we can always consider that it forms a strict monoidal category).

As another application, the algebras of Mon in the 2-category **Cat** (of categories, functors, and natural transformations) are precisely monads.

10.1.6 2-categories admitting a presentation. We should first note that, contrarily to the case of 1-categories (§2.3.14), not every 2-category admits a presentation by a 3-polygraph. Namely, any 2-category C is presented by a 3-polygraph P is a quotient of the free 2-category $P^*_{\leqslant 2}$ by the congruence generated by P_3. Since there is no quotient on 1-cells, the underlying 1-category of C is always free (on the underlying 1-polygraph of P). Therefore, only 2-categories whose underlying 1-category is free may have a presentation.

For instance, consider the category $\mathbb{Z}$ corresponding to the additive monoid of integers is not free, by §2.3.13. Therefore the 2-category with $\mathbb{Z}$ as underlying category and only identity 2-cells admits no presentation by a 3-polygraph. Extensions of the notion of polygraph aimed at addressing this problem have been proposed in [100, 118, 286].

10.1.7 The canonical and standard presentations. Any 2-category C whose underlying category is freely generated by a 1-polygraph Q admits a presentation by a 3-polygraph. The *canonical presentation of* C is the 3-polygraph P with

- Q as underlying 1-polygraph,
- the set $P_2 = C_2$ of all 2-cells of C as 2-generators,
- the subset P_3 of $P_2^* \times P_2^*$ of pairs of parallel 2-cells whose evaluation as 2-cells of C are equal.

An analogous of the standard presentation (§2.3.14) can also be defined for 2-categories, giving rise to slightly smaller presentations.

10.1.8 The category of 3-polygraphs. A *morphism* $f\colon P \to Q$ between 3-polygraphs P and Q consists of a morphism $f\colon P_{\leqslant 2} \to Q_{\leqslant 2}$ between the underlying 2-polygraphs (see §2.2.3) together with a function $f_3\colon P_3 \to Q_3$ such that $s_2 \circ f_3 = f_2 \circ s_2$ and $t_2 \circ f_3 = f_2 \circ t_2$. These compose in the expected way, and we write $\mathbf{Pol}_3$ for the category of 3-polygraphs and their morphisms.

10.2 Rewriting Properties of 3-Polygraphs

A 3-polygraph P can be seen as a 3-dimensional rewriting system: its underlying 2-polygraph generates a 2-category, whose 2-cells are the "terms" which get rewritten by the 3-generators. For this reason, the elements of P_3 are sometimes called *rewriting rules*. We now formalize this point of view.

10.2.1 Occurrences of 2-generators. Let P be a 2-polygraph. Given a 2-cell ϕ in P_2^* and a 2-generator $\alpha \in P_2$, we write $|\phi|_\alpha$ for the number of occurrences of α in ϕ. It can be formally defined as follows.

We write N for the 2-category with one 0-cell $\star$, one 1-cell $1_\star$, $\mathbb{N}$ as set of 2-cells, horizontal and vertical compositions being given by addition and the identity 2-cell by 0. Given a 2-generator $\alpha \in P_2$, there exists a unique 2-functor

$$|-|_\alpha \colon P_2^* \to N$$

such that $|\alpha|_\alpha = 1$ and $|\beta|_\alpha = 0$ for every 2-generator $\beta \in P_2$ with $\beta \neq \alpha$. Given a 2-cell ϕ in P_2^*, the natural number $|\phi|_\alpha$ is called the *number of occurrences* of the generator α in ϕ. We also write

$$|-| \colon P_2^* \to N$$

for the 2-functor such that $|\alpha| = 1$ for every generator $\alpha \in P_2$. Given a morphism ϕ, we have

$$|\phi| = \sum_{\alpha \in P_2} |\phi|_\alpha$$

and this quantity is called the *number of generators* in ϕ or the *size* of ϕ.

10.2.2 Contexts. Let P be a 2-polygraph and $u, v \colon x \to y$ be two parallel 1-cells in P_1^*. We write $P[X]$ for the 2-polygraph with the same 0- and 1-cells as P and with $P_2 \sqcup \{X\}$ as 2-cells, with $s_1(X) = u$ and $t_1(X) = v$. A *context* K of type (u, v) in P is a cell K in $P[X]_2^*$ in which the generator X occurs exactly once, i.e., such that $|K|_X = 1$.

10.2.3 Lemma. *Any context K of type (u, v) can be written in the form*

$$K = \psi *_1 (w *_0 X *_0 w') *_1 \psi'$$

for some 1-cells $w \colon x' \to x$ and $w' \colon y \to y'$ and 2-cells $\phi \colon u' \Rightarrow wuw'$ and $\phi' \colon wvw' \Rightarrow v'$. Graphically, K can be depicted as

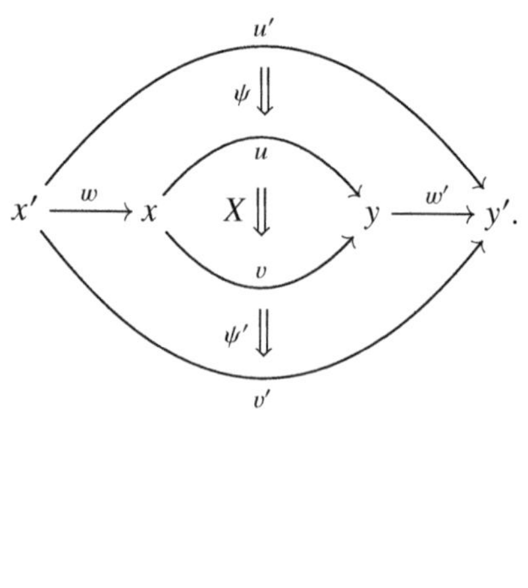

Given a context K as in previous lemma and a 2-cell $\phi\colon u \Rightarrow v$ in P_2^*, we write $C[\phi]$ for the following 2-cell of P_2^*:

$$K[\phi] = \psi *_1 (w *_0 \phi *_0 w') *_1 \psi'.$$

10.2.4 Rewriting steps. Let P be a 3-polygraph. Given a context K in P_2 of type (u, v) as in Lemma 10.2.3 and a 3-cell

$$F\colon \phi \Rrightarrow \phi' : u \Rightarrow v : x \to y$$

in P_3^*, we extend the previous notation and write

$$K[F] = \psi *_1 (w *_0 F *_0 w') *_1 \psi'$$

Graphically,

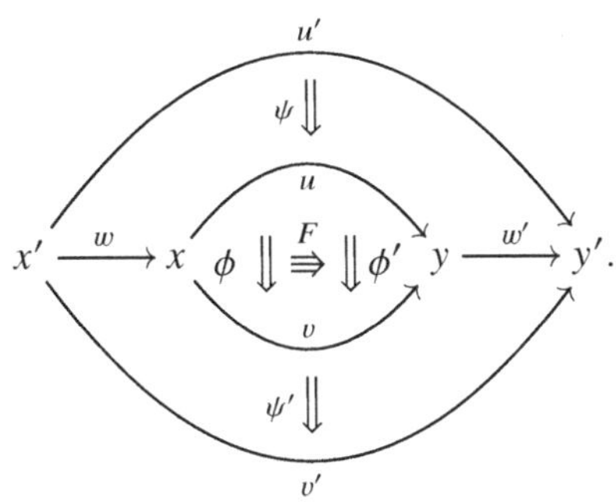

A *rewriting step* is a 3-cell of the form $K[A]$ for some context K and 3-generator $A\colon \phi \Rrightarrow \phi'$. The 2-cells $K[\phi]$ and $K[\phi']$ are respectively called the source and target of the rewriting step. Using the axioms of 3-categories, one shows that every 3-cell of P^* is a composite of rewriting steps:

10.2.5 Lemma. *Any 3-cell F of P^* can be decomposed as*

$$F = F_1 *_2 F_2 *_2 \dots *_2 F_k$$

where the F_i's are rewriting steps. Moreover, the number $k \in \mathbb{N}$ is the same for all such decompositions.

The number k in the previous lemma is called the *length* of F.

10.2.6 Termination and confluence. Given a 3-polygraph P, we write here P^{rs} for its set of rewriting steps and $s_2, t_2\colon P^{\mathrm{rs}} \to P_2^*$ for the functions respectively taking a rewriting step to its source and target. Any 3-polygraph P induces an abstract rewriting system

$$P_2^* \underset{t_2}{\overset{s_2}{\leftleftarrows}} P^{\mathrm{rs}}$$

with 2-cells as vertices and rewriting steps as edges. A 3-polygraph is said to be *terminating*, *Church–Rosser*, *confluent*, *locally confluent*, or *convergent* when the associated abstract rewriting system is, see Section 1.3.

The termination of a 3-polygraph can be shown in a similar way as for 2-polygraphs (Section 4.4) by considering suitable reduction orders: this will be detailed in Chapter 11. In the next section, we study local confluence through critical branchings, generalizing the techniques introduced for 2-polygraphs.

10.2.7 Branchings. Let P be a 3-polygraph. A *branching* is a branching of the underlying abstract rewriting system. It consists of a pair (F_1, F_2) of 3-cells $F_1 \colon \phi \Rrightarrow \psi_1$ and $F_2 \colon \phi \Rrightarrow \psi_2$ in P^* with the same source. We say that the 2-cell ϕ is the *source* of (F_1, F_2). A branching (F_1, F_2) of P is *local* if both F_1 and F_2 are rewriting steps; it is *confluent* if there exist cofinal 3-cells $F_1' \colon \psi_1 \Rrightarrow \chi$ and $F_2' \colon \psi_2 \Rrightarrow \chi$ in P^*.

We say that a local branching (F_1, F_2) of P is *trivial* if $F_1 = F_2$. We say that the branching (F_1, F_2) (resp. (F_2, F_1)) is *orthogonal* if F_1 and F_2 are of the form

$$F_1 = \psi *_1 (w_1 *_0 A_1 *_0 w_1') *_1 \psi' *_1 (w_1 *_0 \phi_2 *_0 w_2') *_1 \psi''$$
$$F_2 = \psi *_1 (w_1 *_0 \phi_1 *_0 w_1') *_1 \psi' *_1 (w_1 *_0 A_2 *_0 w_2') *_1 \psi''$$

where $A_1 \colon \phi_1 \Rrightarrow \phi_1'$ and $A_2 \colon \phi_2 \Rrightarrow \phi_2$ are 3-generators, w_1 , w_1', w_2, and w_2' are 1-cells of P^*, and ψ, ψ', and ψ'' are 2-cells of P^*. The orthogonal situation can be pictured as

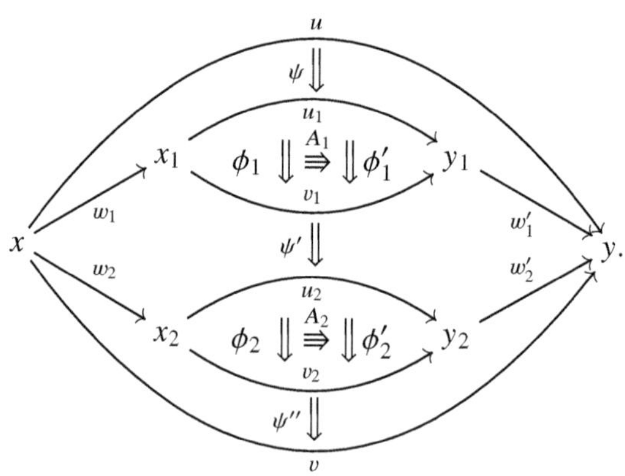

Local branchings are ordered by the relation $\sqsubseteq$ generated by

$$(F_1, F_2) \sqsubseteq (\phi *_i F_1, \phi *_i F_2) \quad \text{and} \quad (F_1, F_2) \sqsubseteq (F_1 *_i \phi, F_2 *_i \phi),$$

where ϕ ranges over the 2-cells of P^* and i over $\{0, 1\}$ such that the involved composites are defined. A branching of P is called *critical* if it is local, orthogonal, and minimal for $\sqsubseteq$. We say that P is *critically confluent* if all its critical branchings are confluent.

As in the case of 2-polygraphs, Lemma 4.3.7, the critical branching lemma holds for 3-polygraphs:

10.2.8 Lemma. *A* 3*-polygraph is locally confluent if and only if all its critical branchings are confluent.*

As a direct corollary of this lemma and Newman's lemma (Lemma 1.3.21), we may state the following proposition, which is used to show the convergence of polygraphs in the vast majority of cases.

10.2.9 Proposition. *A* 3*-polygraph which is terminating and has all its critical branchings confluent is convergent.*

10.2.10 Example. The 3-polygraph Mon of Example 10.1.2 has five critical branchings. All of them are confluent, as shown by the string diagrams below:

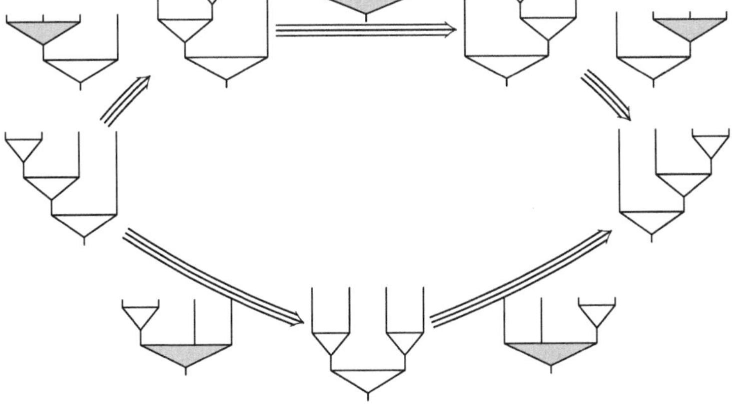

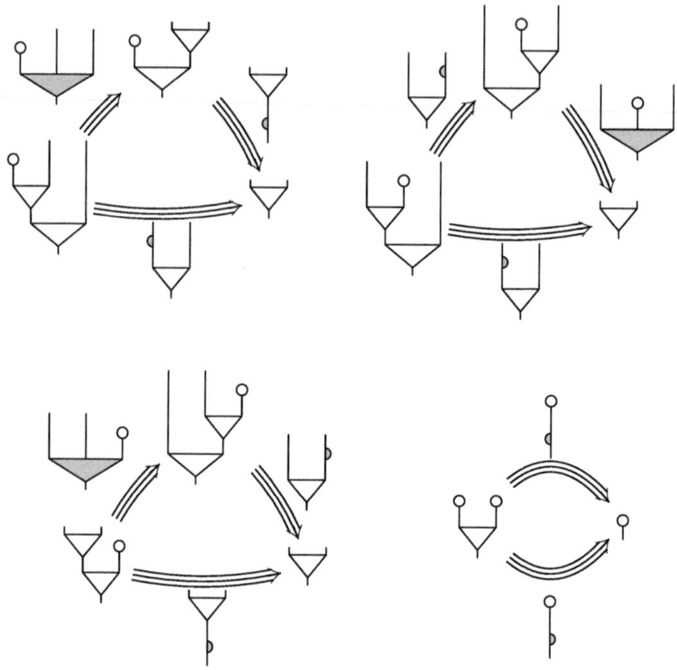

For instance, the source of the first critical branching is

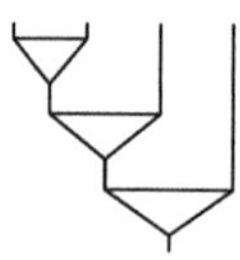 $= (\mu *_0 a *_0 a) *_1 (\mu *_0 a) *_1 \mu$

which can be rewritten by using the 3-generator

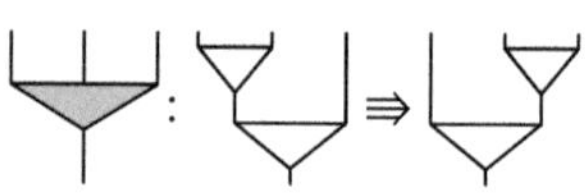

in two ways, yielding the two rewriting steps:

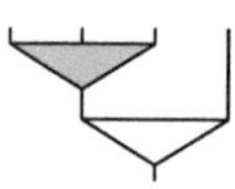 $= (A *_0 a) *_1 \mu$ 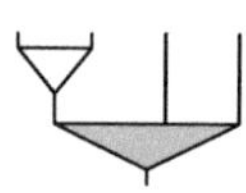 $= (\mu *_0 a *_0 a) *_1 A.$

We will see in Example 11.2.4 that this polygraph is terminating, based on the observation that rewriting either removes 2-generators or moves subtrees to the right. By Proposition 10.2.9, it is thus convergent.

10.2.11 The Knuth–Bendix completion procedure. A Knuth–Bendix completion procedure can be defined for 3-polygraphs. We do not detail it much, because it is very similar to the one for 2-polygraphs given in Section 5.2:

starting with a 3-polygraph P and a total reduction order adapted to the polygraph (Definition 11.1.1), we compute the critical branchings and, for each of those branchings, normalize both members and add a new rule when the normal forms differ, oriented according to the reduction order. As for 2-polygraphs, this procedure might not terminate because we keep on adding new rules. However, there is a new potential source of non-termination for 3-polygraphs: we will see in Section 10.4 that a finite polygraph might give rise to an infinite number of critical branchings, and the completion procedure will have to examine each of them.

10.3 Constructing Presentations

When a 3-polygraph P is convergent, the normal forms provide canonical representatives of equivalence classes of 2-cells in P_2^* modulo the congruence generated by P_3. We explain here that this can be exploited to show that P presents a given 2-category C, by showing that the 2-cells of C are in bijection with the normal forms of the polygraph. The following proposition thus generalizes the method proposed in Section 4.5 to construct presentations of categories.

10.3.1 Proposition. *Let P be a convergent 3-polygraph and C a 2-category whose underlying category is isomorphic to the category freely generated by the underlying 1-polygraph of P, i.e., we have isomorphisms $f_0\colon P_0 \to C_0$ and $f_1\colon P_1^* \to C_1$. Let moreover $f_2\colon P_2 \to C_2$ be a function compatible with source and target, i.e., f_2 sends a 2-generator $\alpha\colon u \Rightarrow v$ in P_2 to a 2-cell $f_2(\alpha)\colon f_1(u) \Rightarrow f_1(v)$. We extend f_2 to the 2-cells in P_2^* by functoriality by*

$$
\begin{aligned}
f_2(\phi *_0 \psi) &= f_2(\phi) *_0 f_2(\psi),\\
f_2(\phi *_1 \psi) &= f_2(\phi) *_1 f_2(\psi),\\
f_2(1_u) &= 1_{f_1(u)}.
\end{aligned}
$$

Suppose finally that

- *for any 3-generator $A\colon \phi \Rrightarrow \psi$ in P_3, we have $f_2(\phi) = f_2(\psi)$,*
- *the function $f_2^*\colon P_2^* \to C_2$ restricts to a bijection between normal forms in P_2^* and C_2.*

Then P is a presentation of C.

Proof. Let us write VC for the 2-polygraph with P as underlying 1-polygraph and whose set of 2-generators is the set C_2 of 2-cells of C. The triple of morphisms (f_0, f_1, f_2) precisely corresponds to a morphism of 2-polygraphs $f\colon P_{\leqslant 2} \to VC$. This morphism induces a functor $f^*\colon P^*_{\leqslant 2} \to C$ from the freely generated 2-category. The first condition ensures that it induces a quotient functor $\overline{f}\colon \overline{P} \to C$, and the second condition ensures that $\overline{f}$ is a bijection. Namely, $\overline{f}$ is a bijection in dimensions 0 and 1 by hypothesis. Moreover, the 2-cells of $\overline{P}$ are in bijection with 2-cells of P^*_2 in normal form because the polygraph is convergent (Proposition 1.3.24), and those are in bijection with the 2-cells of C by hypothesis. □

10.3.2 A presentation of Δ_+. As a detailed example of the above method, we show here that the 3-polygraph Mon of Example 10.1.2 presents the augmented simplicial category Δ_+. This category was introduced in §4.5.6: its objects are natural numbers and morphisms are non-decreasing functions. It is moreover monoidal, with tensor product given on objects by addition (such a monoidal category is called a PRO, see §2.4.10). As such, it can be considered as a 2-category (see §2.4.10), of which we now make an explicit description.

The 2-category Δ_+ has one 0-cell $\star$, the 1-cells are natural numbers and the 2-cells $f\colon m \to n$ are non-decreasing maps from $[m]$ to $[n]$, where $[n] = \{0, \ldots, n-1\}$ for $n \geqslant 0$. The vertical composition of 2-cells is the usual composition of functions, with identities as neutral elements. The horizontal composition of 1-cells is given by addition, with 0 as neutral element, and the horizontal composition of 2-cells $f\colon m \to n$ and $f'\colon m' \to n'$ is given by

$$(f *_0 f')(i) = \begin{cases} f(i) & \text{if } 0 \leqslant i < m, \\ f(i-m)+n & \text{if } m \leqslant i < m+m'. \end{cases} \tag{10.1}$$

We have seen in Example 10.2.10 that the polygraph Mon is convergent. The normal forms in P^*_2 can be characterized as follows. Given $n \in \mathbb{N}$, we define the *right comb* $\mu_n\colon n \Rightarrow 1$ in P^*_2 by induction:

$\mu_0 = \eta =$  $\qquad \mu_1 = a = |$ $\qquad \mu_{n+2} = (a *_0 \mu_{n+1}) *_1 \mu =$ 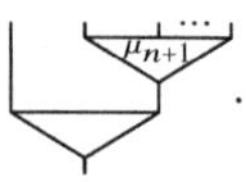 .

A *right forest* is a horizontal composite of right combs, i.e., a morphism of the form

$\mu_{n_1} *_0 \mu_{n_2} *_0 \cdots *_0 \mu_{n_k} =$ 

for some $k \in \mathbb{N}$ called the *width* of the right forest, and $(n_1, \ldots, n_k) \in \mathbb{N}^k$.

10.3.3 Lemma. *Given $n, n' \in \mathbb{N}$, $(\mu_n *_0 \mu_{n'}) *_1 \mu$ rewrites to $\mu_{n+n'}$.*

Proof. By induction on n. For $n = 0$, we have $(\mu_0 *_0 \mu_{n'}) *_1 \mu = \mu_{n'+1}$. Otherwise $(\mu_{n+1} *_0 \mu_{n'}) *_1 \mu$ rewrites in one step to $(\mu_n *_0 \mu_{n'+1}) *_1 \mu$ and we conclude using the induction hypothesis. □

10.3.4 Lemma. *The 2-cells of P_2^* in normal form are precisely the right forests.*

Proof. The right forests are easily checked to be normal forms. We now show that, conversely, every normal form is a right forest by showing that every 2-cell ϕ in P_2^* rewrites to a right forest. The proof is done by induction on the size of ϕ. If the size of ϕ is 0 then it is an identity, which is a right forest. Otherwise it can be decomposed as

$$\phi = \psi *_1 (a^i *_0 \mu *_0 a^j) =$$ 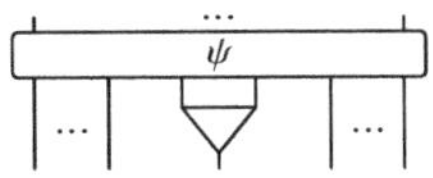

(a^i denoting the horizontal composition of i instances of a) where, by induction, ψ is a right forest

$$\psi = \mu_{n_1} *_0 \mu_{n_2} *_0 \cdots *_0 \mu_{n_{i+1+j}}.$$

By Lemma 10.3.3, ϕ rewrites to the right forest

$$\mu_{n_1} *_0 \cdots *_0 \mu_{n_{i-1}} *_0 \mu_{n_i+n_{i+1}} *_0 \mu_{n_{i+2}} *_0 \cdots *_0 \mu_{n_{i+1+j}}$$

which can be graphically depicted as

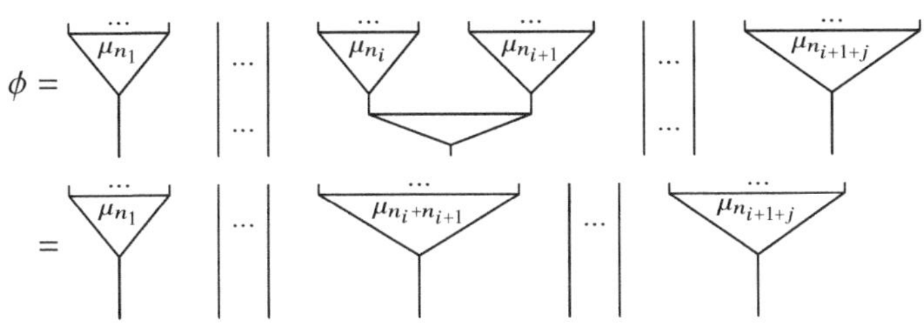

and concludes the proof. □

The underlying 1-category of Δ_+ is the additive monoid $\mathbb{N}$, seen as a category. It is thus the free category on the 1-polygraph $\langle \star \mid a \rangle$, which is the underlying 1-polygraph of Mon. We define a function $f_2\colon \mathrm{Mon}_2 \to (\Delta_+)_2$ where $f_2(\mu)\colon 2 \Rightarrow 1$ and $f_2(\eta)\colon 0 \Rightarrow 1$ are both terminal arrows. Consider the 3-generator

$$A\colon \phi \Rrightarrow \psi\colon a^3 \Rightarrow a\colon \star \to \star$$

with $\phi = (\mu *_0 a) *_1 \mu$ and $\psi = (a *_0 \mu) *_1 \mu$. We necessarily have $f_2(\phi) = f_2(\psi)$ (where f_2 is extended by functoriality, as in Proposition 10.3.1) because both

2-cells are of type $3 \Rightarrow 1$ in Δ_+, and the object 1 is terminal in this category. A similar reasoning can be held for the two other 3-generators L and R.

10.3.5 Lemma. *The function f_2 induces a bijection between the normal forms in P_2^* and the 2-cells of Δ_+.*

Proof. Given a 2-cell $g: n \to k$ in Δ_+, i.e., a non-decreasing function $g: [n] \to [k]$, and $j \in [k]$, we write n_j for the cardinal of the set $f^{-1}(j)$. Note that the function g is entirely determined by the tuple $(n_1, \ldots, n_k) \in \mathbb{N}^k$ since, for $i \in [n]$, $g(i)$ is the unique element of $[k]$ satisfying

$$\sum_{0 \leqslant j < g(i)} n_j \leqslant i < \sum_{0 \leqslant j \leqslant g(i)} n_j$$

and conversely any k-tuple of natural numbers $(n_1, \ldots, n_k)$ determines an increasing function from $\sum_j n_j$ to k in this way. Given a right forest $\mu_{n_1} *_0 \ldots *_0 \mu_{n_k}$, one easily checks that its image under f is the non-decreasing function with associated k-tuple $(n_1, \ldots, n_k)$, thus establishing a bijection between forests of width k and non-decreasing functions with codomain k. $\square$

In order to illustrate the above proof, consider the function $g: 5 \to 3$ whose graph is depicted on the left below:

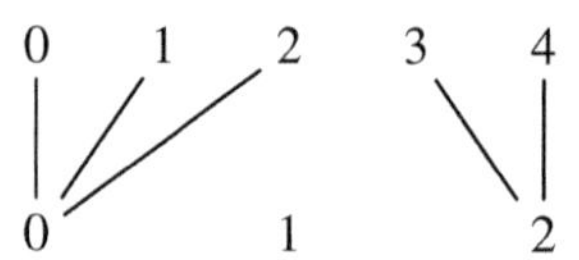

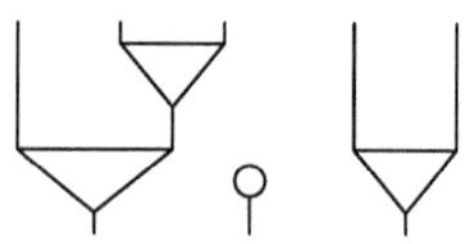

The associated sequence in $\mathbb{N}^3$ is $(3, 0, 2)$ and the associated normal form is the right forest $\mu_3 *_0 \mu_0 *_0 \mu_2$ as pictured on the right. Note the clear correspondence between the two figures.

Let us sum up the results we have obtained in this section for the augmented simplicial category Δ_+. We have

1. constructed two 1-cells $\mu: 2 \to 1$ and $\eta: 0 \to 1$ in the 2-category Δ_+: both are terminal 2-cells,
2. shown that they generate the 2-category: every morphism of Δ_+ can be written as a (horizontal or vertical) composite of those 2-cells,
3. shown that those two 2-cells satisfy the axioms A, L, and R of Example 10.1.2 (reading 3-cells are equalities), expressing that μ is associative and admits η as left and right unit,
4. shown that this set of axioms is complete: if two composites of μ and η give rise to the same 2-cell in Δ_+ then one can show that they are equal using axioms A, L, R, and axioms for 2-categories.

Since the generators and the relations are precisely those of monoids, we can deduce that Δ_+ impersonates the notion of monoid, in the sense that a 2-functor $\Delta_+ \to C$ to some 2-category C determines a monoid, as explained in Example 10.1.5.

10.3.6 Presentations from polygraphs with canonical forms. The observation made for 2-polygraphs in Remark 4.5.4 generalizes to 3-polygraphs. Namely, in the proof of Proposition 10.3.1, we did not fully use the convergence of the polygraph, only the existence of canonical representatives of equivalence classes provided by normal forms. This suggests the following generalization of the above method, which applies in cases where, even though it is difficult to construct a convergent presentation, one can still directly come up with a notion of canonical form. Many applications of this methodology were studied by Lafont [235].

10.3.7 Proposition. *Suppose given*

- *a 3-polygraph P,*
- *a 2-category C whose underlying category is isomorphic to the free category on the underlying 1-polygraph of P,*
- *a function $f_2 \colon P_2 \to C_2$ which is compatible with source and target,*
- *a set $\widehat{P_2^*} \subseteq P_2^*$ of 2-cells called* canonical forms,

such that

1. *for any 2-generator $A \colon \phi \Rrightarrow \psi$ in P_3, we have $f_2(\phi) = f_2(\psi)$,*
2. *every 2-cell in P_2^* is equivalent, with respect to the congruence $\approx^P$ generated by P_3, to a canonical form,*
3. *f_2 restricts to a bijection between canonical forms $u \Rightarrow v$ in P_2^* and 2-cells $f(u) \Rightarrow f(v)$ in C, where f_2 is implicitly extended to 2-cells in P_2^* by functoriality.*

Then P is a presentation of the 2-category C.

Proof. As in the proof of Proposition 10.3.1, by the first condition, f_2 induces a 2-functor $f \colon \overline{P} \to C$, which is bijective on 0- and 1-cells. By surjectivity of f_2 in the third condition, for every 2-cell ϕ in C_2, there is a canonical form $\widehat{\phi}$ such that $f_2(\widehat{\phi}) = \phi$. We have $f(\overline{\widehat{\phi}}) = f_2(\widehat{\phi}) = \phi$ and f is thus surjective on 2-cells. Moreover, consider two parallel 2-cells ϕ and ψ of $\overline{P}_2$, such that $f(\phi) = f(\psi)$. The second condition ensures that there are canonical forms $\widehat{\phi}$ and $\widehat{\psi}$ such that $\phi = \overline{\widehat{\phi}}$ and $\psi = \overline{\widehat{\psi}}$. We have

$$f_2(\widehat{\phi}) = f(\overline{\widehat{\phi}}) = f(\phi) = f(\psi) = f(\overline{\widehat{\psi}}) = f_2(\widehat{\psi}).$$

By injectivity of f_2 in the third condition, we have $\widehat{\phi} = \widehat{\psi}$. Thus, $\phi = \overline{\widehat{\phi}} = \overline{\widehat{\psi}} = \psi$ and f is injective on 2-cells. □

In particular, when P is a convergent 3-polygraph, we can take $\widehat{P_2^*}$ to be the set of 2-cells in normal form, thus recovering Proposition 10.3.1 as a particular case of previous proposition.

10.3.8 Example. In order to show that Mon presents Δ_+, we could have chosen the following alternative definition of right combs:

$$\mu_0 = \eta \qquad\qquad \mu_{n+1} = (a *_0 \mu_n) *_1 \mu$$

Right forests are obtained as horizontal composites of such right combs, and we consider those as canonical forms. The definition is mostly the same as before except that we have

$$\mu_1 = (a *_0 \eta) *_1 \mu =$$ 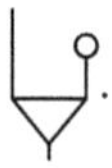 .

It can be shown that every 2-cell is equivalent to a canonical form using a variant of the proof of Lemma 10.3.4, and one can construct a bijection between canonical forms and 2-cells in Δ_+ using a variant of the proof of Lemma 10.3.5. We can thus conclude that Mon is a presentation of Δ_+ by Proposition 10.3.7. Note that the canonical form associated to 1_a is μ_1, so there is no hope to obtain canonical forms as normal forms for some convergent 3-polygraph, because no terminating polygraph can rewrite identities.

10.3.9 Remark. As a variant of the previous example, consider the category $\Delta_+^2 = \Delta_+ \times \Delta_+$. The monoidal structure on Δ_+ induces one on Δ_+^2 given on objects by

$$(m_1, n_1) \otimes (m_2, n_2) = (m_1 \otimes m_2, n_1 \otimes n_2)$$

and similarly on morphisms. The underlying monoid of objects of this category is $\mathbb{N} \times \mathbb{N}$ which is abelian and thus not free (see §2.3.13). Therefore, by §10.1.6, there is no presentation of it (seen as a 2-category induced by the monoidal structure, see §2.4.10) by a 2-polygraph. This situation is detailed in [100]. The same argument applies to most products of 2-categories, but for degenerated cases.

10.3.10 Presenting categories. Let P be a 3-polygraph presenting a monoidal category C, seen as a 2-category: this presentation is of the form

$$P = \langle\, \star \mid P_1 \mid P_2 \mid P_3 \,\rangle.$$

The monoidal category C has an underlying category obtained by forgetting the tensor product and unit object. This category admits, as a category, a presentation by the following 2-polygraph Q constructed from P:

- $Q_0 = P_1^*$ is the set of 1-cells $u\colon \star \to \star$ in P_1^*,
- $Q_1 = P_1^* P_2 P_2^*$ contains a 1-generator

$$u\alpha w\colon uvw \to uv'w$$

for every 1-cells $u, w \in P_1^*$ and 2-generator $\alpha\colon v \Rightarrow v'$ in P_2,
- Q_2 contains a 2-generator

$$uAw\colon u\phi w \Rightarrow u\psi w$$

for every 1-cells $u, w \in P_1^*$ and 3-generator $A\colon \phi \Rrightarrow \psi$, where $u\phi w$ and $u\psi w$ are seen as elements of Q_2^* in the expected functorial way, it also contains a 2-generator

$$X_{u,\alpha,u',\beta,u''}\colon u\alpha u'wu'' * uv'u'\beta u'' \Rightarrow uvu'\beta u'' * u\alpha u'w'u''$$

for every 1-cells $u, u', u'' \in P_1^*$ and 2-generators $\alpha\colon v \Rightarrow v'$ and $\beta\colon w \Rightarrow w'$ in P_2 (which encodes the exchange law).

10.3.11 Example. We have seen above that the augmented simplicial category Δ_+ was presented, as a monoidal category, by the 3-polygraph Mon, defined in Example 10.1.2. We deduce the presentation of Δ_+, as a category, by the 2-polygraph P with

- 0-generators: for $i \in \mathbb{N}$, a generator a^i,
- 1-generators: for $i, j \in \mathbb{N}$,

$$a^i\mu a^j : a^{i+2+j} \to a^{i+1+j} \qquad\qquad a^i\eta a^j : a^{i+j} \to a^{i+1+j}$$

- 2-generators: for $i, j \in \mathbb{N}$,

$$\begin{aligned}
a^iAa^j &: & a^i\mu a^{j+1} * a^i\mu a^j &\Rightarrow a^{i+1}\mu a^j * a^i\mu a^j\\
a^iLa^j &: & a^i\eta a^{j+1} * a^i\mu a^j &\Rightarrow a^{i+j}\\
a^iRa^j &: & a^{i+1}\eta a^j * a^i\mu a^j &\Rightarrow a^{i+j}\\
X_{a^i,\mu,a^j,\mu,a^k} &: & a^i\mu a^{j+2+k} * a^{i+1+j}\mu a^k &\Rightarrow a^{i+2+j}\mu a^k * a^i\mu a^{j+1+k}\\
X_{a^i,\mu,a^j,\eta,a^k} &: & a^i\mu a^{j+k} * a^{i+1+j}\eta a^k &\Rightarrow a^{i+2+j}\eta a^k * a^i\mu a^{j+1+k}\\
X_{a^i,\eta,a^j,\mu,a^k} &: & a^i\eta a^{j+2+k} * a^{i+1+j}\mu a^k &\Rightarrow a^{i+j}\mu a^k * a^i\eta a^{j+1+k}\\
X_{a^i,\eta,a^j,\eta,a^k} &: & a^i\eta a^{j+k} * a^{i+1+j}\eta a^k &\Rightarrow a^{i+j}\eta a^k * a^i\eta a^{j+1+k}
\end{aligned}$$

It can be checked that we precisely recover the presentation for the simplicial category given in §4.5.6, up to renaming the 1-generators $a^i\mu a^j$ to s_i^{i+j+1} and $a^i\eta a^j$ to d_i^{i+j}.

10.4 Indexed Critical Branchings

There is a major difference between rewriting in 3-polygraphs compared to the case of 2-polygraphs studied in previous chapters: contrarily to presentations of categories (see Lemma 4.3.10), the number of critical branchings of a finite 3-polygraph can be infinite. We begin with an example of this phenomenon, originally observed by Lafont [235].

10.4.1 Presenting the theory for symmetries. The category **S** is the category whose objects are natural numbers and a morphism $f\colon m \to n$ is a bijection (also called a *permutation*) from $[m]$ to $[n]$, the ordinals with m and n elements, respectively, with usual compositions and identities. Here, all morphisms are in fact endomorphisms. This category is monoidal with tensor product given by addition on objects (this is a PRO) and as for Δ_+ on morphisms, see (10.1) in §10.3.2.

Starting from the fact that any bijection can be decomposed as a composite of transpositions, we expect that this monoidal category, seen as a 2-category, admits a presentation by the following 3-polygraph P:

$$\langle \star \mid a \mid \gamma\colon a *_0 a \to a *_0 a \mid I, Y \rangle.$$

Here, the 2-generator γ corresponds to the transposition on a set with two elements and is usually pictured as

The two 3-generators express

- the *involutivity* of the transposition:

$$I\colon \gamma *_1 \gamma \Rrightarrow 1_a *_0 1_a$$

  which can be represented as

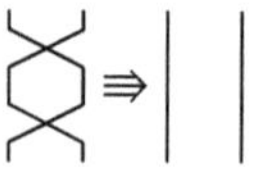

- the *Yang–Baxter* relation Y of type

$$(\gamma *_0 1_a) *_1 (1_a *_0 \gamma) *_1 (\gamma *_0 1_a) \Rrightarrow (1_a *_0 \gamma) *_1 (\gamma *_0 1_a) *_1 (1_a *_0 \gamma)$$

  which can be represented as

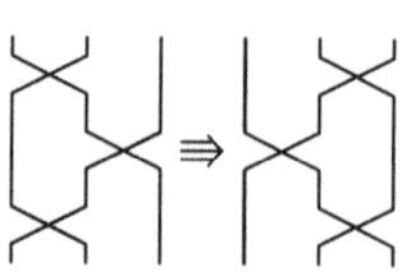

In this polygraph, it can be noted that, for any 2-cell $\phi\colon a^{m+1} \Rightarrow a^{n+1}$, the 2-cell

$$(\gamma *_0 a^{m+1}) *_1 (a *_0 \gamma *_0 a^m) *_1 (\gamma *_0 \phi) *_1 (a *_0 \gamma *_0 a^n) *_1 (\gamma *_0 a^{n+1})$$

of P_2^* can be rewritten in two ways using the rule Y:

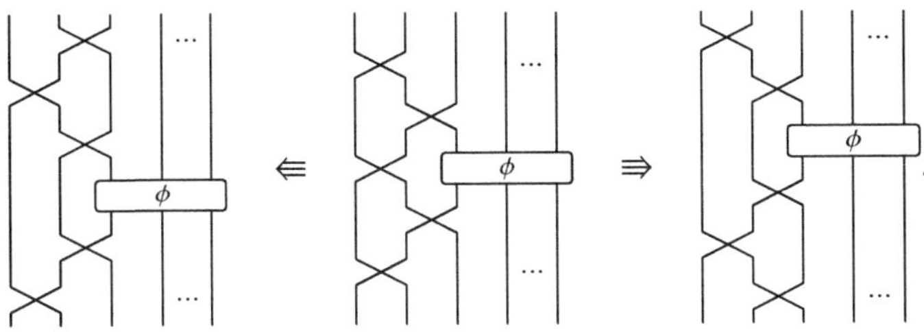

This gives rise to a critical branching when ϕ is either 1_a or of the form γ^{n+1} (the vertical composite of $n+1$ instances of γ). The critical branchings of the rewriting system are thus

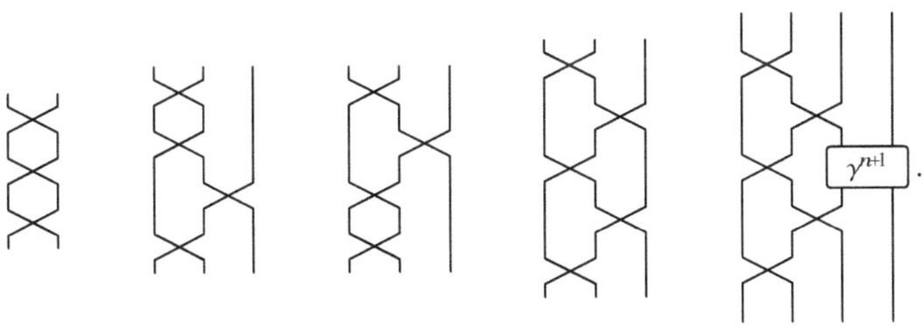

In the first case, the 2-cell can be rewritten by I in two different ways, in the second and third case, the 2-cells can be rewritten both by I and Y. The fourth and fifth cases can be rewritten by Y in two ways, as described above. A finite 3-polygraph can thus give rise to an infinite number of critical branchings. Still, they can be checked to be confluent. In the first four cases, this can be checked directly:

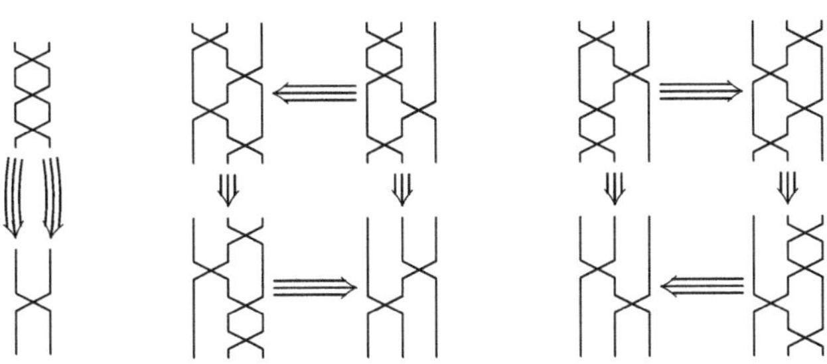

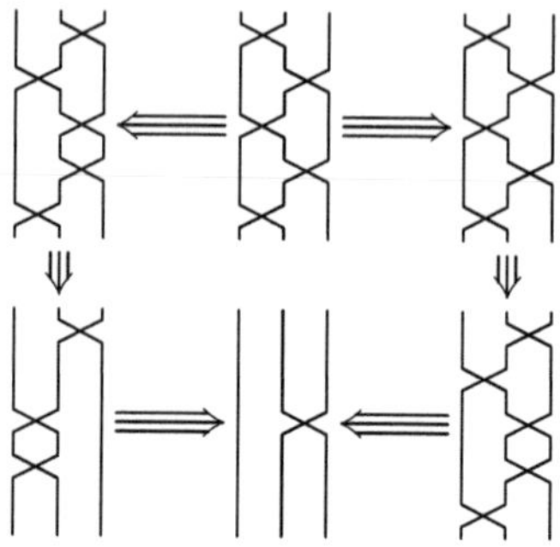

In the fifth case, we have, depending on whether n is odd or even:

$$\gamma^{n+1} \overset{*}{\Rrightarrow} 1_{a^2} \qquad \text{or} \qquad \gamma^{n+1} \overset{*}{\Rrightarrow} \gamma$$

and in the two cases the local branching is confluent:

– if $\gamma^{n+1} \overset{*}{\Rrightarrow} 1_{a^2}$:

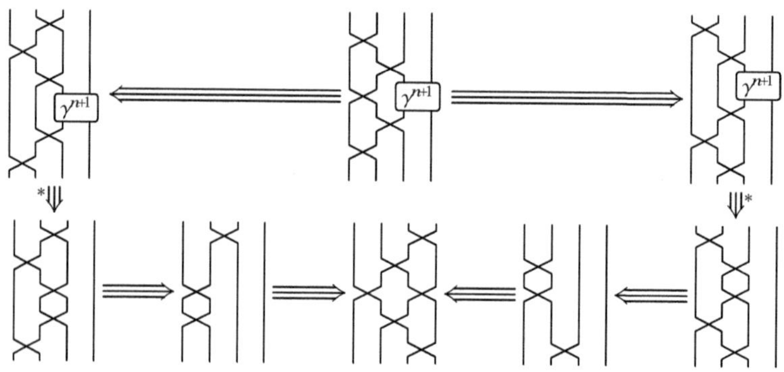

– if $\gamma^{n+1} \overset{*}{\Rrightarrow} \gamma$:

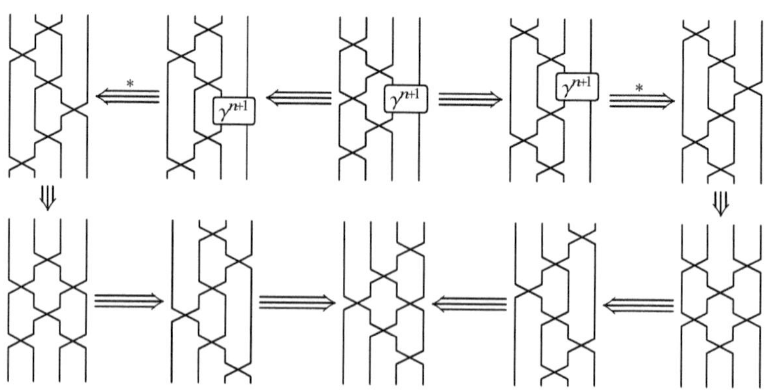

Finally, the rewriting system can be shown to be terminating (Examples 11.2.9 and 11.3.9) and thus convergent.

The normal forms can be characterized as follows. We first define, for every $n \in \mathbb{N}$, a 2-cell

$$\gamma_n : a^{n+1} \to a^{n+1}$$

by

$$\gamma_0 = 1_a \qquad\qquad \gamma_{n+1} = (\gamma_n *_0 a) *_1 (a^n *_0 \gamma).$$

This cell can be seen as a generalization of γ since $\gamma_1 = \gamma$, and γ_n is most naturally pictured as

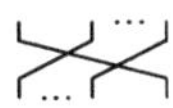

i.e., for low values of n,

$\gamma_0 =$ $\qquad \gamma_1 =$  $\qquad \gamma_2 =$  $\qquad \gamma_3 =$ .

The normal forms of the rewriting system can be characterized as follows, from which we can conclude that this is indeed a presentation of **S**.

10.4.2 Proposition. *A 2-cell ϕ is a normal form if and only if it is either*

- $\phi = 1_{a^0}$, *or*
- *there exists a normal form $\psi : n \to n$ and $m \in \mathbb{N}$ with $0 \leqslant m \leqslant n$ such that ϕ is*

  $$(a *_0 \psi) *_1 (\gamma_m *_0 a^{n-m}) : m + 1 \to m + 1$$

  what we write

  $$\phi = \Gamma_m \psi$$

  i.e., graphically,

  $\phi =$ 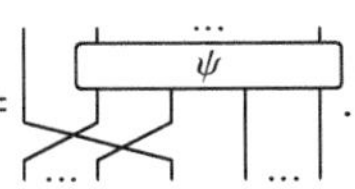.

Proof. We call *canonical forms* the 2-cells of the above form. Those are in normal form: this can be checked directly. Conversely, we show that any 2-cell ϕ rewrites to a canonical form by induction on its size. If the size of ϕ is 0 then ϕ is of the form 1_{a^n} and we have $\phi = \Gamma_0 \ldots \Gamma_0 \Gamma_0 1_{a^0}$ (with n occurrences of Γ_0). Otherwise, ϕ is of the form

$\phi = \phi' *_1 (a^i *_0 \gamma *_0 a^j) =$ 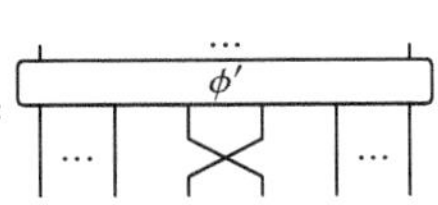

where ϕ' rewrites to a canonical form $\Gamma_m\psi$. Then depending on the respective values of m and i, four generic situations are possible, and in all of them ϕ rewrites to a canonical form (by using the induction hypothesis in the first and in the last case):

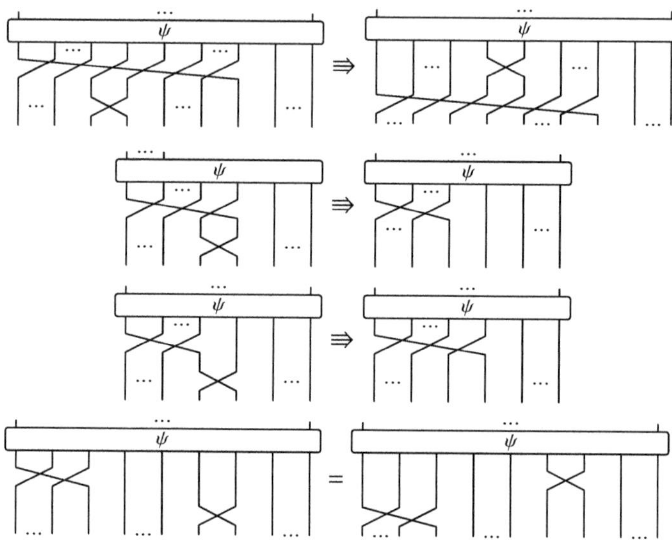

This concluded the proof. $\square$

Given a bijection $f\colon [n] \to [n]$ its *Lehmer code* is a sequence of n natural numbers $(k_0, k_1, \ldots, k_{n-1})$ such that $0 \leqslant k_i < n - i$ for every index i, where k_i is the cardinal of the set $\{j > i \mid f(j) < f(i)\}$. It can be shown that this induces a bijection between permutations of $[n]$ and such sequences, see [240, 248]. For instance, consider the permutation of [4] whose images are (2 0 4 3 1), pictured on the left:

(10.2)

The associated Lehmer code is $(2, 0, 2, 1, 0)$. We can finally conclude that the polygraph P defined in 10.4.1 is a presentation of the PRO of symmetries, see [235] for details.

10.4.3 Theorem. *The polygraph P is a presentation of* **S**.

Proof. Following the method described in Proposition 10.3.1, we interpret the morphism $\gamma\colon 2 \to 2$ as the transposition $[2] \to [2]$, and this interpretation is compatible with the two rules. It is easy to see that the interpretation of a normal form $\Gamma_{k_0}\dots\Gamma_{k_{n-1}}1_{a^0}$ is the bijection whose Lehmer code $(k_0, k_1, \dots, k_n)$. For instance, the normal form associated to the bijection (10.2) is $\Gamma_2\Gamma_0\Gamma_2\Gamma_1\Gamma_0 1_{a^0}$, which is pictured on the right of (10.2). This clearly establishes a bijection between 2-cells in normal form in $P^*_{\leqslant 2}$ and 2-cells of $\mathbf{S}$. □

10.4.4 Classification of critical branchings. Critical branchings in 3-polygraphs are classified in [161]. This case is more difficult than in dimension 2 mainly because, as initially noted in [235] and observed in previous section, a finite 3-polygraph may have an infinite number of critical branchings. However, an analysis of the possible shapes of these critical branchings yields a sufficient condition for confluence that only requires to consider a finite subset of them.

Assume that P is a 3-polygraph. By examination of the different possibilities, the critical branchings of P are classified as follows [161, Section 5.1.1].

1. *Inclusion* critical branchings, with the following source, if χ is the source of a 3-generator of P, and $\phi *_1 u\chi v *_1 \psi$ is the source of another one:

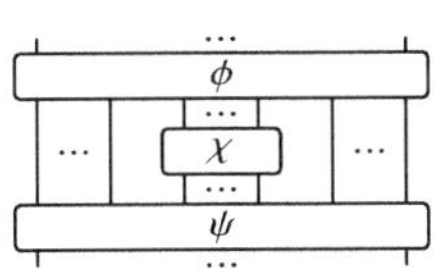

2. *Regular* critical branchings, with the following source, if $\phi *_1 u\chi$ and $\chi v *_1 \psi$ (or $\phi *_1 \chi v$ and $u\chi *_1 \psi$) are the sources of two 3-generators of P:

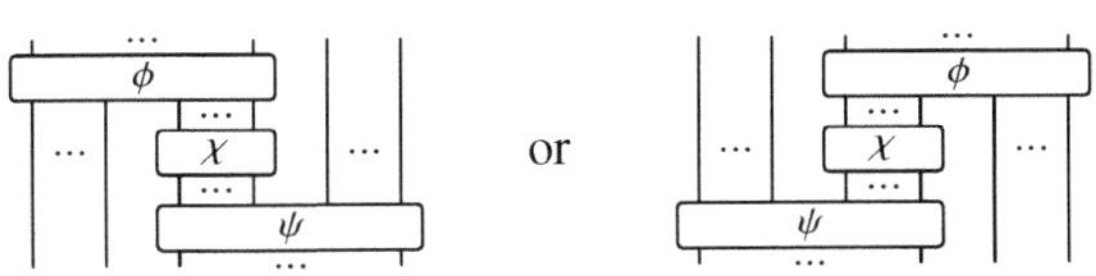

3. Instances of *left-indexed* critical branchings, with the following source, if $\phi *_1 u\chi$ and $v\chi *_1 \psi$ are the sources of two 3-generators of P, and $\zeta : wu \to xv$ is a 2-cell of P^*:

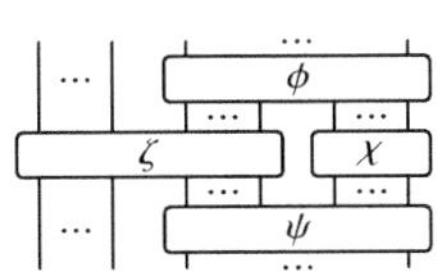

4. Instances of *right-indexed* critical branchings, with the following source, if $\phi *_1 \chi u$ and $\chi v *_1 \psi$ are the sources of 3-generators of P, and $\zeta : uw \Rightarrow vx$ is a 2-cell of P^*:

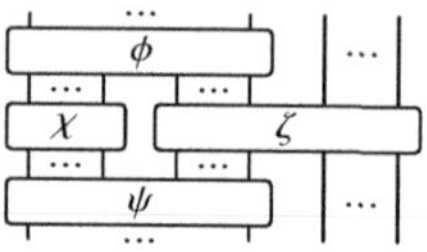

5. Instances of *multi-indexed* critical branchings, in all the other cases: one has a 3-generator with a source of the form

$$\phi *_1 (u_0 *_0 \chi_1 *_0 u_1 *_0 \chi_2 *_0 \cdots *_0 u_{n-1} *_0 \chi_n *_0 u_n)$$

and another 3-generator with a source of the form

$$(v_0 *_0 \chi_1 *_0 v_1 *_0 \chi_2 *_0 \cdots *_0 v_{n-1} *_0 \chi_n *_0 v_n) *_1 \psi$$

so that the source of the branching is of the form

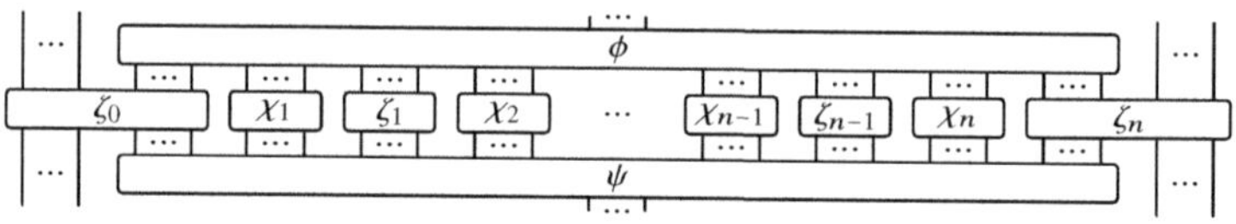

For example, in the presentation of §10.4.1, the four first branchings are regular whereas the generic family of branchings is right-indexed (by γ^{n+1}).

An *instance* of a left- or right-indexed branching is a left- or right-indexed branching as above with a particular value for the 2-cell ζ. It is a *normal instance* when ζ is in normal form.

10.4.5 Indexed polygraphs. We say that a 3-polygraph P is *non-indexed* if it has inclusion or regular critical branchings only, *left-indexed* (resp. *right-indexed*) if it has inclusion, regular or left-indexed (resp. right-indexed) critical branchings only, and *finitely indexed* if each of its indexed critical branchings has a finite number of reduced instances. Then we have the following results, which apply to the presentation of permutations in §10.4.1.

10.4.6 Proposition ([161, Proposition 5.1.3]). *If P has a finite number of 3-cells, then it has a finite number of inclusion and regular critical branchings.*

10.4.7 Proposition ([161, Proposition 5.3.1]). *If P is terminating and left-indexed (resp. right-indexed), then P is confluent if and only if all its inclusion and regular critical branchings, and all the reduced instances of its left-indexed (resp. right-indexed) critical branchings are confluent.*

10.5 Distributive Laws

The notions introduced in Section 3.3 for combining presentations using distributive laws generalize easily to presentations of monoidal categories as we now briefly explain, following [230].

10.5.1 Monoidal categories as monads. We have seen in §3.3.13 that a category corresponds precisely to a monad in **Span**(**Set**), and our aim is to generalize this situation to strict monoidal categories. The main starting point is that, in a strict monoidal category, the set of objects forms a monoid, with tensor as product.

The category **Mon** of monoids and their morphisms has small limits and it therefore makes sense to consider the bicategory **Span**(**Mon**) of spans internals to **Mon** as explained in §3.3.11: a 0-cell of this bicategory is a monoid, a 1-cell from A to B is a diagram of the form

$$A \longleftarrow C \longrightarrow B$$

in **Mon** and composition is given by pullback. It can then be observed that a monad in this bicategory (see §3.3.12) is precisely a monoidal category. In particular, a PRO corresponds to a monad on the monoid $(\mathbb{N}, +, 0)$. This point of view allows us to compose monoidal categories through distributive laws between the corresponding monads. We briefly present it below.

10.5.2 Distributive laws. Given two monoidal categories C and D with the same monoid of objects, a *distributive law* between them is a distributive law between the corresponding monads in **Span**(**Mon**), in the sense of §3.3.12. It consists of a distributive law

$$\ell \colon D \otimes C \to C \otimes D$$

in the sense of §3.3.2, between the underlying categories, which is compatible with tensor product in the sense that, for f_1, f_2, f'_1, f'_2 and g_1, g_2, g'_1, g'_2 morphisms of C and D respectively,

$$\begin{aligned} \ell(g'_1, f'_1) &= (f_1, g_1) \\ \ell(g'_2, f'_2) &= (f_2, g_2) \end{aligned} \qquad \text{implies} \qquad \ell(g'_1 \otimes g'_2, f'_1 \otimes f'_2) = (f_1 \otimes f_2, g_1 \otimes g_2).$$

10.5.3 Factorization systems. Given monoidal categories C, D, and E, with the same monoid of objects, we have $C \otimes_\ell D = E$ precisely when C and D form a *monoidal factorization system* for E, i.e., C and D are monoidal subcategories of E such that every morphism of E factorizes uniquely a morphism of C followed by a morphism of D.

In the case where C (resp. D) admits a presentation by a 3-polygraph P (resp. Q), with $P_0 = Q_0 = \{\star\}$ and $P_1 = Q_1$, the category E admits a presentation by the 3-polygraph R with

$$R_0 = \{\star\} \qquad R_1 = P_1 = Q_1 \qquad R_2 = P_2 \sqcup Q_2 \qquad R_3 = P_3 \sqcup Q_3 \sqcup R_3^\ell$$

where R_3^ℓ presents the distributive law similarly to Theorem 3.3.6.

10.5.4 Composing PROPs. Generalized composition of categories, as described in §3.3.17, also extends to this setting. A typical situation where this is useful is the case of PROPs: those are strict symmetric monoidal categories, with $(\mathbb{N}, +, 0)$ as monoid of objects. When composing two PROPs, one would like to identify the symmetric structures already present in both of them.

The category $\mathbf{S}$ of finite cardinals and bijections is the free PROP (this follows from the presentation constructed in §10.4.1, see also §C.1.4), and a PROP C can thus be seen as a monad on the additive monoid $\mathbb{N}$ in $\mathbf{Span}(\mathbf{Mon})$ equipped with a functor $\mathbf{S} \to C$, i.e., an object on the category on the left of (10.3) below. As explained in §3.3.20, we have an isomorphism

$$\mathbf{S}/\mathbf{Mon}(\mathbf{Span}(\mathbf{Mon})(\mathbb{N}, \mathbb{N})) \simeq \mathbf{Mon}(\mathbf{Mod}(\mathbf{Span}(\mathbf{Mon}))(\mathbf{S}, \mathbf{S})) \qquad (10.3)$$

which thus allows one to consider a PROP as a monad on bimodules of spans of monoids over $\mathbf{S}$, and two PROPs C and D can be composed along a distributive law

$$\ell \colon D \otimes_{\mathbf{S}} C \to C \otimes_{\mathbf{S}} D$$

between the corresponding monads: the composite defined in this way always gives rise to a PROP and identifies the symmetry structure in the composed PROPs as expected [230, 359]. When the two PROPs C and D are presented, one can obtain a presentation of their composite using a direct generalization of §3.3.21.

Similarly, Lawvere theories can be seen as monads in bimodules of spans of monoids over $\mathbf{F}$ and can be composed along distributive laws [84], see also §13.1.18.

10.5.5 Example. We write $\mathbf{F}$ (resp. $\mathbf{F}_\mu$, resp. $\mathbf{F}_\eta$) for the PROP of finite cardinals and functions (resp. surjective functions, resp. injective functions), see §C.2. The PROPs $\mathbf{F}_\mu$ and $\mathbf{F}_\eta$ are subcategories of $\mathbf{F}$, any morphism $h \in \mathbf{F}$ factorizes as $h = g \circ f$ with $f \in \mathbf{F}_\mu$ and $g \in \mathbf{F}_\eta$, and for any other factorization $h = g' \circ f'$ there exists a permutation $w \in \mathbf{S}$ making the following diagram commute:

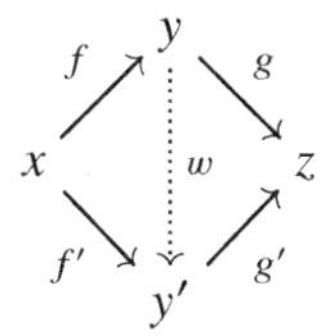

i.e., we have $\mathbf{F} = \mathbf{F}_\mu \otimes_\ell \mathbf{F}_\eta$. The categories $\mathbf{F}_\mu$ and $\mathbf{F}_\eta$ respectively admit presentations by the polygraphs P and Q with generators

$P_0 = Q_0 = \{\star\}$ $\quad P_1 = Q_1 = \{a\}$ $\quad P_2 =$ 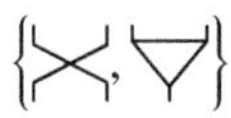 $\quad Q_2 =$ 

where the relations in P_3 are

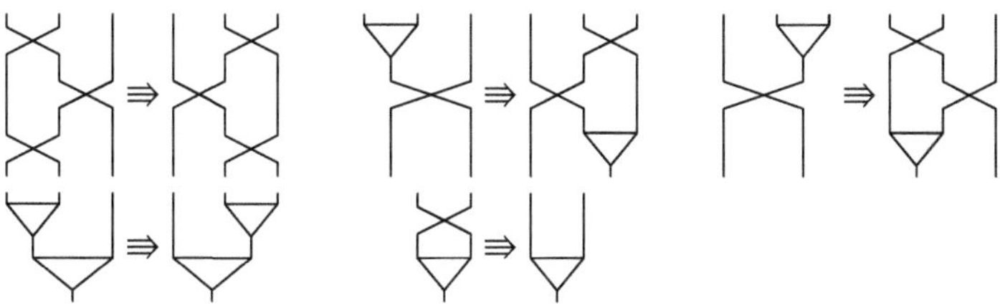

and the relations in Q_3 are

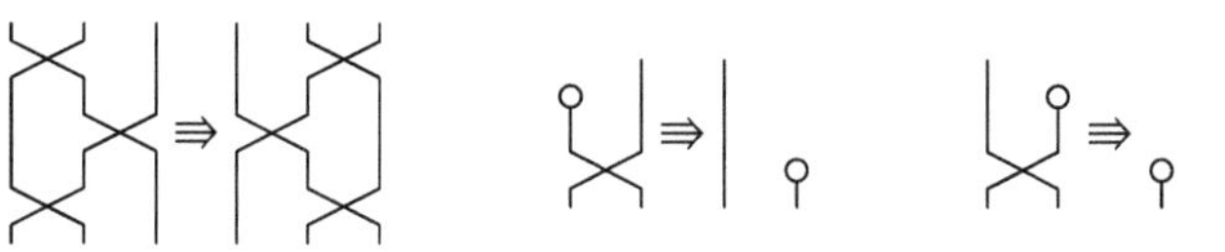

From those, we deduce the presentation of $\mathbf{F}$ by the polygraph R with

$R_0 = \{\star\}$ $\quad R_1 = \{a\}$ $\quad R_2 = P_2 \sqcup Q_2 =$ 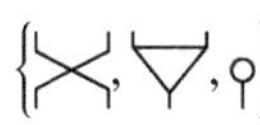

and the relations are

– the common relation for symmetry:

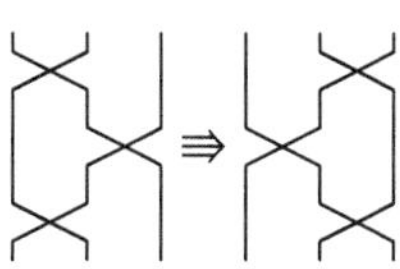

– the relations of P:

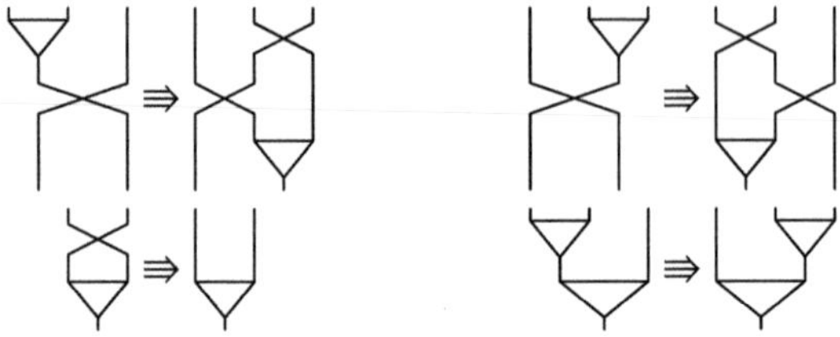

– the relations of Q:

– the relations generated by the distributive law:

11

Termination of 3-Polygraphs

This chapter presents techniques for proving the termination of 3-polygraphs, generalizing those already introduced for 2-polygraphs in Section 4.4. A first method, described in Section 11.1, is based on a certain type of well-founded orders called *reduction orders*. We then turn in Section 11.2 to *functorial interpretations*: these amount to construct a functor from the underlying category to another category which already bears a reduction order. This covers quite a few useful examples. To address more complex cases, we present in Section 11.3 a powerful technique, due to Guiraud [158, 161], based on the construction of a derivation from the polygraph. Here, termination is obtained by specifying quantities on 2-cells which decrease during rewriting, based on information propagated by the 2-cells themselves.

11.1 Reduction and Termination Orders

We begin by extending the notion of reduction order introduced in Section 4.4 from 2-polygraphs to 3-polygraphs.

11.1.1 Definition. Given a 2-category C, a *reduction order* is a partial order $\geqslant$ on pairs of parallel 2-cells which is

- well-founded: every weakly decreasing sequence of 2-cells is eventually stationary,
- compatible with 0-composition: for every 2-cells

$$\phi\colon u \Rightarrow u' : x' \to x \qquad \psi_1, \psi_2\colon v \Rightarrow v' : x \to y \qquad \phi' : w \Rightarrow w' : y \to y'$$

which can be represented as

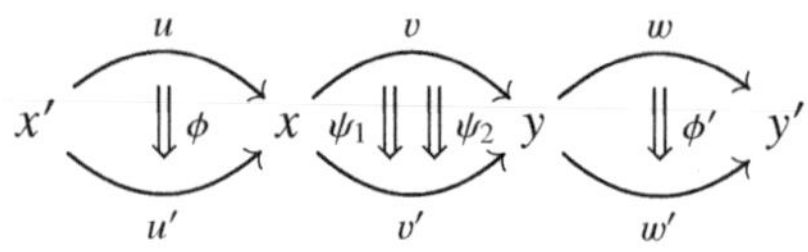

we have that

$$\psi_1 > \psi_2 \qquad \text{implies} \qquad \phi *_0 \psi_1 *_0 \phi' > \phi *_0 \psi_2 *_0 \phi',$$

– compatible with 1-composition: for every 2-cells

$$\phi\colon u' \Rightarrow u\colon x \to y \qquad \psi_1, \psi_2\colon u \Rightarrow v\colon x \to y \qquad \phi'\colon v \Rightarrow v'\colon x \to y$$

which can be represented as

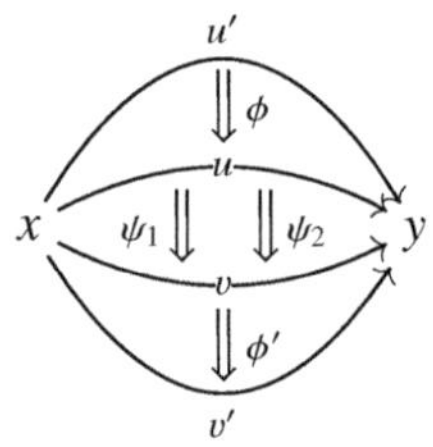

we have that

$$\psi_1 > \psi_2 \qquad \text{implies} \qquad \phi *_1 \psi_1 *_1 \phi' > \phi *_1 \psi_2 *_1 \phi'.$$

Given a 3-polygraph P, a reduction order $\geqslant$ on the 2-category P_2^* is said to be *compatible* with the rules of P when $\phi > \phi'$ for every rule $A\colon \phi \Rrightarrow \phi'$ in P_3. In this case, the order $\geqslant$ is called a *termination order* for P.

In an arbitrary 3-polygraph P, we write $\Rrightarrow^*$ for the relation on parallel 2-cells of P_2^* such that $\phi \Rrightarrow^* \psi$ whenever ϕ rewrites to ψ, or equivalently whenever there is a 3-cell $F\colon \phi \Rrightarrow \psi$ in P_3^*.

11.1.2 Proposition. *Given a 3-polygraph P, the following statements are equivalent.*

1. *The polygraph P is terminating.*
2. *The relation $\Rrightarrow^*$ is a termination order.*
3. *The polygraph P admits a termination order.*

Proof. Similar to the proof of Proposition 4.4.2. □

11.2 Functorial Interpretations

The main criterion for showing the termination of a 3-polygraph is given by constructing a suitable interpretation of its 2-cells in 2-categories for which a reduction order is known. This generalizes the technique of reduction functions introduced in §4.4.5.

11.2.1 Proposition. *Let P be a 3-polygraph. The following statements are equivalent.*

1. *The 3-polygraph P terminates.*
2. *There exists a 2-category C, equipped with a reduction order $\geqslant$, and a 2-functor $[-] : P_2^* \to C$, such that $[\phi] > [\psi]$ for every 3-generator $A\colon \phi \Rrightarrow \psi$ in P_3.*

Proof. $1 \Rightarrow 2$. If the polygraph P terminates, we can take $C = P_2^*$ and $[-] = 1_{P_2^*}$. By Proposition 11.1.2, taking $\Rrightarrow^*$ for $\geqslant$ gives a termination order for P, hence a reduction order on C such that $[\phi] > [\psi]$ for every 3-generator $A\colon \phi \Rrightarrow \psi$ in P_3.

$2 \Rightarrow 1$. Conversely, $[-]$ being a functor, and $\geqslant$ being compatible with the compositions of C imply that, for every rewriting step $F\colon \phi \Rrightarrow \psi$, we have $[f] > [g]$ in C. Now, assume that P does not terminate. Then there exists an infinite sequence of composable rewriting steps in P, yielding an infinite decreasing sequence for $\geqslant$, which is excluded by well-foundedness of $\geqslant$. □

11.2.2 The number of generators. A first very simple situation in which a 3-polygraph terminates is when each rewriting rule (and thus each rewriting step) decreases the number of 2-generators in 2-cells (we recall that the number $|\phi|$ of 2-generators in a 2-cell ϕ was formally defined in §10.2.1). Namely, the usual order $\geqslant$ on natural numbers is a reduction order because it is well-founded and addition is strictly increasing. Thus, the following statement is a direct application of Proposition 11.2.1.

11.2.3 Proposition. *A 3-polygraph P such that $|\phi| > |\phi'|$ for every rewriting rule $A\colon \phi \Rrightarrow \phi'$ in P_3, is terminating.*

11.2.4 Example. Consider the 3-polygraph Mon of monoids described in Examples 10.1.2 and 10.2.10, and write Mon′ for the polygraph obtained from Mon by removing the rewriting rule A corresponding to associativity. For both rewriting rules L and R, the number of generators of the source is 2 and the number of generators of the target is 0. By Proposition 11.2.3, the polygraph Mon′ is thus terminating.

However, in the rule A, the number of generators in the source and in the target are both 2. Therefore the above proposition does not apply to Mon. We see below a more general method which is able to handle this case, see Example 11.2.8.

11.2.5 A 2-category of posets. We define a 2-category **Ord** with one 0-cell $\star$, 1-cells are posets, and 2-cells are weakly increasing functions. The composition of two 1-cells $(X, \leqslant_X)$ and $(Y, \leqslant_Y)$ is given by their cartesian product $(X \times Y, \leqslant_{X\times Y})$, where the product order is such that $(x, y) \leqslant_{X\times Y} (x', y')$ if and only if $x \leqslant_X x'$ and $y \leqslant_Y y'$. Likewise, the horizontal composition of 2-cells is given by their cartesian product, and their vertical composition is the usual composition of functions.

We write $\succcurlyeq$ for the pointwise order on 2-cells: given $f, g\colon X \Rightarrow Y$, we have $f \succ g$ if and only if $f(x) > g(x)$ for every $x \in X$. This order is always compatible with horizontal and vertical composition of 2-cells.

11.2.6 Lemma. *Given posets X and Y such that X is non-empty and Y is well-founded, the order $\succcurlyeq$ on functions $X \Rightarrow Y$ is well-founded.*

Proof. Fix an arbitrary element x in X, which is supposed to be non-empty. An infinite strictly decreasing sequence of functions $f_0 \succ f_1 \succ \ldots$ would induce an infinite strictly decreasing sequence $f_0(x) > f_1(x) > \ldots$ of elements of Y. This is excluded by well-foundedness of Y. □

This provides us with the following technique for showing termination of polygraphs, first considered in [235]:

11.2.7 Proposition. *Let P be a 3-polygraph and $(X, \leqslant)$ a non-empty well-founded poset. Suppose that to each 2-generator $\alpha\colon u \Rightarrow v$ in P_2 whose source has length $m = |u|$ and target has length $n = |v|$ is assigned a strictly increasing function $[\alpha] : X^m \to X^n$, and write $[-] : P_2^* \to$ **Ord** for the induced 2-functor. If, for every rewriting rule $A\colon \phi \Rrightarrow \psi$ in P_3, we have $[\phi] \succ [\psi]$ then the polygraph P terminates.*

Proof. Write C for the full sub-2-category of **Ord** whose 1-cells are the powers X^n of the poset X. The order on X^n is well-founded as a product of well-founded orders, and therefore the induced order $\succcurlyeq$ on 2-cells $X^m \Rightarrow X^n$ is also well-founded by Lemma 11.2.6. For every 2-cell ϕ, we have that $[\phi]$ is a strictly increasing function and thus the order is compatible with composition of 2-cells in the sense of Definition 11.1.1, it is thus a reduction order on C. We conclude with Proposition 11.2.1. □

11.2.8 Example. Consider the 3-polygraph Mon of monoids introduced in Example 10.1.5. Graphically, its rules are

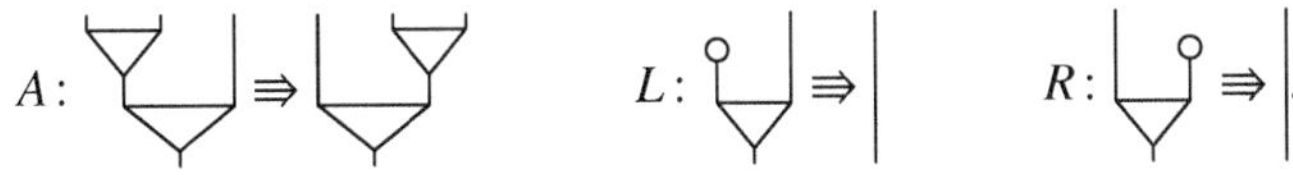

Writing $X = \mathbb{N} \setminus \{0\}$ equipped with the usual order, we define

By Proposition 11.2.7, the following strict inequalities imply that the 3-polygraph Mon terminates:

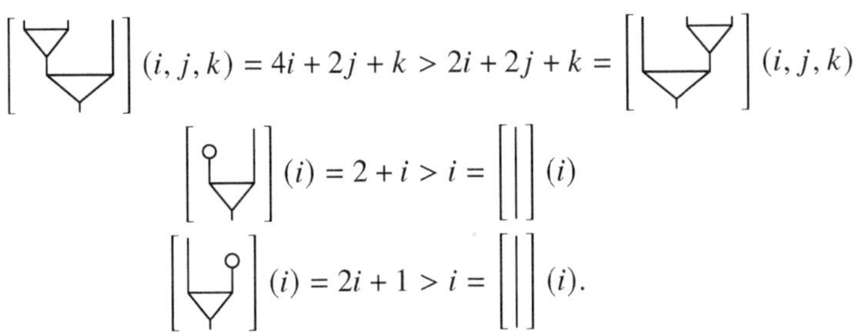

11.2.9 Example. Consider the 2-polygraph of permutations introduced in §10.4.1. It has one 0-generator $\star$, one 1-generator a, one 2-generator

and its rules are

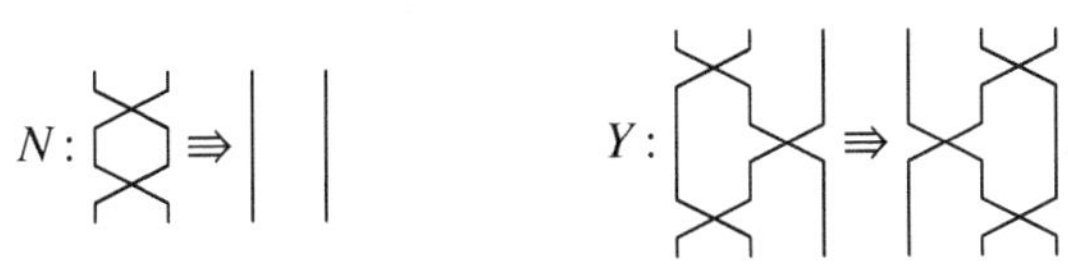

We consider the well-founded poset $X = \mathbb{N} \setminus \{0\}$ equipped with the usual order and consider the interpretation

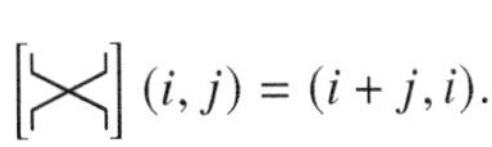

By Proposition 11.2.7, we ensure that the polygraph is terminating by showing that the rewriting rules are strictly decreasing:

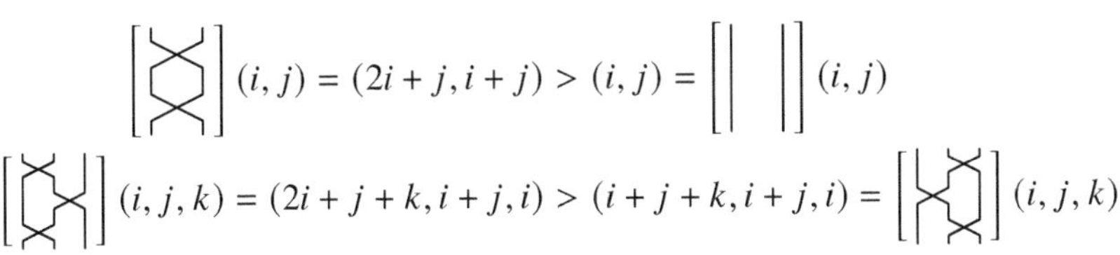

11.2.10 Example. By combining the two previous examples, one can construct a convergent rewriting system corresponding to the theory of commutative monoids (see §C.2.5) which is terminating [235].

11.2.11 Remark. Consider a 3-polygraph P containing a 3-generator

$$A : \phi \Rrightarrow \psi : u \Rightarrow v : x \to y$$

such that $|v| = 0$, i.e., $x = y$ and $v = 1_x$. Then X^0 is reduced to one element and we cannot have $[\phi] > [\psi]$. Therefore, in this case, we cannot use Proposition 11.2.7 to show the termination of the polygraph.

11.2.12 Remark. Proposition 11.2.7 can be generalized by taking a possibly different well-founded poset $X(a)$ for each 1-generator a of the 3-polygraph P. In that case, the interpretation of each 2-generator $\alpha : u \to v$ is replaced by an increasing map $[\alpha] : X(u) \to X(v)$, where X is extended to a 1-functor from the free 1-category P_1^* to the underlying 1-category of **Ord**.

11.3 Termination by Derivations

The above technique is often not applicable for rewriting systems such that some 2-generators have multiple outputs (with the notable exception of Example 11.2.9). We present here another technique for showing the termination of 3-polygraphs, which is due to Guiraud [158, 161] and is based on the following intuition. As suggested by the string diagrammatic representation, we can think of a 2-cell in a polygraph as some kind of electric circuit built from basic components, the 2-generators. For each of those components, we are going to specify how the current is transmitted from inputs to outputs (in both directions, from top to bottom and from bottom to top), as well as how much heat it emits when the current flows through. Finally, if the rewriting rules are such that rewriting a circuit strictly decreases its heat, and the heats are taken in a well-founded order, then we will be able to conclude that the rewriting system is terminating.

11.3.1 The category of contexts. Given a 2-polygraph P, we write $\mathcal{K}_P$ for the *category of contexts* of P. The objects of this category are the 2-cells of P_2^* and a morphism from a 2-cell $\phi : u \Rightarrow v : x \to y$ to a 2-cell $\phi' : u' \Rightarrow v' : x' \to y'$ is a context K of type (u, v), as defined in §10.2.2, such that $K[\phi] = \phi'$.

11.3.2 Natural system. A *natural system* on a 2-polygraph P is a functor $N : \mathcal{K}_P \to \mathbf{Ab}$, associating an abelian group N_ϕ to every 2-cell $\phi \in P_2^*$ and a morphism of groups $N_K : N_\phi \to N_{\phi'}$ to every context K such that $K[\phi] = \phi'$. This notion is a 2-categorical variant of the one already encountered in §8.3.1.

By abuse of notation, given an element $n \in N_\phi$ and a context K of suitable type, we sometimes write $K[n]$ instead of $N_K(n)$. Moreover, given i-composable morphisms ϕ and ϕ', for $i \in \{0, 1\}$, and elements $n \in N_\phi$ and $n' \in N_{\phi'}$, we often write $n *_i \phi'$ (resp. $\phi *_i n'$) instead of $N_{K'}(n)$ (resp. $N_K(n')$) where K' is the context $X *_i \phi'$ (resp. K is the context $\phi *_i X$).

11.3.3 Derivation. Given a natural system N on a 2-polygraph P, a *derivation* d of P into N is a function which to every 2-cell ϕ in P_2^* associates an element of the group N_ϕ, in such a way that

$$d(\phi *_i \psi) = d(\phi) *_i \psi + \phi *_i d(\psi)$$

for suitably i-composable 2-cells ϕ and ψ in P_2^*, with $i \in \{0, 1\}$. Note that such a derivation d is uniquely determined by the images $d(\alpha) \in N_\alpha$ of the 2-generators $\alpha \in P_2$.

11.3.4 Lemma. *Given a derivation* d *as above and a* 1*-cell* $u \in P_1^*$, *we have* $d(1_u) = 0$.

Proof. We have

$$d(1_u) = d(1_u *_1 1_u) = d(1_u) *_1 1_u + 1_u *_1 d(1_u) = d(1_u) + d(1_u)$$

from which we conclude. □

11.3.5 Example. We write $Z : \mathcal{K}_P \to \mathbb{Z}$ for the *trivial natural system* which sends every object to $\mathbb{Z}$ and every morphism to the identity on $\mathbb{Z}$. Fix a 2-generator α in P_2. The operation introduced in §10.2.1, which to a 2-cell ϕ in P_2^* associates the number $|\phi|_\alpha$ of occurrences of α in ϕ, is the derivation of P into the trivial natural system such that $|\alpha|_\alpha = 1$ and $|\beta|_\alpha = 0$ for $\beta \in P_2$ such that $\beta \neq \alpha$.

11.3.6 A natural system of interest. Suppose fixed

- a 2-functor $X : P_2^* \to \mathbf{Ord}$,
- a 2-functor $Y : (P_2^*)^{\mathrm{co}} \to \mathbf{Ord}$, where $(P_2^*)^{\mathrm{co}}$ is the 2-category obtained from P_2^* by formally changing the direction of all 2-cells,
- a commutative monoid $(M, +, 0)$ whose addition is strictly increasing.

We define a natural system $N : \mathcal{K}_P \to \mathbf{Ab}$ as follows.

- To every 2-cell $\phi : u \Rightarrow v$ in P_2^*, N associates the monoid N_ϕ of functions $X_u \times Y_v \to M$ with addition being induced pointwise by the one in M.

– For every 2-cell $\phi\colon u \Rightarrow v\colon x \to y$ in P_2^*, and every pair of 1-cells $w\colon x' \to x$, $w'\colon y \to y'$ as in

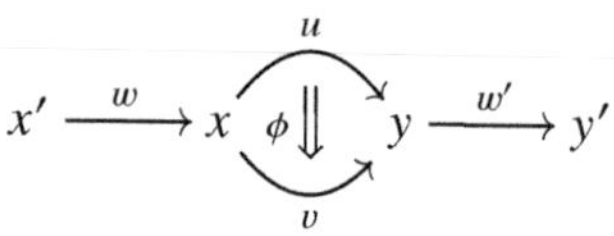

the image of the context $K = w *_0 - *_0 w'$ is the group morphism

$$N_K\colon N_\phi \to N_{w *_0 \phi *_0 w'}$$

which sends a function

$$f\colon X_u \times Y_v \to M$$

to the function

$$N_K(f)\colon X_w \times X_u \times X_{w'} \times Y_w \times Y_v \times Y_{w'} \to M$$

obtained by precomposing f with the canonical projection

$$X_w \times X_u \times X_{w'} \times Y_w \times Y_v \times Y_{w'} \to X_u \times Y_v.$$

– For every 2-cell $\phi\colon u \Rightarrow v$ and 2-cells $\psi\colon u' \Rightarrow u$ and $\psi'\colon v \Rightarrow v'$ as in

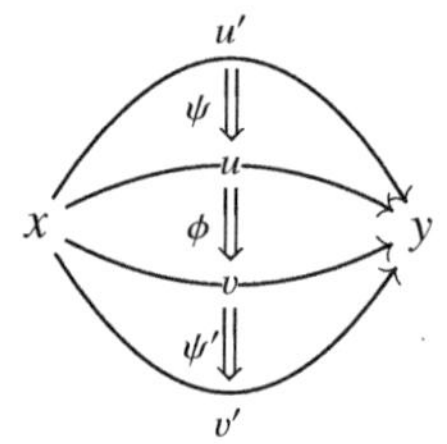

the image of the context $K = \psi *_1 - *_1 \psi'$ is the group morphism

$$N_K\colon N_\phi \to N_{\psi *_1 \phi *_1 \psi'}$$

which sends a function

$$f\colon X_u \times X_v \to M$$

to the function

$$N_K(f)\colon X_{u'} \times Y_{v'} \to M$$

defined by

$$N_K(f) = f \circ (X_\psi \times Y_{\psi'}).$$

The above conditions entirely determine the derivation N, which we often denote by $O(X, Y, M)$ to make clear the dependency on X, Y, and M. Note that such a natural system is entirely determined by the data of

- the posets X_a and Y_a for every a in P_2,
- the functions $X_\alpha : X_u \to X_v$ and $Y_\alpha : Y_v \to Y_u$ for every 2-generator $\alpha : u \Rightarrow v$ in P_2,

where $X_{a_1 \ldots a_n} = X_{a_1} \times \ldots \times X_{a_n}$.

11.3.7 Remark. We could generalize the definition to the case where X and Y are objects in an arbitrary cartesian category and M is a commutative monoid internal to this category.

Given a 2-cell $\phi : u \Rightarrow v$ in P_2^*, the monoid N_ϕ is canonically equipped with the order such that, for elements $f, g : X_u \times Y_v \to M$ of N_ϕ we have

$$f > g \qquad \text{if and only if} \qquad f(x, y) > g(x, y) \text{ for every } (x, y) \in X_u \times Y_v.$$

The above construction was introduced by Guiraud [158, 159, 161]. It is the basis of the following useful termination criterion.

11.3.8 Theorem. *Consider a 3-polygraph P. Suppose given*

- *two 2-functors $X : P_2^* \to$ **Ord** and $Y : (P_2^*)^{\mathrm{co}} \to$ **Ord** such that for every 1-generator $a \in P_1$ the posets X_a and Y_a are non-empty, and $X_\phi \geqslant X_\psi$ and $Y_\phi \geqslant Y_\psi$ for every 3-generator $A : \phi \Rrightarrow \psi$ in P_3,*
- *a well-founded partially ordered commutative monoid $(M, +, 0)$ such that addition is strictly increasing,*
- *a derivation d from the underlying 2-polygraph of P to $O(X, Y, M)$ such that $d(\phi) > d(\psi)$ for every 3-generator $A : \phi \Rrightarrow \psi$ in P_3.*

Then the polygraph P is terminating.

Proof. Let $K[\alpha] : \chi \Rrightarrow \chi'$ be a rewriting step for some 2-generator $\alpha : \phi \Rrightarrow \phi'$ in P_2 and context $K \in \mathcal{K}_P$. By Lemma 10.2.3, the context K can be written in the form

$$K = \psi *_1 (w *_0 - *_0 w') *_1 \psi'$$

for suitably typed 1-cells w and w' in P_1^* and 2-cells ψ and ψ' in P_2^*. Using the definition of derivations, see §4.4.13 and Lemma 11.3.4, we have that $d(K[\chi])$ is equal to

$$d(\psi) *_1 (w *_0 \chi *_0 w') *_1 \psi' + \psi *_1 (w *_0 d(\chi) *_0 w') *_1 \psi' + \psi *_1 (w *_0 \chi *_0 w') *_1 d(\psi')$$

and similarly for $d(K[\chi'])$. By hypothesis, we have $d(\chi) > d(\chi')$, and therefore

$$\psi *_1 (w *_0 d(\chi) *_0 w') *_1 \psi' > \psi *_1 (w *_0 d(\chi) *_0 w') *_1 \psi'.$$

Moreover, since X and Y are decreasing on generators, by functoriality of X and Y we have $X_\chi \geqslant X_{\chi'}$, and thus

$$d(\psi) *_1 (w *_0 \chi *_0 w') *_1 \psi' \geqslant d(\psi) *_1 (w *_0 \chi' *_0 w') *_1 \psi'$$

and similarly

$$\psi *_1 (w *_0 \chi *_0 w') *_1 d(\psi') \geqslant \psi *_1 (w *_0 \chi' *_0 w') *_1 d(\psi').$$

Finally, since addition is strictly increasing, we deduce

$$d(K[\chi]) > d(K[\chi']).$$

An infinite sequence of rewriting steps starting from a 2-cell $\phi\colon u \Rightarrow v$, would thus induce a strictly decreasing sequence

$$f_0 > f_1 > f_2 > \dots$$

of elements of $d(\phi)$, i.e., functions $X_u \times Y_v \to M$. Since X_u and Y_v are supposed to be non-empty, we can pick an element $(x, y) \in X_u \times Y_v$, and we would have a strictly decreasing sequence

$$f_0(x, y) > f_1(x, y) > f_2(x, y) > \dots$$

of elements of M, which is excluded by hypothesis. The polygraph P is thus terminating. □

In order to give some intuition, let us consider a 3-polygraph P. A 2-generator

$$\alpha\colon a_1 a_2 \dots a_m \Rightarrow b_1 b_2 \dots b_n$$

in P_2, where the a_i and b_i are 1-generators in P_1, can be seen as an operation with m inputs and n outputs

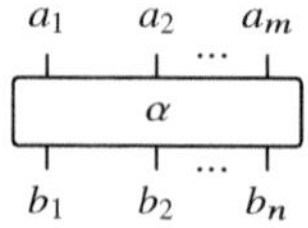

which, as the figure suggests, can be thought of as a building piece of some electrical circuit. The poset X_{a_i} (resp. Y_{a_i}, X_{b_i}, Y_{b_i}) is the set of possible values for the currents flowing into a_i (resp. out from a_i, out from b_i, into b_i). The function

$$X_\alpha\colon X_{a_1} \times \dots \times X_{a_m} \to X_{b_1} \times \dots \times X_{b_n}$$

then indicates, given currents flowing into the inputs a_i, what currents we get from the outputs b_i. Similarly, the function

$$Y_\alpha : Y_{b_1} \times \cdots \times Y_{b_n} \to Y_{a_1} \times \cdots \times Y_{a_m}$$

indicate the current we obtain from the a_i if we use the device "upside down" and flow currents into the b_i. Finally, the monoid M can be thought of as the possible values for "heat" emitted by our electrical circuit and the function

$$d(\alpha) : X_{a_1} \times \cdots \times X_{a_m} \times Y_{b_1} \times \cdots \times Y_{b_n} \to M$$

indicates, given currents flowing into the a_i and into the b_i, the heat that our circuit produces. The fact that it is a derivation amounts to impose that the heat produced by a circuit is the sum of the heat emitted by its components. Finally, the hypotheses of Theorem 11.3.8 ensure that rewriting a circuit will always transform it into a "colder" (i.e., less heat-emitting) circuit.

11.3.9 Example. Following [161, Section 5.4], let us apply the above technique to show that the rewriting system for permutations, already considered in Example 11.2.9, is terminating. We suppose here that

- the poset X_a associated to the 1-generator a is $\mathbb{N}$ equipped with the usual order, and X is defined on the 2-generator by

  X  $(i, j) = (j + 1, i)$,

- $Y_a = \{*\}$ is the terminal poset (reduced to one element $*$), and Y is defined on the 2-generator by

  Y 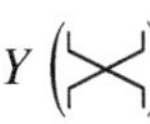 $(*, *) = (*, *)$,

- the monoid M is the additive monoid $\mathbb{N}$,
- the derivation d is defined on the 2-generator by

  d 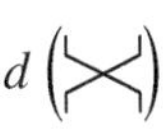 $(i, j) = i$

  (more precisely, the derivation takes four arguments (i, j, k, l), but the two last arguments are necessarily equal to $*$, the only element of Y_a, and are thus omitted).

The rewriting rules make X weakly decrease

X  $(i, j) = (i + 1, j + 1) \geqslant (i, j) = X$ 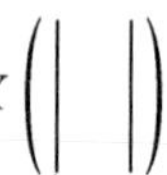 (i, j)

X 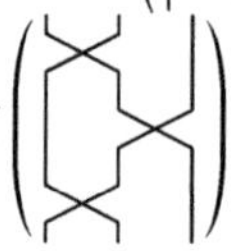 $(i, j, k) = \quad (k + 2, j + 1, i) \quad = X$ 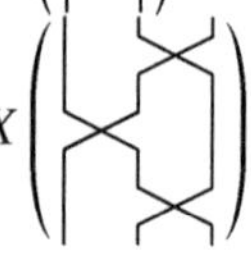 (i, j, k)

as well as obviously Y, and make the derivation strictly decrease

d 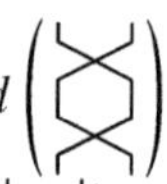 $(i, j) = i + j + 1 > \quad 0 \quad = d$ 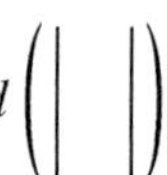 (i, j)

d 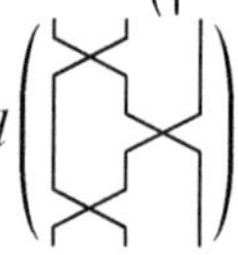 $(i, j, k) = 2i + j + 1 > 2i + j = d$ 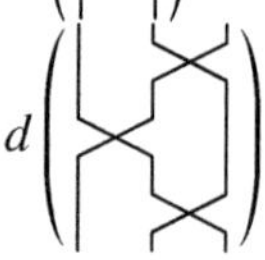 $(i, j, k).$

By Theorem 11.3.8, the 3-polygraph is thus terminating.

A presentation which cannot be handled with the techniques of Section 11.2, see Remark 11.2.11, but can be handled with derivations is given in §12.2.8.

12

Coherent Presentations of 2-Categories

In this chapter, we generalize the definitions and results of Chapter 7 from categories to 2-categories, following [161, 162]. In Section 12.1, we introduce the notion of coherent presentation of a 2-category by a $(4, 2)$-polygraph, where the 4-generators encode the relations among relations. We explain in Section 12.2 that, in the case of convergent polygraphs, we can construct the Squier completion, which is a coherent completion whose 4-generators come from confluence diagrams for the critical branchings. We show in §12.2.8 that, contrarily to the case of categories, a 2-category presented by a finite convergent 3-polygraph is not necessarily of finite derivation type. In Section 12.3, we develop a 3-dimensional generalization of the notion of PRO, for which coherent presentations can be given by $(4, 2)$-polygraphs. This allows us, in Section 12.4, to use the constructions of coherent presentations to obtain coherence results such as Mac Lane's coherence theorem for monoidal categories, as well as generalizations to symmetric and braided monoidal categories in Section 12.5.

12.1 Coherent Presentations of 2-Categories

12.1.1 (4, 2)-Polygraphs. A $(4, 2)$-*polygraph* is a pair (P, P_4) consisting of a 3-polygraph P, and a cellular extension P_4 of the free 3-category $P^\top$ over P. It thus consists of a diagram of sets and functions

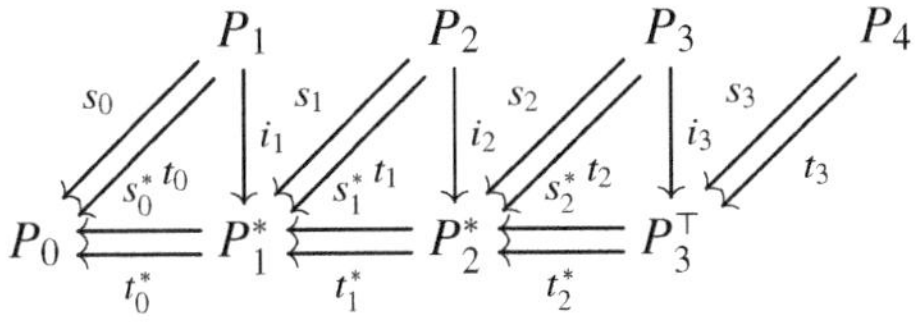

together with the compositions and identities of the underlying $(3,2)$-category

$$P_0 \underset{t_0^*}{\overset{s_0^*}{\Longleftarrow}} P_1^* \underset{t_1^*}{\overset{s_1^*}{\Longleftarrow}} P_2^* \underset{t_2^*}{\overset{s_2^*}{\Longleftarrow}} P_3^{\top},$$

whose *source* and *target maps* s_i and t_i satisfy the globular relations

$$s_i^* \circ s_{i+1} = s_i^* \circ t_{i+1} \qquad \text{and} \qquad t_i^* \circ s_{i+1} = t_i^* \circ t_{i+1},$$

for every $0 \leqslant i \leqslant 2$. The elements of the cellular extension P_4 are called the 4-*generators* of the polygraph P. We write $\Lambda\colon F \Rrightarrow G$ for a 4-generator Λ in P_4 such that $s_2(\Lambda) = F$ and $t_2(\Lambda) = G$. Given a 4-polygraph P, we write $P_{\leqslant 3}$ for its underlying 3-polygraph.

12.1.2 Coherence. A $(4,2)$-polygraph P is *coherent* when for any parallel 3-cells $F, G\colon \phi \Rrightarrow \psi$ of $P_{\leqslant 3}^{\top}$, the free $(3,2)$-category generated by the underlying 3-polygraph of P, we have $F \simeq^P G$, i.e., F and G are related by the congruence generated by the cells in P_4.

Given a $(3,2)$-category C a cellular extension X of C is *acyclic* when $F \simeq^X G$ holds for every pair of parallel 3-cells F and G of C. Here $\simeq^X$ is the congruence generated by X (which is defined in the expected way, generalizing the definition of §7.1.2). Given a $(3,2)$-polygraph P, a $(4,2)$-polygraph (P, P_4) is thus coherent precisely when P_4 is an acyclic extension of $P^{\top}$.

12.1.3 Polygraphs of finite derivation type. The property of finite derivation type defined in Chapter 8 for 2-polygraphs is extended to 3-polygraphs as follows. One says that a 3-polygraph P has *finite derivation type* when it is finite and when the free $(3,2)$-category $P^{\top}$ admits a finite acyclic cellular extension. As in the case of presentation of 1-categories, given two presentations of the same 2-category by finite 3-polygraphs, the following result proves that both have finite derivation type or none at all.

12.1.4 Theorem. *Let P and Q be two Tietze equivalent 3-polygraphs such that P_2 and Q_2 are finite. Then P has finite derivation type if and only if Q has finite derivation type.*

Proof. The proof is similar to the one given in the case of 1-categories by Theorem 8.1.2. □

As a consequence of Theorem 12.1.4, one can say that a 2-category has *finite derivation type* when it admits a presentation by a 3-polygraph having finite derivation type. The property of having finite derivation type is invariant by Tietze equivalence for finite 3-polygraphs. This is not the case for infinite ones as shown by the following example [161, Section 4.3.10].

12.1.5 Example. Consider the 3-polygraph P with one 0-generator, one 1-generator, three 2-generators  and the following two 3-generators:

A: 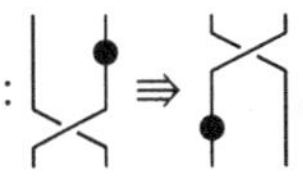 , B: 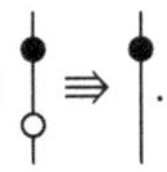 .

We prove that the $(3,2)$-category $P^{\top}$ admits an empty acyclic extension and thus has finite derivation type.

The 3-polygraph P is Tietze equivalent to the 3-polygraph Q defined the same way as P except for the orientation of the 3-cell A:

A: 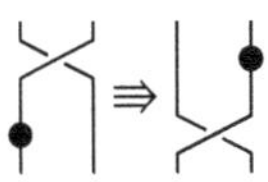 , B: 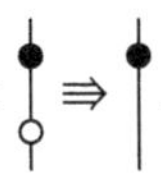 .

In this polygraph, we introduce the notation  for the 2-cell defined by induction on the natural number k by

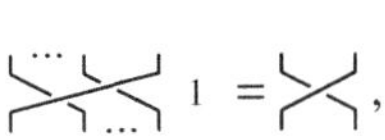

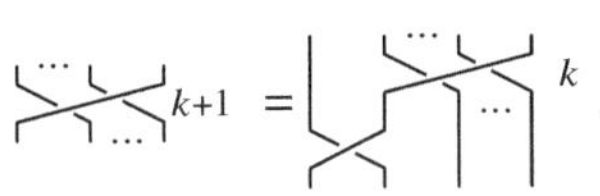

The polygraph Q is not convergent, but we can complete it into the infinite 3-polygraph $Q_\infty = Q \sqcup \{B_k \mid k \geqslant 1\}$, where B_0 is B and B_k is the following 3-cell:

B_k: 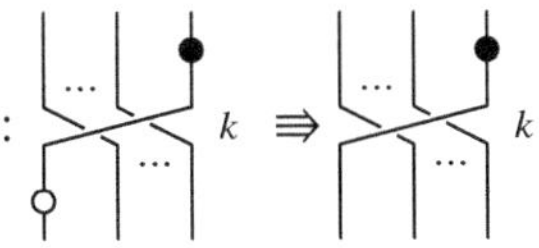 .

It can be shown that the 3-polygraph Q_∞ does not have finite derivation type. In particular the $(3,2)$-category $Q_\infty^{\top}$ has an infinite acyclic extension $\{\Lambda_k \mid k \in \mathbb{N}\}$ with

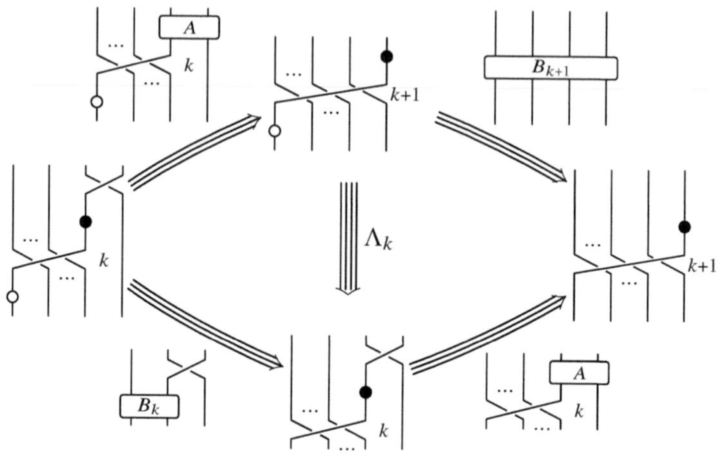

which cannot be reduced to a finite one.

12.1.6 Generating confluences. Theorems 2.5.10 and 7.3.5 state that the set of critical branchings of a convergent n-polygraph P generates an acyclic extension of the $(n,n-1)$-category $P^{\top}$ when $n \leqslant 2$. The proof of this result can be extended to 3-polygraphs as follows. Given a convergent 3-polygraph P, a *family of generating confluences of P* is a cellular extension of the free $(3,2)$-category $P^{\top}$ that contains exactly one 4-cell Λ of the form

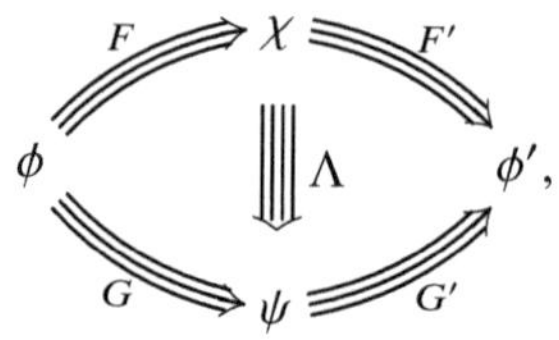

for every critical branching (F, G) of P. We define the *Squier completion* of the 3-polygraph P as the $(4,2)$-polygraph denoted by $\mathrm{Sq}(P)$ and defined by $\mathrm{Sq}(P) = (P, P_4)$, where P_4 is a chosen family of generating confluences of P. As in the case were $n = 2$, see Theorem 7.3.5, we have [161, Proposition 4.3.4]:

12.1.7 Theorem. *Given a convergent presentation of a 2-category C by a 3-polygraph P, any Squier completion of P is coherent presentation of C.*

As a consequence of Theorem 12.1.7, a finite convergent 3-polygraph with a finite set of critical branchings has finite derivation type. In particular, a terminating polygraph with no critical branching has finite derivation type. However, this result fails to generalize to n-categories when $n \geqslant 2$, see Section 16.8 and [161]. A counterexample to show this for $n = 3$ is developed in §12.2.8.

12.2 Squier's Completion of 3-Polygraphs

12.2.1 Non-indexed 3-polygraphs. Recall from §10.4.5 that a 3-polygraph is non-indexed when each of its critical branchings is an inclusion one or a regular one. It can be proved that a 3-polygraph with a finite set of 3-cells has a finite number of inclusion and regular critical branchings [161, Proposition 5.1.3]. As a consequence, we have the following finiteness condition in the non-indexed case [161, Theorem 5.1.4]:

12.2.2 Theorem. *A finite, convergent, and non-indexed* 3*-polygraph has finite derivation type.*

12.2.3 Confluence in indexed 3-polygraphs. Now let us consider the problem of finite convergence for finitely indexed 3-polygraphs (those for which each indexed critical branching has a finite number of normal instances). The situation is more complicated than in the non-indexed case. However, we have the following confluence result [161, Proposition 5.3.1]:

12.2.4 Proposition. *Let P be a terminating right-indexed (resp. left-indexed)* 3*-polygraph. Then P is confluent if and only if every inclusion critical branching, every regular critical branching, and every instance of every right-indexed (resp. left-indexed) critical branching is confluent.*

Proof. Suppose that P is a terminating right-indexed 3-polygraph (the left-indexed case is similar) such that all its inclusion critical branchings, regular critical branchings, and all the normal instances of its right-indexed critical branchings are confluent. It is sufficient to prove that every non-normal instance of its right-indexed critical branchings is confluent. Let us consider an instance of right-indexed critical branching. With the notations of §10.4.4, it is of the form

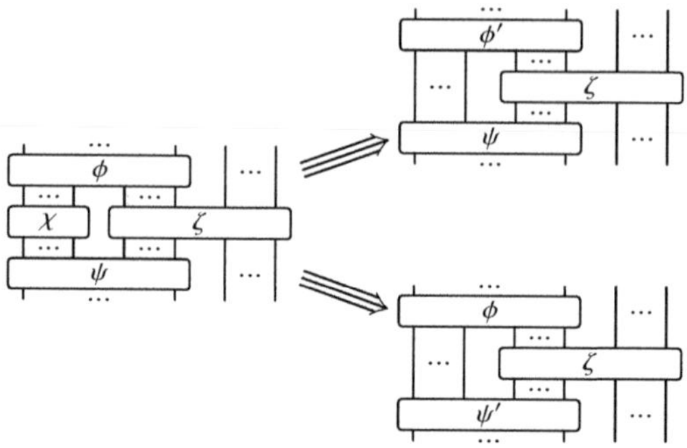

for some 2-cell ζ in P_2^*. If ζ is not a normal form, it admits a normal form $\widehat{\zeta}$, because P terminates. There is another instance of the above critical branching with $\widehat{\zeta}$ in place of ζ. Since $\widehat{\zeta}$ is a normal form, this is a normal instance, so that, by hypothesis, it is confluent. This ensures the confluence of the original branching as follows:

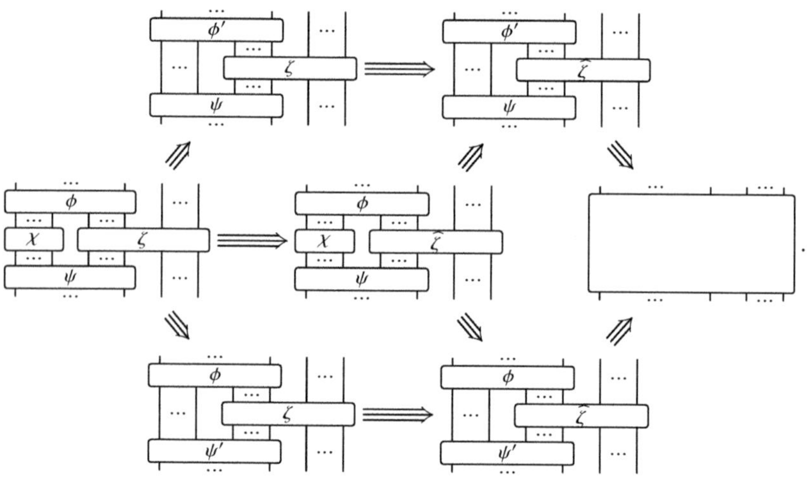

In this way, we prove that the polygraph P is confluent. □

12.2.5 Acyclic extensions of indexed 3-polygraphs. Let P be a locally confluent and right-indexed (resp. left-indexed) 3-polygraph. Suppose that a confluence has been chosen for each inclusion and regular critical branching and each normal instance of each right-indexed (resp. left-indexed) critical branching. Let P_4 be the cellular extension of the $(3,2)$-category $P^{\top}$ corresponding to these confluence diagrams. We can prove that if P is convergent and right-indexed (resp. left-indexed), then P_4 forms an acyclic extension of the $(3,2)$-category $P^{\top}$, i.e., the $(4,2)$-polygraph (P, P_4) is coherent [161, Proposition 5.3.3]. The proof follows the same scheme as the proof given for Theorems 7.3.5 and 12.1.7. It is the same for trivial, inclusion, and regular critical branchings. For right-indexed (resp. left-indexed) critical branchings, we follow a reasoning similar to the proof of Proposition 12.2.4. We thus have the following result [161, Theorem 5.3.4]:

12.2.6 Theorem. *A finite, convergent, and finitely indexed* 3*-polygraph has finite derivation type.*

In the next section, we present an illustration of this result with a 3-polygraph which is finite, convergent, and right-indexed, and thus has an infinite number of critical branchings. Yet, the polygraph has finite derivation type thanks to finite indexation.

12.2.7 Example: the 3-polygraph of permutations. Consider the 3-polygraph P presenting the PRO **S** of whose morphisms are permutations, which is introduced in §10.4.1. This polygraph has one 0-cell, one 1-cell, one 2-cell $\times$, and the following two 3-cells:

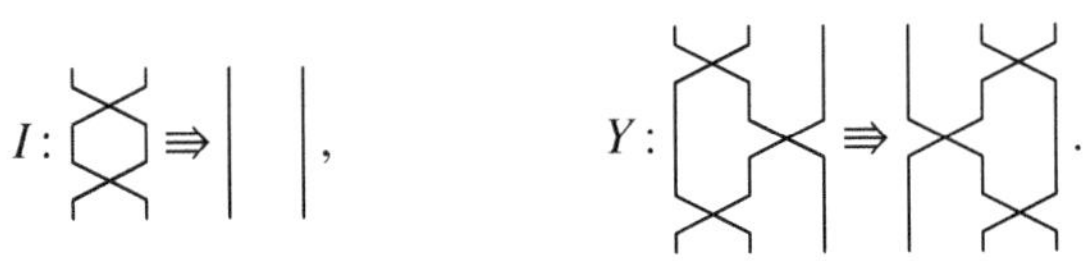

This polygraph was shown to be terminating in Examples 11.2.9 and 11.3.9. We have seen in §10.4.1 that it has three regular and one right-indexed critical branchings, with the following sources:

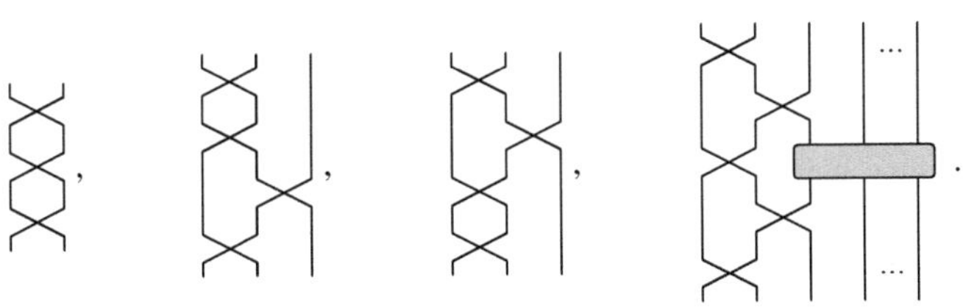

From Proposition 12.2.4, we know that, to show the confluence of the polygraph, it is sufficient to prove that the three regular critical branchings are confluent and that each normal instance of the right-indexed one is. This is, in fact, what we have been doing in §10.4.1: we only briefly recall here those confluence diagrams. First, the three regular critical branchings are confluent:

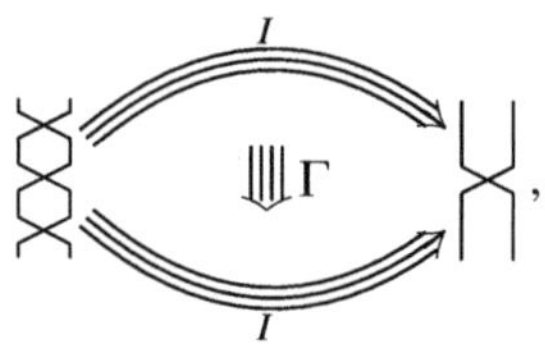

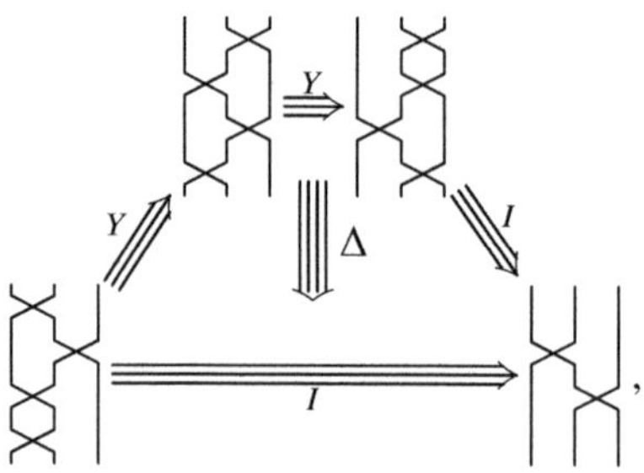

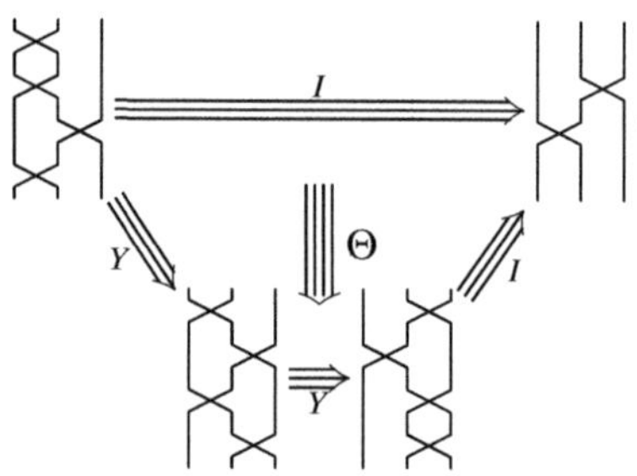

From the characterization of the set of normal forms given in §10.4.1, we deduce that there are two normal instances of the right-indexed critical branching: for $k = |$ and $k = \times$. We check that both are confluent. For $k = |$, we have:

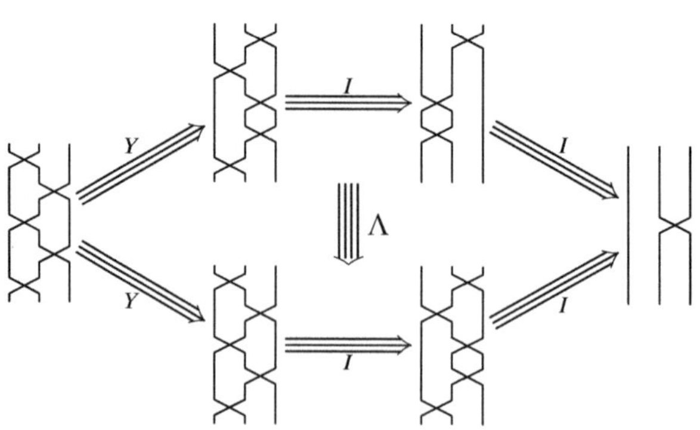

For $k = \times$, we have:

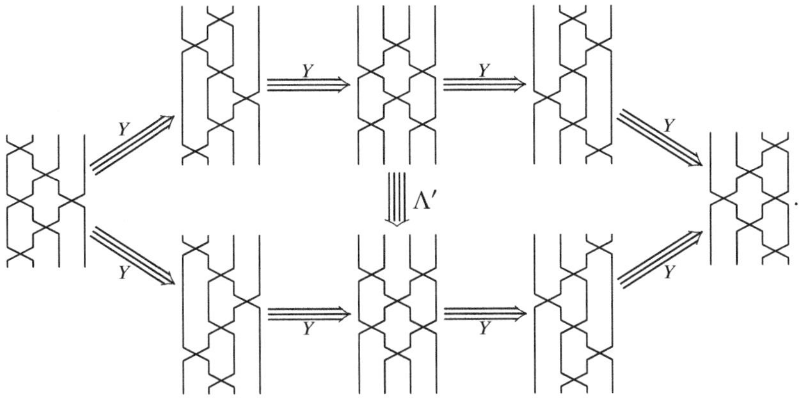

The 3-polygraph P is finite, convergent, and finitely indexed, by Theorem 12.2.6, it follows that it has finite derivation type. More precisely, the five 4-cells Γ, Δ, Θ, Λ, and Λ' form an acyclic extension of the $(3,2)$-category $P^\top$.

12.2.8 Main counterexample: the polygraph of pearls. Let us mention a 3-polygraph, studied in [161], which illustrates the fact that, without finite indexation, finiteness, and convergence are not sufficient to ensure finiteness of derivation type. We consider the 3-polygraph P of *pearls* with one 0-cell $\star$, one 1-cell a, three 2-cells $\bullet$, $\cap$, and $\cup$, and the following four 3-cells:

We define by induction on the natural number k the 2-cell $\bullet^k$ as follows:

Let us show that the polygraph is terminating. The rules C and D make the number of 2-generators strictly decrease, while this number is invariant by the rules A and B, so that we only have to show that the rules A and B are terminating. In order to show this, we apply Theorem 11.3.8 and use the notations of this theorem in the following. The posets associated to the 1-generator are $X_a = Y_a = \mathbb{N}$ equipped with the usual order. The interpretations of the 2-generators are

$$X(\bullet)(i) = i+1, \qquad X(\cap)() = (0,0), \qquad X(\cup)(i,j) = (),$$
$$Y(\bullet)(i) = i+1, \qquad X(\cup)() = (0,0), \qquad X(\cap)(i,j) = ().$$

We take M to be the monoid $(\mathbb{N}, +, 0)$ and define the derivation d by

$$d(\dot{\vert})(i,j) = 0, \qquad d(\cap)(i,j) = i, \qquad d(\cup)(i,j) = i.$$

For the rule A, we have, using the properties of derivation,

$$\begin{aligned} d(s_2^*(A)) &= d(\dot{\vert}\cap \text{ composite}) = d(\cap) *_1 \dot{\vert}\,\vert + \cap *_1 d(\dot{\vert}\,\vert), \\ d(t_2^*(A)) &= d(\cap\dot{\vert} \text{ composite}) = d(\cap) *_1 \vert\,\dot{\vert} + \cap *_1 d(\vert\,\dot{\vert}), \end{aligned}$$

so that

$$\begin{aligned} d(s_2^*(A))(i,j) &= d(\cap)(i+1,j) + d(\dot{\vert})(0,i) \\ &= (i+1) + 0 \\ &> i + 0 \\ &= d(\cap)(i, j+1) + d(\dot{\vert})(0,j) \\ &= d(t_2^*(A))(i,j) \end{aligned}$$

and therefore $d(s_2^*(A)) > d(t_2^*(A))$. Similarly, $d(s_2^*(B)) > d(t_2^*(B))$. By Theorem 11.3.8, we thus deduce that the polygraph is terminating.

The 3-polygraph P has four regular critical branchings whose sources are

It also has one right-indexed critical branching, generated by the 3-cells A and B, with source

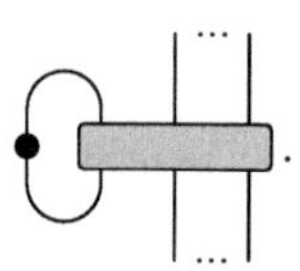

Thus P is a terminating and right-indexed 3-polygraph. By application of Proposition 12.2.4, the confluence of P can be shown by proving that its four regular critical branchings and all normal instances of its right-indexed critical branchings are confluent. For the regular ones, we have the following confluence diagrams:

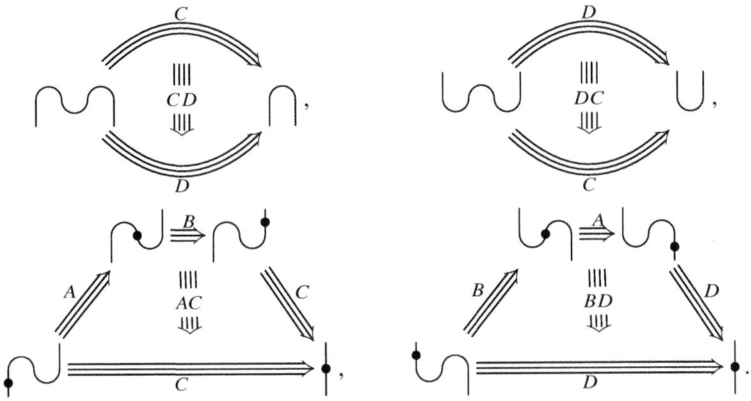

From the characterization of normal forms of the polygraph given in [161, Section 5.5.2], the normal instances of the right-indexed critical branching AB 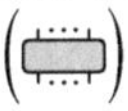 are the instances corresponding to the following 2-cells

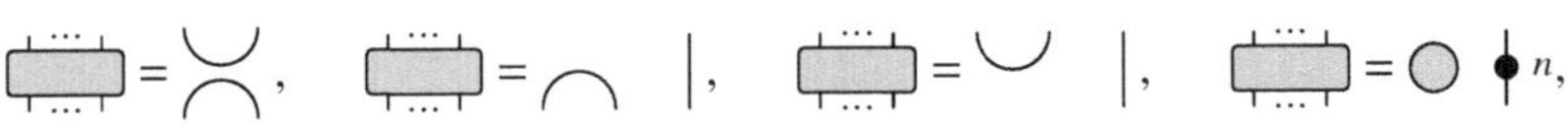

where, in the latter, $n \in \mathbb{N}$ and ○ ranges over the set N_0, the subset of P_2^* consisting of normal forms of P with degenerate source and target, which are characterized by the following two construction rules:

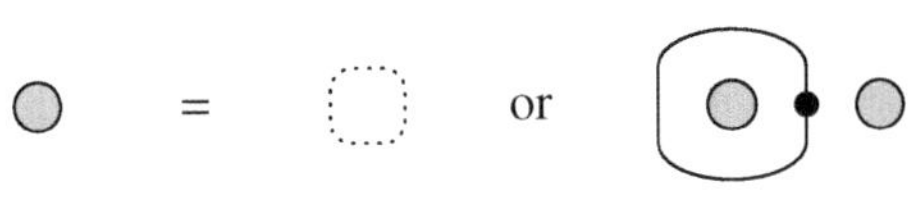

(on the left, this is the empty diagram). Now we check that, for each one of these 2-cells, the corresponding critical branching AB 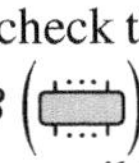 is confluent. Let us note that, for the first three cases, there are several possible confluence diagrams, because they also contain regular critical branchings of P.

– For = , we choose the following one:

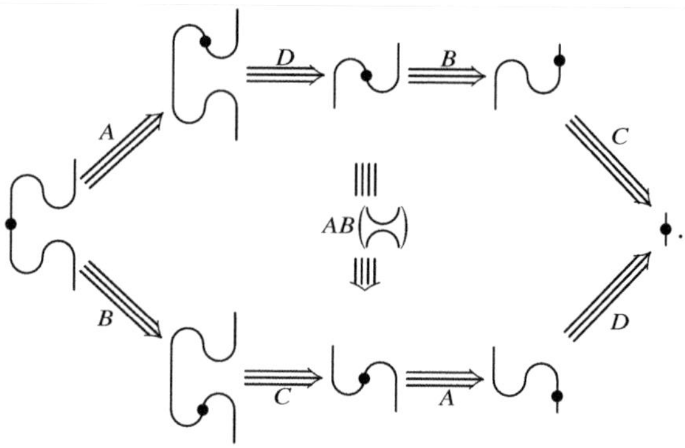

– For = :

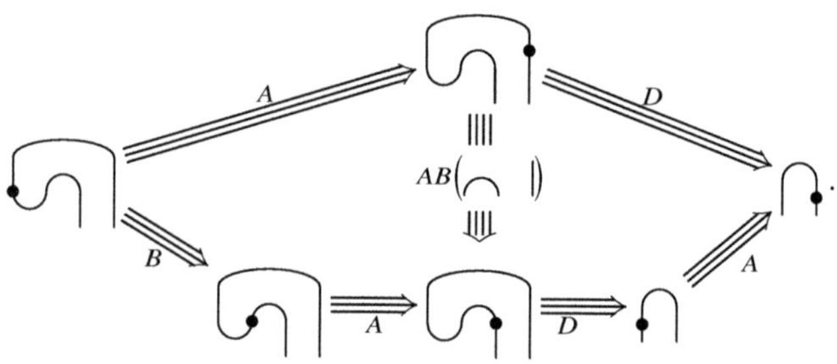

– For = :

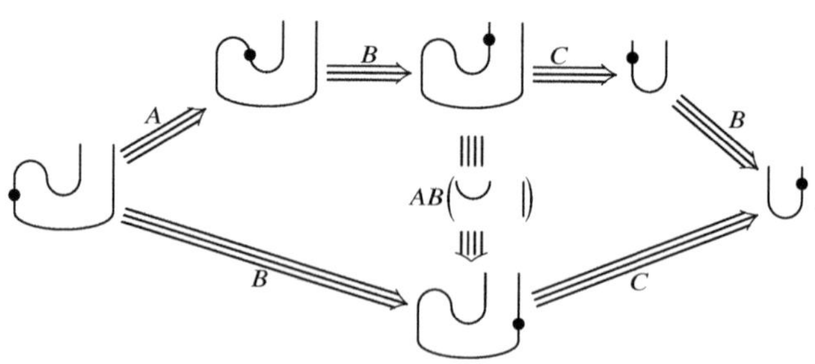

– Finally, for = ○ ⫯n :

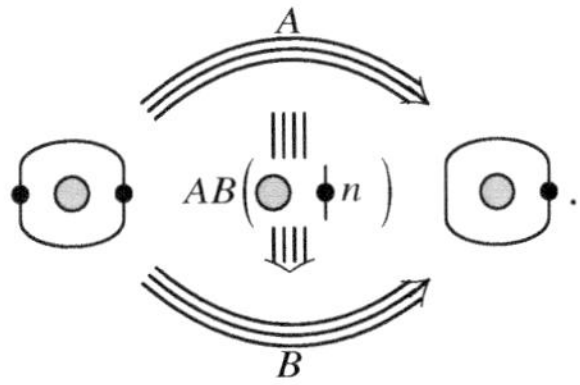

It follows that the 3-polygraph P is convergent and right-indexed, and the following 4-cells form an acyclic extension of $P^\top$:

$$CD,\ DC,\ AC,\ BD,\ AB\,(⋈),\ AB\,(⌢\ |),\ AB\,(⌣\ |),\ AB\,(○\ ⫯^{n}),$$

where ○ is in N_0 and n is in $\mathbb{N}$. It can be observed that the 4-cells $AB\,(⋈)$, $AB\,(⌢\ |)$ and $AB\,(⌣\ |)$ are superfluous. Namely, the 3-spheres forming their boundaries are also the boundaries of 4-cells of $Q^\top$ where Q is the $(4,2)$-polygraph obtained from the 3-polygraph P by adding the 4-generators AC and BD.

Let us denote by X_0 the family made of the 4-cells CD, DC, AC, and BD. Then, for every natural number n, one defines:

$$X_{n+1} = X_n \sqcup \left\{ AB\,(○\ ⫯^{n}) \mid ○ \in N_0 \right\}.$$

Thus, the following set of 4-cells forms an acyclic extension of the $(3,2)$-category $P^\top$:

$$X = \bigcup_{n\in\mathbb{N}} X_n.$$

It can be shown that this infinite number of confluence diagrams cannot be filled by a finite cellular extension and thus that the 3-polygraph P does not have finite derivation type [161, Theorem 5.5.7]:

12.2.9 Theorem. *The above* 3*-polygraph P does not have finite derivation type.*

We will see in §16.8.4 a generalization of this result to n-polygraphs with $n \geqslant 3$.

12.3 (3, 2)-PROs

We generalize, in dimension 3, the notion of PRO introduced in §2.4.10, as well as introduce symmetric and coherent variants. This will be used in subsequent sections to show coherence theorems such as Mac Lane's coherence theorem for monoidal categories.

12.3.1 3-PROs. A 3-*monoid* is a 3-category C where there is exactly one 0-cell $\star$. In such a 3-category, the set C_1 is canonically a monoid when equipped with 0-composition as multiplication and $1_\star$ as unit. A 3-*PRO* is a 3-monoid whose monoid of 1-cells is the additive monoid $\mathbb{N}$. Note that the underlying 2-category of a 3-PRO is always a PRO, as defined in §2.4.10, thus the name. A 3-PRO which is also a $(3,2)$-category is called a $(3,2)$-*PRO*: this is a 3-PRO in which every 3-cell is invertible with respect to composition $*_2$.

12.3.2 3-PROPs. A *symmetry* on a 3-monoid C is an invertible transformation

$$\gamma_{a,b} \colon a *_0 b \to b *_0 a,$$

indexed by 1-cells $a, b \in C_1$, which is natural in both components and makes the following diagrams commute:

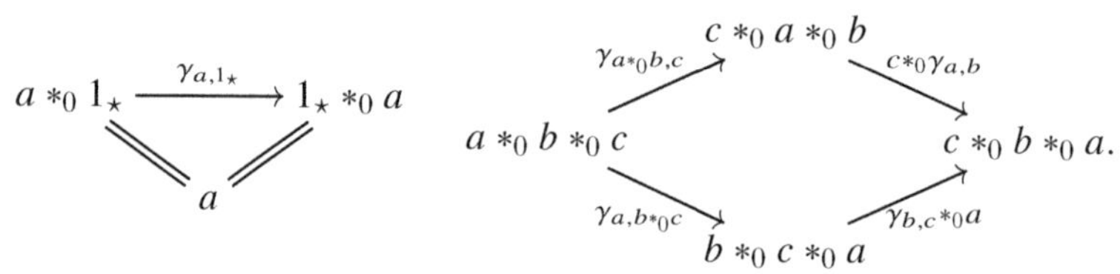

A 3-*PROP* (resp. $(3,2)$-*PROP*) is a 3-PRO (resp. $(3,2)$-PRO) equipped with a symmetry.

12.3.3 Algebras over 3-PRO(P)s. The 2-category **Cat** of categories, functors, and natural transformations is monoidal when equipped with the cartesian product as tensor product. By Mac Lane's coherence theorem, it can be considered as a strict monoidal category or, equivalently, as a 3-PRO with categories as 1-cells, functors as 2-cells, natural transformations as 3-cells, cartesian product as 0-composition, composition of functors as 1-composition, vertical composition of natural transformations as 2-composition.

If C is a 3-PRO, a C-*algebra* is a 3-functor from C to **Cat**. If C is a 3-PROP, we moreover require that this 3-functor preserves the symmetry. Given a C-algebra A, we often write $A_\star$ instead $A(\star)$. If A and B are C-algebras, a *morphism of C-algebras* from A to B is a natural transformation from A to B, i.e., a pair (F, ϕ) where $F \colon A_\star \to B_\star$ is a functor and ϕ is a map sending every 2-cell $f \colon m \Rightarrow n$ in C to a natural isomorphism with the following shape:

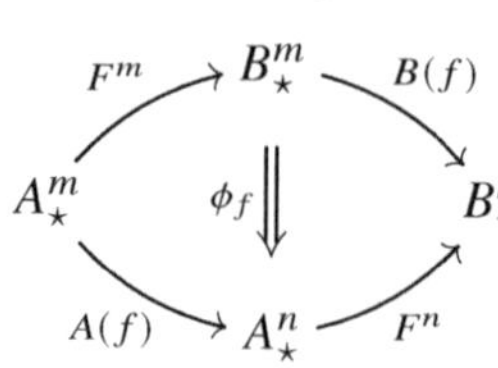

such that the following relations hold:

– for every 2-cells $f: m \Rightarrow n$ and $g: p \Rightarrow q$ of C, we have

$$\phi_{f *_0 g} = \phi_f \times \phi_g,$$

i.e., graphically,

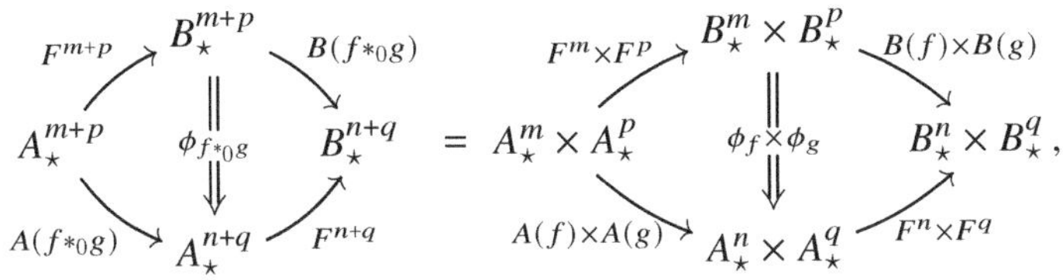

– for every 2-cells $f: m \Rightarrow n$ and $g: n \Rightarrow p$ in C, we have

$$\phi_{f *_1 g} = (\phi_f *_1 B(g)) *_2 (A(f) *_1 \phi_g),$$

i.e., graphically,

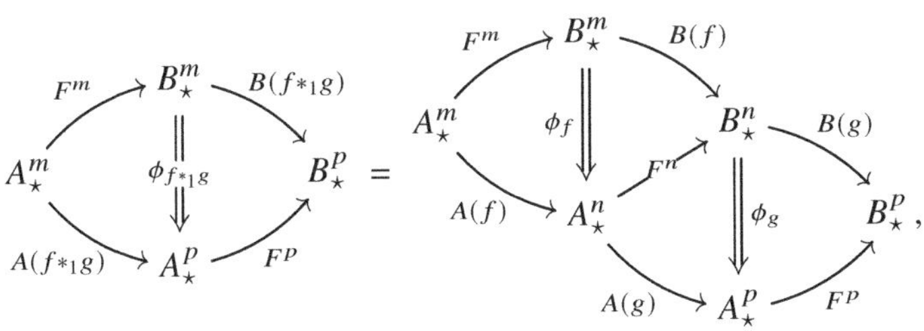

– for every 3-cell $\alpha: f \Rrightarrow g: m \Rightarrow n$ in C, we have

$$\phi_f *_2 (A(\alpha) *_1 F^n) = (F^m *_1 B(\alpha)) *_2 \phi_g,$$

i.e., graphically,

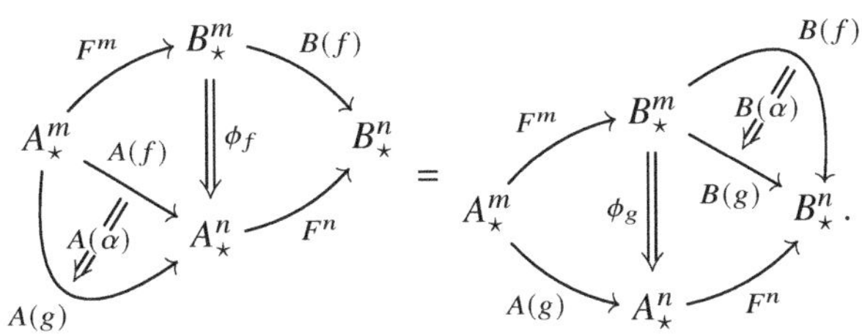

The C-algebras and their morphisms form a category denoted by $\mathbf{Alg}(C)$.

12.3.4 The coherence problem for algebras over a 3-PRO(P). Let C be a 3-PRO(P) and A be a C-algebra. A C-*diagram* in A is the image $A(\Delta)$ of a 3-sphere Δ in C, i.e., a pair (α, β) of 3-cells with the same source, and with the same target, where α (resp. β) is the *source* (resp. *target*) of the 3-sphere and is denoted by $s(\Delta)$ (resp. $t(\Delta)$). A C-diagram $A(\Delta)$ in A *commutes* if the relation

$$A(s(\Delta)) = A(t(\Delta))$$

is satisfied in **Cat**.

The *coherence problem* for algebras over a 3-PRO(P) is the following question:

Given a 3-PRO(P) C, does every C-diagram commute in every C-algebra?

A 3-PRO is *aspherical* when there is at most one 3-cell between two given 2-cells, i.e., any two parallel 3-cells are equal. As a consequence of this definition, we have the following sufficient condition for giving a positive answer to the coherence problem:

12.3.5 Proposition. *If C is an aspherical 3-PRO(P), then every C-diagram commutes in every C-algebra.*

12.3.6 Presentations of (3, 2)-PROs. A *presentation* of a $(3,2)$-PRO C is a $(4,2)$-polygraph P such that $C \simeq P^{\top}_{\leqslant 3}/P_4$, i.e., C is isomorphic to the $(3,2)$-category generated by the underlying 3-polygraph $P_{\leqslant 3}$, quotiented by the congruence generated by the 4-generators. By definition of a (3, 2)-PRO, in the case where we have a presentation as above, the 3-polygraph P necessarily has exactly one 0-cell and one 1-cell. A presentation P of C is called *coherently convergent* rather than *convergent* when P is a convergent 3-polygraph and P_4 is a cellular extension of generating confluences of P, see [162, Section 2.1.1].

12.3.7 Example. Consider the $(4,2)$-polygraph

$$P = \langle\, \star \mid a \mid \mu\colon aa \Rightarrow a \mid A\colon (\mu *_0 a) *_1 \mu \Rrightarrow (a *_0 \mu) *_1 \mu \mid \Gamma \,\rangle.$$

The 3-generator μ is often pictured as $\triangledown$ and A as

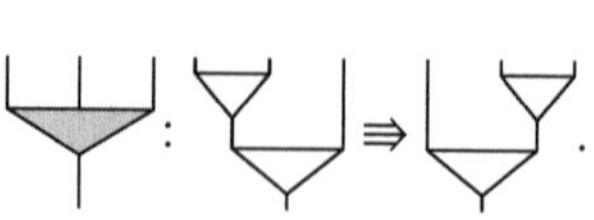

Similarly, the 4-generator Γ is pictured as  and its boundary is given by

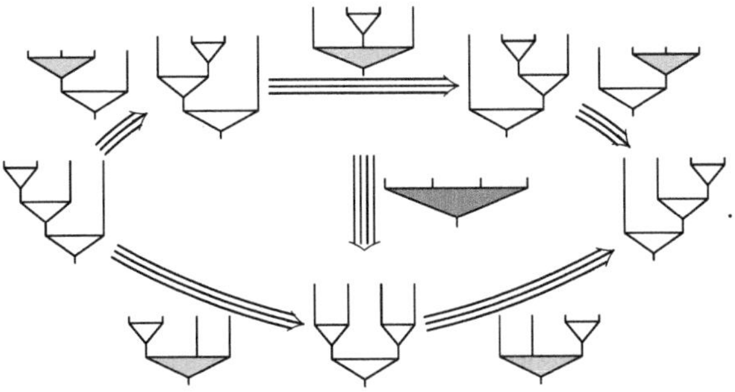

Consider the category Δ_μ whose objects are natural numbers and morphisms from m to n are surjective increasing functions $[m] \to [n]$ where $[n]$ denotes the set $\{0, \ldots, n-1\}$. This is a monoidal subcategory of the augmented simplicial category Δ_+ (see §§4.5.6 and 10.3.2), already encountered in Example 3.3.10. As a variant of the presentation of Δ_+, see §10.3.2, the underlying 3-polygraph of P can be shown to present the monoidal category Δ_μ. Moreover, the rewriting system is convergent and the boundary of the 4-generator Γ shown above is a confluence diagram for the only critical branching of the rewriting system so that, by Theorem 12.1.7, P_4 forms an acyclic extension of the $(3,2)$-category $P^{\top}_{\leqslant 3}$, i.e., P is a coherent presentation of Δ_μ.

We write AsCat for the $(3,2)$-PRO presented by P, i.e., $\mathrm{AsCat} = P^{\top}_{\leqslant 3}/P_4$. The category of its algebras $\mathbf{Alg}(\mathrm{AsCat})$ is isomorphic to the category of associative categories: we recall that an *associative category* is a category C equipped with a bifunctor $\otimes\colon C \times C \to C$ and a natural transformation $\alpha_{a,b,c}\colon (a \otimes b) \otimes c \to a \otimes (b \otimes c)$ satisfying the usual coherence law (see §12.4.1). Namely, the correspondence between an associative category $(C, \otimes, \alpha)$ and a 3-functor $A\colon \mathrm{AsCat} \to \mathbf{Cat}$ is given by

$A\Big($  $\Big) = C,$ $\qquad A\Big($  $\Big) = \otimes,$ $\qquad A\Big($ 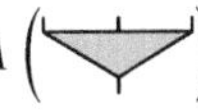 $\Big) = \alpha.$

This correspondence is well-defined since the coherence diagram satisfied by associative categories corresponds to the 4-cell .

Since $P_4 = \{$$\}$ is an acyclic extension of $\mathrm{As}^{\top}_{\leqslant 3}$, we have that AsCat is an aspherical $(3,2)$-PRO. As a consequence, in every associative category C, every AsCat-diagram is commutative. This fact can be informally restated as: every diagram built in C from the functor $\otimes$ and the natural transformation α is commutative.

12.3.8 Coherence in algebras over (3, 2)-PROs. By definition, a 3-PRO C is aspherical if, for every presentation P of C, the cellular extension P_4 of $P^{\top}_{\leqslant 3}$

is acyclic. The latter condition is satisfied by any convergent presentation of C yielding the following sufficient condition for giving a positive answer to the coherence problem for C-algebras [162, Theorem 2.1.2]:

12.3.9 Theorem. *If a $(3,2)$-PRO C admits a convergent presentation then every C-diagram commutes in every C-algebra.*

12.4 Coherence in Monoidal Categories

The coherence problems in monoidal categories can be formulated in terms of asphericity problems for $(3,2)$-categories. This section briefly reviews this approach in the case of monoidal categories. Symmetric and braided monoidal categories are handled in the next section.

12.4.1 Monoidal categories. A *monoidal category* is a category C, equipped with two functors

$$\otimes\colon C\times C\to C, \qquad\qquad e\colon 1\to C,$$

and three natural isomorphisms

$$\alpha_{x,y,z}\colon (x\otimes y)\otimes z\to x\otimes(y\otimes z),\quad \lambda_x\colon e\otimes x\to x,\quad \rho_x\colon x\otimes e\to x,$$

such that the following two diagrams commute in C:

$$\begin{array}{ccccc}
 & & (x\otimes y)\otimes(z\otimes t) & & \\
 & \overset{\alpha_{x\otimes y,z,t}}{\nearrow} & & \overset{\alpha_{x,y,z\otimes t}}{\searrow} & \\
((x\otimes y)\otimes z)\otimes t & & & & x\otimes(y\otimes(z\otimes t)) \\
 & \underset{\alpha_{x,y,z}\otimes t}{\searrow} & & \underset{x\otimes\alpha_{y,z,t}}{\nearrow} & \\
 & (x\otimes(y\otimes z))\otimes t & \xrightarrow[\alpha_{x,y\otimes z,t}]{} & x\otimes((y\otimes z)\otimes t) &
\end{array}$$

$$\begin{array}{ccc}
(x\otimes e)\otimes y & \xrightarrow{\rho_x\otimes y} & x\otimes(e\otimes y) \\
 & \underset{\alpha_{x,e,y}}{\searrow}\quad\underset{x\otimes\lambda_y}{\nearrow} & \\
 & x\otimes(e\otimes y). &
\end{array}$$

12.4.2 The 3-PRO of monoidal categories. Consider the $(4,2)$-polygraph

$$P=\langle\,\star\mid a\mid \mu\colon aa\Rightarrow a,\eta\colon 1_\star\to a\mid A,L,R\mid\Gamma,\Delta\,\rangle,$$

whose 2-generators μ and η are respectively pictured as  and , whose 3-generators A, L, and R are respectively pictured as

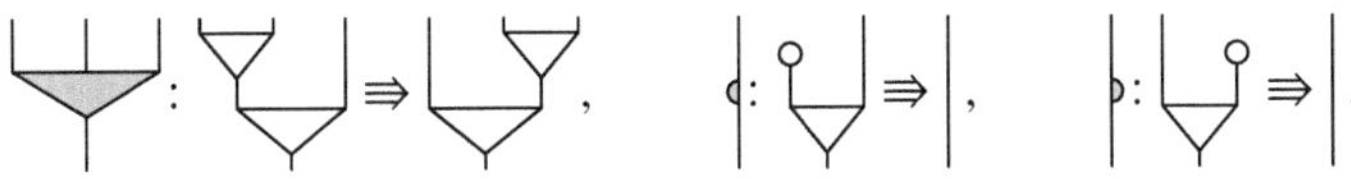

whose 4-generators Γ and Δ are respectively

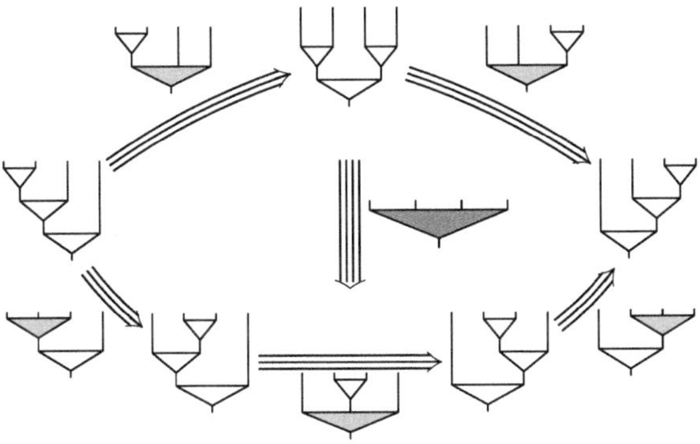

, and

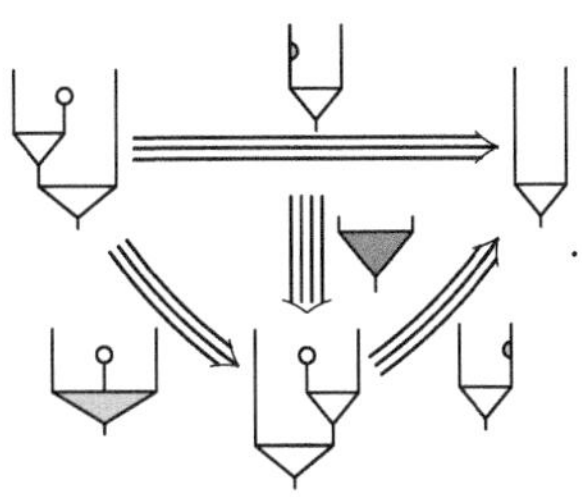

We denote by MonCat the $(3,2)$-PRO presented by this polygraph. It is easily seen that the category of small monoidal categories and monoidal functors is isomorphic to the category **Alg**(MonCat) [162, Lemma 2.3.2].

Note that the underlying 3-polygraph $P_{\leqslant 3}$ of P is the polygraph of monoids defined in Example 10.1.2, which was shown to be terminating in Example 11.2.4. Its five critical branchings are computed in Example 10.2.10 and shown to be confluent. Consider the cellular extension X of $P^{\top}_{\leqslant 3}$ with five 4-cells: the 4-generators Γ and Δ, as well as

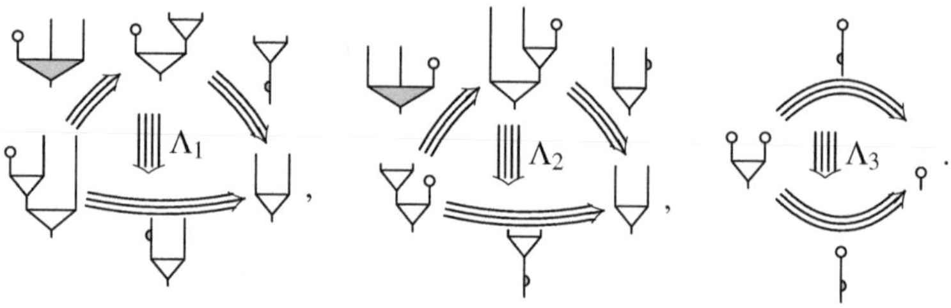

By Theorem 12.1.7, X forms an acyclic extension of $P^{\top}_{\leqslant 3}$ since its elements are a choice of confluence diagrams for the five critical branchings. It can be shown that Λ_1, Λ_2, and Λ_3 are superfluous in this cellular extension, i.e., for each 4-cell Λ_i, we have $\overline{s(\Lambda_i)} = \overline{t(\Lambda_i)}$ in MonCat [162, Section 2.3.3]. Therefore $\{\Gamma, \Delta\}$ is still an acyclic extension, i.e., the polygraph P is coherent. We have thus proved [161, Theorem 5.2.2], [162, Proposition 2.3.3]:

12.4.3 Proposition. *The above* $(4,2)$*-polygraph* P *is coherent.*

Mac Lane's coherence theorem [261, Theorem VII.2.1] states that, in a monoidal category, every diagram whose arrows are built up from instances of $\otimes$, α, λ, and ρ commute. From Proposition 12.4.3, we can deduce this theorem, which can be reformulated as follows:

12.4.4 Theorem. *The* 3*-PRO* MonCat *is aspherical.*

12.5 Coherence in Symmetric and Braided Monoidal Categories

12.5.1 Symmetric monoidal categories. A *symmetric monoidal category* is a monoidal category $(C, \otimes, e, \alpha, \lambda, \rho)$ equipped with a natural isomorphism

$$\gamma_{x,y} : x \otimes y \longrightarrow y \otimes x,$$

called the *symmetry* and such the following two diagrams commute in C:

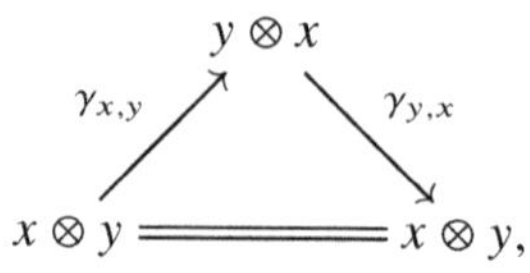

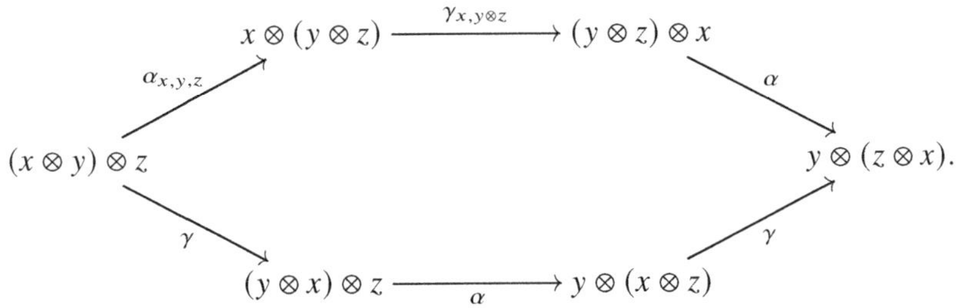

12.5.2 PROPs. Recall from §2.4.10 that a PRO is a strict monoidal category whose monoid of objects is $(\mathbb{N}, +, 0)$. We now introduce the following symmetric variant. A *PROP* is a strict symmetric monoidal category whose monoid of objects is $(\mathbb{N}, +, 0)$. In the following, we consider PROs and PROPs as 2-categories with one 0-cell. In particular, the underlying 2-category of a 3-PROP, as defined in §12.3.2, is a PROP.

12.5.3 PROPs as PROs. PROPs can be characterized among PROs as follows, see §C.1.4 and [157, Proposition A.3 and Corollary A4]. A PRO C is a PROP if and only if it contains a 2-cell $\gamma\colon 2 \Rightarrow 2$, represented by ⤫, such that the following relations hold:

- *involutivity* of the symmetry

$$\gamma *_1 \gamma = 1_2,$$

  which can be pictured as

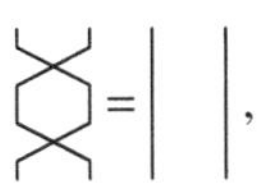

- the *Yang–Baxter relation*

$$(\gamma *_0 1) *_1 (1 *_0 \gamma) *_1 (\gamma *_0 1) = (1 *_0 \gamma) *_1 (\gamma *_0 1) *_1 (1 *_0 \gamma),$$

  which can be pictured as

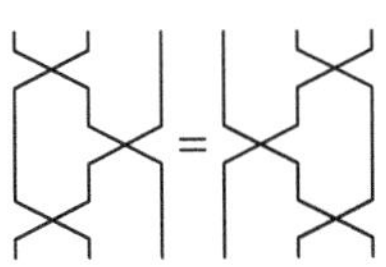

- for every 2-cell $\phi\colon m \Rightarrow n$ of C, the *left and right naturality relations* for ϕ

$$(\phi *_0 1) *_1 \gamma_{n,1} = \gamma_{m,1} *_1 (1 *_0 \phi),$$
$$(1 *_0 \phi) *_1 \gamma_{1,n} = \gamma_{1,m} *_1 (\phi *_0 1),$$

with the inductively defined notations:

$$\gamma_{0,1} = \gamma_{1,0} = 1_1, \qquad \gamma_{n+1,1} = (n *_0 \gamma) *_1 (\gamma_{n,1} *_0 1),$$
$$\gamma_{1,n+1} = (\gamma *_0 n) *_1 (1 *_0 \gamma_{1,n}).$$

If we represent ϕ by , any $\gamma_{n,1}$ by 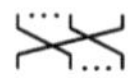, and any $\gamma_{1,n}$ by 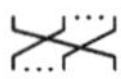, the naturality relations for ϕ are

12.5.4 The PROP of permutations. The initial PROP is the PROP of *permutations*, denoted by **S** and introduced in §10.4.1, whose 2-cells from n to n are the permutations of $\{0, \ldots, n-1\}$ and with no 2-cell from m to n if $m \neq n$. The 2-PROP **S** is presented by the 3-polygraph P of permutations defined in §10.4.1 whose 3-cells correspond to the involutivity and Yang–Baxter relations:

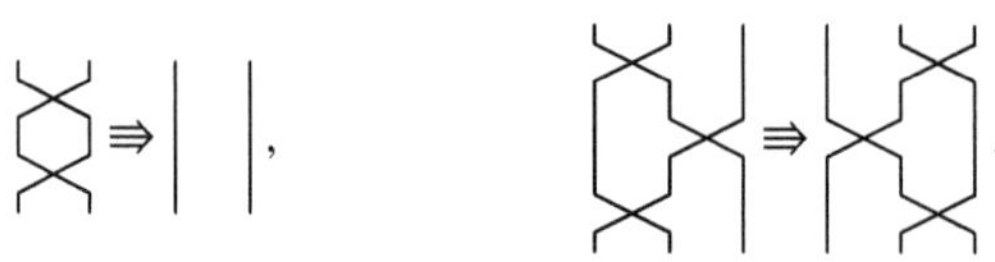 (12.1)

There is an isomorphism between the category of small categories and functors and the category **Alg**(Sym).

12.5.5 Presentations of PROPs. Let P be a 2-polygraph with one 0-cell and one 1-cell. We denote by SP the 3-polygraph obtained from P by adjoining a 2-cell : $2 \Rightarrow 2$ and the following 3-cells:

- the symmetry 3-cell and the Yang–Baxter 3-cell (12.1),
- two 3-cells for every 2-cell ϕ = 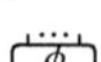 of P, corresponding to the naturality relations for f:

The *free PROP generated by P* is the 2-category, denoted by P^S, presented by the 3-polygraph SP, see also §C.1.3. We define a *presentation of a PROP C* as a pair (P, P_3) made of a 2-polygraph P with one 0-cell and one 1-cell and a cellular extension P_3 of the free 2-PROP P^S, such that $C \simeq P_2^S/P_3$.

12.5.6 Presentations of (3, 2)-PROPs. Let (P, P_3) be a presentation of a PROP. We denote by Q the $(4,2)$-polygraph obtained from the 3-polygraph SP by adjoining the 3-cells of P_3 and a cellular extension made of the following two 4-cells for each 3-generator $A\colon \phi \Rrightarrow \psi$ in P_3, corresponding to the naturality relations for A:

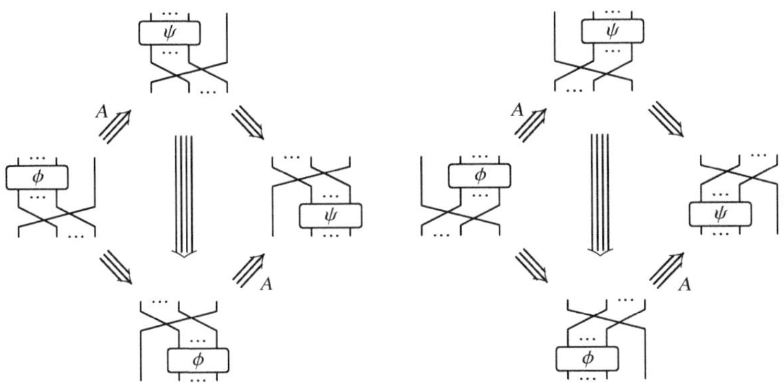

The *free* $(3,2)$*-PROP generated by* P is the $(3,2)$-category, denoted by P^S, presented by the $(4,2)$-polygraph Q defined above: $P^S = Q^{\top}_{\leqslant 3}/Q_4$.

We define a *presentation of a* $(3,2)$*-PROP* C as a pair (P, P_4), where P is a presentation of a PROP and P_4 is a cellular extension of the free $(3,2)$-PROP P^S generated by P, such that $C \simeq P^S/P_4$. A presentation P of a $(3,2)$-PROP is called *convergent* when the 3-polygraph SP is convergent.

12.5.7 Application to symmetric monoidal categories. Let **SCat** be the $(3,2)$-PROP presented by the polygraph P given as follows.

- $P_0 = \{\star\}$, $P_1 = \{a\}$,
- P_2 is the 2-polygraph, containing two 2-cells ▽ and ♀ ,
- P_3 is the cellular extension of the free 2-PROP P_2^S generated by P_2 containing the three 3-cells

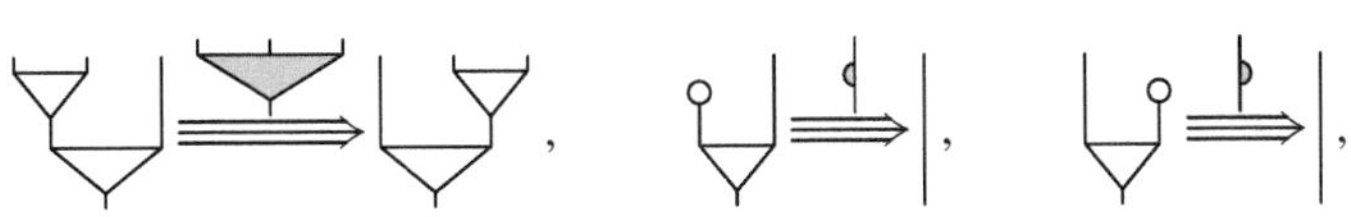

plus the following extra 3-cell:

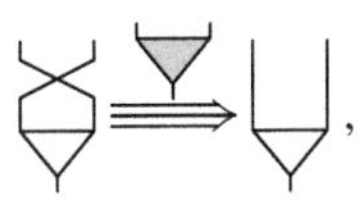

– P_4 is the cellular extension of the free $(3,2)$-PROP P_3^S generated by P_3 containing the two 4-cells

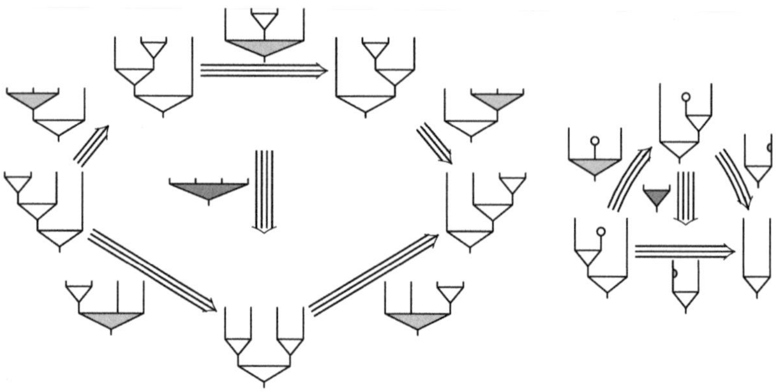

plus the following two extra 4-cells:

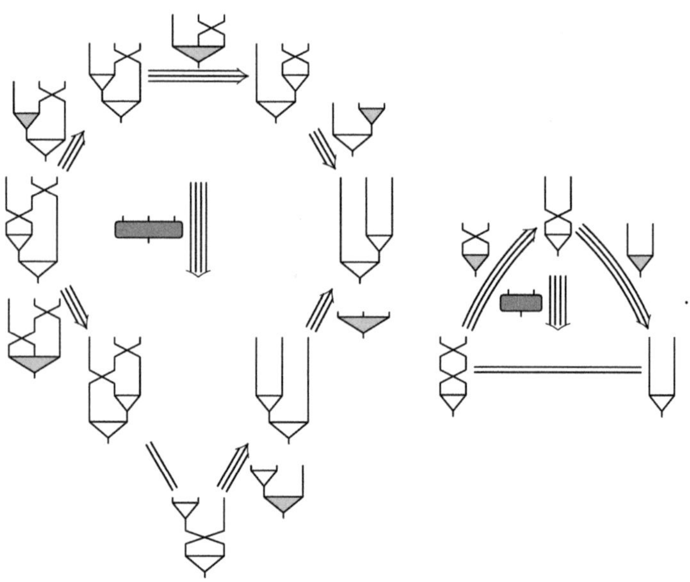

The category of small symmetric monoidal categories and symmetric monoidal functors is isomorphic to the category **Alg**(**SCat**). A convergent presentation of the $(3,2)$-PROP **SCat** is constructed in [162, Section 3.2]. The coherence theorem for symmetric monoidal categories [259] can be deduced from this construction: the $(3,2)$-PROP **SCat** is aspherical [162, Corollary 3.3.6].

12.5.8 Braided monoidal categories. A *braided monoidal category* is a monoidal category $(C, \otimes, e, \alpha, \lambda, \rho)$ equipped with a natural isomorphism

$$\beta_{x,y} \colon x \otimes y \longrightarrow y \otimes x,$$

called the *braiding* and such that the following diagrams commute in C:

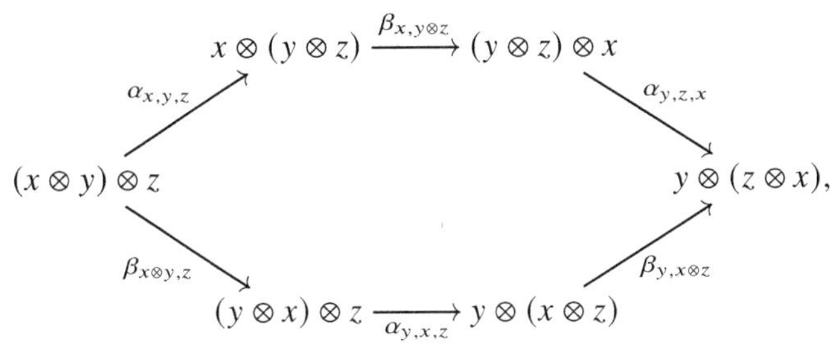

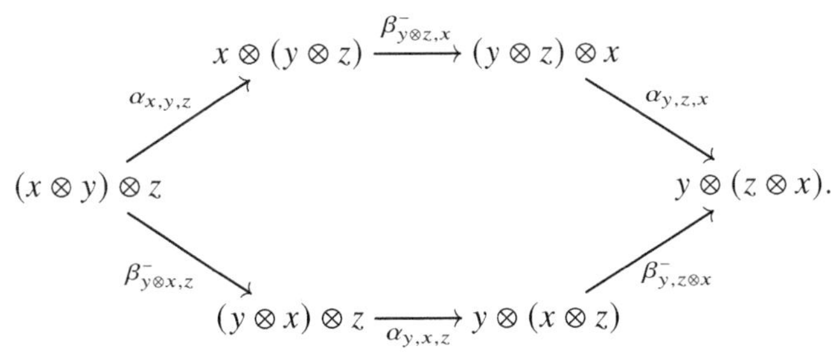

12.5.9 Generalized coherence theorems. Contrarily to the case of monoidal and symmetric monoidal categories, we do not have that every diagram commutes in a braided monoidal category. For instance, the morphisms $\beta_{x,y}$ and $\beta^-_{y,x}$, from $x \otimes y$ to $y \otimes x$, have no reason to be equal. In fact, they are equal if and only if β is a symmetry, hence if and only if all diagrams commute. As a consequence, the coherence problem for braided monoidal categories requires a generalized version of the coherence problem we have considered so far. The *generalized coherence problem* is the following one:

Given a $(3,2)$-PROP C, decide, for any 3-sphere α of C, whether or not the diagram $A(\alpha)$ commutes in every C-algebra A.

A solution for the generalized coherence problem is a decision procedure for the equality of 3-cells of C. For the coherence problems considered so far, this decision procedure answers yes for every 3-sphere. A method to study the generalized coherence theorem of 3-PROPs is given in [162, Section 4] and illustrated on the $(3, 2)$-PROP of braided monoidal categories. In this way, we recover the coherence result of Joyal and Street [209]: a diagram Δ commutes if and only if both $s(\Delta)$ and $t(\Delta)$ have the same associated braid.

13
Term Rewriting Systems

The study of *universal algebra*, that is, the description of algebraic structures by means of symbolic expressions subject to equations, dates back to the end of the 19th century [355]. It was motivated by the large number of fundamental mathematical structures fitting into this framework: groups, rings, lattices, and so on. From the 1970s on, the algorithmic aspect became prominent and led to the notion of term rewriting system. This chapter briefly revisits these ideas from a polygraphic viewpoint, introducing only what is strictly necessary for understanding. We refer the reader to standard textbooks such as [20, 342] for a proper study of this vast topic.

In Section 13.1, we begin by introducing term rewriting systems as presentations of Lawvere theories, which are particular cartesian categories. Some classical results on Lawvere theories are recalled in Section 13.2. The theory of rewriting in this context is explored in Section 13.3 by defining rewriting steps, critical branchings, and the original Knuth–Bendix completion procedure [218]. In Section 13.4, we show that a term rewriting system can also be described by a 3-polygraph in which variables are handled explicitly, i.e., by taking into account their duplication and erasure. Finally, in Section 13.5, we give a precise meaning to the statement that term rewriting systems are "cartesian polygraphs".

13.1 Presentations of Lawvere Theories

13.1.1 Signatures. In the context of term rewriting systems, a *signature* P consists of

- a set P_0 of *sorts*,
- a set P_1 of *operations* together with functions $s_0 \colon P_1 \to P_0^*$ and $t_0 \colon P_1 \to P_0$

respectively associating to each operation the sorts of its inputs and of its output, where P_0^* denotes the free monoid over P_0.

A *morphism* $f\colon P \to Q$ between signatures P and Q consists of two functions $f_0\colon P_0 \to Q_0$ and $f_1\colon P_1 \to Q_1$ such that $s_0 \circ f_1 = f_0^* \circ s_0$ and $t_0 \circ f_1 = f_0 \circ t_0$ (where $f_0^*\colon P_0^* \to Q_0^*$ is the extension of f_0 as a morphism of monoids). We write $\mathbf{Pol}_1^\times$ for the resulting category (this notation will be justified in Section 13.5).

Given an operation α, we write $\alpha\colon a_1 \ldots a_n \to a$ to indicate that its source is $s_0(\alpha) = a_1 \ldots a_n$ and target is $t_0(\alpha) = a$. The natural number n is called the *arity* of α. A signature is *mono-sorted* when P_0 is reduced to one element: in this case, $P_0^* = \mathbb{N}$ and t_0 is the terminal function.

13.1.2 Terms. Given a signature P, a *variable* is a symbol of the form x_i^u with $u \in P_0^*$ and $i \in \mathbb{N}$. A *term* on a signature P is a "well-typed" tree whose nodes are decorated in operations and leaves are decorated in variables. Formally, the family of sets $P_1^*(u, a)$ of *terms* from $u \in P_0^*$ to $a \in P_0$ is the smallest family, indexed by u and a, such that

– given $a_1, \ldots, a_n \in P_0$ and $1 \leqslant i \leqslant n$, we have a variable term

$$x_i^{a_1 \ldots a_n} \quad \in \quad P_1^*(a_1 \ldots a_n, a_i),$$

– given an operation $\alpha\colon a_1 \ldots a_n \to a$ in P_1, $u \in P_0^*$, and terms $\phi_i \in P_1^*(u, a_i)$ for $1 \leqslant i \leqslant n$, we have a composite term

$$\alpha(\phi_1, \ldots, \phi_n) \quad \in \quad P_1^*(u, a).$$

We write $\phi\colon u \to a$ to indicate that ϕ is a term in $P_1^*(u, a)$. In the following, we sometimes omit writing the superscripts from variables.

13.1.3 Substitutions. Given sorts $u \in P_0^*$ and $a_1, \ldots, a_n \in P_0$, a *substitution* $\sigma\colon u \to a_1 \ldots a_n$ is an n-tuple of terms $\sigma = \langle \sigma_1, \ldots, \sigma_n \rangle$, where the $\sigma_i\colon u \to a_i$ are terms with $1 \leqslant i \leqslant n$. Given a term $\phi\colon a_1 \ldots a_n \to a$, we write

$$\phi \cdot \sigma\colon u \to a$$

for the term obtained from ϕ by replacing each variable x_i by σ_i: this term is defined inductively by

$$x_i \cdot \sigma = \sigma_i, \qquad \alpha(\phi_1, \ldots, \phi_m) \cdot \sigma = \alpha(\phi_1 \cdot \sigma, \ldots, \phi_m \cdot \sigma).$$

13.1.4 The generated category. Given a signature P, we write P^* for the category with P_0^* as objects and substitutions $\sigma\colon u \to v$ as morphisms. Given two substitutions $\sigma\colon u \to v$ and $\tau\colon v \to w$ with $\tau = \langle \tau_1, \dots, \tau_n \rangle$, their composite is the substitution

$$\tau \circ \sigma = \langle \tau_1 \cdot \sigma, \dots, \tau_n \cdot \sigma \rangle,$$

and given an object $u = a_1 \dots a_n$ the identity on u is $\langle x_1^u, \dots, x_n^u \rangle$. Note that, given a term $\phi\colon w \to a$, we have

$$\phi \cdot (\tau \circ \sigma) = (\phi \cdot \tau) \cdot \sigma, \qquad\qquad \phi \cdot \langle x_1^u, \dots, x_n^u \rangle = \phi.$$

We write P_1^* for the set of all morphisms of P^* and $s_0^*, t_0^*\colon P_1^* \to P_0$ for the source and target functions.

13.1.5 Cartesian categories. In a category C, a *cartesian product* of two objects u and v is an object, usually noted $u \times v$, together with morphisms $\pi_1\colon u \times v \to u$ and $\pi_2\colon u \times v \to v$, called *projections*, such that for every object w and morphisms $\phi\colon w \to u$ and $\psi\colon w \to v$, there exists a unique morphism $\langle \phi, \psi \rangle\colon w \to u \times v$ satisfying $\pi_1 \circ \langle \phi, \psi \rangle = \phi$ and $\pi_2 \circ \langle \phi, \psi \rangle = \psi$:

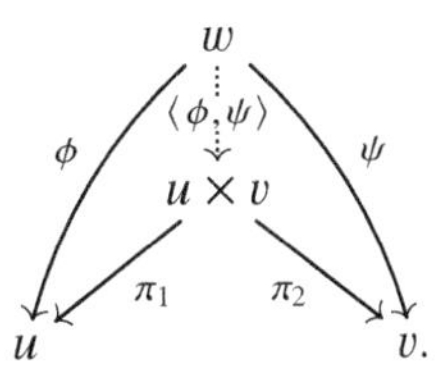

An object 1 is *terminal* in a category when for every object u there exists a unique morphism $u \to 1$. A category is *cartesian* when it has a terminal object and every pair of objects admits a cartesian product. In the following, for simplicity, we suppose fixed a choice of a product for any pair of objects in C, which we can suppose to be strictly associative and unital by Mac Lane's coherence theorem (Theorem 12.4.4).

A *morphism* $f\colon C \to D$ of cartesian categories, also called a *cartesian functor*, is a functor which preserves cartesian products and the terminal object. Here, we only consider functors for which this preservation is strict, by which we mean that $f(u \times v) = f(u) \times f(v)$ and $f(1) = 1$. We write **Cart**, or $\mathbf{Cart}_1$, for the category of cartesian categories.

13.1.6 Lemma. *The category P^* is cartesian.*

Proof. The product of two objects $u = a_1 \dots a_p$ and $v = b_1 \dots b_q$ is given by their concatenation uv, and the canonical projections are

$$\langle x_1^{uv}, \ldots, x_p^{uv} \rangle : uv \to u, \qquad\qquad \langle x_{p+1}^{uv}, \ldots, x_{p+q}^{uv} \rangle : uv \to v.$$

Finally, given two morphisms

$$\sigma = \langle \sigma_1, \ldots, \sigma_p \rangle : w \to u, \qquad\qquad \tau = \langle \tau_1, \ldots, \tau_q \rangle : w \to v,$$

the associated universal morphism is

$$\langle \sigma, \tau \rangle = \langle \sigma_1, \ldots, \sigma_p, \tau_1, \ldots, \tau_q \rangle : w \to uv. \qquad \square$$

Let us describe an important case of the above construction. We write $\mathbf{F}$ for the category whose objects are natural numbers and morphisms $m \to n$ are functions $[m] \to [n]$ where $[n] = \{0, \ldots, n-1\}$ is a set with n elements, see also Section C.2. We write $I \colon \mathbf{F} \to \mathbf{Set}$ for the canonical inclusion functor. Given a set P_0, we (abusively) write $\mathbf{F}/P_0$ for the comma category $I \downarrow P_0$ of I over the set P_0.

13.1.7 Lemma. *Given a signature P such that $P_1 = \emptyset$, we have $P^* \simeq (\mathbf{F}/P_0)^{\mathrm{op}}$. In particular, if $P_0 = \{\star\}$ then $P^* \simeq \mathbf{F}^{\mathrm{op}}$.*

We now describe the universal property satisfied by the construction P^*.

13.1.8 Lawvere theories. Suppose fixed a set P_0 of sorts. A *P_0-sorted Lawvere theory* (or *algebraic theory*) is a cartesian category C equipped with functor

$$(\mathbf{F}/P_0)^{\mathrm{op}} \to C,$$

which preserves finite products and is the identity on objects. A morphism between two Lawvere theories C and D is a functor $f \colon C \to D$ making the following diagram commute:

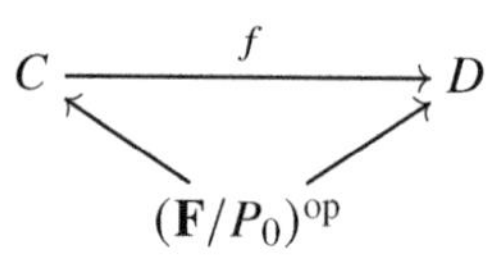

We write $\mathbf{Law}_{P_0}$ for the resulting category.

13.1.9 The free Lawvere theory. Consider the subcategory $\mathbf{S}_{P_0}$ of the category $\mathbf{Pol}_1^{\times}$ of signatures, where objects are the signatures having P_0 as sorts, and morphisms are those which are identity on sorts. There is a forgetful functor

$$W_0 \colon \mathbf{Law}_{P_0} \to \mathbf{S}_{P_0}$$

sending a Lawvere theory C to the signature P with

$$P_1 = \coprod_{uP_0^*, a \in P_0} C(u,a)$$

as operations, with source and target being respectively given by the indices u and a of the coproduct. The cartesian category generated by a signature introduced in §13.1.4 can be shown to be freely generated in the following sense.

13.1.10 Proposition. *The functor W_0 admits a left adjoint*

$$L_0 \colon \mathbf{S}_{P_0} \to \mathbf{Law}_{P_0},$$

such that the image of a signature P is the Lawvere theory P^*.

13.1.11 Congruence. A *congruence* on a Lawvere theory C is a relation $\approx$ on parallel morphisms, such that

– given morphisms $f \colon u' \to u$, $h \colon v \to v'$, and $g, g' \colon u \to v$,

$$g \approx g' \qquad \text{implies} \qquad f * g * h \approx f * g' * h,$$

– given morphisms $f, f' \colon w \to u$ and $g, g' \colon w \to v$,

$$f \approx f' \quad \text{and} \quad g \approx g' \qquad \text{implies} \qquad \langle f, g \rangle \approx \langle f', g' \rangle.$$

Given such a congruence, we write $C/\!\approx$ for the associated *quotient* Lawvere theory, obtained from C by quotienting morphisms under $\approx$.

13.1.12 Term rewriting systems. A *term rewriting system* P consists of a signature (P_0, s_0, t_0, P_1) together with a set P_2 of *rewriting rules* (or *relations*) equipped with *source* and *target* functions $s_1, t_1 \colon P_2 \to P_1^*$ such that the associated morphisms are parallel, i.e., $s_0^* \circ s_1 = s_0^* \circ t_1$ and $t_0^* \circ s_1 = t_0^* \circ t_1$. A rewriting rule A with source (resp. target) $\phi \colon u \to a$ (resp. $\psi \colon u \to a$) is often denoted

$$A \colon \phi \Rightarrow \psi \colon u \to a.$$

The Lawvere theory *presented* by a term rewriting system P is $\overline{P} = P^*/P_2$, i.e., the category obtained from P^* by quotienting morphisms under the congruence $\approx^P$ generated by P_2, and we say that P is a *presentation* of $\overline{P}$.

A morphism $P \to Q$ between term rewriting systems consists of a morphism between the underlying signatures together with a function $P_2 \to Q_2$ which is compatible with source and target. We write $\mathbf{Pol}_2^\times$ for the category of term rewriting systems.

13.1.13 Models. A *model* of a Lawvere theory C is a functor $C \to \mathbf{Set}$ which preserves finite products. In the case where C admits a presentation P, a model amounts to the data of

- a set $[\![a]\!]$ for every sort $a \in P_0$,
- a function $[\![\alpha]\!] : [\![a_1]\!] \times \ldots \times [\![a_n]\!] \to [\![a]\!]$ for every operation $\alpha \colon a_1 \ldots a_n \to a$ in P_1,

such that $[\![\phi]\!] = [\![\psi]\!]$ for every relation $A \colon \phi \Rightarrow \psi$, where $[\![-]\!]$ is extended to terms by

$$[\![\alpha(\phi_1, \ldots, \phi_n)]\!] = [\![\alpha]\!] \circ \langle [\![\phi_1]\!], \ldots, [\![\phi_n]\!] \rangle$$

and $[\![x_i^{a_1 \ldots a_n}]\!] : [\![a_1]\!] \times \cdots \times [\![a_n]\!] \to [\![a_i]\!]$ is the canonical projection. We sometimes abusively speak of a model of a signature (resp. term rewriting system) to mean a model of the generated (resp. presented) Lawvere theory.

13.1.14 Example. The theory of groups is presented by P with

$$P_0 = \{a\}, \qquad P_1 = \{\mu \colon 2 \to 1, \eta \colon 0 \to 1, \iota \colon 1 \to 1\},$$

and rewriting rules

$$\mu(\eta, x_1) \Rightarrow x_1, \qquad \mu(x_1, \eta) \Rightarrow x_1, \quad \mu(\mu(x_1, x_2), x_3) \Rightarrow \mu(x_1, \mu(x_2, x_3)),$$
$$\mu(\iota(x_1), x_1) \Rightarrow \eta, \quad \mu(x_1, \iota(x_1)) \Rightarrow \eta,$$

where, given $n \in \mathbb{N}$, we write n instead of a^n for an element of P_0^*. A model for this theory is a group.

Of course, as a variation of the previous example, usual algebraic structures have an associated Lawvere theory: groups, rings, modules, vector spaces, algebras, lattices, etc. Most are mono-sorted, apart from the theory for modules (as well as the one of vector spaces) which has two sorts: one corresponding to the ring of scalars and one to the abelian group. As a notable exception, there is no Lawvere theory corresponding to fields: intuitively, this is because the inverse operation is only partially defined (0 is not invertible). Below, we give an example of a Lawvere theory of more computational nature.

13.1.15 Example. Combinatory logic was introduced by Schönfinkel [321] and Curry [101] as an algebraic way of capturing binding and substitution. It can be presented by the mono-sorted term rewriting system P with operations

$$\alpha \colon 2 \to 1, \qquad \sigma \colon 0 \to 1, \qquad \kappa \colon 0 \to 1, \qquad \iota \colon 0 \to 1$$

(α should be read as an "application", and the constants σ, κ, and ι are usually respectively denoted S, K, and I) and relations

$$\begin{aligned}\alpha(\alpha(\alpha(\sigma,x_1),x_2),x_3) &\Rightarrow \alpha(\alpha(x_1,x_2),\alpha(x_1,x_3)),\\ \alpha(\alpha(\kappa,x_1),x_2) &\Rightarrow x_1,\\ \alpha(\iota,x_1) &\Rightarrow x_1.\end{aligned}$$

A combinatory term ϕ in P_1^* can be interpreted as a λ-term $[\![\phi]\!]$ by

$$[\![\alpha(\phi,\psi)]\!] = [\![\phi]\!]\,[\![\psi]\!], \quad [\![\sigma]\!] = \lambda xyz.(xz)(yz), \quad [\![\kappa]\!] = \lambda xy.x, \quad [\![\iota]\!] = \lambda x.x,$$

and conversely every λ-term can be interpreted as a combinatory logic term, giving rise to a correspondence between the morphisms in the presented category and λ-terms modulo β-reduction, although the details are subtle, see [322] for a survey on the subject. A model of this Lawvere theory is called a *combinatory algebra*.

13.1.16 Tietze transformations. The *elementary Tietze transformations* consist, starting from a presentation P, in

(T1) *adding a superfluous operation*: given a term $\phi\colon u \to a$ in P_1^*, we construct the presentation P' such that

$$P'_0 = P_0, \qquad P'_1 = P_1 \sqcup \{\alpha\colon u \Rightarrow a\}, \qquad P'_2 = P_2 \sqcup \{A\colon \phi \Rightarrow \alpha\},$$

(T2) *adding a superfluous relation*: given two terms ϕ and ψ such that $\phi \approx^P \psi$, we construct the presentation P' such that

$$P'_0 = P_0, \qquad P'_1 = P_1, \qquad P'_2 = P_2 \sqcup \{A\colon \phi \Rightarrow \psi\}.$$

The *Tietze equivalence* is the smallest equivalence relation on presentations such that P is Tietze equivalent to P' whenever there exists a Tietze transformation from P to P'. The proof of the following theorem carries over as in the case of polygraphs (see Theorem 5.1.2).

13.1.17 Theorem. *Two presentations with finite sets of operations and relations present isomorphic categories if and only if they are Tietze equivalent.*

13.1.18 Composing presentations. In Section 3.3, we have seen that we could compose presented categories, when given a distributive law between them, and this was extended to presentations of monoidal categories in Section 10.5. We mention here that this also generalizes to Lawvere theories: the corresponding notion of distributive law is studied in [84].

13.2 More on Models

In this section, we briefly recall some of the classical theory of Lawvere theories, as initiated by Lawvere in his PhD thesis [245], see [2, 3] for an in-depth presentation. For the sake of simplicity, we only handle here the mono-sorted case.

13.2.1 Models. Given a Lawvere theory C, we have seen in §13.1.13 that a model is a functor $C \to \mathbf{Set}$ which preserves finite limits. A morphism between models is a natural transformation and we write $\mathbf{Mod}(C)$ for the category of models.

13.2.2 Free models. Given a Lawvere theory C, there is a forgetful functor

$$U \colon \mathbf{Mod}(C) \to \mathbf{Set}$$

which to a model $M \colon C \to \mathbf{Set}$ associates $M(1)$.

13.2.3 Theorem. *The functor U is monadic: it admits a left adjoint and the category* $\mathbf{Mod}(C)$ *is equivalent to the category of T-algebras, where T is the monad associated to the adjunction.*

13.2.4 Monads. By the above theorem, every Lawvere theory C canonically induces a monad T on **Set**. An explicit description of this monad can be given by the following coend formula:

$$TX = \int^{n} C(n,1) \times X^{n}.$$

Not every monad arises in this way, and those which do can be characterized as being *finitary*, i.e., preserving filtered colimits. Writing **Mnd** for the category of monads on **Set**, we have the following equivalence of categories.

13.2.5 Theorem. *The category* **Law** *of Lawvere theories is equivalent to the full subcategory of* **Mnd** *whose objects are finitary monads.*

We have already explained above how to associate a finitary monad to a Lawvere theory. Conversely, given such a monad T, the opposite category of the Kleisli category is always a Lawvere theory. These constructions give rise to the equivalence stated in the theorem.

13.2.6 The Birkhoff theorem. We now turn to a different approach to models of Lawvere theory. Fix a signature P. In the context of model theory, its models are sometimes called *structures*. Given a presentation Q on this signature (i.e., $Q_0 = P_0$ and $Q_1 = P_1$), we have a quotient functor

$$P^* \to \overline{Q} = P^*/Q_2,$$

which induces, by precomposition, a functor

$$\mathbf{Mod}(Q) \to \mathbf{Mod}(P)$$

between the categories of models. The functor $P \to Q$ being surjective on objects and full, the induced functor between models is full and faithful, and we can thus consider $\mathbf{Mod}(Q)$ as a full subcategory of $\mathbf{Mod}(P)$. Conversely, given a full subcategory C of $\mathbf{Mod}(P)$, one may wonder whether there is a set P_2 of relations such that C is precisely the category of models of the Lawvere theory presented by (P, P_2). The following theorem, due to Birkhoff [43], see [3], and sometimes called the *HSP theorem*, gives a characterization of those situations.

13.2.7 Theorem. *Given a signature P and a full subcategory C of* $\mathbf{Mod}(P)$, *C is the category of models of a term rewriting system Q on the signature P if and only if it is closed under*

(H) homomorphic images*: given a regular epimorphism $f\colon M \to N$ in* $\mathbf{Mod}(P)$ *(i.e., f is an epi which can be obtained as a coequalizer) with $M \in C$, the object N also belongs to C,*

(S) subalgebras*: given a monomorphism $f\colon M \to N$ in* $\mathbf{Mod}(P)$ *with $N \in C$, the object M also belongs to C, and*

(P) products*: given $M, N \in C$, their product $M \times N$ in* $\mathbf{Mod}(P)$ *also belongs to C.*

13.3 Term Rewriting

Up to now, we have been using term rewriting systems as a notion of presentation, for which the orientation of the rules does not really matter. We now introduce the rewriting structure, following what we have done for 1-polygraphs (Section 1.3), 2-polygraphs (Chapter 4), and 3-polygraphs (Section 10.2).

13.3.1 Occurrences. Given a term $t\colon a_1 \dots a_n \to a$ and an index i, with $1 \leqslant i \leqslant n$, the number $o_i(t)$ of *occurrences* of the i-th variable x_i into t is defined by induction on t by

$$o_i(x_i) = 1, \qquad o_i(x_j) = 0, \qquad o_i(\alpha(\phi_1, \dots, \phi_k)) = \sum_{i=1}^{k} o_i(\phi_i),$$

for $j \neq i$. A variable x_i is *linear* in a term t when it occurs exactly once, i.e., $o_i(t) = 1$.

13.3.2 Contexts. A *context* $\kappa\colon a_1 \dots a_n a \to b$ is a term such that the variable x_{n+1} (of type a) is linear in C. Given a term $\phi\colon a_1 \dots a_n \to a$, we write

$$\kappa \cdot \phi\colon a_1 \dots a_n \to b$$

for the term obtained from κ by substituting ϕ for x_{n+1}, i.e.,

$$\kappa \cdot \phi = \kappa \cdot \langle x_1, \dots, x_n, \phi \rangle,$$

with the notations of §13.1.3.

Given two contexts $\kappa\colon a_1 \dots a_n a \to b$ and $\rho\colon a_1 \dots a_n b \to c$, their *composite* is the context

$$\rho \circ \kappa\colon a_1 \dots a_n a \to c,$$

obtained from ρ by replacing the variable x_{n+1} (of type b) by κ, i.e.,

$$\rho \circ \kappa = \rho \cdot \kappa = \rho \cdot \langle x_1, \dots, x_n, \kappa \rangle$$

and the *identity context* is

$$x_{n+1}^{a_1 \dots a_n a}\colon a_1 \dots a_n a \to a.$$

Note that given a term $\phi\colon a_1 \dots a_n \to a$, we have

$$(\rho \circ \kappa) \cdot \phi = \rho \cdot (\kappa \cdot \phi), \qquad\qquad x_{n+1}^{a_1 \dots a_n a} \cdot \phi = \phi.$$

Given a fixed $u \in P_0^*$, we can thus build a category with P_0 as set of objects, a morphism $\kappa\colon a \to b$ being a context $\kappa\colon ua \to b$.

It is also useful to introduce a *base change* operation on contexts. Given a context $\kappa\colon a_1 \dots a_n a \to b$ and a substitution $\sigma\colon u \to a_1 \dots a_n$, we write

$$\sigma^*(\kappa)\colon ua \to b$$

for the context defined by

$$\sigma^*(\kappa) = \kappa \cdot \langle \sigma_1, \dots, \sigma_n, x_{n+1}^{ua} \rangle.$$

13.3.3 Contexts and substitutions. Given a context κ, a term ϕ and a substitution σ of appropriate type, an expression of the form $\kappa \cdot \phi \cdot \sigma$ is always implicitly bracketed as $\kappa \cdot (\phi \cdot \sigma)$. These operations are compatible with the categorical structures in the sense that, for suitably typed contexts and substitutions, we have

$$\rho \cdot (\kappa \cdot \phi \cdot \tau) \cdot \sigma = (\rho \circ \sigma^*(\kappa)) \cdot \phi \cdot (\tau \circ \sigma), \qquad\qquad 1 \cdot \phi \cdot 1 = \phi.$$

Graphically, it is sometimes convenient to depict the term $\kappa \cdot \phi \cdot \sigma$ as

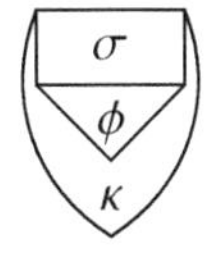

.

13.3.4 k-ary contexts. Generalizing the construction of §13.3.2, a *k-ary context*, for $k \in \mathbb{N}$, is a term

$$\kappa \colon a_1 \dots a_n a'_1 \dots a'_k \to a,$$

such that the variables $x_{n+1}, \dots, x_{n+k}$ are linear in κ. Given terms

$$\phi_i \colon a_1 \dots a_n \to a'_i,$$

with $1 \leqslant i \leqslant k$, we write

$$\kappa \cdot (\phi_1, \dots, \phi_k) = \kappa\langle x_1, \dots, x_n, \phi_1, \dots, \phi_k\rangle.$$

In the following, we will only use binary contexts, i.e., the case where $k = 2$.

13.3.5 Rewriting steps. A *rewriting step*

$$\kappa \cdot A \cdot \sigma \colon \kappa \cdot \phi \cdot \sigma \Rightarrow \kappa \cdot \psi \cdot \sigma \colon u' \to a'$$

consists of a rewriting rule

$$A \colon \phi \Rightarrow \psi \colon u \to a,$$

together with a substitution and a context

$$\sigma \colon u' \to u \qquad \text{and} \qquad \kappa \colon u'a \to a'.$$

In this case, we say that the term $\kappa \cdot \phi \cdot \sigma$ *rewrites* in one step to the term $\kappa \cdot \psi \cdot \sigma$. A *rewriting path* is a sequence of composable rewriting steps.

13.3.6 Rewriting. A term rewriting system P induces an abstract rewriting system with the terms in P_1^* as vertices and rewriting steps as edges. The rewriting system is *terminating*, (*locally*) *confluent*, etc., when the associated abstract rewriting system is.

13.3.7 Critical branchings. A *branching* is a pair

$$\kappa_1 \cdot \psi_1 \cdot \sigma_1 \xLeftarrow{\kappa_1 \cdot A_1 \cdot \sigma_1} \kappa_1 \cdot \phi_1 \cdot \sigma_1 = \kappa_2 \cdot \phi_2 \cdot \sigma_2 \xRightarrow{\kappa_2 \cdot A_2 \cdot \sigma_2} \kappa_2 \cdot \psi_2 \cdot \sigma_2 \qquad (13.1)$$

of coinitial rewriting steps. We can identify the following families of branchings.

1. A branching is *trivial* when it is of the form

$$\psi \xLeftarrow{\kappa\cdot A\cdot\sigma} \phi \xRightarrow{\kappa\cdot A\cdot\sigma} \psi.$$

   Such a branching is clearly confluent.
2. A branching is *parallel orthogonal* when it is of the form

$$\phi_1 \xLeftarrow{\kappa\cdot(A_1\cdot\sigma_1,\psi_2\cdot\sigma_2)} \phi \xRightarrow{\kappa\cdot(\psi_1\cdot\sigma_1,A_2\cdot\sigma_2)} \phi_2,$$

   for some suitably typed binary context κ, rewriting rules $A_1\colon \psi_1 \Rightarrow \psi_1'$ and $A_2\colon \psi_2 \Rightarrow \psi_2'$, and substitutions σ_1 and σ_2.
3. A branching is *inclusion orthogonal* when it is of the form

$$\phi_1 \xLeftarrow{\kappa\cdot A_1\cdot\langle\sigma_1,\ldots,\sigma_n\rangle} \phi \xRightarrow{\kappa\cdot\psi_1\cdot\langle\sigma_1,\ldots,\kappa'\cdot A_2\cdot\sigma',\ldots,\sigma_n\rangle} \phi_2,$$

   for some suitably typed contexts κ and κ', rewriting rules $A_1\colon \psi_1 \Rightarrow \psi_1'$ and $A_2\colon \psi_2 \Rightarrow \psi_2'$, and substitutions $\sigma = \langle\sigma_1,\ldots,\sigma_n\rangle$ and σ', such that $\sigma_i = \kappa' \cdot \psi_2 \cdot \sigma'$ for some index i with $1 \leqslant i \leqslant n$.
4. A branching is *non-minimal* when it is of the form

$$\phi_1 \xLeftarrow{\kappa\cdot(\kappa_1\cdot A_1\cdot\sigma_1)\cdot\sigma} \phi \xRightarrow{\kappa\cdot(\kappa_1\cdot A_1\cdot\sigma_1)\cdot\sigma} \phi_2,$$

   for some suitably typed context κ and substitution σ, not both identities, and of suitable types.
5. A branching is *critical* when it is not of any of the above forms.

The above definition of critical branching makes it easy to show the critical branchings lemma, stated below. Moreover, those can be efficiently computed, see [20, Section 6.2] for a presentation of the classical algorithms.

13.3.8 Lemma. *A term rewriting system is locally confluent if and only if all its critical branchings are.*

13.3.9 Example. The theory of monoids, with 0-generators $P_0 = \{a\}$, 1-generators $P_1 = \{\mu\colon 2 \to 1, \eta\colon 0 \to 1\}$ and 2-generators

$$\mu(\eta,x_1) \Rightarrow x_1, \quad \mu(x_1,\eta) \Rightarrow x_1, \quad \mu(\mu(x_1,x_2),x_3) \Rightarrow \mu(x_1,\mu(x_2,x_3)),$$

is locally confluent since its five critical branchings, whose source is shown below, are confluent:

$$\mu(\mu(\mu(x_1,x_2),x_3),x_4), \qquad \mu(\mu(\eta,x_1),x_2), \qquad \mu(\mu(x_1,\eta),x_2),$$
$$\mu(\mu(x_1,x_2),\eta), \qquad \mu(\eta,\eta).$$

The rewriting system can be shown to be terminating and is thus convergent.

13.3.10 Reduction orders. Given a signature P, a *reduction order* $>$ is an order on P_1^* which is

- well-founded,
- closed under application: for every $\alpha \in P_1$ of arity n and every term $\phi_1, \dots, \phi_n \in P_1^*$ and $\phi_i' \in P_1^*$, we have that $\phi_i > \phi_i'$ implies

$$\alpha(\phi_1, \dots, \phi_{i-1}, \phi_i, \phi_{i+1}, \dots, \phi_n) > \alpha(\phi_1, \dots, \phi_{i-1}, \phi_i', \phi_{i+1}, \dots, \phi_n),$$

- closed under substitution: given terms $\phi, \phi' \in P_1^*$ and substitution σ, we have $\phi > \phi'$ implies $\phi \cdot \sigma > \phi' \cdot \sigma$.

Given a rewriting system (P, P_2) on a signature P, a *termination* order $>$ is a reduction order on P such that $\phi > \psi$ for every rewriting rule $A \colon \phi \Rightarrow \psi$ in P. As in the case of 2-polygraphs (Proposition 4.4.2), we have [20, Theorem 5.2.3]:

13.3.11 Proposition. *A rewriting system is terminating if and only if it admits a termination order.*

13.3.12 Completion. Given a rewriting system P equipped with a termination order $>$, we can turn it into a convergent rewriting system by the following Knuth–Bendix completion procedure [218]. It is very similar to the one already given in Section 5.2 and consists in iteratively applying the following steps.

- For every critical branching $\phi_1 \Leftarrow \phi \Rightarrow \phi_2$, compute normal forms $\widehat{\phi_1}$ and $\widehat{\phi_2}$ for ϕ_1 and ϕ_2, respectively.
- If $\widehat{\phi_1} = \widehat{\phi_2}$ for every possible branching, the procedure halts.
- Otherwise, there is a critical branching with $\widehat{\phi_1} \neq \widehat{\phi_2}$:
  - if $\widehat{\phi_1} > \widehat{\phi_2}$, we add the rule $\widehat{\phi_1} \Rightarrow \widehat{\phi_2}$ to P,
  - if $\widehat{\phi_2} > \widehat{\phi_1}$, we add the rule $\widehat{\phi_2} \Rightarrow \widehat{\phi_1}$ to P.

As in the case of 2-polygraphs, the procedure is not guaranteed to stop. In the case it does, the resulting rewriting system is a convergent presentation of the Lawvere theory $\overline{P}$. When the procedure does not terminate, the above steps produce an infinite sequence of rewriting systems P^i with $P^0 = P$, by iteratively adding rules, and the inductive limit $\bigcup_i P^i$ is always a convergent presentation of $\overline{P}$.

13.3.13 Example. The presentation of groups given in Example 13.1.14 is not locally confluent. By applying the Knuth–Bendix completion procedure, one can arrive at the following convergent presentation with the same 0- and 1-generators and whose relations are those of Example 13.1.14 together with

$$\mu(\iota(x_1,\mu(x_1,x_2))) \Rightarrow x_2, \qquad \iota(\eta) \Rightarrow \eta, \quad \iota(\mu(x_1,x_2)) \Rightarrow \mu(\iota(x_2),\iota(x_1)),$$
$$\mu(x_1,\mu(\iota(x_1,x_2))) \Rightarrow x_2, \quad \iota(\iota(x_1)) \Rightarrow x_1,$$

see [218, Example 1].

13.3.14 Non-linearity and confluence. Because of the presence of variables which can potentially duplicate terms when substituted, one should be careful when the rewriting system is not terminating. For instance, contrary to the case of polygraphs studied in previous chapters, it is not true that a rewriting system without critical branchings is always confluent. Namely, consider the mono-sorted rewriting system due to Huet [188], with generators

$$\tau\colon 0 \to 1, \qquad \phi\colon 0 \to 1, \qquad \omega\colon 0 \to 1, \qquad \sigma\colon 1 \to 1, \qquad \varepsilon\colon 2 \to 1,$$

(which should respectively be read as “true”, “false”, “infinity”, “successor”, and “equality”) and relations

$$\omega \Rightarrow \sigma(\omega), \qquad \varepsilon(x_1,x_1) \Rightarrow \tau, \qquad \varepsilon(x_1,\sigma(x_1)) \Rightarrow \phi.$$

The first rule clearly makes the rewriting system non-terminating. There is no critical branching, yet the system is not confluent:

$$\tau \Longleftarrow \varepsilon(\omega,\omega) \Longrightarrow \varepsilon(\omega,\sigma(\omega)) \Longrightarrow \phi.$$

It can however be shown that a rewriting system which is *left-linear* (i.e., where no variable occurs twice in the left member of a rewriting rule) and without critical branchings is always confluent [20, Section 6.4].

13.4 Term Rewriting Systems and 3-Polygraphs

We now explain that presentations of Lawvere theories can be seen as particular 3-polygraphs [72, 136]. This is based on the idea, familiar to people working on linear logic, that a cartesian category is a monoidal category in which every object can be duplicated and erased, see for instance [277, Section 6]. For the sake of simplicity, we consider only strict monoidal categories here, as justified by Mac Lane’s coherence theorem (Theorem 12.4.4) although many results extend seamlessly to the general case.

13.4.1 Underlying monoidal category. Suppose given a cartesian category C. It can be equipped with a structure of symmetric monoidal category. The unit object is the terminal object. The tensor product of two objects is their cartesian product, and the tensor product of two morphisms $\phi\colon u \to u'$ and $\psi\colon v \to v'$ is the morphism

$$\phi \times \psi\colon u \times v \to u' \times v',$$

obtained by the universal property of the product:

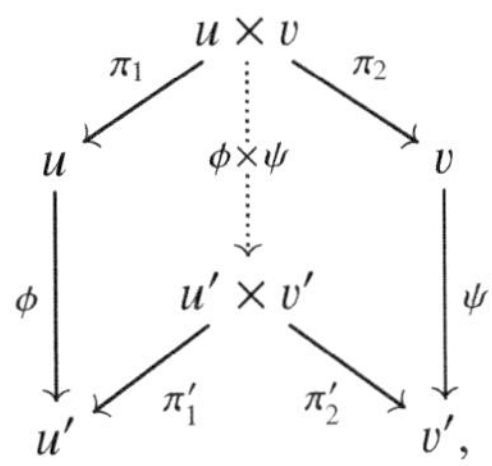

where the morphisms $\pi_1, \pi_2, \pi'_1, \pi'_2$ are the projections. Finally, the symmetry

$$\gamma_{u,v}\colon v \times u \to u \times v$$

is defined by

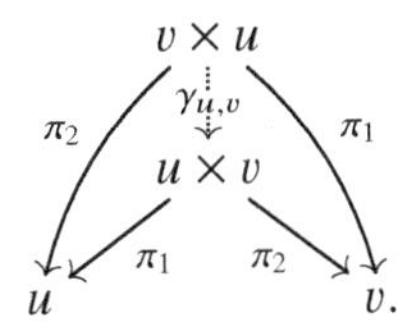

In general, this monoidal structure is not strict, but Mac Lane's coherence theorem ensures that there is no harm in considering it to be strict up to monoidal equivalence of categories.

A cartesian monoidal category is a symmetric monoidal category which is induced by a cartesian category as above. We now show that those can be characterized among symmetric monoidal categories as being those in which every object is equipped with a structure of commutative comonoid in a natural way.

13.4.2 Comonoids. Suppose given a strict monoidal category $(C, \otimes, i)$. A *comonoid* (u, δ, ε) in C consists of an object u together with morphisms

$$\delta\colon u \to u \otimes u, \qquad\qquad \varepsilon\colon i \to u,$$

satisfying the usual associativity and unitality axioms

$$(\delta \otimes 1_u) \circ \delta = (1_u \otimes \delta) \circ \delta, \quad (\varepsilon \otimes 1_u) \circ \delta = 1_u, \quad (1_u \otimes \varepsilon) \circ \delta = 1_u.$$

This structure is dual to the one of monoid (see Example 10.1.5). In the case where the monoidal category is equipped with a symmetry gamma, the comonoid is *commutative* when it satisfies

$$\gamma_{u,u} \circ \delta = \delta.$$

The following theorem is detailed in various places, e.g., [277]:

13.4.3 Theorem. *In a symmetric monoidal category* $(C, \otimes, i)$, *the tensor product is a cartesian product if and only if there are natural transformations of components*

$$\delta_u : u \to u \otimes u, \qquad \varepsilon_u : u \to i,$$

which are monoidal, i.e., for every objects $u, v \in C$ *we have*

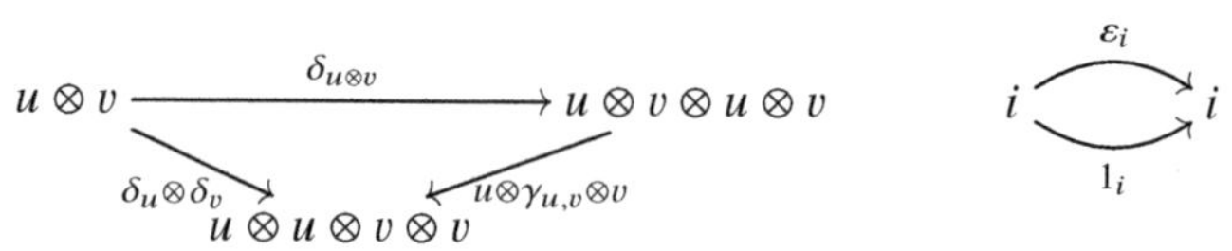

and such that $(u, \delta_u, \varepsilon_u)$ *is a commutative comonoid for every object u.*

Proof. Suppose that C is a cartesian category. Given an object u, the comonoid morphisms δ_u are induced by the universal property of the product:

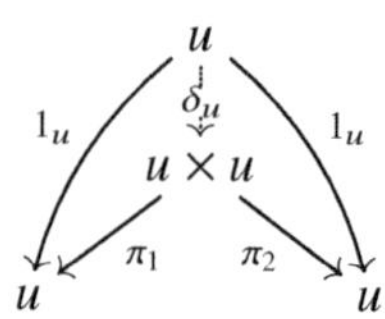

and $\varepsilon_u : u \to 1$ is the terminal morphism. The verification of axioms of commutative comonoids and naturality is left to the reader.

Conversely, suppose that C is a symmetric monoidal category equipped with natural transformations δ and ε as in the statement of the theorem. For any pair of objects u and v, we claim that their cartesian product is $u \otimes v$ equipped with projections

$$1_u \otimes \varepsilon_v : u \otimes v \to u, \qquad \varepsilon_v \otimes 1_v : u \otimes v \to v.$$

Given morphisms $\phi : w \to u$ and $\psi : w \to v$, we claim that the universal morphism is $\chi = (\phi \otimes \psi) \circ \delta_w$:

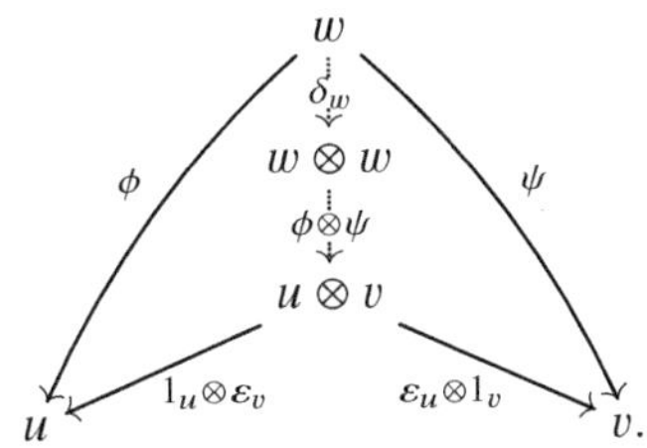

Namely, we have

$$
\begin{aligned}
(1_u \otimes \varepsilon_v) \circ (\phi \otimes \psi) \circ \delta_w &= (1_u \circ \phi) \otimes (\varepsilon_v \circ \psi) \circ \delta_w, && \text{(interchange law)}\\
&= (\phi \otimes \varepsilon_w) \circ \delta_w, && \text{(naturality of } \varepsilon\text{)}\\
&= \phi \circ (1_w \otimes \varepsilon_w) \circ \delta_w, && \text{(interchange law)}\\
&= \phi, && \text{(axiom of comonoids)}
\end{aligned}
$$

so that the triangle on the left commutes, and similarly for the one on the right. Conversely, given a morphism $\chi : w \to u \otimes v$ such that $\pi_1 \circ \chi = \phi$ and $\pi_2 \circ \chi = \psi$, we have

$$
\begin{aligned}
\chi &= (((1_u \otimes \varepsilon_u) \circ \delta_u) \otimes ((\varepsilon_v \otimes 1_v) \circ \delta_v)) \circ \chi, && \text{(axiom of comonoids)}\\
&= (1_u \otimes \varepsilon_u \otimes \varepsilon_v \otimes 1_v) \circ (\delta_u \otimes \delta_v) \circ \chi, && \text{(interchange)}\\
&= (1_u \otimes \varepsilon_u \otimes \varepsilon_v \otimes 1_v) \circ (1_u \otimes \gamma_{u,v} \otimes 1_v) \circ \delta_{u \otimes v} \circ \chi, && (\delta \text{ is monoidal})\\
&= (1_u \otimes \varepsilon_v \otimes \varepsilon_u \otimes 1_v) \circ \delta_{u \otimes v} \circ \chi, && (\gamma \text{ natural}).
\end{aligned}
$$

This concludes the proof. □

13.4.4 Remark. In a more general way, it can be shown that the forgetful functor **Cart** → **MonCat** from cartesian categories to monoidal categories admits a right adjoint

$$\text{Comon} : \mathbf{Cart} \to \mathbf{MonCat}$$

which associates to a monoidal category C, the category of comonoids in C. This has been rediscovered many times and can be traced back to Fox [136].

As a consequence of this theorem, the free cartesian category on a presented symmetric monoidal category can be presented as follows [72]:

13.4.5 Theorem. *Let P be a 3-polygraph presenting a symmetric monoidal category $\overline{P}$ (in particular, $P_0 = \{\star\}$ is reduced to one element). The free cartesian category on $\overline{P}$ is presented by the 3-polygraph Q such that*

$$Q_0 = P_0,$$
$$Q_1 = P_1,$$
$$Q_2 = P_2 \sqcup \{\delta_a : a \to aa, \varepsilon_a : a \to 1 \mid a \in P_1\},$$
$$Q_3 = P_3 \sqcup Q_3',$$

where Q_3' consists of the generators

$$A_a : \delta_a * \delta_a a \Rrightarrow \delta_a * a\delta_a, \quad L_a : \delta_a * \varepsilon_a a \Rrightarrow 1_a, \quad D_\alpha : \alpha * \delta_v \Rrightarrow \delta_u * \alpha\alpha,$$
$$R_a : \delta_a * a\varepsilon_a \Rrightarrow 1_a, \quad E_\alpha : \alpha * \varepsilon_v \Rrightarrow \varepsilon_u,$$

indexed by 1-generators a in P_1 and 2-generators $\alpha : u \Rightarrow v$ in P_2. Here, given $u \in P_1^*$, the 2-cells δ_u and ε_u are defined by induction on u by

$$\delta_\star = 1_\star, \qquad \delta_{au} = \delta_a\delta_u * a\gamma_{u,a} * u, \qquad \varepsilon_\star = 1_\star, \qquad \varepsilon_{au} = \varepsilon_a\varepsilon_u.$$

Graphically, the morphisms δ_u and ε_u can be respectively depicted as

and satisfy

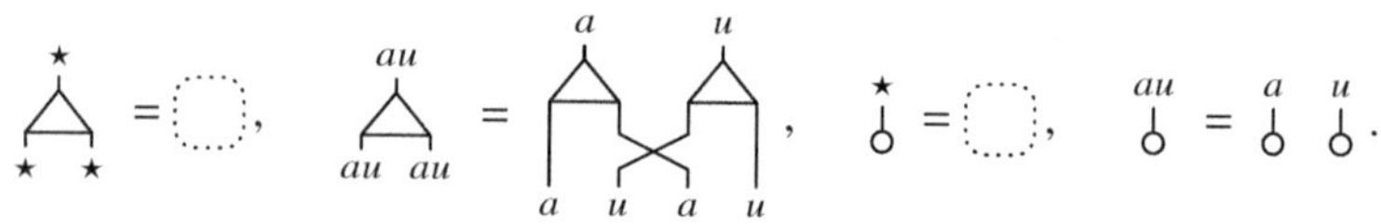

The relations are

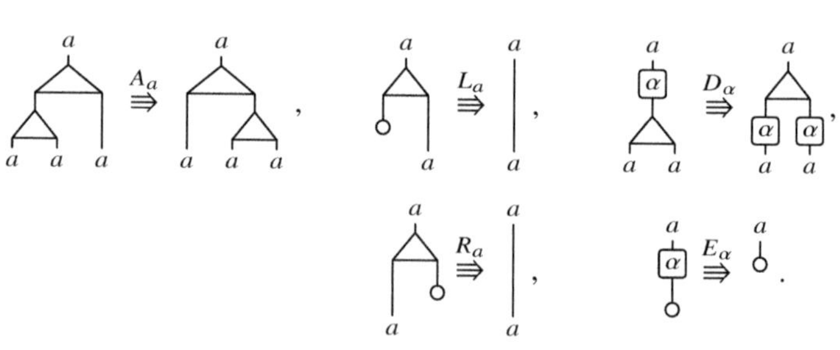

More generally, the free cartesian category on a presented monoidal category C, can be obtained by first presenting the free symmetric monoidal category on C, see §12.5.5, and then applying the above construction.

13.4.6 Presentations of Lawvere theories by polygraphs. The above construction can be used to translate a term rewriting system presenting a Lawvere theory C to a polygraph presenting the underlying monoidal category of C. First, we can define a functor

$$U\colon \mathbf{Pol}_1^{\times} \to \mathbf{Pol}_2,$$

which to every signature P associates the 2-polygraph UP defined by

$$(UP)_0 = \{\star\}, \qquad (UP)_1 = P_0, \qquad (UP)_2 = P_1.$$

This functor induces an isomorphism between $\mathbf{Pol}_1^{\times}$ and the full subcategory of $\mathbf{Pol}_2$ whose objects are polygraphs with $\star$ as only 0-generator. However, note that, given a signature P, the categories P^* and $(UP)^*$ are generally not isomorphic: the former is cartesian whereas the latter is generally only monoidal. In order to address this discrepancy, Theorem 13.4.3 suggests that we consider the 2-polygraph obtained from UP by formally adding a symmetry, see §12.5.5 and Theorem C.1.5, and a natural structure of commutative comonoid for every object, see Theorem 13.4.5. We thus define a functor

$$L\colon \mathbf{Pol}_1^{\times} \to \mathbf{Pol}_3,$$

where LP is the polygraph obtained from UP (seen as a 3-polygraph by the canonical inclusion $\mathbf{Pol}_2 \to \mathbf{Pol}_3$ adding an empty set of 3-generators) by performing those constructions.

13.4.7 Proposition. *Given a signature $P \in \mathbf{Pol}_1^{\times}$, the monoidal category $\overline{LP}$ presented 3-polygraph LP is the cartesian category P^* generated by P.*

As a variant of the above construction, one can show [72]:

13.4.8 Theorem. *For every term rewriting system P, there is a 3-polygraph Q such that $\overline{P}$ is isomorphic to $\overline{Q}$ (as monoidal categories). Moreover, when P is finite, the polygraph Q can also be chosen to be finite.*

13.4.9 Example. We have described, in Example 13.1.14, a term rewriting system corresponding to the theory of groups. By applying the above construction, we obtain the following 3-polygraph P which presents the same Lawvere theory, considered as a monoidal category. We have $P_0 = \{\star\}$, $P_1 = \{a\}$ (thus $P_1^* \simeq \mathbb{N}$), the 2-generators are those coming from the original term rewriting system

$$\mu\colon 2 \to 1, \qquad \eta\colon 0 \to 1, \qquad \iota\colon 1 \to 1,$$

respectively pictured as

, , ,

as well as those corresponding to the cartesian structure

$$\delta\colon 1 \to 2, \qquad \varepsilon\colon 1 \to 0, \qquad \gamma\colon 2 \to 2,$$

respectively pictured as

and the relations are those coming from the term rewriting system

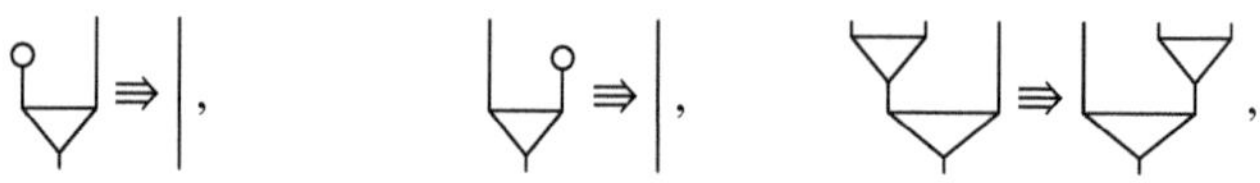

$$\mu a * \mu \Rrightarrow 1_a, \qquad a\mu * \mu \Rrightarrow 1_a, \qquad \mu a * \mu \Rrightarrow a\mu * \mu,$$

$$\delta * \iota a * \mu \Rrightarrow \varepsilon, \qquad \delta * a\iota * \mu \Rrightarrow \varepsilon,$$

in addition to those corresponding to the cartesian structure (omitted here). Note that the use of δ in the two last relations is due to the fact that the variable x_1 is used twice in the corresponding relation on terms. It turns out that this is the polygraph for cocommutative Hopf algebras, see §C.4.6.

13.4.10 Example. The 3-polygraph P corresponding to the Lawvere theory of commutative monoids is the polygraph of bicommutative bialgebras, see §C.4.3.

13.5 Cartesian Polygraphs

As the notations used in Section 13.1 are meant to suggest, term rewriting systems can be seen as particular instances of a notion of *cartesian polygraph* adapted to presenting cartesian categories [268]. We briefly review this notion here.

13.5.1 Cartesian 0-polygraphs. The category $\mathbf{Pol}_0^\times$ of cartesian 0-polygraphs is the category of sets (as for regular polygraphs, see Section 1.1).

13.5.2 Cartesian 1-polygraphs. The category $\mathbf{Pol}_1^\times$ of cartesian 1-polygraphs is the category of signatures, see §13.1.1.

13.5.3 Cartesian 2-polygraphs. The category $\mathbf{Pol}_2^\times$ of cartesian 2-polygraphs is the category of term rewriting systems, see §13.1.12. Given a 2-polygraph P, we write $P_{\leqslant 1}$ for the underlying 1-polygraph.

13.5.4 Cartesian 2-categories. In order to define cartesian 3-polygraphs, we first need to introduce the following notion. A 2-category C is *cartesian* if its underlying category is cartesian and for every pair of 2-cells

$$F\colon \phi \Rightarrow \phi' : w \to u, \qquad\qquad G\colon \psi \Rightarrow \psi' : w \to v,$$

with same 0-source (resp. 0-target), there exists a unique morphism

$$\langle F, G\rangle\colon \langle \phi, \phi'\rangle : w \to u \times v,$$

such that $\langle F, G\rangle *_0 \pi_1 = F$ and $\langle F, G\rangle *_0 \pi_2 = G$. Graphically,

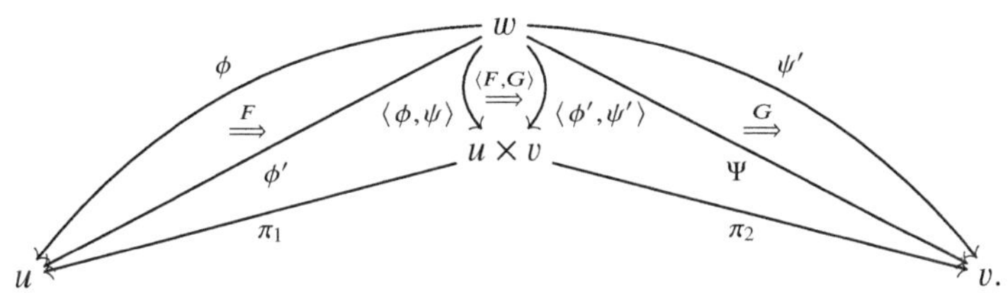

We write $\mathbf{Cart}_2$ for the category of cartesian 2-categories, morphisms being 2-functors whose underlying functor is cartesian.

13.5.5 Lawvere 2-theories. Given a set P_0, the cartesian category $(\mathbf{F}/P_0)^{\mathrm{op}}$ can canonically be seen as cartesian 2-category with only identity 2-cells. A *Lawvere 2-theory* is a cartesian 2-category equipped with a cartesian 2-functor

$$(\mathbf{F}/P_0)^{\mathrm{op}} \to C,$$

which preserves finite products and is the identity on objects [358].

13.5.6 The generated cartesian (2, 1)-category. Given a cartesian 2-polygraph P, we write $P^\top$ for the cartesian $(2,1)$-category it generates. It has the category $P^*_{\leqslant 1}$ freely generated by the underlying signature (i.e., 1-polygraph) as underlying category and its 2-cells are generated under composition and inverses by the elements of P_2, with source and target indicated by s_1 and t_1. We write $P_2^\top$ for the set of 2-cells of $P^\top$. The cartesian $(2,1)$-category $P^\top$ is canonically a Lawvere 2-theory, with P_0 as sorts.

13.5.7 Cartesian (3, 1)-polygraphs. A *cartesian* $(3,1)$*-polygraph* consists of

– a 2-polygraph P,

– a set P_3 of 3-generators together with functions $s_2, t_2 \colon P_3 \longrightarrow P_2^{\top}$ such that $s_1^* \circ s_2 = t_1^* \circ s_2$ and $s_1^* \circ t_2 = t_1^* \circ t_2$.

A *morphism* $f \colon P \to Q$ of cartesian $(3,1)$-polygraphs consists of a morphism $P_{\leqslant 2} \to Q_{\leqslant 2}$ between the underlying 2-polygraphs together with a function $P_3 \to Q_3$ which commutes with source and with target.

13.5.8 Congruence. A *congruence* $\approx$ on a cartesian 2-category C is a congruence on the underlying 2-category such that, for every 1-cells

$$F, F' \colon \phi \Rightarrow \phi' \colon w \to u, \qquad\qquad G, G' \colon \psi \Rightarrow \psi' \colon w \to v,$$

we have that

$$F \approx F' \quad \text{and} \quad G \approx G' \qquad \text{implies} \qquad \langle F, G \rangle \approx \langle F', G' \rangle.$$

Given a 3-polygraph P, the *P-congruence* $\approx^P$ is the smallest congruence such that $F \approx^P G$, for every 3-generator $\Lambda \colon F \Rrightarrow G$.

13.5.9 Coherent presentation. A $(3,1)$-polygraph P is a *coherent presentation* of a cartesian category C when C is the cartesian category presented by the underlying 2-polygraph, i.e., $\overline{P_{\leqslant 2}} = C$, and for every parallel 2-cells $F, G \colon \phi \Rightarrow \psi$ in $P_2^{\top}$ one has $F \approx^P G$.

13.5.10 Cartesian Squier homotopical theorem. An analogous of Squier's homotopical theorem (Theorems 7.3.5 and 12.1.7) can be formulated in this context: the cartesian category presented by a convergent term rewriting system P admits a coherent presentation by the $(3,1)$-polygraph (P, P_3), where P_3 consists of a confluence diagram for every critical branching of P. This can be used to recover various coherence results (such as Mac Lane's coherence theorem for monoidal categories) through term rewriting systems [38, 92, 286, 358].

13.5.11 Homological invariants. Finally, we shall briefly mention that the homological tools developed for 2-polygraphs in Chapter 9 can be adapted to term rewriting systems [266, 267]. In particular, the homology of a Lawvere theory can be used in order to obtain lower bounds on generators and relations that any presentation should have [195, 196, 267].

PART IV

POLYGRAPHS

14
Higher Categories

The remaining chapters of this book present a general theoretical background underlying all constructions encountered so far. Thus, from this point on, we shall assume that the reader is well acquainted with the basics of category theory, as developed in [261].

Among the many existing notions of higher categories, the notion of strict globular n-category that we shall describe is, in some sense, the most basic one. The earliest published reference to the concept appears to be [63], where it is motivated by the study of higher homotopies. Precisely, one looks here for higher-dimensional analogues of the fundamental groupoid of a space. Grothendieck soon afterward realized the need for a weak version of infinity-groupoids to fulfill this purpose, see [153, 270]. The theory has been further developed in the highly influential paper [334], advocating the use of strict higher categories as coefficients for non-abelian cohomology. The same paper introduces the notion of *freely generated ω-category over a computad*—here called a polygraph—which is central in the present work, together with the definition of oriented simplices or orientals. Orientals yield a nerve functor from strict ω-categories to simplicial sets, turning the former into models of homotopy types, as developed in [13, 144].

The present work stresses yet another aspect of strict ω-categories, especially the free ones, as higher-dimensional rewriting "spaces" in the spirit of [163].

In this chapter, we set the essential definitions and notations. Starting with a description of the basic "shapes", that is, the presheaf category $\mathbf{Glob}_\omega$ of globular sets, we define a family of operations endowing a globular set with a structure of ω-category. We then prove that the category $\mathbf{Cat}_\omega$ of strict ω-categories is exactly the category of algebras of the monad induced by the forgetful functor from $\mathbf{Cat}_\omega$ to $\mathbf{Glob}_\omega$. We finally define important subcategories of $\mathbf{Cat}_\omega$ obtained by requiring cells to be invertible above a given dimension.

14.1 Globular Sets

14.1.1 Globes. We first define the small category $\mathbb{O}$ of *globes*: its objects are the integers $0, 1, \ldots$ and its morphisms are generated by a double sequence

$$\sigma_n, \tau_n \colon n \to n+1,$$

where $n \in \mathbb{N}$,

$$0 \overset{\sigma_0}{\underset{\tau_0}{\rightrightarrows}} 1 \overset{\sigma_1}{\underset{\tau_1}{\rightrightarrows}} \cdots \overset{\sigma_{n-1}}{\underset{\tau_{n-1}}{\rightrightarrows}} n \overset{\sigma_n}{\underset{\tau_n}{\rightrightarrows}} n+1 \overset{\sigma_{n+1}}{\underset{\tau_{n+1}}{\rightrightarrows}} \cdots,$$

quotiented by the equations

$$\sigma_{n+1} \circ \sigma_n = \tau_{n+1} \circ \sigma_n, \tag{14.1}$$
$$\sigma_{n+1} \circ \tau_n = \tau_{n+1} \circ \tau_n. \tag{14.2}$$

As a consequence of these equations, whenever $0 \leqslant m < n$, the hom-set $\mathbb{O}(m, n)$ contains exactly two morphisms:

$$\sigma_m^n = \sigma_{n-1} \circ \cdots \circ \sigma_m,$$
$$\tau_m^n = \tau_{n-1} \circ \cdots \circ \tau_m.$$

A *globular set* is then a presheaf on $\mathbb{O}$, that is, a functor $X \colon \mathbb{O}^{\mathrm{op}} \to \mathbf{Set}$. Thus, a globular set X amounts to a sequence of sets $X(n)$ of n-dimensional globes, for each $n \geqslant 0$, together with source and target maps

$$X(\sigma_n), X(\tau_n) \colon X(n+1) \to X(n)$$

satisfying the *globular relations*, dual to (14.1) and (14.2). Let us denote $X(\sigma_n)$ by s_n and $X(\tau_n)$ by t_n. Whenever $m \leq n$, we set

$$s_m^n = s_m \circ \cdots \circ s_{n-1},$$
$$t_m^n = t_m \circ \cdots \circ t_{n-1},$$

thus $s_m^n, t_m^n \colon X(n) \to X(m)$.

Globular sets and natural transformations between them define a category denoted by $\mathbf{Glob}_\omega$. For any globular set X and integer n, we shall denote $X(n)$ by X_n. The elements of X_n are called *n-cells*. For any n-cell x and $m \leq n$, the notations $s_m(x)$ and $t_m(x)$ will stand for $s_m^n(x)$ and $t_m^n(x)$, respectively. The m-cell $s_m(x)$ will be called the *m-source* of x (or simply the *source* if $m = n-1$) and the m-cell $t_m(x)$ the *m-target* of x (or simply the *target* if $m = n-1$). Two n-cells x, y are called *parallel* if either $n = 0$ or $s_{n-1}(x) = s_{n-1}(y)$ and $t_{n-1}(x) = t_{n-1}(y)$ otherwise.

Let us denote $\mathbb{O}^{(n)}$ the full subcategory of $\mathbb{O}$ whose objects are $0, \ldots, n$. Then the category $\mathbf{Glob}_n$ of *n-globular sets* is by definition the category of presheaves on $\mathbb{O}^{(n)}$. If $0 \leqslant m < n$, the canonical inclusion

$$\mathbb{O}^{(m)} \to \mathbb{O}^{(n)}$$

gives rise, by precomposition, to a truncation functor

$$U_m^n \colon \mathbf{Glob}_n \to \mathbf{Glob}_m.$$

Also, for each $n \geqslant 0$, the canonical inclusion $\mathbb{O}^{(n)} \to \mathbb{O}$ gives rise to a truncation functor

$$U_n \colon \mathbf{Glob}_\omega \to \mathbf{Glob}_n$$

making all the following triangles commute:

$$\begin{array}{ccc} \mathbf{Glob}_\omega & & \\ {\scriptstyle U_n}\downarrow & \searrow^{U_m} & \\ \mathbf{Glob}_n & \xrightarrow[U_m^n]{} & \mathbf{Glob}_m. \end{array}$$

For any n-globular set X, and $m < n$, the notation $U_m(X)$ will stand for $U_m^n(X)$. Remark that $\mathbf{Glob}_\omega$ is the projective limit of the diagram

$$\mathbf{Glob}_0 \xleftarrow{U_0^1} \mathbf{Glob}_1 \longleftarrow \cdots \longleftarrow \mathbf{Glob}_{n-1} \xleftarrow{U_{n-1}^n} \mathbf{Glob}_n \longleftarrow \cdots.$$

14.1.2 Globes and spheres. As for any presheaf category, we get a Yoneda embedding

$$\mathrm{Y} \colon \mathbb{O} \to \mathbf{Glob}_\omega$$

defined on objects by $Y(m)(n) = \mathbb{O}(m, n)$. For each $n \geqslant 0$, we call *n-globe* and denote by $\mathbb{O}_n$ the representable globular set $\mathrm{Y}(n)$. The n-globe has exactly two i-cells in dimensions $0 \leqslant i < n$, one n-cell, and no i-cell for $i > n$. The sub-globular set of $\mathbb{O}_n$ having the same cells as $\mathbb{O}_n$ in all dimensions $i \neq n$ and no n-cell will be called the *n-sphere*, and denoted by $\partial\mathbb{O}_n$. Remark that $\partial\mathbb{O}_0$ is the initial globular set with no cells at all. Globes and spheres come with a family of canonical inclusion morphisms

$$\mathrm{i}_n \colon \partial\mathbb{O}_n \to \mathbb{O}_n$$

which we shall encounter in numerous occasions. For example, the case $n = 2$ may be pictured as

14.2 Strict n-Categories

14.2.1 Definition. A *strict ω-category* is given by a globular set C together with a family of partial binary composition operations $(*_i)_{i\in\mathbb{N}}$ and identity operations $(1^i)_{i\in\mathbb{N}\setminus\{0\}}$ subject to the following conditions:

- if $0 \leqslant i < k$ and x, y are k-cells such that $t_i(x) = s_i(y)$ (in which case we say that x and y are *i-composable*) there is a k-cell $x *_i y$,
- if $k > 0$ and x is a $(k-1)$-cell, there is a k-cell 1^k_x, and more generally, if $i \geq 0$ and x is an i-cell, we may define recursively on $k > i$ a k-cell 1^k_x by $1^k_x = 1^k_{1^{k-1}_x}$.

Compositions and units are subject to:

1. positional conditions prescribing the source and target of composites and units, namely
   - if $0 \leqslant i < j$, then $s_j(x *_i y) = s_j(x) *_i s_j(y)$ and $t_j(x *_i y) = t_j(x) *_i t_j(y)$,
   $$s_j(x *_i y) = s_j(x) *_i s_j(y) \quad \text{and} \quad t_j(x *_i y) = t_j(x) *_i t_j(y),$$
   - if $0 \leqslant j \leqslant i$, then
   $$s_j(x *_i y) = s_j(x) \quad \text{and} \quad t_j(x *_i y) = t_j(y),$$
   - if $0 \leqslant i < k$ and x is an i-cell, then
   $$s_i(1^k_x) = x = t_i(1^k_x),$$
2. computational conditions of
   - associativity: if $i < k$ and x, y, z are k-cells such that $t_i(x) = s_i(y)$ and $t_i(y) = s_i(z)$, then
   $$(x *_i y) *_i z = x *_i (y *_i z),$$
   - neutrality of units: if $0 \leqslant i < k$ and x is a k-cell, then
   $$1^k_{s_i(x)} *_i x = x *_i 1^k_{t_i(x)} = x,$$
   - exchange: if $i < j < k$ and x, y, z, v are k-cells such that $t_j(x) = s_j(y)$, $t_j(z) = s_j(v)$ and $t_i(x) = s_i(z)$, then also $t_j(y) = s_j(v)$, and
   $$(x *_j y) *_i (z *_j v) = (x *_i z) *_j (y *_i v),$$
   - compatibility of units: if $0 \leqslant i < j < k$ and x, y are i-composable j-cells, then
   $$1^k_{x *_i y} = 1^k_x *_i 1^k_y.$$

Let C, D be two strict ω-categories. An *ω-functor* $f: C \to D$ is a morphism of the underlying globular sets which preserves the compositions and units. Strict ω-categories and strict ω-functors build a (large) category we denote by $\mathbf{Cat}_\omega$. If we restrict the above construction to cells of dimension at most n, we get the category of strict n-categories, denoted by $\mathbf{Cat}_n$.

From now on we will drop the adjective "strict" and we will speak of "ω-categories" and "n-categories" when we mean "strict ω-categories" and "strict n-categories".

14.2.2 Remark. The structure of n-category is sometimes presented by the alternative set of operations and axioms described in Appendix D.

14.2.3 ω-categories as models of a projective sketch. The above axioms for ω-categories can be presented in diagrammatic form as follows.

For any globular set X and $0 \leqslant i < n$, there is a pullback square in $\mathbf{Set}$:

$$\begin{array}{ccc} X_n \underset{X_i}{\times} X_n & \xrightarrow{r_i^n} & X_n \\ {\scriptstyle l_i^n}\downarrow & & \downarrow{\scriptstyle s_i^n} \\ X_n & \xrightarrow[t_i^n]{} & X_i . \end{array}$$

The operations of compositions and units become maps:

$$*_i : X_n \underset{X_i}{\times} X_n \to X_n$$

and

$$1^n : X_i \to X_n$$

for $0 \leqslant i < n$.

The positional conditions for compositions amount to the commutation of the following diagrams:

$$\begin{array}{ccc} X_n \underset{X_i}{\times} X_n & \xrightarrow{*_i} & X_n \\ {\scriptstyle s_j^n \times_{X_i} s_j^n}\downarrow & & \downarrow{\scriptstyle s_j^n} \\ X_j \underset{X_i}{\times} X_j & \xrightarrow[*_i]{} & X_i \end{array} \qquad \begin{array}{ccc} X_n \underset{X_i}{\times} X_n & \xrightarrow{*_i} & X_n \\ {\scriptstyle t_j^n \times_{X_i} t_j^n}\downarrow & & \downarrow{\scriptstyle t_j^n} \\ X_j \underset{X_i}{\times} X_j & \xrightarrow[*_i]{} & X_i \end{array}$$

for $0 \leqslant i < j < n$ and

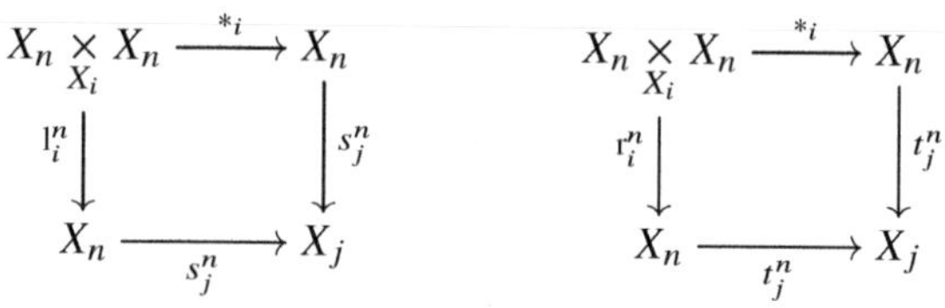

for $0 \leqslant j \leqslant i < n$.

As for units, if $0 \leqslant i < n$, the positional conditions amount to the commutations of

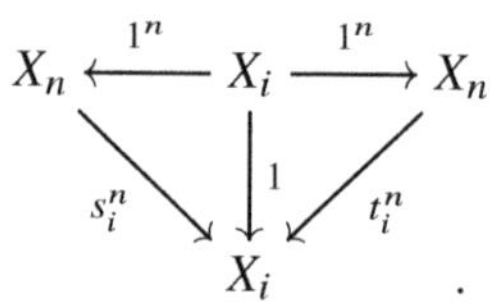

Now, each axiom is expressed by the commutation of a diagram in **Set** involving arrows derived from the source, target, compositions, and unit arrows by means of universal constructions.

– Associativity of compositions amounts to the commutation of

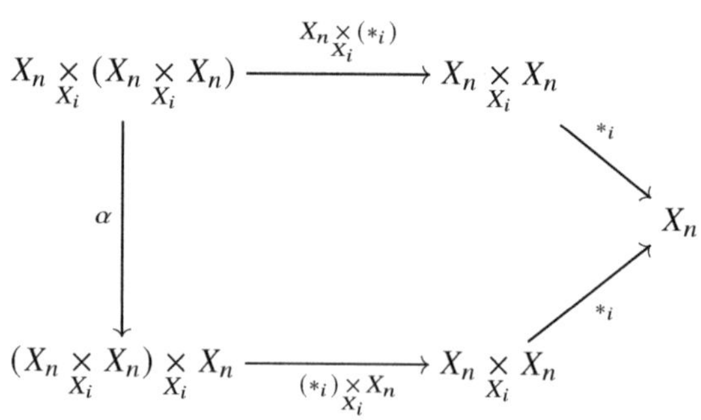

where α is the canonical bijection between both pullbacks.

– Let $0 \leqslant i < j < n$. We build the diagram

$$\begin{array}{ccccc}
(X_n \underset{X_j}{\times} X_n) \underset{X_i}{\times} (X_n \underset{X_j}{\times} X_n) & & \xrightarrow{\quad \mathrm{r} \quad} & & X_n \underset{X_j}{\times} X_n \\
 & \overset{\lambda}{\searrow} & & & \downarrow{\scriptstyle \mathrm{l}_j^n} \\
\downarrow{\scriptstyle 1} & & X_n \underset{X_i}{\times} X_n & \xrightarrow{\mathrm{r}_i^n} & X_n \\
 & & \downarrow{\scriptstyle \mathrm{l}_i^n} & & \downarrow{\scriptstyle s_n^i} \\
X_n \underset{X_j}{\times} X_n & \xrightarrow[\mathrm{l}_j^n]{} & X_n & \xrightarrow[t_n^i]{} & X_i
\end{array}$$

where both solid squares are pullbacks. There is a unique universal arrow λ making the whole diagram commute. Similarly, by replacing the left projection l_j^n with the right projection r_j^n in the above diagram, we get a universal arrow

$$\rho\colon (X_n \underset{X_j}{\times} X_n) \underset{X_i}{\times} (X_n \underset{X_j}{\times} X_n) \to X_n \underset{X_i}{\times} X_n.$$

Consider now the diagram

$$\begin{array}{ccccc}
(X_n \underset{X_j}{\times} X_n) \underset{X_i}{\times} (X_n \underset{X_j}{\times} X_n) & & \xrightarrow{\quad \rho \quad} & & X_n \underset{X_i}{\times} X_n \\
 & \overset{\theta}{\searrow} & & & \downarrow{\scriptstyle *_i} \\
\downarrow{\scriptstyle \lambda} & & X_n \underset{X_j}{\times} X_n & \xrightarrow{\mathrm{r}_j^n} & X_n \\
 & & \downarrow{\scriptstyle \mathrm{l}_j^n} & & \downarrow{\scriptstyle s_n^j} \\
X_n \underset{X_i}{\times} X_n & \xrightarrow[*_i]{} & X_n & \xrightarrow[t_n^j]{} & X_j.
\end{array}$$

The positional conditions on compositions and the globular relations ensure that the outer square commutes, whereas the small solid square is a pullback by definition. Therefore, we get a unique universal arrow θ as shown in the diagram.

Now the exchange rule amounts to the commutation of

$$\begin{array}{ccc}
(X_n \underset{X_j}{\times} X_n) \underset{X_i}{\times} (X_n \underset{X_j}{\times} X_n) & \xrightarrow{\theta} & X_n \underset{X_j}{\times} X_n \\
\downarrow{\scriptstyle (*_j) \underset{X_i}{\times} (*_j)} & & \downarrow{\scriptstyle *_j} \\
X_n \underset{X_i}{\times} X_n & \xrightarrow[*_i]{} & X_n.
\end{array}$$

- Let $0 \leqslant i < n$. The positional conditions on units imply that the following diagram of solid arrows commutes:

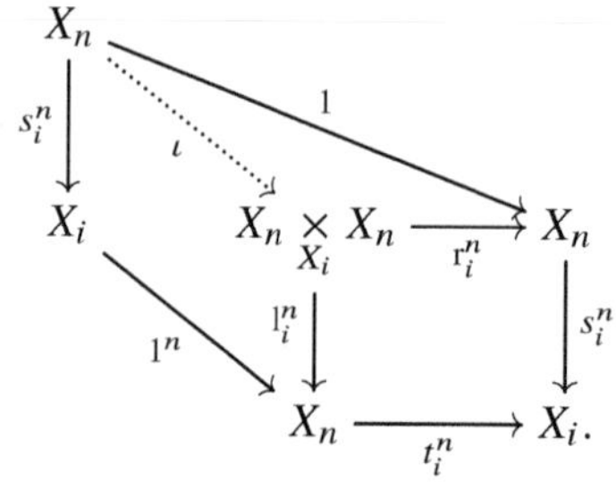

  Therefore, there is a unique universal arrow ι making the whole diagram commute. Now the first axiom for left units amounts to the fact that ι equalizes the pair $(\mathrm{r}_i^n, *_i)$, that is, the commutation of

$$X_n \xrightarrow{\ \iota\ } X_n \underset{X_i}{\times} X_n \underset{*_i}{\overset{\mathrm{r}_i^n}{\rightrightarrows}} X_n.$$

  The first axiom for right units is treated similarly.
- Finally, the compatibility of compositions with units amounts to the commutation of the following diagram

$$\begin{array}{ccc} X_m \underset{X_i}{\times} X_m & \xrightarrow{\ *_i\ } & X_m \\ {\scriptstyle 1^n \underset{X_i}{\times} 1^n}\big\downarrow & & \big\downarrow{\scriptstyle 1^n} \\ X_n \underset{X_i}{\times} X_n & \xrightarrow{\ *_i\ } & X_n \end{array}$$

  whenever $0 \leqslant i < m < n$.

14.2.4 Proposition. *The category* $\mathbf{Cat}_\omega$ *is a category of models of a projective sketch.*

Proof. The above diagrammatic presentation of the axioms of ω-categories defines a sketch whose models are actual ω-categories. □

14.2.5 Corollary. *The category* $\mathbf{Cat}_\omega$ *is complete and cocomplete.*

Proof. This follows from the previous result by using Proposition G.1.10. □

14.2.6 Truncation functors. As for globular sets, whenever $0 \leqslant m < n$, we get *truncation functors*

$$U_m^n \colon \mathbf{Cat}_n \to \mathbf{Cat}_m$$

and

$$U_n \colon \mathbf{Cat}_\omega \to \mathbf{Cat}_n$$

making all triangles

$$\begin{array}{ccc} \mathbf{Cat}_\omega & & \\ {\scriptstyle U_n}\downarrow & \searrow^{U_m} & \\ \mathbf{Cat}_n & \xrightarrow[U_m^n]{} & \mathbf{Cat}_m \end{array}$$

commute.

Here again, $\mathbf{Cat}_\omega$ appears as the projective limit of the diagram

$$\mathbf{Cat}_0 \xleftarrow{U_0^1} \mathbf{Cat}_1 \longleftarrow \cdots \longleftarrow \mathbf{Cat}_n \xleftarrow{U_n^{n+1}} \mathbf{Cat}_{n+1} \longleftarrow \cdots.$$

Remark that, by abuse of language, we use the same notation for truncation functors among n-categories and among n-globular sets.

By construction, ω-categories are globular sets with structure, whence a forgetful functor

$$V \colon \mathbf{Cat}_\omega \to \mathbf{Glob}_\omega$$

which restricts for each $n \in \mathbb{N}$ to

$$V_n \colon \mathbf{Cat}_n \to \mathbf{Glob}_n.$$

These forgetful functors commute with the above truncation functors, that is, the following diagram commutes whenever $0 \leqslant m < n$:

$$\begin{array}{ccc} \mathbf{Cat}_n & \xrightarrow{U_m^n} & \mathbf{Cat}_m \\ {\scriptstyle V_n}\downarrow & & \downarrow{\scriptstyle V_m} \\ \mathbf{Glob}_n & \xrightarrow[U_m^n]{} & \mathbf{Glob}_m. \end{array} \tag{14.3}$$

14.2.7 Proposition. *The forgetful functor* $V \colon \mathbf{Cat}_\omega \to \mathbf{Glob}_\omega$ *admits a left adjoint* $F \colon \mathbf{Glob}_\omega \to \mathbf{Cat}_\omega$*. Likewise, for each* $n \in \mathbb{N}$, $V_n \colon \mathbf{Cat}_n \to \mathbf{Glob}_n$ *admits a left adjoint* $F_n \colon \mathbf{Glob}_n \to \mathbf{Cat}_n$*.*

Proof. The categories $\mathbf{Glob}_\omega$ and $\mathbf{Cat}_\omega$ are categories of models of projective sketches S and S', respectively, whereas the functor V is the one induced on models by the inclusion morphism $S \hookrightarrow S'$. Therefore V admits a left adjoint (see Theorem G.1.8). The same arguments hold for $\mathbf{Glob}_n$ and $\mathbf{Cat}_n$, where $n \in \mathbb{N}$. □

14.2.8 Proposition. *The forgetful functor $V\colon \mathbf{Cat}_\omega \to \mathbf{Glob}_\omega$ preserves filtered colimits.*

Proof. This is a general property of functors induced by a morphism of projective sketches involving only finite cones (see [26, Chapter 4, Theorem 4.4]). Concretely, the colimit C of a filtered diagram in $\mathbf{Cat}_\omega$ is obtained by taking the colimit X of the underlying diagram in $\mathbf{Glob}_\omega$ and defining a structure of ω-category on X in the obvious way, so that $V(C) = X$. Thus, the stronger statement that V *creates* filtered colimits holds. □

14.2.9 Globes and spheres. By abuse of notation, the free ω-category $F(\mathbb{O}_n)$ generated by $\mathbb{O}_n$ will be still denoted by $\mathbb{O}_n$ and called the *n-globe*. Likewise we denote by $\partial\mathbb{O}_n$ the free ω-category $F(\partial\mathbb{O}_n)$, and call it the *n-sphere*. Remark that, in the case of globes and spheres, the free functor only adds new identity cells to the ones already present in the globular globes and spheres.

For any ω-category C, the set of *n-globes of C* is the hom-set $\mathbf{Cat}_\omega(\mathbb{O}_n, C)$, which amounts to the set C_n of n-cells in C. Likewise, the set of *n-spheres of C* is the hom-set $\mathbf{Cat}_\omega(\partial\mathbb{O}_{n+1}, C)$, which amounts to the set of pairs of parallel n-cells in C, that is, parallel in the underlying globular set.

14.3 Basic Examples

Let us first mention a few immediate examples of ω-categories.

- Sets: as $\mathbf{Cat}_0 = \mathbf{Glob}_0 = \mathbf{Set}$ and $\mathbf{Cat}_n$ naturally embeds in $\mathbf{Cat}_\omega$ (see §14.4.6), any set S can be viewed as an ω-category, precisely the ω-category C whose 0-cells are the elements of S and whose n-cells, $n > 0$, are all of the form 1^n_x for $x \in S$.
- Monoids: any monoid M, being a 1-category with a unique object, can be seen as an ω-category whose n-cells are identities for all $n \geqslant 2$.
- Commutative monoids: to any *commutative* monoid $(A, +)$ we may associate the ω-category C defined by $C_0 = \{\star\}$, $C_1 = \{1^1_\star\}$, $C_2 = A$ and having only identity cells in higher dimensions. The source and target maps are uniquely determined, whereas compositions are defined for any pair (u, v) of 2-cells by $u *_0 v = u *_1 v = u + v$. The axioms of ω-categories are easily checked. Conversely, for any ω-category C such that $C_0 = \{\star\}$, $C_1 = \{1^1_\star\}$ and C_n has only identity cells for $n > 2$, the $*_0$ and $*_1$ compositions on C_2 coincide and are commutative operations, so that $(C_2, *_0)$ (or $(C_2, *_1)$) becomes an abelian monoid.

14.4 More Properties of $\mathbf{Cat}_\omega$

14.4.1 Local presentability. By definition, the category $\mathbf{Glob}_\omega$ of globular sets is a category of presheaves, so that limits and colimits are computed pointwise. On the other hand, we already noticed that $\mathbf{Cat}_\omega$ is the category of models of a projective sketch and is hence complete and cocomplete (see Corollary 14.2.5). The forgetful functor $V\colon \mathbf{Cat}_\omega \to \mathbf{Glob}_\omega$, being a right adjoint, preserves limits. Thus limits in $\mathbf{Cat}_\omega$ are computed as in $\mathbf{Glob}_\omega$. Colimits however are hard to compute, even in $\mathbf{Cat}_1$.

14.4.2 Enrichment. For each $n \geqslant 0$, the category $\mathbf{Cat}_n$ has a monoidal structure defined by its cartesian product and terminal object. Thus, the notion of $\mathbf{Cat}_n$-enriched category makes sense: in fact, $\mathbf{Cat}_n$-enriched categories are just $(n+1)$-categories. As for $n = \omega$, it turns out that $\mathbf{Cat}_\omega$ is enriched over itself.

14.4.3 Proposition. *The forgetful functor* $V\colon \mathbf{Cat}_\omega \to \mathbf{Glob}_\omega$ *is monadic.*

Proof. First remark that, for any $0 \leqslant i < n$, the correspondence

$$X \mapsto X_n \underset{X_i}{\times} X_n$$

is functorial from $\mathbf{Glob}_\omega$ to $\mathbf{Set}$. Indeed, the Yoneda embedding yields a pushout

$$\begin{array}{ccc} \mathbb{O}_i & \xrightarrow{\mathrm{Y}(\tau_i^n)} & \mathbb{O}_n \\ {\scriptstyle \mathrm{Y}(\sigma_i^n)}\downarrow & & \downarrow \\ \mathbb{O}_n & \longrightarrow & \mathbb{O}_n \underset{\mathbb{O}_i}{+} \mathbb{O}_n \end{array}$$

in $\mathbf{Glob}_\omega$ and we get a natural bijection

$$X_n \underset{X_i}{\times} X_n \simeq \mathbf{Glob}_\omega(\mathbb{O}_n \underset{\mathbb{O}_i}{+} \mathbb{O}_n, X)$$

so that our correspondence is the object part of the (representable) functor

$$\mathbf{Glob}_\omega(\mathbb{O}_n \underset{\mathbb{O}_i}{+} \mathbb{O}_n, -)\colon \mathbf{Glob}_\omega \to \mathbf{Set}.$$

Now, by Proposition 14.2.7, the functor V admits a left adjoint F. By Beck's monadicity theorem [261], it is sufficient to prove that V creates absolute coequalizers. Thus, let C, D be ω-categories, $f, g\colon C \to D$ a pair of morphisms, $X = VC$, $Y = VD$ the underlying globular sets. Suppose $w\colon Y \to Z$ is an absolute coequalizer of the pair $u = Vf$, $v = Vg$. We must prove the existence of a unique ω-category E, and a unique morphism $h\colon D \to E$ such that

– $VE = Z$,

– $Vh = w$

and check that h is the coequalizer of the pair (f, g) in $\mathbf{Cat}_\omega$.

Let us first define a structure of ω-category on the globular set Z. By hypothesis, the diagram

$$X \underset{v}{\overset{u}{\rightrightarrows}} Y \xrightarrow{w} Z \tag{14.4}$$

is an absolute coequalizer. Thus, by applying the functor $\mathbf{Glob}_\omega(\mathbb{O}_n \underset{\mathbb{O}_i}{+} \mathbb{O}_n, -)$ defined in the preliminary remark, we still get a coequalizer, now in **Set**:

$$X_n \underset{X_i}{\times} X_n \underset{v'}{\overset{u'}{\rightrightarrows}} Y_n \underset{Y_i}{\times} Y_n \xrightarrow{w'} Z_n \underset{Z_i}{\times} Z_n.$$

Likewise by applying the functor $\mathbf{Glob}_\omega(\mathbb{O}_n, -)$, we get another coequalizer diagram

$$X_n \underset{v_n}{\overset{u_n}{\rightrightarrows}} Y_n \xrightarrow{w_n} Z_n.$$

Consider now the following diagram:

$$\begin{array}{ccccc}
X_n \underset{X_i}{\times} X_n & \underset{v'}{\overset{u'}{\rightrightarrows}} & Y_n \underset{Y_i}{\times} Y_n & \xrightarrow{w'} & Z_n \underset{Z_i}{\times} Z_n \\
{\scriptstyle *_i}\downarrow & & {\scriptstyle *_i}\downarrow & & \vdots ? \\
X_n & \underset{v_n}{\overset{u_n}{\rightrightarrows}} & Y_n & \xrightarrow{w_n} & Z_n.
\end{array}$$

As u and v come from morphisms in $\mathbf{Cat}_\omega$, the squares on u', u_n, and v', v_n commute. Therefore, $w_n \circ (*_i)$ coequalizes the pair (u', v'), and there exists a unique map from $Z_n \underset{Z_i}{\times} Z_n$ to Z_n making the right-hand square commute. This map defines the i-composition among n-cells in Z, still denoted by $*_i$. By a similar argument we may define the unit map $1^n : Z_{n-1} \to Z_n$.

It remains to check that the compositions and units just defined on Z satisfy the axioms of ω-categories. This amounts to check the commutation of all diagrams expressing these axioms. We shall treat the axiom of associativity in detail and leave the remaining axioms as exercises. By applying appropriate functors to the coequalizer diagram (14.4), we get the following diagram in **Set**:

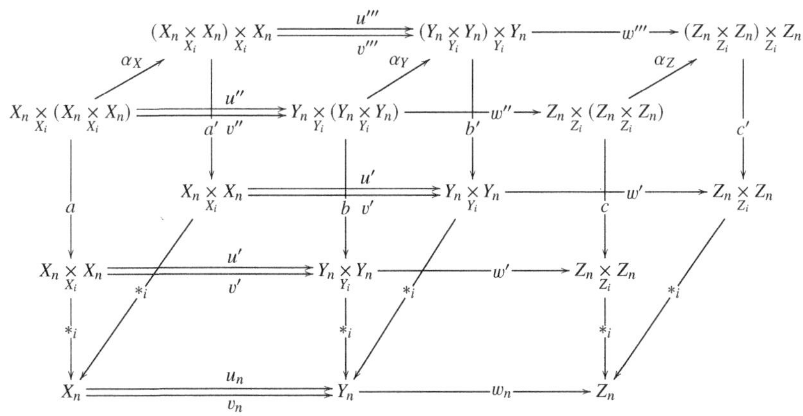

where $a = X_n \underset{X_i}{\times} (*_i)$, $a' = (*_i) \underset{X_i}{\times} X_n$, and b, b', c, c' are defined accordingly.

Because the coequalizer (14.4) is absolute, all horizontal lines are also coequalizer diagrams. Now $w_n \circ (*_i) \circ b$ coequalizes the pair (u'', v''), therefore

$$(*_i) \circ c \circ w'' = w_n \circ (*_i) \circ b. \tag{14.5}$$

On the other hand, by naturality of α,

$$\begin{aligned}(*_i) \circ c' \circ \alpha_Z \circ w'' &= (*_i) \circ c' \circ w''' \circ \alpha_Y \\ &= (*_i) \circ w' \circ b' \circ \alpha_Y \\ &= w_n \circ (*_i) \circ b' \circ \alpha_Y\end{aligned}$$

and by the associativity of $*_i$ on Y, the latter expression is equal to $w_n \circ (*_i) \circ b$ so that we get

$$(*_i) \circ c' \circ \alpha_Z \circ w'' = w_n \circ (*_i) \circ b. \tag{14.6}$$

Now as w'' is a coequalizer, (14.5) and (14.6) imply

$$(*_i) \circ c' \circ \alpha_Z = (*_i) \circ c.$$

This is the commutation of the rightmost pentagon and $*_i$ is associative on Z. The other axioms are proved in a similar way. Thus we have defined an ω-category E with underlying globular set $Z = VE$.

Now the commutation of the square involving w_n and w' in the above diagram expresses the preservation of the composition $*_i$ by w. The preservation of units holds by a similar argument. Therefore, we get a unique morphism $h\colon D \to E$ in $\mathbf{Cat}_\omega$ such that $Vh = w$.

Finally, we must show that the diagram

$$C \underset{g}{\overset{f}{\rightrightarrows}} D \xrightarrow{h} E$$

is itself a coequalizer in $\mathbf{Cat}_\omega$. Thus, let K be an ω-category and $k\colon D \to K$ a morphism such that $kf = kg$. We have to prove the existence of a unique morphism $\ell\colon E \to K$ such that $\ell h = k$. Now, if $T = VF$ and $t = Vk$, there is a unique morphism $s\colon Z \to T$ such that $sw = t$:

$$\begin{array}{ccccc} X & \underset{v}{\overset{u}{\rightrightarrows}} & Y & \xrightarrow{w} & Z \\ & & & \searrow^{t} & \downarrow{\scriptstyle s} \\ & & & & T. \end{array}$$

We look for an ℓ such that $V\ell = s$. Uniqueness is obvious, and existence reduces to the observation that s preserves compositions and units. As for compositions, consider the following diagram:

$$\begin{array}{ccccc} Y_n \underset{Y_i}{\times} Y_n & \xrightarrow{w'} & Z_n \underset{Z_i}{\times} Z_n & \xrightarrow{s'} & T_n \underset{T_i}{\times} T_n \\ {\scriptstyle *_i}\downarrow & & {\scriptstyle *_i}\downarrow & & {\scriptstyle *_i}\downarrow \\ Y_n & \xrightarrow[w_n]{} & Z_n & \xrightarrow[s_n]{} & T_n. \end{array}$$

The left-hand square commutes because $w = Vh$, and the outer square commutes because $sw = t$ and $t = Vk$. But w' is a coequalizer map, whence the right-hand square also commutes: this shows that s preserves compositions, as required. A similar argument proves the preservation of units, and we get the unique morphism $\ell\colon E \to K$ such that $s = V\ell$. □

14.4.4 Remark. Instead of using Beck's criterion to prove the monadicity of $\mathbf{Cat}_\omega$ over $\mathbf{Glob}_\omega$, we could have used a less known criterion in terms of sketches due to Lair. Precisely, globular sets are models of a projective sketch S with underlying category $\mathbb{O}^{\mathrm{op}}$ and no cones, whereas ω-categories are models of a projective sketch S', with an obvious sketch inclusion $S \hookrightarrow S'$, inducing the forgetful functor $V\colon \mathbf{Cat}_\omega \to \mathbf{Glob}_\omega$ between the corresponding categories of models. Now (i) the base of each cone of S' already belongs to S and (ii) each object of S' not in S is the tip of at least one cone of S'. By Theorem G.1.11, these two conditions ensure the monadicity of V.

14.4.5 Remark. Recall that monadicity is not transitive. For example, the forgetful functor $\mathbf{Cat} \to \mathbf{Graph}$ is monadic, as well as the forgetful functor $\mathbf{Graph} \to \mathbf{Set}^2$ taking the graph $X_0 \leftleftarrows X_1$ to the pair (X_0, X_1). However the composite $\mathbf{Cat} \to \mathbf{Set}^2$ is *not* monadic. Consider in fact the following categories:

- the category C, freely generated on the graph having a set of five vertices $V_C = \{0, 1, 2, 3, 4\}$ and a set of two edges $E_C = \{a\colon 0 \to 1; b\colon 3 \to 4\}$,
- the subcategory D of C obtained by removing the isolated vertex 2.

Define two morphisms $u, v\colon C \to D$ such that u and v are both retractions of the inclusion $D \to C$, $u(2) = 1$ and $v(2) = 3$. We leave it as an exercise to check that the forgetful functor $\mathbf{Cat} \to \mathbf{Set}^2$ takes the pair (u, v) to a pair (f, g) whose coequalizer e in $\mathbf{Set}^2$ is split, but the coequalizer w of (u, v) in $\mathbf{Cat}$ is not sent to e. The generating graphs for C, D and of the coequalizer E of the pair (u, v) are represented in the picture below:

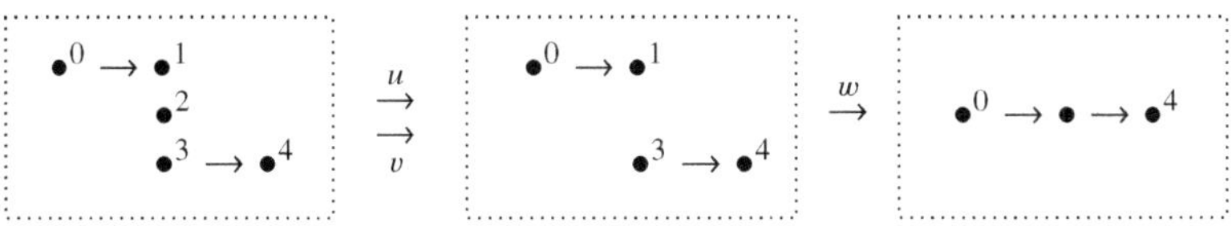

C D E

14.4.6 Adjunctions. If $0 \leqslant m < n$, there is a canonical inclusion functor

$$I_n^m\colon \mathbf{Cat}_m \to \mathbf{Cat}_n$$

taking an m-category C to the n-category $D = I_n^m(C)$ such that $U_m^n(D) = C$ and all i-cells of D are units for $m < i \leqslant n$. Likewise, we get a canonical inclusion

$$I^m\colon \mathbf{Cat}_m \to \mathbf{Cat}_\omega.$$

Therefore, each m-category may be naturally identified with an n-category for any $m < n \leqslant \omega$.

The functor I_n^m (resp. I^m) has a right adjoint, namely the truncation functor U_m^n (resp. U_m).

Now I_n^m also admits a left adjoint

$$\overline{U}_m^n\colon \mathbf{Cat}_n \to \mathbf{Cat}_m.$$

Let C be an n-category and $D = \overline{U}_m^n(C)$. Up to dimension $m - 1$, D coincides with $U_{m-1}^n(C)$, whereas D_m is the quotient of C_m modulo the congruence generated by C_{m+1}. Precisely, two parallel m-cells x, y in C_m are congruent modulo C_{m+1} if and only if there is a sequence $x_0 = x, x_1, \dots, x_p = y$ of m-cells of C_m and a sequence $z_1, \dots, z_p$ of $(m + 1)$-cells in C_{m+1} such that, for each $i = 1, \dots, p$, either $z_i = x_{i-1} \to x_i$ or $z_i\colon x_i \to x_{i-1}$. Note that the source and target maps, as well as compositions on C_m are compatible with the congruence relation. Therefore, D is a well-defined m-category, as expected. Also, the action of $\overline{U}^n_m$ on morphisms is immediate, and clearly functorial. Likewise, I^m admits a left adjoint $\overline{U}_m\colon \mathbf{Cat}_\omega \to \mathbf{Cat}_m$.

Let us finally remark that the truncation functor U^n_m (resp. U_n) also admits a right adjoint $\underline{I}^m_n\colon \mathbf{Cat}_m \to \mathbf{Cat}_n$ (resp. $\underline{I}^m\colon \mathbf{Cat}_m \to \mathbf{Cat}_\omega$): let C be an m-category, $D = \underline{I}^m_n(C)$ is the n-category such that (i) $U^n_m(D) = C$; (ii) for each pair x, y of parallel m-cells in D_m, there is exactly one $(m+1)$-cell $z\colon x \to y$ in D_{m+1}; and (iii) all i-cells of D are units whenever $i > m+1$. The ω-category $\underline{I}^m(C)$ is defined accordingly. To sum up, omitting the indices, we get a series of adjunctions between inclusions and truncation functors:

$$\overline{U} \quad \dashv \quad I \quad \dashv \quad U \quad \dashv \quad \underline{I}.$$

14.5 (*n*, *p*)-Categories

14.5.1 Invertible cells. Let $0 \leqslant i < k \leqslant n \leqslant \omega$. Let C be an n-category and u a k-cell of C. A k-cell v of C is a *$*_i$-inverse to u* if v is left and right i-composable with u and $u *_i v = 1^k_{s_i(u)}$ and $v *_i u = 1^k_{t_i(u)}$. If such a k-cell v exists, it is necessarily unique. In that case, we call u an *$*_i$-invertible* cell. A k-cell u is called simply *invertible* if it is $*_{k-1}$-invertible.

14.5.2 Lemma. *If a 2-cell is $*_0$-invertible, then it is also $*_1$-invertible.*

Proof. Let u be a $*_0$-invertible 2-cell and v its $*_0$-inverse, so that $u *_0 v = 1^2_{s_0(u)}$ and $v *_0 u = 1^2_{t_0(u)}$. This implies that the 1-cells $s_1(u)$ and $t_1(u)$ are $*_0$-invertible, with $s_1(v)$ and $t_1(v)$ as respective $*_0$-inverses. Let

$$v' = 1^2_{t_1(u)} *_0 v *_0 1^2_{s_1(u)}.$$

We claim that v' is a $*_1$-inverse to u. In fact,

$$s_1(v') = t_1(u) *_0 s_1(v) *_0 s_1(u) = t_1(u)$$

so that u and v' are $*_1$-composable. Moreover, by using the exchange rule

$$\begin{aligned}
u *_1 v' &= \big(u *_0 1^2_{s_1(v)} *_0 1^2_{s_1(u)}\big) *_1 \big(1^2_{t_1(u)} *_0 v *_0 1^2_{s_1(u)}\big) \\
&= \big(u *_1 1^2_{t_1(u)}\big) *_0 \big(1^2_{s_1(v)} *_1 v\big) *_0 1^2_{s_1(u)} \\
&= u *_0 v *_0 1^2_{s_1(u)} \\
&= 1^2_{s_1(u)}.
\end{aligned}$$

Likewise, one checks that $v' *_1 u = 1^2_{t_1(u)}$. □

14.5.3 Corollary. *Let $i \leqslant j < k$. Each $*_i$-invertible k-cell is also $*_j$-invertible.*

Proof. Let $i \leqslant j < k \leqslant n$, and let C be an n-category. We define a 2-category D by $D_0 = C_i$, $D_1 = C_j$, $D_2 = C_k$ with the obvious source and target maps, units and compositions induced by C. The statement then immediately follows by applying Lemma 14.5.2 to the 2-cells of D. □

14.5.4 Definition. Let $0 \leqslant p \leqslant n \leqslant \omega$. The category $\mathbf{Cat}_{n,p}$ is the full subcategory of $\mathbf{Cat}_n$ having as objects the n-categories whose k-cells are invertible for all $k > p$. These objects are called *(n, p)-categories*. In particular, the objects of $\mathbf{Cat}_{n,0}$ are the *n-groupoids*, where k-cells are invertible for all $k > 0$. We also denote $\mathbf{Cat}_{n,0}$ by $\mathbf{Gpd}_n$.

14.5.5 Proposition. *Let $p \leqslant i < k \leqslant n$ and let C be an (n, p)-category. Each k-cell of C is $*_i$-invertible.*

Proof. As in the proof of Corollary 14.5.3, the statement reduces to the fact that, in each 2-category C all whose 1-cells are $*_0$-invertible, any 2-cell u is $*_1$-invertible if and only if it is $*_0$-invertible. The "if" direction follows from Lemma 14.5.2. Conversely, suppose that all 1-cells are invertible, and let u be a $*_1$-invertible 2-cell, with $*_1$-inverse v. By hypothesis, $s_1(u)$ and $t_1(u)$ have $*_0$-inverses v_1^- and v_1^+, respectively. One easily checks that the required $*_0$-inverse to u is given by

$$v' = 1^2_{v_1^+} *_0 v *_0 1^2_{v_1^-}$$

(see also [16, §1.3]). □

14.5.6 Proposition. *For each n and p such that $0 \leqslant p \leqslant n \leqslant \omega$, the category $\mathbf{Cat}_{n,p}$ is complete, cocomplete, and monadic over $\mathbf{Glob}_n$. Moreover, the inclusion functor $\mathbf{Cat}_{n,p} \to \mathbf{Cat}_n$ admits a left adjoint.*

Proof. Like $\mathbf{Cat}_n$, the category $\mathbf{Cat}_{n,p}$ is the category of models of a projective sketch $S_{n,p}$, hence it is complete and cocomplete. The inclusion functor $\mathbf{Cat}_{n,p} \to \mathbf{Cat}_n$ is induced by the morphism of corresponding sketches $S_n \to S_{n,p}$, hence admits a left adjoint. Finally the monadicity of $\mathbf{Cat}_{n,p}$ over $\mathbf{Glob}_n$ is proved as in Proposition 14.4.3. □

15
Polygraphs

The notion of 2-polygraph, already introduced in Chapter 2, first appears in [333] under the name of *computad* as an essential tool in proving the existence of limits in 2-categories. Although its relevance to rewriting theory was recognized by Eilenberg and Street from the very beginning [120], this point of view is not explicitly mentioned in the literature until the early 1990s. The general notion of n-computad explicitly appears in [304], and independently in [71] and [72] under the name of polygraph. We adopt here Burroni's presentation and terminology. The source of Burroni's approach can be traced back in his work on graphical algebras [70], where he presents a "concept of dimension in formal languages". Let us mention that [28] introduces a wide generalization of the notion of n-computad attached to a finitary monad T on globular sets, presented in more detail in Chapter 18. We deal in this chapter with the particular case where the monad T comes from the adjunction between $\mathbf{Cat}_\omega$ and $\mathbf{Glob}_\omega$.

15.1 Main Definitions

Throughout this section, we denote by n a natural number.

15.1.1 Cellular extensions. Given an n-category C, a *cellular extension of* C is a family

$$(X_i : \partial\mathbb{O}_{n+1} \to C)_{i \in I}$$

of n-spheres in C indexed by a set I. This amounts to a family of pairs of parallel n-cells in C. Equivalently, it can also be seen as an ω-functor

$$X : \coprod_{i \in I} \partial\mathbb{O}_{n+1} \to C.$$

Note that, in order for X to make sense, we identify the n-category C with its image in $\mathbf{Cat}_\omega$ by the inclusion functor defined in §14.4.6. A morphism

$$f\colon (C, (X_i)_{i\in I}) \to (D, (Y_j)_{j\in J})$$

between two cellular extensions of n-categories consists of a pair (g, h), where $g\colon C \to D$ is a morphism in $\mathbf{Cat}_n$ and $h\colon I \to J$ is a map such that, for each $i \in I$, $g \circ X_i = Y_{h(i)}$. We write $\mathbf{Cat}_n^+$ for the resulting category. More abstractly, the category $\mathbf{Cat}_n^+$ is the pullback of $\mathbf{Cat}_n$ and $\mathbf{Glob}_{n+1}$ over $\mathbf{Glob}_n$ in $\mathbf{CAT}$ (which denotes the category of possibly large categories and functors):

$$\begin{array}{ccc} \mathbf{Cat}_n^+ & \dashrightarrow & \mathbf{Glob}_{n+1} \\ \vdots & & \downarrow{\scriptstyle U_n^{n+1}} \\ \mathbf{Cat}_n & \xrightarrow[V_n]{} & \mathbf{Glob}_n. \end{array} \tag{15.1}$$

In the above diagram, the forgetful functor V_n and the truncation functor U_n^{n+1} are those defined in Chapter 14, and the forgetful functor from $\mathbf{Cat}_n^+$ to $\mathbf{Cat}_n$ takes a cellular extension (C, X) to the n-category C. Finally, the horizontal dotted arrow takes a cellular extension (C, X) to the $(n + 1)$-globular set extending $V_n(C)$ with the set of $(n + 1)$-cells determined by X.

15.1.2 Freely generated category. Consider now the forgetful functor

$$W_n\colon \mathbf{Cat}_{n+1} \to \mathbf{Cat}_n^+$$

which to an $(n+1)$-category C associates the pair $(U_n^{n+1}(C), (X_x)_{x\in C_{n+1}})$ where for each cell $x \in C_{n+1}$, X_x is the n-sphere $(s_n(x), t_n(x))$. This functor W_n is in fact the universal arrow from $\mathbf{Cat}_{n+1}$ to $\mathbf{Cat}_n^+$ resulting from the commutation of the diagram

$$\begin{array}{ccc} \mathbf{Cat}_{n+1} & \xrightarrow{V_{n+1}} & \mathbf{Glob}_{n+1} \\ {\scriptstyle U_n^{n+1}}\downarrow & & \downarrow{\scriptstyle U_n^{n+1}} \\ \mathbf{Cat}_n & \xrightarrow[V_n]{} & \mathbf{Glob}_n \end{array}$$

and the pullback property of (15.1).

15.1.3 Proposition. *The functor W_n admits a left adjoint L_n:*

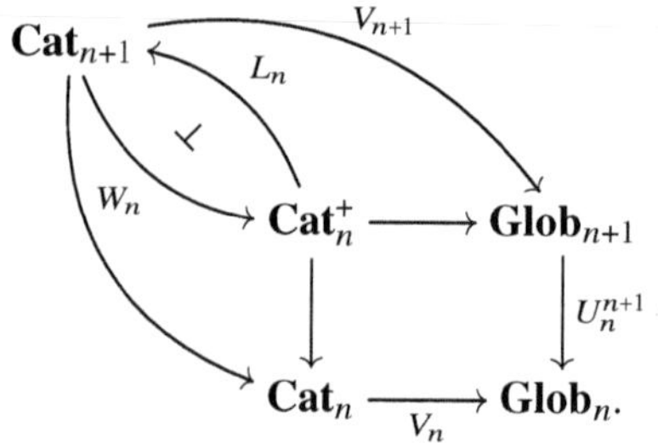

Proof. Let C be an n-category and $(X_i : \partial\mathbb{O}_{n+1} \to C)_{i\in I}$ a cellular extension, so that (C, X) is an object of $\mathbf{Cat}_n^+$. By considering C, $\partial\mathbb{O}_{n+1}$ and $\mathbb{O}_{n+1}$ as $(n + 1)$-categories, and remembering that $\mathbf{Cat}_{n+1}$ is cocomplete, we define $L_n(C, X)$ as the pushout given by the following diagram in $\mathbf{Cat}_{n+1}$:

$$\begin{array}{ccc} \coprod_{i\in I} \partial\mathbb{O}_{n+1} & \xrightarrow{X} & C \\ {\scriptstyle \coprod_{i\in I} \mathrm{i}_{n+1}}\downarrow & & \downarrow \\ \coprod_{i\in I} \mathbb{O}_{n+1} & \longrightarrow & L_n(C, X). \end{array}$$

This construction yields a functor $L_n : \mathbf{Cat}_n^+ \to \mathbf{Cat}_{n+1}$. We claim that L_n is left adjoint to W_n. Consider in fact (C, X) an object of $\mathbf{Cat}_n^+$, D an $(n+1)$-category, and $f : (C, X) \to W_n(D)$ a morphism in $\mathbf{Cat}_n^+$. Recall that f is a pair (g, h) where $g : C \to U_n(D)$ is an n-functor and $h : X \to D_{n+1}$ is a map preserving the globular structure. This amounts to a commutative diagram in $\mathbf{Cat}_{n+1}$ of the form

$$\begin{array}{ccc} \coprod_{i\in I} \partial\mathbb{O}_{n+1} & \xrightarrow{X} & C \\ {\scriptstyle \coprod_{i\in I} \mathrm{i}_{n+1}}\downarrow & & \downarrow{\scriptstyle \tilde{g}} \\ \coprod_{i\in I} \mathbb{O}_{n+1} & \xrightarrow[\tilde{h}]{} & D, \end{array}$$

where $\tilde{g}$ and $\tilde{h}$ are the $(n + 1)$-functors built from g and h, respectively. The pushout property then gives a unique morphism $f^* : L_n(C, X) \to D$ making the following diagram commutative:

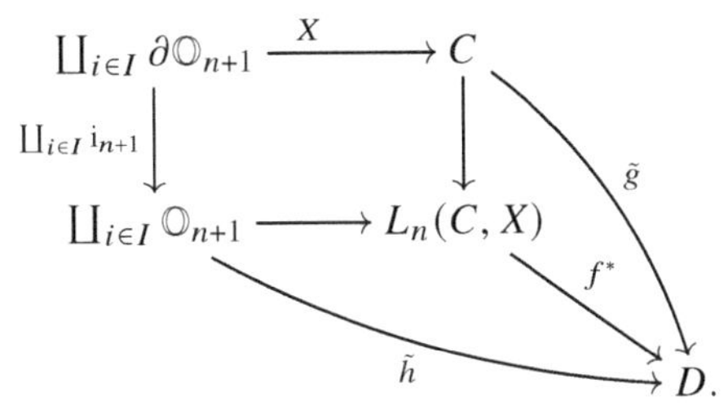

The correspondence $f \mapsto f^*$ is then a natural isomorphism

$$\mathbf{Cat}_n^+((C,X), W_n(D)) \xrightarrow{\ \cong\ } \mathbf{Cat}_{n+1}(L_n(C,X), D),$$

which ends the proof. □

15.1.4 Remark. There are several approaches to the construction of the above functor L_n. On the abstract side, its existence comes from the fact that W_n is a limit and filtered colimit preserving functor between locally presentable categories, see [142, 14.6], [26, Theorem 4.1], or [2, 1.66]. More concretely, a purely syntactic construction of L_n based on a type system is given in [279]. An alternative construction is given in Appendix D.

Given a cellular extension (C,X) of an n-category C, we call $L_n(C,X)$ the *freely generated* $(n+1)$-*category* on this extension and denote it by $C[X]$. In practice, the universal property of $C[X]$ will be used by applying the following lemma.

15.1.5 Lemma. *Let $(C,(X_i)_{i\in I})$ be a cellular extension of an n-category C. For every $(n+1)$-category D, every morphism $g\colon C \to U_n^{n+1}(D)$ of* $\mathbf{Cat}_n$ *and every map $h\colon I \to D_{n+1}$ making (g,h) a morphism of cellular extensions, there exists a unique morphism $\overline{g}\colon C[X] \to D$ in* $\mathbf{Cat}_{n+1}$ *such that $W_n(\overline{g}) = (g,h)$.*

Proof. The statement of the lemma is a mere rephrasing of the fact that L_n is left adjoint to W_n. □

15.1.6 Quotient category. For any cellular extension (C,X) of an n-category C, the *quotient* n-category C/X is the n-category obtained from C by identifying n-cells under the smallest congruence (with respect to compositions and identities) containing all spheres (x,y) in X. More precisely, C/X is nothing but $\overline{U}_n^{n+1}(C[X])$ were $\overline{U}$ is the truncation functor already described in §14.4.6.

15.1.7 n-Polygraphs. We already gave an explicit description of n-polygraphs for $0 \leqslant n \leqslant 3$ in previous chapters. We now turn to the general definition. Thus, the category $\mathbf{Pol}_n$ of n-polygraphs is defined by induction on n, together with a functor $F_n\colon \mathbf{Pol}_n \to \mathbf{Cat}_n$.

- The category $\mathbf{Pol}_0$ is $\mathbf{Set} = \mathbf{Cat}_0$ and $F_0 = 1$.
- Given $\mathbf{Pol}_n$ and $F_n\colon \mathbf{Pol}_n \to \mathbf{Cat}_n$, the category $\mathbf{Pol}_{n+1}$ is defined by the following pullback in $\mathbf{CAT}$

$$\begin{array}{ccc} \mathbf{Pol}_{n+1} & \xrightarrow{J_n} & \mathbf{Cat}_n^+ \\ {\scriptstyle U_n^{n+1}}\downarrow & & \downarrow \\ \mathbf{Pol}_n & \xrightarrow[F_n]{} & \mathbf{Cat}_n, \end{array}$$

whereas F_{n+1} is $L_n J_n$:

$$\mathbf{Pol}_{n+1} \xrightarrow[J_n]{} \mathbf{Cat}_n^+ \xrightarrow[L_n]{} \mathbf{Cat}_{n+1}, \qquad F_{n+1} = L_n J_n .$$

More explicitly, an $(n+1)$-polygraph $P^{(n+1)}$ is a pair $(P^{(n)}, X)$ where $P^{(n)}$ is an n-polygraph and X is a cellular extension of the n-category $F_n(P^{(n)})$. The first projection of the pullback yields a truncation functor $\mathbf{Pol}_{n+1} \to \mathbf{Pol}_n$ we still denote by U_n^{n+1} as in the case of n-globular sets and n-categories. Note also that the following square commutes:

$$\begin{array}{ccc} \mathbf{Pol}_n & \xleftarrow{U_n^{n+1}} & \mathbf{Pol}_{n+1} \\ {\scriptstyle F_n}\downarrow & & \downarrow{\scriptstyle F_{n+1}} \\ \mathbf{Cat}_n & \xleftarrow[U_n^{n+1}]{} & \mathbf{Cat}_{n+1}. \end{array}$$

For each n-polygraph P and $k < n$, the polygraph $U_k^{k+1} \circ \cdots \circ U_{n-1}^n(P)$ will be denoted by $P_{\leqslant k}$. Thus, the data defining an n-polygraph P may be displayed in the following diagram in $\mathbf{Set}$:

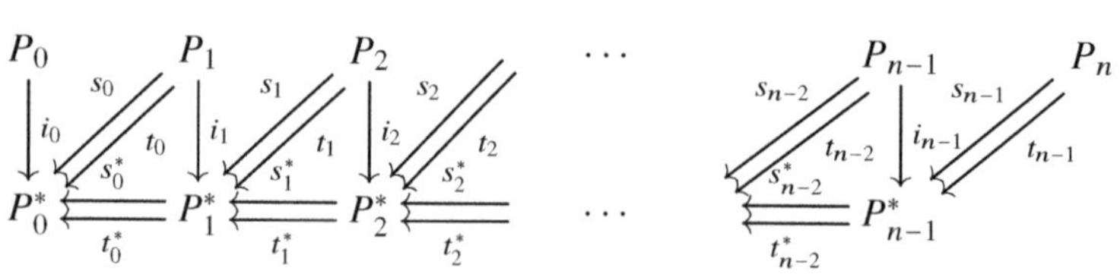

where, for each $0 \leqslant k < n-1$, $s_k^* \circ i_{k+1} = s_k$ and $t_k^* \circ i_{k+1} = t_k$. Also, at each level $k \leqslant n$, $k > 0$, we add a new set P_k of *k-generators* together with source and target maps

$$s_{k-1}, t_{k-1}\colon P_k \to P_{k-1}^*$$

such that

$$s^*_{k-2} \circ s_{k-1} = s^*_{k-2} \circ t_{k-1} \qquad t^*_{k-2} \circ t_{k-1} = t^*_{k-2} \circ t_{k-1} \tag{15.2}$$

for $k > 2$, thus defining a k-cellular extension of the $(k-1)$-category $P^*_{\leqslant k-1}$

$$P^*_0 \underset{t^*_0}{\overset{s^*_0}{\Longleftarrow}} P^*_1 \underset{t^*_1}{\overset{s^*_1}{\Longleftarrow}} P^*_2 \underset{t^*_2}{\overset{s^*_2}{\Longleftarrow}} \cdots \underset{t^*_{k-2}}{\overset{s^*_{k-2}}{\Longleftarrow}} P^*_{k-1}$$

already defined at the previous level. The vertical map i_k is the natural inclusion of the set P_k of k-generators into the set P^*_k of k-cells of the free k-category on the cellular extension of $P^*_{\leqslant k-1}$ by $s_{k-1}, t_{k-1} \colon P_k \to P^*_{k-1}$. Formally, i_k is the k-dimensional component of the unit of the monad $W_{k-1}L_{k-1}$ on this cellular extension. Concretely, the cells in P^*_k are all formal compositions of k-generators in P_k and units on cells in P^*_{k-1}, quotiented by the axioms of k-categories.

15.1.8 Proposition. *Let P be an n-polygraph and let $0 < k < n$. Suppose that $A \subseteq P^*_k$ is a subset of P^*_k such that*

- *$i_k(P_n) \subseteq A$,*
- *A contains all cells of the form 1^k_u for $u \in P^*_{k-1}$,*
- *for each $i \in \{0, \dots, k-1\}$ and each pair $u, v \in A$ of $*_i$-composable cells, A contains their $*_i$-composition $u *_i v$.*

Then $A = P^*_k$.

Proof. Let C be the k-category whose $(k-1)$-truncation is $P^*_{\leqslant k-1}$ and such that $C_k = A$, with source and target maps $A \to P^*_{k-1}$ given by the restriction of $s_{k-1}, t_{k-1} \colon P^*_k \to P^*_{k-1}$. By hypothesis, C is a sub-k-category of $P^*_{\leqslant k}$, and there is an inclusion morphism $j \colon C \to P^*_{\leqslant k}$. On the other hand, the inclusion $i_k \colon P_k \to A$ determines a unique morphism $f \colon P^*_{\leqslant k} \to C$ by Lemma 15.1.5. Now $j \circ f \colon P^*_{\leqslant k} \to P^*_{\leqslant k}$ is the identity on $P^*_{\leqslant k}$, by the uniqueness property from Lemma 15.1.5. This implies that $j_k \circ f_k \colon P^*_k \to P^*_k$ is the identity on P^*_k, whence j_k is surjective. Therefore $A = P^*_k$. □

15.1.9 Structural induction. Proposition 15.1.8 allows reasoning by structural induction on the cells of freely generated ω-categories. Precisely, in order to prove that a certain property A holds for all cells in P^*_k, it suffices to check that

- A holds for all units 1_v where $v \in P^*_{k-1}$,
- A holds for all generating cells of the form $i_k(a)$ for $a \in P_k$,

– whenever A holds for two i-composable cells u and v, then A holds for $w = u *_i v$.

15.1.10 ω-Polygraphs. The category $\mathbf{Pol}_\omega$ of ω-polygraphs (or simply polygraphs) is the projective limit of the following diagram in **CAT**:

$$\mathbf{Pol}_0 \xleftarrow{U_0^1} \mathbf{Pol}_1 \longleftarrow \cdots \longleftarrow \mathbf{Pol}_n \xleftarrow{U_n^{n+1}} \mathbf{Pol}_{n+1} \longleftarrow \cdots .$$

Thus, as in the case of globular sets and ω-categories, we get a family of truncation functors $U_n\colon \mathbf{Pol}_\omega \to \mathbf{Pol}_n$. Also, keeping the above notations, for each ω-polygraph P and integer n, the n-polygraph $U_n(P)$ will be denoted by $P_{\leqslant n}$. Likewise, for each n, any morphism $f\colon P \to Q$ in $\mathbf{Pol}_\omega$ gives rise by truncation to a morphism $f_{\leqslant n}\colon P_{\leqslant n} \to Q_{\leqslant n}$.

15.2 Three Adjunctions

This section investigates some fundamental adjunctions between the categories $\mathbf{Pol}_\omega$, $\mathbf{Cat}_\omega$, and $\mathbf{Glob}_\omega$. We have already examined the monadic adjunction between $\mathbf{Cat}_\omega$ and $\mathbf{Glob}_\omega$ (see Propositions 14.2.7 and 14.4.3). We now turn to two further important pairs of adjoint functors.

15.2.1 For each $n \in \mathbb{N}$, we first define a functor $G_n\colon \mathbf{Cat}_n \to \mathbf{Pol}_n$ together with a natural transformation $\varepsilon\colon F_nG_n \to 1$.

Thus, let C be an n-category. The n-polygraph $P = G_n(C)$, as well as ε_C, are defined dimensionwise as follows:

- The set $P_0 = (G_n(C))_0$ is just $C_0 = (F_n(P))_0$, and $(\varepsilon_C)_0$ is the identity.
- Suppose P and ε_C have been defined up to dimension $k < n$. The set P_{k+1} of $(k+1)$-generators then consists in triples $p = (z, x, y)$ where x, y are parallel k-cells in $P_k^* = (F_nG_n(C))_k$ and z is a $(k+1)$-cell in C_{k+1} of the form $z\colon \varepsilon_C(x) \to \varepsilon_C(y)$. The source and target maps

  $$s_k, t_k\colon P_{k+1} \to P_k^*$$

  are given by $s_k(p) = x$ and $t_k(p) = y$, and ε_C extends to P_{k+1} by $\varepsilon_C(p) = z$. Thus P is now defined up to dimension $k+1$, and by applying Lemma 15.1.5, the map ε_C extends to P_{k+1}^*, yielding a $(k+1)$-functor.

Likewise, G_n is defined on morphisms. Functoriality of G_n and naturality of ε immediately follow from the construction.

15.2.2 Lemma. *For each $n \in \mathbb{N}$, the functor G_n is right adjoint to F_n.*

Proof. It is sufficient to check that the natural transformation ε, which becomes of course the counit of the adjunction, satisfies the following universal property: for any n-functor $f\colon F_n(P) \to C$, where P is an n-polygraph and C an n-category, there is a unique morphism $g\colon P \to G_n(C)$ in $\mathbf{Pol}_n$ such that the following triangle commutes:

$$\begin{array}{ccc} & & F_nG_n(C) \\ & \overset{F_n(g)}{\nearrow} & \downarrow{\scriptstyle \varepsilon_C} \\ F_n(P) & \xrightarrow[f]{} & C. \end{array}$$

Here again g is built by induction on all dimensions $0 \leqslant k \leqslant n$. For $k = 0$, we must have $g_0 = f_0$. Suppose now that g has been defined up to dimension $k < n$, satisfying the commutation condition. Let $p \in P_{k+1}$ be a $(k+1)$-generator of P, $u = s_k(p)$, and $v = t_k(p)$ in P^*_k. The induction hypothesis and the definition of ε imply that $g(p) = (f(p), u, v)$. Now Lemma 15.1.5 applies, and we may extend $F_n(g)$ up to a $(k+1)$-functor still satisfying the commutation condition. □

15.2.3 Consider now the diagram

$$\begin{array}{ccccccccc} \mathbf{Pol}_0 & \xleftarrow{U_0^1} & \mathbf{Pol}_1 & \longleftarrow \cdots \longleftarrow & \mathbf{Pol}_n & \xleftarrow{U_n^{n+1}} & \mathbf{Pol}_{n+1} & \longleftarrow & \cdots \\ {\scriptstyle F_0}\downarrow\uparrow{\scriptstyle G_0} & & {\scriptstyle F_1}\downarrow\uparrow{\scriptstyle G_1} & & {\scriptstyle F_n}\downarrow\uparrow{\scriptstyle G_n} & & {\scriptstyle F_{n+1}}\downarrow\uparrow{\scriptstyle G_{n+1}} & & \\ \mathbf{Cat}_0 & \xleftarrow{U_0^1} & \mathbf{Cat}_1 & \longleftarrow \cdots \longleftarrow & \mathbf{Cat}_n & \xleftarrow{U_n^{n+1}} & \mathbf{Cat}_{n+1} & \longleftarrow & \cdots \end{array}$$

where the squares involving F and the squares involving G commute. The projective limit of the top row is the category $\mathbf{Pol}_\omega$ and the projective limit of the bottom row is the category $\mathbf{Cat}_\omega$. Therefore we get a pair of functors $F\colon \mathbf{Pol}_\omega \to \mathbf{Cat}_\omega$ and $G\colon \mathbf{Cat}_\omega \to \mathbf{Pol}_\omega$ such that, for each n, $F_n \circ U_n = U_n \circ F$ and $G_n \circ U_n = U_n \circ G$. Moreover F is left adjoint to G. For any polygraph P, $F(P)$ is the *freely generated ω-category* on P and will be denoted by P^*.

15.2.4 Let us now define a pair of adjoint functors between categories $\mathbf{Pol}_\omega$ and $\mathbf{Glob}_\omega$. Consider first the functor $N\colon \mathbf{Pol}_\omega \to \mathbf{Glob}_\omega$ which takes a polygraph P to a globular set X by keeping only "hereditary globular" generators. Precisely, let P be a polygraph, we define the globular set $X = N(P)$ dimensionwise, such that for each $n \in \mathbb{N}$, $X_n \subseteq P_n$; recall that $i_k\colon P_k \to P^*_k$ denotes the canonical insertion of k-generators into k-cells:

– For $n = 0$, $X_0 = P_0$.

- Let $n > 0$ and suppose we have defined $X_k \subseteq P_k$ for all $k < n$, together with source and target maps building an $(n-1)$-globular set. Let $X_n \subseteq P_n$ be the set of n-generators a of P such that $s_{n-1}(a)$ and $t_{n-1}(a)$ belong to $i_{n-1}(X_{n-1})$ and define source and target maps $s^X_{n-1}, t^X_{n-1} : X_n \to X_{n-1}$ as the unique maps such that $i_{n-1}s^X_{n-1}(a) = s_{n-1}(a)$ and $i_{n-1}t^X_{n-1}(a) = t_{n-1}(a)$ for each $a \in X_n$. This extends X to an n-globular set, as shown in the following diagram:

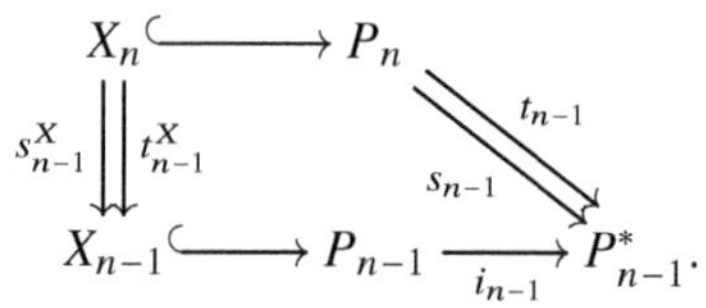

The previous construction is clearly functorial and defines the required functor N. Remark that N admits a left adjoint $M : \mathbf{Glob}_\omega \to \mathbf{Pol}_\omega$ which takes the globular set X to a polygraph P such that $P_n = X_n$, in other words, M defines a natural inclusion of $\mathbf{Glob}_\omega$ into $\mathbf{Pol}_\omega$.

15.2.5 Lemma. *There is a natural isomorphism $\phi : NG \to V$, that is, the diagram*

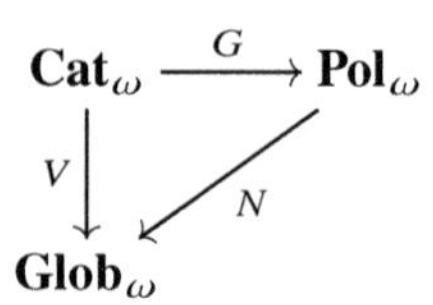

commutes up to a natural isomorphism.

Proof. Let C be an ω-category, and $X = NG(C)$. For each $n \in \mathbb{N}$, let $\phi^C_n : X_n \to C_n$ be the composition of the following maps

$$X_n \hookrightarrow G(C)_n \xrightarrow{i_n} FG(C)_n \xrightarrow{(\varepsilon_C)_n} C_n.$$

The family $(\phi^C_n)_{n\in\mathbb{N}}$ defines a globular morphism $\phi^C : NG(C) \to V(C)$, natural in C. Thus we get a natural transformation $\phi : NG \to V$.

Let us now define $\chi^C_n : C_n \to X_n$ by induction on n such that $\phi^C_n \circ \chi^C_n = 1_{C_n}$:

- For $n = 0$, $X_0 = C_0$ and $\phi^C_0 = 1_{C_0} = 1_{X_0}$, so that $\chi^C_0 : C_0 \to X_0$ is also $1_{C_0} = 1_{X_0}$.
- Suppose $n > 0$ and χ^C_k has been defined up to $k = n - 1$, and let $z \in C_n$. Let $u = s_{n-1}(z)$ and $v = t_{n-1}(z)$ in C_{n-1}. By induction hypothesis, $\chi^C_{n-1}(u)$ and $\chi^C_{n-1}(v)$ belong to X_{n-1}. Let $x = i_{n-1}\chi^C_{n-1}(u)$, $y = i_{n-1}\chi^C_{n-1}(v)$ in $FG(C)_{n-1}$ and define $a = \chi^C_n(z) = (z, x, y)$. By construction $a \in X_n$ and $\phi^C_n(a) = z$.

It remains to prove that ϕ_n^C is injective. We reason again by induction on n.

- For $n = 0$, ϕ_0^C is an identity, hence injective.
- Suppose $n > 0$ and ϕ_{n-1}^C injective. Let $a_i = (z_i, x_i, y_i) \in X_n$ for $i = 0, 1$ such that $\phi_n^C(a_0) = \phi_n^C(a_1)$. Thus $z_0 = z_1$. Also

$$\begin{aligned}\phi_{n-1}^C(s_{n-1}^X(a_0)) &= s_{n-1}(\phi_n^C(a_0))\\ &= s_{n-1}(\phi_n^C(a_1))\\ &= \phi_{n-1}^C(s_{n-1}^X(a_1))\end{aligned}$$

and because ϕ_{n-1}^C is injective,

$$s_{n-1}^X(a_0) = s_{n-1}^X(a_1).$$

Now

$$\begin{aligned}x_0 &= s_{n-1}(a_0)\\ &= i_{n-1}s_{n-1}^X(a_0)\\ &= i_{n-1}s_{n-1}^X(a_1)\\ &= s_{n-1}(a_1)\\ &= x_1.\end{aligned}$$

Likewise $y_0 = y_1$, and we get $a_0 = a_1$. Hence ϕ_n^C is injective, and we are done. □

15.3 (*n*, *p*)-Polygraphs

In the same way as an n-polygraph generates an n-category, we may define, for each $n \geqslant p$, a notion of *(n, p)-polygraph* generating an (n, p)-category. The particular case where $n = 3$ and $p = 1$ has been already used to define coherent presentations of categories in Chapter 7. Remark that the construction of the left adjoint in Proposition 15.1.3 applies to the case where $\mathbf{Cat}_n$ is replaced by $\mathbf{Cat}_{n,p}$, as in the following diagram:

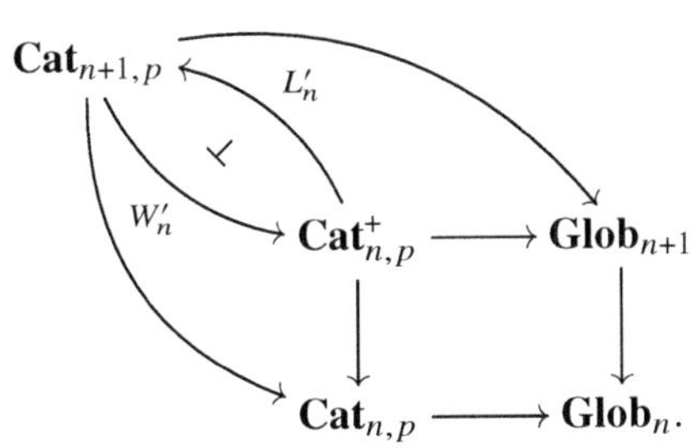

Here $\mathbf{Cat}^+_{n,p}$ is defined as above by the pullback square of the obvious forgetful functors. We may now define the category $\mathbf{Pol}_{n,p}$ of (n, p)-polygraphs, together with a functor $F'_n : \mathbf{Pol}_{n,p} \to \mathbf{Cat}_{n,p}$ taking an (n, p)-polygraph to the (n, p)-category it generates. The definition is by induction on $n \geqslant p$.

- For $n = p$, $\mathbf{Pol}_{n,p}$ is just $\mathbf{Pol}_p$, and F'_n is F_p, as $\mathbf{Cat}_{n,p}$ is just $\mathbf{Cat}_p$.
- Given $\mathbf{Pol}_{n,p}$ and $F'_n : \mathbf{Pol}_{n,p} \to \mathbf{Cat}_{n,p}$, the category $\mathbf{Pol}_{n+1,p}$ is defined by the following pullback in $\mathbf{CAT}$:

$$\begin{array}{ccc} \mathbf{Pol}_{n+1,p} & \xrightarrow{J'_n} & \mathbf{Cat}^+_{n,p} \\ {\scriptstyle U'_{n+1,n}}\downarrow & & \downarrow \\ \mathbf{Pol}_{n,p} & \xrightarrow[F'_n]{} & \mathbf{Cat}_{n,p}, \end{array}$$

whereas F'_{n+1} is $L'_n J'_n$:

$$\mathbf{Pol}_{n+1,p} \xrightarrow[J'_n]{} \mathbf{Cat}^+_{n,p} \xrightarrow[L'_n]{} \mathbf{Cat}_{n+1,p}. \qquad (F'_{n+1} \text{ the composite})$$

Concretely, let $(P_k)_{0 \leqslant k \leqslant n}$ the sequence of k-dimensional generators of an (n, p)-polygraph P, and C be the (n, p)-category generated by P, we shall denote C_k by P^*_k for $0 \leqslant k \leqslant p$ and by $P^\top_k$ for $p < k \leqslant n$. Note that in the latter case, $P^\top_k$ contains all composites *and inverses* of the generators in P_k. As for n-polygraphs in §15.1.7, the data defining an (n, p)-polygraph P may be displayed in the following diagram in $\mathbf{Set}$:

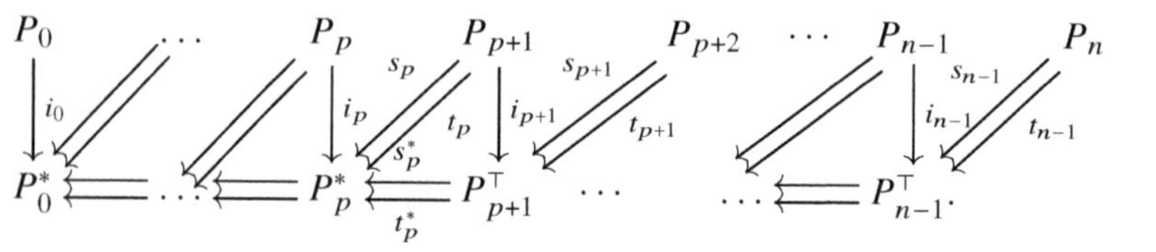

Finally, as in the case of (plain) polygraphs, we may define a category $\mathbf{Pol}_{\omega,p}$ of (ω, p)-polygraphs as the projective limit of the system $(\mathbf{Pol}_{n,p}, U_{n+1,n})_{n \geqslant p}$.

16
Properties of the Category of n-Polygraphs

In this chapter, we establish the main properties of the category $\mathbf{Pol}_n$ of n-polygraphs. We first show how to compute limits and colimits and prove that $\mathbf{Pol}_n$ is complete and cocomplete for any $n \geqslant 0$. The behavior of the cartesian product deserves a special attention in that it does *not* correspond to the product of generators. The monomorphisms (resp. epimorphisms) in $\mathbf{Pol}_n$ are then characterized as injective (resp. surjective) maps between generators. The linearization of polygraphic expressions plays a central role in proving these facts. Whereas $\mathbf{Pol}_n$ is a presheaf category for $n \in \{0, 1, 2\}$, it already fails to be cartesian closed for $n \geqslant 3$, as proved in [265], the culprit for this defect being as usual the Eckmann-Hilton phenomenon. The categories $\mathbf{Pol}_n$ are however locally presentable for all $n \in \mathbb{N} \cup \{\omega\}$. We introduce the technical notion of *context*, in relation with n-dimensional rewriting, and use it to prove that if an ω-category is freely generated by a polygraph then this polygraph is unique up to isomorphism. Finally, we show how to define rewriting properties of n-polygraphs and to prove coherence results by rewriting on $(n-1)$-categories presented by convergent n-polygraphs.

16.1 Limits and Colimits

16.1.1 Terminal object. The category $\mathbf{Pol}_\omega$ has a terminal object $\mathbf{1}^{\mathrm{pol}}$, defined as the image of the terminal ω-category $\mathbf{1}^{\mathrm{cat}}$ by the right adjoint functor $G\colon \mathbf{Cat}_\omega \to \mathbf{Pol}_\omega$. Concretely, $(\mathbf{1}^{\mathrm{pol}})_0$ consists in a single 0-cell, whereas for each $n > 0$, the set $(\mathbf{1}^{\mathrm{pol}})_n$ of n-generators consists in all pairs (u, v) of parallel cells in $(\mathbf{1}^{\mathrm{pol}})^*_{n-1}$. Thus $\mathbf{1}^{\mathrm{pol}}$ has exactly one 0-generator, one 1-generator, and infinitely many n-generators for each $n \geqslant 2$. A detailed description in the case $n = 2$ can be found in [338].

16.1.2 Products. Let P, Q be a pair of polygraphs, and $!^P : P \to \mathbf{1}^{\mathrm{pol}}$, $!^Q : Q \to \mathbf{1}^{\mathrm{pol}}$ the canonical morphisms from P, Q to the terminal object. We define a polygraph R, together with morphisms $p : R \to P$, $q : R \to Q$, by induction on the dimension.

- For $n = 0$, $R_0 = P_0 \times Q_0$ and $p_0 : R_0 \to P_0$, $q_0 : R_0 \to Q_0$ are the usual projection maps.
- Let $n > 0$ and suppose R, p, q have been defined up to dimension $n - 1$. The set R_n of n-generators consists of quadruples

$$c = (a, b, u, v) \in P_n \times Q_n \times R^*_{n-1} \times R^*_{n-1}$$

  such that $!^P_n(a) = !^Q_n(b)$ and u, v are parallel cells satisfying the equations $p^*_{n-1}u = s_{n-1}(a)$, $q^*_{n-1}u = s_{n-1}(b)$, $p^*_{n-1}v = t_{n-1}(a)$, and $q^*_{n-1}v = t_{n-1}(b)$. The projection maps are defined by $p_n c = a$ and $q_n c = b$. The source and target maps $s_{n-1}, t_{n-1} : R_n \to R^*_{n-1}$ are defined by $s_{n-1}(r) = u$ and $t_{n-1}(r) = v$.

Now the following square is a pullback in $\mathbf{Pol}_\omega$:

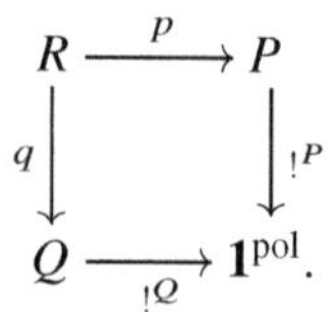

In fact, let S be a polygraph and $f : S \to P$, $g : S \to Q$ morphisms such that the following diagram commutes:

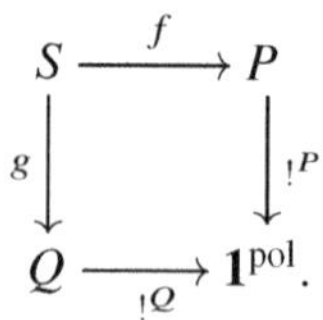

(16.1)

We show that there is a unique morphism $h : S \to R$ making the following diagram commute, for each $n \geqslant 0$:

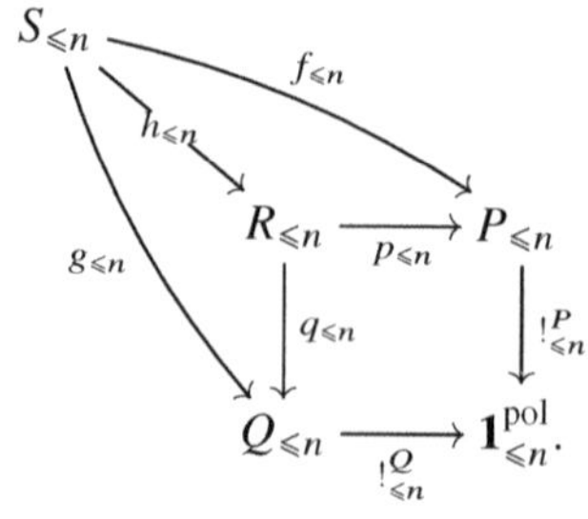

(16.2)

Let us build h and prove its uniqueness by induction on n.

- If $n = 0$, $R_0 = P_0 \times Q_0$, $h_0 : S_0 \to R_0$ takes $c \in S_0$ to $(f_0 c, g_0 c)$ and this choice is unique.
- Let $n \geqslant 0$ and suppose that h has been defined up to dimension n such that (16.2) commutes. We extend h to dimension $n + 1$ as follows: given $c \in S_{n+1}$, we define $h_{n+1}c = (a, b, u, v)$ where

$$a = f_{n+1}c, \tag{16.3}$$

$$b = g_{n+1}c, \tag{16.4}$$

$$u = h_n^* s_n(c), \tag{16.5}$$

$$v = h_n^* t_n(c). \tag{16.6}$$

  The equations (16.3) to (16.6) ensure that $h_{n+1}c \in R_n$ and that the globular relations $s_n(h_{n+1}c) = h_n^* s_n(c)$ and $t_n(h_{n+1}c) = h_n^* t_n(c)$ hold, so that h extends to a morphism up to dimension $n+1$. Moreover, this extension now satisfies the required commutation conditions $f = ph$ and $g = qh$ up to dimension $n+1$. Conversely, the commutation conditions imply (16.3) and (16.4), and the requirement that h be a morphism implies (16.5) and (16.6). Hence, the above choice for $h_{n+1}c$ is unique and we are done.

Therefore R is the cartesian product $P \times Q$ in $\mathbf{Pol}_\omega$. It should be emphasized that generally, from $n = 2$ on, the map $(a, b, u, v) \mapsto (a, b)$ from R_n to $P_n \times Q_n$ is not surjective, and from $n = 3$ on, not injective either. For example, if

$$P = \langle\, \star_1, \star_2 \mid a, b : \star_1 \to \star_2 \mid f : a \to b \,\rangle,$$

$$Q = \langle\, \star_1, \star_2, \star_3 \mid a : \star_1 \to \star_2, b : \star_2 \to \star_3, c : \star_1 \to \star_3 \mid g : a *_0 b \to c \,\rangle$$

and $R = P \times Q$, then $R_2 = \emptyset$ and $P_2 \times Q_2 = \{(f, g)\}$, thus showing that $R_2 \to P_2 \times Q_2$ is not surjective. Another example of non-injectivity is given in 16.3.2. A similar proof shows that $\mathbf{Pol}_\omega$ has all small products $\prod_{i \in I} P^i$.

16.1.3 Equalizers. Let P, Q be polygraphs and $f, g : P \to Q$ be two morphisms in $\mathbf{Pol}_\omega$. For each $n \in \mathbb{N}$, we define a subset of P_n by

$$R_n = \{a \in P_n \mid f(a) = g(a)\}.$$

As the inclusions $j_n : R_n \to P_n$ commute with the source and target maps, this defines a polygraph R together with a morphism $j : R \to P$ in $\mathbf{Pol}_\omega$. Now $j : R \to P$ is clearly an equalizer of the pair (f, g) in $\mathbf{Pol}_\omega$.

16.1.4 Coproducts. The coproducts in $\mathbf{Pol}_\omega$ are built by taking the coproducts of the corresponding sets of n-generators in each dimension n. Thus, if $(P^i)_{i \in I}$ is a family of polygraphs, we have for each dimension n, $(\coprod_{i \in I} P^i)_n = \coprod_{i \in I} P^i_n$.

16.1.5 Coequalizers. Let P, Q be two polygraphs and $f, g\colon P \to Q$ be a pair of morphisms in $\mathbf{Pol}_\omega$. We first build, for each $n \in \mathbb{N}$, a coequalizer $k_n\colon Q_n \to R_n$ of the pair (f_n, g_n) in **Set**:

$$P_n \underset{g_n}{\overset{f_n}{\rightrightarrows}} Q_n \xrightarrow{k_n} R_n.$$

Concretely, there is a binary relation on Q_n defined by $b \rightsquigarrow_n b'$ if and only if there is an $a \in P_n$ such that $b = f_n(a)$ and $b' = g_n(a)$. If $\sim_n$ is the smallest equivalence relation on Q_n containing $\rightsquigarrow_n$, then R_n is the set $Q_n/{\sim_n}$ of equivalence classes of $\sim_n$ and k_n is the canonical surjection.

We now define, by induction on n, a polygraph R with R_n as the set of n-generators, together with a morphism $k\colon Q \to R$ in $\mathbf{Pol}_\omega$ whose n-dimensional component is the above k_n.

- For $n = 0$, we just take R_0 and k_0 as above, and there is nothing to prove.
- Let $n > 0$ and suppose that the polygraph R has been constructed up to dimension $n-1$, together with the morphism $k\colon Q \to R$ and that, for all $0 \leqslant i < n$, $k_i^* f_i^* = k_i^* g_i^*$. Taking R_n as above, we have to define source and target maps $s_{n-1}, t_{n-1}\colon R_n \to R_{n-1}^*$. Consider the following diagram

$$\begin{array}{ccccc}
P_n & \underset{g_n}{\overset{f_n}{\rightrightarrows}} & Q_n & \xrightarrow{k_n} & R_n \\
{\scriptstyle s_{n-1}}\downarrow & & {\scriptstyle s_{n-1}}\downarrow & & \vdots\,{\scriptstyle s_{n-1}} \\
P_{n-1}^* & \underset{g_{n-1}^*}{\overset{f_{n-1}^*}{\rightrightarrows}} & Q_{n-1}^* & \xrightarrow[k_{n-1}^*]{} & R_{n-1}^*
\end{array}$$

in which all solid arrows are already given, and let $c \in R_n$. Let b be a representative of c in Q_n, that is, $c = k_n(b)$ and set $s_{n-1}(c) = k_{n-1}^* s_{n-1}(b) \in R_{n-1}^*$. We must show that $s_{n-1}(c)$ so defined does not depend on the choice of the representative b. Thus let $b' \in Q_n$ such that $k_n(b') = c$. By definition, $b \sim_n b'$. In order to show that $k_{n-1}^* s_{n-1}(b) = k_{n-1}^* s_{n-1}(b')$, it is sufficient to check it in the base case where $b \rightsquigarrow_n b'$. In this case, there is a generator $a \in P_n$ such that $f_n(a) = b$ and $g_n(a) = b'$, therefore

$$\begin{aligned}
k_{n-1}^* s_{n-1}(b) &= k_{n-1}^* s_{n-1} f_n(a) \\
&= k_{n-1}^* f_{n-1}^* s_{n-1}(a) \\
&= k_{n-1}^* g_{n-1}^* s_{n-1}(a) \\
&= k_{n-1}^* s_{n-1} g_n(a) \\
&= k_{n-1}^* s_{n-1}(b')
\end{aligned}$$

so that the source map on R_n is well-defined. Of course the target map t_{n-1} is defined accordingly. As for the globular relations, remark that if $n \geqslant 2$, by induction hypothesis,

$$\begin{aligned} t_{n-2}s_{n-1}k_n &= t_{n-2}k^*_{n-1}s_{n-1} \\ &= k^*_{n-2}t_{n-2}s_{n-1} \\ &= k^*_{n-2}t_{n-2}t_{n-1} \\ &= t_{n-2}k^*_{n-1}t_{n-1} \\ &= t_{n-2}t_{n-1}k_n \end{aligned}$$

and because k_n is surjective, this implies $t_{n-2}s_{n-1} = t_{n-2}t_{n-1}$. Likewise $s_{n-2}s_{n-1} = s_{n-2}t_{n-1}$. Thus, the polygraph R is now defined up to dimension n. By using Lemma 15.1.5, we finally extend k to a morphism of polygraphs up to dimension n, so that $k^*_n f^*_n = k^*_n g^*_n$.

Having defined $k \colon Q \to R$, a similar induction process shows that k is indeed a coequalizer of the pair $(f,)g$ in $\mathbf{Pol}_\omega$.

16.1.6 Proposition. *The category* $\mathbf{Pol}_\omega$ *is complete and cocomplete.*

Proof. The category $\mathbf{Pol}_\omega$ has all small products and equalizers of all pairs of morphisms, hence has all small limits (see [261, Chapter V, Theorem 1]). Dually, $\mathbf{Pol}_\omega$ has coproducts and coequalizers, hence has all small colimits. □

16.1.7 Remark. To each polygraph P we may associate a set

$$|P| = \coprod_{k \in \mathbb{N}} P_k$$

consisting of all generators of P. This correspondence is the object part of a functor $|-| \colon \mathbf{Pol}_\omega \to \mathbf{Set}$. In fact, a morphism $f \colon P \to Q$ in $\mathbf{Pol}_\omega$ takes, for each $k \in \mathbb{N}$, the set P_k to the set Q_k, whence a map $|f| \colon |P| \to |Q|$. Now, the above construction of the colimits in $\mathbf{Pol}_\omega$ shows that this functor $|-|$ preserves all small colimits. As a morphism f of polygraphs is entirely determined by its components f_n, the functor $|-|$ is faithful, making $\mathbf{Pol}_\omega$ a *concrete* category.

16.2 Morphisms in $\mathbf{Pol}_\omega$

We need a few technical preliminaries before characterizing monomorphisms and epimorphisms in $\mathbf{Pol}_\omega$. First, let C be an ω-category and $X \subseteq C_n$ a set of n-cells of C: we say that X is *closed under divisors* if, for any i-composable n-cells $u, v \in C_n$ such that $u *_i v \in X$, then $u \in X$ and $v \in X$. We may then state the following crucial property of cellular extensions:

16.2.1 Lemma. *Let $n \in \mathbb{N}$, and $f\colon (C,X) \to (D,Y)$ a morphism in $\mathbf{Cat}_n^+$, where $f = (g,h)$, with $g\colon C \to D$ a morphism in $\mathbf{Cat}_n$ and $h\colon X \to Y$ a source and target preserving map. Let $f^*\colon L_n(C,X) \to L_n(D,Y)$ be the induced morphism in $\mathbf{Cat}_{n+1}$. Suppose that the maps $g_n\colon C_n \to D_n$ and $h\colon X \to Y$ are injective and that the image of g_n is closed under divisors. Then f^*_{n+1} is injective and its image is closed under divisors.*

Proof. We restrict here to the general line of reasoning and refer to [255, Section 2.3] for a complete proof. Note also that [264] contains a thorough analysis of the shape of freely generated cells, essentially encompassing the present material. Thus, let (C,X) and (D,Y) as in the above statement. Let C_{n+1} (resp. D_{n+1}) the set of $(n+1)$-cells in $L_n(C,X)$ (resp. $L_n(D,Y)$). According to [279, Section 4.1], there are sets of well-typed formal expressions E^C and E^D endowed with binary relations $\sim^C$ and $\sim^D$ generating congruence relations $\simeq^C$ and $\simeq^D$ such that $C_{n+1} = E^C/\simeq^C$ and $D_{n+1} = E^D/\simeq^D$. Moreover, f induces a map $\overline{f}\colon E^C \to E^D$ such that the following diagram commutes, the vertical maps being the canonical surjections:

$$\begin{array}{ccc} E^C & \xrightarrow{\overline{f}} & E^D \\ \downarrow & & \downarrow \\ C_{n+1} & \xrightarrow[f^*_{n+1}]{} & D_{n+1}. \end{array}$$

Then structural induction on formal expressions shows that for each $a, b \in E^C$ such that $\overline{f}(a) \sim^D \overline{f}(b)$, then $a \sim^C b$. Thus, whenever $\overline{f}(a) \simeq^D \overline{f}(b)$, $a \simeq^C b$. Therefore f^*_{n+1} is injective. Moreover, the image of f^*_{n+1} is still closed by divisors. □

16.2.2 Proposition. *Let P, Q be polygraphs, $f\colon P \to Q$ be a morphism in $\mathbf{Pol}_\omega$ and $n \in \mathbb{N}$. If for all $k \leqslant n$, $f_k\colon P_k \to Q_k$ is injective, then $f_n^*\colon P_n^* \to Q_n^*$ is injective.*

Proof. Suppose $f\colon P \to Q$ is a morphism in $\mathbf{Pol}_\omega$ such that, for all $k \leqslant n$, the map $f_k\colon P_k \to Q_k$ is injective. Then, by applying Lemma 16.2.1 dimension-wise, we check the two following properties by induction on $k \in \{0, ..., n-1\}$.

- The maps $f_k^*\colon P_k^* \to Q_k^*$ and $f_{k+1}\colon P_{k+1} \to Q_{k+1}$ are injective.
- The image of f_k^* is closed by divisors.

Therefore, by applying once more Lemma 16.2.1, f_n^* is injective. □

Let (C,X) be a cellular extension of an n-category C by X and $C[X]$ the freely generated $(n+1)$-category on this extension. As all categories $\mathbf{Cat}_m$

canonically embed into $\mathbf{Cat}_\omega$, we may view C and $C[X]$ as ω-categories, and consider the canonical ω-functor $j\colon C \to C[X]$. Let us call an ω-functor *injective* if its underlying globular map is a monomorphism in $\mathbf{Glob}_\omega$, that is, injective in each dimension. Then the following result follows immediately as a particular case of Lemma 16.2.1.

16.2.3 Proposition. *For any cellular extension (C, X), the canonical ω-functor $j\colon C \to C[X]$ is injective.*

16.2.4 Remark. In fact, Proposition 16.2.3 holds for a more general interpretation of the notion of cellular extension, namely one when C is *any* ω-category and X is a set of cells of *any dimensions* freely attached to C. We refer to [264, Section 4, p. 36] for a complete proof of this generalized statement. When translated into the present language, Makkai's theorem states precisely the following: for any ω-category C and any set X, together with a family of morphisms $f_x\colon \partial\mathbb{O}_{n_x} \to C$, $x \in X$, the morphism $j\colon C \to C[X]$ in the pushout square

$$\begin{array}{ccc} \coprod_{x\in X} \partial\mathbb{O}_{n_x} & \xrightarrow{[f_x]_{x\in X}} & C \\ {\scriptstyle \coprod_{x\in X} \mathrm{i}_{n_x}}\downarrow & & \downarrow{\scriptstyle j} \\ \coprod_{x\in X} \mathbb{O}_{n_x} & \longrightarrow & C[X] \end{array}$$

is injective. The proof of Makkai goes along the same lines as the one of Lemma 16.2.1 and involves a precise analysis of the formal expressions denoting the cells of $C[X]$.

16.2.5 Linearization. To each pair (X, n) such that X is a set and $n \in \mathbb{N}$ we may associate an ω-category $C(X, n)$ whose only non-trivial cells are in dimension n, where

- $C(X, n)_n = X$ if $n = 0$,
- $C(X, n)_n = \mathbb{N}[X]$, the free abelian monoid generated by X if $n > 0$.

Note that $\mathbb{N}[X]$ consists of linear combinations of the form $u = \sum_{x\in X} n_x x$ where $n_x \in \mathbb{N}$ and $n_x = 0$ for all but a finite number of indices x. Moreover, for all $i < n$, the i-composition of such n-cells is given by $u *_i v = u + v$. Remark that, *except for* $n = 1$, the category $C(X, n)$ is freely generated by a polygraph $P(X, n)$ whose generators are given by $P(X, n)_n = X$, $P(X, n)_0 = \{*\}$ if $n > 0$ and $P(X, n)_i = \emptyset$ otherwise, the source and target maps being uniquely determined by these data.

Let now P be a polygraph and $n \in \mathbb{N}$. To any n-generator $a \in P_n$ and any n-cell $u \in P_n^*$ we may unambiguously attach a natural number $\mathrm{w}_a(u)$, the *weight* of a in u, measuring the number of occurrences of a in u. Consider indeed the n-polygraph $P_{\leqslant n}$ which coincides with P up to dimension n, and let $X = P_n$: the above defined ω-category $C(X, n)$ can be seen as an n-category. Now, by Lemma 15.1.5, the unique morphism in $\mathbf{Cat}_{n-1}$ taking $(P_{\leqslant n-1})^*$ to the terminal object extends uniquely to a morphism $\lambda\colon (P_{\leqslant n})^* \to C(X, n)$ whose restriction to P_n is the identity $P_n \to X$. We may therefore define the natural numbers $\mathrm{w}_a(u)$ by

$$\lambda_n(u) = \sum_{a \in P_n} \mathrm{w}_a(u) a$$

for any n-cell $u \in P_n^*$. Note that $\lambda_n(u)$ may be seen as a *multiset* $\mathrm{supp}^\sharp(u)$ on P_n mapping each generator $a \in P_n$ to $\mathrm{w}_a(u)$.

16.2.6 Support. The above notion of weight leads to the technical notion of *support*. Let P be a polygraph and u an n-cell in P_n^*. We first define the set of n-generators actually occurring in u by

$$\mathrm{supp}_n(u) = \{a \in P_n \mid \mathrm{w}_a(u) \neq 0\}\,.$$

Now the *total support* of u is defined by induction on the dimension of u:

- For $n = 0$, $u \in P_0^* = P_0$ and $\mathrm{supp}(u) = \{u\}$.
- For $n > 0$, $\mathrm{supp}(u) = U \cup S \cup T \cup S' \cup T'$ where

$$\begin{aligned} U &= \mathrm{supp}_n(u), \\ S &= \mathrm{supp}(s_{n-1}(u)), \\ T &= \mathrm{supp}(t_{n-1}(u)) \\ S' &= \cup_{a \in \mathrm{supp}_n(u)} \mathrm{supp}(s_{n-1}(a)), \\ T' &= \cup_{a \in \mathrm{supp}_n(u)} \mathrm{supp}(t_{n-1}(a)). \end{aligned}$$

In other words, the total support of u consists in all generators in $\cup_{0 \leqslant i \leqslant n} P_i$ needed to express u. Likewise, for any subset $A \subseteq \coprod_{k \in \mathbb{N}} P_k$ of generators of P, we define the *support of* A by $\mathrm{supp}(A) = \cup_{a \in A} \mathrm{supp}(a^*)$.

16.2.7 Subpolygraph. Let P be a polygraph and $a \in P_n$ an n-generator. Let a^* denote the corresponding n-cell in P_n^*. The *subpolygraph of P generated by a*, denoted $Q = \langle a \rangle^P$, is defined as follows: for each $k \leqslant n$, the set of k-generators of Q is $Q_k = P_k \cap \mathrm{supp}(a^*)$, where $\mathrm{supp}(a^*)$ is the total support of a^* defined in §16.2.6, whereas the source and target maps $s_{k-1}, t_{k-1}\colon Q_k \to Q_{k-1}^*$ are obtained by restriction of $s_{k-1}, t_{k-1}\colon P_k \to P_{k-1}^*$ for $k > 0$. Of course $Q_k = \emptyset$ for $k > n$. Thus, we get a canonical morphism $\kappa_a\colon \langle a \rangle^P \to P$ such that $Q_n = \{a\}$ and $\kappa_a(a) = a$. More generally, for any subset $A \subseteq \coprod_{k \in \mathbb{N}} P_k$ of generators of

P, the *subpolygraph of P generated by A* is the polygraph $Q = \langle A\rangle^P$ such that, for each $k \in \mathbb{N}$, $Q_k = P_k \cap \operatorname{supp}(A)$, and built dimensionwise, together with a canonical morphism $k_A : \langle A\rangle^P \to P$ by using at each level the appropriate restriction to Q of the source and target maps of P.

16.2.8 Proposition. *A morphism $f: Q \to P$ in $\mathbf{Pol}_\omega$ is a monomorphism if and only if $f_n : Q_n \to P_n$ is injective in each dimension n.*

Proof. In one direction, notice that the functor $\mathbf{Pol}_\omega \to \mathbf{Set}^{\mathbb{N}}$ taking a polygraph P to the sequence $(P_n)_{n\in\mathbb{N}}$ is faithful by construction, hence reflects monomorphisms. Therefore, if f_n is injective for all n, then f is a monomorphism.

Conversely, suppose that $f: Q \to P$ is a monomorphism. Suppose also that there is a $k \in \mathbb{N}$ such that f_k is not injective, and let n be the smallest such integer. By hypothesis, there are two distinct n-generators $a, a' \in Q_n$ such that $f_n(a) = f_n(a')$. Hence

$$f^*_{n-1}(s_{n-1}(a)) = s_{n-1}(f_n(a)) = s_{n-1}(f_n(a')) = f^*_{n-1}(s_{n-1}(a')).$$

But as f_{n-1} is injective, so is f^*_{n-1} by Lemma 16.2.2. Hence $s_{n-1}(a) = s_{n-1}(a')$. Likewise $t_{n-1}(a) = t_{n-1}(a')$. As a consequence, there is a unique morphism $h: \langle a\rangle^Q \to \langle a'\rangle^Q$ taking a to a'. Let $k = k_a$ and $k' = k_b \circ h$. We now have a diagram

$$\langle a\rangle^Q \underset{k'}{\overset{k}{\rightrightarrows}} Q \xrightarrow{\;f\;} P$$

such that $fk = fk'$ but $k \neq k'$. This contradicts the hypothesis, and we are done. □

16.2.9 Corollary. *The functor $F: \mathbf{Pol}_\omega \to \mathbf{Cat}_\omega$ preserves monomorphisms.*

Proof. Suppose that $f: Q \to P$ is a monomorphism in $\mathbf{Pol}_\omega$. By Proposition 16.2.8, f_n is injective in each dimension n, and by Lemma 16.2.2, so is f^*_n. Now the functor $C \mapsto (C_n)_{n\in\mathbb{N}}$ from $\mathbf{Cat}_\omega$ to $\mathbf{Set}^{\mathbb{N}}$ is faithful and reflects monomorphisms. Hence $f^*: Q^* \to P^*$ is a monomorphism. □

16.2.10 Remark. Proposition 16.2.8 shows that for each subset A of generators of a polygraph P, the canonical morphism $k_A : \langle A\rangle^P \to P$ is a monomorphism. Conversely, any monomorphism $f: Q \to P$ in $\mathbf{Pol}_\omega$ factorizes as $f = k_A \circ h$, where $A = \{f(q) \mid q \in Q_k, k \in \mathbb{N}\}$ and $h: Q \to \langle A\rangle^P$ is an isomorphism.

16.2.11 Proposition. *A morphism $f: P \to Q$ in $\mathbf{Pol}_\omega$ is an epimorphism if and only if f_n is surjective in each dimension n.*

Proof. As above, the functor $P \mapsto (P_n)_{n\in\mathbb{N}}$ is faithful and reflects epimorphisms. Therefore if f_n is surjective in each dimension, then f is an epimorphism. Conversely, suppose that $f\colon P \to Q$ is an epimorphism in $\mathbf{Pol}_\omega$. Consider the abelianization functor $\mathrm{Ab}\colon \mathbf{Cat}_\omega \to \mathbf{Ch}_{\mathbb{Z},\geqslant 0}$ (see Chapter 22). By Proposition 22.1.2, Ab is a left adjoint. Recall that $F\colon \mathbf{Pol}_\omega \to \mathbf{Cat}_\omega$ is also a left adjoint. Therefore the composition $\mathrm{Ab}\circ F\colon \mathbf{Pol}_\omega \to \mathbf{Ch}_{\mathbb{Z},\geqslant 0}$ is a left adjoint and preserves epimorphisms. As a consequence, the map $(\mathrm{Ab}(f^*))_n\colon \mathbb{Z}[P_n] \to \mathbb{Z}[Q_n]$ is surjective for each n, but $(\mathrm{Ab}(f^*))_n$ is nothing but the linearization of f_n, hence f_n itself is surjective. □

16.2.12 Proposition. *A morphism $f\colon P \to Q$ in $\mathbf{Pol}_\omega$ is an isomorphism if and only if, for each $n \geqslant 0$, it induces a bijection $f_n\colon P_n \to Q_n$.*

Proof. In one direction, if f is an isomorphism, so is $|f|$ by functoriality, whence also all maps f_n for $n \geqslant 0$. Conversely, suppose f_n is a bijection in all dimensions n. We define $g\colon Q \to P$ inverse to f in $\mathbf{Pol}_\omega$ by induction on the dimension.

- For $n = 0$, let $g_0\colon Q_0 \to P_0$ be the inverse map $(f_0)^{-1}$ of f_0.
- Let $n \geqslant 0$ and suppose g has been defined up to dimension n such that $g_k \circ f_k = (1_P)_k$ and $f_k \circ g_k = (1_Q)_k$ for all $0 \leqslant k \leqslant n$. Define

  $$g_{n+1} = (f_{n+1})^{-1}\colon Q_{n+1} \to P_{n+1}.$$

  Let $a \in Q_{n+1}$. We have to check that $s_n(g_{n+1}(a)) = g_n^*(s_n(a))$. Now

  $$f_n^*(s_n(g_{n+1}(a)) = s_n(f_{n+1}g_{n+1}(a)) = s_n(a) = f_n^*(g_n^*(s_n(a)).$$

  By induction hypothesis, f_n^* is a bijection, whence the desired equality.

Likewise $t_n(g_{n+1}(a)) = g_n^*(t_n(a))$. Therefore, g is defined and is inverse to f up to dimension $n+1$, whence the result. □

16.2.13 Remark. As an immediate consequence of the above results, the isomorphisms in $\mathbf{Pol}_\omega$ are exactly the morphisms which are monomorphisms and epimorphisms. Moreover the functor $|{-}|$ from Remark 16.1.7 preserves and reflects monomorphisms, epimorphisms, and isomorphisms.

16.2.14 Subobject classifier. The category $\mathbf{Pol}_\omega$ has a *subobject classifier*, that is, an object Ω together with a monomorphism $\mathsf{true}\colon \mathbf{1}^{\mathrm{pol}} \to \Omega$ such that for every monomorphism $f\colon P \to Q$ in $\mathbf{Pol}_\omega$, there is a unique morphism $\chi_f\colon Q \to \Omega$ making the following diagram a pullback square:

$$\begin{array}{ccc} P & \xrightarrow{!^P} & \mathbf{1}^{\mathrm{pol}} \\ {\scriptstyle f}\downarrow & & \downarrow{\scriptstyle \mathsf{true}} \\ Q & \xrightarrow[\chi_f]{} & \Omega. \end{array} \tag{16.7}$$

As usual, these conditions determine Ω up to isomorphism. Let us now build $\mathsf{true}\colon \mathbf{1}^{\mathrm{pol}} \to \Omega$ by induction on the dimension. To avoid overloaded notation, we denote $\mathbf{1}^{\mathrm{pol}}$ by T throughout the construction.

- For $n = 0$, T_0 is a singleton, whereas $\Omega_0 = T_0 + T_0$ elements and true_0 is the left inclusion $T_0 \to T_0 + T_0$.
- Let $n \geqslant 0$ and suppose Ω and true have been defined up to dimension n. Recall from §16.1.1 that the set of $(n+1)$-generators of T is

  $$T_{n+1} = \left\{(u, v) \;\middle|\; u \in T_n^*, v \in T_n^*, s_{n-1}(u) = s_{n-1}(v), t_{n-1}(u) = t_{n-1}(v)\right\}$$

  with source and target maps defined by $s_n(u, v) = u$ and $t_n(u, v) = v$. Consider now the set

  $$S_{n+1} = \left\{(u, v) \;\middle|\; u \in \Omega_n^*, v \in \Omega_n^*, s_{n-1}(u) = s_{n-1}(v), t_{n-1}(u) = t_{n-1}(v)\right\}$$

  and the following two subsets of S_{n+1}:

  $$\begin{aligned} S^0_{n+1} &= \left\{(\mathsf{true}^*_n u, \mathsf{true}^*_n v) \;\middle|\; (u, v) \in T_{n+1}\right\}, \\ S^1_{n+1} &= S_{n+1} \setminus S^0_{n+1}. \end{aligned}$$

  The set of $(n+1)$-generators of Ω is then

  $$\Omega_{n+1} = S^0_{n+1} + S^0_{n+1} + S^1_{n+1} \tag{16.8}$$

  and true_{n+1} sends T_{n+1} to the first copy of S^0_{n+1} by

  $$\mathsf{true}_{n+1}(u, v) = (\mathsf{true}^*_n u, \mathsf{true}^*_n v).$$

  The source and target maps $s_n, t_n\colon \Omega_{n+1} \to \Omega^*_n$ are naturally given by $s_n(u, v) = u$ and $t_n(u, v) = v$, making true a morphism of $(n+1)$-polygraphs.

Suppose now that $f\colon P \to Q$ is a monomorphism. We define a morphism $\chi_f\colon Q \to \Omega$ by induction on the dimension.

- If $n = 0$, $(\chi_f)_0\colon Q_0 \to \Omega_0 = T_0 + T_0$ sends a 0-cells u of Q to the left component if and only if $a \in \operatorname{im} f_0$.
- Suppose χ_f has been defined up to dimension n, and let $a \in Q_{n+1}$, with $s_n(a) = u$ and $t_n(a) = v$. The pair $c = ((\chi^*_f)_n(u), (\chi^*_f)_n(v))$ is by induction a pair of parallel n-cells of Ω^*_n and three cases are possible:

- If $c \in S^0_{n+1}$ and $a \in \operatorname{im} f_{n+1}$, $(\chi_f)_{n+1}$ sends a to c in the first S^0_{n+1} component of (16.8).
- If $c \in S^0_{n+1}$ and $a \notin \operatorname{im} f_{n+1}$, $(\chi_f)_{n+1}$ sends a to c in the second S^0_{n+1} component of (16.8).
- If $c \in S^1_{n+1}$, $(\chi_f)_{n+1}$ sends a to c in the S^0_{n+1} component of (16.8).

This defines χ_f as a morphism of $(n+1)$-polygraphs. We easily check that χ_f so defined is the unique morphism such that (16.7) is a pullback square.

16.2.15 A counterexample. Let us end this review of morphisms in $\mathbf{Pol}_\omega$ by the following small observation. To each ω-category C corresponds an ω-groupoid $\widehat{C}$ in $\mathbf{Gpd}_\omega$ obtained by formally inverting all n-cells, $n > 0$, in C. The correspondence $C \mapsto \widehat{C}$ is in fact the object part of the left adjoint to the inclusion $\mathbf{Gpd}_\omega \to \mathbf{Cat}_\omega$ and the unit of the associated monad yields a morphism $\eta_C \colon C \to \widehat{C}$. In case $C = P^*$ is freely generated by a polygraph P, one could expect η_C to be injective. However, this is proved wrong by the following counterexample. Let

$$P = \langle \star \mid x\colon \star \to \star \mid a, b\colon x \to 1_\star, c\colon 1_\star \to x \rangle$$

and $C = P^*$. Consider the 2-cells $u, v \in C_2$ given by $u = a *_1 c *_1 b$, $v = b *_1 c *_1 a$ and define $u' = \eta_C u$, $v' = \eta_C v$. Whereas $u \neq v$ in C, the presence of a strict inverse a^{-1} for a in $\widehat{C}$ implies that $u' = v'$, as

$$\begin{aligned} u' &= a *_1 c *_1 b = a *_1 (c *_1 a) *_1 (a^{-1} *_1 b) \\ &= a *_1 (a^{-1} *_1 b) *_1 (c *_1 a) = b *_1 c *_1 a = v'. \end{aligned}$$

16.3 Is $\mathbf{Pol}_n$ a Topos?

As the category of n-polygraphs is complete and cocomplete and has a subobject classifier, it is natural to ask if $\mathbf{Pol}_n$ is a topos. In fact, $\mathbf{Pol}_0$ is the category of sets and $\mathbf{Pol}_1$ is the category of graphs, hence both are presheaf categories and thus are topoi. As for $n = 2$, Carboni and Johnstone [75] (corrected in [76]) have proved that $\mathbf{Pol}_2$ is also a presheaf category and thus a topos. However, from $n = 3$ on, $\mathbf{Pol}_n$ fails to be cartesian *closed*, as shown by Makkai and Zawadowski [265] (see also [264, Section 6, p. 57]), and therefore is not even an elementary topos. Let us finally mention that Batanin [30] defines a notion of T-computad for each monad T on globular sets and gives sufficient conditions on T for the category of T-computads to be a presheaf category. Of course, [265] implies that these conditions do not hold when T is the monad of strict ω-categories on globular sets.

16.3.1 The category $\mathbf{Pol}_2$ is a presheaf category. The category $\mathbf{Pol}_2$ is a category of presheaves over the small category whose objects are p_0, p_1, and $p_{m,n}$ for $m, n \in \mathbb{N}$, morphisms are generated by $s: p_0 \to p_1$, $t: p_0 \to p_1$ and

$$\begin{aligned} s^i_{m,n}&: p_1 \to p_{m,n} && \text{for } m, n \in \mathbb{N} \text{ with } 0 \leqslant i < m,\\ t^i_{m,n}&: p_1 \to p_{m,n} && \text{for } m, n \in \mathbb{N} \text{ with } 0 \leqslant i < n,\\ \sigma^i_{m,n}&: p_0 \to p_{m,n} && \text{for } m, n \in \mathbb{N} \text{ with } 0 \leqslant i \leqslant m,\\ \tau^i_{m,n}&: p_0 \to p_{m,n} && \text{for } m, n \in \mathbb{N} \text{ with } 0 \leqslant i \leqslant n, \end{aligned}$$

subject to the relations

$$\begin{aligned} \sigma^i_{m,n} &= s^i_{m,n} \circ s & \sigma^{i+1}_{m,n} &= s^i_{m,n} \circ t & \sigma^0_{m,n} &= \tau^0_{m,n}\\ \tau^i_{m,n} &= t^i_{m,n} \circ s & \tau^{i+1}_{m,n} &= t^i_{m,n} \circ t & \sigma^m_{m,n} &= \tau^n_{m,n} \end{aligned}$$

for all indices such that the morphisms are defined. A presheaf P over this category then corresponds to a polygraph with $P(p_0)$ as 0-cells, $P(p_1)$ as 1-cells

$$a: x \to y$$

with $x = P(s(a))$ and $y = P(t(a))$, and $P(p_{m,n})$ as 2-cells

$$\alpha: a_1 \ldots a_m \to b_1 \ldots b_n$$

with $a_i = P(s^i_{m,n}(\alpha))$ and $b_i = P(t^i_{m,n}(\alpha))$.

16.3.2 The category $\mathbf{Pol}_3$ is not cartesian closed. The original argument by Makkai and Zawadowski being quite intricate, we give here the simpler proof due to Cheng [83], based on an explicit counterexample: we shall describe 3-polygraphs P, Q, R, and S such that the diagram

$$P \underset{f_2}{\overset{f_1}{\rightrightarrows}} Q \xrightarrow{\ g\ } R$$

is a coequalizer in $\mathbf{Pol}_3$ not preserved by the functor $- \times S: \mathbf{Pol}_3 \to \mathbf{Pol}_3$. Therefore $- \times S$ does not preserve colimits, hence admits no right adjoint and $\mathbf{Pol}_3$ is not cartesian closed. Now P, Q, R, S are as follows:

- $P_0 = \{\star\}$, $P_1 = \emptyset$, $P_2 = \{a\}$, and $P_3 = \emptyset$ so that there is no choice for the source and target maps,
- $Q_0 = \{\star\}$, $Q_1 = \emptyset$, $Q_2 = \{b_1, b_2\}$ where $f_i a = b_i$ for $i \in \{1,2\}$, and $Q_3 = \{b_3\}$ where $b_3: b_1 *_1 b_2 \to 1_2(\star)$,
- $R_0 = \{\star\}$, $R_1 = \emptyset$, $R_2 = \{c\}$ where $c = gb_1 = gb_2$, and $R_3 = \{d\}$ where $gb_3 = d$ and $d: c *_1 c \to 1_2(\star)$,
- $S = Q$.

Let $P' = P \times S$, $Q' = Q \times S$, and $R' = R \times S$. By the construction of products described in §16.1.2, P', Q', and R' have a single 0-generator $\star' = (\star, \star)$ and no 1-generator. As for generators in dimensions 2 and 3,

- P'_2 has two elements $a'_i = (a, b_i, 1_1(\star'), 1_1(\star'))$ for $i \in \{1,2\}$ and $P'_3 = \emptyset$,
- Q'_2 has four elements $b'_{ij} = (b_i, b_j, 1_1(\star'), 1_1(\star'))$ for $(i,j) \in \{1,2\} \times \{1,2\}$,
- Q'_3 has *two* elements

$$b'_{31} = (b_3, b_3, b'_{12} *_1 b'_{21}, 1_2(\star')),$$
$$b'_{32} = (b_3, b_3, b'_{11} *_1 b'_{22}, 1_2(\star')),$$

- R'_2 has two elements $c'_i = (c, b_i, 1_1(\star'), 1_1(\star'))$ for $i \in \{1,2\}$,
- R'_3 has a single element $d' = (d, b_3, c'_1 *_1 c'_2, 1_2(\star'))$.

Consider now the coequalizer diagram

$$P' \underset{f'_2}{\overset{f'_1}{\rightrightarrows}} Q' \xrightarrow{g'} R''$$

where $f'_i = f_i \times 1_S$. As $P'_3 = \emptyset$ and Q'_3 has two elements, R''_3 also has two elements, whereas R'_3 only has one. Therefore R' is not isomorphic to R'' and we are done. Note that the key of the above argument is the Eckmann-Hilton phenomenon, according to which $b_1 *_1 b_2 = b_2 *_1 b_1$: this in turn implies that b'_{31} and b'_{32} are defined in such a way that both projections of $b'_{12} *_1 b'_{21}$ and $b'_{11} *_1 b'_{22}$ match the actual source of b_3.

16.3.3 Presheaf subcategories of polygraphs. Although $\mathbf{Pol}_\omega$ is not a presheaf category, there are several interesting full subcategories of $\mathbf{Pol}_\omega$ which are presheaf categories, as shown in [176, 177]. The general principle is to restrict the shape of generators to sufficiently "regular" ones. Important examples are *many-to-one* polygraphs, where the target of each generator is a generator, or *non-unital* polygraphs, where the source and target of each generator cannot be identities. An alternative approach is taken in [169–171], where a category of *regular polygraphs* is *defined* as the presheaf category on a small category of certain globular shapes, then shown to be equivalent to a full subcategory of $\mathbf{Pol}_\omega$.

16.4 Local Presentability

This section closely follows [264, Section 5]. Appendix G recalls everything we need about locally presentable categories in the present section. Let us call a polygraph P *finite* when $|P| = \coprod_k P_k$ is a finite set. Given any polygraph Q, we may then consider the set of finite subpolygraphs of Q, in the sense of §16.2.7. Let I_Q be the subcategory of $\mathbf{Pol}_\omega$ whose objects are the finite subpolygraphs of Q, the morphisms are the canonical inclusions $P \to P'$ for all P, P' such that $|P| \subseteq |P'|$ and $D\colon I_Q \to \mathbf{Pol}_\omega$ the corresponding inclusion functor. Let P, P' be objects in I_Q, together with their canonical inclusion morphisms $f\colon P \to Q$ and $f'\colon P' \to Q$. Consider the finite subpolygraph P'' of Q generated by $|P| \cup |P'|$ and $h\colon P'' \to Q$ the corresponding inclusion morphism. Clearly f and f' factor through f'' as in the diagram

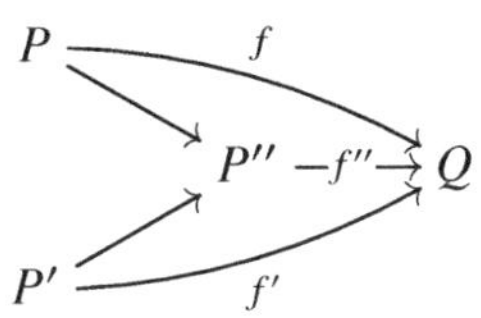

so that I_Q is a directed poset. Now, by taking the colimit of the sets $|P|$ for $P \in I_Q$ in **Set**, we get

$$\varinjlim_{P \in I_Q} |D(P)| = \bigcup_{F \in I_Q} |P| = |Q|$$

but the functor $|-|$ preserves small colimits, as remarked in Remark 16.1.7, so that

$$|\varinjlim_{P \in I_Q} D(P)| = |Q|.$$

Therefore the canonical map $j\colon \varinjlim_{P \in I_Q} D(P) \to Q$ is such that $|j|$ is an identity. By Remark 16.2.13, the functor $|-|$ reflects isomorphisms, hence j is an isomorphism in $\mathbf{Pol}_\omega$. Thus we have proved the following statement:

16.4.1 Proposition. *Any polygraph is a canonical colimit of its finite subpolygraphs.*

Let now P be a finite polygraph and $(Q^{(i)}, f_{ij})$ a filtered system in $\mathbf{Pol}_\omega$ indexed by a small category I, with colimit $Q = \varinjlim_i Q^{(i)}$ and $f_i\colon Q^{(i)} \to Q$ the canonical morphisms for $i \in I$. Let $f\colon P \to Q$ be a morphism in $\mathbf{Pol}_\omega$. Because the functor $|-|$ preserves colimits and $|P|$ is finite, there is an $i \in I$ and a map $h\colon |P| \to |Q^{(i)}|$ such that

- $|f|$ factors as in the following diagram

$$\begin{array}{ccc} & & |Q^{(i)}| \\ & \overset{h}{\nearrow} & \downarrow{\scriptstyle |f_i|} \\ |P| & \xrightarrow[|f|]{} & |Q|, \end{array}$$

- for all $a \in |P|$, $b \in |Q^{(i)}|$, and $b' \in |Q^{(i)}|$ such that

$$|f_i|(b) = |f_i|(b') = |f|(a),$$

we have $b = b'$.

A double induction on the pair (n, p), where n is the highest dimension k such that $P_k \neq \emptyset$ and p is the cardinal of P_n, shows with the help of Lemma 15.1.5 that there is a morphism $g \colon P \to Q^{(i)}$ is $\mathbf{Pol}_\omega$ factoring f as in the diagram

$$\begin{array}{ccc} & & Q^{(i)} \\ & \overset{g}{\nearrow} & \downarrow{\scriptstyle f_i} \\ P & \xrightarrow[f]{} & Q. \end{array}$$

Thus

$$\mathbf{Pol}_\omega(P, \varinjlim_i Q^{(i)}) \simeq \varinjlim_i \mathbf{Pol}_\omega(P, Q^{(i)})$$

and we have proved the following result:

16.4.2 Proposition. *A finite polygraph is a finitely presentable object in* $\mathbf{Pol}_\omega$.

One easily checks that the isomorphism classes of finite polygraphs form a set. As a consequence of Propositions 16.4.1 and 16.4.2 we get the following theorem:

16.4.3 Theorem. *The category* $\mathbf{Pol}_\omega$ *is locally finitely presentable.*

16.5 Contexts

We recall here very briefly the notion of *context*, based on the presentation of [279, Section 5, p. 191], and refer to this article for detailed proofs. Let $n \geqslant 1$, P a polygraph and $x = (u, v)$ an ordered pair of parallel $(n-1)$-cells in P^*. We call here such a pair x an *n-type* (a convenient alternative terminology for "n-sphere" as defined in §15.1.1). The polygraph P may be extended to a polygraph $P[x]$ by adjoining a new generator x in dimension n, such that $s_{n-1}x = u$ and $t_{n-1}x = v$. We call the n-cell $\mathbf{x} = x^*$ (where x^* is short for $i_n(x)$) of $P[x]^*$ an *n-indeterminate of type x over P*.

16.5.1 Definition. Let $n \geqslant 1$, P a polygraph and x an n-type of P. An n-cell $u \in P[x]^*$ is a *context* of type x if $\mathrm{w}_x(u) = 1$.

A context u of type x will be denoted $u = c[\mathbf{x}]$, where $\mathbf{x} = x^*$. An n-context $u = c[\mathbf{x}]$ of type x is *thin* whenever $\mathrm{w}_a(u) = 0$ for all n-generators $a \in P[x]_n \setminus \{x\}$, that is, all generators in $\mathrm{supp}(u)$ are of dimension $< n$ but x itself, and $c[\mathbf{x}]$ is *trivial* if $c[\mathbf{x}] = \mathbf{x}$. There is a well-defined operation of *substitution* in n-contexts: let z be an n-cell of P^* of type $x = (u, v)$, that is, such that $s_{n-1}z = u$ and $t_{n-1}z = v$, and $c[\mathbf{x}]$ an n-context of type x over P. From Lemma 15.1.5, we get a morphism

$$\mathrm{sub}_z : P[x]^* \to P^*$$

in $\mathbf{Cat}_\omega$ taking $\mathbf{x}$ to z and leaving other generators unchanged. Then the substitution of $\mathbf{x}$ by z in c, noted $c[z]$ may be defined as $\mathrm{sub}_z(c[\mathbf{x}])$.

Recall from [279] that thin n-contexts can always be expressed in the (non-unique) form:

$$c[\mathbf{x}] = u_{n-1} *_{n-2} (\dots *_1 (u_1 *_0 \mathbf{x} *_0 v_1) *_1 \dots) *_{n-2} v_{n-1} \tag{16.9}$$

where u_i and v_i are identities over i-dimensional cells. This remark leads to the following technical result:

16.5.2 Lemma. *For each $n > 1$ there is a map ∂ taking each thin n-context $c[\mathbf{x}]$ to an $(n-1)$-context $\partial c[\mathbf{x}']$ of type $x' = (s_{n-2}\mathbf{x}, t_{n-2}\mathbf{x})$ such that*

- *for each n-cell z of type x, $s_{n-1}c[z] = \partial c[s_{n-1}z]$ and $t_{n-1}c[z] = \partial c[t_{n-1}z]$,*
- *if $\partial c[\mathbf{x}']$ is trivial, then so is $c[\mathbf{x}]$.*

Proof. See [279, p. 193]. □

16.5.3 Remark. Conversely, let $c[\mathbf{y}]$ be an $(n-1)$-context of type $y = (u, v)$, where u, v are parallel $(n-2)$-cells, and $\mathbf{x}$ be an n-indeterminate such that $s_{n-2}(\mathbf{x}) = u$ and $t_{n-2}(\mathbf{x}) = v$, there is a unique thin n-context $\overline{c}[\mathbf{x}]$ such that $\partial\overline{c}[\mathbf{y}] = c[\mathbf{y}]$.

16.5.4 Composition. Let now $u = c[\mathbf{x}]$ be an n-context of type x over P, and $d[\mathbf{y}]$ be an n-context of type $y = (s_{n-1}u, t_{n-1}u)$ over $P[x]$. The previously defined substitution process yields an n-context $d[c[\mathbf{x}]]$ of type x over P.

16.5.5 Remark. Let $u = c[\mathbf{x}]$ be an $(n-1)$-context of type x, $d[\mathbf{y}]$ be an $(n-1)$-context of type $y = (s_{n-2}u, t_{n-2}u)$, and $e[\mathbf{x}] = d[c[\mathbf{x}]]$ the composed context as above. For each n-indeterminate $\mathbf{x}$ such that

$$(s_{n-2}\mathbf{x}, t_{n-2}\mathbf{x}) = (s_{n-2}u, t_{n-2}u),$$

the thin n-context defined in Remark 16.5.3 satisfies the equation

$$\overline{e}[\mathbf{x}] = \overline{d}[\overline{c}[\mathbf{x}]]. \tag{16.10}$$

16.5.6 Lemma. *For any n-cell z of type x, if $d[c[z]] = z$ then both contexts $c[\mathbf{x}]$ and $d[\mathbf{y}]$ are trivial.*

Proof. We reason by induction on n. If $n = 1$, by computing the weights on both sides of the equality $d[c[z]] = z$, we see that $c[\mathbf{x}]$ and $d[\mathbf{y}]$ are thin, but thin 1-contexts are trivial. Let $n > 1$ and suppose that the result holds in dimension $n - 1$. Let $c[\mathbf{x}]$ and $d[\mathbf{y}]$ be n-contexts and z an n-cell as in the statement, such that $d[c[z]] = z$. By computing the weights on both sides of this equality, we see that $c[\mathbf{x}]$ and $d[\mathbf{y}]$ are thin contexts. Thus, Lemma 16.5.2 yields $(n-1)$-contexts $\partial c[\mathbf{x}']$ and $\partial d[\mathbf{y}']$ such that, for $z' = s_{n-1}z$, $z' = \partial d[\partial c[z']]$. By induction hypothesis, $\partial c[\mathbf{x}']$ and $\partial d[\mathbf{y}']$ are trivial, and so are $c[\mathbf{x}]$ and $d[\mathbf{y}]$ by Lemma 16.5.2. □

16.6 Basis Uniqueness

Let P be a polygraph and $C = P^*$ the free ω-category it generates. We shall prove that the generators of P are entirely determined by C (see also [264, Section 4.(8.3)]).

16.6.1 Definition. Let P be a polygraph and $n > 0$. An n-cell $u \in P^*_n$ is *irreducible* if it is not a unit and whenever $u = v *_i w$, then either $u = v$ and $w = 1^n_{t_i(v)}$ or $u = w$ and $v = 1^n_{s_i(w)}$.

16.6.2 Lemma. *An n-cell $u \in P^*_n$ is irreducible if and only if it is a generating cell of the form $u = a^*$ for $a \in P_n$.*

Proof. Suppose that u is irreducible. By structural induction on u (see §15.1.9), either $u = 1^n_v$, which contradicts the hypothesis, or $u = a^*$ with $a \in P^*_n$, in which case we get the result, or $u = v *_i w$. By definition, $u = v$ or $u = w$, so that the induction hypothesis applies and u is a generating cell.

Conversely, suppose that $u = a^*$ and $u = v *_i w$. By computing the weights on both sides, we may suppose without loss of generality that $\mathrm{w}_a(v) = 1$ and $\mathrm{w}_a(w) = 0$. Then, there is a context $c[\mathbf{x}]$ of type $x = (s_{n-1}a^*, t_{n-1}a^*)$ such that $v = c[a^*]$. Let $y = (s_{n-1}v, t_{n-1}v)$ and denote by $d[\mathbf{y}]$ the context $\mathbf{y} *_i w$ of type y. By substitution, $d[c[a^*]] = u = a^*$. By Lemma 16.5.6, both contexts $c[\mathbf{x}]$ and $d[\mathbf{y}]$ are trivial. In particular, $\mathbf{y} *_i w = \mathbf{y}$. By repeated applications of Lemma 16.5.2, we see that $s_k(w)$ must be a unit cell for all $k > i$, which implies $w = 1^n_{t_i(\mathbf{y})} = 1^n_{t_i(v)}$. Therefore $u = v *_i w = v$ and we are done. □

We may now state the main result of this section:

16.6.3 Proposition. *Let P, Q two polygraphs such that there is an isomorphism $f\colon P^* \to Q^*$ in $\mathbf{Cat}_\omega$. Then there is a morphism $g\colon P \to Q$ in $\mathbf{Pol}_\omega$ such that, for each $n \in \mathbb{N}$, $g_n\colon P_n \to Q_n$ is a bijection, and $f = g^*$.*

Proof. Let $a \in P_n$ be an n-generator. By Lemma 16.6.2, a^* is irreducible, and as $f\colon P^* \to Q^*$ is an isomorphism, so is $f(a^*)$. Therefore, by Lemma 16.6.2 again, there is a $b = g_n(a) \in Q_n$ such that $f(a^*) = b^*$. This defines the required map $g_n\colon P_n \to Q_n$. By construction, $f = g^*$. □

16.7 Rewriting Properties of n-Polygraphs

The theory of rewriting developed in Chapters 1, 4, and 10 extends to n-polygraphs in a seamless way. Throughout this section, we fix an n-polygraph P, with $n \geqslant 2$. Recall from §15.1.7, that, for any $k \leqslant n$, we denote by $P_{\leqslant k}$ the k-polygraph obtained by truncating P to dimension k. By definition, the set P_n of n-generators is a cellular extension of the category $P^*_{\leqslant n-1}$ freely generated by the $(n-1)$-polygraph $P_{\leqslant n-1}$ and we think of P as a presentation of the $(n-1)$-category

$$\overline{P} = P^*_{\leqslant n-1}/P_n$$

defined as the quotient $(n-1)$-category, in the sense of §15.1.6. In this setting, the elements of P_n are called *rewriting rules of P*.

16.7.1 Rewriting step. Let x, y be two parallel $(n-1)$-cells in $P^*_{\leqslant n-1}$. A *rewriting step of P* from x to y is an n-cell $w \in P^*_n$ with source x and target y of the form $w = c[a]$ where a is a rewriting rule of P and $c[\mathbf{w}]$ is a thin n-context of type $(s_{n-1}(a), t_{n-1}(a))$. From (16.9), each rewriting step can be expressed in a (non-unique) form as

$$c[a] = u_{n-1} *_{n-2} (\dots *_1 (u_1 *_0 a *_0 v_1) *_1 \dots) *_{n-2} v_{n-1}$$

where u_i and v_i are identities over i-dimensional cells, for every $1 \leqslant i \leqslant n-1$. We denote by P_n^{steps} the set of rewriting steps of the polygraph P.

16.7.2 Rewriting path. A *rewriting path of P* is a sequence

$$\phi = w_1, \ldots, w_k \tag{16.11}$$

of rewriting steps of P such that $t_{n-1}(w_i) = s_{n-1}(w_{i+1})$, for every $1 \leqslant i \leqslant k-1$. Therefore, a rewriting path ϕ of type (16.11) yields an n-cell of P^*_n

$$\overline{\phi} = w_1 *_{n-1} \ldots *_{n-1} w_k. \tag{16.12}$$

In the case where $k = 0$, the rewriting path ϕ is said to be *empty*, and the corresponding n-cell $\overline{\phi}$ is of the form 1_u for some $u \in P^*_{n-1}$. Note that for any rewriting path $\phi = w_1, \ldots, w_k$, the $(n-1)$-cells $x = s_{n-1}(w_1)$ and $y = t_{n-1}(w_k)$ are respectively the source and target of $\overline{\phi}$, so that there is no ambiguity in writing $\phi\colon x \to y$. As in §4.1.3, if $\phi\colon x \to y$ and $\psi\colon y \to z$ are rewriting paths, we denote their concatenation by $\phi * \psi$. In particular, the sequence (16.11) may be denoted

$$\phi = w_1 * \ldots * w_k. \tag{16.13}$$

The following result is now an easy consequence of Proposition 15.1.8.

16.7.3 Proposition. *For each n-cell z in P^*_n, there is a rewriting path ϕ of P such that $z = \overline{\phi}$.*

16.7.4 Remark. In Proposition 16.7.3, the rewriting path ϕ is not uniquely determined by z. However, if $\phi = w_1 * \ldots * w_k$ and $\phi' = w'_1 * \ldots * w'_{k'}$ are two rewriting paths such that $\overline{\phi} = \overline{\phi'}$, then $k' = k$. Therefore, the *length* of a rewriting path ϕ only depends on $\overline{\phi}$. Moreover, given ϕ and ϕ' as above, there is a permutation σ of $\{1, \ldots, k\}$ such that for each $i \in \{1, \ldots, k\}$, w_i and $w'_{\sigma(i)}$ are thin contexts $w_i = c[a]$ and $w'_{\sigma(i)} = c'[a]$ over the *same* generator $a \in P_{n-1}$.

16.7.5 Rewriting properties. Let P be an n-polygraph P. Its two top dimensions build a 1-polygraph $(P^*_{n-1}, P^{\text{steps}}_n)$, i.e., an abstract rewriting system, whose 0-generators are the $(n-1)$-cells of P^*_{n-1} and 1-generators are rewriting steps of P. We thus extend the rewriting notions defined on 1-polygraphs in Section 1.3 to n-polygraphs as follows. We say that the n-polygraph P is

terminating / quasi-terminating / Church–Rosser /
confluent / locally confluent / decreasing
convergent / quasi-convergent

when the 1-polygraph $(P^*_{n-1}, P^{\text{steps}}_n)$ is. In particular, the following general result still applies here, see Lemma 1.3.21:

16.7.6 Proposition. *A terminating polygraph P is confluent if and only if it is locally confluent.*

The above proposition naturally leads us to investigate local branchings.

16.7.7 Classification of local branchings. A *branching of P* is a pair (ϕ, ψ) of rewriting paths $\phi\colon x \to y$, $\psi\colon x \to z$ of P with common source x. Such a branching is *local* if both ϕ and ψ are of length 1, that is, are rewriting *steps* of P. A local branching (ϕ, ψ) is

- *trivial* if $\psi = \phi$,
- *orthogonal* if there are rewriting steps $w\colon u \to v$ and $w'\colon u' \to v'$ such that $\overline{\phi} = w *_{n-2} 1_{u'}$ and $\overline{\psi} = 1_u *_{n-2} w'$ (or $\overline{\phi} = 1_{u'} *_{n-2} w$ and $\overline{\psi} = w' *_{n-2} 1_u$),
- *overlapping* if it is neither trivial nor orthogonal.

16.7.8 Minimal branching. The notion of *minimal branching* extends in arbitrary dimension $n \geqslant 2$ as follows. There is a binary relation $\sqsubseteq$ on the set of local branchings defined by $(\phi, \psi) \sqsubseteq (\phi', \psi')$ if and only if there is an $(n-1)$-context $c[\mathbf{y}]$ of type $(s_{n-2}(\overline{\phi}), t_{n-2}(\overline{\phi}))$ (which is the same as $(s_{n-2}(\overline{\psi}), t_{n-2}(\overline{\psi})))$ such that $\overline{\phi'} = \overline{c}[\overline{\phi}]$ and $\overline{\psi'} = \overline{c}[\overline{\psi}]$, where $\overline{c}[\mathbf{x}]$ is the thin n-context defined in Remark 16.5.3. The relation $\sqsubseteq$ is clearly reflexive. Antisymmetry follows from Lemma 16.5.6, whereas transitivity is a consequence of (16.10). Therefore $\sqsubseteq$ is a partial order and we may define a minimal branching as a minimal element of the set of local branchings with respect to this order.

16.7.9 Critical branching. A branching is *critical* when it is overlapping and minimal.

16.7.10 Lemma. *Given $n \geqslant 2$, an n-polygraph is locally confluent if and only if all its critical branchings are confluent.*

Proof. Same proof as for Lemma 4.3.7. □

16.8 Polygraphs With Finite Derivation Type

The property of finite derivation type has been studied in Chapter 8 for 2-polygraphs and in Chapter 12 for 3-polygraphs. It can be extended to any n-polygraphs as follows. An n-polygraph P has *finite derivation type* if it is finite and if the free $(n, n-1)$-category $P^{\top}$ admits a finite acyclic cellular extension, that is, a finite cellular extension generating all n-spheres of $P^{\top}$. An $(n-1)$-category has *finite derivation type* when it admits a presentation by an n-polygraph with finite derivation type. As in the case of 1- and 2-categories, given two presentations of the same $(n-1)$-category by finite n-polygraphs, both are of finite derivation type or neither is. The proof of this result given in [161, Proposition 3.3.4] is similar to the proof in the case of 1- and 2-categories given in Theorems 8.1.2 and 12.1.4, respectively.

16.8.1 Proposition. *Let P and Q be Tietze equivalent finite n-polygraphs. Then the polygraph P has finite derivation type if and only if Q has.*

16.8.2 Squier's coherence theorem. For $n \leqslant 3$, we have shown that, for a convergent n-polygraph P, the set of critical branchings generate a homotopy basis of the $(n, n-1)$-category $P^\top$, see Theorems 2.5.10, 7.3.5, and 12.1.7. The proofs of these results extend to higher-dimensional polygraphs as follows, see [161, Proposition 4.3.4].

16.8.3 Theorem. *Let P be a convergent n-polygraph, and P_{n+1} be a cellular extension of the free $(n, n-1)$-category $P^\top$. If P_{n+1} contains, for every critical branching (ϕ, ψ) of P, one $(n+1)$-generator of the form*

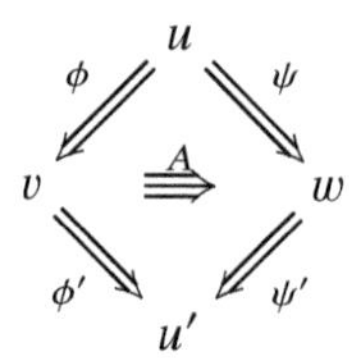

where ϕ' and ψ' are n-cells in P_n^*, then the $(n+1, n-1)$-polygraph (P, P_{n+1}) is coherent.

Following Theorem 16.8.3, for every $n \geqslant 1$, a finite convergent n-polygraph with a finite set of critical branchings has finite derivation type. A 1-category having a finite convergent presentation therefore has finite derivation type, see Theorem 8.2.1. Note, however, that this result fails to generalize to n-categories for $n \geqslant 2$ as shown with the following counterexample.

16.8.4 A counterexample. We have constructed in §12.2.8 a finite convergent 3-polygraph Pearl_3 which does not have finite derivation type. By shifting dimensions on the polygraph Pearl_3, we obtain an n-polygraph Pearl_n, for any $n \geqslant 3$. It has exactly the same cells and compositions in dimensions $n-3$, $n-2$, $n-1$, and n as Pearl_3 has in dimensions 0, 1, 2, and 3; on top of that, it has one cell in each dimension up to $n-4$ and no other possible compositions, except with degenerate cells. By construction, the polygraph Pearl_n is finite and convergent, yet it still fails to have finite derivation type. We have thus proved that, for every natural number $n \geqslant 2$, there exists an n-category which does not have finite derivation type and admits a presentation by a finite convergent $(n+1)$-polygraph. We refer the reader to [161] for more details on this result.

17
A Catalogue of n-Polygraphs

We have already presented a wealth of low-dimensional examples of n-polygraphs in Parts I, II, and III, and many more will be found in Appendices A and C. The present chapter concentrates on some useful families of n-polygraphs based on familiar shapes: cylinders, cubes, and simplices, namely Street's *orientals* defined in the seminal paper [334]. These families are crucial in the development of a homotopy theory of ω-categories.

We shall explain two methods for generating the above families. The first one is based on a direct definition of the cylinder polygraph $\mathbb{O}_1 \otimes P$ of a polygraph P. The second is based on Steiner's theory of augmented directed complexes [330], which is a very powerful tool to build polygraphs using chain complexes. In particular, it allows to define a tensor product for polygraphs (or even ω-categories) from which we can recover the cylinder polygraph, as well as a join operation.

17.1 First Examples of n-Polygraphs

17.1.1 Monoids. Of particular importance is the 2-polygraph associated to a presentation of a monoid by generators and rewriting rules (see Chapter 4 and Appendix A). Recall that given M a monoid presented by a set of generators P_1 and a set P_2 of rewriting rules of the form $\alpha : w \Rightarrow w'$ where $w, w' \in P_1^*$, we get a 2-polygraph $P = \langle\, P_0 \mid P_1 \mid P_2 \,\rangle$, where $P_0 = \{\star\}$.

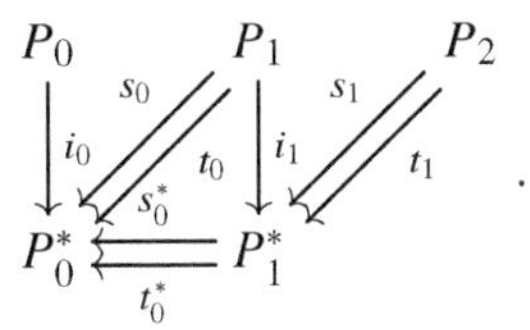

The source and target maps are defined by $s_0(a) = t_0(a) = \star$ for each $a \in P_1$, and $s_1(\alpha) = w$, $t_1(\alpha) = w'$ for each rewrite rule $\alpha\colon w \Rightarrow w'$.

Remark that in this case we recover M as the quotient category C/X where C is the free category $P_0^* \leftleftarrows P_1^*$ and $X = P_2$.

17.1.2 Terminal polygraph. The functor $G\colon \mathbf{Cat}_\omega \to \mathbf{Pol}_\omega$ from 15.1.10 also provides many natural examples of polygraphs. In particular $\mathbf{1}^{\mathrm{pol}} = G(\mathbf{1}^{\mathrm{cat}})$ where $\mathbf{1}^{\mathrm{cat}}$ is the terminal ω-category has infinitely many generators in all dimensions $k \geqslant 2$. It is in fact a terminal object in the category $\mathbf{Pol}_\omega$, as immediately implied by the adjunction.

17.1.3 Globular sets. Each n-globular set X may be seen as an n-polygraph. Precisely, for each $n \in \mathbb{N}$ there is a natural inclusion functor $\mathbf{Glob}_n \to \mathbf{Pol}_n$ taking the globular set X to the polygraph P whose set of k-generators is just $P_k = X_k$ for $0 \leqslant k \leqslant n$. For example, the n-globe $\mathbb{O}_n$ is a polygraph with exactly two generators in dimensions $0 \leqslant k < n$ and a unique generator in dimension n. Likewise, the n-spheres $\partial\mathbb{O}_n$ provide another example of a series of n-polygraphs.

17.2 Tensoring Polygraphs by $\mathbb{O}_1$

17.2.1 Before we turn to more complex examples of polygraphs, we need to describe a useful general construction, namely the *tensor product* of a polygraph by the 1-globe $\mathbb{O}_1$. In order to check the coherence of the following construction, we shall rely on the existence and the properties of the functor $\Gamma\colon \mathbf{Cat}_\omega \to \mathbf{Cat}_\omega$ taking an ω-category C to the ω-category ΓC of "small cylinders internal to C". We refer to Chapter 20 for a detailed account of this functor Γ. As for the present argument, we just need to know that Γ comes equipped with natural transformations

$$\overline{\pi}, \underline{\pi}\colon \Gamma \to 1_{\mathbf{Cat}_\omega}$$

and maps $|-|\colon (\Gamma C)_n \to C_{n+1}$ taking an n-cylinder $u \in (\Gamma C)_n$ to its *principal cell* $|u| \in C_{n+1}$, such that, for any n-cylinder $u \in (\Gamma C)_n$, the source and target of $|u|$ are given by

$$|u|\colon \overline{\pi}_C(u){*_0}|t_0(u)|{*_1}\cdots{*_{n-1}}|t_{n-1}(u)| \to |s_{n-1}(u)|{*_{n-1}}\cdots{*_1}|s_0(u)|{*_0}\underline{\pi}_C(u).$$

Let us first introduce a few notations to be used throughout the construction.

- The 0-generators of $\mathbb{O}_1$ will be denoted by $\mathbf{0}^-$ and $\mathbf{0}^+$.
- The 1-generator of $\mathbb{O}_1$ will be denoted by $\mathbf{1}$, so that $\mathbf{1}\colon \mathbf{0}^- \to \mathbf{0}^+$.
- For any set A of symbols, we denote by $\mathbf{0}^- \otimes A$ (resp. $\mathbf{0}^+ \otimes A$, $\mathbf{1} \otimes A$) the set of all symbols of the form $\mathbf{0}^- \otimes a$ (resp. $\mathbf{0}^+ \otimes a$, $\mathbf{1} \otimes a$), where $a \in A$.
- Given a symbol a, we define the following formal expressions:

$$\begin{aligned}
\mathrm{S}_0(a) &= \mathbf{0}^- \otimes a, \\
\mathrm{T}_0(a) &= \mathbf{0}^+ \otimes a, \\
\mathrm{S}_1(a) &= (\mathbf{0}^- \otimes a) *_0 (\mathbf{1} \otimes t_0(a)), \\
\mathrm{T}_1(a) &= (\mathbf{1} \otimes s_0(a)) *_0 (\mathbf{0}^+ \otimes a);
\end{aligned}$$

and more generally, for each integer $i > 1$:

$$\begin{aligned}
\mathrm{S}_i(a) &= (\mathbf{0}^- \otimes a) *_0 (\mathbf{1} \otimes t_0(a)) *_1 \cdots *_{i-1} (\mathbf{1} \otimes t_{i-1}(a)), \\
\mathrm{T}_i(a) &= (\mathbf{1} \otimes s_{i-1}(a)) *_{i-1} \cdots *_1 (\mathbf{1} \otimes s_0(a)) *_0 (\mathbf{0}^+ \otimes a).
\end{aligned}$$

These expressions will eventually denote actual cells whenever the indeterminate a is interpreted appropriately.

17.2.2 Let now P be a polygraph. We shall build a new polygraph $Q = \mathbb{O}_1 \otimes P$ endowed with a morphism $h : P^* \to \Gamma Q^*$ giving rise to the following families of maps:

- morphisms $\mathsf{top}, \mathsf{bot}\colon P \to Q$ in $\mathbf{Pol}_\omega$ respectively taking an n-cell $u \in P_n^*$ to $\mathsf{top}(u) = \mathbf{0}^- \otimes u = \overline{\pi}_{Q^*}(h(u))$ and $\mathsf{bot}(u) = \mathbf{0}^+ \otimes u = \underline{\pi}_{Q^*}(h(u))$,
- maps $\mathbf{1} \otimes -\colon P_n^* \to Q_{n+1}^*$ defined as the composite

$$P_n^* \xrightarrow{h_n} \Gamma Q^*{}_n \xrightarrow{|-|} Q_{n+1}^*$$

such that

$$\mathbf{1} \otimes u\colon \mathrm{S}_n(u) \to \mathrm{T}_n(u) \tag{17.1}$$

for any n-cell $u \in P_n^*$.

17.2.3 We now define Q, bot, top, and $\mathbf{1} \otimes -$ by simultaneous induction on the dimension.

- For $n = 0$, the set of 0-generators of Q is defined by $Q_0 = \mathbf{0}^- \otimes P_0 \sqcup \mathbf{0}^+ \otimes P_0$ whereas top_0 and bot_0 are given by $\mathsf{top}_0(a) = \mathbf{0}^- \otimes a$ and $\mathsf{bot}_0(a) = \mathbf{0}^+ \otimes a$ for $a \in P_0$.
- The set of 1-generators of Q is defined by $Q_1 = \mathbf{0}^- \otimes P_1 \sqcup \mathbf{0}^+ \otimes P_1 \sqcup \mathbf{1} \otimes P_0$, whereas $\mathsf{top}_1, \mathsf{bot}_1 : P_1 \to Q_1$ and $\mathbf{1} \otimes - : P_0 \to Q_1$ are given by the obvious canonical inclusions. The source and target maps $s_0^Q, t_0^Q : Q_1 \to Q_0^*$ are given, for each $a \in P_1$ by $s_0^Q(\mathbf{0}^- \otimes a) = \mathbf{0}^- \otimes s_0^P(a)$, $s_0^Q(\mathbf{0}^+ \otimes a) = \mathbf{0}^+ \otimes s_0^P(a)$, $t_0^Q(\mathbf{0}^- \otimes a) = \mathbf{0}^- \otimes t_0^P(a)$ and $t_0^Q(\mathbf{0}^+ \otimes a) = \mathbf{0}^+ \otimes t_0^P(a)$, so that top and bot are as expected morphisms up to dimension 1. As for $a \in P_0$, we define $s_0^Q(\mathbf{1} \otimes a) = \mathbf{0}^- \otimes a$ and $t_0^Q(\mathbf{1} \otimes a) = \mathbf{0}^+ \otimes a$. Because $P_0^* = P_0$ there is nothing more to verify here, and (17.1) holds.
- Suppose now Q has been defined up to dimension n, as well as morphisms $\mathsf{top}, \mathsf{bot} : P \to Q$ in $\mathbf{Pol}_\omega$ up to dimension n, a morphism $h : P^* \to \Gamma Q^*$ in $\mathbf{Cat}_\omega$ up to dimension $n-1$ and corresponding maps $\mathbf{1} \otimes - : P_k^* \to Q_{k+1}^*$ satisfying equations (17.1) for all $k \in \{0, \ldots, n-1\}$. The set of $(n+1)$-generators of Q is defined by $Q_{n+1} = \mathbf{0}^- \otimes P_{n+1} \sqcup \mathbf{0}^+ \otimes P_{n+1} \sqcup \mathbf{1} \otimes P_n$ whereas $\mathsf{top}_{n+1}, \mathsf{bot}_{n+1} : P_{n+1} \to Q_{n+1}$ and $\mathbf{1} \otimes - : P_n \to Q_{n+1}$ are given by the obvious canonical inclusions. The source and target maps are defined as above for the generators of the form $\mathbf{0}^- \otimes a$ and $\mathbf{0}^+ \otimes a$, namely:

$$
\begin{aligned}
s_n^Q(\mathbf{0}^- \otimes a) &= \mathbf{0}^- \otimes s_n^P(a), \\
s_n^Q(\mathbf{0}^+ \otimes a) &= \mathbf{0}^+ \otimes s_n^P(a), \\
t_n^Q(\mathbf{0}^- \otimes a) &= \mathbf{0}^- \otimes t_n^P(a), \\
t_n^Q(\mathbf{0}^+ \otimes a) &= \mathbf{0}^+ \otimes t_n^P(a).
\end{aligned}
$$

  As for generators of the form $\mathbf{1} \otimes a$, with $a \in P_n$, the induction hypothesis implies that $\mathrm{S}_n(a)$ and $\mathrm{T}_n(a)$ denote well-defined cells in Q_n^* and moreover that these two cells are parallel. Hence, the n-source and n-target of $\mathbf{1} \otimes a$ may be defined by $s_n^Q(\mathbf{1} \otimes a) = \mathrm{S}_n(a)$ and $t_n^Q(\mathbf{1} \otimes a) = \mathrm{T}_n(a)$, whence $\mathbf{1} \otimes a : \mathrm{S}_n(a) \to \mathrm{T}_n(a)$. It follows that the polygraph Q is now defined up to dimension $n+1$, together with the morphisms $\mathsf{top}, \mathsf{bot} : P \to Q$, and that (17.1) holds for the $(n+1)$-*generators* of P. This implies that the morphism h can be extended in dimension n by a map $h_n : P_n \to (\Gamma Q^*)_n$ commuting to source and target maps. By the universal property of polygraphs (Lemma 15.1.5), h extends as a morphism $P^* \to \Gamma Q^*$ up to dimension n, and by composing with the principal cell map $|-| : (\Gamma Q^*)_n \to Q_{n+1}^*$, we get a map

$$\mathbf{1} \otimes -: P_n^* \to Q_{n+1}^*$$

satisfying (17.1) for all cells $u \in P_n^*$. Thus, the induction is complete.

17.3 Families of Polygraphs

17.3.1 Cylinders. Let $n \geqslant 0$. The *free-standing n-cylinder* is by definition the polygraph

$$\mathrm{Cyl}_n = \mathbb{O}_1 \otimes \mathbb{O}_n.$$

Now recall that the n-globe $\mathbb{O}_n$ has $2n + 1$ generators, namely the only n-generator $\mathbf{n}$ together with the $2n$ generators of the form $\mathbf{i}^- = s_i(\mathbf{n})$ and $\mathbf{i}^+ = t_i(\mathbf{n})$ for $i \in \{0, ..., n-1\}$. Thus, Cyl_0 is just $\mathbb{O}_1$, whereas for $n > 0$ the generators of Cyl_n are listed below.

- There are four 0-generators, namely $\mathbf{0}^- \otimes \mathbf{0}^-$, $\mathbf{0}^- \otimes \mathbf{0}^+$, $\mathbf{0}^+ \otimes \mathbf{0}^-$, and $\mathbf{0}^+ \otimes \mathbf{0}^+$.
- For $0 < i \leqslant n-1$, there are six i-generators, namely $\mathbf{0}^- \otimes \mathbf{i}^-$, $\mathbf{0}^- \otimes \mathbf{i}^+$, $\mathbf{0}^+ \otimes \mathbf{i}^-$, $\mathbf{0}^+ \otimes \mathbf{i}^+$, $\mathbf{1} \otimes (\mathbf{i{-}1})^-$, and $\mathbf{1} \otimes (\mathbf{i{-}1})^+$.
- There are four n-generators, namely $\mathbf{0}^- \otimes \mathbf{n}$, $\mathbf{0}^+ \otimes \mathbf{n}$, $\mathbf{1} \otimes (\mathbf{n{-}1})^-$, and $1 \otimes (n-1)^+$.
- There is only one $(n+1)$-generator $\mathbf{1} \otimes \mathbf{n}$.

Therefore Cyl_n has exactly $6n + 3$ generators. Moreover, the source and target of these generators are given by the formulas (17.1). Here are pictures of Cyl_0, Cyl_1, and Cyl_2:

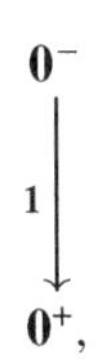

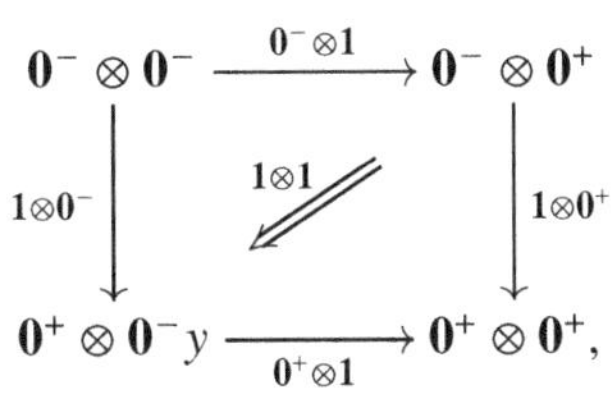

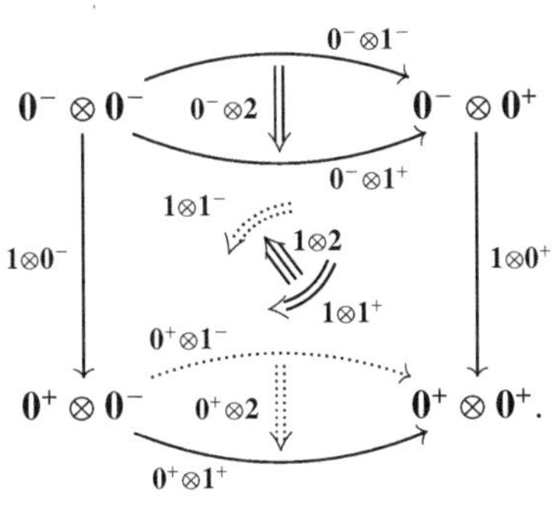

17.3.2 Cubes. The tensor product construction above also leads to the definition of the *polygraphic n-cubes*. Precisely, this family $(\mathrm{Cub}_n)_{n\in\mathbb{N}}$ of polygraphs is defined by

- $\mathrm{Cub}_0 = \mathbb{O}_0$,
- $\mathrm{Cub}_{n+1} = \mathbb{O}_1 \otimes \mathrm{Cub}_n$.

It follows from the above construction that, for each $i \in \{0, \ldots, n\}$, the set of i-generators of Cub_n has $\binom{n}{i}2^{n-i}$ elements, and therefore the total number of generators in Cub_n is

$$\sum_{i=0}^{n}\binom{n}{i}2^{n-i} = 3^n.$$

A convenient way to encode these generators is by labeling them by words on the alphabet $\{-, 1, +\}$, the i-generators being those with exactly i occurrences of the letter 1. For example, Cub_2 looks like

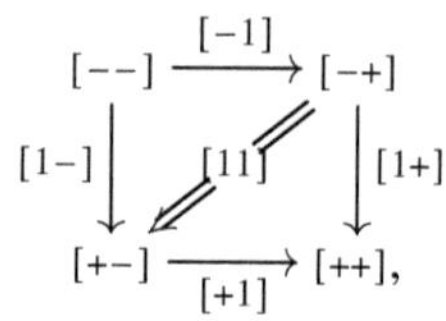

whereas Cub_3 looks like

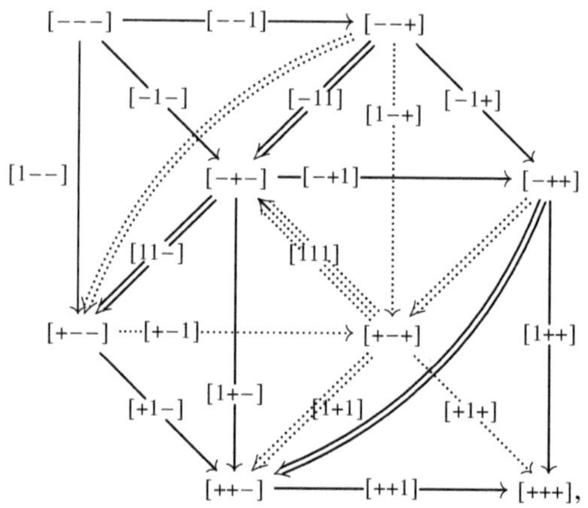

where the 3-cell $[111]$ goes from the composition of the front faces

$$s_2[111] = ([-11] *_0 [1{+}{+}]) *_1 ([-1-] *_0 [1{+}1]) *_1 ([11-] *_0 [{+}{+}1])$$

to the composition of the back faces

$$t_2[111] = ([--1] *_0 [11{+}]) *_1 ([1-1] *_0 [{+}1{+}]) *_1 ([1--] *_0 [{+}11]).$$

17.3.3 Simplices. The correspondence $P \mapsto \mathbb{O}_1 \otimes P$ from $\mathbf{Pol}_\omega$ to $\mathbf{Pol}_\omega$ is easily seen to be functorial. Now, for each polygraph P, we can form the following pushout in $\mathbf{Cat}_\omega$:

$$\begin{array}{ccc} P^* & \xrightarrow{\mathsf{top}} & (\mathbb{O}_1 \otimes P)^* \\ {\scriptstyle !}\downarrow & & \downarrow \\ \mathbf{1}^{\mathrm{cat}} & \longrightarrow & C(P). \end{array}$$

This defines a functor $P \mapsto C(P)$ from $\mathbf{Pol}_\omega$ to $\mathbf{Cat}_\omega$. Because top is a cofibration of $\mathbf{Cat}_\omega$ (see §19.2.1), $\mathbf{1}^{\mathrm{cat}} \to C(P)$ is also a cofibration and as $\mathbf{1}^{\mathrm{cat}}$ is cofibrant, so is $C(P)$. By Theorem 21.1.6, there is a polygraph $S(P)$ such that $C(P) = S(P)^*$. It is now possible to define a family $(O_n)_{n \in \mathbb{N}}$ of polygraphs by

- $O_0 = \mathbb{O}_0$,
- $O_{n+1} = S(O_n)$.

It turns out that O_n is precisely the polygraphic n-th simplex, or *n-th oriental*, first introduced in [334]. For each $0 \leqslant i \leqslant n$ the set of i-generators ("i-faces") of O_n has $\binom{n+1}{i+1}$ elements. The i-generators of O_n may be conveniently encoded by strictly increasing sequences of integers $\langle n_0, ..., n_i \rangle$, where $0 \leqslant n_0 < n_1 < \cdots < n_i \leqslant n$. For example, O_2 and O_3 may be pictured as follows:

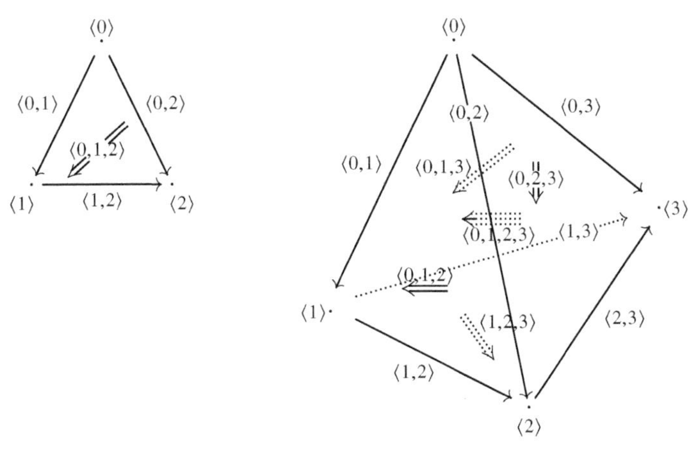

17.4 Construction of Polygraphs via Steiner's Theory

We shall now present Steiner's theory of augmented directed complexes [330] and use it to construct some polygraphs. Other similar formalisms include Street's *parity complexes* [335, 336] and Johnson's *pasting schemes* [206].

17.4.1 Augmented directed complexes. In this section, by "chain complex" we will always mean "chain complex of abelian groups in non-negative degree" (see §E.2.1). Recall that an *augmented chain complex* (K, d, e) is a chain complex (K, d) endowed with an *augmentation* e, that is, with a map of abelian groups $e\colon K_0 \to \mathbb{Z}$ such that $ed_1 = 0$. A morphism from an augmented chain complex (K, d_K, e_K) to a second augmented chain complex (L, d_L, e_L) is a morphism of chain complex f from (K, d_K) to (L, d_L) such that $e_L f_0 = e_K$.

An *augmented directed complex* is an augmented chain complex (K, d, e) equipped with, for every $n \geq 0$, a submonoid K_n^+ of *positive chains* of the abelian group K_n. A morphism $f\colon K \to L$ between two augmented directed complexes consists of a morphism of the underlying augmented chain complexes that respects the positive chains in the sense that $f(K_n^+) \subseteq L_n^+$ for every $n \geq 0$. We will denote by **ADC** the resulting category.

17.4.2 From ω-categories to augmented directed complexes. We define a functor $\lambda\colon \mathbf{Cat}_\omega \to \mathbf{ADC}$ by sending an ω-category C to the following augmented directed complex $\lambda(C)$:

- For every $n \geq 0$, the abelian group $(\lambda(C))_n$ is generated by elements $[x]$, for x an n-cell of C, subject to the relations

$$[x *_i y] = [x] + [y]$$

  for pairs of i-composable n-cells x and y.
- For every $n \geq 1$ and every n-cell x of C, we set

$$d_n([x]) = [t_{n-1}(x)] - [s_{n-1}(x)].$$

- If x is a 0-cell of C, we set

$$e([x]) = 1.$$

- Finally, for every $n \geq 0$, the submonoid $(\lambda(C))_n^+$ is the submonoid of $(\lambda(C))_n$ generated by the elements $[x]$ for $x \in C_n$.

The globular relations easily imply that d is indeed a differential so that $\lambda(C)$ is indeed an augmented directed complex.

Given an ω-functor $f\colon C \to D$, the morphism of augmented directed complexes $\lambda(f)\colon \lambda(C) \to \lambda(D)$ is defined on generators by

$$(\lambda(f))([x]) = [f(x)].$$

17.4.3 From chain complexes to ω-categories. We now define a functor $\mu\colon \mathbf{Ch}_{\mathbb{Z},\geqslant 0} \to \mathbf{Cat}_\omega$, where $\mathbf{Ch}_{\mathbb{Z},\geqslant 0}$ denotes the category of chain complexes. Given a chain complex K, the associated ω-category $\mu(K)$ is defined as follows:

- For $n \geq 0$, an n-cell of $\mu(K)$ consists of a table

$$x = \begin{pmatrix} x_0^- & \cdots & x_n^- \\ x_0^+ & \cdots & x_n^+ \end{pmatrix},$$

  where

  - x_i^ε, for $0 \leq i \leq n$ and $\varepsilon = \pm$, belongs to K_i,
  - $x_n^- = x_n^+$,
  - $d(x_i^\varepsilon) = x_{i-1}^+ - x_{i-1}^-$ for $1 \leq i \leq n$ and $\varepsilon = \pm$.
- For $n \geq 1$, the source and target of such a cell x are given by

$$s(x) = \begin{pmatrix} x_0^- & \cdots & x_{n-1}^- \\ x_0^+ & \cdots & x_{n-1}^- \end{pmatrix} \quad \text{and} \quad t(x) = \begin{pmatrix} x_0^- & \cdots & x_{n-1}^+ \\ x_0^+ & \cdots & x_{n-1}^+ \end{pmatrix}.$$

- For $n \geq 0$, the unit cell of such a cell x is given by

$$1_x = \begin{pmatrix} x_0^- & \cdots & x_n^- & 0 \\ x_0^+ & \cdots & x_n^+ & 0 \end{pmatrix}.$$

- Finally, if

$$x = \begin{pmatrix} x_0^- & \cdots & x_n^- \\ x_0^+ & \cdots & x_n^+ \end{pmatrix} \quad \text{and} \quad y = \begin{pmatrix} y_0^- & \cdots & y_n^- \\ y_0^+ & \cdots & y_n^+ \end{pmatrix}$$

  are two n-cells of $\mu(K)$ such that the $t_i(x) = s_i(y)$ for some $0 \leq i < n$, then

$$x *_i y = \begin{pmatrix} x_0^- & \cdots & x_i^- & x_{i+1}^- + y_{i+1}^- & \cdots & x_n^- + y_n^- \\ y_0^+ & \cdots & y_i^+ & x_{i+1}^- + y_{i+1}^- & \cdots & x_n^- + y_n^- \end{pmatrix}.$$

One checks that $\mu(K)$ is indeed an ω-category.

If $f\colon K \to L$ is a morphism of chain complexes, then the action on n-cells of the ω-functor $\mu(f)$ is defined by

$$\begin{pmatrix} x_0^- & \cdots & x_n^- \\ x_0^+ & \cdots & x_n^+ \end{pmatrix} \mapsto \begin{pmatrix} f(x_0^-) & \cdots & f(x_n^-) \\ f(x_0^+) & \cdots & f(x_n^+) \end{pmatrix}.$$

17.4.4 Remark. The construction μ of the previous paragraph actually lands into the category $\mathbf{Cat}_\omega(\mathbf{Ab})$ of ω-categories internal to abelian groups, that is, of ω-categories whose set of n-cells is endowed with a structure of abelian group, and whose operations (sources, target, units, and compositions) are compatible with these structures of abelian groups on cells. More precisely, the functor μ naturally lifts to a functor $\mathbf{Ch}_{\mathbb{Z},\geqslant 0} \to \mathbf{Cat}_\omega(\mathbf{Ab})$. A theorem of Bourn [54] states that this functor is an equivalence of categories. This is sometimes called the *globular Dold-Kan correspondence*.

17.4.5 From augmented directed complexes to ω-categories. We now define a functor $\nu\colon \mathbf{ADC} \to \mathbf{Cat}_\omega$ as a subfunctor of the functor

$$\mathbf{ADC} \xrightarrow{U} \mathbf{Ch}_{\mathbb{Z},\geqslant 0} \xrightarrow{\mu} \mathbf{Cat}_\omega,$$

where U denotes the obvious forgetful functor. If K is an augmented directed complex, an n-cell

$$\begin{pmatrix} x_0^- & \cdots & x_n^- \\ x_0^+ & \cdots & x_n^+ \end{pmatrix}$$

of $\mu(K)$ belongs to the sub-ω-category $\nu(K)$ if

- x_i^ε, for $0 \leq i \leq n$ and $\varepsilon = \pm$, is a positive chain (that is, is in K_i^+),
- $e(x_0^\varepsilon) = 1$ for $\varepsilon = \pm$.

One checks that the operations of the ω-category $\mu(K)$ restrict to $\nu(K)$ and that if $f\colon K \to L$ is a morphism of augmented directed complexes, then the ω-functor $\mu(f)\colon \mu(K) \to \mu(L)$ restricts to an ω-functor $\nu(f)\colon \nu(K) \to \nu(L)$.

17.4.6 Proposition. *The functors*

$$\lambda\colon \mathbf{Cat}_\omega \to \mathbf{ADC} \qquad\qquad \nu\colon \mathbf{ADC} \to \mathbf{Cat}_\omega$$

define a pair of adjoint functors.

Proof. We will only define the components of the adjunction morphisms and leave the verification to the reader. If C is an ω-category, the unit of the adjunction at C is the ω-functor $\eta_C\colon C \to \nu(\lambda(C))$ defined on n-cells by

$$x \mapsto \begin{pmatrix} [s_0(x)] & \cdots & [s_n(x)] \\ [t_0(x)] & \cdots & [t_n(x)] \end{pmatrix}.$$

If K is an augmented chain complex, the counit of the adjunction K is the morphism $\varepsilon_K\colon \lambda(\nu(K)) \to K$ defined on generating n-chains by

$$\left[\begin{pmatrix} x_0^- & \cdots & x_n^- \\ x_0^+ & \cdots & x_n^+ \end{pmatrix}\right] \mapsto x_n,$$

where x_n denotes the n-chain $x_n^- = x_n^+$. □

17.4.7 Remark. Similarly, the functors

$$U\lambda\colon \mathbf{Cat}_\omega \to \mathbf{Ch}_{\mathbb{Z},\geqslant 0} \qquad\qquad \mu\colon \mathbf{Ch}_{\mathbb{Z},\geqslant 0} \to \mathbf{Cat}_\omega,$$

where $U\colon \mathbf{ADC} \to \mathbf{Ch}_{\mathbb{Z},\geqslant 0}$ denotes the forgetful functor, define a pair of adjoint functors.

17.4.8 Augmented directed complexes with basis. A *basis* B of an augmented directed complex is a sequence of subsets $B_n \subseteq K_n$, indexed by $n \geq 0$, such that

- B_n is a basis of the $\mathbb{Z}$-module K_n,
- B_n generates K_n^+ as a submonoid of K_n.

The data of such a basis gives, for every $n \geq 0$, an isomorphism of abelian groups between K_n and $\mathbb{Z}^{(B_n)}$ restricting to an isomorphism of monoids between K_n^+ and $\mathbb{N}^{(B_n)}$. The elements of B_n can be recovered as the minimal non-zero elements of K_n^+ for the order on K_n defined by $x \leq y$ if $y - x \in K_n^+$. This shows that if such a basis exists it is unique.

17.4.9 Proposition. *Let P be a polygraph. Then, for each $n \geqslant 0$, the family $([x])_{x\in P_n}$ forms a basis of $\lambda(P^*)$.*

Proof. As any n-cell of P^* can be expressed as a composition of units and n-generators (see Proposition 15.1.8), the set P_n generates the monoid $\lambda(P^*)_n^+$. Let us show that it is a $\mathbb{Z}$-basis of $\lambda(P^*)_n$. We have to prove that the morphism $\beta\colon \mathbb{Z}[P_n] \to \lambda(P^*)_n$ which, given x in P_n, sends $[x]$ in $\mathbb{Z}[P_n]$ to $[x]$ in $\lambda(P^*)_n$ is an isomorphism. By 16.2.5, we have a map $\gamma\colon P_n^* \to \mathbb{Z}[P_n]$ sending an n-generator x of P to $[x]$ and compositions to sums. We thus get a morphism $\bar\gamma\colon \lambda(P^*)_n \to \mathbb{Z}[P_n]$. We claim that $\bar\gamma$ is an inverse of β. As $\mathbb{Z}[P_n]$ is generated by the $[x]$, it suffices to check equality $\bar\gamma \circ \beta = 1$ on these elements, which is true by definition. As $\lambda(P^*)_n$ is generated by the $[x]$, we similarly get the equality $\beta \circ \bar\gamma = 1$. □

17.4.10 Let K be an augmented directed complex with a basis B. Let $n \geq 0$ and let x be an n-chain of K. We can write

$$x = \sum_{b\in B_n} n_b b,$$

where the n_b are integers, in a unique way. The *support* $\operatorname{supp}(x)$ of x is the set of b such that n_b is non-null. The *negative support* $\operatorname{supp}^-(x)$ and *positive support* $\operatorname{supp}^+(x)$ of x are the sets

$$\mathrm{supp}^-(x) = \{b \mid n_b < 0\} \qquad \text{and} \qquad \mathrm{supp}^+(x) = \{b \mid n_b > 0\}.$$

We define

$$x^- = \sum_{b\in\mathrm{supp}^-(x)} (-n_b)b \qquad \text{and} \qquad x^+ = \sum_{b\in\mathrm{supp}^+(x)} n_b b.$$

We have

$$x = x^+ - x^-.$$

Actually, x^- and x^+ are the only positive n-chains y and z with disjoint support such that $x = z - y$.

If now $n \geq 1$ and x is still an n-chain, we define the positive $(n-1)$-chains $d^-(x)$ and $d^+(x)$ to be

$$d^-(x) = (d(x))^- \qquad \text{and} \qquad d^+(x) = (d(x))^+.$$

More generally, if $0 \leq i \leq n$, we define two positive i-chains $d_i^-(x)$ and $d_i^+(x)$ by

$$d_i^-(x) = (d^-)^{n-i}(x) \qquad \text{and} \qquad d_i^+(x) = (d^+)^{n-i}(x).$$

17.4.11 Unital augmented directed complexes. Let K be an augmented directed complex with basis B. For every element of the basis B, we define a table

$$\langle x\rangle = \begin{pmatrix} d_0^-(x) & \cdots & d_{n-1}^-(x) & x \\ d_0^+(x) & \cdots & d_{n-1}^+(x) & x \end{pmatrix}.$$

This table defines an n-cell of $\nu(K)$ if and only if $e(d_0^\varepsilon(x)) = 1$ for $\varepsilon = \pm$.

This motivates the following definition. An augmented directed complex with basis is said to be *unital* if for every $n \geq 0$ and every n-chain x of the basis, we have $e(d_0^-(x)) = 1$ and $e(d_0^+(x)) = 1$. If K is a unital augmented directed complex, the cells of the form $\langle x\rangle$, for x in the basis of K, are called *atoms*.

17.4.12 Loop-free augmented directed complexes. Let K be an augmented directed complex K with basis B. We say that K is *loop-free* if there exists a partial order on $\coprod_{n\geq 0} B_n$ such that, for every $n \geq 1$, every x in B_n and every $0 \leq i < n$, any element of the support of $d_i^-(x)$ is smaller than any element of the support of $d_i^+(x)$.

Similarly, we say that K is *strongly loop-free* if there exists a partial order $\preccurlyeq$ on $\coprod_{n\geq 0} B_n$ such that, for every $n \geq 1$, every x in B_n, every y in the support of d^-x and every z in the support of d^+x, one has

$$y \preccurlyeq x \preccurlyeq z.$$

As the terminology suggests, one can show that a strongly loop-free augmented directed complex is loop-free.

17.4.13 Remark. The definition of "loop-free" given in the previous paragraph is not the one from [330] but the one used in [331]. The two definitions can be shown to be equivalent.

17.4.14 Steiner complexes. We will say that an augmented directed complex is a *Steiner complex* if it is unital and loop-free. Similarly, we will say that it is a *strong Steiner complex* if it is unital and strongly loop-free.

17.4.15 Loop-free polygraphs. Let P be a polygraph. We say that P is *loop-free* if there exists a partial order on the set $\coprod_{n\geq 0} P_n$ such that, for every $n \geq 1$, every x in P_n and every $0 \leq i < n$, any element of the support of $s_i(x)$ is strictly smaller than any element of the support of $t_i(x)$.

Similarly, we will say that P is *strongly loop-free* if there exists a partial order $\preccurlyeq$ on $\coprod_{n\geq 0} P_n$ such that, for every $n \geq 1$, every x in P_n, every $0 \leq i < n$, every y in the support of $s_i(x)$ and every z in the support of $t_i(x)$, one has $y \preccurlyeq x \preccurlyeq z$.

We will say that an ω-category is a *Steiner ω-category* (resp. a *strong Steiner ω-category*) if it is generated by loop-free polygraph (resp. by a strongly loop-free polygraph).

We can now state a reformulation of the main theorems of [330]:

17.4.16 Theorem. *The adjunction*

$$\mathbf{Cat}_\omega \underset{\nu}{\overset{\lambda}{\rightleftarrows}} \mathbf{ADC} \quad (\lambda \dashv \nu)$$

restricts to an equivalence of categories between

- *the full subcategory of* $\mathbf{Cat}_\omega$ *spanned by Steiner's ω-categories (resp. by strong Steiner's ω-categories) and*
- *the full subcategory of* **ADC** *on Steiner complexes (resp. by strong Steiner complexes).*

Moreover, if K is a Steiner complex, the ω-category $\nu(K)$ is freely generated in the sense of polygraphs by its atoms.

17.4.17 Remark. The description of the full subcategories of $\mathbf{Cat}_\omega$ appearing in the previous theorem is different in [330]. Nevertheless, it is shown in [11] that they are equivalent.

We will now use Steiner's theory to construct some polygraphs and in particular recover the ones defined in the previous section.

17.4.18 Orientals. Consider the functor

$$c\colon \widehat{\Delta} \to \mathbf{Ch}_{\mathbb{Z},\geqslant 0}$$

sending a simplicial set (see §F.1.1) to its normalized chain complex (see §F.1.2). This functor naturally lifts to a functor

$$c_{\mathrm{ad}}\colon \widehat{\Delta} \to \mathbf{ADC}.$$

Indeed, if X is a simplicial set, then the chain complex $c(X)$ can be equipped with the following structure of augmented directed complex.

- The augmentation $e\colon \mathbb{Z}[X_0] \to \mathbb{Z}$ sends $[x]$, for x a 0-simplex, to 1.
- The submonoid $c(X)_n^+$, for $n \geq 0$, is generated by canonical basis of $c(X)_n$.

Steiner showed that the composite

$$\Delta \hookrightarrow \widehat{\Delta} \xrightarrow{c_{\mathrm{ad}}} \mathbf{ADC},$$

where the first functor denotes the Yoneda embedding, lands into strong Steiner complexes. In particular, by post-composing by $\nu\colon \mathbf{ADC} \to \mathbf{Cat}_\omega$, we get a functor

$$O\colon \Delta \to \mathbf{Cat}_\omega$$

landing into ω-categories generated by polygraphs. This functor is the so-called *cosimplicial object of orientals*. In particular, for $n \geq 0$, we recover the n-th oriental O_n as defined in §17.3.3.

More generally, it is shown in [14] that if the simplicial set X is a simplicial complex (this means, first, that the $(n+1)$-faces of any n-simplex are distinct and, second, that for any set E of $n+1$ 0-simplices, there is at most one non-degenerate n-simplex whose set of 0-simplices is E), then $c_{\mathrm{ad}}(X)$ is a strong Steiner complex. Under this condition, we thus get an ω-category $\nu(c_{\mathrm{ad}}(X))$ generated by a polygraph, which deserves to be called the *oriental* associated to X.

17.4.19 Globes. The n-category $\mathbb{O}_n$ is freely generated by a globular set. In particular, it is generated by a polygraph P whose generators are the cells of this globular set. If we denote by x the principal n-cell of $\mathbb{O}_n$, then

$$s_0(x) \preccurlyeq \cdots \preccurlyeq s_{n-1}(x) \preccurlyeq s_n(x) = x = t_n(x) \preccurlyeq t_{n-1}(x) \preccurlyeq \cdots \preccurlyeq t_0(x)$$

is a total order on generators of P showing that P is strongly loop-free. This shows that $\mathbb{O}_n$ is a strong Steiner ω-category, so that we have

$$\mathbb{O}_n \simeq \nu(\lambda(\mathbb{O}_n)).$$

Let us describe explicitly the augmented directed complex $\lambda(\mathbb{O}_n)$.

- The chains are defined by

$$\lambda(\mathbb{O}_n)_i = \begin{cases} \mathbb{Z}[\{s_i(x), t_i(x)\}] & \text{if } 0 \leq i < n, \\ \mathbb{Z}[\{x\}] & \text{if } i = n, \\ 0 & \text{if } i > n. \end{cases}$$

- If $0 < i < n$, then

$$d(s_i(x)) = t_{i-1}(x) - s_{i-1}(x) \qquad \text{and} \qquad d(t_i(x)) = t_{i-1}(x) - s_{i-1}(x),$$

  and

$$d(x) = t_{n-1}(x) - s_{n-1}(x).$$

- The augmentation is defined by

$$e(s_0(x)) = 1 \qquad \text{and} \qquad e(t_0(x)) = 1.$$

- The monoids of positive chains are given by

$$\lambda(\mathbb{O}_n)_i^+ = \begin{cases} \mathbb{N}[\{s_i(x), t_i(x)\}] & \text{if } 0 \leq i < n, \\ \mathbb{N}[\{x\}] & \text{if } i = n, \\ 0 & \text{if } i > n. \end{cases}$$

  One can easily check that $\nu(\lambda(\mathbb{O}_n))$ is indeed isomorphic to $\mathbb{O}_n$ without invoking Steiner's theorem.

17.4.20 Tensor product of augmented directed complexes. Let K and L be two augmented directed complexes. We define their *tensor product* $K \otimes L$ in the following way:

- For $n \geq 0$,

$$(K \otimes L)_n = \bigoplus_{\substack{i+j=n \\ i\geq 0,\, j\geq 0}} K_i \otimes L_j.$$

- If x is in K_i and y is in K_j with $i + j > 0$, then

$$d(x \otimes y) = d(x) \otimes y + (-1)^i x \otimes d(y),$$

  where by convention $d(z) = 0$ is z is in K_0 or L_0.

– If x is in K_0 and y is in L_0, then

$$e(x \otimes y) = e(x)e(y).$$

– For $n \geq 0$, the submonoid $(K \otimes L)_n^+$ is generated by the chains of the form $x \otimes y$ with x in K_i^+ and y in L_j^+ with $i + j = n$.

This tensor product defines a (non-symmetric) monoidal structure on **ADC** which is biclosed in the sense that, for X an object, the functors $X \otimes -$ and $- \otimes X$ both admit a right adjoint. The unit of this tensor product is $\lambda(\mathbb{O}_0)$.

17.4.21 Proposition. *The tensor product of two strong Steiner complexes is a strong Steiner complex.*

Proof. This is [330, Example 3.10]. □

17.4.22 Tensor product of strong Steiner ω-categories. Let C and D be two strong Steiner ω-categories. We define their tensor product by

$$C \otimes D = \nu(\lambda(C) \otimes \lambda(D)).$$

Steiner's theory and the previous proposition imply that $C \otimes D$ is still a strong Steiner ω-category. Moreover, if C is generated by a polygraph P and D is generated by a polygraph Q, the ω-category $C \otimes D$ is generated by a polygraph $P \otimes Q$ whose n-generators are given by the formula

$$(P \otimes Q)_n = \coprod_{i+j=n} P_i \times Q_j.$$

17.4.23 Theorem. *There exists a unique (up to unique isomorphism) biclosed monoidal structure on* $\mathbf{Cat}_\omega$ *that extends the monoidal structure given by the tensor product of Steiner ω-categories.*

Proof. This is stated in [330, Section 7]. A detailed proof can be found in [15, Appendix A]. □

17.4.24 Remark. The tensor product given by the previous theorem is called the *Gray tensor product*. It was defined for 2-categories by Gray in [151]. It was first extended to ω-categories by Al-Agl and Steiner [5]. Crans then gave alternate descriptions [96].

17.4.25 Proposition. *The tensor product of two ω-categories generated by polygraphs is generated by a polygraph.*

Proof. See [168, Theorem 1.35] or [256, Proposition 5.1.2.7]. □

17.4.26 Remark. This proposition shows that it makes sense to talk of the tensor product of two polygraphs. In particular, identifying $\mathbb{O}_1$ with its generating polygraph, for any polygraph P, we get a polygraph $\mathbb{O}_1 \otimes P$. One can show that this polygraph is the same as the one defined in Section 17.2.

17.4.27 Cylinders and cubes. It follows from the previous remark that one has

$$(\mathrm{Cyl}_n)^* \simeq \mathbb{O}_1 \otimes \mathbb{O}_n \simeq \nu(\lambda(\mathbb{O}_1) \otimes \lambda(\mathbb{O}_n))$$

and

$$(\mathrm{Cub}_n)^* \simeq \mathbb{O}_1 \otimes \cdots \otimes \mathbb{O}_1 \simeq \nu(\lambda(\mathbb{O}_1) \otimes \cdots \otimes \lambda(\mathbb{O}_1)),$$

where $\mathbb{O}_1$ and $\lambda(\mathbb{O}_1)$ both appear n times.

17.4.28 Join of augmented directed complexes. Let K and L be two augmented directed complexes. We define their *join* $K \star L$ in the following way:

- For $n \geq 0$,

$$(K \star L)_n = \bigoplus_{\substack{i+1+j=n \\ i \geq -1,\, j \geq -1}} K_i \otimes L_j,$$

  where by convention $K_{-1} = \mathbb{Z}$ and $L_{-1} = \mathbb{Z}$.
- If x is in K_i and y is in K_j with $i + 1 + j > 0$, then

$$d(x \otimes y) = d(x) \otimes y + (-1)^{i+1} x \otimes d(y),$$

  where by convention $d(z) = e(z)$ if z is in K_0 or L_0, and $d(n) = 0$ for n in K_{-1} or L_{-1}.
- If x is in K_0 and y is in L_0, then

$$e(x \otimes 1) = e(x) \qquad \text{and} \qquad e(1 \otimes y) = e(y).$$

- For $n \geq 0$, the submonoid $(K \star L)_n^+$ is generated by the chains of the form $x \otimes y$ with x in K_i^+ and y in L_j^+ with $i + 1 + j = n$.

The join defines a (non-symmetric) monoidal structure on **ADC**. The unit is the initial augmented directed complex, the null complex. This monoidal structure is not biclosed but only locally biclosed in some appropriate sense (see [15, paragraph 5.7]). This locally biclosedness is equivalent to the fact that the joint commutes with (non-empty) connected colimits in each variable.

17.4.29 Proposition. *The join of two strong Steiner complexes is a strong Steiner complex.*

Proof. This is [15, Corollary 6.21]. □

17.4.30 Join of strong Steiner ω-categories. Let C and D be two strong Steiner ω-categories. We define their *join* by

$$C \star D = \nu(\lambda(C) \star \lambda(D)).$$

Steiner's theory and the previous proposition imply that $C \star D$ is still a strong Steiner ω-category. Moreover, if C is generated by a polygraph P and D is generated by a polygraph Q, the ω-category $C \star D$ is generated by a polygraph $P \star Q$ whose n-generators are given by the formula

$$(P \star Q)_n = \coprod_{\substack{i+1+j=n \\ i \geq -1,\, j \geq -1}} P_i \times Q_j,$$

where by convention P_{-1} and Q_{-1} are both singletons.

17.4.31 Theorem. *There exists a unique (up to unique isomorphism) monoidal structure on* $\mathbf{Cat}_\omega$*, called the* join, *that extends the monoidal structure given by the join of Steiner ω-categories and whose monoidal product commutes with (non-empty) connected colimits in each variable.*

Proof. This is [15, Theorem 6.29]. □

17.4.32 Proposition. *The join of two ω-categories generated by polygraphs is generated by a polygraph.*

Proof. This is [12, Corollary 7.6]. □

17.4.33 Remark. This proposition shows that it makes sense to talk of the join of two polygraphs. In particular, identifying $\mathbb{O}_0$ with its generating polygraph, for any polygraph P, we get a polygraph $\mathbb{O}_0 \star P$. One can show that it is canonically isomorphic to the polygraph $S(P)$ defined in §17.3.3.

17.4.34 Orientals. It follows from the previous paragraph that one has

$$(O_n)^* \simeq \mathbb{O}_0 \star \cdots \star \mathbb{O}_0 \simeq \nu(\lambda(\mathbb{O}_0) \star \cdots \star \lambda(\mathbb{O}_0)),$$

where $\mathbb{O}_0$ and $\lambda(\mathbb{O}_0)$ both appear n times. Note that the full cosimplicial object $O\colon \Delta \to \mathbf{Cat}_\omega$ can be recovered from this definition of the orientals. Indeed, as $\mathbb{O}_0$ is a terminal object in $\mathbf{Cat}_\omega$, it is canonically endowed with a monoid structure for the monoidal structure given by the join on $\mathbf{Cat}_\omega$. By the universal property of the augmented simplicial category Δ_+ (see [261, Chapter VII, Section 5, Proposition 1]), this monoid structure induces a functor $\Delta_+ \to \mathbf{Cat}_\omega$, whose restriction to Δ gives back the cosimplicial object of orientals.

18
Generalized Polygraphs

For each $n \in \mathbb{N} \cup \{\omega\}$, strict n-categories are the algebras of the monad $T = V_n F_n$ induced by the forgetful functor $V_n : \mathbf{Cat}_n \to \mathbf{Glob}_n$ and its left adjoint F_n, as shown in Chapter 14. However, the notion of *strict* n-category is sometimes too restrictive, whence the need for a notion of *weak* n-category. One proposal, due to Penon [296], defines weak n-categories as algebras of another monad P on $\mathbf{Glob}_n$, which in some sense "relaxes" the above monad T. In the same vein, Batanin [31] describes a general process consisting in replacing equalities with coherence cells, of which Penon's construction is a typical instance. In [28], Batanin generalizes the notion of polygraph to a notion of T*-polygraph* (that he calls T*-computad*), where T is any finitary monad on globular sets.

This chapter starts with our presentation of Batanin's ideas, the main point being the fairly general adjunction result of §18.1.5. This immediately applies to (n, p)-polygraphs as a straightforward generalization of n-polygraphs.

We then turn to two key examples of this general setting: the monad of weak n-categories (Section 18.2), which was the initial motivation for the general construction, and the monad associated to linear polygraphs (§18.3.1), of special interest in the present book.

18.1 T-Polygraphs

The definition of polygraphs presented in Chapter 15 strongly relies on Proposition 15.1.3, which asserts the existence of a left adjoint to a certain functor

$$W_n : \mathbf{Cat}_{n+1} \to \mathbf{Cat}_n^+.$$

This functor W_n is in turn based on the monad of strict ω-categories on globular sets. Following Batanin's original idea [30], we shall briefly explain how this construction adapts to an arbitrary finitary monad T on globular sets, giving rise to the general notion of "T-polygraph".

18.1.1 Globular algebras. Let $m \in \mathbb{N} \cup \{\omega\}$ and T_m be a finitary monad on $\mathbf{Glob}_m$, that is, a monad whose underlying endofunctor preserves filtered colimits. Let $\mathbf{Glob}_m^T$ denote the category of T_m-algebras. The category $\mathbf{Glob}_m^T$ comes with a forgetful functor $V_m\colon \mathbf{Glob}_m^T \to \mathbf{Glob}_m$, right adjoint to a functor $E_m\colon \mathbf{Glob}_m \to \mathbf{Glob}_m^T$. Now for each $n < m$, there is a truncation functor

$$U_n^m\colon \mathbf{Glob}_m \to \mathbf{Glob}_n$$

right adjoint to the canonical inclusion

$$I_n^m\colon \mathbf{Glob}_n \to \mathbf{Glob}_m.$$

Therefore, we get a pair of adjoint functors $U_n^m V_m\colon \mathbf{Glob}_m^T \to \mathbf{Glob}_n$ and $E_m I_n^m\colon \mathbf{Glob}_n \to \mathbf{Glob}_m^T$ whose composition gives a monad $T_n = U_n^m V_m E_m I_n^m$ on $\mathbf{Glob}_n$. Let again $\mathbf{Glob}_n^T$ denote the category of T_n-algebras and V_n, E_n the corresponding adjoint functors. There is then a unique comparison functor $K\colon \mathbf{Glob}_m^T \to \mathbf{Glob}_n^T$ making the following diagrams commute:

$$\begin{array}{ccc} \mathbf{Glob}_m^T & \xrightarrow{V_m} & \mathbf{Glob}_m \\ {\scriptstyle K}\downarrow & & \downarrow{\scriptstyle U_n^m} \\ \mathbf{Glob}_n^T & \xrightarrow[V_n]{} & \mathbf{Glob}_n, \end{array} \qquad \begin{array}{ccc} \mathbf{Glob}_m & \xrightarrow{E_m} & \mathbf{Glob}_m^T \\ {\scriptstyle I_n^m}\uparrow & & \downarrow{\scriptstyle K} \\ \mathbf{Glob}_n & \xrightarrow[E_n]{} & \mathbf{Glob}_n^T. \end{array}$$

Thinking of K as a truncation functor, we shall denote it from now on by the same letter U_n^m as the corresponding truncation between globular sets.

18.1.2 Freely adjoining cells. Following the pattern of Chapter 15, we turn to the special case of the above situation where $m = n + 1$, and consider the category $\mathbf{Glob}_n^{T+}$ defined by the following pullback square in $\mathbf{CAT}$:

$$\begin{array}{ccc} \mathbf{Glob}_n^{T+} & \longrightarrow & \mathbf{Glob}_{n+1} \\ \downarrow & & \downarrow{\scriptstyle U_n^{n+1}} \\ \mathbf{Glob}_n^T & \xrightarrow[V_n]{} & \mathbf{Glob}_n. \end{array}$$

As the square

$$\begin{CD} \mathbf{Glob}^T_{n+1} @>{V_{n+1}}>> \mathbf{Glob}_{n+1} \\ @V{U^{n+1}_n}VV @VV{U^{n+1}_n}V \\ \mathbf{Glob}^T_n @>>{V_n}> \mathbf{Glob}_n \end{CD}$$

commutes, there is a unique functor $W_n\colon \mathbf{Glob}^T_{n+1} \to \mathbf{Glob}^{T+}_n$ such that the following diagram commutes:

$$\begin{tikzcd} \mathbf{Glob}^T_{n+1} \arrow[dr, "W_n"] \arrow[drr, bend left, "V_{n+1}"] \arrow[ddr, bend right, "U^{n+1}_n"'] & & \\ & \mathbf{Glob}^{T+}_n \arrow[r] \arrow[d] & \mathbf{Glob}_{n+1} \arrow[d, "U^{n+1}_n"] \\ & \mathbf{Glob}^T_n \arrow[r, "V_n"'] & \mathbf{Glob}_n . \end{tikzcd} \tag{18.1}$$

As in the definition of "ordinary" polygraphs, the crucial step will be the construction of a left adjoint L_n for W_n. Before proving this fact, we shall need a few preliminary results on pullbacks in **CAT**.

18.1.3 Remark. Pullbacks in **CAT**, sometimes called *strict pullbacks*, are generally badly behaved and do not preserve equivalences of categories. Therefore, given categories **A**, **B**, **C** and functors $U\colon \mathbf{A} \to \mathbf{C}$, $V\colon \mathbf{B} \to \mathbf{C}$, one usually defines the *pseudo-pullback*

$$\begin{CD} \mathbf{P} @>>> \mathbf{A} \\ @VVV @VV{U}V \\ \mathbf{B} @>>{V}> \mathbf{C} \end{CD}$$

by taking for objects of **P** the triples (a, b, ϕ) where a is an object of **A**, b an object of **B** and ϕ an isomorphism from Ua to Vb. Note that the above square only commutes up to a canonical isomorphism. However, in case U is an *isofibration*, that is, if for any object a in **A** and any isomorphism $g\colon Ua \to c$ in **C** there is an isomorphism $f\colon a \to a'$ such that $Uf = g$, it turns out that such a **P** is equivalent to the strict pullback (exercise!).

18.1.4 Definition of $\mathbf{Glob}^{T+}_n$. In the present setting, the truncation functor $U^{n+1}_n\colon \mathbf{Glob}_{n+1} \to \mathbf{Glob}_n$ is in fact an isofibration, and we shall slightly depart here from Batanin's presentation by defining $\mathbf{Glob}_n^{T+}$ as the *strict* pullback of U_n^{n+1} and V_n. Thus, $\mathbf{Glob}_n^{T+}$ has objects pairs (X, A) where X is an $(n+1)$-globular set and A a T_n-algebra such that $U_n^{n+1}X = V_n A$. Morphisms are defined accordingly.

18.1.5 Pullbacks of monadic functors. Let $U\colon \mathbf{A} \to \mathbf{C}$ and $V\colon \mathbf{B} \to \mathbf{C}$ be two functors, and consider their strict pullback in **CAT**:

$$\begin{array}{ccc} \mathbf{P} & \xrightarrow{\widetilde{V}} & \mathbf{A} \\ {\scriptstyle \widetilde{U}}\downarrow & & \downarrow{\scriptstyle U} \\ \mathbf{B} & \xrightarrow[V]{} & \mathbf{C}. \end{array}$$

Suppose in addition that

- V is strictly monadic, with left adjoint E, meaning that $\mathbf{B}$ is isomorphic to the category of algebras of the associated monad VE,
- $\mathbf{A}$ is cocomplete,
- U is an isofibration,
- U admits a left adjoint I such that $UI = 1_{\mathbf{C}}$ and for each object c, the unit $\eta_c\colon c \to UIc$ is 1_c, whence also, for each object a in $\mathbf{A}$, $U(\varepsilon_a) = 1_a$,
- U has a right adjoint.

Then the functor $\widetilde{V}$ is also monadic. We shall denote by η' and ε' the unit and counit of the adjunction between V and E.

Before proving the statement, let us point out that the above hypotheses are immediately satisfied in case $V = V_n$ and $U = U_n^{n+1}$ as defined in 18.1.2.

As for the existence of a left adjoint for $\widetilde{V}$, let a be an object of $\mathbf{A}$ and define a pair $p = (a^+, b)$, where a^+ is an object of $\mathbf{A}$ and b an object of $\mathbf{B}$ as follows: taking first $b = EUa$, in order for p to be an object of $\mathbf{P}$, we need to define a^+ such that $Ua^+ = Vb$. Consider first the following pushout square in $\mathbf{A}$, which exists because of the cocompleteness assumption:

$$\begin{array}{ccc} IUa & \xrightarrow{\varepsilon_a} & a \\ {\scriptstyle I(\eta'_{Ua})}\downarrow & & \downarrow \\ IVEUa & \longrightarrow & a'. \end{array} \tag{18.2}$$

Now U being left adjoint, it preserves pushouts and moreover $U(\varepsilon_a) = 1_a$. Therefore, by applying U to the above square, the bottom arrow becomes an isomorphism $\phi\colon Vb \to Ua'$. Now U being an isofibration, we may choose an object a^+ and an isomorphism $\psi\colon a^+ \to a'$ such that $U\psi = \phi$. We therefore get a pushout square

$$\begin{array}{ccc} IUa & \xrightarrow{\varepsilon_a} & a \\ {\scriptstyle I(\eta'_{Ua})}\big\downarrow & & \big\downarrow \\ IVEUa & \longrightarrow & a^+ \end{array} \tag{18.3}$$

whose top and bottom arrows are taken to identities by U. The construction $\widetilde{E}: a \mapsto (a^+, EUa)$ is clearly functorial.

It remains to show that $\widetilde{E}$ is in fact left adjoint to $\widetilde{V}$. Let us define two natural transformations

$$\tilde{\eta}: 1_{\mathbf{A}} \to \widetilde{V}\widetilde{E}$$

and

$$\tilde{\varepsilon}: \widetilde{E}\widetilde{V} \to 1_{\mathbf{P}}$$

as follows.

For an object a in $\mathbf{A}$, as $\widetilde{V}\widetilde{E}a = a^+$ we may define $\tilde{\eta}_a$ as the right vertical arrow in the pushout square (18.3). This clearly defines a natural transformation from $1_{\mathbf{A}}$ to $\widetilde{V}\widetilde{E}$.

Let now $p = (a, b)$ be an object of $\mathbf{P}$, that is, $Ua = Vb$. By definition, $\widetilde{E}\widetilde{V}p = (a^+, EVb)$ and we thus look for a morphism

$$\tilde{\varepsilon}_p: (a^+, EVb) \to (a, b).$$

The second component is immediately given by the counit $\varepsilon'_b: EVb \to b$. As for the first component, we need to build a morphism

$$t_a: a^+ \to a$$

in $\mathbf{A}$. Note first that the following triangle

$$\begin{array}{ccc} Ua & & \\ {\scriptstyle \eta'_{Ua}}\big\downarrow & \searrow^{1_{Ua}} & \\ VEUa & \xrightarrow[V(\varepsilon'_b)]{} & Ua \end{array}$$

commutes because $Ua = Vb$ and the triangular identity between η' and ε'. By applying the functor I to the above triangle, we get the commutation of

$$\begin{array}{ccc} IUa & & \\ {\scriptstyle I(\eta'_{Ua})}\big\downarrow & \searrow^{1_{IUa}} & \\ IVEUa & \xrightarrow[IV(\varepsilon'_b)]{} & IUa, \end{array}$$

therefore also the commutation of

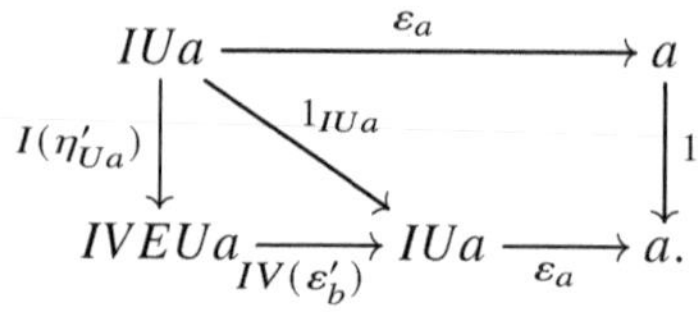

As the outer square above commutes, the universal property of the pushout yields a unique morphism $t_a \colon a^+ \to a$ such that the following diagram commutes:

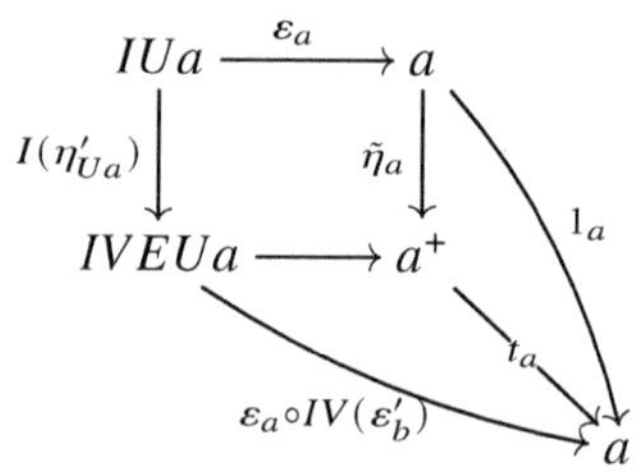

(18.4)

Now, by applying U to the above diagram, one gets $U(t_a) = V(\varepsilon'_b)$, whence $\tilde{\varepsilon}_p$ is indeed a morphism in **P**. The naturality of the construction is immediate. It remains to establish the triangular identities between $\tilde{\eta}$ and $\tilde{\varepsilon}$. Let first $p = (a, b)$ be an object of **P**, so that $a = \widetilde{V}p$. The commutation of the right triangle in (18.4) reads

$$\widetilde{V}(\tilde{\varepsilon}_p) \circ \tilde{\eta}_{\widetilde{V}p} = 1_{\widetilde{V}p}$$

which gives the first identity. Finally, let a be an object of **A** and apply $\widetilde{E}$ to the right triangle in (18.4), one obtains

$$\widetilde{E}(t_a) \circ \widetilde{E}(\tilde{\eta}_a) = 1_{\widetilde{E}a}$$

and we have to show that

$$\widetilde{E}(t_a) = \tilde{\varepsilon}_{\widetilde{E}a}$$

which reduces to the equality between both components:

$$\widetilde{U}\widetilde{E}(t_a) = \widetilde{U}(\tilde{\varepsilon}_{\widetilde{E}a}) \quad \text{and} \quad \widetilde{V}\widetilde{E}(t_a) = \widetilde{V}(\tilde{\varepsilon}_{\widetilde{E}a}).$$

The first one comes from the fact that $\widetilde{U}\widetilde{E} = EU$ and by applying this last functor to the bottom triangle in (18.4), and the second from the definition of $\widetilde{E}$ and $\tilde{\varepsilon}$ according to which $\widetilde{V}\widetilde{E}(t_a) = t_{a^+} = \widetilde{V}(\tilde{\varepsilon}_{\widetilde{E}a})$. Therefore

$$\tilde{\varepsilon}_{\widetilde{E}a} \circ \widetilde{E}(\tilde{\eta}_a) = 1_{\widetilde{E}a}$$

and the second triangular identity is proved. Thus $\widetilde{E}$ is left adjoint to $\widetilde{V}$.

The monadicity of $\widetilde{V}$ now follows from Beck's criterion. Let $p = (a, b)$, $p' = (a', b')$ in **P** and $u, v\colon p \to p'$ be a pair of parallel morphisms such that

$$\widetilde{V}p \underset{\widetilde{V}v}{\overset{\widetilde{V}u}{\rightrightarrows}} \widetilde{V}p' \xrightarrow{\ e\ } a'' \tag{18.5}$$

is an absolute coequalizer in **A**. It follows that

$$Vb \underset{U\widetilde{V}v}{\overset{U\widetilde{V}u}{\rightrightarrows}} Vb' \xrightarrow{\ Ue\ } Ua'' \tag{18.6}$$

is also an absolute coequalizer in **C**. As V is strictly monadic, it creates such coequalizers. There is therefore a unique morphism $f\colon b' \to b''$ in **B** such that $Vf = Ue$ and

$$b \underset{\widetilde{U}v}{\overset{\widetilde{U}u}{\rightrightarrows}} b' \xrightarrow{\ f\ } b''$$

is a coequalizer in **B**. Clearly (a'', b'') is in **P** and the morphism

$$(e, f)\colon (a', b') \to (a'', b'')$$

is the coequalizer of u and v we are looking for.

The existence of a left adjoint for the functor W_n of (18.1) is then a consequence of the following result in general category theory:

18.1.6 Proposition. *Let* $X\colon \mathbf{A} \to \mathbf{B}$ *and* $Y\colon \mathbf{B} \to \mathbf{C}$ *be two functors and* $Z = YX$, *and suppose that*

- **A** *is cocomplete,*
- Z *has a left adjoint,*
- Y *is monadic.*

Then the functor X *admits a left adjoint.*

Proof. This is essentially a simpler, less general version of Dubuc's adjoint triangle theorem [116]. □

18.1.7 Existence of the functor L_n. This above statement immediately applies to the triangle

$$\begin{array}{ccc} \mathbf{Glob}^T_{n+1} & & \\ & \overset{W_n}{\searrow} \quad \overset{V_{n+1}}{\searrow\searrow} & \\ & \mathbf{Glob}^{T+}_n & \xrightarrow[\widetilde{V}_n]{} \mathbf{Glob}_{n+1}. \end{array}$$

In fact, the cocompleteness of $\mathbf{Glob}^T_{n+1}$ comes from the finitary assumption on the monad T, the functor V_{n+1}, being the forgetful functor from $(n+1)$-globular algebras to $(n+1)$-globular sets admits a left adjoint, and we have just shown that $\widetilde{V}_n$ is monadic. Therefore W_n admits a left adjoint L_n.

18.1.8 Definition of T-polygraphs. We may now define by induction on $n \geqslant 0$ a category $\mathbf{Pol}^T_n$, together with a functor

$$F_n\colon \mathbf{Pol}^T_n \to \mathbf{Glob}^T_n,$$

following the pattern of §15.1.7.

- For $n = 0$, $\mathbf{Pol}^T_0 = \mathbf{Glob}_0 = \mathbf{Set} = \mathbf{Glob}^T_0$ and F_0 is the identity functor.
- Let $n \geqslant 0$ and suppose we have defined

  $$F_n\colon \mathbf{Pol}^T_n \to \mathbf{Glob}^T_n.$$

  The category $\mathbf{Pol}^T_{n+1}$ is then given by the following pullback square in $\mathbf{CAT}$:

  $$\begin{array}{ccc} \mathbf{Pol}^T_{n+1} & \xrightarrow{J_n} & \mathbf{Glob}^{T+}_n \\ \downarrow & & \downarrow \\ \mathbf{Pol}^T_n & \xrightarrow[F_n]{} & \mathbf{Glob}^T_n; \end{array}$$

  and the functor F_{n+1} is the composite

  $$L_nJ_n\colon \mathbf{Pol}^T_{n+1} \to \mathbf{Glob}^T_{n+1}$$

  where L_n is the above defined left adjoint to W_n.

If we start with a finitary monad T on $\mathbf{Glob}_\omega$, we get by construction a sequence of canonical functors in $\mathbf{CAT}$

$$\mathbf{Pol}^T_0 \longleftarrow \mathbf{Pol}^T_1 \longleftarrow \mathbf{Pol}^T_2 \longleftarrow \cdots$$

whose projective limit $\mathbf{Pol}^T_\omega$ is by definition the category of ω-polygraphs with respect to the monad T.

Finally, for each $n \geqslant 0$, there is a functor

$$G_n\colon \mathbf{Glob}^T_n \to \mathbf{Pol}^T_n$$

and a natural transformation

$$\varepsilon\colon F_nG_n \to 1$$

such that F_n is left adjoint to G_n and ε is the counit of this adjunction.

The construction of G_n in the general case being essentially the same as the one explained in Section 15.2 for the particular case where $\mathbf{Glob}_n^T = \mathbf{Cat}_n$, we shall only very briefly sketch the induction step yielding G_{n+1} from G_n.

Thus, suppose we have defined G_n as a right adjoint to F_n, with counit ε, and let C be an object of $\mathbf{Glob}_{n+1}^T$, with underlying $(n+1)$-globular set X. Let $C' = U_n^{n+1}(C)$. We define $G_{n+1}(C)$ as a pair (P, D^+) where P is in $\mathbf{Pol}_n^T$ and D^+ in $\mathbf{Glob}_n^{T+}$ by taking $P = G_n(C')$ and $D^+ = (D, Z)$, with $D = F_n(P)$ and Z the $(n+1)$-globular set defined as follows. Up to dimension n, Z coincides with the underlying n-globular set of $F_n(P)$, whereas Z_{n+1} consists of triples $(z, x, y) \in X_{n+1} \times Z_n \times Z_n$ such that $z\colon \varepsilon_n^{C'}(x) \to \varepsilon_n^{C'}(y)$.

Now $\varepsilon^{C'} : F_n G_n(C') \to C'$ extends to a natural transformation

$$\theta\colon D^+ \to W_n(C)$$

by sending the generator (z, x, y) to z.

Thus, by adjunction, we get $\theta^*\colon L_n(D^+) \to C$, but $L_n(D^+)$ is precisely $F_{n+1}G_{n+1}(C)$, so that θ^* defines the counit $\varepsilon^C : F_{n+1}G_{n+1}(C) \to C$.

18.1.9 Basic examples. Besides the basic case of strict ω-categories, an immediate example of the above construction is given by (n, p)-polygraphs introduced in Section 15.3. We just remark here that, again, these polygraphs are particular instances of T-polygraphs, where the monad T is the one induced by the forgetful functor

$$V\colon \mathbf{Cat}_{n,p} \to \mathbf{Glob}_n.$$

18.2 Polygraphs for Weak n-Categories

The general construction of Section 18.1 applies to Penon's monad on n-globular sets from [296]. This section basically follows Penon's approach, but for a correction pointed out by Cheng and Makkai in [85]: whereas Penon works over *reflexive* globular sets, it turns out that the category of plain globular sets, that is, our category $\mathbf{Glob}_n$, yield more examples of weak n-categories. In particular, braided monoidal categories fit in the latter setting, but not in the former.

Given $n \in \mathbb{N} \cup \{\omega\}$, we first consider the category $\mathbf{Mag}_n$ of *n-magmas*, whose objects are n-globular sets endowed with the same family of binary composition operations as (strict) n-categories, satisfying the same "positional" conditions with respect to source and target maps, but without requiring identities, associativity and exchange. The morphisms of $\mathbf{Mag}_n$ are the globular maps preserving all compositions. Of course any n-category is a particular n-magma. Now, let M be an n-magma and C be an n-category seen as an n-magma, a morphism

$$f\colon M \to C$$

in $\mathbf{Mag}_n$ is called a *categorical stretching* — "étirement catégorique" in Penon's terminology, or "trivial fibration" in Batanin's — if for any $k < n$ and any pair (a, b) of parallel k-cells in M such that $f(a) = f(b)$, there is a $(k+1)$-cell c in M such that $f(c) = 1^{k+1}_{f(a)}$, and if moreover, in case $n \neq \omega$, f_n is injective. A *trivialization* of a categorical stretching f is a map $[,]_f$ choosing a cell $c = [a, b]_f$ with $f(c) = 1^{k+1}_{f(a)}$ for each pair (a, b) as above. There is now a category $\mathbf{Q}$ whose objects are pairs $(f, [,]_f)$, where f is a categorical stretching and $[,]_f$ a trivialization of f, and whose morphisms are commutative squares

$$\begin{array}{ccc} M & \xrightarrow{u} & M' \\ {\scriptstyle f}\downarrow & & \downarrow{\scriptstyle f'} \\ C & \xrightarrow[v]{} & C' \end{array}$$

in $\mathbf{Mag}_n$ such that $u[a, b]_f = [va, vb]_{f'}$ for all parallel k-cells a, b, such that $f(a) = f(b)$. As each n-magma M has an underlying n-globular set, the correspondence taking $f\colon M \to C$ to M induces a forgetful functor

$$U\colon \mathbf{Q} \to \mathbf{Glob}_n.$$

Now, as shown in [296], this functor U admits a left adjoint F, thus defining a monad $P = UF$ on $\mathbf{Glob}_n$. The algebras of this monad P are precisely the weak n-categories we were looking for.

Finally, this monad P satisfies the hypotheses of Section 18.1 and therefore produces an appropriate notion of P-polygraphs.

18.3 Linear Polygraphs

18.3.1 Linear polygraphs as T-polygraphs. In Chapter 6 we introduced the notion of 1-dimensional linear polygraphs as rewriting systems for associative unital algebras over a field $\Bbbk$. These are again special cases of the general construction 18.1, for a certain monad on n-globular sets we now briefly describe. Let $n \in \mathbb{N} \cup \{\omega\}$ and $\Bbbk$ be a field. We denote by $\mathbf{Alg}$ the category of unital and associative $\Bbbk$-algebras. The category $\mathbf{Alg}_n$ of n-algebras is the category of n-categories internal to $\mathbf{Alg}$. Consider the forgetful functor

$$U\colon \mathbf{Alg}_n \to \mathbf{Glob}_n$$

obtained by composing both functors $\mathbf{Alg}_n \to \mathbf{Cat}_n$ and $\mathbf{Cat}_n \to \mathbf{Glob}_n$. The categories $\mathbf{Glob}_n$, $\mathbf{Cat}_n$, **Set**, and **Alg** are models of projective sketches S, S', T, and T', respectively, where T is of course the trivial sketch on the terminal category. Moreover, there are inclusion morphisms of sketches $i\colon S \to S'$ and $j\colon T \to T'$, and both of them satisfy Lair's conditions of Theorem G.1.11. These induce a morphism

$$i \otimes j\colon S \otimes T \to S' \otimes T'$$

still satisfying Lair's conditions (see [4, p. 18] about tensoring sketches). Now the models of $S' \otimes T'$ in **Set** are also the models of S' in $\mathrm{Mod}(T') = \mathbf{Alg}$, in other words

$$\mathrm{Mod}(S' \otimes T') = \mathbf{Alg}_n$$

whereas $S \otimes T \simeq S$. Therefore our functor U is precisely the one induced by $i \otimes j$ on models:

$$U = \mathrm{Mod}(i \otimes j)\colon\ \mathrm{Mod}(S' \otimes T') \to \mathrm{Mod}(S).$$

As a consequence of Theorem G.1.11, the functor U is monadic. It is also easily seen to preserve filtered colimits, whence inducing a finitary monad L on $\mathbf{Glob}_n$. Thus the machinery of 18.1 applies, and produces the category of L-polygraphs, that is, the *linear polygraphs* we wanted.

18.3.2 Bimodules. The description of n-algebras and linear n-polygraphs can be made more explicit by introducing the following notion of globular bimodules. For an algebra A, we denote by $\mathbf{Bimod}(A)$ the category of bimodules over A and their morphisms. We consider the category $\mathbf{Glob}(\mathbf{Bimod})(A)$ of *globular A-bimodules*, that is, functors

$$X\colon \mathbb{O}^{\mathrm{op}} \to \mathbf{Bimod}(A),$$

and their morphisms. Let us define the category $\mathbf{Bimod}_G$ whose objects are pairs (A, M) made of an algebra A and a globular A-bimodule M, and whose morphisms from (A, M) to (B, N) are pairs (F, G) made of a morphism $F\colon A \to B$ of algebras and a morphism $G\colon M \to N$ of bimodules, that is,

$$G(ama') = F(a)G(m)F(a')$$

holds for all a and a' in A and m in M.

The following result gives a characterization of the category of ω-algebras.

18.3.3 Theorem ([160, Theorem 1.3.3]). *The category* $\mathbf{Alg}_\omega$ *is isomorphic to the full subcategory of* $\mathbf{Bimod}_G$ *whose objects are the pairs* (A, M) *such* *that M_0 is equal to A, with its canonical A-bimodule structure, and that satisfy, for all n-cells a and b of M, the relation*

$$as_0(b) + t_0(a)b - t_0(a)s_0(b) = s_0(a)b + at_0(b) - s_0(a)t_0(b). \qquad (18.7)$$

18.3.4 Explicit construction. The key step in the explicit inductive construction of T-polygraphs at level n is the concrete description of the left adjoint L_n to the forgetful functor $W_n\colon \mathbf{Glob}^T_{n+1} \to \mathbf{Glob}^{T+}_n$ of (18.1). In the present case of linear polygraphs, this forgetful functor is

$$W_n\colon \mathbf{Alg}_{n+1} \to \mathbf{Alg}^+_n$$

and its left adjoint L_n takes an extended n-algebra (A, X) to the $(n+1)$-algebra $A[X]$ constructed as follows. First, we consider the A_0-bimodule

$$M = \left(A_0 \otimes \Bbbk[X] \otimes A_0\right) \oplus A_n$$

obtained by the direct sum of the free A_0-bimodule with basis X and of a copy of A_n, equipped with its canonical A_0-bimodule structure. Thus M contains linear combinations of elements axb, for a and b in A_0 and x in X, and of an n-cell c of A. We define the source, target, and identity maps

$$M \underset{t}{\overset{s}{\rightrightarrows}} A_n \qquad i\colon A_n \to M$$

by

$$\begin{aligned} s(axb) &= as(x)b, & s(c) &= c, & i(c) &= c,\\ t(axb) &= at(x)b, & t(c) &= c, \end{aligned}$$

for all x in X, a and b in A_0, and c in A_{n-1}. Then we define the A_0-bimodule $A[X]_{n+1}$ as the quotient of M by the A_0-bimodule ideal generated by all the elements

$$\left(as_0(b) + t_0(a)b - t_0(a)s_0(b)\right) - \left(s_0(a)b + at_0(b) - s_0(a)t_0(b)\right),$$

where a and b range over $A_0 \otimes \Bbbk X \otimes A_0$. We prove that the source and target maps are compatible with the quotient, so that, by Theorem 18.3.3, the A_0-bimodule $A[X]_{n+1}$ extends A into a uniquely defined $(n+1)$-algebra $A[X]$.

PART V

HOMOTOPY THEORY OF POLYGRAPHS

19

Polygraphic Resolutions

The purpose of this chapter is to introduce the notion of a polygraphic resolution of an ω-category. This notion was introduced by Métayer [278] to define a homology theory for ω-categories, that is now known as the *polygraphic homology*. It was then showed by himself and Lafont [236] that this homology recovers the classical homology of monoids for ω-categories coming from monoids. It is now known by the work of Lafont, Métayer, and Worytkiewicz [237] that these polygraphic resolutions are resolutions in the sense of a model category structure on $\mathbf{Cat}_\omega$, the so-called *folk model structure*, that we will present in the next chapters.

Roughly speaking, a polygraphic resolution is a non-abelian version of a resolution of a module in the sense of homological algebra. More precisely, if C is an ω-category, a polygraphic resolution of C is a polygraph P endowed with an ω-functor $P^* \to C$ that is a *trivial fibration*, meaning in a nutshell that it is surjective with source and target fixed at all levels. Technically, we will define these trivial fibrations by a right lifting property with respect to inclusions of spheres into disks in a very similar way as trivial fibrations of topological spaces are defined.

The chapter is organized as follows. In the first section, we introduce the notion of a weak factorization system in a general category. We explain the *small object argument*. In the second section, we define *cofibrations* and *trivial cofibrations* of ω-categories as the classes appearing in the weak factorization system generated by the set $\mathcal{I}$ of inclusions of spheres into disks. We also introduce the class of *relative polygraphs*, which are cell complexes generated by $\mathcal{I}$, and we explain why polygraphs almost tautologically correspond to relative polygraphs of the form $\varnothing \to C$. In the third section, we introduce the notion of a polygraphic resolution. We show that every ω-category admits such a resolution and we give the example of the so-called *canonical resolution*. Finally, in the last section, we study the uniqueness of these resolutions, showing that there is always a map between two such resolutions and postponing to a later chapter the fact that two such maps are homotopic in some appropriate sense.

19.1 Weak Factorization Systems

The purpose of this section is to introduce some basic results on weak factorization systems, which will be applied in the next section to a factorization system giving rise to "polygraphic resolutions". This factorization system has the additional property of being generated by morphisms between finitely presentable objects. This additional hypothesis will allow us to avoid the use of ordinals and cardinals.

In the section, we fix a category C. Our case of interest is $C = \mathbf{Cat}_\omega$.

19.1.1 Lifting properties. Let $f\colon X \to Y$ and $g\colon Z \to T$ be two morphisms of C. One says that f *has the left lifting property* with respect to g or that g *has the right lifting property* with respect to f if for every commutative square

$$\begin{array}{ccc} X & \longrightarrow & Z \\ {\scriptstyle f}\downarrow & & \downarrow{\scriptstyle g} \\ Y & \longrightarrow & T \end{array}$$

there exists a *lift*, that is, a morphism $h\colon Y \to Z$ making the two triangles

$$\begin{array}{ccc} X & \longrightarrow & Z \\ {\scriptstyle f}\downarrow & \overset{h}{\nearrow} & \downarrow{\scriptstyle g} \\ Y & \longrightarrow & T \end{array}$$

commute. More generally, one says that f has the left lifting property with respect to a class of maps $\mathcal{I}$ if it has the left lifting property with respect to every morphism in $\mathcal{I}$, and similarly for the right lifting property. We will denote by $l(\mathcal{I})$ and $r(\mathcal{I})$ the class of maps having the left or right lifting property with respect to a class $\mathcal{I}$.

19.1.2 Stability properties of $l(\mathcal{I})$. Suppose C is cocomplete and let $\mathcal{I}$ be a class of morphisms of C. One checks that $l(\mathcal{I})$ contains the class of isomorphisms and is stable under

1. sums: if $(i_k\colon X_k \to Y_k)_{k\in K}$ is a (small) family of elements of $l(\mathcal{I})$, then the sum

$$\coprod_{k\in K} i_k\colon \coprod_{k\in K} X_k \to \coprod_{k\in K} Y_k$$

belongs to $l(\mathcal{I})$,

2. pushouts: if

$$\begin{array}{ccc} X & \longrightarrow & Z \\ {\scriptstyle i}\downarrow & & \downarrow{\scriptstyle j} \\ Y & \longrightarrow & T \end{array}$$

is a pushout square and i belongs to $l(\mathcal{I})$, then so does j,

3. countable compositions: if

$$X_0 \xrightarrow{i_1} X_1 \xrightarrow{i_2} \cdots \xrightarrow{i_n} X_n \xrightarrow{i_{n+1}} \cdots$$

is a diagram of elements of $l(\mathcal{I})$, then the morphism

$$X_0 \to \varinjlim_{n \geq 0} X_n$$

belongs to $l(\mathcal{I})$,

4. retracts: if

$$\begin{array}{ccccc} & & 1_X & & \\ X & \longrightarrow & Z & \longrightarrow & X \\ {\scriptstyle i}\downarrow & & \downarrow{\scriptstyle j} & & \downarrow{\scriptstyle i} \\ Y & \longrightarrow & T & \longrightarrow & Y \\ & & 1_Y & & \end{array}$$

is a commutative diagram and j belongs to $l(\mathcal{I})$, then so does i.

19.1.3 Remark. More generally, under the same assumption and with the same notation as in the paragraph above, the class $l(\mathcal{I})$ is stable under transfinite compositions (see [187, Definition 2.1.1]). Note that being stable by transfinite compositions and pushouts implies being stable by sums.

19.1.4 Weak factorization systems. A *weak factorization system* on C is a pair $(\mathcal{L}, \mathcal{R})$ of classes of morphisms of C satisfying the following conditions:

1. Every morphism f of C factors as $f = pi$, where p is in $\mathcal{R}$ and i is in $\mathcal{L}$.
2. We have

$$\mathcal{L} = l(\mathcal{R}) \qquad \text{and} \qquad \mathcal{R} = r(\mathcal{L}).$$

Recall that an object A of a category is *finitely presentable* when the functor $C(A,-): C \to \mathbf{Set}$ preserves filtered colimits. In practice, we are mostly interested in preservation of colimits of diagrams in C of the form

$$X_0 \longrightarrow X_1 \longrightarrow X_2 \longrightarrow \cdots$$

In this case, the fact that the canonical arrow $\varinjlim_n C(A, X_n) \to C(A, \varinjlim_n X_n)$ is an isomorphism means in particular that any morphism $f\colon A \to \varinjlim_n X_n$ factors through some X_n. For instance, the finitely presentable sets are the finite sets.

19.1.5 Proposition (Small object argument: first version)**.** *Suppose C is cocomplete and let I be a* set *of morphisms of C whose sources are finitely presentable objects. Then $(lr(I), r(I))$ is a weak factorization system on C.*

Proof. The second condition of the definition of a weak factorization system follows from the general equality $rlr(I) = r(I)$. Let us sketch a proof for the first one. Let $f\colon X \to Y$ be a morphism of C. Let S be the set of commutative squares of the form

$$\begin{array}{ccc} A_s & \longrightarrow & X \\ {\scriptstyle i_s}\big\downarrow & & \big\downarrow{\scriptstyle f} \\ B_s & \longrightarrow & Y, \end{array}$$

where i_s is an element of I. Summing all these squares, we get a commutative square

$$\begin{array}{ccc} \coprod_{s\in S} A_s & \longrightarrow & X \\ {\scriptstyle \coprod_{s\in S} i_s}\big\downarrow & & \big\downarrow{\scriptstyle f} \\ \coprod_{s\in S} B_s & \longrightarrow & Y. \end{array}$$

Taking the pushout of the top left corner, we get a factorization of f:

$$\begin{array}{ccccc} \coprod_{s\in S} A_s & \longrightarrow & X & & \\ {\scriptstyle \coprod_{s\in S} i_s}\big\downarrow & & \big\downarrow{\scriptstyle j_1} & \searrow{\scriptstyle f} & \\ \coprod_{s\in S} B_s & \longrightarrow & X_1 & & \\ & \searrow & & \searrow{\scriptstyle p_1} & \\ & & & & Y. \end{array}$$

By §19.1.2, the morphism j_1 belongs to $lr(I)$, but there is no reason for p_1 to belong to $r(I)$: we thus apply the same procedure to p_1, obtaining a factorization

$p_1 = p_2 j_2$, where $j_2 \colon X_1 \to X_2$ belongs to $lr(I)$. Going on by induction, we get a sequence of morphisms

$$X \xrightarrow{j_1} X_1 \xrightarrow{j_2} \cdots \xrightarrow{j_n} X_n \xrightarrow{j_{n+1}} \cdots$$

belonging to $lr(I)$ and hence a morphism

$$j_\infty \colon X \longrightarrow X_\infty = \varinjlim_n X_n$$

in $lr(I)$ giving rise to a factorization

$$f = p_\infty j_\infty,$$

where p_∞ is the morphisms induced by the p_n's. To conclude the proof, it suffices to show that p_∞ belongs to $r(I)$. Consider a commutative square

$$\begin{array}{ccc} A & \xrightarrow{k} & X_\infty \\ {\scriptstyle i}\downarrow & & \downarrow{\scriptstyle p_\infty} \\ B & \xrightarrow[l]{} & Y, \end{array}$$

where i is in I. Since by our additional assumption, the object A is finitely presentable, the morphism k factors through some X_n and the diagram factors as

$$\begin{array}{ccccc} A & \longrightarrow & X_n & \longrightarrow & X_\infty \\ {\scriptstyle i}\downarrow & & & & \downarrow{\scriptstyle p_\infty} \\ B & & \xrightarrow[l]{} & & Y, \end{array}$$

where $X_n \to X_\infty$ is the canonical morphism. The composite from X_n to Y is p_n, and this diagram defines a square appearing in the definition of X_{n+1}. This means that there exists a lift

$$\begin{array}{ccccccc} A & \longrightarrow & X_n & \xrightarrow{j_{n+1}} & X_{n+1} & \longrightarrow & X_\infty \\ {\scriptstyle i}\downarrow & & & \nearrow & & & \downarrow{\scriptstyle p_\infty} \\ B & & & \xrightarrow[l]{} & & & Y, \end{array}$$

thereby proving the result. □

19.1.6 Remark. The same conclusion holds with the weaker hypothesis that there exists a regular cardinal κ such that the sources of the morphisms in I are κ-presentable (see Section G.2), the above statement being the case $\kappa = \aleph_0$. This condition is automatic if C is locally presentable. See also [184, Section 10.5] for weaker assumptions for the small object argument. The proof of these more general results is basically the same except that one has to proceed by transfinite induction.

19.1.7 $\mathcal{I}$-cells. Suppose C is cocomplete and let $\mathcal{I}$ be a class of morphisms of C. The class $\mathrm{cell}_\omega(\mathcal{I})$ of *countable $\mathcal{I}$-cellular extensions* is the smallest class of morphisms of C containing $\mathcal{I}$ and stable under sums, pushouts, and countable compositions. Every element of $\mathrm{cell}_\omega(\mathcal{I})$ can be obtained as a countable composition of pushouts of sums of elements of $\mathcal{I}$.

19.1.8 Remark. When $\mathcal{I}$ is a set and the source of the elements of $\mathcal{I}$ are finitely presentable, by [270, Proposition A.6], the class $\mathrm{cell}_\omega(\mathcal{I})$ is equal to the more classical class $\mathrm{cell}(\mathcal{I})$ of $\mathcal{I}$-cellular extensions, which is defined as $\mathrm{cell}_\omega(\mathcal{I})$ but asking also for stability under transfinite compositions.

19.1.9 Proposition (Small object argument: refined version)**.** *Suppose C is cocomplete and let I be a* set *of morphisms of C whose sources are finitely presentable. Then every morphism of C factors as $f = pi$, where p is in $r(I)$ and i is in* $\mathrm{cell}_\omega(I)$. *Moreover, every element of $lr(I)$ is a retract of an element of* $\mathrm{cell}_\omega(I)$.

Proof. The first point was actually proven in the proof of Proposition 19.1.5. The second point follows from the following lemma, the so-called "retract lemma". □

19.1.10 Lemma (Retract lemma)**.** *Suppose we have a factorization $f = pi$, in a category C, where f has the left lifting property with respect to p. Then f is a retract of i.*

Proof. Denote $i\colon X \to Y$ and $p\colon Y \to Z$. By hypothesis, there exists $h\colon Z \to Y$ making the square

$$
\begin{array}{ccc}
X & \xrightarrow{\;i\;} & Y \\
{\scriptstyle f}\downarrow & \overset{h}{\nearrow} & \downarrow{\scriptstyle p} \\
Z & = & Z
\end{array}
$$

commute. The commutative diagram

$$
\begin{array}{ccccc}
X & = & X & = & X \\
{\scriptstyle f}\downarrow & & \downarrow{\scriptstyle i} & & \downarrow{\scriptstyle f} \\
Z & \xrightarrow[h]{} & Y & \xrightarrow[p]{} & Z
\end{array}
$$

proves the result. □

19.1.11 Remark. Proposition 19.1.9 holds under the weaker hypothesis described in Remark 19.1.6 if one replaces $\mathrm{cell}_\omega(I)$ with $\mathrm{cell}(I)$.

19.2 Cofibrations and Trivial Fibrations

We will now apply the theory of weak factorization systems described in the previous section to the category $\mathbf{Cat}_\omega$ and the set of inclusions of boundaries of globes. This will lead to a notion of cofibrations and trivial fibrations of ω-categories.

19.2.1 Cofibrations, trivial fibrations, and relative polygraphs. Recall that for every $n \geq 0$ we denote by $\mathrm{i}_n : \partial\mathbb{O}_n \to \mathbb{O}_n$ the inclusion of the $(n-1)$-sphere into the n-globe. We set

$$\mathcal{I} \quad = \quad \{\mathrm{i}_n \mid n \geq 0\} .$$

From this set $\mathcal{I}$, we obtain three classes of ω-functors:

- the class $lr(\mathcal{I})$ of *cofibrations*,
- the class $r(\mathcal{I})$ of *trivial fibrations*, and
- the class $\mathrm{cell}_\omega(\mathcal{I})$ of *relative polygraphs*.

The category $\mathbf{Cat}_\omega$ being locally presentable (see §14.4.1) and the spheres being finitely presentable ω-categories, we can apply the small object argument (Propositions 19.1.5 and 19.1.9) and we obtain:

19.2.2 Proposition. *The following assertions hold:*

1. *The pair ({cofibrations}, {trivial fibrations}) is a weak factorization system on* $\mathbf{Cat}_\omega$.
2. *Every ω-functor f factors as $f = pi$, where i is a relative polygraph and p is a trivial fibration.*
3. *The cofibrations are the retracts of relative polygraphs.*

We will now describe more concretely trivial fibrations and relative polygraphs.

19.2.3 Trivial fibrations. By definition, an ω-functor $f : C \to D$ is a trivial fibration if, for every $n \geq 0$, it has the right lifting property with respect to $\mathrm{i}_n : \partial\mathbb{O}_n \to \mathbb{O}_n$. The data of a commutative diagram

$$\begin{array}{ccc} \partial\mathbb{O}_n & \longrightarrow & C \\ {\scriptstyle \mathrm{i}_n}\downarrow & & \downarrow{\scriptstyle f} \\ \mathbb{O}_n & \longrightarrow & D \end{array}$$

is equivalent to the data of a pair (x, y) of parallel $(n-1)$-cells of C and of an n-cell $b\colon f(x) \to f(y)$ in D. (To make sense of this when $n = 0$, one has to consider that every ω-category has a unique cell of dimension -1.) A lift for such a square is given by an n-cell $a\colon x \to y$ such that $f(a) = b$.

In other words, $f\colon C \to D$ is a trivial fibration if and only if the following conditions hold:

1. f is surjective on objects,
2. for every $n \geq 0$, every pair x, y of parallel n-cells of C and every $(n+1)$-cells $b\colon f(x) \to f(y)$ in D, there exists an $(n+1)$-cell $a\colon x \to y$ in C such that $f(a) = b$.

19.2.4 Relative polygraphs. By definition, an ω-functor $f\colon C \to D$ is a relative polygraph if f can be obtained as a countable composition

$$C = C_0 \xrightarrow{i_1} C_1 \xrightarrow{i_2} \cdots \xrightarrow{i_n} C_n \xrightarrow{i_{n+1}} \cdots,$$

i.e., if f is the canonical morphism $C_0 \to \varinjlim_n C_n$, where i_n, for $n \geq 1$, is part of a pushout square

$$\begin{array}{ccc} \coprod_{k\geq 0} E_k \times \partial\mathbb{O}_k & \longrightarrow & C_{n-1} \\ {\scriptstyle \coprod_{k\geq 0} E_k\times \mathrm{i}_n}\downarrow & & \downarrow{\scriptstyle i_n} \\ \coprod_{k\geq 0} E_k \times \mathbb{O}_k & \longrightarrow & C_n, \end{array}$$

the E_k's being sets (depending on n). In other words, C_n is obtained from C_{n-1} by freely adding cells (of any dimension) and $i_n\colon C_{n-1} \to C_n$ is the canonical ω-functor. This means that D is obtained from C by adding cells (in any order) in several steps. Using the fact that the forgetful functor from $\mathbf{Cat}_m$ to $\mathbf{Cat}_n$, when $m > n$, respects colimits (see §14.4.6) and in particular pushouts, one gets that the new cells can always be attached dimension by dimension. This means that we can suppose that i_n, for $n \geq 1$, is part of a pushout square of the simpler form

$$\begin{array}{ccc} \coprod_{F_n} \partial \mathbb{O}_n & \longrightarrow & C_{n-1} \\ {\scriptstyle \coprod_{F_n} \mathrm{i}_n} \big\downarrow & & \big\downarrow {\scriptstyle i_n} \\ \coprod_{F_n} \mathbb{O}_n & \longrightarrow & C_n, \end{array}$$

where F_n is any set. In other words, i_n is the canonical ω-functor associated to the cellular extension (C_{n-1}, F_n). In particular, we get:

19.2.5 Proposition. *An ω-category C is generated by a polygraph if and only if the unique functor from the initial ω-category to C is a relative polygraph.*

19.2.6 Remark. We will see in Chapter 21 that the previous proposition can be strengthened by saying that C is generated by a polygraph if and only if the unique ω-functor from the initial ω-category to C is a cofibration. In other words, the "cofibrant objects" are exactly the ω-categories generated by polygraphs.

19.2.7 Proposition. *Every cofibration is a monomorphism.*

Proof. Cofibrations are retracts of countable compositions of canonical ω-functors associated to cellular extensions. As retracts and countable compositions of monomorphic ω-functors are monomorphisms, the result follows from the fact that the canonical ω-functor associated to a cellular extension is a monomorphism (Proposition 16.2.3). □

19.3 Polygraphic Resolutions

Polygraphic resolutions are the ω-categorical version of the free resolutions of homological algebra, see §E.3.1. We will see shortly that every ω-category admits a polygraphic resolution and that such a resolution is in some sense unique up to homotopy.

19.3.1 Polygraphic resolutions. A *polygraphic resolution* of an ω-category C is a pair (P, p), where P is a polygraph and $p\colon P^* \to C$ is a trivial fibration.

19.3.2 Proposition. *Every ω-category admits a polygraphic resolution.*

Proof. Let C be an ω-category. Consider the unique ω-functor $\varnothing_C\colon \varnothing \to C$ from the initial ω-category $\varnothing$ to C. By Proposition 19.2.2, this ω-functor factors as $p \circ \varnothing_{P^*}$, where P is a polygraph, $\varnothing_{P^*}\colon \varnothing \to P^*$ is the unique such morphism and p is a trivial fibration, thereby proving the result. □

The previous proposition shows the existence of a polygraphic resolution for any ω-category by abstract non-sense. Here is a canonical choice of such a resolution:

19.3.3 The canonical resolution. Let C be an ω-category. The counit of the adjunction between the categories of polygraphs and ω-categories gives an ω-functor $\varepsilon_C \colon G(C)^* \to C$, where $G \colon \mathbf{Cat}_\omega \to \mathbf{Pol}_\omega$ denotes the functor described in §15.1.10. This ω-functor is a trivial fibration, essentially by definition. Indeed, it is bijective on objects by definition and, if x and y are two parallel k-cells of $G(C)^*$, for some $k \geq 0$, and z is a $(k+1)$-cell of $G(C)^*$ from $\varepsilon_C(x)$ to $\varepsilon_C(y)$, then, by definition of $G(C)$, there is a generating $(k+1)$-cell $p = (z, x, y)$ in $G(C)$ from x to y such that $\varepsilon_C(p) = z$. This means that $(G(C), \varepsilon_C)$ is a polygraphic resolution of C. This polygraphic resolution is called the *canonical resolution* of C. Note that it is functorial in C.

19.4 Uniqueness of Polygraphic Resolutions

In this section, we prove that polygraphic resolutions are unique up to a non-canonical homotopy in an appropriate sense.

19.4.1 Proposition. *Let $f \colon C \to D$ be an ω-functor, let (P, p) and (Q, q) two polygraphic resolutions of C and D, respectively. There exists a (non-canonical) ω-functor $g \colon P^* \to Q^*$ making the square*

$$\begin{array}{ccc} P^* & \xrightarrow{\;g\;} & Q^* \\ {\scriptstyle p}\downarrow & & \downarrow{\scriptstyle q} \\ C & \xrightarrow[\;f\;]{} & D \end{array}$$

commute.

Proof. Consider the commutative square

$$\begin{array}{ccc} \varnothing & \longrightarrow & Q^* \\ \downarrow & & \downarrow{\scriptstyle q} \\ P^* & \xrightarrow[\;p\;]{} C \xrightarrow[\;f\;]{} & D, \end{array}$$

where the unlabeled arrows are the unique such ω-functors. By Proposition 19.3.2, the ω-functor $\varnothing \to P^*$ is a cofibration. Since by definition, the ω-functor q is a trivial fibration, the square admits a lift f, thereby proving the result. $\square$

19.4.2 Remark. In the particular case where $C = D$ and f is the identity ω-functor, we get the existence of an ω-functor

$$\begin{array}{ccc} P^* & \xrightarrow{g} & Q^* \\ & {}_{p}\searrow \quad \swarrow_{q} & \\ & C & \end{array}$$

between any two polygraphic resolutions of C.

19.4.3 Remark. We will prove in Chapter 21, using the "folk model structure" on $\mathbf{Cat}_\omega$, that the ω-functor g of the previous proposition is unique up to some appropriate notion of homotopy (see Proposition 21.2.9).

20

Toward the Folk Model Structure: ω-Equivalences

In this chapter, we introduce all the notions and tools that will allow us to define and establish the existence, in the next chapter, of the folk model category structure on $\mathbf{Cat}_\omega$. We particularly focus on the concept of ω-equivalences, that will be the weak equivalences of this model structure. Essentially all the material of this chapter is extracted from Lafont, Métayer, and Worytkiewicz [237], although cylinders first appeared in work by Métayer [279].

The class of ω-equivalences is the appropriate generalization to ω-categories of the class of equivalences of ordinary categories. In particular, an ω-equivalence between 1-categories is nothing but an equivalence of categories. To define this notion, we need to generalize the concept of an invertible cell (or isomorphism). This leads to the notion of a reversible cell, which is, in intuitive terms, a cell admitting an inverse up to cells admitting inverses, up to cells admitting inverses, etc. Another fundamental tool is the ω-category $\widetilde{\Gamma}(C)$ of reversible cylinders in an ω-category C that we will lead to a sensible notion of homotopy.

The first section is devoted to ω-equivalences. We start by defining the notion of a *reversible cell* in an ω-category. Following [237], we define it by coinduction. We then introduce the class of *ω-equivalences*. We prove some basic stability properties of this class. We observe that trivial fibrations in the sense of the previous chapter are ω-equivalences. In the second section, we introduce the ω-category $\Gamma(C)$ of *cylinders* in an ω-category C. The purpose of this ω-category is to allow the definition of ω-functors playing the role of homotopies. This leads to the definition of an *oplax transformations*, generalizing the 1-categorical natural transformations and the 2-categorical oplax transformations. The third section is about the sub-ω-category $\widetilde{\Gamma}(C) \subseteq \Gamma(C)$ of *reversible cylinders*. This is a fundamental tool to establish properties of ω-equivalences. We show that this ω-category behaves as a homotopical path object, and we use its properties to show that the class of ω-equivalences satisfies the 2-out-of-3 property, a result that turns out to be non-trivial. In the fourth section, we introduce a "coherent" version of reversible cells leading to the notion of a *fibration* of ω-categories. We show that an ω-functor is a trivial fibration if and only if it is both an ω-equivalence and a fibration. Finally, in the last section, we study the class of ω-functors having the left lifting properties with respect to fibrations (that could be called trivial cofibrations). We show that it consists of ω-functors being both an ω-equivalence and a cofibration. To do so, we introduce the notion of an *immersion*, which is a kind of strong deformation retract with respect to the "path object" of reversible cylinders.

20.1 ω-Equivalences

The notion of an ω-equivalence is the higher-dimensional generalization of the notion of an equivalence of categories. In particular, ω-equivalent ω-categories can be considered as being close to be equal. The definition is a bit involved and requires the introduction of the auxiliary concept of a reversible cell.

20.1.1 Reversible cells. Given an ω-category C, the notion of a reversible cell of C is defined by *coinduction* in the following way. An n-cell $u\colon x \to y$ of C, for some $n > 0$, is *reversible* if there exists an n-cell $\overline{u}\colon y \to x$ in C and $(n+1)$-cells

$$\alpha\colon 1_x \to u *_{n-1} \overline{u} \qquad \text{and} \qquad \beta\colon \overline{u} *_{n-1} u \to 1_y$$

in C that are both reversible.

Concretely, this means that an n-cell $u\colon x \to y$ of C is reversible if and only if it belongs to a set X of cells of C having the following property: for every $n' > 0$ and every n'-cell $u'\colon x' \to y'$ in X, there exists an n'-cell $\overline{u'}\colon y' \to x'$ and $(n'+1)$-cells $\alpha'\colon 1_{x'} \to u' *_{n'-1} \overline{u'}$ and $\beta'\colon \overline{u'} *_{n'-1} u' \to 1_{y'}$ in X.

20.1.2 Remark. Strictly speaking, coinduction allows to define algebraic structures. What we have really defined in the previous paragraph is a "reversibility structure" where all the "there exists" are replaced by actual choices. This structure can somehow be flattened into a kind of tree. Then one can define a reversible cell as a cell that can be endowed with a reversibility structure.

We will often say that "we reason by coinduction". This basically means that we are defining a "reversibility structure" according to its actual definition, producing choices of $\overline{u}$, α, and β as in the definition. In particular, if R is a set of cells of an ω-category C, "proving by coinduction" that cells of R are reversible will consist in producing, for every n-cell u of R, a formula giving $\overline{u}$,

α, and β as in the definition of a reversible cell *assuming* that the $(n+1)$-cells of R are reversible.

20.1.3 Weak inverses. Let $n \geq 0$. If an n-cell $u\colon x \to y$ in some ω-category is reversible, then any $\overline{u}\colon y \to x$ as in the definition will be called a *weak inverse* of u.

If C is a 1-category, a reversible 1-cell of C is the same thing as an isomorphism of C and a weak inverse is an inverse. This follows from the fact that units are reversible (see the next lemma). Similarly, if C is a 2-category, a 1-cell $u\colon x \to y$ of C is reversible if and only if it is an equivalence, that is, if and only if there exists a 1-cell $\overline{u}\colon y \to x$ and 2-cells $\alpha\colon 1_x \to u *_0 \overline{u}$ and $\beta\colon \overline{u} *_0 u \to 1_y$ that are isomorphisms.

20.1.4 Lemma. *Let C be an ω-category.*

1. *If x is a cell of C, then 1_x is reversible.*
2. *If $u\colon x \to y$ is a reversible cell of C, then any weak inverse $\overline{u}\colon y \to x$ of u is reversible and has u as a weak inverse.*
3. *If $u\colon x \to y$ and $v\colon y \to z$ are two reversible n-cells of C for an $n > 0$, then $u *_{n-1} v$ is reversible.*
4. *More generally, if u and v are two reversible n-cells of C such that the composition $u *_i v$ is defined for some $0 \leq i < n$, then this composition is reversible.*

Proof. We proceed by coinduction.

1. Let x be an n-cell of C. Set $u = \overline{u} = 1_x$. Then $u *_n \overline{u} = \overline{u} *_n u = 1_{1_x}$ is reversible by coinduction, and so is 1_x by definition.
2. This is immediate by the symmetry in the definition of a reversible cell.
3. By definition, there exist cells $\overline{u}\colon y \to x$, $\overline{v}\colon z \to y$ and reversible cells $\alpha\colon 1_x \to u *_{n-1} \overline{u}$, $\beta\colon \overline{u} *_{n-1} u \to 1_y$, $\gamma\colon 1_y \to v *_{n-1} \overline{v}$, $\delta\colon \overline{v} *_{n-1} v \to 1_z$. We get cells
$$\alpha *_n (u *_{n-1} \gamma *_{n-1} \overline{u})\colon 1_x \to u *_{n-1} v *_{n-1} \overline{v} *_{n-1} \overline{u}$$
$$(\overline{v} *_{n-1} \beta *_{n-1} v) *_n \delta\colon \overline{v} *_{n-1} \overline{u} *_{n-1} u *_{n-1} v \to 1_z.$$
By coinduction, it suffices to show that the cells $\alpha *_n (u *_{n-1} \gamma *_{n-1} \overline{u})$ and $(\overline{v} *_{n-1} \beta *_{n-1} v) *_n \delta$ are reversible. It thus suffices to show that the whiskering of a reversible cell by any cell is reversible. This can be shown by a new coinduction but also follows from the next proposition.
4. The case where $i = n-1$ is the previous assertion. If $i < n-1$, then, using the exchange law, this follows from the case $i = n-1$ and the case of a whiskering (see the proof of the previous assertion). □

20.1.5 Proposition. *Let $f: C \to D$ be an ω-functor. If u is a reversible cell of C, then $f(u)$ is a reversible cell of D.*

Proof. Let $u: x \to y$ be a reversible n-cell of C. By definition, there exists a cell $\overline{u}: y \to x$ and two reversible cells $\alpha: 1_x \to u *_{n-1} \overline{u}$ and $\beta: \overline{u} *_{n-1} u \to 1_x$. By coinduction, the cells $f(\alpha): 1_{f(x)} \to f(u) *_{n-1} f(\overline{u})$ and $f(\beta): f(\overline{u}) *_{n-1} f(u) \to 1_{f(y)}$ are reversible. This means that $f(u)$ is indeed reversible. □

20.1.6 ω-equivalent cells. Let $n \geq 0$. Two n-cells x and y of an ω-category C are *ω-equivalent* if there exists a reversible cell $u: x \to y$ in C. We denote this relation by $x \sim y$.

We chose to define the notion of being ω-equivalent in terms of the notion of a reversible cell but we could also have defined this notion directly using coinduction: two parallel n-cells x and y are ω-equivalent if there exists $(n+1)$-cells $u: x \to y$ and $\overline{u}: y \to x$ such that 1_x and $u *_n \overline{u}$ are ω-equivalent, and $\overline{u} *_n u$ and 1_y are ω-equivalent.

It follows from §20.1.3 that ω-equivalent objects in a 1-category are isomorphic objects and that ω-equivalent objects in a 2-category are equivalent objects.

20.1.7 Proposition. *Let C be an ω-category. The relation "being ω-equivalent" is a congruence relation on the set of cells of C in the sense that:*

1. *This relation is an equivalence relation.*
2. *This equivalence relation is compatible with compositions: if u and u', and v and v' are ω-equivalent cells, then $u *_i v$ is ω-equivalent to $u' *_i v'$ when these compositions make sense.*

Proof. This is a direct consequence of Lemma 20.1.4. □

20.1.8 Proposition. *Let $f: C \to D$ be an ω-functor. If two cells x and y of C are ω-equivalent, then so are $f(x)$ and $f(y)$.*

Proof. This follows from Proposition 20.1.5 □

20.1.9 We now turn to a result of paramount importance in the construction of the folk model structure, namely the "division Lemma" ([237, Lemma 4.6]). As the proof is quite intricate, it will be convenient to introduce the following terminology: for any property $\mathcal{P}$ applying to n-cells, we say that there is a *weakly unique* n-cell satisfying $\mathcal{P}$ whenever any two n-cells satisfying $\mathcal{P}$ are ω-equivalent.

20.1.10 Lemma (Division lemma). *Any reversible* 1*-cell* $u\colon x \longrightarrow y$ *satisfies the following* left division property*:*

- *For any* 1*-cell* $w\colon x \longrightarrow z$*, there is a weakly unique* 1*-cell* $v\colon y \longrightarrow z$ *such that* $u *_0 v \sim w$*.*
- *For any pair of parallel* 1*-cells* $a, b\colon y \longrightarrow z$ *and any* 2*-cell* $w\colon u{*_0}a \longrightarrow u{*_0}b$*, there is a weakly unique* 2*-cell* $v\colon a \longrightarrow b$ *such that* $u *_0 v \sim w$*.*
- *More generally, for any* $n > 0$*, any pair of parallel* n*-cells* a, b *such that* $s_0(a) = s_0(b) = y$*,* $t_0(a) = t_0(b) = z$ *and any* $(n+1)$*-cell* $w\colon u{*_0}a \longrightarrow u{*_0}b$*, there is a weakly unique* $(n+1)$*-cell* $v\colon a \longrightarrow b$ *such that* $u *_0 v \sim w$*.*

Likewise, reversible 1*-cells satisfy the corresponding right division property.*

Proof. Let $u\colon x \longrightarrow y$ be a reversible 1-cell. By definition there is a weak inverse $\overline{u}\colon y \longrightarrow x$ together with a reversible 2-cell $\beta\colon \overline{u} *_0 u \longrightarrow 1_y$, as well as a reversible 2-cell $\overline{\beta}\colon 1_y \longrightarrow \overline{u} *_0 u$ such that $\beta *_1 \overline{\beta} \sim 1_{\overline{u}*_0 u}$.

- In the first case, define $v = \overline{u}{*_0}w\colon y \longrightarrow z$. Then $u{*_0}v = u{*_0}\overline{u}{*_0}w \sim w$. Moreover, for any $v'\colon y \longrightarrow z$ such that $u *_0 v' \sim w$, we get $\overline{u} *_0 u *_0 v' \sim \overline{u} *_0 w = v$, whence $v' \sim v$ and weak uniqueness holds. Likewise, the right division property holds in the first case.
- Let $a, b\colon y \longrightarrow z$ be 1-cells and $w\colon u *_0 a \longrightarrow u *_0 b$. Suppose there is a 2-cell $v\colon a \longrightarrow b$ such that $u{*_0}v \sim w$. Then, by applying exchange and compatibility of ω-equivalence with compositions, one gets

$$\begin{aligned}(\overline{\beta} *_0 a) *_1 (\overline{\beta} *_0 w) *_1 (\beta *_0 b) &\sim (\overline{\beta} *_0 a) *_1 (u *_0 v) *_1 (\beta *_0 b)\\ &\sim (\overline{\beta} *_1 \beta) *_0 v\\ &\sim v,\end{aligned}$$

which implies weak uniqueness for v. As for existence, define

$$v = (\overline{\beta} *_0 a) *_1 (\overline{\beta} *_0 w) *_1 (\beta *_0 b).$$

By definition, $v\colon a \longrightarrow b$. Consider now

$$v' = (\beta *_0 a) *_1 v *_1 (\overline{\beta} *_0 b).$$

Again, exchange and compatibility yield

$$v' \sim \overline{u} *_0 w$$

but also

$$v' \sim \overline{u} *_0 (u *_0 v).$$

By applying weak uniqueness to left division by $\overline{u}$, one gets $w \sim u *_0 v$, whence the result. Right division is proved accordingly in this case.

– The general case is proved by induction on n. The case $n = 1$ has been just proved above. Let now $n > 1$ and suppose that the property of left and right division holds for any reversible 1-cell in any ω-category C up to dimension $n - 1$. Let a, b be parallel n-cells such that $s_0(a) = s_0(b) = y$, $t_0(a) = t_0(b) = z$ and $w\colon u *_0 a \to u *_0 b$ an $(n + 1)$-cell in C. Consider

$$w' = (\overline{u} *_0 w) *_1 (\beta *_0 t_1(a)).$$

We have

$$s_n(w') = (\beta *_0 s_1(a)) *_1 a \quad \text{and} \quad t_n(w') = (\beta *_0 s_1(a)) *_1 b.$$

Now $\beta *_0 s_1(a)$ may be seen as a reversible 1-cell in the ω-category $C(y, z)$, so that the induction hypothesis applies and yields a weakly unique $(n+1)$-cell $v\colon a \to b$ such that

$$(\beta *_0 s_1(a)) *_1 v \sim w'. \tag{20.1}$$

One then shows that any $(n+1)$-cell $v'\colon a \to b$ such that $u *_0 v' \sim w$ satisfies the equation (20.1). Therefore, by induction, $v' \sim v$ and weak uniqueness is proved. It remains to check that the above $(n + 1)$-cell v satisfies $u *_0 v \sim w$. Rewriting (20.1) using the exchange law, we get

$$(\overline{u} *_0 u *_0 v) *_1 (\beta *_0 t_1(a)) \sim (\overline{u} *_0 w) *_1 (\beta *_0 t_1(a)).$$

By induction applied to the reversible 1-cell $\beta *_0 t_1(a)$ of $C(y, z)$, weak uniqueness for right division implies

$$\overline{u} *_0 u *_0 v \sim \overline{u} *_0 w,$$

and finally, by weak uniqueness applied to $\overline{u}$, $u *_0 v \sim w$, as required. □

20.1.11 ω-Equivalences. An ω-functor $f\colon C \to D$ is an *ω-equivalence* if the following conditions are satisfied:

1. For every object y of D, there exists an object x of C such that $f(x)$ is ω-equivalent to y.
2. For every $n \geq 0$, every pair of parallel n-cells x and y of C and every $(n + 1)$-cells $v\colon f(x) \to f(y)$, there exists an $(n + 1)$-cells $u\colon x \to y$ such that $f(u)$ is ω-equivalent to v.

If $f\colon C \to D$ is a 1-functor between 1-categories, then f is an ω-equivalence if and only if it is an equivalence of categories. Similarly, if $f\colon C \to D$ is a 2-functor, then f is an ω-equivalence if and only if f is a biequivalence. Note that an ω-equivalence between two 2-categories is not the same thing as a biequivalence as an ω-equivalence is required to be a strict 2-functor as opposed to a bifunctor.

20.1.12 Remark. In general, an ω-equivalence does *not* admit an inverse, in any reasonable sense, that is a *strict* ω-equivalence. Morally, it admits a *weak* ω-functor as a weak inverse in some sense. For instance, the obvious ω-functor from the "pseudo-2-triangle"

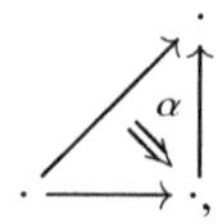

where α is a reversible cell, to the commutative triangle, is easily seen to be an ω-equivalence, but there is no ω-equivalence from the commutative triangle to the pseudo-2-triangle.

20.1.13 Proposition. *Trivial fibrations are ω-equivalences.*

Proof. This follows immediately from the characterization of trivial fibrations given at the end of §19.2.3 and the fact that identities are reversible cells. □

An ω-equivalence is injective up to ω-equivalence of cells in the following sense:

20.1.14 Proposition. *Let $f: C \to D$ be an ω-equivalence and let x and y be two parallel cells of C. If $f(x)$ and $f(y)$ are ω-equivalent, then x and y are ω-equivalent.*

Proof. By definition, there exists a cell $v: f(y) \to f(x)$ such that $1_{f(x)}$ and $f(u) *_{n-1} v$, and $v *_{n-1} f(u)$ and $1_{f(y)}$ are ω-equivalent cells. Since f is an ω-equivalence, there exists a cell $\overline{u}: y \to x$ such that $f(\overline{u})$ is ω-equivalent to v. Using Proposition 20.1.7, we get that $f(1_x)$ and $f(u *_{n-1} \overline{u})$, and $f(\overline{u} *_{n-1} u)$ and $f(1_y)$ are ω-equivalent cells. The result thus follows by coinduction. □

20.1.15 Proposition. *The composition of two ω-equivalences is an ω-equivalence.*

Proof. Let $f: C \to D$ and $g: D \to E$ be two ω-equivalences. Let us prove that $gf: C \to E$ is an ω-equivalence.

1. Let z be an object of E. Since g is an ω-equivalence, there exists an object y of D such that $g(y)$ and z are ω-equivalent. Similarly, since f is an ω-equivalence, there exists an object x of C such that $f(x)$ and y are ω-equivalent. It follows from Proposition 20.1.5 that $g(f(x))$ and $g(y)$ are ω-equivalent. Hence, by transitivity of the relation of ω-equivalence, $gf(x)$ and z are ω-equivalent.

2. Let $n \geq 0$, let x and y be two parallel n-cells of C, and let $w\colon gf(x) \to gf(y)$ be an $(n+1)$-cell of E. Since g is an ω-equivalence, there exists an $(n+1)$-cell $v\colon f(x) \to f(y)$ of D such that $g(v)$ and w are ω-equivalent. Similarly, since f is an ω-equivalence, there exists an $(n+1)$-cell $u\colon x \to y$ such that $f(u)$ and v are ω-equivalent. It follows from Proposition 20.1.8 that $gf(u)$ and $g(v)$ are ω-equivalent, and hence that $gf(u)$ and w are ω-equivalent, thereby proving the result. □

20.1.16 2-out-of-3 Property. Recall that a class of maps $\mathcal{W}$ in a category C is said to satisfy the *2-out-of-3 property* if for any commutative triangle

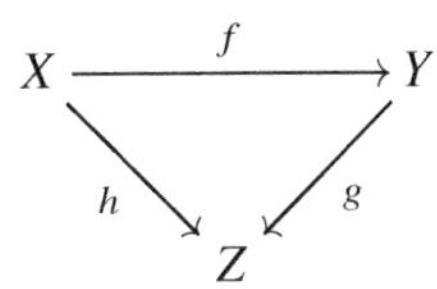

in C, if two morphisms among f, g, and h are in $\mathcal{W}$, then so is the third one. For instance, isomorphisms in a category satisfy the 2-out-of-3 property. More generally, any reasonable notion of "equivalence" in a category should satisfy this property.

The 2-out-of-3 property is made of three different properties, depending on which of two morphisms among f, g, and h are assumed to be in $\mathcal{W}$. If $\mathcal{W}$ is the class of ω-equivalences, the previous proposition gives the case where these two morphisms are f and g. The following proposition is the case where they are g and h:

20.1.17 Proposition. *Let $f\colon C \to D$ and $g\colon D \to E$ be two ω-functors. If g and gf are ω-equivalences, then so is f.*

Proof. Let us prove that f is an ω-equivalence.

1. Let y be an object of D. Consider the object $g(y)$ of E. Since gf is an ω-equivalence, there exists an object x of C such that $gf(x)$ is ω-equivalent to $g(y)$. By Proposition 20.1.14, this implies that $f(x)$ and y are ω-equivalent.
2. Let $n \geq 0$, let x and y be two parallel n-cells of C and let $v\colon f(x) \to f(y)$ be an $(n+1)$-cell of D. Consider the $(n+1)$-cell $g(v)\colon gf(x) \to gf(y)$ of E. Since gf is an ω-equivalence, there exists an $(n+1)$-cell $u\colon x \to y$ such that $gf(u)$ and $g(v)$ are ω-equivalent. By Proposition 20.1.14, this implies that $f(u)$ and v are ω-equivalent. □

The remaining case of the 2-out-of-3 property is much harder to prove and requires the introduction of the ω-category of cylinders, which is the topic of the next section.

We end the section with two easy stability conditions of the class of ω-equivalences.

20.1.18 Proposition. *A retract of an ω-equivalence is an ω-equivalence.*

Proof. Consider a commutative diagram

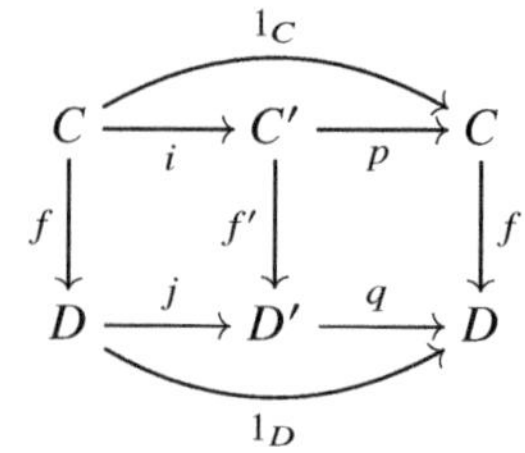

where f' is an ω-equivalence. Let us prove that f is an ω-equivalence.

1. Let y be an object of D. Since f is an ω-equivalence, there exists a cell x' in C' such that $f'(x') = y'$, where $y' = j(y)$. We thus get an object $x = p(x')$ in C. As $f(x) = fp(x') = qf'(x)$, the cell $f(x)$ is ω-equivalent to $qy' = qj(y) = y$.
2. Let x and y be parallel cells of C and let $v\colon f(x) \to f(y)$ be a cell of D. Consider the cell $v' = j(v)\colon jf(x) \to jf(y)$. We have $v'\colon f'i(x) \to f'i(y)$ and, since f' is an ω-equivalence, there exists a cell $u'\colon i(x) \to i(y)$ such that $f(u')$ is ω-equivalent to v'. Setting $u = pf(u')$, we get as in the previous point that $f(u)$ is ω-equivalent to v. □

20.1.19 Proposition. *A filtered colimit of ω-equivalences is an ω-equivalence. In particular, a countable composition (or more generally a transfinite composition) of ω-equivalences is an ω-equivalence.*

Proof. Let I be a filtered category, let $F, G\colon I \to \mathbf{Cat}_\omega$ be two functors, and let $f\colon F \to G$ a natural transformation such that for every i in I, $f_i\colon F(i) \to G(i)$ is an ω-equivalence. Let us prove that $f_\infty = \varinjlim f\colon \varinjlim F \to \varinjlim G$ is an ω-equivalence. Recall (see Proposition 14.2.8) that the forgetful functor from ω-categories to globular sets respects filtered colimits. This means that for every $n \geq 0$, the set of n-cells of $\varinjlim F$ is the colimit of the sets of n-cells of the $F(i)$. If x is an n-cell of $F(i)$, we will denote by $[x]$ the corresponding n-cell of $\varinjlim F$.

1. Let $[y]$ be an object of $\varinjlim G$ coming from an object y of $G(i)$ for some i in I. Since $f_i\colon F(i) \to G(i)$ is an ω-equivalence, there exists an object x of $F(i)$ such that $\alpha_i(x)$ is ω-equivalence to y. By definition, we have $f_\infty([x]) = [y]$.

Moreover, since the canonical ω-functor $G(i) \to \varinjlim G$, as any ω-functor, sends ω-equivalent cells to ω-equivalent cells, $[x]$ and $[y]$ are ω-equivalent.

2. Let $n \geq 0$ and let $[x]$ and $[y]$ be two parallel n-cells of $\varinjlim F$, where x is an n-cell of $F(i)$ and y is an n-cell of $F(j)$. Let $[u]: f_\infty([x]) \to f_\infty([y])$ be an $(n+1)$-cell of $\varinjlim G$, where u is an $(n+1)$-cell of $G(k)$. Using the fact that I is filtered, one can suppose that $i = j = k$ and $u: f_i(x) \to f_i(y)$. Since f_i is an ω-equivalence, there exists $v: x \to y$ such that $f_i(u)$ is ω-equivalent to u. This implies that $[u]: [x] \to [y]$ and that $f_\infty([u])$ is ω-equivalent to $[v]$. □

20.2 The ω-Category of Cylinders

In this section, we fix an ω-category C.

20.2.1 Cylinders. For $n \geq 0$, an *n-cylinder in C* is given by two n-cells x and y of C, and a sequence

$$\alpha_0^-, \alpha_0^+, \ldots, \alpha_{n-1}^-, \alpha_{n-1}^+, \alpha_n^- = \alpha_n = \alpha_n^+$$

where α_j^ε, for $0 \leq j \leq n$ and $\varepsilon = \pm$, is a $(j+1)$-cell whose source and target are

$$\alpha_j^-: s_j(x) *_0 \alpha_0^+ *_1 \cdots *_{j-1} \alpha_{j-1}^+ \to \alpha_{j-1}^- *_{j-1} \cdots *_1 \alpha_0^- *_0 s_j(y),$$
$$\alpha_j^+: t_j(x) *_0 \alpha_0^+ *_1 \cdots *_{j-1} \alpha_{j-1}^+ \to \alpha_{j-1}^- *_{j-1} \cdots *_1 \alpha_0^- *_0 t_j(y).$$

Note that for $n = j$, we get

$$\alpha_n: x *_0 \alpha_0^+ *_1 \cdots *_{n-1} \alpha_{n-1}^+ \to \alpha_{n-1}^- *_{n-1} \cdots *_1 \alpha_0^- *_0 y.$$

We say that such a cylinder is a *cylinder from x to y* and we write $\alpha: x \curvearrowright y$.

Here are pictures of a 0-cylinder, a 1-cylinder, and a 2-cylinder in C:

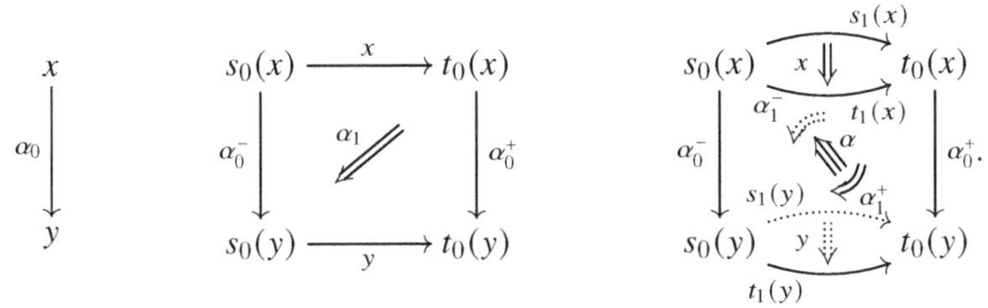

20.2.2 Inductive definition of cylinders. Alternatively, the notion of an n-cylinder $\alpha\colon x \curvearrowright y$, where x and y are n-cells of C, can defined inductively in the following way. A 0-cylinder $\alpha\colon x \to y$ is a 1-cell $\alpha_0\colon x \to y$. For $n > 0$, an n-cylinder $\alpha\colon x \curvearrowright y$ consists of two 1-cells

$$\alpha_0^-\colon s_0(x) \to t_0(x) \qquad \text{and} \qquad \alpha_0^+\colon t_0(x) \to t_0(x)$$

and an $(n-1)$-cylinder

$$[\alpha]\colon [x *_0 \alpha_0^+] \curvearrowright [\alpha_0^- *_0 y]$$

in the ω-category $C(s_0(x), t_0(y))$, where $[x *_0 \alpha_0^+]$ and $[\alpha_0^- *_0 y]$ denote the n-cells $x *_0 \alpha_0^+$ and $\alpha_0^- *_0 y$ of C seen as $(n-1)$-cells of $C(s_0(x), t_0(y))$.

The equivalence between the two definitions easily follows by induction. This second definition is the one used [237]. Its main advantage is to allow inductive arguments on cylinders.

We will see that the cylinders in C organize themselves in an ω-category. We now define the associated operations.

20.2.3 Source and target of a cylinder. Let x and y be two n-cells for some $n > 0$, and let $\alpha\colon x \curvearrowright y$ be an n-cylinder. We define the *source* of this cylinder to be the $(n-1)$-cylinder

$$s_{n-1}(\alpha)\colon s_{n-1}(x) \curvearrowright s_{n-1}(y)$$

given by the cells

$$\alpha_0^-, \alpha_0^+, \ldots, \alpha_{n-2}^-, \alpha_{n-2}^+, \alpha_{n-1}^-.$$

Similarly, the *target* of such a cylinder is the $(n-1)$-cylinder

$$t_{n-1}(\alpha)\colon t_{n-1}(x) \curvearrowright t_{n-1}(y)$$

defined by the cells

$$\alpha_0^-, \alpha_0^+, \ldots, \alpha_{n-2}^-, \alpha_{n-2}^+, \alpha_{n-1}^+.$$

Geometrically, the source of a cylinder is the "back face" of this cylinder and the target is the "front face". For instance, the source and target of a 2-cylinder

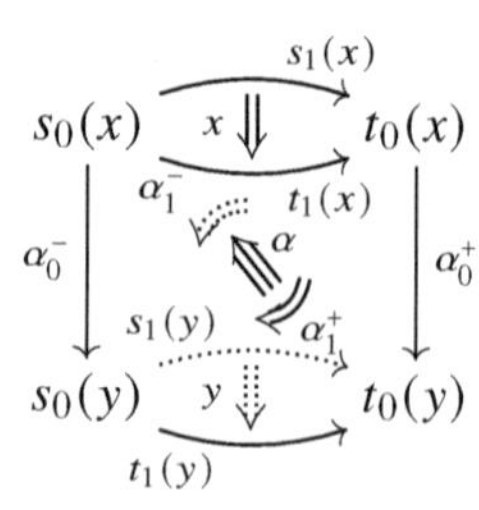

are the 1-cylinders

$$\begin{array}{ccc} s_0(x) & \xrightarrow{s_1(x)} & t_0(x) \\ {\scriptstyle \alpha_0^-}\downarrow & \overset{\alpha_1^-}{\swarrow} & \downarrow{\scriptstyle \alpha_0^+} \\ s_0(y) & \xrightarrow[s_1(y)]{} & t_0(y) \end{array} \qquad \text{and} \qquad \begin{array}{ccc} s_0(x) & \xrightarrow{t_1(x)} & t_0(x) \\ {\scriptstyle \alpha_0^-}\downarrow & \overset{\alpha_1^+}{\swarrow} & \downarrow{\scriptstyle \alpha_0^+} \\ s_0(y) & \xrightarrow[t_1(y)]{} & t_0(y). \end{array}$$

20.2.4 Unit of a cylinder. Let $\alpha\colon x \curvearrowright y$ be an n-cylinder in C. We define the *unit* of this cylinder to be the $(n+1)$-cylinder

$$1_\alpha\colon 1_x \curvearrowright 1_y$$

defined by the cells

$$\alpha_0^-, \alpha_0^+, \dots, \alpha_{n-1}^-, \alpha_{n-1}^+, \alpha_n, \alpha_n, 1_{\alpha_n}.$$

Geometrically, this cylinder is obtained by gluing two copies of α and putting a trivial cell in the middle. For instance, the unit of the 1-cylinder

$$\begin{array}{ccc} s_0(x) & \xrightarrow{x} & t_0(x) \\ {\scriptstyle \alpha_0^-}\downarrow & \overset{\alpha_1}{\swarrow} & \downarrow{\scriptstyle \alpha_0^+} \\ s_0(y) & \xrightarrow[y]{} & t_0(y) \end{array}$$

is the 2-cylinder

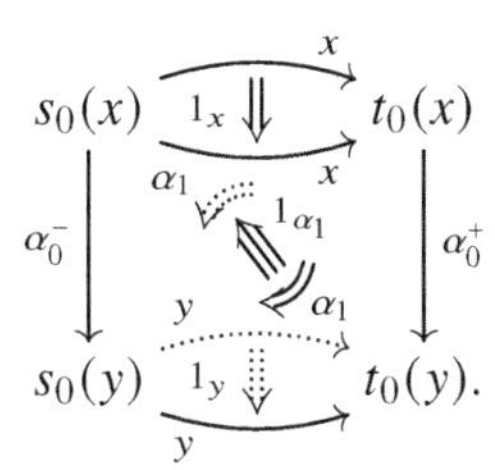

20.2.5 Composition of cylinders. Let $\alpha\colon x \curvearrowright y$ and $\beta\colon z \curvearrowright t$ be two n-cylinders in C for some $n > 0$. Let $0 \le i < n$ and suppose that α and β are composable in dimension i, that is, that we have

$$t_i(\alpha)\colon t_i(x) \curvearrowright t_i(y) = s_i(\beta)\colon s_i(x) \curvearrowright s_i(y).$$

We define the *composition in dimension i* of these two n-cylinders to be the n-cylinder

$$\gamma = \alpha *_i \beta\colon x *_i z \curvearrowright y *_i t$$

defined by, for $j < i$,

$$\begin{aligned}
\gamma_j^\varepsilon &= \alpha_j^\varepsilon = \beta_j^\varepsilon \\
\gamma_i^- &= \alpha_i^- \\
\gamma_i^+ &= \beta_i^+ \\
\gamma_{i+1}^- &= \left(s_{i+1}(x) *_0 \beta_0^+ *_1 \cdots *_{i-1} \beta_{i-1}^+ *_i \beta_{i+1}^\varepsilon\right) \\
&\qquad *_{i+1} \left(\alpha_{i+1}^\varepsilon *_i \alpha_{i-1}^- *_{i-1} \cdots *_1 \alpha_0^- *_0 s_{i+1}(t)\right) \\
\gamma_{i+1}^+ &= \left(t_{i+1}(x) *_0 \beta_0^+ *_1 \cdots *_{i-1} \beta_{i-1}^+ *_i \beta_{i+1}^\varepsilon\right) \\
&\qquad *_{i+1} \left(\alpha_{i+1}^\varepsilon *_i \alpha_{i-1}^- *_{i-1} \cdots *_1 \alpha_0^- *_0 t_{i+1}(t)\right)
\end{aligned}$$

if $i + 1 < n$, and

$$\gamma_j^\varepsilon = \left(s_{i+1}(x) *_0 \beta_0^+ *_1 \cdots *_{i-1} \beta_{i-1}^+ *_i \beta_j^\varepsilon\right) *_{i+1} \left(\alpha_j^\varepsilon *_i \alpha_{i-1}^- *_{i-1} \cdots *_1 \alpha_0^- *_0 t_{i+1}(t)\right)$$

for $i + 1 < j \leq n$. In particular, for $j = n$, we have

$$\gamma_n^\varepsilon = \left(s_{i+1}(x) *_0 \beta_0^+ *_1 \cdots *_{i-1} \beta_{i-1}^+ *_i \beta_n\right) *_{i+1} \left(\alpha_n *_i \alpha_{i-1}^- *_{i-1} \cdots *_1 \alpha_0^- *_0 t_{i+1}(t)\right).$$

Here are some pictures of composable cylinders in low dimensions:

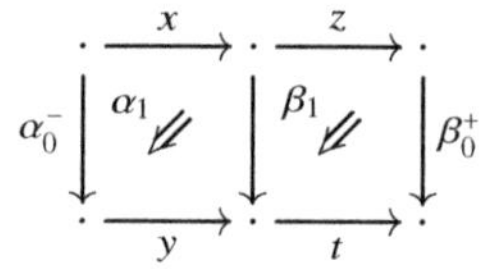

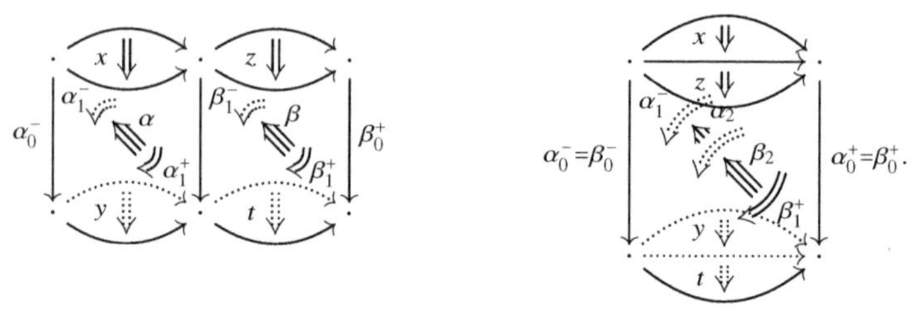

20.2.6 The ω-category of cylinders. We define the ω-category $\Gamma(C)$ of *cylinders in C* to be the ω-category whose n-cells are n-cylinders in C and whose sources, targets, units, and compositions are given according to the formulas given in the previous paragraphs. Tedious calculations show that:

20.2.7 Theorem. *If C is an ω-category, then $\Gamma(C)$ is indeed an ω-category.*

Proof. See [278, Appendix A]. ∎

20.2.8 The free-standing n-cylinder. Theorem 20.2.7 is a key ingredient in the construction $\mathbb{O}_1 \otimes (-)$ presented in Section 17.2. In particular, we get for each $n \geq 0$ a polygraph $\mathrm{Cyl}_n = \mathbb{O}_1 \otimes \mathbb{O}_n$. The associated ω-category Cyl_n^* is the free-standing n-cylinder, and by construction, for any ω-category C, the set of n-cylinders in C is just $\mathbf{Cat}_\omega(\mathrm{Cyl}_n^*, C)$.

20.2.9 Remark. The ω-category $\Gamma(C)$ can be defined more conceptually using the Gray tensor product of ω-categories. Indeed, the Gray tensor product is a biclosed monoidal structure on ω-category. This means that we have functors

$$\underline{\mathrm{Hom}}_{\mathrm{oplax}} \colon \mathbf{Cat}_\omega^{\mathrm{op}} \times \mathbf{Cat}_\omega \to \mathbf{Cat}_\omega \quad \text{and} \quad \underline{\mathrm{Hom}}_{\mathrm{lax}} \colon \mathbf{Cat}_\omega^{\mathrm{op}} \times \mathbf{Cat}_\omega \to \mathbf{Cat}_\omega$$

and isomorphisms

$$\mathbf{Cat}_\omega(C \otimes D, E) \simeq \mathbf{Cat}_\omega(C, \underline{\mathrm{Hom}}_{\mathrm{oplax}}(D, E))$$

and

$$\mathbf{Cat}_\omega(C \otimes D, E) \simeq \mathbf{Cat}_\omega(D, \underline{\mathrm{Hom}}_{\mathrm{lax}}(C, E)),$$

natural in C, D, and E in $\mathbf{Cat}_\omega$. (See for instance [15, Appendix A] for definitions.) It can be shown that the ω-category $\Gamma(C)$ is canonically isomorphic to $\underline{\mathrm{Hom}}_{\mathrm{lax}}(\mathbb{O}_1, C)$ (see [15, Section B.1]).

20.2.10 Functoriality of the cylinder ω-category. Let $f \colon C \to D$ be an ω-functor. If $\alpha \colon x \curvearrowright y$ is an n-cylinder in C, we define an n-cylinder $f(\alpha) \colon f(x) \curvearrowright f(y)$ in D by applying f to the components of α, that is, by the cells

$$f(\alpha_0^-), f(\alpha_0^+), \cdots f(\alpha_n^-), f(\alpha_n^+), f(\alpha_{n+1}).$$

One checks that this assignment defines an ω-functor $\Gamma(f) \colon \Gamma(C) \to \Gamma(D)$. Moreover, this assignment is functorial in f. In other words, we have defined an endofunctor

$$\Gamma \colon \mathbf{Cat}_\omega \to \mathbf{Cat}_\omega.$$

20.2.11 Projections. Let $\alpha \colon x \curvearrowright y$ be a cylinder in C. We set

$$\overline{\pi}(\alpha) = x \qquad \text{and} \qquad \underline{\pi}(\alpha) = y.$$

Geometrically, $\overline{\pi}(\alpha)$ and $\underline{\pi}(\alpha)$ are, respectively, the top face and the bottom face of the cylinder α. The formulas defining the structure of the ω-category $\Gamma(C)$ make transparent the fact that these assignments define ω-functors

$$\overline{\pi}_C, \underline{\pi}_C \colon \Gamma(C) \to C.$$

Moreover, these ω-functors are natural in C and we have defined natural transformations

$$\overline{\pi}, \underline{\pi} \colon \Gamma \to 1_{\mathbf{Cat}_\omega}.$$

20.2.12 Trivial cylinders. Let x be an n-cell of C. The *trivial cylinder* on x is the n-cylinder $\iota(x) \colon x \curvearrowright x$ defined by the cells

$$1_{s_0(x)}, 1_{t_0(x)}, \dots, 1_{s_{n-1}(x)}, 1_{t_{n-1}(x)}, 1_x.$$

For instance, the trivial cylinder on a 2-cell

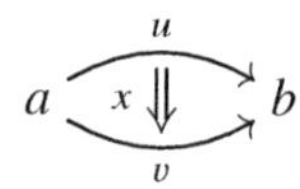

can be pictured as

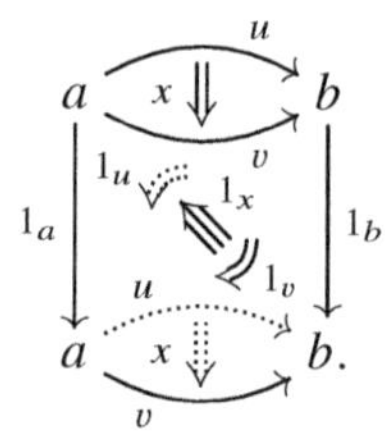

One checks that this assignment defines an ω-functor $\iota_C \colon C \to \Gamma(C)$ and this ω-functor is natural in C. In other words, we have defined a natural transformation

$$\iota \colon 1_{\mathbf{Cat}_\omega} \to \Gamma.$$

20.2.13 Oplax transformations. If C is an ω-category, then the diagonal ω-functor $C \to C \times C$ factors as

$$C \xrightarrow{\iota_C} \Gamma(C) \xrightarrow{(\overline{\pi}_C, \underline{\pi}_C)} C \times C.$$

As every factorization of the diagonal, this factorization induces some kind of notion of homotopies. These homotopies will be called *oplax transformations*. In other words, if $f, g \colon C \to D$ are two ω-functors, an oplax transformation α from f to g, denoted by $\alpha \colon f \Rightarrow g$, is an ω-functor $\alpha \colon C \to \Gamma(D)$ making the diagram

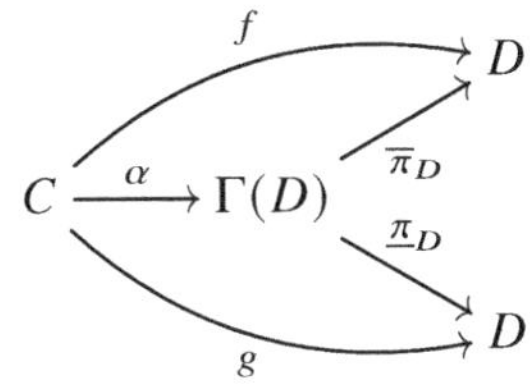

commute.

20.2.14 Algebraic description of oplax transformations. Let $\alpha\colon f \Rightarrow g$ be an oplax transformation. If x is an n-cell of C, we will denote by α_x the principal cell of the n-cylinder $\alpha(x)$. Thus, α_x is an $(n+1)$-cell of D. One can show that the cells α_x of D, where x varies among the cells of C, fully determine α. Better, an oplax transformation can be fully defined in terms of these α_x.

More precisely, if $f, g\colon C \to D$ are two ω-functors, the data of an oplax transformation $\alpha\colon f \Rightarrow g$ is equivalent to the data of, for every n-cell x of C, an $(n+1)$-cell

$$\alpha_x\colon f(x) *_0 \alpha_{t_0(x)} *_1 \cdots *_{n-1} \alpha_{t_{n-1}(x)} \to \alpha_{s_{n-1}(x)} *_{n-1} \cdots *_1 \alpha_{s_0(x)} *_0 g(x)$$

such that the following relations are satisfied:

- for every cell x of C, we have
$$\alpha_{1_x} = 1_{\alpha_x},$$
- for every $n > i \geq 0$ and every pair of n-cells x and y such that $x *_i y$ is well-defined, we have
$$\begin{aligned}\alpha_{x *_i y} = \big(f(s_{i+1}(x)) *_0 \alpha_{t_0(y)} *_1 \cdots *_{i-1} \alpha_{t_{i-1}(y)} *_i \alpha_y\big) *_{i+1} \\ \big(\alpha_x *_i \alpha_{s_{i-1}(x)} *_{i-1} \cdots *_1 \alpha_{s_0(x)} *_0 g(t_{i+1}(y))\big).\end{aligned}$$

20.2.15 Some operations on oplax transformations. Let $f\colon C \to C$ be an ω-functor. The ω-functor $\iota_C\colon C \to \Gamma(C)$ defines an oplax transformation from f to f that we will call the *unit oplax transformation* of f and that we will denote by $1_f\colon f \Rightarrow f$. If x is a cell of C, we have

$$(1_f)_x = 1_x.$$

If $\alpha\colon f \Rightarrow f'$ is an oplax transformation between ω-functors from C to D and $g\colon D \to E$ is an ω-functor, we define an oplax transformation $g\alpha\colon gf \Rightarrow gf'$ by the composition

$$C \xrightarrow{\alpha} \Gamma(D) \xrightarrow{\Gamma(g)} \Gamma(E).$$

If x is a cell of C, we have

$$(g\alpha)_x = g(\alpha_x).$$

Similarly, if $f\colon C \to D$ is an ω-functor and $\alpha\colon g \Rightarrow g'$ is an oplax transformation between ω-functors from D to E, we define an oplax transformation $\alpha f\colon gf \Rightarrow g'f$ by the composition

$$C \xrightarrow{f} D \xrightarrow{\alpha} \Gamma(E)$$

and, for x a cell of C, we have

$$(\alpha f)_x = \alpha_{f(x)}.$$

20.2.16 Remark. If $f, g, h\colon C \to D$ are three ω-functors and $\alpha\colon f \Rightarrow g$ and $\beta\colon g \Rightarrow h$ are oplax transformations, one can define in a natural way a composite $\beta \circ \alpha\colon f \Rightarrow h$. We will only need the existence of this composition and therefore we do not give its precise definition. One can show that ω-categories, ω-functors, and oplax transformations, with the operations defined in the previous paragraph and this operation $\circ$, form a sesquicategory (see §4.1.5 for a definition). However, this sesquicategory is *not* a 2-category. We refer the reader to [15, Appendix C] for more details.

20.3 The ω-Category of Reversible Cylinders

20.3.1 Reversible cylinders. Let C be an ω-category. An n-cylinder $\alpha\colon x \curvearrowright y$ is said to be *reversible* if the cells

$$\alpha_0^-, \alpha_0^+, \dots, \alpha_{n-1}^-, \alpha_{n-1}^+, \alpha_n$$

are reversible (see §20.1.1).

20.3.2 The ω-category of reversible cylinders. Let C be an ω-category and let $\widetilde{\Gamma}(C)_n$, for $n \geq 0$, be the set of reversible n-cylinders in C. By definition, we have an inclusion $\widetilde{\Gamma}(C)_n \subseteq \Gamma(C)_n$. One immediately checks that these $\widetilde{\Gamma}(C)_n$ define a subcategory $\widetilde{\Gamma}(C)$ of $\Gamma(C)$. Moreover, as the image of a reversible cell by an ω-functor is reversible, the functor $\Gamma\colon \mathbf{Cat}_\omega \to \mathbf{Cat}_\omega$ induces a functor

$$\widetilde{\Gamma}\colon \mathbf{Cat}_\omega \to \mathbf{Cat}_\omega.$$

Since units are reversible cells, the trivial cylinder on a cell is reversible and the natural transformation $\iota\colon 1_{\mathbf{Cat}_\omega} \to \Gamma$ factors by a natural transformation

$$\iota\colon 1_{\mathbf{Cat}_\omega} \to \widetilde{\Gamma}.$$

Finally, by restriction, the natural transformations $\overline{\pi}, \underline{\pi}\colon \Gamma \to 1_{\mathbf{Cat}_\omega}$ define natural transformations

$$\overline{\pi}, \underline{\pi}\colon \widetilde{\Gamma} \to 1_{\mathbf{Cat}_\omega}.$$

20.3.3 Reversible oplax transformations. Let $f, g\colon C \to D$ be two ω-functors. We say that an oplax transformation $\alpha\colon f \Rightarrow g$ is *reversible* if the ω-functor $\alpha\colon C \to \Gamma(D)$ factors through the inclusion $\widetilde{\Gamma}(D) \subseteq \Gamma(D)$. In others words, we have a factorization

$$D \xrightarrow{\iota_D} \widetilde{\Gamma}(D) \xrightarrow{(\overline{\pi}_D, \underline{\pi}_D)} D \times D$$

of the diagonal of D and a reversible oplax transformation from f to g is an ω-functor $\alpha\colon C \to \widetilde{\Gamma}(D)$ making the diagram

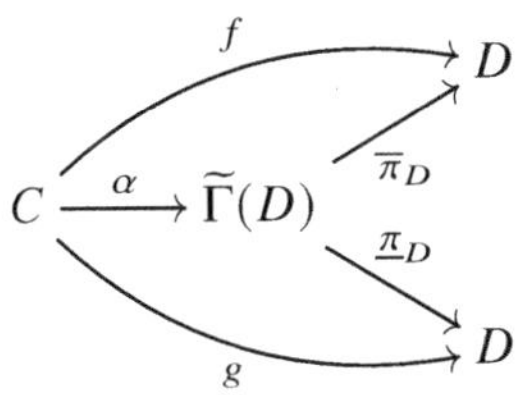

commute.

In the algebraic definition of an oplax transformation given in §20.2.14, an oplax transformation α is reversible if and only if α_x is reversible for every cell x.

One immediately checks that the various operations on oplax transformations defined in §20.2.15 restrict to reversible oplax transformations. In other words, the unit oplax transformation of an ω-functor is reversible and if we can compose horizontally a reversible oplax transformation and an ω-functor (in both directions) we get a reversible oplax transformation.

We now introduce some terminology to state the "transport lemma" which is of crucial importance to prove the existence of the folk model structure.

20.3.4 Incomplete cylinders. Let C be an ω-category. For $n \geq 0$, a *bottom-incomplete n-cylinder* in C consists of the same data as an n-cylinder $\alpha\colon x \curvearrowright y$ except that y and α_n are not given. More formally, it consists of an object x of C if $n = 0$ and, if $n > 0$, of an n-cell x, two parallel $(n-1)$-cells y^- and y^+ and a sequence

$$\alpha_0^-, \alpha_0^+, \ldots, \alpha_{n-1}^-, \alpha_{n-1}^+$$

where α_j^ε is a $(j+1)$-cell whose source and target are

$$\alpha_j^- : s_j(x) *_0 \alpha_0^+ *_1 \cdots *_{j-1} \alpha_{j-1}^+ \to \alpha_{j-1}^- *_{j-1} \cdots *_1 \alpha_0^- *_0 s_j(y^-),$$
$$\alpha_j^+ : t_j(x) *_0 \alpha_0^+ *_1 \cdots *_{j-1} \alpha_{j-1}^+ \to \alpha_{j-1}^- *_{j-1} \cdots *_1 \alpha_0^- *_0 t_j(y^+).$$

Here are pictures of bottom-incomplete cylinders in dimensions 0, 1, and 2:

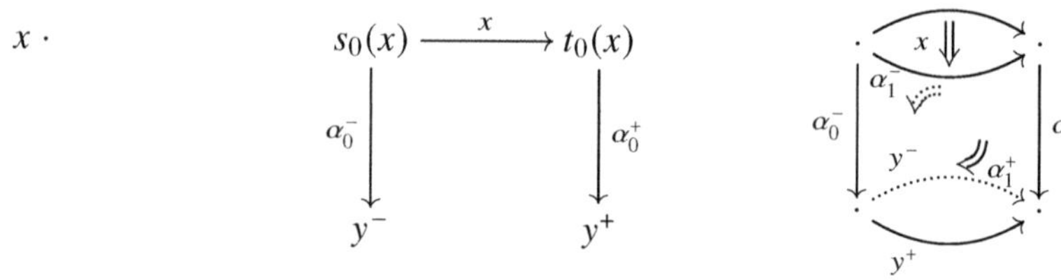

We say that such a bottom-incomplete n-cylinder is *reversible* if the cells

$$\alpha_0^-, \alpha_0^+, \ldots, \alpha_{n-2}^-, \alpha_{n-2}^+, \alpha_{n-1}^-$$

are reversible.

We define similarly the notion of a *top-incomplete n-cylinder* and of *reversible top-incomplete n-cylinder.*

20.3.5 Lemma (Transport lemma). *Let C be an ω-category.*

1. *Any reversible bottom-incomplete n-cylinder extends (non-uniquely) to a reversible n-cylinder.*
2. *Consider a reversible bottom-incomplete n-cylinder.*

   a.b *If* $\alpha : x \curvearrowright y$ *and* $\alpha' : x \curvearrowright y'$ *are two extensions as in 1, then y and y′ are ω-equivalent.*

   b.b *If* $\alpha : x \curvearrowright y$ *is an extension as in 1 and y′ is an n-cell ω-equivalent to y, then there exists an extension* $\alpha : x \curvearrowright y'$ *as in 1.*

Proof. The case $n = 0$ is obvious. For $n = 1$, consider the incomplete 1-cylinder defined by $(x, y^-, y^+, \alpha_0^-, \alpha_0^+)$ and let $\overline{\alpha_0^-}$ be a weak inverse of α_0^-. The 1-cell

$$y = \overline{\alpha_0^-} *_0 x *_0 \alpha_0^+$$

satisfies the relation $x *_0 \alpha_0^+ \sim \alpha_0^- *_0 y$ where $\sim$ denotes ω-equivalence as defined in §20.1.6. This is witnessed by a reversible 2-cell $\alpha_1 : x *_0 \alpha_0^+ \to \alpha_0^- *_0 y$, which gives the desired extension. Conditions 2(a) and 2(b) immediately follow from the remark that $y \sim y'$ if and only if $\alpha_0^- *_0 y \sim \alpha_0^- *_0 y'$.

Let us now suppose $n > 1$ and let

$$(x, y^-, y^+, \alpha_0^-, \alpha_0^+, \ldots, \alpha_{n-2}^-, \alpha_{n-2}^+, \alpha_{n-1}^-)$$

be a reversible bottom-incomplete n-cylinder. We have to prove that there is a weakly unique n-cell y such that

$$x *_0 \alpha_0^+ *_1 \cdots *_{n-1} \alpha_{n-1}^+ \quad \sim \quad \alpha_{n-1}^- *_{n-1} \cdots *_1 \alpha_0^- *_0 y, \tag{20.2}$$

where $y \colon y^- \to y^+$. Let $w = x *_0 \alpha_0^+ *_1 \cdots *_{n-1} \alpha_{n-1}^+$ denote the left member of (20.2). We prove by induction on k such that $1 \le k \le n$ that there is a weakly unique n-cell v^k in C such that

$$w \quad \sim \quad \alpha_{n-1}^- *_{n-1} \cdots *_1 \alpha_{n-k}^- *_{n-k} v^k. \tag{20.3}$$

For $k = 1$, this equation reads

$$w \quad \sim \quad \alpha_{n-1}^- *_{n-1} v^1. \tag{20.4}$$

It can be seen as an equation on 1-cells in the ω-category C' of $(n-2)$-cells of C. Therefore the division lemma (Lemma 20.1.10) applies and yields a weakly unique 1-cell v^1 of C', which is also an n-cell of C satisfying (20.4). Let k such that $1 \le k < n$ and suppose that (20.3) has a weakly unique solution v^k. By applying again the division lemma, there is a weakly unique n-cell v^{k+1} solution of

$$v^k \quad \sim \quad \alpha_{n-k-1}^- *_{n-k-1} v^{k+1}. \tag{20.5}$$

Therefore, substituting v^k by the second member of (20.5) in (20.3) one gets

$$w \quad \sim \quad \alpha_{n-1}^- *_{n-1} \cdots *_1 \alpha_{n-k-1}^- *_{n-k-1} v^{k+1},$$

which completes the induction. Now $v^n = y$ provides a weakly unique solution of (20.2), as required. □

20.3.6 Remark. A similar result holds for reversible top-incomplete cylinders.

20.3.7 Alternative description of the transport lemma. In practice, the transport lemma will be used in the following way. Let $\alpha \colon x_0 \curvearrowright x_1, \beta \colon y_0 \curvearrowright y_1$ be two parallel n-cylinders and let $u \colon x_0 \to x_1$ be an $(n+1)$-cell in an ω-category C. The data of α, β, and u is equivalent to the data of a bottom-incomplete $(n+1)$-cylinder (informally, α, β, and u corresponds respectively to the back face, the front face, and the top face of the bottom-incomplete cylinder). Moreover, this bottom-incomplete cylinder is reversible if and only if the cylinders α and β are reversible. In this case, an extension, as in the transport lemma, corresponds to an $(n+1)$-cell $v \colon y_0 \to y_1$ and a reversible $(n+1)$-cylinder $\Lambda \colon u \curvearrowright v$, whose source and target cylinders are α and β.

We will often write $\Lambda\colon u \curvearrowright v\colon \alpha \to \beta$ in such a situation. The existence and uniqueness properties of the second part of the transport lemma apply to v.

20.3.8 Proposition. *Let C be an ω-category.*

1. *The projections $\overline{\pi}_C, \underline{\pi}_C\colon \widetilde{\Gamma}(C) \to C$ are trivial fibrations.*
2. *The ω-functor $\iota_C\colon C \to \widetilde{\Gamma}(C)$ is an ω-equivalence.*

Proof. 1. Let us prove the result for $\overline{\pi}$, the other proof being analogous. The surjectivity on objects is clear. Let $n \geq 0$ and let $\alpha\colon x \curvearrowright y$ and $\beta\colon z \curvearrowright t$ be two parallel reversible n-cylinders. Suppose we have an n-cell between their images by the ω-functor $\overline{\pi}$, i.e., an n-cell $u\colon x \to z$. We are precisely in the situation of the previous paragraph and the transport lemma gives a reversible $(n+1)$-cylinder $\Lambda\colon \alpha \to \beta\colon u \curvearrowright v$, showing that $\overline{\pi}$ is indeed a trivial fibration.

2. Since the composite

$$C \xrightarrow{\iota_C} \widetilde{\Gamma}(C) \xrightarrow{\overline{\pi}_C} C$$

is the identity and hence an ω-equivalence, and that the trivial fibration $\overline{\pi}_C$ is an ω-equivalence (see Proposition 20.1.13), it follows from one of the known cases of the 2-out-of-3 property (Proposition 20.1.17) that ι_C is an ω-equivalence. □

20.3.9 Mapping path space factorization. Let $f\colon C \to D$ be an ω-functor. We define an ω-category $\widetilde{\Gamma}_f$ by the pullback square

$$\begin{array}{ccc} \widetilde{\Gamma}_f & \xrightarrow{\pi_2} & \widetilde{\Gamma}(D) \\ {\scriptstyle \pi_1}\big\downarrow & & \big\downarrow{\scriptstyle \overline{\pi}} \\ C & \xrightarrow[f]{} & D. \end{array}$$

Explicitly, an n-cell of $\widetilde{\Gamma}_f$ is a pair $(x, \gamma\colon f(x) \curvearrowright y)$, where x and y are n-cells of C and D, and γ is a reversible n-cylinder. Note that since trivial fibrations are stable under pullback, by the previous proposition, the projection $\pi_1\colon \widetilde{\Gamma}_f \to C$ is a trivial fibration.

The ω-functor $f\colon C \to D$ factors as

$$C \xrightarrow{i_f} \widetilde{\Gamma}_f \xrightarrow{p_f} D,$$

where i_f is defined by

$$i_f(x) = (x, \iota(f(x))\colon f(x) \curvearrowright f(x))$$

and p_f by

$$p_f(x, \alpha\colon f(x) \curvearrowright y) = y.$$

As the triangle

$$\begin{array}{ccc} C & \xrightarrow{i_f} & \widetilde{\Gamma}_f \\ & {\scriptstyle 1_C}\searrow & \downarrow{\scriptstyle \pi_1} \\ & & C \end{array}$$

is commutative and π_1 is a trivial fibration, by one of the known cases of the 2-out-of-3 property (Proposition 20.1.17), i_f is an ω-equivalence.

20.3.10 Proposition. *An ω-functor $f\colon C \to D$ is an ω-equivalence if and only if the $p_f\colon \widetilde{\Gamma}_f \to D$ is a trivial fibration.*

Proof. If p_f is a trivial fibration and hence an ω-equivalence, then, since i_f is always an ω-equivalence, so is $f = p_f i_f$.

Suppose conversely that f is an ω-equivalence.

1. Let y be an object of D. As f is an ω-equivalence, there exists an object x of C and a reversible 1-cell $u\colon f(x) \to y$. This 1-cell defines a reversible 0-cylinder $u\colon f(x) \curvearrowright y$ sent to y by p_f.
2. Let $n > 0$ and let $(x, \alpha\colon f(x) \curvearrowright y)$ and $(x', \alpha'\colon f(x') \curvearrowright y')$ be two parallel n-cells of $\widetilde{\Gamma}_f$. Suppose that we have a cell between their image by p_f, i.e., an n-cell $v\colon y \to y'$. By the transport lemma (Lemma 20.3.5), there exist an $(n+1)$-cell $v'\colon f(x) \to f(x')$ and an $(n+1)$-cylinder $\Lambda\colon v' \curvearrowright v\colon \alpha \to \alpha'$. Since f is an ω-equivalence, there exists $u\colon x \to x'$ such that $f(u)$ is ω-equivalent to v'. Using the transport lemma again, we get a reversible $(n+1)$-cylinder $\Lambda'\colon f(u) \curvearrowright v\colon \alpha \to \alpha'$. This means that $(u, \Lambda'\colon f(u) \curvearrowright v)$ is an $(n+1)$-cell of $\widetilde{\Gamma}_f$ sent to v by p_f, thereby proving the result. □

20.3.11 Theorem. *The class of ω-equivalences satisfies the 2-out-of-3 property.*

Proof. The only remaining case is the following one (see Propositions 20.1.15 and 20.1.17 for the other ones). Let $f\colon C \to D$ and $g\colon D \to E$ be two ω-functors. Suppose f and gf are ω-equivalences. Let us prove that g is an ω-equivalence.

1. Let z be an object of E. Since gf is an ω-equivalence, we get an object x of C such that $gf(x)$ and z are ω-equivalent and $y = f(x)$ shows that g is surjective up to ω-equivalence.

2. Let y_0 and y_1 be two parallel n-cells of D and let $w\colon g(y_0) \to g(y_1)$ be an $(n+1)$-cell. By the previous proposition, since f is an ω-equivalence, $p_f\colon \widetilde{\Gamma}_f \to D$ is a trivial fibration. We can thus lift the pair of parallel cells y_0, y_1 to a pair of parallel cells

$$(x_0, \alpha\colon f(x_0) \curvearrowright y_0) \qquad \text{and} \qquad (x_1, \beta\colon f(x_1) \curvearrowright y_1)$$

of $\widetilde{\Gamma}_f$. By applying g, we get a pair of parallel cells

$$(x_0, g(\alpha)\colon gf(x_0) \curvearrowright g(y_0)) \qquad \text{and} \qquad (x_1, g(\beta)\colon gf(x_1) \curvearrowright g(y_1))$$

of $\widetilde{\Gamma}_{gf}$. Since gf is an ω-equivalence, by the previous proposition, the projection $p_{gf}\colon \widetilde{\Gamma}_{gf} \to E$ is a trivial fibration. We can thus lift the cell $w\colon g(y_0) \to g(y_1)$ to a cell

$$(u\colon x_0 \to x_1, \Lambda\colon gf(u) \curvearrowright w\colon g(\alpha) \to g(\beta))$$

of $\widetilde{\Gamma}_{gf}$. Consider the bottom-incomplete cylinder defined by $\alpha\colon f(x_0) \curvearrowright y_0$, $\beta\colon f(x_1) \curvearrowright y_1$ and $f(u)\colon f(x_0) \to f(x_1)$. By the transport lemma (Lemma 20.3.5), it can be extended and we get a cell $v\colon y_0 \to y_1$ and a cylinder $\Delta\colon f(u) \curvearrowright v\colon \alpha \to \beta$. Applying the uniqueness part of the transport lemma to the cylinders

$$g(\Delta)\colon gf(u) \curvearrowright g(v)\colon g(\alpha) \to g(\beta)$$

and

$$\Lambda\colon gf(u) \curvearrowright w\colon g(\alpha) \to g(\beta),$$

we get that $g(v)$ is ω-equivalent to w, thereby proving the result. □

20.4 Coherent Reversible Cells and Fibrations

20.4.1 The free-standing reversible cell. We define an ω-category R_1 generated by a polygraph in the following way. The ω-category R_1 has two objects, 0 and 1. There are two generating 1-cells

$$r\colon 0 \to 1 \qquad \text{and} \qquad \overline{r}\colon 1 \to 0$$

and four generating 2-cells

$$r_-\colon 1_0 \to r *_0 \overline{r}, \quad \overline{r}_-\colon r *_0 \overline{r} \to 1_0, \quad r_+\colon \overline{r} *_0 r \to 1_1, \quad \overline{r}_+\colon 1_1 \to \overline{r} *_0 r.$$

More generally, for $j \geq 2$, there are 2^j generating j-cells

$$r_{l_1,\ldots,l_{j-1}} \qquad \text{and} \qquad \overline{r}_{l_1,\ldots,l_{j-1}},$$

where for $1 \leq k \leq j-1$, $l_k = \pm$. The source and target of the generators are given by

$$\begin{aligned}
r_{l_1,\dots,l_{j-2},-} &: & 1_{s_{j-2}(r_{l_1,\dots,l_{j-2}})} &\to r_{l_1,\dots,l_{j-2}} *_{j-2} \overline{r}_{l_1,\dots,l_{j-2}},\\
\overline{r}_{l_1,\dots,l_{j-2},-} &: & r_{l_1,\dots,l_{j-2}} *_{j-2} \overline{r}_{l_1,\dots,l_{j-2}} &\to 1_{s_{j-2}(r_{l_1,\dots,l_{j-2}})},\\
r_{l_1,\dots,l_{j-2},+} &: & \overline{r}_{l_1,\dots,l_{j-2}} *_{j-2} r_{l_1,\dots,l_{j-2}} &\to 1_{t_{j-2}(r_{l_1,\dots,l_{j-2}})},\\
\overline{r}_{l_1,\dots,l_{j-2},+} &: & 1_{t_{j-2}(r_{l_1,\dots,l_{j-2}})} &\to \overline{r}_{l_1,\dots,l_{j-2}} *_{j-2} r_{l_1,\dots,l_{j-2}}.
\end{aligned}$$

This definition was made so that R_1 is, in some sense, freely generated by a reversible 1-cell r. In particular, R_1 comes with a *canonical inclusion* $\mathbb{O}_1 \hookrightarrow R_1$ corresponding to the 1-cell r and, if C is an ω-category, a 1-cell u of C is reversible if and only if there exists a dotted arrow making the triangle

$$\begin{array}{ccc}
\mathbb{O}_1 & \xrightarrow{u} & C \\
\big\downarrow & \nearrow & \\
R_1 & &
\end{array}$$

commute. In particular, two objects x and y of C are ω-equivalent if and only if there exists a dotted arrow making the triangle

$$\begin{array}{ccc}
\partial\mathbb{O}_1 & \xrightarrow{(x,y)} & C \\
\big\downarrow & \nearrow & \\
R_1 & &
\end{array},$$

where the vertical inclusion is the composite $\partial\mathbb{O}_1 \hookrightarrow \mathbb{O}_1 \hookrightarrow R_1$, commute.

Similarly, for every $n \geq 1$, one can define an ω-category R_n, endowed with a *canonical inclusion* $\mathbb{O}_n \hookrightarrow R_n$ corresponding to an n-cell r, modeled on the definition of a reversible n-cell. This ω-category has, by definition, the following properties: an n-cell u of an ω-category C is reversible if and only if there exists a dotted arrow making the triangle

$$\begin{array}{ccc}
\mathbb{O}_n & \xrightarrow{u} & C \\
\big\downarrow & \nearrow & \\
R_n & &
\end{array}$$

commute; in particular, two parallel $(n-1)$-cells x and y of C are ω-equivalent if and only if there exists a dotted arrow making the triangle

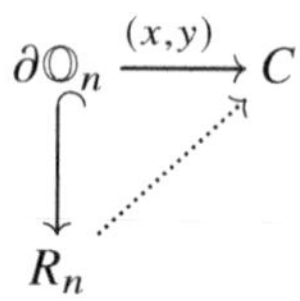

commute.

20.4.2 Incoherence. If x and y are two objects of a 2-category, it is well known that x and y are equivalent if and only if there exists an adjoint equivalence between x and y. In other words, if there exist 1-cells $u\colon x \to y$, $v\colon y \to x$ and 2-cells $\eta\colon 1_x \to uv$ and $\varepsilon\colon vu \to 1_y$ that are isomorphisms, then we can choose these cells such that the triangular identities

$$(\eta * u)(u * \varepsilon) = 1_u \quad \text{and} \quad (v * \eta)(\varepsilon * v) = 1_v$$

hold. A 4-tuple $(u, v, \eta, \varepsilon)$ as above is a witness of the fact that x and y are ω-equivalent. We can think of this witness as *incoherent* if the triangular identities do not hold and as *coherent* if they hold.

Similarly, if x and y are two objects of an ω-category C, the data of an ω-functor $R_1 \to C$ making the triangle

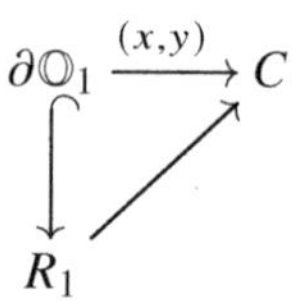

commute is witness of the fact that x and y are ω-equivalent. This witness is *incoherent* as, for instance, it does not satisfy the triangular identities up to ω-equivalence. Technically, this boils down to the fact that R_1 is not contractible (see §22.4.1).

20.4.3 Coherence. Let $n \geq 0$. Consider the ω-functor

$$\partial\mathbb{O}_{n+1} \to \mathbb{O}_n$$

corresponding to the collapsing of the two non-trivial n-cells of $\partial\mathbb{O}_{n+1}$. By Proposition 19.2.2, this ω-functor factors (non-uniquely) as a cofibration followed by a trivial fibration. Let us fix such a factorization

$$\partial\mathbb{O}_{n+1} \xrightarrow{\ j\ } J_{n+1} \xrightarrow{\ p\ } \mathbb{O}_n.$$

Let x and y be two parallel n-cells of an ω-category C. The fact that p is a trivial fibration makes it reasonable to think of a dotted arrow making the triangle

$$\begin{array}{ccc} \partial\mathbb{O}_{n+1} & \xrightarrow{(x,y)} & C \\ {\scriptstyle j}\downarrow & \nearrow & \\ J_{n+1} & & \end{array}$$

commute as a *coherent* witness of the fact that x and y are ω-equivalent.

20.4.4 Remark. As already mentioned, the ω-category J_{n+1} introduced in the previous paragraph is not uniquely defined. Some choices are wiser than others. For instance, we can choose J_{n+1} so that its underlying $(n+1)$-category is obtained from $\mathbb{O}_{n+1}$ by adding an $(n+1)$-cell $v\colon y \to x$ in the other direction than the principal cell $u\colon x \to y$ of $\mathbb{O}_{n+1}$. We can even fix its underlying $(n+2)$-category by saying that it is generated by two $(n+2)$-cells $\eta\colon 1_x \to u *_n v$ and $\varepsilon\colon v *_n u \to 1_y$. We do not assume that these additional properties hold in the remaining of the text. (We will use the existence of these better choices in Section 21.3.)

The following proposition shows that, as in the case of 2-categories, there exist coherent witnesses if and only if there exist incoherent witnesses.

20.4.5 Proposition. *Let $n \geq 0$ and let x and y be two parallel n-cells of an ω-category C. The following assertions are equivalent:*

1. *The cells x and y are ω-equivalent.*
2. *There exists a dotted arrow making the triangle*

$$\begin{array}{ccc} \partial\mathbb{O}_{n+1} & \xrightarrow{(x,y)} & C \\ \downarrow & \nearrow & \\ R_{n+1} & & \end{array}$$

   commute.
3. *There exists a dotted arrow making the triangle*

$$\begin{array}{ccc} \partial\mathbb{O}_{n+1} & \xrightarrow{(x,y)} & C \\ {\scriptstyle j}\downarrow & \nearrow & \\ J_{n+1} & & \end{array}$$

   commute.

4. *There exists a factorization*

$$\partial\mathbb{O}_{n+1} \xrightarrow{k} K \xrightarrow{q} \mathbb{O}_n$$

of the ω-functor $\partial\mathbb{O}_{n+1} \to \mathbb{O}_n$ of §20.4.3 with q a trivial fibration, and a dotted arrow making the triangle

$$\begin{array}{ccc} \partial\mathbb{O}_{n+1} & \xrightarrow{(x,y)} & C \\ {\scriptstyle k}\downarrow & \nearrow & \\ K & & \end{array}$$

commute.

Proof. The equivalence between 1 and 2 is true by the definition of R_{n+1}.

By definition, 3 implies 4. The converse follows by considering the solid commutative square

$$\begin{array}{ccc} \partial\mathbb{O}_{n+1} & \xrightarrow{k} & K \\ {\scriptstyle j}\downarrow & \nearrow & \downarrow{\scriptstyle q} \\ J_{n+1} & \xrightarrow[p]{} & \mathbb{O}_n \end{array}$$

that admits a lift as j is a cofibration and q is a trivial fibration.

Let us prove that 4 implies 1. The ω-functor $k\colon \partial\mathbb{O}_{n+1} \to K$ defines a pair of parallel n-cells in K. These two cells are sent to the principal cell of $\partial\mathbb{O}_n$ by $q\colon K \to \mathbb{O}_n$. Since this ω-functor q is a trivial fibration, by Proposition 20.1.14, there exists a reversible $(n+1)$-cell r in K between these two parallel n-cells. By commutativity of the triangle of the hypothesis, we have $f(r)\colon x \to y$, showing that x and y are ω-equivalent.

To conclude the proof, let us prove that 1 implies 4. Consider the ω-functor $x\colon \mathbb{O}_n \to C$ corresponding to the n-cell x and the associated ω-category $\widetilde{\Gamma}_x$ (see §20.3.9). We get a factorization

$$\partial\mathbb{O}_{n+1} \xrightarrow{(k_1,k_2)} \widetilde{\Gamma}_x \xrightarrow{\pi_1} \mathbb{O}_n$$

of the ω-functor $\partial\mathbb{O}_{n+1} \to \mathbb{O}_n$ of the assertion by choosing two parallel n-cells k_1 and k_2 of $\widetilde{\Gamma}_x$ in the following way. Denote by p the principal cell of $\mathbb{O}_n$. We set $k_1 = (p, \iota(x)\colon x \curvearrowright x)$. As for the cell k_2, we choose a reversible $(n+1)$-cell $u\colon x \to y$. Using this u, one easily defines a reversible cylinder $\alpha_u\colon x \curvearrowright y$ and we set $k_2 = (p, \alpha_u\colon x \curvearrowright y)$. This ends the definition of the desired factorization. Moreover, by definition, the triangle

$$\begin{tikzcd} \partial\mathbb{O}_{n+1} \arrow[r, "{(x,y)}"] \arrow[d, "{(k_1,k_2)}"'] & C \\ \widetilde{\Gamma}_x \arrow[ur, "p_x"'] & \end{tikzcd}$$

commutes, thereby proving the result. □

20.4.6 Proposition. *Let $f\colon C \to D$ be an ω-functor. The following assertions are equivalent:*

1. *The ω-functor f is an ω-equivalence.*
2. *For every $n \geq 0$, every commutative square*

$$\begin{tikzcd} \partial\mathbb{O}_n \arrow[r] \arrow[d, "\mathrm{i}_n"'] & C \arrow[d, "f"] \\ \mathbb{O}_n \arrow[r] & D \end{tikzcd}$$

factors as

$$\begin{tikzcd} \partial\mathbb{O}_n \arrow[r, "\mathrm{i}_n"] \arrow[d, "\mathrm{i}_n"'] & \mathbb{O}_n \arrow[r] \arrow[d] & C \arrow[d, "f"] \\ \mathbb{O}_n \arrow[r] & R_{n+1} \arrow[r] & D, \end{tikzcd}$$

where the two ω-functors $\mathbb{O}_n \to R_n$ are the two components of the inclusion $\partial\mathbb{O}_{n+1} \hookrightarrow \mathbb{O}_{n+1} \hookrightarrow R_{n+1}$.

3. *For every $n \geq 0$, every commutative square*

$$\begin{tikzcd} \partial\mathbb{O}_n \arrow[r] \arrow[d, "\mathrm{i}_n"'] & C \arrow[d, "f"] \\ \mathbb{O}_n \arrow[r] & D \end{tikzcd}$$

factors as

$$\begin{tikzcd} \partial\mathbb{O}_n \arrow[r, "\mathrm{i}_n"] \arrow[d, "\mathrm{i}_n"'] & \mathbb{O}_n \arrow[r] \arrow[d] & C \arrow[d, "f"] \\ \mathbb{O}_n \arrow[r] & J_{n+1} \arrow[r] & D \end{tikzcd}$$

where the two ω-functors $\mathbb{O}_n \to J_n$ are the two components of the ω-functor $j\colon \partial\mathbb{O}_{n+1} \to J_{n+1}$.

Proof. Using the defining property of R_{n+1}, the second assertion translates directly to the definition of an ω-equivalence. Indeed, if $n = 0$, the data of a square as in the assertion boils down to the data of an object y of D and the data of a factorization boils down to the data of an object x in C and of an incoherent witness of the fact that $f(x)$ and y are ω-equivalent. Similarly, if $n > 0$, the data of the square boils down to the data of two $(n-1)$-cells x and x' of C and an $(n+1)$-cell $v\colon f(x) \to f(x')$ in D, and the data of the factorization boils down to the data of an $(n+1)$-cell $u\colon x \to y$ of C and an incoherent witness that $f(u)$ and v are ω-equivalent. The third assertion translates similarly replacing incoherent witnesses by coherent witnesses and the result follows from the previous proposition. □

20.4.7 We will denote by

$$\mathrm{j}_n\colon \mathbb{O}_n \to J_{n+1}$$

the composite

$$\mathbb{O}_n \xhookrightarrow{\iota_1} \partial\mathbb{O}_{n+1} \xrightarrow{j} J_{n+1},$$

where $\iota_1\colon \mathbb{O}_n \hookrightarrow \partial\mathbb{O}_{n+1}$ denotes the ω-functor corresponding to "source n-cell" of $\partial\mathbb{O}_{n+1}$. Note that j_n is defined as a composite of two cofibrations and is hence a cofibration. Moreover, it is an ω-equivalence. Indeed, we have a commutative triangle

$$\begin{array}{ccc} \mathbb{O}_n & \xrightarrow{\mathrm{j}_n} & J_{n+1} \\ & {\scriptstyle 1_{\mathbb{O}_n}}\searrow & \downarrow{\scriptstyle p} \\ & & \mathbb{O}_n, \end{array}$$

where p is trivial fibration and hence an ω-equivalence, and this follows from the 2-out-of-3 property (and more precisely, the easy case of Proposition 20.1.17). We set

$$\mathcal{J} = \{\mathrm{j}_n \mid n \geq 0\}.$$

An ω-functor $f\colon C \to D$ is a *fibration* if it has the right lifting property with respect to $\mathcal{J}$. Concretely, this means that f is a fibration if, for every $n \geq 0$, every n-cell u of C, every n-cell v of D, every coherent witness that $f(u)$ and v are ω-equivalent can be lifted to a coherent witness that u and some n-cell u' such that $f(u') = v$ are ω-equivalent.

20.4.8 Theorem. *An ω-functor f is a trivial fibration if and only if it is both an ω-equivalence and a fibration.*

Proof. As elements of $\mathcal{J}$ are cofibrations, i.e., elements of $lr(\mathcal{I})$, we have

$$r(\mathcal{I}) = rlr(\mathcal{I}) \subseteq r(\mathcal{J}),$$

meaning that trivial fibrations are fibrations. As we already know that trivial fibrations are ω-equivalence (Proposition 20.1.13), this establishes one implication.

Conversely, let $f: C \to D$ be an ω-functor being both an ω-equivalence and a fibration. Fix $n \geq 0$ and consider a commutative square

$$\begin{array}{ccc} \partial\mathbb{O}_n & \longrightarrow & C \\ {\scriptstyle \mathrm{i}_n}\downarrow & & \downarrow{\scriptstyle f} \\ \mathbb{O}_n & \longrightarrow & D. \end{array}$$

As f is an ω-equivalence, by Proposition 20.4.6 this square factors as

$$\begin{array}{ccccc} \partial\mathbb{O}_n & \xrightarrow{\mathrm{i}_n} & \mathbb{O}_n & \longrightarrow & C \\ {\scriptstyle \mathrm{i}_n}\downarrow & & {\scriptstyle \mathrm{j}_n}\downarrow & & \downarrow{\scriptstyle f} \\ \mathbb{O}_n & \longrightarrow & J_{n+1} & \longrightarrow & D. \end{array}$$

But as f is a fibration, the right square of the factorization admits a lift, showing that the initial square admits a lift and hence that f is a trivial fibration. □

20.5 Immersions

20.5.1 An ω-functor $i: C \to D$ is an *immersion* if it admits a retraction $r: D \to C$, so that we have $ri = 1_C$, and a reversible oplax transformation $\alpha: ir \Rightarrow 1_D$ such that $\alpha * i = 1_i$ (see §20.2.15). Diagrammatically, we have the following commutative diagrams:

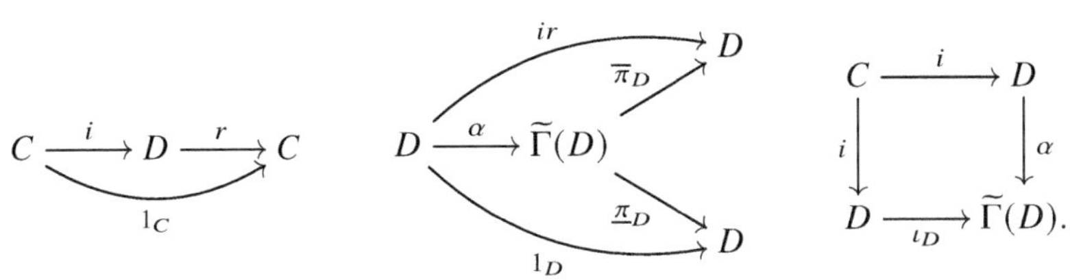

The last square can also be written

$$\begin{array}{ccc} C & \xrightarrow{\;\; i \;\;} & D \\ {\scriptstyle \iota_C}\Big\downarrow & & \Big\downarrow{\scriptstyle \alpha} \\ \widetilde{\Gamma}(C) & \xrightarrow[\widetilde{\Gamma}(i)]{} & \widetilde{\Gamma}(D). \end{array}$$

This notion of an immersion is an ω-categorical version of the notion of a "strong deformation retracts" in topology.

20.5.2 Proposition. *An ω-functor $f\colon C \to D$ is an immersion if and only if the commutative square*

$$\begin{array}{ccc} C & \xrightarrow{\;\; i_f \;\;} & \widetilde{\Gamma}_f \\ {\scriptstyle f}\Big\downarrow & & \Big\downarrow{\scriptstyle p_f} \\ D & \xrightarrow[1_D]{} & D \end{array}$$

admits a lift.

Proof. By definition, we have a pullback square

$$\begin{array}{ccc} \widetilde{\Gamma}_f & \xrightarrow{\;\; \pi_2 \;\;} & \widetilde{\Gamma}(D) \\ {\scriptstyle \pi_1}\Big\downarrow & & \Big\downarrow{\scriptstyle \overline{\pi}} \\ C & \xrightarrow[f]{} & D. \end{array}$$

This means that an ω-functor $h\colon D \to \widetilde{\Gamma}_f$ corresponds to the data of an ω-functor $g\colon D \to C$ and an ω-functor $\alpha\colon D \to \widetilde{\Gamma}(D)$ such that $\overline{\pi}\alpha = fg$, i.e., a reversible oplax transformation whose source is fg. The commutativity of the right-lower triangle of

$$\begin{array}{ccc} C & \xrightarrow{\;\; i_f \;\;} & \widetilde{\Gamma}_f \\ {\scriptstyle f}\Big\downarrow & \overset{h}{\nearrow} & \Big\downarrow{\scriptstyle p_f} \\ D & \xrightarrow[1_D]{} & D, \end{array}$$

using the equality $p_f = \underline{\pi}\pi_2$, means that $\underline{\pi}\alpha = 1_D$, i.e., that the target of α is 1_D. Finally, the commutativity of the left-upper triangle, using the fact that $i_f = (1_C, \iota_D f)\colon C \to C \times_D \widetilde{\Gamma}(D)$, means that $(g, \alpha)f = (1_C, \iota_D f)$, i.e., that $gf = 1_C$ and that $\alpha * f = 1_f$, thereby proving the result. □

20.5.3 Corollary. *If an ω-functor is both an ω-equivalence and a cofibration, then it is an immersion.*

Proof. If $f\colon C \to D$ is an ω-equivalence, then, by Proposition 20.3.10, $p_f\colon \widetilde{\Gamma}_f \to D$ is a trivial fibration. In particular, if f is additionally a cofibration, then the square of the previous proposition admits a lift, showing that f is indeed an immersion. □

20.5.4 Proposition. *Immersions are ω-equivalences.*

Proof. Let $i\colon C \to D$ be an immersion, and let r and α be as in the definition of §20.5.1.

1. Let y be an object of D. The reversible cell $\alpha_y\colon ir(y) \to y$ shows that, for $x = r(y)$, $i(x)$ and y are ω-equivalent.
2. Let x and x' be two parallel n-cells of D and let $v\colon i(x) \to i(y)$ be an $(n+1)$-cell of D. Consider the cell $r(v)$. As $ri = 1_C$, we have $r(v)\colon x \to y$. Consider now the reversible cell α_v. A priori, its source and target are given by

$$\alpha_v\colon ir(v) *_0 \alpha_{t_0(v)} *_1 \cdots *_n \alpha_{t_n(v)} \to \alpha_{s_n(v)} *_n \cdots *_1 \alpha_{s_0(v)} *_0 v.$$

As the cells $s_0(v), t_0(v), \ldots, s_n(v), t_n(v)$ are in the image of i, it follows from the relation $\alpha * i = 1_i$ that the cells $\alpha_{s_0(v)}, \alpha_{t_0(v)}, \ldots, \alpha_{s_n(v)}, \alpha_{t_n(v)}$ are units, so that we have $\alpha_v\colon ir(v) \to v$. This shows that $i(u)$, where $u = r(v)$, and v are ω-equivalent. □

20.5.5 Proposition. *The class of immersions is closed under pushouts.*

Proof. Let $i\colon C \to D$ be an immersion, and let r and α be as in the definition of §20.5.1. Consider a pushout square

$$\begin{array}{ccc} C & \xrightarrow{f} & C' \\ {\scriptstyle i}\downarrow & & \downarrow{\scriptstyle i'} \\ D & \xrightarrow[g]{} & D'. \end{array}$$

Using the universal property of this square, we get a retraction r' of i':

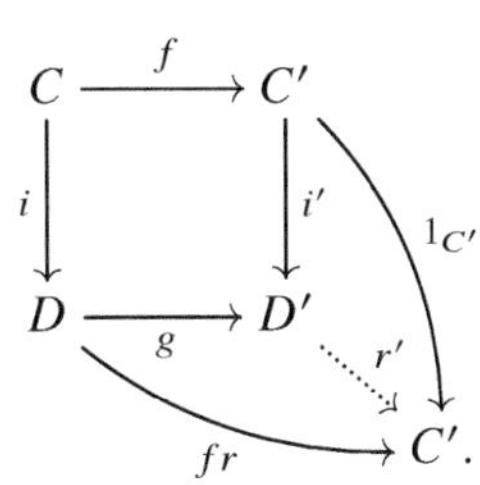

Similarly, we get a reversible oplax transformation α':

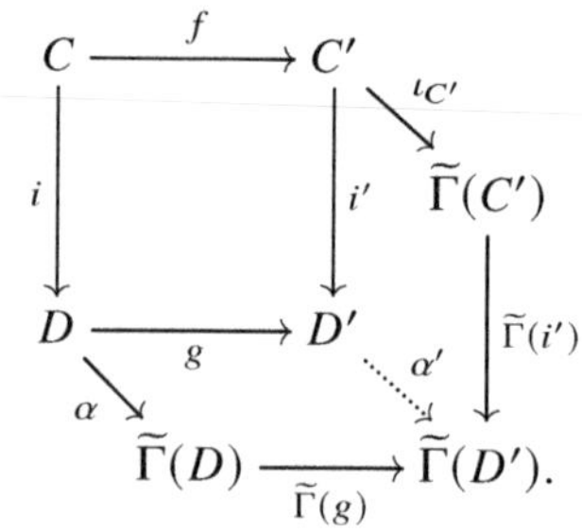

Indeed, the outer diagram commutes as

$$\widetilde{\Gamma}(i')\iota'_C f = \widetilde{\Gamma}(i')\widetilde{\Gamma}(f)\iota_C = \widetilde{\Gamma}(g)\widetilde{\Gamma}(i)\iota_C = \widetilde{\Gamma}(g)\alpha i.$$

Using the uniqueness part of the same universal property, one checks that $\overline{\pi}\alpha = i'r'$ and $\underline{\pi}\alpha = 1_{D'}$, i.e., that $\alpha\colon i'r' \Rightarrow 1_{D'}$. The fact that $\alpha' * i' = 1_{i'}$ is expressed in one of the commutative squares defining α'. This proves that $i'\colon C' \to D'$ is indeed an immersion. □

20.5.6 Theorem. *The class of ω-functors that are both ω-equivalences and cofibrations is closed under pushouts.*

Proof. The class of cofibrations being closed under pushouts (see §19.1.2), all we have to show is that the pushout of an ω-functor f being both an ω-equivalence and a cofibration is an ω-equivalence. But by Corollary 20.5.3, such an f is an immersion. Hence, by the previous proposition, a pushout of f is still an immersion, and hence an ω-equivalence by Proposition 20.5.4. □

20.5.7 Corollary. *Every element of $lr(\mathcal{J})$ is both an ω-equivalence and a cofibration.*

Proof. Denote by $\mathcal{W}$ the class of ω-equivalences and by *Cof* the class of cofibrations. By Propositions 20.1.18 and 20.1.19, the class $\mathcal{W}$ is closed under retracts and countable compositions. The same properties hold for the class $\mathit{Cof} = lr(\mathcal{I})$ (see §19.1.2). Using the previous proposition, we get that the class $\mathcal{W} \cap \mathit{Cof}$ is closed under retracts, pushouts, and countable compositions. It follows, using the small object argument (Proposition 19.1.9), that to show the inclusion $lr(\mathcal{J}) \subseteq \mathcal{W} \cap \mathit{Cof}$, it suffices to show the inclusion $\mathcal{J} \subseteq \mathcal{W} \cap \mathit{Cof}$. But we already noted in §20.4.7 that for every n, the ω-functor $\mathrm{j}_n\colon \mathbb{O}_n \to J_{n+1}$ is both an ω-equivalence and a cofibration. □

20.5.8 Remark. We will see in the proof of Theorem 21.1.2 that it follows formally from what we proved so far that, conversely, an ω-functor being both an ω-equivalence and a cofibration is in $lr(\mathcal{J})$.

21

The Folk Model Structure

This chapter is about proving the existence of the so-called folk model category structure on $\mathbf{Cat}_\omega$, following Lafont, Métayer, and Worytkiewicz [237]. This model category structure is a generalization of a model category structure on **Cat** whose weak equivalences are the equivalences of categories, a folklore result, whence the name. The analogous result for 2-categories was proved by Lack [229, 231].

The *folk model category structure* is a model category structure on $\mathbf{Cat}_\omega$ whose weak equivalences are the ω-equivalences and whose cofibrant resolutions are the polygraphic resolutions. It is the natural homotopical framework in which the notion of polygraphic resolutions lives. As a convincing evidence of this, we will see in the next chapter that Métayer's polygraphic homology can be expressed as a derived functor with respect to the folk model category structure.

The chapter is organized as follows. The first section is devoted to the proof of the existence of the folk model category structure. The hard work was done in the previous chapter, and we are basically just assembling various results. Our proof differs in one point from the original one: we avoid the use of Smith's theorem. In the second section, we prove, still according to [237], that the ω-category $\widetilde{\Gamma}(C)$ of reversible cylinders in an ω-category C forms a path object for C in the sense of the folk model category structure. We deduce from this fact that polygraphic resolutions are unique in a stronger sense than the one proved before. In the last section, we transfer the folk model structure on $\mathbf{Cat}_\omega$ to the category of (n, p)-categories. We give explicit descriptions of the resulting structures for various special cases: n-categories with particular small values of n, ω-groupoids, and $(\omega, 1)$-categories.

21.1 The Folk Model Structure on $\mathbf{Cat}_\omega$

The purpose of this section is to prove the existence of the so-called "folk model structure" on $\mathbf{Cat}_\omega$. We start by recalling the definition of a model category, see Appendix H for a more detailed presentation.

21.1.1 A *model category* is a category $\mathcal{M}$ endowed with three classes of maps: the *weak equivalences*, the *cofibrations*, and the *fibrations*; these data are required to satisfy the following axioms:

1. the category $\mathcal{M}$ is finitely complete and finitely cocomplete,
2. the class of weak equivalences satisfies the 2-out-of-3 property,
3. the class of weak equivalences, cofibrations, and fibrations are closed under retracts,
4. cofibrations have the left lifting property with respect to *trivial fibrations* (that is, maps that are both a fibration and a weak equivalence); trivial cofibrations (that is, maps that are both a cofibration and a weak equivalence) have the left lifting property with respect to fibrations, and
5. every map of $\mathcal{M}$ factors as a cofibration followed by a trivial fibration, and as a trivial cofibration followed by a fibration.

21.1.2 Theorem. *The classes of ω-equivalences (see §20.1.11), cofibrations (see §19.2.1), and fibrations (see §20.4.7) define a model structure on* $\mathbf{Cat}_\omega$ *known as the "folk model structure".*

Proof. Let us denote by $\mathcal{W}$ the class of weak equivalences, by $\mathcal{C}of$ the class of cofibrations and by $\mathcal{F}ib$ the class of fibrations. By definition, we have

$$\mathcal{C}of = lr(\mathcal{I}) \quad \text{and} \quad \mathcal{F}ib = r(\mathcal{J})$$

(see §§19.2.1 and 20.4.7). We showed that

1. $\mathcal{W}$ satisfies the 2-out-of-3 property (Theorem 20.3.11) and is closed under retracts (Proposition 20.1.18),
2. $lr(\mathcal{J}) \subseteq \mathcal{W} \cap \mathcal{C}of$ (Corollary 20.5.7),
3. $\mathcal{W} \cap \mathcal{F}ib = r(\mathcal{I})$ (Theorem 20.4.8).

We will now see that these properties formally imply the theorem.

Let us first prove that the inclusion in 2 is actually an equality. Let f be an ω-functor in $\mathcal{W} \cap \mathcal{C}of$. Applying the small object argument (Proposition 19.1.9) to f and the set $\mathcal{J}$ yields a factorization $f = pi$, where p is in $\mathcal{F}ib = r(\mathcal{J})$ and i in $lr(\mathcal{J})$. By 2, i is in $\mathcal{W}$, and so is p by 1. It follows from 3 that p is in $r(\mathcal{I})$. As $f = pi$ is in $\mathcal{C}of = lr(\mathcal{I})$, it has the left lifting property with respect to p

and, by the retract lemma (Lemma 19.1.10), f is a retract of i. As i is in $lr(\mathcal{J})$, so is f, thereby proving the desired equality.

Let us now check that we indeed have a model structure. The category $\mathbf{Cat}_\omega$ is complete and cocomplete. We have already proved that the class of ω-equivalences satisfies the 2-out-of-3 property and is closed under retracts. The fact that the classes of cofibrations and fibrations can defined by lifting conditions implies that they are closed under retracts as well. The equalities

$$\mathcal{C}of = lr(\mathcal{I}), \quad \mathcal{F}ib = r(\mathcal{J}), \quad \mathcal{W} \cap \mathcal{C}of = lr(\mathcal{J}), \quad \text{and} \quad \mathcal{W} \cap \mathcal{F}ib = r(\mathcal{I})$$

show that the required lifting properties are fulfilled and, by applying the small object argument, that the required factorization properties are fulfilled as well. □

21.1.3 Remark. The folk model structure is what is called a "combinatorial model structure": a model category C is said to be *combinatorial* if, first, the category C is locally presentable (see Appendix G) and, second, there exist *sets* (by opposition to classes) of morphisms $\mathcal{I}$ and $\mathcal{J}$ such that the class of cofibration is $lr(\mathcal{I})$ and the class of trivial cofibrations is $lr(\mathcal{J})$. The sets $\mathcal{I}$ and $\mathcal{J}$ (which are not unique) are then said to *generate* the model category C.

21.1.4 Let C be a model category. An object X of C is said to be *cofibrant* if the unique morphism from the initial object of C to X is a cofibration; similarly, X is said to be *fibrant* if the unique morphism from X to the terminal object of C is a fibration.

21.1.5 Proposition. *Every ω-category is fibrant in the folk model structure.*

Proof. Let C be an ω-category. Consider a diagram

$$\begin{array}{ccc} D & \xrightarrow{f} & C \\ {\scriptstyle i}\downarrow & & \\ E & & \end{array},$$

where i is a trivial cofibration. By Corollary 20.5.3, i admits a retraction r and the triangle

$$\begin{array}{ccc} D & \xrightarrow{f} & C \\ {\scriptstyle i}\downarrow & \nearrow_{fr} & \\ E & & \end{array}$$

is commutative, thereby proving that C is fibrant. □

21.1.6 Theorem. *The cofibrant objects of the folk model structure are exactly the ω-categories generated by polygraphs.*

Proof. Let C be a cofibrant ω-category. By Proposition 19.2.2, the ω-functor $\emptyset \to C$ is a retract of a relative polygraph $D \to E$. In particular, $\emptyset$ is a retract of D and C is a retract of E. This means that $D \simeq \emptyset$ and hence that D is generated by a polygraph. This shows that C is a retract of an ω-category generated by a polygraph.

To conclude the proof, it suffices to show that a retract of an ω-category generated by a polygraph is also generated by a polygraph. This immediately follows from [279, Theorem 7.1], stating that the full subcategory of $\mathbf{Cat}_\omega$ consisting of those ω-categories that are generated by polygraphs is Cauchy-complete. We briefly describe the general idea of the argument, and refer to [279] for a complete proof. Thus, let P be a polygraph and $h\colon P^* \to P^*$ be an idempotent morphism in $\mathbf{Cat}_\omega$. We have to build a polygraph Q, together with morphisms $r\colon P^* \to Q^*$ and $s\colon Q^* \to P^*$ such that $rs = 1_{Q^*}$ and $sr = h$. The polygraph Q and the maps r, s are defined by simultaneous induction on the dimension. In dimension 0, $Q_0 = \{h(x) \mid x \in P_0^* = P_0\}$, the map s is given by the obvious inclusion of Q_0 into P_0 and $r(x) = h(x)$ for all $x \in P_0$. Let now $n > 0$ and suppose we have defined Q, r, and s up to dimension $n-1$, satisfying the above equations. We must now define the set Q_n of n-generators in Q, together with source and target maps $s_{n-1}, t_{n-1}\colon Q_n \to Q_{n-1}^*$. The crucial observation is that h induces a partition of P_n in three subsets:

$$P_n = P_n^0 \sqcup P_n^1 \sqcup P_n^2$$

where P_n^0 is the set of those n-generators whose image by h is an identity, P_n^1 is the set of those generators $a \in P_n$ such that $h(a^*) = c[a^*]$ where $c[\mathbf{x}]$ is a thin n-context (see Section 16.5), and P_n^2 is the set of remaining n-generators. We may now define

$$Q_n = \{h(a^*) \mid a \in P_n^1\}$$

which gives an obvious inclusion $i\colon Q_n \to P_n^*$. By induction hypothesis, the morphism r is defined up to dimension $n-1$, so that we can define

$$s_{n-1}, t_{n-1}\colon Q_n \to Q_{n-1}^*$$

by requiring the commutation of the following square

$$\begin{array}{ccc} Q_n & \xrightarrow{s_{n-1},t_{n-1}} & Q_{n-1}^* \\ {\scriptstyle i}\downarrow & & \uparrow{\scriptstyle r} \\ P_n^* & \xrightarrow{s_{n-1},t_{n-1}} & P_{n-1}^* \end{array}$$

for source and target maps. Thus the ω-category Q^* is now defined up to dimension n and one easily checks that $i\colon Q_n \to P_n^*$ induces an extension of s up to dimension n from Q^* to P^* satisfying $hs = s$.

Finally, we must extend r up to dimension n. The main difficulty is to find where the generators belonging to S_n^2 should be sent. Therefore, we first consider an auxiliary subpolygraph R of P, identical to P up to dimension $n-1$, and whose set of n-generators is just $R_n = P_n^0 \sqcup P_n^1$. The inclusion $j\colon R \to P$ now induces a morphism j^* of n-categories from $R^* \to P^*$, and one shows the existence of n-morphisms $h'\colon R^* \to R^*$ and $k\colon P^* \to R^*$ such that the following square commutes:

$$\begin{array}{ccc} R^* & \xrightarrow{h'} & R^* \\ {\scriptstyle j^*}\big\downarrow & \nearrow^{k} & \big\downarrow{\scriptstyle j^*} \\ P^* & \xrightarrow[h]{} & P^* \end{array}$$

The purpose of the previous step was precisely to get rid of S_n^2. Now one builds $r'\colon R^* \to Q^*$ and $s'\colon Q^* \to R^*$, extending r, s up to dimension n is such a way that $s'r' = h'$ and $r's' = 1_{Q^*}$. Note that the proof of this last equation relies on the fine properties of contexts, as presented in Section 16.5. We obtain the desired retraction r up to dimension n by taking $r = r'k$. □

21.1.7 Remark. The previous theorem easily implies that the folk model structure is not "cartesian" (meaning that the cartesian product functor is not a Quillen bifunctor in the sense of [187, Definition 4.2.1]). Indeed, in a cartesian model structure, the product of two cofibrant objects is cofibrant. But it is not true that the product of two ω-categories generated by a polygraph is generated by a polygraph. For instance, the product $\mathbb{O}_1 \times \mathbb{O}_1$ is not generated by a polygraph.

21.2 The Path Objects of Cylinders

The goal of this section is to prove that, for every ω-category C, the ω-category $\widetilde{\Gamma}C$ (see §20.3.2), endowed with the maps

$$C \xrightarrow{\iota_C} \widetilde{\Gamma}(C) \xrightarrow{(\overline{\pi},\underline{\pi})} C \times C,$$

is a path object for C in the folk model structure in the following sense:

21.2.1 Path objects. Let $\mathcal{M}$ be a model category and let X be an object of $\mathcal{M}$. A *path object* for X in $\mathcal{M}$ is an object P of $\mathcal{M}$, endowed with a factorization

$$X \xrightarrow{r} P \xrightarrow{p} X \times X$$

of the diagonal of X, where r a weak equivalence of $\mathcal{M}$ and p a fibration of $\mathcal{M}$.

As we already know that $\iota_C \colon C \to \widetilde{\Gamma}(C)$ is an ω-equivalence (Proposition 20.3.8), all we have to prove is that $(\overline{\pi}, \underline{\pi}) \colon \widetilde{\Gamma}(C) \to C \times C$ is a fibration. To do so, we will use the following variation on the notion of an immersion:

21.2.2 Strong immersions. An ω-functor $i \colon C \to D$ is a *strong immersion* if it admits a retraction $r \colon D \to C$ and reversible oplax transformations $\alpha \colon ir \Rightarrow 1_D$ and $\alpha' \colon 1_D \Rightarrow ir$ such that $\alpha * i = 1_i$ and $\alpha' * i = 1_i$.

21.2.3 Proposition. *Let C be an ω-category. Then the ω-functor*

$$(\overline{\pi}, \underline{\pi}) \colon \widetilde{\Gamma}(C) \to C \times C$$

has the right lifting properties with respect to strong immersions.

Proof. Let $i \colon D \to E$ be a strong immersion and let r, α, and α' be as in §21.2.2. Consider a commutative square

$$\begin{array}{ccc} D & \xrightarrow{\gamma} & \widetilde{\Gamma}(C) \\ {\scriptstyle i}\downarrow & & \downarrow{\scriptstyle (\overline{\pi}, \underline{\pi})} \\ E & \xrightarrow[(f,g)]{} & C \times C. \end{array}$$

In other words, we have two ω-functors $f, g \colon E \to C$ and a reversible oplax transformation $\gamma \colon fi \Rightarrow gi$. A lift of such a square amounts to a reversible oplax transformation $\delta \colon f \Rightarrow g$ such that $\delta * i = \gamma$. One defines δ as the composite

$$f \xRightarrow{f * \alpha'} fir \xRightarrow{\gamma * r} gir \xRightarrow{g * \alpha} g$$

(see Remark 20.2.16). We have

$$\begin{aligned} \delta * i &= \big((g * \alpha)(\gamma * r)(f * \alpha')\big) * i \\ &= (g * \alpha * i)(\gamma * ri)(f * \alpha' * i) \\ &= 1_{gi}(\gamma * 1_D)1_{fi} = \gamma, \end{aligned}$$

thereby proving the result. □

21.2.4 Proposition. *For every $n \geq 0$, the ω-functor $\mathrm{j}_n \colon \mathbb{O}_n \to J_{n+1}$ of §20.4.3 is a strong immersion.*

Proof. Let $n \geq 0$. By Corollary 20.5.3, every trivial cofibration is an immersion. This means that there exists a section $r\colon J_{n+1} \to \mathbb{O}_n$ and a reversible oplax transformation $\alpha\colon ir \Rightarrow 1_D$. A dual proof shows that there exists a section $r'\colon J_{n+1} \to \mathbb{O}_n$ and a reversible oplax transformation $\alpha'\colon 1_D \Rightarrow ir'$. (More formally, one can apply Corollary 20.5.3 to $(\mathrm{j}_n)^{\mathrm{o}}$, where $-^{\mathrm{o}}\colon \mathbf{Cat}_\omega \to \mathbf{Cat}_\omega$ denotes the duality of $\mathbf{Cat}_\omega$ consisting in reverting the orientation of n-cells for every $n \geq 1$.) To conclude the proof it suffices to check that $r = r'$. Consider the reversible oplax transformation $r * \alpha'\colon r \Rightarrow rir' = r'$. As in $\mathbb{O}_n$, the only reversible cells are the identities, this oplax transformation has to be the identity, showing that $r = r'$. □

21.2.5 Corollary. *For every ω-category C, the ω-functor*

$$(\overline{\pi}, \underline{\pi})\colon \widetilde{\Gamma}(C) \to C \times C$$

is a fibration.

Proof. This is an immediate consequence of the two previous propositions. □

21.2.6 Remark. The same proof shows that $(\overline{\pi}, \underline{\pi})\colon \Gamma(C) \to C \times C$ is also a fibration.

We thus have showed:

21.2.7 Theorem. *For every ω-category C, the ω-category $\widetilde{\Gamma}(C)$ (endowed with the ω-functors described before) is a path object for C in the folk model structure.*

21.2.8 Remark. Although this is not related to the goal of the section, let us mention that one can prove in a very similar way that for every ω-functor $f\colon C \to D$, the ω-functor $p_f\colon \widetilde{\Gamma}_f \to D$ is a fibration. Indeed, one checks that it has the right lifting property with respect to (not necessarily strong) immersions and the assertion thus follows from Corollary 20.5.3.

21.2.9 Proposition. *Let $f\colon C \to D$ be an ω-functor and let (P, p) and (Q, q) be polygraphic resolutions (see §19.3.1) of C and D, respectively. If $g, g'\colon P^* \to Q^*$ are two ω-functors such that the two squares*

$$\begin{array}{ccc} P^* & \overset{g}{\underset{g'}{\rightrightarrows}} & Q^* \\ {\scriptstyle p}\downarrow & & \downarrow{\scriptstyle q} \\ C & \xrightarrow[f]{} & D \end{array}$$

commute, then there exists a reversible oplax transformation $\alpha\colon g \Rightarrow g'$.

Proof. The fact that $\widetilde{\Gamma}(A)$ is a path object for any ω-category A implies that $u\colon A \to B$ is "right homotopic" to $v\colon A \to B$ in the sense of model categories if and only if there exists a reversible oplax transformation from u to v. The result thus follows from general properties of cofibrant resolutions in model categories (see for instance [184, Proposition 8.1.25]). □

21.2.10 Remark. In particular, in the case where $C = D$ and f is the identity ω-functor, we get a diagram

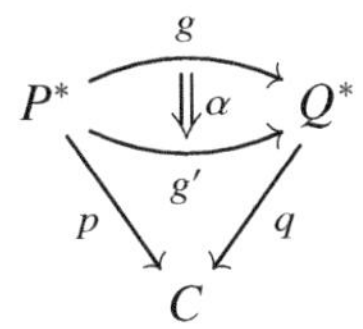

from which one can deduce that the morphism given in Remark 19.4.2 between any two resolutions is unique up to a (non-canonical) reversible oplax transformation.

21.3 The Folk Model Structure on $\mathbf{Cat}_n$ and $\mathbf{Cat}_{n,p}$

The purpose of this section is to transfer the folk model structure on $\mathbf{Cat}_\omega$ to subcategories of $\mathbf{Cat}_\omega$ such as the $\mathbf{Cat}_n$ or more generally $\mathbf{Cat}_{n,p}$. To do so, we will use the following classical transfer lemma:

21.3.1 Lemma. *Let $(\mathcal{M}, \mathcal{W}, \mathcal{C}of, \mathcal{F}ib)$ be a combinatorial model category generated by I and J. Let C be a locally presentable category, and let*

$$F\colon \mathcal{M} \to C, \quad G\colon C \to \mathcal{M}$$

be a pair of adjoint functors. Suppose that

$$G(lr(F(J))) \subseteq \mathcal{W},$$

where $\mathcal{W}$ denotes the class of weak equivalences of $\mathcal{M}$. Then $F(I)$ and $F(J)$ generate a combinatorial model structure on C, whose class of weak equivalences is $G^{-1}(\mathcal{W})$ and whose class of fibrations is $G^{-1}(\mathcal{F}ib)$.

Proof. See for instance [97, Theorem 3.3]. □

21.3.2 Proposition. *Let C be a reflective subcategory of $\mathbf{Cat}_\omega$ closed under pushouts and filtered colimits. Suppose that, for every ω-category C in C, the ω-category $\Gamma(C)$ is still in C. Then there exists a model structure on C whose weak equivalences are the ω-equivalences between objects of C and whose* *fibrations are the "folk" fibrations (i.e., the one defined in §20.4.7) between objects of C. This model structure is generated by $F(\mathcal{I})$ and $F(\mathcal{J})$, where $F\colon \mathbf{Cat}_\omega \to C$ denotes the left adjoint of the inclusion functor.*

Proof. As C is a reflective subcategory closed under filtered colimits of a locally presentable category, the category C is itself locally presentable. Let us check the hypothesis of the previous lemma: we have to show that $lr(F(\mathcal{J}))$ is included in the class of ω-equivalences. By the small object argument applied to $F(\mathcal{J})$ in C, elements of $lr(F(\mathcal{J}))$ are transfinite compositions of pushouts of elements of $F(\mathcal{J})$, these colimits being taken in C. By hypothesis, the inclusion functors to $\mathbf{Cat}_\omega$ preserves these colimits and they can be computed in $\mathbf{Cat}_\omega$. By the following lemma, the elements of $F(\mathcal{J})$ are immersions. By Proposition 20.5.5, a pushout of such an immersion is an immersion, and hence is an ω-equivalence by Proposition 20.5.4. As, by Proposition 20.1.19, ω-equivalences are stable under transfinite compositions, this concludes the proof. □

21.3.3 Lemma. *Under the hypothesis of the previous proposition, if i is an immersion, so is $F(i)$.*

Proof. Let D be an ω-category. By hypothesis, $\widetilde{\Gamma}F(D)$ is in C. For E an ω-category, denote by $\varepsilon_E\colon E \to F(E)$ the canonical ω-functor. The universal property of F gives a dotted arrow

$$\begin{array}{ccc} \widetilde{\Gamma}(D) & \xrightarrow{\widetilde{\Gamma}(\varepsilon_D)} & \widetilde{\Gamma}F(D) \\ {\scriptstyle \varepsilon_{\widetilde{\Gamma}(D)}}\downarrow & \nearrow_{\tau_D} & \\ F\widetilde{\Gamma}(D) & & \end{array}$$

making the triangle commute. If $f, g\colon C \to D$ are two ω-functors and $\alpha\colon f \Rightarrow g$ is a reversible oplax transformation, then

$$F(C) \xrightarrow{F(\alpha)} F\widetilde{\Gamma}(D) \xrightarrow{\tau_D} \widetilde{\Gamma}F(D)$$

defines a reversible oplax transformation. By abuse of notation, we will denote it by $F(\alpha)$. One checks that we have $F(\alpha)\colon F(f) \Rightarrow F(g)$ as expected.

Let now $i\colon C \to D$ be an immersion and let r and α be as in the definition of §20.5.1. By functoriality, $F(r)$ is a retraction of $F(i)$. The previous paragraph gives a reversible oplax transformation $F(\alpha)\colon F(i)F(r) \Rightarrow 1_{F(D)}$. One checks that $F(\alpha) * F(i) = 1_{F(i)}$, thereby proving the result. (We refer the reader to [237, Lemma 6.2] for a more detailed proof.) □

From now on, we fix n to be either an integer or ω, and $p \leq n$.

21.3.4 Proposition. *If C is an (n,p)-category, then so is $\widetilde{\Gamma}(C)$.*

Proof. Let $k > n$ and let $\alpha\colon x \curvearrowright y$ be a k-cell of $\widetilde{\Gamma}(C)$. Such a cell is a unit in $\widetilde{\Gamma}(C)$ if and only if x, y, and α_k are units in C and $\alpha^-_{k-1} = \alpha^+_{k-1}$. As C is an n-category, the k-cells x, y, α^-_{k-1}, α^+_{k-1} and the $(k+1)$-cell α_k are units. As

$$\begin{aligned} s_{k-1}(\alpha^-_{k-1}) &= s_{k-1}(x) *_0 \alpha^+_0 *_1 \cdots *_{k-2} \alpha^+_{k-2} \\ &= t_{k-1}(x) *_0 \alpha^+_0 *_1 \cdots *_{k-2} \alpha^+_{k-2} = s_{k-1}(\alpha^+_{k-1}), \end{aligned}$$

we have $\alpha^-_{k-1} = \alpha^+_{k-1}$, showing that $\alpha\colon x \curvearrowright y$ is indeed a unit.

Let now $k > p$ and let $\alpha\colon x \curvearrowright y$ be a k-cell of $\widetilde{\Gamma}(C)$. As C is an (n,p)-category, the k-cells x and y are invertible. We now define an inverse $\beta\colon x^{-1} \curvearrowright y^{-1}$ of $\alpha\colon x \curvearrowright y$ in $\widetilde{\Gamma}(C)$ in the following way:

$$\beta^\varepsilon_l = \alpha^\varepsilon_l \quad \text{for } l \leq k-2,$$

$$\beta^-_{k-1} = \alpha^+_{k-1}, \qquad \beta^+_{k-1} = \alpha^-_{k-1},$$

$$\begin{aligned} \beta_k = {} & (x^{-1} *_0 \alpha^+_0 *_1 \cdots *_{k-2} \alpha^+_{k-2}) \\ & *_{k-1} \alpha_k^{-1} *_{k-1} (\alpha^-_{k-2} *_{k-2} \cdots *_1 \alpha^-_0 *_0 y^{-1}), \end{aligned}$$

the inverse of the $(k+1)$-cells α_k existing by hypothesis. One checks that this indeed defines a revertible cylinder and that this cylinder is an inverse of $\alpha\colon x \curvearrowright y$. □

21.3.5 Theorem. *There exists a model structure on $\mathbf{Cat}_{n,p}$, known as "the folk model structure", whose weak equivalences are the ω-equivalences between (n,p)-categories and whose fibrations are the "folk" fibrations (i.e., the one defined in §20.4.7) between (n,p)-categories. This model structure is generated by $F(\mathcal{I})$ and $F(\mathcal{J})$, where $F\colon \mathbf{Cat}_\omega \to \mathbf{Cat}_{n,p}$ denotes the left adjoint of the inclusion functor.*

Proof. The result will be a consequence of Proposition 21.3.2 once we check that the hypotheses hold. The previous proposition gives us one of the hypotheses and we are left to check that the subcategory $\mathbf{Cat}_{n,p}$ of $\mathbf{Cat}_\omega$ is stable under pushout and filtered colimits. But it is actually stable under any colimits as the inclusion functor $\mathbf{Cat}_{n,p} \hookrightarrow \mathbf{Cat}_\omega$ admits a right adjoint, namely the functor taking an ω-category C to the (n,p)-category obtained from C by throwing out k-cells for $k > n$, and non-invertible l-cells for $l > p$. □

Let us now describe more precisely this model structure in some specific cases.

21.3.6 The folk model structure on $\mathbf{Cat}_n$. The ω-category $\mathbf{Cat}_n$, which is nothing but $\mathbf{Cat}_{n,n}$, is endowed with a folk model structure by the previous theorem. Its weak equivalences, fibrations and hence trivial fibrations are inherited from the folk model structure on $\mathbf{Cat}_\omega$. Let us describe its cofibrations. By the previous theorem, the class of cofibrations is $lr(F(\mathcal{I}))$, where $F\colon \mathbf{Cat}_\omega \to \mathbf{Cat}_n$ denotes the left adjoint to the inclusion functor. For $k \geq 0$, we have

$$F(\mathbb{O}_k) = \begin{cases} \mathbb{O}_k & \text{if } k \leq n, \\ \mathbb{O}_n & \text{if } k > n, \end{cases}$$

and, similarly,

$$F(\partial\mathbb{O}_k) = \begin{cases} \partial\mathbb{O}_k & \text{if } k \leq n+1, \\ \mathbb{O}_n & \text{if } k > n+1. \end{cases}$$

This implies that

$$F(\mathrm{i}_k) \begin{cases} \mathrm{i}_k & \text{if } k \leq n, \\ \bar{\mathrm{i}}_{n+1} & \text{if } k \leq n+1, \\ 1_{\mathbb{O}_n} & \text{if } k > n+1, \end{cases}$$

where $\bar{\mathrm{i}}_{n+1}\colon \partial\mathbb{O}_{n+1} \to \mathbb{O}_n$ corresponds to the collapsing of the two non-trivial n-cells of $\partial\mathbb{O}_{n+1}$. This shows that the class of cofibrations is generated by

$$\mathrm{i}_0, \ldots, \mathrm{i}_n, \bar{\mathrm{i}}_{n+1}.$$

If C is an n-category, then the ω-functor from the empty n-category to C is a cellular extension for these generators if and only if C is presented by an $(n+1)$-polygraph, that is, if and only if its underlying $(n-1)$-category is freely generated by a polygraph. In particular, any n-category admits such an n-category as a fibrant replacement in $\mathbf{Cat}_n$. Actually, one can show that an n-category is cofibrant if and only if it is presented by an $(n+1)$-polygraph.

Similarly, if one chooses carefully the J_{k+1} (see Remark 20.4.4), one gets that $F(\mathrm{j}_k) = 1_{\mathbb{O}_n}$ if $k \geq n$ so that trivial cofibrations of $\mathbf{Cat}_n$ are generated by

$$\bar{\mathrm{j}}_1 = F(\mathrm{j}_1), \ldots, \bar{\mathrm{j}}_{n-1} = F(\mathrm{j}_{n-1}).$$

21.3.7 The folk model structure on $\mathbf{Cat}_0$, $\mathbf{Cat}_1$, and $\mathbf{Cat}_2$. The weak equivalence of the folk model structure on $\mathbf{Set} = \mathbf{Cat}_0$ are the ω-equivalences between sets, that is, the bijections. It follows from the discussion in the previous paragraph that its class of cofibrations is generated by

$$\varnothing \to \{x\} \quad \text{and} \quad \{x, y\} \to \{z\}.$$

This implies that any map is cofibration. In particular, every set is cofibrant. Similarly, its class of trivial cofibrations is generated by an empty set of generators, showing that any map is fibration.

The weak equivalences of the folk model structure on **Cat** = **Cat**$_1$ are the ω-equivalences between categories, that is, the equivalences of categories. Its class of cofibrations is generated by the obvious functors of the form

$$\varnothing \hookrightarrow \cdot, \qquad \{\cdot \quad \cdot\} \hookrightarrow \{\cdot \to \cdot\}, \quad \{\cdot \rightrightarrows \cdot\} \to \{\cdot \longrightarrow \cdot\}.$$

This implies that cofibrations are exactly the functors injective on objects. In particular, every category is cofibrant. Similarly, the class of trivial cofibrations is generated by $\bar{\mathrm{j}}_1$. If one chooses J_1 according to Remark 20.4.4, one gets that $\bar{\mathrm{j}}_1$ is the inclusion functor

$$\{x\} \hookrightarrow \{\, x \overset{\sim}{\longrightarrow} y \,\},$$

where the symbol $\sim$ denotes an isomorphism. This means that an ω-functor $f: C \to D$ is a fibration in **Cat** if and only if it is an *iso-fibration*, that is, if and only if for any object x of C and any isomorphism $v: f(x) \to y$ of D, there exists an isomorphism $u: x \to x'$ in C such that $f(u) = v$. We have thus recovered the classical folk model structure on **Cat**, as described for instance in [312].

Let us now move on to **Cat**$_2$. The weak equivalences are the 2-equivalences of 2-categories. Its class of cofibrations is generated by the obvious 2-functors of the form

$$\varnothing \hookrightarrow \cdot, \qquad \{\cdot \quad \cdot\} \hookrightarrow \{\cdot \to \cdot\}, \quad \{\cdot \rightrightarrows \cdot\} \hookrightarrow \{\cdot \overset{\Downarrow}{\rightrightarrows} \cdot\}.$$

and

$$\{\cdot \overset{\Downarrow\,\Downarrow}{\rightrightarrows} \cdot\} \to \{\cdot \overset{\Downarrow}{\rightrightarrows} \cdot\}.$$

Its class of trivial cofibrations is generated by

$$\{x\} \hookrightarrow F(J_1) \quad \text{and} \quad \{\cdot \overset{u}{\frown} \cdot\} \hookrightarrow \{\cdot \underset{v}{\overset{u}{\Downarrow\wr}} \cdot\}.$$

Choosing J_1 according to Remark 20.4.4, one gets that $F(J_1)$ is the free-standing adjoint equivalence: it has two objects x and y, two generating 1-cells $u: 0 \to 1$ and $v: 1 \to 0$, two generating 2-cells $\eta: 1_x \to uv$ and $\varepsilon: vu \to 1_y$ and these 2-cells satisfy the triangular identities: $(\eta * u)(u * \varepsilon) = 1_u$ and $(v * \eta)(\varepsilon * v) = 1_v$. We have thus recovered the folk model structure introduced by Lack in [229], with a correction in [231].

21.3.8 The folk model structure on $\mathbf{Gpd}_\omega$. The category $\mathbf{Gpd}_\omega$, which is nothing but $\mathbf{Cat}_{\omega,0}$, is endowed with a folk model structure. The weak equivalences are the ω-equivalences between ω-groupoids. They can be characterized as the ω-functors inducing bijections on connected components and homotopy groups but we will not enter into that. Denote by $F\colon \mathbf{Cat}_\omega \to \mathbf{Gpd}_\omega$ the left adjoint to the inclusion functor. This functor sends an ω-category C to the ω-groupoid obtained by formally inverting every cell of C. Set

$$\widetilde{\mathbb{O}}_n = F(\mathbb{O}_n).$$

The class of cofibrations of $\mathbf{Gpd}_\omega$ is generated by the inclusion

$$F(\mathrm{i}_n)\colon \partial\widetilde{\mathbb{O}}_n \hookrightarrow \widetilde{\mathbb{O}}_n,$$

where $\partial\widetilde{\mathbb{O}}_n$ denotes the underlying $(n-1)$-groupoid of $\widetilde{\mathbb{O}}_n$. Less trivially, the class of trivial cofibrations is generated by the set $\widetilde{\mathcal{J}}$ of ω-functors

$$\widetilde{\mathrm{j}}_n\colon \widetilde{\mathbb{O}}_n \hookrightarrow \widetilde{\mathbb{O}}_{n+1},$$

where $n \geq 1$, corresponding to the source of the principal cell of $\widetilde{\mathbb{O}}_{n+1}$. (Note that these ω-functors are *not* the $F(\mathrm{j}_n)$.) To prove this, one can proceed as follows. First, one notes that these ω-functors are trivial cofibrations, so that the class $\widetilde{\mathcal{J}}$ is included in the class of trivial cofibration. Second, one checks that an ω-equivalence between ω-groupoids having the right lifting property with respect to the $\widetilde{\mathrm{j}}_n$ is a trivial fibration. Third, one concludes using a similar argument as in the proof of Theorem 21.1.2. Let f be a trivial cofibration in $\mathbf{Gpd}_\omega$. Using the small object argument, one can factor it as $f = pi$, where i is in $lr(\widetilde{\mathcal{J}})$ and p is in $r(\widetilde{\mathcal{J}})$. By the first point, i is a trivial cofibration and so p is a weak equivalence by the 2-out-of-3 property. This implies that p is a trivial fibration by the second point. Thus, f has the left lifting property with respect to p and, by the retract lemma, f is a retract of i, showing that f is in $lr(\widetilde{\mathcal{J}})$.

We refer the reader to [16] for more details on the folk model structure on $\mathbf{Gpd}_\omega$. In this paper, this model structure is called the Brown-Golasiński model structure as it coincides, through the equivalence of categories between ω-groupoids and crossed complexes, with the model structure on crossed complexes introduced by Brown and Golasiński in [62].

21.3.9 The folk model structure on $\mathbf{Cat}_{\omega,1}$. The description of the folk model structure on $\mathbf{Cat}_{\omega,1}$ is very close to the one of $\mathbf{Gpd}_\omega$. Its weak equivalences are the ω-equivalences between $(\omega,1)$-categories. Setting

$$\widetilde{\mathbb{O}}_n = F(\mathbb{O}_n),$$

where $F\colon \mathbf{Cat}_\omega \to \mathbf{Cat}_{\omega,1}$ denotes the left adjoint to the inclusion functor, the class of cofibrations of $\mathbf{Cat}_{\omega,1}$ is generated by the inclusions

$$F(\mathrm{i}_n)\colon \partial\widetilde{\mathbb{O}}_n \hookrightarrow \widetilde{\mathbb{O}}_n,$$

where $\partial\widetilde{\mathbb{O}}_n$ denotes the underlying $(n-1,1)$-category of $\widetilde{\mathbb{O}}_n$. The class of trivial cofibrations can be shown to be generated by

$$\widetilde{\mathrm{j}}_1 = F(\mathrm{j}_1)\colon \widetilde{\mathbb{O}}_0 \to F(J_1)$$

and by the

$$\widetilde{\mathrm{j}}_n\colon \widetilde{\mathbb{O}}_n \hookrightarrow \widetilde{\mathbb{O}}_{n+1},$$

for $n \geq 1$, corresponding to the source of the principal cell of $\widetilde{\mathbb{O}}_{n+1}$.

22
Homology of ω-Categories

This chapter is about Métayer's polygraphic homology of ω-categories [278]. This homology theory was first defined in the following way: the polygraphic homology of an ω-category is the homology of the abelianization of any of its polygraphic replacements. Métayer then showed with Lafont that for every monoid, considered as an ω-category, its polygraphic homology coincides with its classical homology as a monoid [236]. This result was then generalized to 1-categories by Guetta [155].

In this chapter, we prove that the polygraphic homology is the left derived functor of a linearization functor from $\mathbf{Cat}_\omega$ to the category $\mathbf{Ch}_{\mathbb{Z},\geqslant 0}$ of chain complexes, where $\mathbf{Cat}_\omega$ is endowed with ω-equivalences and $\mathbf{Ch}_{\mathbb{Z},\geqslant 0}$ with quasi-isomorphisms.

In the first section of the chapter, we define the abelianization functor, and we prove that an oplax transformation induces a chain homotopy after abelianization. In the second section, we show that the abelianization functor is a left Quillen functor for the folk model category structure on $\mathbf{Cat}_\omega$ and the projective structure on $\mathbf{Ch}_{\mathbb{Z},\geqslant 0}$. We define polygraphic homology using this derived functor. The fact that polygraphic resolutions are cofibrant resolutions for the folk model category structure proves that this homology is indeed Métayer's polygraphic homology. In the third section, we show that polygraphic homology generalizes monoid homology. The proof we present is based on a conceptual reinterpretation of Lafont and Métayer's proof by Guetta. Finally, in the last section, we compute some examples of polygraphic homology.

In the whole chapter, we assume some familiarity with homological algebra, and in particular homology of monoids. We refer the reader to Appendix E for a quick introduction.

22.1 The Abelianization Functor

22.1.1 Abelianization of an ω-category. We now define the *abelianization functor*

$$\mathrm{Ab}\colon \mathbf{Cat}_\omega \to \mathbf{Ch}_{\mathbb{Z},\geqslant 0},$$

where $\mathbf{Ch}_{\mathbb{Z},\geqslant 0}$ denotes the category of chain complexes of abelian groups in non-negative degree, see §E.2.1. Using the notation of Section 17.4, this functor is just the composite

$$\mathbf{Cat}_\omega \xrightarrow{\lambda} \mathbf{ADC} \xrightarrow{U} \mathbf{Ch}_{\mathbb{Z},\geqslant 0}.$$

Nevertheless, we will give a direct definition.

Let C be an ω-category. We define the chain complex $\mathrm{Ab}(C)$ in the following way. For every $n \geq 0$, the group $\mathrm{Ab}(C)_n$ is generated by elements $[x]$, for every n-cell x of C, subject to the relations $[x *_i y] = [x] + [y]$, for every pair of n-cells x and y such that $x *_i y$ is defined. It follows that if z is an m-cell for $m < n$, then $[1_z] = 0$, where 1_z denotes the iterated identity in dimension n. For $n \geq 1$, we define d_n by

$$d_n([x]) = [t_{n-1}(x)] - [s_{n-1}(x)],$$

where x is an n-cell of C. The axioms of ω-categories giving the sources and targets of compositions imply that these maps are well-defined and the globular relations that they define a chain complex.

If $f\colon C \to D$ is an ω-functor, then we define a chain map

$$\mathrm{Ab}(f)\colon \mathrm{Ab}(C) \to \mathrm{Ab}(D)$$

by setting

$$\mathrm{Ab}(f)_n([x]) = [f(x)]$$

for $n \geq 0$ and x an n-cell of C. One checks that this is well-defined, that $\mathrm{Ab}(f)$ is a chain map and that Ab is indeed a functor.

22.1.2 Proposition. *The functor* Ab *is a left adjoint. In particular, it preserves colimits.*

Proof. We saw in Remark 17.4.7 that Ab, which is $U\lambda$ with the notation of this remark, admits the functor $\nu\colon \mathbf{Ch}_{\mathbb{Z},\geqslant 0} \to \mathbf{Cat}_\omega$ of §17.4.5 as a left adjoint. □

22.1.3 Proposition. *Let P be a polygraph. For every $n \geq 0$, the abelian group* $\mathrm{Ab}(P^*)_n$ *is free with basis the set of the elements of the form $[x]$, with x is in P_n.*

Proof. This is part of Proposition 17.4.9. □

22.1.4 Proposition. *Let $f, g\colon C \to D$ be two ω-functors and let $\alpha\colon f \Rightarrow g$ be an oplax transformation. Then, setting*

$$\mathrm{Ab}(\alpha)_n([x]) = [\alpha_x]$$

for $n \geq 0$ and x in C_n, we obtain a chain homotopy (see §E.2.4)

$$\mathrm{Ab}(\alpha)\colon\ \mathrm{Ab}(f) \Rightarrow \mathrm{Ab}(g).$$

Proof. Let us first check that the map $\mathrm{Ab}(\alpha)_n$ is well-defined. Let x and y be two n-cells such that $x *_i y$ is defined for some $i < n$. We have to check the equality $\mathrm{Ab}(\alpha)_n([x *_i y]) = \mathrm{Ab}(\alpha)_n([x]) + \mathrm{Ab}(\alpha)_n([y])$, that is, $[\alpha_{x *_i y}] = [\alpha_x] + [\alpha_y]$. But this is exactly what we get by linearizing the formula

$$\begin{aligned}\alpha_{x *_i y} = \big(f(s_{i+1}(x)) *_0 \alpha_{t_0(y)} *_1 \cdots *_{i-1} \alpha_{t_{i-1}(y)} *_i \alpha_y\big) *_{i+1} \\ \big(\alpha_x *_i \alpha_{s_{i-1}(x)} *_{i-1} \cdots *_1 \alpha_{s_0(x)} *_0 g(t_{i+1}(y))\big),\end{aligned}$$

keeping in mind that, as we are linearizing in dimension $n + 1$, we have $[z] = 0$ for z an m-cell with $m < n + 1$.

Let us now prove that $\mathrm{Ab}(\alpha)$ is indeed a chain homotopy. If x is 0-cell of C, we have $\alpha_x\colon f(x) \to g(x)$ and so $d_1[\alpha_x] = [g(x)] - [f(x)]$, showing that $d_1(\mathrm{Ab}(\alpha)_0([x])) = \mathrm{Ab}(g)_0([x]) - \mathrm{Ab}(f)_0([x])$. Similarly, if x is an n-cell of C_n for $n > 0$, we have

$$\alpha_x\colon f(x) *_0 \alpha_{t_0(x)} *_1 \cdots *_{n-1} \alpha_{t_{n-1}(x)} \to \alpha_{s_{n-1}(x)} *_{n-1} \cdots *_1 \alpha_{s_0(x)} *_0 g(x)$$

and, by linearizing,

$$d_{n+1}[\alpha_x] = [\alpha_{s_{n-1}(x)}] + [g(x)] - [f(x)] - [\alpha_{t_{n-1}(x)}].$$

As

$$\begin{aligned}[\alpha_{t_{n-1}(x)}] - [\alpha_{s_{n-1}(x)}] &= \mathrm{Ab}(\alpha)_n([t_{n-1}(x)]) - \mathrm{Ab}(\alpha)_n([s_{n-1}(x)]) \\ &= \mathrm{Ab}(\alpha)_n([t_{n-1}(x)] - [s_{n-1}(x)]) \\ &= \mathrm{Ab}(\alpha)_n(d_n[x]),\end{aligned}$$

we get

$$d_{n+1}(\mathrm{Ab}(\alpha)_n([x])) + \mathrm{Ab}(\alpha)_{n-1}(d_n[x]) = \mathrm{Ab}(g)_n([x]) - \mathrm{Ab}(f)_n([x]),$$

thereby showing the result. □

22.2 Deriving the Abelianization Functor

In this section, we will show that the abelianization functor

$$\mathrm{Ab}\colon \mathbf{Cat}_\omega \to \mathbf{Ch}_{\mathbb{Z},\geqslant 0}$$

can be derived, thus producing a homology theory for ω-categories.

We start with some preliminaries on derived functors.

22.2.1 Homotopy category. A *localizer* (also called *relative category*) is a category C endowed with a class $\mathcal{W}$ of morphisms called *weak equivalences*. The *homotopy category* of a localizer $(C, \mathcal{W})$ is the category $C[\mathcal{W}^{-1}]$ obtained from C by formally inverting arrows in $\mathcal{W}$. We will also denote this category by $\mathrm{Ho}(C)$, making implicit the class $\mathcal{W}$. There is a canonical functor $p\colon C \to \mathrm{Ho}(C)$.

In particular, any model category $\mathcal{M}$ has an underlying localizer $(\mathcal{M}, \mathcal{W})$ and thus a homotopy category $\mathrm{Ho}(\mathcal{M})$.

22.2.2 Left derived functors. Let $(C, \mathcal{W}_C)$ and $(\mathcal{D}, \mathcal{W}_\mathcal{D})$ be two localizers, and let $F\colon C \to \mathcal{D}$ be a functor. The *(total) left derived functor* of F, if it exists, is the universal pair consisting of a functor

$$\mathbb{L}F\colon \mathrm{Ho}(C) \to \mathrm{Ho}(\mathcal{D})$$

and a natural transformation

$$\begin{array}{ccc} C & \xrightarrow{F} & \mathcal{D} \\ {\scriptstyle p_C}\downarrow & \overset{\lambda}{\nearrow\!\!\!\nearrow} & \downarrow{\scriptstyle p_\mathcal{D}} \\ \mathrm{Ho}(C) & \underset{\mathbb{L}F}{\dashrightarrow} & \mathrm{Ho}(\mathcal{D}). \end{array}$$

By abuse of language, one often refers to $\mathbb{L}F$ as the left derived functor of F.

One important use of model categories is to provide tools to prove the existence of derived functors. In particular, the so-called left Quillen functors can be left derived.

22.2.3 Left Quillen functors. Let $\mathcal{M}$ and $\mathcal{N}$ be two model categories. A left adjoint functor $F\colon \mathcal{M} \to \mathcal{N}$ is called a *left Quillen functor* if it sends cofibrations of $\mathcal{M}$ to cofibrations of $\mathcal{N}$ and trivial cofibration of $\mathcal{M}$ to trivial cofibrations of $\mathcal{N}$.

Similarly, a right adjoint functor $G\colon \mathcal{N} \to \mathcal{M}$ is said to be a *right Quillen functor* if it sends fibrations to fibrations and trivial fibrations to trivial fibrations.

If

$$F\colon \mathcal{M} \to \mathcal{N} \qquad\qquad G\colon \mathcal{N} \to \mathcal{M}$$

is a pair of adjoint functors, then F is a left Quillen functor if and only if G is a right Quillen functor. The pair (F, G) is then called a *Quillen pair* or a *Quillen adjunction.*

22.2.4 Theorem (Quillen). *A left Quillen functor* $F\colon \mathcal{M} \to \mathcal{N}$ *admits a left derived functor* $\mathbb{L}F\colon \operatorname{Ho}(\mathcal{M}) \to \operatorname{Ho}(\mathcal{N})$. *Moreover, if* X *is an object of* $\mathcal{M}$, *then* $\mathbb{L}F(p_{\mathcal{M}}(X))$ *is canonically isomorphic to* $p_{\mathcal{N}}(F(Q))$, *where* $(Q, Q \to X)$ *is a cofibrant replacement, i.e., a cofibrant object* Q *endowed with a weak equivalence* $Q \to X$.

Similarly, a right Quillen functor admits a right derived functor that can be computed using fibrant replacement.

22.2.5 Remark. One actually only needs F to send trivial cofibrations between cofibrant objects to weak equivalences for the previous theorem to apply.

We will now apply Quillen's result to the abelianization functor

$$\operatorname{Ab}\colon \mathbf{Cat}_\omega \to \mathbf{Ch}_{\mathbb{Z},\geqslant 0}.$$

We first introduce a model category structure on $\mathbf{Ch}_{\mathbb{Z},\geqslant 0}$.

22.2.6 The projective model structure on chain complexes. The category of chain complexes $\mathbf{Ch}_{\mathbb{Z},\geqslant 0}$ can be endowed with the so-called *projective model structure* (see [307, Chapter II, Section 4, page 11]):

- the weak equivalences are the quasi-isomorphisms (see §E.3.5),
- the cofibrations are the monomorphisms f such that, for every $n \geq 0$, the cokernel of f_n is projective,
- the fibrations are the morphisms f such that, for every $n > 0$, f_n is an epimorphism.

In particular, every chain complex is fibrant for this model structure and the cofibrant chain complexes are exactly the ones that are projective in every degree.

22.2.7 Theorem. *The functor* $\operatorname{Ab}\colon \mathbf{Cat}_\omega \to \mathbf{Ch}_{\mathbb{Z},\geqslant 0}$ *is a left Quillen functor, where* $\mathbf{Cat}_\omega$ *is endowed with the folk model structure and* $\mathbf{Ch}_{\mathbb{Z},\geqslant 0}$ *is endowed with the projective model structure.*

Proof. We have to show that Ab preserves cofibrations and trivial cofibrations. Since the cofibrations and trivial cofibrations of $\mathbf{Cat}_\omega$ are generated by sets $\mathcal{I}$ and $\mathcal{J}$, it suffices to prove that these sets are sent to cofibrations and weak equivalences, respectively.

Let $\mathrm{i}_n\colon \partial\mathbb{O}_n \to \mathbb{O}_n$ be an element of $\mathcal{I}$. Proposition 22.1.3 gives a concrete description of the morphism $\operatorname{Ab}(\mathrm{i}_n)\colon \operatorname{Ab}(\partial\mathbb{O}_n) \to \operatorname{Ab}(\mathbb{O}_n)$: for k such that $0 \leq k < n$, the morphism $\operatorname{Ab}(\mathrm{i}_n)_k$ can be identified with the identity of $\mathbb{Z}^2$, for $k = n$, it can be identified with the unique morphism $0 \to \mathbb{Z}$ and, for $k > n$, with the identity of 0. All these morphisms are monomorphisms of cokernel either 0 or $\mathbb{Z}$. This means that $\mathrm{Ab}(\mathrm{i}_n)$ is a cofibration.

Let now $\mathrm{j}_n : \mathbb{O}_n \to J_{n+1}$ be an element of $\mathcal{J}$. By Corollary 20.5.3, this ω-functor admits an inverse up to reversible oplax transformations. (Now that we have the folk model structure, this also follows from the so-called "Whitehead Theorem", see for instance [184, Theorem 7.5.10], as j_n is a weak equivalence between cofibrant and fibrant objects.) But since by Proposition 22.1.4, oplax transformations are sent to chain homotopies, the morphism $\mathrm{Ab}(\mathrm{j}_n)$ is a homotopy equivalence (see §E.3.5) and thus a quasi-isomorphism, thereby proving the result. □

22.2.8 Polygraphic homology of ω-categories. By the previous theorem, the functor

$$\mathrm{Ab}\colon \mathbf{Cat}_\omega \to \mathbf{Ch}_{\mathbb{Z},\geqslant 0}$$

can be derived to a functor

$$\mathbb{L}\,\mathrm{Ab}\colon \mathrm{Ho}(\mathbf{Cat}_\omega) \to \mathrm{Ho}(\mathbf{Ch}_{\mathbb{Z},\geqslant 0}).$$

In particular, for $n \geq 0$, by post-composing by the functor

$$H_n\colon \mathrm{Ho}(\mathbf{Ch}_{\mathbb{Z},\geqslant 0}) \to \mathbf{Ab}$$

(induced by the $H_n\colon \mathbf{Ch}_{\mathbb{Z},\geqslant 0} \to \mathbf{Ab}$ functor of §E.2.2), we get a functor

$$H_n^{\mathrm{Pol}}\colon \mathrm{Ho}(\mathbf{Cat}_\omega) \to \mathbf{Ab}.$$

If C is an ω-category, by definition, the *n-th polygraphic homology group* of C is the abelian group $H_n^{\mathrm{Pol}}(C)$.

Concretely, by Theorem 22.2.4, the group $H_n^{\mathrm{Pol}}(C)$ is computed in the following way:

$$H_n^{\mathrm{Pol}}(C) = H_n(\mathrm{Ab}(P^*)),$$

where $(P, P^* \to C)$ is any polygraphic resolution of C.

22.3 Comparison With Homology of Monoids

In the previous section, we defined the polygraphic homology of an ω-category C. In this section, we will show, following [236], that when M is a monoid, the polygraphic homology of M, seen as an ω-category, coincides with the homology of the monoid M. A similar result holds for 1-categories, the homology of 1-categories being defined using simplicial sets (see Section F.1), as proved in [155].

22.3.1 Theorem. *Let M be a monoid. The polygraphic homology of M seen as an ω-category coincides with the homology of the monoid M.*

Proof. Let $p\colon P^* \to M$ be a polygraphic resolution of M, considered as a 1-category, such that P^0 consists of a unique element that we will denote by $*$. For instance, one could take the standard resolution. By definition, the polygraphic homology of M is the homology of the abelianization of P^*. We will now associate to this polygraphic resolution a resolution by free left $\mathbb{Z}M$-modules of $\mathbb{Z}$ (endowed with the trivial action).

Denote by e the unit element of M and by $e\backslash M$ the coslice category. Explicitly, an object of $e\backslash M$ is an element of M, and if m and m' are two objects, a morphism from m to m' is an element n of M such that $nm = m'$. Denote by $e\backslash P^*$ the ω-category defined by the pullback square

$$\begin{array}{ccc} e\backslash P^* & \longrightarrow & P^* \\ {\scriptstyle q}\big\downarrow & & \big\downarrow{\scriptstyle p} \\ e\backslash M & \longrightarrow & M, \end{array}$$

where the bottom horizontal morphism is the forgetful functor. Let us describe the cells of this ω-category. An object of $e\backslash P^*$ consists of an element m of M that we will denote by $(m, *)$. A 1-cell from an object m to an object m' consists of a 1-cell x such that $p(x)m = m'$. A 1-cell is thus uniquely determined by a pair (m, x), where m is in M and x is a 1-cell of P^*. The source of such a 1-cell is $p(x)m$ and its target is m'. Similarly, an i-cell for $i > 1$ can be described as a pair (m, x), where m is in M and x is an i-cell of P^*. Its source is $(m, s_{i-1}(x))$ and its target $(m, t_{i-1}(x))$.

One can show that the ω-category $e\backslash P^*$ is freely generated in the sense of polygraphs by its cells of the form (m, a), where m is in M and a is in P_i for $i \geq 0$. This implies that the linearization of this ω-category in degree i is

$$\mathrm{Ab}(e\backslash P^*)_i \simeq \mathbb{Z}[M \times P_i] \simeq \mathbb{Z}M[P_i].$$

In particular, it has a natural structure of left $\mathbb{Z}M$-module. If x is a cell, setting $[x] = [(e, x)]$, we have $[(m, x)] = m[x]$. By definition, if m is in M and x is an i-cell for $i > 0$, we have

$$d_i(m[x]) = \begin{cases} m[*] - mp(x)[*] & \text{if } i = 1, \\ m[t_{i-1}(x)] - m[s_{i-1}(x)] & \text{if } i > 1, \end{cases}$$

and the map d_i is thus $\mathbb{Z}M$-linear. This shows that $\mathrm{Ab}(e\backslash P^*)$ is a complex of free left $\mathbb{Z}M$-modules.

The unique ω-functor from $e\backslash P^*$ to the terminal ω-category induces a morphism of complexes $\mathrm{Ab}(e\backslash P^*)$ to $\mathbb{Z}$. In other words, by sending $m[*]$ in $\mathrm{Ab}(e\backslash P^*)_0$ to 1 in $\mathbb{Z}$, we get an augmented complex of left $\mathbb{Z}M$-modules

$$\mathbb{Z} \longleftarrow \mathbb{Z}M[P_0] \longleftarrow \mathbb{Z}M[P_1] \longleftarrow \mathbb{Z}M[P_2] \longleftarrow \cdots.$$

We will see that this complex is exact. Assuming this, let us end the proof. As the complex is exact, we have built, as announced, a resolution of $\mathbb{Z}$ by free left $\mathbb{Z}M$-module. The homology of the monoid M is thus the homology of the complex

$$\mathbb{Z} \otimes_{\mathbb{Z}M} \mathbb{Z}M[P_0] \longleftarrow \mathbb{Z} \otimes_{\mathbb{Z}M} \mathbb{Z}M[P_1] \longleftarrow \mathbb{Z} \otimes_{\mathbb{Z}M} \mathbb{Z}M[P_2] \longleftarrow \cdots,$$

which is canonically isomorphic to the complex

$$\mathbb{Z}[P_0] \longleftarrow \mathbb{Z}[P_1] \longleftarrow \mathbb{Z}[P_2] \longleftarrow \cdots,$$

which is nothing but $\mathrm{Ab}(P^*)$. The homology of the monoid M is thus the homology of $\mathrm{Ab}(P^*)$, that is, the polygraphic homology of M.

To end the proof, it thus suffices to show that the augmented complex introduced at the beginning of the previous paragraph is exact. We have a unique ω-functor $r\colon e\backslash P^* \to \mathbf{1}$, where $\mathbf{1}$ denotes the terminal ω-category. We have to show that $\mathrm{Ab}(r)$ is a quasi-isomorphism. Consider the ω-functor $e\colon \mathbf{1} \to e\backslash P^*$ corresponding to the object $(e, *)$ of $e\backslash P^*$. The composition pe is the identity of $\mathbf{1}$. We will construct an oplax transformation $\alpha\colon pe \Rightarrow 1_{e\backslash P^*}$. Using Proposition 22.1.4, we will get that $\mathrm{Ab}(r)$ is a homotopy equivalence and thus a quasi-isomorphism. Let us construct this oplax transformation α. First, note that there exists an oplax transformation $\beta\colon qpe \Rightarrow q$, where $q\colon e\backslash P^* \to e\backslash M$ is the ω-functor introduced at the beginning of the proof. (Note that qpe is the constant ω-functor of value e.) Indeed, as $e\backslash M$ admits e as an initial object, we have a natural transformation $\gamma\colon e \Rightarrow 1_{e\backslash M}$, where e denotes the constant endofunctor of $e\backslash M$ of value e. The oplax transformation β is thus $\gamma * q$. Second, note that q is a trivial fibration, as it is defined by pulling back the trivial fibration p. The existence of the transformation α thus follows from the following lemma that concludes the proof. □

22.3.2 Lemma. *Let $p\colon C \to D$ and $f, g\colon B \to C$ be three ω-functors and let $\beta\colon pf \Rightarrow pg$ be an oplax transformation. If p is a trivial fibration and T is a cofibrant ω-category, then there exists an oplax transformation $\alpha\colon f \Rightarrow g$ such that $\beta = p * \alpha$.*

Proof. The oplax transformation β corresponds to an ω-functor $B \to \Gamma(D)$. Consider the pullback square

$$\begin{array}{ccc} \Gamma(D) \times_{D\times D} C \times C & \longrightarrow & \Gamma(D) \\ \downarrow & & \downarrow{\scriptstyle (\overline{\pi},\underline{\pi})} \\ C \times C & \xrightarrow[p\times p]{} & D \times D. \end{array}$$

As the source of β is pf and its target is pg, we get an ω-functor

$$(\beta, f \times g)\colon B \to \Gamma(D) \times_{D\times D} C \times C.$$

The naturality square

$$\begin{array}{ccc} \Gamma(C) & \xrightarrow{\Gamma(p)} & \Gamma(D) \\ {\scriptstyle (\overline{\pi},\underline{\pi})}\downarrow & & \downarrow{\scriptstyle (\overline{\pi},\underline{\pi})} \\ C \times C & \xrightarrow[p\times p]{} & D \times D \end{array}$$

induces an ω-functor

$$q\colon \Gamma(C) \to \Gamma(D) \times_{D\times D} C \times C.$$

An oplax transformation α as in the statement exactly corresponds to a dotted arrow making the triangle

$$\begin{array}{ccc} & & \Gamma(C) \\ & \nearrow & \downarrow{\scriptstyle q} \\ B & \xrightarrow[(\beta,f\times g)]{} & \Gamma(D) \times_{D\times D} C \times C \end{array}$$

commute. As B is cofibrant, to get the result it suffices to show that q is a trivial fibration.

Let us prove this:

1. An object of $\Gamma(D) \times_{D\times D} C \times C$ is given by two objects x and x' of C and a 1-cell $v\colon p(x) \to p(x')$. As p is a trivial fibration, there exists $u\colon x \to x'$ such that $p(u) = v$, showing that q is surjective on objects.
2. Let $n \geq 1$ and let $\gamma\colon x \curvearrowright y$ and $\delta\colon z \curvearrowright t$ be two parallel n-cylinders of C. Suppose we have an $(n+1)$-cell from $q(\gamma)$ to $q(\delta)$. This means that we have two n-cells $u\colon x \to z$ and $v\colon y \to t$ of C and an $(n+1)$-cylinder $\Gamma\colon p(\gamma) \to p(\delta)\colon p(u) \curvearrowright p(v)$. As p is a trivial fibration, the $(n+2)$-cell of the $(n+1)$-cylinder Γ can be lifted to C yielding an $(n+1)$-cylinder $\Lambda\colon \gamma \to \delta\colon u \curvearrowright v$ of C such that $p(\Lambda) = \Gamma$. This shows that q is a trivial fibration, thereby proving the lemma. □

22.4 Examples

22.4.1 Polygraphic homology of R_1. We will compute the polygraphic homology of the free-standing reversible cell R_1, introduced in §20.4.1. We will see that its homology is non-trivial. This implies that R_1 is not weakly contractible.

Since R_1 is freely generated by a polygraph, its homology is the homology of its linearization $C = \mathbf{Ab}(R_1)$. We have

$$C_0 = \mathbb{Z}[0,1]$$

and

$$C_j = \mathbb{Z}[\{r_{l_1,\dots,l_{j-1}}, \overline{r}_{l_1,\dots,l_{j-1}} \mid l_k = \pm, 1 \le k < j\}],$$

with

$$d_1([r]) = [t_0(r)] - [s_0(r)] = [1] - [0],$$

and similarly,

$$d_1([\overline{r}]) = [0] - [1] = -d([r]),$$

$$\begin{aligned} d_j([r_{l_1,\dots,l_{j-2},-}]) &= [t_{j-1}(r_{l_1,\dots,l_{j-2},-})] - [s_{j-1}(r_{l_1,\dots,l_{j-1}},-)] \\ &= [r_{l_1,\dots,l_{j-2}} *_{j-2} \overline{r}_{l_1,\dots,l_{j-2}}] - [1_{s_{j-2}(r_{l_1,\dots,l_{j-2}})}] \\ &= [r_{l_1,\dots,l_{j-2}}] + [\overline{r}_{l_1,\dots,l_{j-2}}], \end{aligned}$$

and similarly,

$$\begin{aligned} d_j([\overline{r}_{l_1,\dots,l_{j-2},-}]) &= -d_j([r_{l_1,\dots,l_{j-2},-}]), \\ d_j([r_{l_1,\dots,l_{j-2},+}]) &= -d_j([r_{l_1,\dots,l_{j-2},-}]), \\ d_j([\overline{r}_{l_1,\dots,l_{j-2},+}]) &= d_j([r_{l_1,\dots,l_{j-2},-}]). \end{aligned}$$

Let us compute the homology of this chain complex. One has

$$H_0^{\mathrm{Pol}}(R_1) = \mathbb{Z}[0,1]/\langle [1] - [0] \rangle \simeq \mathbb{Z}.$$

A 1-chain $a[r] + b[\overline{r}]$ is a cycle if and only if

$$d_1(a[r] + b[\overline{r}]) = (a-b)[r] = 0,$$

that is, if and only if $a = b$. But $[r] + [\overline{r}]$ is a boundary. Hence

$$H_1^{\mathrm{Pol}}(R_1) = 0.$$

Similarly, a 2-chain $a[r_-] + b[\overline{r}_-] + c[r_+] + d[\overline{r}_+]$ is a cycle if and only if one has $a - b - c + d = 0$. The abelian group of 2-cycles is thus isomorphic to $\mathbb{Z}^3$. As for the abelian group of 2-boundaries, it is spanned by $[r_-] + [\overline{r}_-]$

and $[r_+] + [\overline{r}_+]$. It is thus isomorphic to $\mathbb{Z}^2$. Computing the quotient, one gets that

$$H_2^{\mathrm{Pol}}(R_1) \simeq \mathbb{Z}.$$

This already proves that R_1 is not weakly contractible. More generally, for $n \geq 2$, one checks that

$$H_n^{\mathrm{Pol}}(R_1) \simeq \mathbb{Z}^{(2^n-2^{n-2})-2^{n-1}} = \mathbb{Z}^{2^{n-2}}.$$

22.4.2 Thomason homology and the homology of $K(\mathbb{N}, 2)$. In this chapter, we defined a homology theory for ω-categories: the polygraphic homology. There is a second homology theory for ω-categories, that we will call the Thomason homology, which can be morally defined in the following way. Let C be an ω-category. Consider the weak ω-groupoid obtained by weakly inverting all the cells of C. This weak ω-groupoid corresponds to a homotopy type and, morally, the Thomason homology of C is the homology of this homotopy type. One way to give a precise definition of this homology is to use Street nerve $N\colon \mathbf{Cat}_\omega \to \widehat{\Delta}$, introduced in [334], which extends the usual nerve functor (see §F.1.3) to ω-categories. If C is an ω-category, for every $n \geq 0$, we set

$$H_n^{\mathrm{Th}}(C) = H_n(NC).$$

This is the *Thomason homology* of C. When C is a category, we recover the classical homology of categories (generalizing the classical homology of monoids) recalled in Section F.1. The main result of [155] says that the polygraphic homology and the Thomason homology of a category agree. It is tempting to think that these two homologies always agree. This is not the case!

Here is a counter-example. Recall (see Section 14.3) that to any abelian monoid M, one can associate a 2-category with only one object, one 1-cell (the identity of the unique object), and M as the set of 2-cells. Let us denote this 2-category by $K(M, 2)$. By [10, Theorem 4.7], if M is an abelian group A, then the Thomason homology of $K(A, 2)$ is the homology of the corresponding Eilenberg–Mac Lane space, that is, of any CW-complex whose homotopy groups are trivial except the second one which is A. In particular, by classical results, the Thomason homology of $K(\mathbb{Z}, 2)$ is the homology of $\mathbb{C}\mathrm{P}^\infty$, the infinite-dimensional complex projective space, and is hence $\mathbb{Z}$ in even degree and null in odd degree. Moreover, by [10, Theorem 4.9], the inclusion 2-functor $K(\mathbb{N}, 2) \hookrightarrow K(\mathbb{Z}, 2)$ induces an isomorphism in homology. We thus have

$$H_n^{\mathrm{Th}}(K(\mathbb{N}, 2)) = \begin{cases} \mathbb{Z} & \text{if } n \text{ is even,} \\ 0 & \text{if } n \text{ is odd.} \end{cases}$$

But $K(\mathbb{N}, 2)$ is freely generated by the unique 2-polygraph P with

$$P_0 = \{\star\}, \quad P_1 = \emptyset, \quad P_2 = \{\alpha\}.$$

Its polygraphic homology is hence the homology of its linearization

$$\mathbb{Z} \longleftarrow 0 \longleftarrow \mathbb{Z} \longleftarrow 0 \longleftarrow \cdots$$

and we have

$$H_n^{\mathrm{Pol}}(K(\mathbb{N}, 2)) = \begin{cases} \mathbb{Z} & \text{if } n = 0 \text{ or } n = 2, \\ 0 & \text{otherwise.} \end{cases}$$

In particular, $H_4^{\mathrm{Pol}}(K(\mathbb{N}, 2)) \not\simeq H_4^{\mathrm{Th}}(K(\mathbb{N}, 2))$.

23

Resolutions by $(\omega, 1)$-Polygraphs

Anick and Green constructed the first explicit free resolutions for algebras from a presentation of relations by non-commutative Gröbner bases [7–9, 152]. Their constructions provide resolutions to compute homological invariants, such as homology groups, Hilbert and Poincaré series of algebras presented by generators and relations given by a Gröbner basis. The chains of these resolutions are defined by iterated overlaps of the leading terms of the Gröbner basis and the differentials are constructed by noetherian induction. Similar methods for calculating free resolutions for monoids and algebras, inspired by string rewriting mechanisms, have been developed in numerous works [60, 154, 219, 221]. A purely polygraphic approach to the construction of these resolutions by rewriting has been developed in [163] using the notion of $(\omega, 1)$-polygraphic resolution, where the mechanism for proving the acyclicity of the resolution relies on the construction of a normalization strategy extended in all dimensions. The construction of polygraphic resolutions by rewriting has also been applied to the case of associative algebras in [160] and shuffle operads in [269], introducing in each case a notion of polygraph adapted to the algebraic structure.

In this chapter, we show how to construct a polygraphic resolution of a category from a convergent presentation of that category and how to deduce an abelian version of such a resolution. The notion of polygraphic resolution of an ω-category was introduced in Section 19.3: it consists of a polygraph which is weakly equivalent to the category. We consider here a variant of this notion adapted to $(\omega, 1)$-categories, related to the folk model structure on the category $\mathbf{Cat}_{\omega,1}$ of $(\omega, 1)$-categories. The chapter is organized as follows. In Section 23.1, we introduce the notion of contraction with respect to a unital section, which we often use to show that an $(\omega, 1)$-polygraph is acyclic. In Section 23.2, we show how to compute a cofibrant replacement of a category in the category $\mathbf{Cat}_{\omega,1}$ from one of its convergent presentations. This construction extends to higher dimensions the one given in low dimensions in Chapter 7 in terms of coherent presentations. In Section 23.3, we explain how to deduce an abelian resolution from a resolution by an $(\omega, 1)$-polygraph, thus again extending the constructions given in low dimensions in Chapter 9. We deduce from this resolution several homological and homotopical finiteness conditions for finite convergence. In Section 23.4, we extend the results about finite derivation type presented in Chapter 8.

23.1 Polygraphic Resolutions and Contractions

In this section, we consider the folk model structure on $\mathbf{Cat}_{\omega,1}$ constructed in Theorem 21.3.5 and described in §21.3.9, and the notion of oplax transformation between ω-functors as defined in §20.2.14.

23.1.1 Polygraphic resolution of an $(\omega, 1)$-category. A *polygraphic resolution of a category C in* $\mathbf{Cat}_{\omega,1}$ is a pair (P, p) made of an $(\omega, 1)$-polygraph P and a trivial fibration $p\colon P^\top \to C$, where $P^\top$ is the free $(\omega, 1)$-category generated by P. Expanding the definition, P is a polygraphic resolution of C if and only if it presents C and, for every $n \geqslant 2$, the extension P_{n+1} of $P^\top_{\leqslant n}$ is acyclic.

23.1.2 Unital sections and essential cells. Let P be an $(\omega, 1)$-polygraph. For u a 1-cell of the quotient category $\overline{P}$, we denote by $P^\top_u$ the corresponding fiber of the canonical projection $\pi\colon P^\top \twoheadrightarrow \overline{P}$. By definition, $P^\top_u$ is an ω-groupoid, whose 0-cells are the representatives of u in $P^\top$. To avoid confusion, we keep the dimensions of the $(\omega, 1)$-category $P^\top$ when talking about the cells and compositions of $P^\top_u$.

A *unital section of P* is a family

$$\iota = \left(\iota_u\colon \mathbf{1} \to P^\top_u\right)_{u \in \overline{P}}$$

of ω-functors, satisfying $\iota_{1_x} = 1_{1_x}$ for every 0-generator x of P. Such a family of functors assigns to every 1-cell u of $\overline{P}$ a representative 1-cell ι_u in $P^\top$, in such a way that identities are mapped to identities. A unital section of P is almost a functorial section of the canonical projection $\pi\colon P^\top \to \overline{P}$, except that it is not defined in dimension 0 and no specific compatibility with the 0-composition is required.

Fix a unital section ι of P. If ϕ is an n-cell of $P^\top$, we will write $\widehat{\phi}$ for $\iota\pi(\phi)$ when no confusion occurs. Note that $\widehat{\phi}$ is an identity if $n \geqslant 2$. A 1-cell u of $P^\top$ is

ι*-reduced* if $u = \widehat{u}$ holds. A non-ι-reduced 1-cell u of $P^\top$ is ι*-essential* if $u = av$, with a a 1-generator of P and v an ι-reduced 1-cell of $P^\top$.

23.1.3 Contractions. Let P be an $(\omega, 1)$-polygraph, and ι be a unital section of P. An ι*-contraction of* P is a family

$$\sigma = \left(\begin{array}{ccc} P_u^\top & \xrightarrow{\;1_{P_u^\top}\;} & P_u^\top \\ & \Downarrow \sigma_u & \\ & P_u^\top \xrightarrow{\varepsilon} \mathbf{1} \xrightarrow{\iota} P_u^\top & \end{array} \right)_{u \in \overline{P}}$$

of oplax transformations such that $\sigma_{\sigma_\phi} = 1_{\sigma_\phi}$ and $\sigma_{\iota_u} = 1_{\iota_u}$ for every cell ϕ in $P^\top$ and 1-cell u in $\overline{P}$, where σ_ψ is a short notation for $(\sigma_{\widehat{\psi}})_\psi$. An ι-contraction is thus almost an oplax transformation from $1_{P^\top}$ to $\iota\varepsilon$, but, like $\iota\varepsilon$, it is not defined on 0-cells and no specific compatibility with the 0-composition is required.

Fix an ι-contraction σ of P. By definition of σ, for all $n \geqslant 1$, n-cell ϕ of $P^\top$, and $1 \leqslant k < n$,

$$s_k(\sigma_\phi) = \phi *_1 \sigma_{t_1(\phi)} *_2 \cdots *_k \sigma_{t_k(\phi)} \quad \text{and} \quad t_k(\sigma_\phi) = \begin{cases} \widehat{\phi} & \text{if } k = 1, \\ \sigma_{s_k(\phi)} & \text{otherwise.} \end{cases}$$

An n-cell ϕ of $P^\top$ is σ*-reduced* if it is an identity or in the image of σ.

23.1.4 Sided contractions. We say that an ι-contraction σ is *right* if, for all $n \geqslant 1$ and n-cells ϕ, ψ of $P^\top$ of respective 1-sources u and v, it satisfies

$$\sigma_{\phi\psi} = u\sigma_\psi *_1 \sigma_{\phi\widehat{v}}. \tag{23.1}$$

Symmetrically, an ι-contraction is *left* if for all $n \geqslant 1$ and n-cells ϕ, ψ of $P^\top$ of respective 1-sources u and v, it satisfies

$$\sigma_{\phi\psi} = \sigma_\phi v *_1 \sigma_{\widehat{u}\psi}. \tag{23.2}$$

In the sequel, we will consider right ι-contractions, however, the definitions and results admit a left version.

If σ is a right ι-contraction of P, and $n \geqslant 1$, a non-σ-reduced n-cell ϕ of $P^\top$ is σ*-essential* if there exist an n-generator α of P and an ι-reduced 1-cell u of $P^\top$ such that $\phi = \alpha u$.

23.1.5 Lemma ([163, Corollary 3.3.5])**.** *Let P be an $(\omega, 1)$-polygraph, and ι be a unital section of P. A right ι-contraction σ of P is uniquely and entirely determined by its values on the ι-essential 1-cells of $P^\top$ and, for every $n \geqslant 1$, on the σ-essential n-cells of $P^\top$.*

Proof. If σ is a right ι-contraction, then its values are prescribed on every cell of $P^\top$ that is not ι-essential or σ-essential. Now, the values of σ on ι-essential and σ-essential cells of $P^\top$ can be chosen freely (with correct source and target), provided that these values make σ compatible with all the defining relations of the structure of $(\omega, 1)$-category, and in particular with exchange relations between the 0-composition and the other compositions. It turns out that (23.1) imposes compatibility with these exchange relations. □

23.1.6 Theorem. *Let P be an $(\omega, 1)$-polygraph, and ι be a unital section of P. The canonical projection $\pi\colon P^\top \to \overline{P}$ is a trivial fibration in* $\mathbf{Cat}_{\omega,1}$ *if and only if P admits a right ι-contraction.*

Proof. Assume that $\pi\colon P^\top \to \overline{P}$ is a trivial fibration. Let us define a right ι-contraction σ of P thanks to Lemma 23.1.5. If au is an ι-essential 1-cell of the free $(\omega, 1)$-category $P^\top$, then $\pi(au) = \pi(\widehat{au})$, so that, by definition of $\overline{P}$, there exists a 2-cell

$$\sigma_{au}\colon au \to \widehat{au}$$

in $P^\top$. Assume that σ is defined on the n-cells of $P^\top$, for $n \geqslant 1$, and let αu be a σ-essential n-cell of $P^\top$. The n-cells $s(\sigma_{\alpha u})$ and $t(\sigma_{\alpha u})$ are parallel, so, by hypothesis, there exists an $(n{+}1)$-cell

$$\sigma_{\alpha u}\colon s(\sigma_{\alpha u}) \to t(\sigma_{\alpha u})$$

in $P^\top$.

Conversely, let σ be an ι-contraction of P, and ϕ, ψ be parallel n-cells of $P^\top$, for $n \geqslant 1$. We have $t(\sigma_\phi) = \sigma_{s(\phi)} = \sigma_{s(\psi)} = t(\sigma_\psi)$ by hypothesis, so that the $(n{+}1)$-cell $\sigma_\phi *_n \sigma_\psi^-$ is well defined, with source $s(\sigma_\phi)$ and target $s(\sigma_\psi)$. The fact that $t_k(\phi) = t_k(\psi)$ holds for every $0 \leqslant k < n$ implies that

$$(\sigma_\phi *_n \sigma_\psi)^- *_{n-1} \sigma^-_{t_{n-1}(\phi)} *_{n-2} \cdots *_0 \sigma^-_{t_0(\phi)}$$

is a well-defined $(n{+}1)$-cell of $P^\top$, with source ϕ and target ψ, thus proving that P_{n+1} is acyclic. □

This theorem shows that to prove that an $(\omega, 1)$-polygraph P is a polygraphic resolution of the category $\overline{P}$, it suffices to provide it with a right ι-contraction. In the next section, we show how to construct such a contraction from a convergent presentation of the category $\overline{P}$.

23.2 Polygraphic Resolutions From Convergence

23.2.1 The cells of the Squier polygraphic resolution. Assume that P is a 2-polygraph. Define $\mathrm{Sq}(P)$ as the graded set $(\mathrm{Sq}_n(P))_{n \geqslant 0}$ where

1. $\mathrm{Sq}_0(P) = P_0$ and $\mathrm{Sq}_1(P) = P_1$,
2. for $n \geqslant 2$, $\mathrm{Sq}_n(P)$ is the set of tuples $(u_1, \ldots, u_n)$, written $u_1|\cdots|u_n$, of non-identity reduced 1-cells of P^* such that

   – u_1 is a 1-generator of P,

   – for every $1 \leqslant i < n$, the 1-cell $u_i u_{i+1}$ is not reduced,

   – for every $1 \leqslant i < n$, every proper left-factor of $u_i u_{i+1}$ is reduced.

23.2.2 Interpretation in the reduced case. Assume that P is a reduced 2-polygraph. Then $u_1|u_2$ is a 2-generator of $\mathrm{Sq}(P)$ if and only if u_1 is a 1-generator of P and $u_1 u_2$ is the source of a 2-generator of P.

From the classification of critical branchings of a 2-polygraph given in §4.3.9, the critical branchings of the reduced polygraph P are of the form

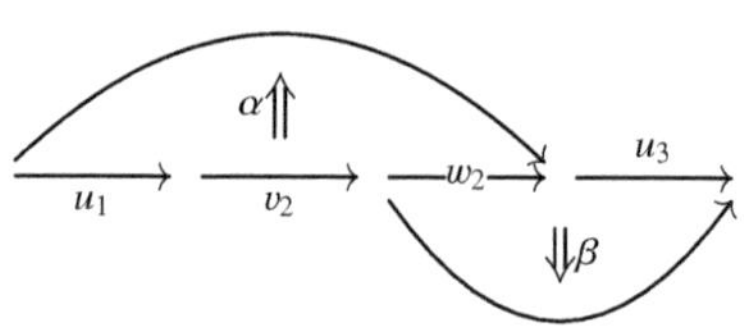

where α and β are 2-generators of P, and $u_1 v_2$, w_2, and u_3 are reduced non-identity 1-cells of P^*, with u_1 a 1-generator of P. Putting $u_2 = v_2 w_2$ induces a one-to-one correspondence between the 3-generators $u_1|u_2|u_3$ of $\mathrm{Sq}_3(P)$ and the critical branchings of P whose source is $u_1 u_2 u_3$.

For $n \geqslant 3$, define the *critical n-branchings of P* as the non-ordered families $(\alpha_1, \ldots, \alpha_n)$ of rewriting steps of P with the same source, overlapping in a non-trivial and minimal way. Conducting a similar analysis as in §4.3.9 shows that the critical 3-branchings of P fall in one of the two cases

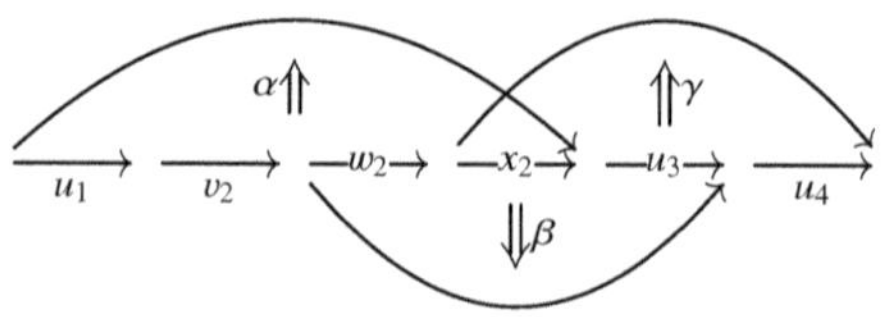

(23.3)

or

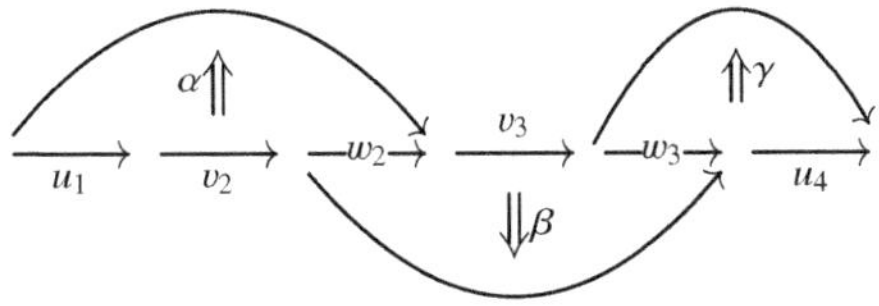 (23.4)

where α, β, and γ are 2-generators of P, and u_1v_2, w_2, u_3, and u_4 in (23.3) or u_1v_2, w_2, v_3w_3, and u_4 in (23.4) are reduced non-identity 1-cells of P^*, with u_1 a 1-generator of P. Putting $u_2 = v_2w_2x_2$, in (23.3), or $u_2 = v_2w_2$ and $u_3 = v_3w_3$, in (23.4), induces a one-to-one correspondence between the 4-generators $u_1|u_2|u_3|u_4$ of $\mathrm{Sq}_4(P)$ and the critical 3-branchings of P. This observation generalizes to establish a bijection between the $(n+1)$-generators of $\mathrm{Sq}(P)$ and the critical n-branchings of P.

23.2.3 Theorem. *Let P be a convergent 2-polygraph. There exists a unique structure of $(\omega, 1)$-polygraph on $\mathrm{Sq}(P)$, and unique unital section ι and right ι-contraction σ of $\mathrm{Sq}(P)$, that satisfy $\iota_u = \widehat{u}$, for every 1-cell u of P^*, and*

$$\sigma_{(u_1|\cdots|u_n)u_{n+1}} = \begin{cases} u_1|\cdots|u_{n+1} & \text{if } u_1|\cdots|u_{n+1} \in \mathrm{Sq}_{n+1}(P), \\ 1_{(u_1|\cdots|u_n)u_{n+1}} & \text{if } u_nu_{n+1} \text{ is reduced,} \end{cases} \tag{23.5}$$

for all n-generator $u_1|\cdots|u_n$ of $\mathrm{Sq}_n(P)$ with $n \geqslant 1$, and reduced 1-cell u_{n+1} of P^*. Moreover, this structure makes $\mathrm{Sq}(P)$ a polygraphic resolution of the category $\overline{P}$.

The polygraphic resolution $\mathrm{Sq}(P)$ thus constructed is called the *Squier polygraphic resolution* of the category $\overline{P}$ with respect to the presentation P.

Proof. If the condition (23.5) is satisfied, then the source and target maps of $\mathrm{Sq}(P)$ are imposed by the first case and the definition of an ι-contraction. Indeed, writing $\underline{u} = u_1|\cdots|u_{n-1}$, we must have

$$s(u_1|\cdots|u_n) = s(\sigma_{\underline{u}u_n}) = \underline{u}u_n *_1 \sigma_{t_1(\underline{u})u_n} *_2 \cdots *_{n-1} \sigma_{t_{n-1}(\underline{u})u_n},$$

and

$$t(u_1|\cdots|u_n) = t(\sigma_{\underline{u}u_n}) = \begin{cases} \widehat{u_1u_2} & \text{if } n = 2, \\ \sigma_{s(\underline{u})u_n} & \text{otherwise.} \end{cases}$$

Then we prove, using the definition of the source and target of an ι-contraction, that these source and target maps satisfy the globular relations. Next, according to Lemma 23.1.5, it is necessary and sufficient to define σ on the ι-essential and σ-essential cells of $\mathrm{Sq}(P)^\top$.

The ι-essential 1-cells are the 1-cells u_1u_2, where u_1 is a 1-generator of P, u_2 is a reduced 1-cell of P^*, and u_1u_2 is not reduced. If $u_1|u_2$ is a 2-generator of $\mathrm{Sq}(P)$, then (23.5) imposes $\sigma_{u_1u_2} = u_1|u_2$. Otherwise, there exists a proper factorization $u_2 = v_2w_2$ such that $u_1|v_2$ is a 2-generator of $\mathrm{Sq}(P)$, and (23.5) reads $\sigma_{(u_1|v_2)w_2} = 1_{(u_1|v_2)w_2}$. This last equality imposes that the source and target of $\sigma_{(u_1|v_2)w_2}$ must be equal, giving the value of σ on u_1u_2:

$$\sigma_{u_1u_2} = t(\sigma_{(u_1|v_2)w_2}) = s(\sigma_{(u_1|v_2)w_2}) = (u_1|v_2)w_2 *_1 \sigma_{\widehat{u_1v_2}w_2}.$$

Now, fix $n \geqslant 2$. The σ-essential n-cells of $\mathrm{Sq}(P)^\top$ are the $\underline{u}u_{n+1}$, where $\underline{u} = u_1|\cdots|u_n$ is an n-cell of $\mathrm{Sq}(P)$, and u_{n+1} is a reduced 1-cell of P^*. We distinguish three cases. First, if $\underline{u}|u_{n+1}$ is an $(n+1)$-generator of $\mathrm{Sq}(P)$, then (23.5) imposes $\sigma_{\underline{u}u_{n+1}} = \underline{u}|u_{n+1}$. Second, if u_nu_{n+1} is reduced, then (23.5) gives $\sigma_{\underline{u}u_{n+1}} = 1_{\underline{u}u_{n+1}}$. Otherwise, there exists a proper factorization $u_{n+1} = v_{n+1}w_{n+1}$ such that $\underline{u}|v_{n+1}$ is an $(n+1)$-generator of $\mathrm{Sq}(P)$. In that case, (23.5) implies that the source and the target of $\sigma_{(\underline{u}|v_{n+1})w_{n+1}}$ are equal. On the one hand, we have

$$s(\sigma_{(\underline{u}|v_{n+1})w_{n+1}}) = (\underline{u}|v_{n+1})w_{n+1} *_1 \sigma_{t_1(\underline{u}|v_{n+1})w_{n+1}} *_2 \cdots *_n \sigma_{t_n(\underline{u}|v_{n+1})w_{n+1}},$$

and, on the other hand, we obtain

$$\begin{aligned} t(\sigma_{(\underline{u}|v_{n+1})w_{n+1}}) &= \sigma_{s(\underline{u}|v_{n+1})w_{n+1}} = \sigma_{s(\sigma_{\underline{u}v_{n+1}})w_{n+1}} \\ &= \sigma_{\underline{u}u_{n+1} *_1 \sigma_{t_1(\underline{u})v_{n+1}} w_{n+1} *_2 \cdots *_n \sigma_{t_n(\underline{u})v_{n+1}} w_{n+1}}. \end{aligned}$$

Using the compatibility of σ with the compositions $*_i$ for $1 \leqslant i \leqslant n$, we develop the latter expression, by induction on n, to obtain a composite $(n+1)$-cell containing $\sigma_{\underline{u}u_{n+1}}, \sigma_{\sigma_{t_n(\underline{u})v_{n+1}} w_{n+1}}$, and lower-dimensional invertible cells. Thus, we obtain a relation between two composite $(n+1)$-cells that defines $\sigma_{\underline{u}u_{n+1}}$ in terms of the other involved cells.

Finally, we apply Theorem 23.1.6 to conclude that $\mathrm{Sq}(P)$ is a polygraphic resolution of the category C. □

23.2.4 Interpretation in the reduced case. Assume that P is a reduced convergent 2-polygraph, and let us examine the first dimensions of $\mathrm{Sq}(P)$.

The 2-generators $a|u$ of $\mathrm{Sq}_2(P)$, for a a 1-generator of P and u a reduced 1-cell of P^* such that au is not reduced, have the shape

$$a|u\colon au \Rightarrow \widehat{au}.$$

The ι-contraction σ is given, on a 1-cell au of P^* with $a \in P_1$ and u reduced, by

$$\sigma_{au} = \begin{cases} a|u & \text{if } a|u \in \mathrm{Sq}_2(P), \\ 1_{au} & \text{if } au \text{ is reduced}, \\ (a|v)w *_1 \sigma_{\widehat{av}w} & \text{if } u = vw \text{ with } a|v \in \mathrm{Sq}_2(P). \end{cases}$$

On more general 1-cells, σ is defined by the fact that it is a right ι-contraction, and the relation

$$\sigma_{uv} = u\sigma_v *_1 \sigma_{\widehat{u\widehat{v}}}.$$

By construction, the 3-generators of $\mathrm{Sq}_3(P)$ have the shape

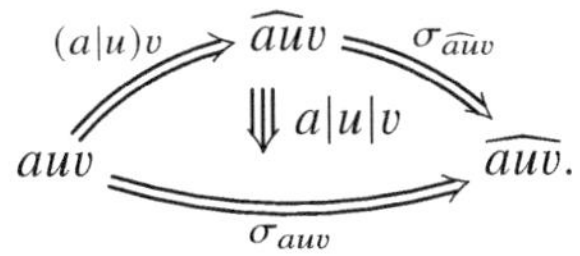

The ι-contraction σ is defined on the 2-generators $(a|u)v$ by (23.5). The simple cases are $\sigma_{(a|u)v} = a|u|v$, if the latter belongs to $\mathrm{Sq}_3(P)$, and $\sigma_{(a|u)v} = 1_{(a|u)v}$ if uv is reduced. The more complicated case is the definition of $\sigma_{(a|u)vw}$ when $a|u|v$ belongs to $\mathrm{Sq}_3(P)$. In this situation, the relation $\sigma_{(a|u|v)w} = 1_{(a|u|v)w}$ implies $s(\sigma_{(a|u|v)w}) = t(\sigma_{(a|u|v)w})$, which develops into

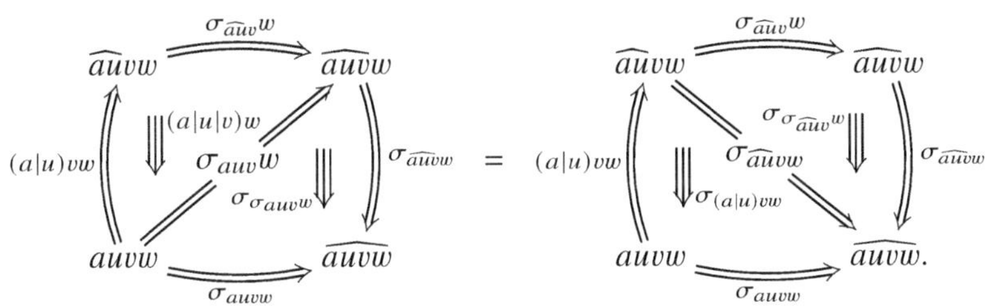

Finally, the 4-generators of $\mathrm{Sq}_4(P)$ have the same shape as this last defining equation, but in the case where vw is not reduced and with all proper left-factors reduced:

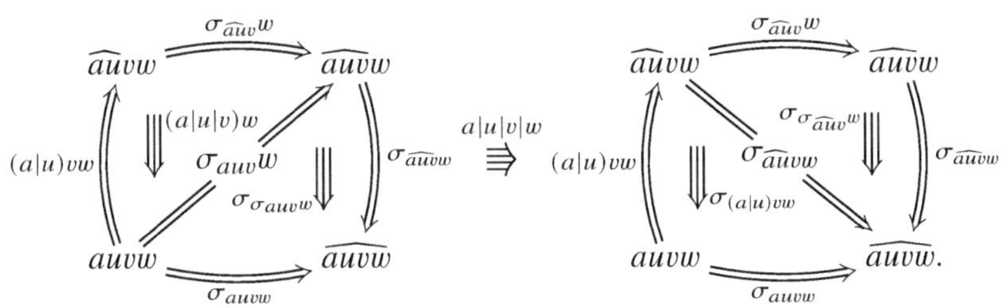

23.3 Abelianization of Polygraphic Resolutions

In Section 22.3, we saw that the polygraphic homology of a monoid, seen as an ω-category, coincides with its integral homology. In the same spirit, in this section we show how to deduce the homology of a small category with coefficients in natural systems from one of its resolutions in $\mathbf{Cat}_{\omega,1}$. In particular, we show

how to associate a resolution of natural systems to a polygraphic resolution of a category in $\mathbf{Cat}_{\omega,1}$, and we illustrate this construction with examples of polygraphic resolutions calculated from convergent polygraphs.

23.3.1 Free natural systems. A natural system on a category C is a functor from the category FC of factorization of C and with values in the category **Ab**, see §F.2.2 for details. For a family X of 1-cells of C, we denote by $F_C[X]$ the free natural system on C generated by X and given by

$$F_C[X] = \bigoplus_{x \in X} \mathrm{F}C(x, -).$$

Fix an $(\omega, 1)$-polygraph P presenting C. We consider the free natural system $F_C[P_0]$ generated by the identity 1-cells 1_x, for $x \in P_0$. If u is a 1-cell of C, then $F_C[P_0]_u$ is the free abelian group generated by the pairs (v, w) of 1-cells of C such that $t(v) = s(w) = x$ and $vw = u$.

We also consider, for every natural number $n \geqslant 1$, the free natural system $F_C[P_n]$ generated by one copy of the 1-cell $\overline{\alpha}$ of C for each n-generator α of P. If u is a 1-cell of C, then $F_C[P_n]_u$ is the free abelian group generated by the triples (v, α, w), denoted by $v[\alpha]w$, made of an n-generator α of P, and 1-cells v and w of C, such that the composite $v\overline{\alpha}w$ is well defined in C and equal to u

The mapping of every 1-generator x of P to the element $[x]$ of $F_C[P_1]_{\overline{x}}$ is extended into a derivation of P_1^* into $F_C[P_1]$ by putting

$$[1_u] = 0 \qquad \text{and} \qquad [uv] = [u]\overline{v} + \overline{u}[v],$$

for all composable 1-cells u and v in C. Here, the natural system $F_C[P_1]$ on C is seen as a natural system on P_1^* by composition with the canonical projection $p\colon P_1^* \to \overline{P}$.

For $n > 1$, the mapping of every n-generator α of P to the element $[\alpha]$ of $F_C[P_n]_{\overline{\alpha}}$ is extended to associate to every n-cell ϕ of $P^\top$ the element $[\phi]$ of $F_C[P_n]_{\overline{\phi}}$, defined by induction on the size of ϕ as follows:

$$[1_\phi] = 0, \qquad [\phi^-] = -[\phi], \qquad [\phi *_k \psi] = \begin{cases} [\phi]\overline{\psi} + \overline{\phi}[\psi] & \text{if } k = 0, \\ [\phi] + [\psi] & \text{otherwise.} \end{cases}$$

23.3.2 Abelianization of polygraphic resolutions. Let P be an $(\omega, 1)$-polygraph. We denote by $F_{\overline{P}}[P]$ the complex

$$\cdots \longrightarrow F_{\overline{P}}[P_n] \xrightarrow{d_n} F_{\overline{P}}[P_{n-1}] \longrightarrow \cdots \xrightarrow{d_1} F_{\overline{P}}[P_0] \xrightarrow{\varepsilon} \mathbb{Z} \longrightarrow 0$$

of natural systems on $\overline{P}$, whose boundary maps are defined as follows. The augmentation morphism ε is defined, on every pair (u, v) of composable 1-cells of $\overline{P}$, by

$$\varepsilon(u, v) = 1.$$

For $n \geqslant 1$, the morphism d_n of natural systems on $\overline{P}$ is given, on the generator $[\alpha]$ corresponding to an n-cell α of P, by

$$d_n[\alpha] = \begin{cases} (1, \overline{\alpha}) - (\overline{\alpha}, 1) & \text{if } n = 1, \\ [t(\alpha)] - [s(\alpha)] & \text{otherwise.} \end{cases}$$

By induction on the size of cells of $P^{\top}$, we prove, for every n-generator ϕ in $P^{\top}$, with $n \geqslant 1$, that

$$d_n[\phi] = \begin{cases} (1, \overline{\phi}) - (\overline{\phi}, 1) & \text{if } n = 1, \\ [t(\phi)] - [s(\phi)] & \text{otherwise.} \end{cases}$$

As a consequence, we have $\varepsilon d_1 = 0$ and $d_n d_{n+1} = 0$, for every $n \geqslant 1$, proving that $F_{\overline{P}}[P]$ is indeed a chain complex.

23.3.3 Theorem. *If P is a polygraphic resolution of a category C, then $F_C[P]$ is a resolution by free natural systems on C of the constant natural system $\mathbb{Z}$.*

Proof. Let ι be a unital section of P. By Theorem 23.1.6, P admits a right ι-contraction σ. Let us consider the following families of morphisms of $\mathbb{Z}$-modules, indexed by a 1-cell w of C:

$$i_{-1} : \mathbb{Z} \to F_C[P_0]_w \qquad\qquad i_n : F_C[P_n]_u \to F_C[P_{n+1}]_w$$

defined by

- $i_{-1}(1) = (w, 1)$,
- $i_0(u, v) = u[\widehat{v}]$, for all 1-cells u, v of C such that $w = uv$,
- $i_n(u[\alpha]v) = u[\sigma_{\alpha\widehat{v}}]$, for all $n \geqslant 1$, n-generator α in P, and 1-cells u, v of C such that $w = u\overline{\alpha}v$.

By induction on the size of the n-cells of the free $(\omega, 1)$-category $P^{\top}$, using the properties of a right ι-contraction, we prove that

$$i_n(u[\phi]v) = u[\sigma_{\phi\widehat{v}}]$$

holds for all $n \geqslant 1$, n-cell ϕ of $P^{\top}$, and 1-cells u, v of C such that the composite $u\overline{\phi}v$ is well defined. We deduce that $(\sigma_n)_{n \geqslant 1}$ is a contracting homotopy for the complex $F_C[P]$. □

23.3.4 Homological syzygies. For $n \geqslant 1$, the kernel of the differential d_n defined in §23.3.2 is the natural system on $\overline{P}$ defined pointwise by

$$(h_n)_w = \ker \left(F_{\overline{P}}[P_n]_w \xrightarrow{d_n} F_{\overline{P}}[P_{n-1}]_w \right),$$

for every 1-cell w in $\overline{P}$. It is denoted by $h_n(P)$, and its elements are called the *homological n-syzygies of P*. As a consequence of Theorems 23.2.3 and 23.3.3, we obtain the following result.

23.3.5 Theorem. *Let C be a category and P a convergent presentation of C. Then, for every $n \geqslant 2$, the natural system $h_n(P)$ is generated by the elements*

$$d_n[u_1|\cdots|u_n] = [u_1|\cdots|u_{n-1}]\overline{u}_n + [\sigma_{t(u_1|\cdots|u_{n-1})u_n}] - [\sigma_{s(u_1|\cdots|u_{n-1})u_n}]$$

where $u_1|\cdots|u_n$ ranges over the n-generators of $\mathrm{Sq}(P)$*, and σ is the right ι-contraction associated to* $\mathrm{Sq}(P)$.

We now state a consequence of Theorems 23.2.3 and 23.3.3 for reduced presentations without critical 3-branchings. In §§23.2.2 and 23.2.4 we give an interpretation of the n-generators of the resolution $\mathrm{Sq}(P)$ when P is a reduced convergent 2-polygraph. In particular, there is a one-to-one correspondence between the 2-generators of the resolution $\mathrm{Sq}(P)$ and the critical branchings on the one hand and between the 3-generators and the critical 3-branchings on the other. Moreover, when P admits no critical 3-branchings, it also has no critical n-branchings for $n \geqslant 3$, and so $\mathrm{Sq}_n(P)$ is empty for $n \geqslant 3$. The result is thus as follows.

23.3.6 Corollary. *Let C be a category, and P a reduced convergent presentation of C without critical 3-branchings. Then the sequence*

$$0 \longrightarrow F_C[\mathrm{Sq}_2(P)] \xrightarrow{d_3} F_C[P_2] \xrightarrow{d_2} F_C[P_1] \xrightarrow{d_1} F_C[P_0] \xrightarrow{\varepsilon} \mathbb{Z} \longrightarrow 0$$

is a partial resolution of length 4 of the trivial natural system $\mathbb{Z}$.

This result was proved by Squier for presentation of monoids by string rewriting systems using a direct method based on the characterization of critical 3-branchings [326, Theorem 3.2]. It is useful for proving examples of categories or monoids of finite type that do not admit a finite convergent presentation, as we did in Chapter 9.

23.3.7 Convergence and homology of categories. The construction given in this section allows us to calculate the homology of a category C from a presentation of this category by a convergent 2-polygraph P. Indeed, by Theorem 23.2.3, $\mathrm{Sq}(P)$ is a polygraphic resolution of the category C, and by

Theorem 23.3.3, $F_C[\mathrm{Sq}(P)]$ is a resolution by free natural systems on C of the constant natural system $\mathbb{Z}$. The homology of C with coefficient in a contravariant natural system D on C, as defined in §F.2.3, is thus given by

$$H_*(C, D) = \mathrm{Tor}^{FC}(D, \mathbb{Z}) = \mathrm{H}_*(D \otimes_{FC} \mathrm{Sq}(P)).$$

23.3.8 Example: the reduced standard polygraphic resolution. Let C be a category. To simplify the example, assume that, if a composite morphism fg of C is an identity, then so are f and g. The *reduced standard presentation of* C is the 2-polygraph $\overline{\mathrm{Std}}_2(C)$ whose 0-generators are objects of C, with one 1-generator $\widehat{f}$ for each non-identity morphism f of C, and one 2-generator

$$f|g : \widehat{f}\widehat{g} \Rightarrow \widehat{fg}$$

for each pair (f, g) of composable non-identity morphisms in C. Without the simplifying hypothesis on C, the target of $f|g$ is replaced by 1_x if $fg = 1_x$ in C.

The 2-polygraph $\overline{\mathrm{Std}}_2(C)$ is reduced and convergent, and applying Theorem 23.2.3 extends it into a polygraphic resolution of C, denoted by $\overline{\mathrm{Std}}(C)$ and called the *reduced standard polygraphic resolution of* C. For $n \geqslant 2$, the n-generators of Std(C) are the $f_1|\cdots|f_n$, such that each f_i is a non-identity morphism of C and each (f_i, f_{i+1}) is composable.

The source and target of the 3-generators of $\overline{\mathrm{Std}}(C)$ are given by

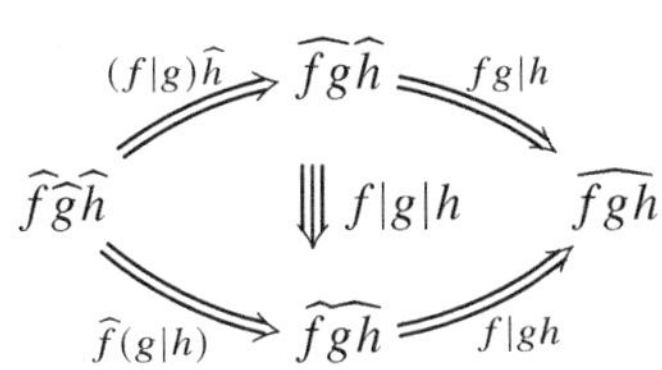

and a 4-generator $a|b|c|d$ has source

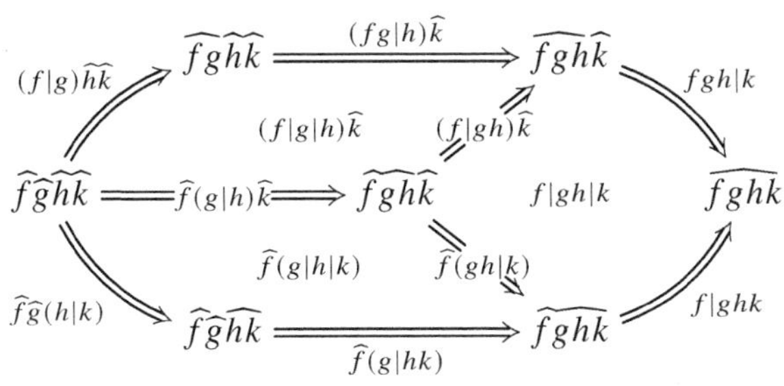

and target

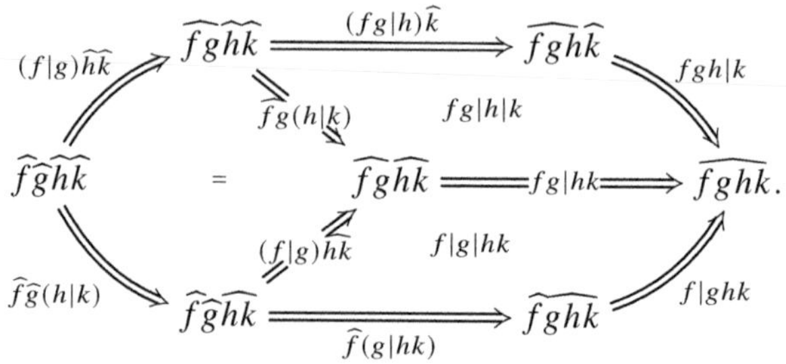

(with the arrows of 3-cells removed for clarity).

For n-cells, $n \geqslant 2$, we prove, by induction on n, that the source and target of n-generators are composites of the $(n-1)$-cells

$$d_i(f_1|\cdots|f_n) = \begin{cases} \widehat{f_1}(f_2|\cdots|f_n) & \text{if } i = 0, \\ f_1|\cdots|f_i f_{i+1}|\cdots|f_n & \text{if } 1 \leqslant i \leqslant n-1, \\ (f_1|\cdots|f_{n-1})\widehat{f_n} & \text{if } i = n, \end{cases}$$

with k-cells, for $1 < k < n-1$. More precisely, the source of $f_1|\cdots|f_n$ contains one copy of each $d_i(f_1|\cdots|f_n)$ for $n-i$ even, and its target, one copy of each $d_i(f_1|\cdots|f_n)$ for $n-i$ odd.

Theorem 23.3.3 applied to $\overline{\mathrm{Std}}(C)$ gives a free resolution

$$\cdots \longrightarrow F_C[\overline{\mathrm{Std}}_n(C)] \xrightarrow{d_n} F_C[\overline{\mathrm{Std}}_{n-1}(C)] \longrightarrow \cdots \longrightarrow F_C[C_0] \xrightarrow{\varepsilon} \mathbb{Z} \longrightarrow 0$$

with differential defined by

$$\begin{aligned} d_n[f_1|\cdots|f_n] &= (-1)^n f_1[f_2|\cdots|f_n] + \\ &\quad \sum_{i=1}^{n-1}(-1)^{n-i}[d_i(f_1|\cdots|f_n)] + [f_1|\cdots|f_{n-1}]f_n. \end{aligned}$$

23.3.9 The associative polygraphic resolution. Let A be the monoid with one non-trivial idempotent element, that is presented by the following 2-polygraph:

$$\mathrm{As}_2 = \langle\, a_0 \mid a_1 \mid a_2\colon a_1 a_1 \Rightarrow a_1 \,\rangle.$$

This polygraph is reduced and convergent, see Example 4.3.12, with one critical n-branching for every $n \geqslant 2$. Thus, the reduced standard polygraphic resolution

$\mathrm{As}_\omega = \mathrm{Sq}(\mathrm{As}_2)$ of A, given by Theorem 23.2.3, has one n-generator a_n for every $n \geqslant 0$, corresponding to the product $a_1|\cdots|a_1$ of n copies of a_1. The 3-generator a_3 of As_ω is given in classical notation and in string diagrams respectively as follows:

$$a_2 a_1 *_1 a_2 \overset{a_3}{\Rrightarrow} a_1 a_2 *_1 a_2$$

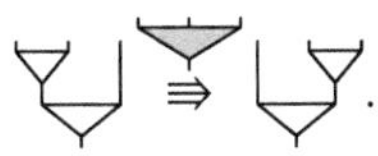

The 4-generator a_4 of As_ω is

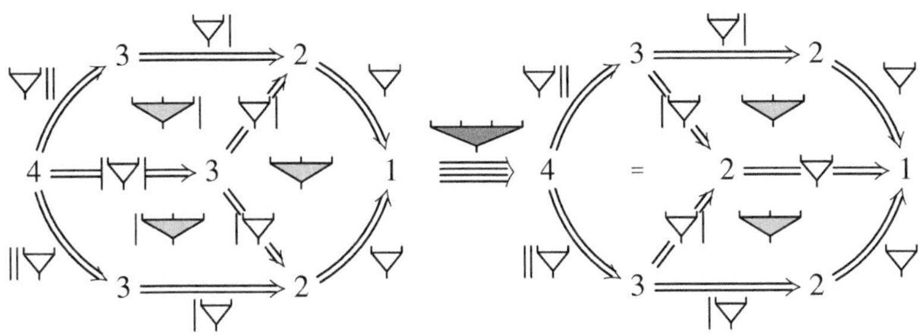

which, contracting by one dimension, can also be pictured as Mac Lane's pentagon or Stasheff's polytope K_4:

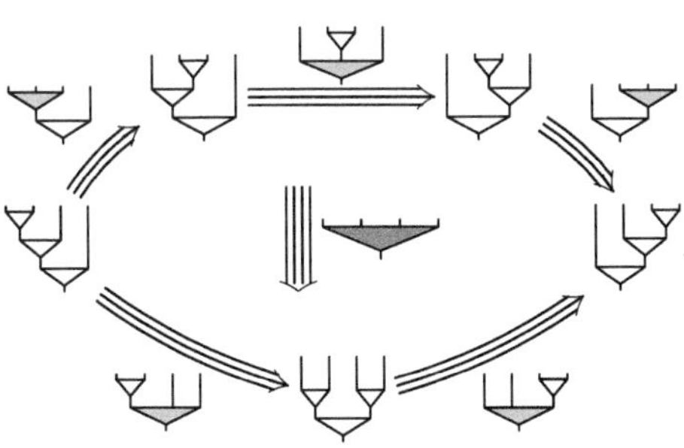

Finally, the 5-generator a_5 of As_ω has the shape of Stasheff's polytope K_5, its source being

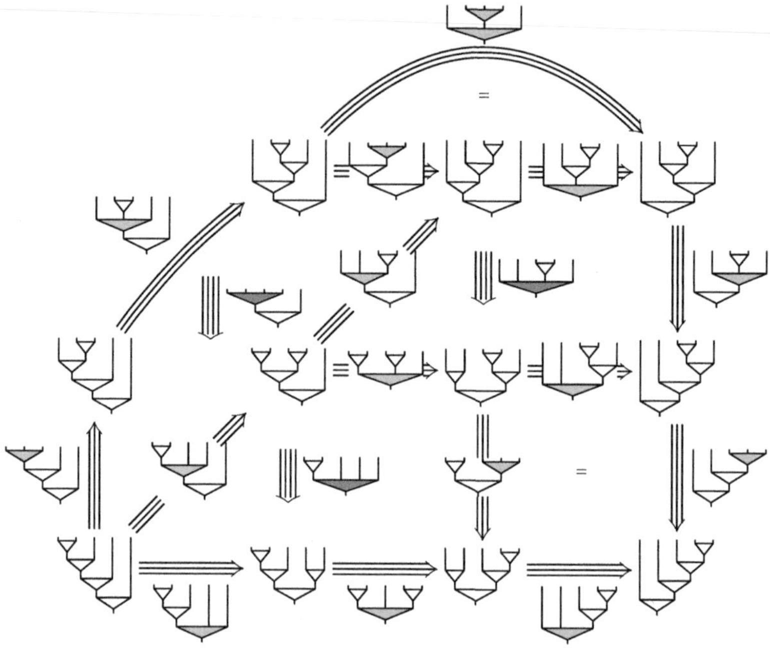

and its target being given by a symmetric composite 4-cell, see [163, Section 6.1]. Theorem 23.3.3, applied to As_ω, yields a resolution

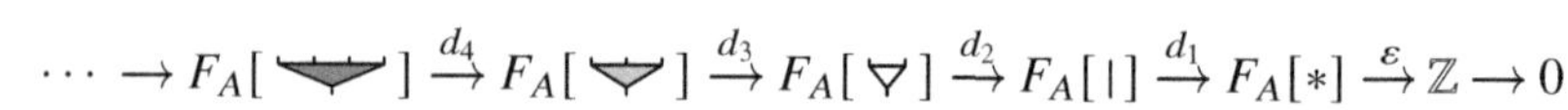

of $\mathbb{Z}$ by free natural systems on A. Computing this differential on each n-cell of As_ω gives generators of the natural systems of homological n-syzygies of As_ω. For example, $h_2(\mathrm{As})$ is generated by

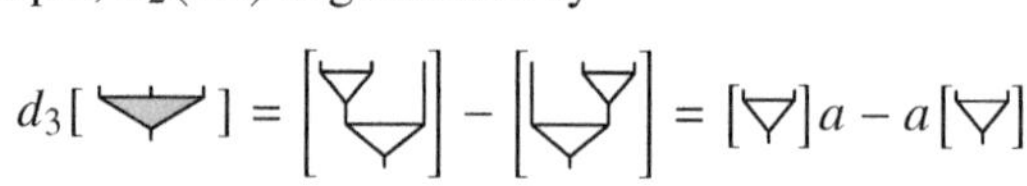

while $h_3(\mathrm{As})$ is generated by

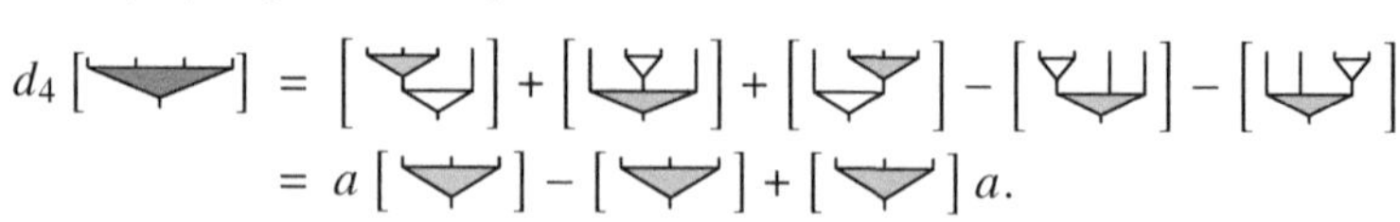

Similarly, $h_4(\mathrm{As})$ is generated by $d_5[a_5]$, which is equal, by definition, to

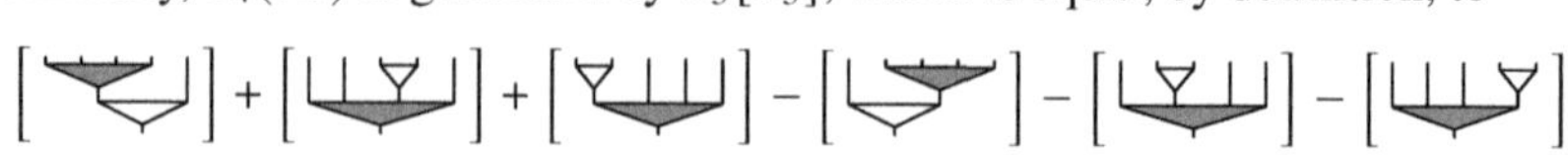

and reduces to

$$d_5[a_5] = [$$  $$]\, a - a\, [$$  $$].$$

23.3.10 The category of monotone surjections. We denote by Δ_μ the subcategory of the simplicial category whose objects are the natural numbers and whose morphisms from m to n are the monotone surjections from $\{0, \ldots, m\}$ to $\{0, \ldots, n\}$. This category, studied in [251], see also [261, Section 7.5, Exercise 3.(a)], admits a presentation by the 2-polygraph P with the natural numbers as 0-generators, with one 1-generators $x_i^n : n+1 \to n$ for all natural numbers $1 \leqslant i \leqslant n$, and one 2-generator

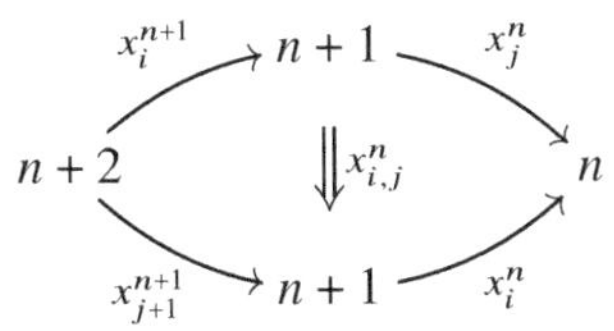

for all natural numbers $0 \leqslant i \leqslant j \leqslant n+1$. The 1-cell x_i^n represents the map

$$x_i^n(j) \;=\; \begin{cases} j & \text{if } j \leqslant i, \\ j-1 & \text{if } j > i. \end{cases}$$

This is a variant of the presentation constructed in §4.5.6. Thereafter, we drop the exponents of the 1-generators and 2-generators of P, simply writing x_i and $x_{i,j}$.

The 2-polygraph P is convergent. Indeed, for termination, given a 1-cell $u = x_{i_1} \ldots x_{i_k}$ of P^*, we define the natural number $\nu(u)$ as the number of pairs (i_p, i_q) such that $i_p \leqslant i_q$, with $1 \leqslant p < q \leqslant k$. In particular, we have $\nu(x_i x_j) = 1$ and $\nu(x_{j+1} x_i) = 0$ when $i \leqslant j$, giving $\nu(s(x_{i,j})) > \nu(t(x_{i,j}))$. Moreover, we have $\nu(wuw') > \nu(wvw')$ when $\nu(u) > \nu(v)$ holds. Thus, for every non-identity 2-cell $a \colon u \Rightarrow v$ of P^*, the strict inequality $\nu(u) > \nu(v)$ is satisfied, giving termination. Moreover, the 2-polygraph P has one critical branching $(x_{i,j} x_k, x_i x_{j,k})$ for all possible $0 \leqslant i \leqslant j \leqslant k \leqslant n+2$, which is confluent.

Theorem 23.2.3, applied to P, gives a polygraphic resolution $\mathrm{Sq}(P)$ of Δ_μ, whose 3-cells are given, in classical notation and in string diagrams (with $x_i = |_i$ and $x_{i,j} = \rtimes_{i,j}$) respectively, by

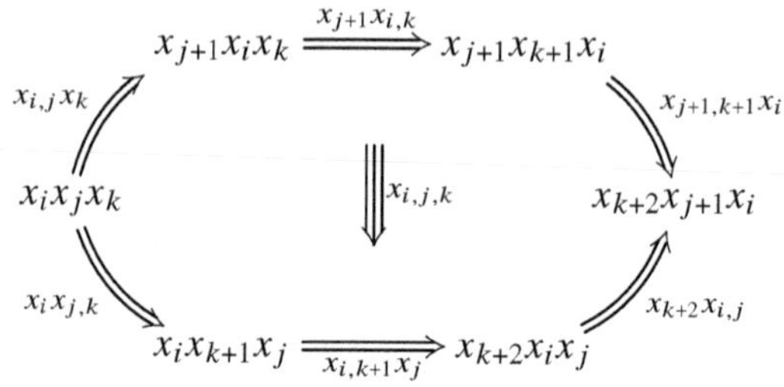

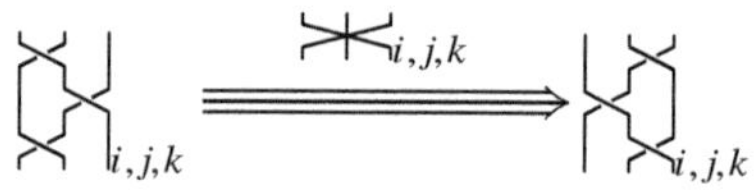

The $(\omega, 1)$-polygraph $\mathrm{Sq}(P)$ has one 4-generator $x_{i,j,k,l}$ for all possible indices $0 \leqslant i \leqslant j \leqslant k \leqslant l \leqslant n+3$, given in string diagrams and omitting the subscripts, by the following diagrams:

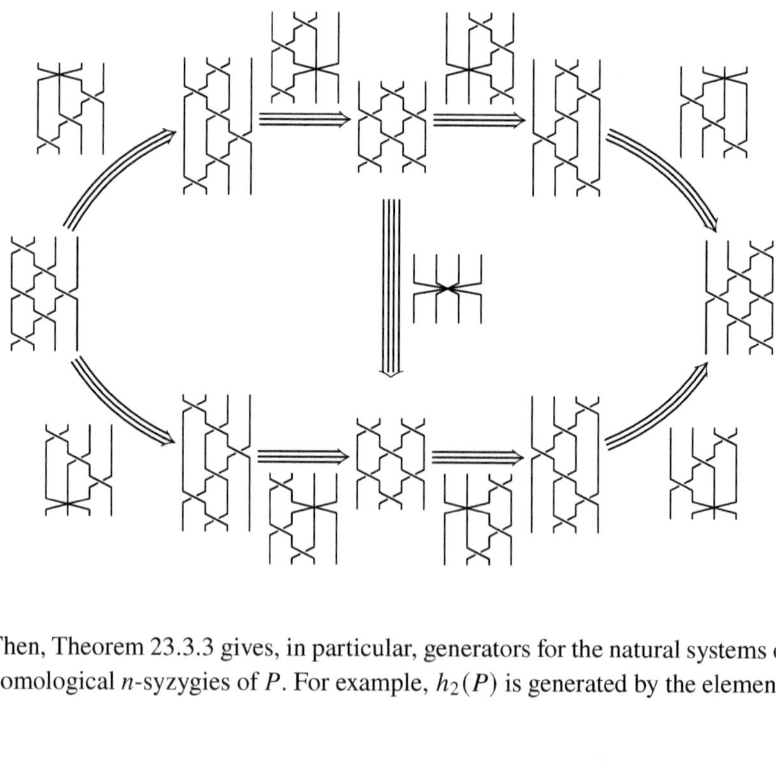

Then, Theorem 23.3.3 gives, in particular, generators for the natural systems of homological n-syzygies of P. For example, $h_2(P)$ is generated by the elements

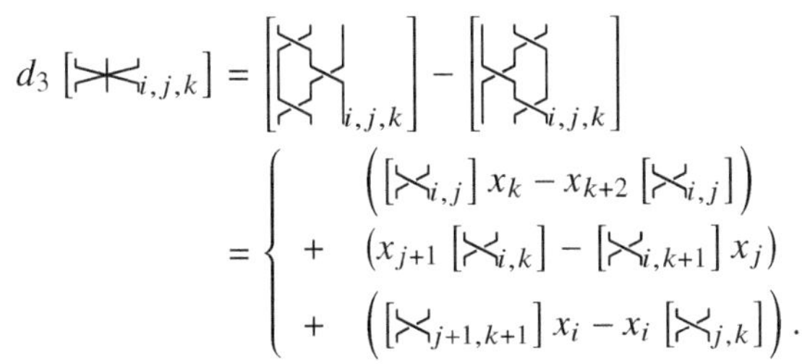

23.4 Categories of Finite Homological Type

In Chapter 8, we introduced the notion of a finite derivation type for categories. In this section, we show how to refine this notion in higher dimensions and how to relate it with finite homological type.

23.4.1 Higher-dimensional finite derivation type. Let C be a category and $n \in \mathbb{N} \cup \{\infty\}$. We say that C has *finite n-derivation type*, FDT_n for short, if it admits a polygraphic resolution P in $\mathbf{Cat}_{\omega,1}$ such that P_k is finite for every $k \leqslant n$. In particular, C has FDT_1 if it is finitely generated, FDT_2 if it is finitely presented, and FDT_3 if it has finite derivation type as defined in Section 8.1. By definition, FDT_∞ implies FDT_n, and FDT_{n+1} implies FDT_n, for every $n \geqslant 0$.

As an immediate consequence of Theorem 23.2.3, we deduce the following condition for finite convergence.

23.4.2 Theorem. *A category with a finite convergent presentation has* FDT_∞.

23.4.3 Categories of finite homological type. A category C is of *homological type* FP_n if the constant natural system $\mathbb{Z}$ on C is of homological type FP_n, see §F.3 for a summary on the notion of finite homological type for categories. As a consequence of Theorems 23.2.3 and 23.3.3, we obtain the following implications.

23.4.4 Theorem. *Let C be a category and $n \in \mathbb{N} \cup \{\infty\}$. If C has* FDT_n, *then it is of homological type* FP_n. *In particular, if C has a finite convergent presentation, then it is of homological type* FP_∞.

This result generalizes [98, Theorem 3.2] and [233, Theorem 3], stating that if a monoid has FDT, then it is FP_3 (see also [306]). It also generalizes Squier's homological theorem [326, Theorem 4.1], that says that a monoid admitting a finite convergent presentation is FP_3, and the extensions of Squier's result in [60, 154, 219] that prove that such a monoid is FP_∞.

23.4.5 Example. The 2-polygraph As_2 defined in §23.3.9 extends to a polygraphic resolution As_ω having one n-generator for every $n \geqslant 0$. Hence the polygraph As_2 and the presented monoid have FDT_∞.

23.5 Homological Syzygies and Identities Among Relations

In this section we establish an isomorphism between the natural systems of homological 2-syzygies and identities among relations for a category presented by a 2-polygraph. This is an extension to category presentations of a Brown-Huebschmann theorem in group theory that states an isomorphism between the modules of identities among relations and homological 2-syzygies for group presentations [64].

In this section, P denotes a 2-polygraph. We aim to build an isomorphism between the natural system $\Pi(P)$ of identities among relations of P defined in §8.3.2 and the natural system $h_2(P)$ of its homological 2-syzygies defined in §23.3.4.

23.5.1 Lemma. *Let P be a 2-polygraph. For every 2-loop ψ of $P^\top$, we have $[\psi] = 0$ in $F_{\overline{P}}[P_2]$ if and only if $\lfloor \psi \rfloor = 0$ holds in $\Pi(P)$.*

Proof. To prove that $\lfloor \psi \rfloor = 0$ implies $[\psi] = 0$, we check that the relations (8.2) and (8.3) defining $\Pi(P)$ are also satisfied in $F_{\overline{P}}[P_2]$. The first relation is given by the definition of the map $[-]$. The second relation is checked as follows:

$$[\psi *_1 \phi] = [\psi] + [\phi] = [\phi] + [\psi] = [\phi *_1 \psi].$$

Conversely, let us consider a 2-loop ψ in $P^\top$ with source w such that $[\psi] = 0$. We decompose ψ into

$$\psi = u_1 \alpha_1^{\epsilon_1} v_1 *_1 \cdots *_1 u_p \alpha_p^{\epsilon_p} v_p,$$

where α_i is a 2-generator, u_i, v_i are 1-cells of $P^\top$, and $\epsilon_i \in \{-, +\}$. Then we get

$$0 = [\psi] = \epsilon_1 \overline{u}_1 [\alpha_1] \overline{v}_1 + \ldots + \epsilon_p \overline{u}_p [\alpha_p] \overline{v}_p.$$

Since the natural system $F_{\overline{P}}[P_2]$ is freely generated by the elements $[\alpha]$ of $F_{\overline{P}}[P_2]_{\overline{\alpha}}$, for α a 2-generator, this implies the existence of a self-inverse permutation τ of $\{1, \ldots, p\}$ such that the following relations are satisfied:

$$\alpha_i = \alpha_{\tau(i)}, \qquad \overline{u}_i = \overline{u}_{\tau(i)}, \qquad \overline{v}_i = \overline{v}_{\tau(i)}, \qquad \epsilon_i = -\epsilon_{\tau(i)}.$$

Let us denote, for every $1 \leqslant i \leqslant p$, the source and target of $\alpha_i^{\epsilon_i}$ by w_i and w_i', respectively. They satisfy $\overline{w}_i = \overline{w}_i'$. We also fix a section $\widehat{\cdot}$ and a left strategy σ for the 2-polygraph P, so that $\widehat{u} = \widehat{v}$ for all 1-cells u and v such that $\overline{u} = \overline{v}$.

For every $1 \leqslant i \leqslant p$, we denote by ψ_i the following 2-cell of $P^\top$:

$$\psi_i = \sigma^-_{u_i w_i v_i} \star_1 u_i \alpha_i^{\epsilon_i} v_i \star_1 \sigma_{u_i w_i' v_i}.$$

Using the facts that $w = u_1 w_1 v_1 = u_p w'_p v_p$ and that $u_i w_i' v_i = u_{i+1} w_{i+1} v_{i+1}$ for every $1 \leqslant i < p$, we can write the 2-loop ψ as the following composite:

$$\psi = \sigma_w \star_1 \psi_1 \star_1 \cdots \star_1 \psi_p \star_1 \sigma^-_{w}.$$

As a consequence, we get

$$\lfloor \psi \rfloor = \left\lfloor \sigma^-_{w} \star_1 \psi \star_1 \sigma_w \right\rfloor = \lfloor \psi_1 \rfloor + \ldots + \lfloor \psi_p \rfloor .$$

In order to conclude, we prove that the equality $\left\lfloor \psi_{\tau(i)} \right\rfloor = - \lfloor \psi_i \rfloor$ holds for every $1 \leqslant i \leqslant p$. Since σ is a left strategy, we have

$$\sigma_{u_i w_i v_i} = \sigma_{u_i} w_i v_i \star_1 \sigma_{\widehat{u}_i w_i} v_i \star_1 \sigma_{\widehat{u_i w_i} v_i}$$

and, using the fact that $\widehat{u_i w_i} = \widehat{u_i w_i'}$, we have

$$\sigma_{u_i w_i' v_i} = \sigma_{u_i} w_i' v_i \star_1 \sigma_{\widehat{u}_i w_i'} v_i \star_1 \sigma_{\widehat{u_i w_i} v_i}.$$

This gives

$$\begin{aligned}
\lfloor \psi_i \rfloor &= \left\lfloor \sigma^-_{\widehat{u_i w_i} v_i} \star_1 \sigma^-_{\widehat{u}_i w_i} v_i \star_1 \sigma^-_{u_i} w_i v_i \star_1 u_i \alpha_i^{\epsilon_i} v_i \star_1 \sigma_{u_i} w_i' v_i \star_1 \sigma_{\widehat{u}_i w_i'} v_i \star_1 \sigma_{\widehat{u_i w_i} v_i} \right\rfloor \\
&= \left\lfloor \sigma^-_{\widehat{u}_i w_i} v_i \star_1 \sigma^-_{u_i} w_i v_i \star_1 u_i \alpha_i^{\epsilon_i} v_i \star_1 \sigma_{u_i} w_i' v_i \star_1 \sigma_{\widehat{u}_i w_i'} v_i \right\rfloor \\
&= \left\lfloor \sigma^-_{\widehat{u}_i w_i} \star_1 \widehat{u}_i \alpha_i^{\epsilon_i} \star_1 \sigma_{\widehat{u}_i w_i'} \right\rfloor \overline{v}_i.
\end{aligned}$$

Now, let us compute $\left\lfloor \psi_{\tau(i)} \right\rfloor$. We already know that $\alpha_{\tau(i)} = \alpha_i$ and $\epsilon_{\tau(i)} = -\epsilon_i$. As a consequence, we get $w_{\tau(i)} = w_i'$ and $w'_{\tau(i)} = w_i$. Moreover, we have $\widehat{u}_{\tau(i)} = \widehat{u}_i$, so that we have:

$$\begin{aligned}
\left\lfloor \psi_{\tau(i)} \right\rfloor = & \Big\lfloor \sigma^-_{\widehat{u_i w_i} v_{\tau(i)}} \star_1 \sigma^-_{\widehat{u}_i w_i'} v_{\tau(i)} \star_1 \sigma^-_{u_i} w_i' v_{\tau(i)} \\
& \star_1 u_{\tau(i)} \alpha_i^{-\epsilon_i} v_{\tau(i)} \star_1 \sigma_{u_i} w_i v_{\tau(i)} \star_1 \sigma_{\widehat{u}_i w_i} v_{\tau(i)} \star_1 \sigma_{\widehat{u_i w_i} v_{\tau(i)}} \Big\rfloor \\
= & \left\lfloor \sigma^-_{\widehat{u}_i w_i'} \star_1 \widehat{u}_i \alpha_i^{-\epsilon_i} \star_1 \sigma_{\widehat{u}_i w_i} \right\rfloor \overline{v}_i = - \lfloor \psi_i \rfloor .
\end{aligned}$$

This implies $\lfloor \psi \rfloor = 0$, thus concluding the proof. □

23.5.2 Lemma. *For every element a in $h_2(P)$, there exists a 2-loop ψ in $P^\top$ such that $a = [\psi]$ holds.*

Proof. Let w be the 1-cell of $\overline{P}$ such that a belongs to $F_{\overline{P}}[P_2]_w$ and let P_3 be an acyclic extension of the (2, 1)-category $P^\top$. Since $d_2(a) = 0$, by acyclicity of P_3 and Theorem 23.3.3, there exists b in $F_{\overline{P}}[P_3]_w$ such that $a = d_3(b)$. By definition of $F_{\overline{P}}[P_3]_w$, we can write

$$b = \epsilon_1 u_1 [\alpha_1] v_1 + \ldots + \epsilon_p u_p [\alpha_p] v_p,$$

with, for every $1 \leqslant i \leqslant p$, $\alpha_i \in P_3$, $u_i, v_i \in \overline{P}$ and $\epsilon_i \in \{-, +\}$ such that $u_i \overline{\alpha}_i v_i = w$ holds. We fix a section $\widehat{\cdot}$ of P and we choose 2-cells

$$\phi_i : \widehat{w} \Rightarrow \widehat{u}_i s_1(\alpha_i^{\epsilon_i}) \widehat{v}_i \qquad \text{and} \qquad \psi_i : \widehat{u}_i t_1(\alpha_i^{\epsilon_i}) \widehat{v}_i \Rightarrow \widehat{w}.$$

Let A be 3-cell of $P_3^\top$ defined by

$$A = \left(\phi_1 *_1 \widehat{u}_1 \alpha_1^{\epsilon_1} \widehat{v}_1 *_1 \psi_1\right) *_1 \cdots *_1 \left(\phi_k *_1 \widehat{u}_k \alpha_k^{\epsilon_k} \widehat{v}_k *_1 \psi_k\right).$$

By definition of $[\cdot]$ on 3-cells, we have

$$[A] = \sum_{i=1}^{p} \left[\phi_i *_1 \widehat{u}_i \alpha_i^{\epsilon_i} \widehat{v}_i *_1 \psi_i\right] = \sum_{i=1}^{p} \left([1_{\phi_i}] + \epsilon_i u_i [\alpha_i] v_i + [1_{\psi_i}]\right) = b.$$

Finally, we get

$$a = d_3[A] = [s(A)] - [t(A)] = [s(A) *_1 t(A)^-].$$

Hence $\psi = s(A) *_1 t(A)^-$ is a 2-loop of $P^\top$ that satisfies $a = [\psi]$. □

When P is a convergent 2-polygraph, we have seen that the natural systems $h_2(P)$ and $\Pi(P)$ on $\overline{P}$ are generated by a family of generating confluences of P. The following result from [163, Theorem 5.6.5] states that, more generally, the natural systems $h_2(P)$ and $\Pi(P)$ are isomorphic, as proved by Brown-Huebschmann for presentations of groups in [64].

23.5.3 Theorem. *Let P be a 2-polygraph. The natural systems $\Pi(P)$ and $h_2(P)$ are isomorphic.*

Proof. We define a morphism of natural systems $\Phi \colon \Pi(P) \to h_2(P)$ by setting $\Phi \lfloor \phi \rfloor = [\phi]$ for every identity ϕ of P. This definition is correct, since the defining relations of $\Pi(P)$ also hold in $F_{\overline{P}}[P_2]$, and thus in $h_2(P)$. Moreover, Φ is a morphism of natural systems, since we have

$$\Phi(u \lfloor \phi \rfloor v) = \Phi(\lfloor \widehat{u} \phi \widehat{v} \rfloor) = [\widehat{u} \phi \widehat{v}] = u[\phi] v = u \Phi(\lfloor \phi \rfloor) v,$$

for every 2-loop ϕ in $P^\top$ and 1-cells u, v in $\overline{P}$ such that $\widehat{u} \phi \widehat{v}$ is defined.

Now, let us define a morphism of natural systems $\Psi \colon h_2(P) \to \Pi(P)$. Let a be an element of $h_2(P)_w$. By Lemma 23.5.2, there exists a 2-loop $\psi : u \Rightarrow u$ in $P^\top$ such that $a = [\psi]$ and $w = \overline{u}$. We define $\Psi(a) = \lfloor \psi \rfloor$. This definition does not depend on the choice of ψ. Indeed, let us assume that $\phi\colon v \Rightarrow v$ is a 2-loop such that $a = [\phi]$. It follows that $\overline{v} = \overline{u}$, and we can choose a 2-cell $\xi\colon u \Rightarrow v$ in $P^\top$. Then we have

$$a = [\psi] = [\phi] = [\xi *_1 \phi *_1 \xi^-].$$

As a consequence, we get

$$[\psi *_1 \xi^- *_1 \phi^- *_1 \xi] = [\psi] - [\xi *_1 \phi *_1 \xi^-] = 0.$$

Thus

$$0 = \lfloor \psi *_1 \xi^- *_1 \phi^- *_1 \xi \rfloor = \lfloor \psi \rfloor - \lfloor \xi *_1 \phi *_1 \xi^- \rfloor = \lfloor \psi \rfloor - \lfloor \phi \rfloor .$$

Finally, the relations $\Psi\Phi = 1_{\Pi(P)}$ and $\Phi\Psi = 1_{h_2(P)}$ are direct consequences of the definitions of Φ and Ψ. □

The following result relates the low-dimensional finiteness properties seen in this chapter and in §8.3.5 for the property $\mathrm{FDT}_{\mathrm{ab}}$.

23.5.4 Theorem. *Let P be a finite 2-polygraph. The following conditions are equivalent.*

1. *The category $\overline{P}$ is of homological type* FP_3.
2. *The natural system $h_2(P)$ on $\overline{P}$ is finitely generated.*
3. *The natural system $\Pi(P)$ on $\overline{P}$ is finitely generated.*
4. *The category $\overline{P}$ has* $\mathrm{FDT}_{\mathrm{ab}}$.

Proof. The equivalence between 1 and 2 comes from the definition of the property FP_3. The equivalence between 2 and 3 is a consequence of Theorem 23.5.3. The equivalence between 3 and 4 is given by Proposition 8.3.7. □

Note that, following Theorem 23.4.4, the property FDT_3 implies FP_3. We expect the reverse implication to be false in general, which amounts to proving that $\mathrm{FDT}_{\mathrm{ab}}$ does not imply FDT_3, since FP_3 is equivalent to $\mathrm{FDT}_{\mathrm{ab}}$ for finitely presented categories. This question is still open.

23.5.5 Identities among relations for higher polygraphs. We conclude this chapter by mentioning some results on identities among relations for n-polygraphs for any $n \geqslant 0$. As in §8.3.5 for the case $n = 2$, we call an $(n, n-1)$-category C *abelian* if, for every $(n-1)$-cell u of C, the group Aut_u^C of n-loops of C with source u is abelian. For C an $(n, n-1)$-category, its *abelianization* C_{ab} is the quotient of C by the cellular extension that contains one n-sphere $\phi *_{n-1} \psi \to \psi *_{n-1} \phi$ for every n-loops ϕ and ψ of C with the same source.

Identities among relations for 2-polygraphs, as defined in §8.3.2, were extended to the structure of an n-polygraph P in [164] to form a natural system $\Pi(P)$ on the $(n-1)$-category $\overline{P}$. This definition is based on a generalization of a result proved by Baues and Jibladze, see [33] for the case $n = 2$, stating that an $(n, n-1)$-category is abelian if and only if it is linear, where $(n, n-1)$-categories correspond to a notion of "globular crossed module" for n-categories. When the polygraph P is convergent, the natural system $\Pi(P)$ is generated by the generating confluences of P [164, Proposition 2.4.2].

A notion of *abelian finite derivation type*, $\mathrm{FDT}_{\mathrm{ab}}$ for short, can also be defined for n-polygraphs. An n-polygraph P has $\mathrm{FDT}_{\mathrm{ab}}$ when the abelian $(n, n-1)$-category $P^{\top}_{\mathrm{ab}}$ admits a finite acyclic cellular extension. As for case $n = 2$ in Proposition 8.3.7, we prove that an n-polygraph is $\mathrm{FDT}_{\mathrm{ab}}$ if and only if the natural system $\Pi(P)$ of identities among relations of P is finitely generated [163, Proposition 5.7.2].

APPENDICES

Appendix A
A Catalogue of 2-Polygraphs

In this section, we list some examples of presentations of monoids and categories by a 2-polygraph.

A.1 Presentations of Monoids

The richest source of presentations of categories can be found in presentations of monoids (and groups), which are seen here as a particular instance of a category.

A.1.1 Monoid. A *monoid* $(M, \times, 1)$ is a set equipped with a binary operation $\times$ which is associative and admits 1 as neutral element. A morphism between two monoids is a function between the underlying sets which preserves multiplication and neutral element.

The following lemma shows that one can always see a monoid as a particular case of a category with only one element conventionally denoted $\star$.

A.1.2 Lemma. *The category* **Mon** *of monoids is isomorphic to the full subcategory of* **Cat** *whose objects are the categories with $\star$ as only object.*

Proof. To a monoid $(M, \times, 1)$ we can associate a category BM, sometimes called the *delooping* of M, with $\star$ as only object, the elements $a \in M$ as morphisms $a\colon \star \to \star$, composition being given by $b \circ a = a \times b$, with 1 as identity on $\star$. This construction is easily extended as a functor which is an isomorphism of categories. □

A presentation of a monoid consists of a 2-polygraph P whose set of 0-generators is reduced to one element $P_0 = \{\star\}$. A group being a monoid, presentations of groups provide many such examples [94]; we mostly restrict here to those which are "really monoids", meaning that the presence of inverses does not play a crucial role in the presentation.

A.1.3 Notations. In a polygraph $P = \langle \star \mid P_1 \mid P_2 \rangle$, we generally omit mentioning the source and target of 1-generators since they are necessarily $\star$. Moreover, as far as we are concerned with the presented category, the names of the 2-generators, as well as their orientation, will not be relevant. The 2-polygraph

$$\langle \star \mid a\colon \star \to \star \mid \alpha\colon aa \Rightarrow 1 \rangle$$

will thus often be simply noted

$$\langle \star \mid a \mid aa = 1 \rangle. \tag{A.1}$$

Traditionally, the set P_0 is even omitted, i.e., the above presentation is noted $\langle a \mid aa = 1 \rangle$, we will however refrain from doing so in order to avoid a possible confusion with a 1-polygraph. We sometimes write the indices at the bottom right of the presentation. For instance

$$\left\langle \star \,\middle|\, a_i \,\middle|\, a_j a_i = a_i a_j \right\rangle_{i,j \in \mathbb{N}}$$

denotes a presentation with

$$P_1 = \{a_i \mid i \in \mathbb{N}\} \qquad \text{and} \qquad P_2 = \left\{a_j a_i = a_i a_j \,\middle|\, i, j \in \mathbb{N}\right\}.$$

A.1.4 Natural numbers. The additive monoid $\mathbb{N}$ of natural numbers admits the presentation

$$\langle \star \mid a \mid \ \rangle$$

where a corresponds to the natural number 1, and more generally a^n to $n \in \mathbb{N}$.

A.1.5 Cyclic monoids. The additive monoid of booleans $\mathbb{N}/2\mathbb{N}$ admits the presentation (A.1) above. More generally, given $n \in \mathbb{N}$, the additive monoid $\mathbb{N}/n\mathbb{N}$ admits the presentation

$$\langle \star \mid a \mid a^n = 1 \rangle.$$

A.1.6 Booleans. The set $\mathbb{B}$ of booleans respectively consists of the two elements $\bot$ (standing for false) and $\top$ (standing for true). We respectively write $\wedge$, $\vee$, and $\times$ for conjunction, disjunction, and exclusive disjunction. The monoid $(\mathbb{B}, \times, \bot)$ is isomorphic to $\mathbb{N}/2\mathbb{N}$ and thus admits (A.1) as presentation. The monoids $(\mathbb{B}, \vee, \bot)$ and $(\mathbb{B}, \wedge, \top)$ are isomorphic and admit the presentation

$$\langle \star \mid a \mid aa = a \rangle.$$

A.1.7 Free monoids. Fix a set X. A *word* u over X is a finite sequence $u = a_1 \dots a_n$ of elements a_i of X. Given two words $u = a_1 \dots a_m$ and $v = b_1 \dots b_n$, their *concatenation* is the word $uv = a_1 \dots a_m b_1 \dots b_n$. The *free monoid* X^* over X can be described as the monoid of words over X, with concatenation as composition and empty word as unit. It admits the presentation

$$\langle \star \mid X \mid \ \rangle.$$

A.1.8 Free commutative monoids. Fix a set X. Recall from §1.4.1 that a *multiset* over X is a function $\mu\colon X \to \mathbb{N}$, assigning a *multiplicity* to an element a of X, such that every element a of X has a null multiplicity, excepting for a finite number. Given two multisets μ and ν, we write $\mu \sqcup \nu$ for their pointwise sum. The constant function equal to 0 is called the *empty multiset*. The *free commutative monoid* over X can be described as the monoid of multisets with $\sqcup$ as composition and empty multiset as unit. It admits the presentation

$$\langle \star \mid X \mid ab = ba \rangle_{a,b \in X}.$$

For instance, the multiplicative monoid $\mathbb{N} \setminus \{0\}$ is the free commutative monoid over the set $\mathbb{N}$, since an element of this monoid can be interpreted as the multiset of its prime factors. It admits the presentation

$$\big\langle \star \,\big|\, a_i \,\big|\, a_i a_j = a_j a_i \big\rangle_{i,j \in \mathbb{N}}$$

where the 1-generator a_i corresponds to the i-th prime number, i.e., $a_0 = 2$, $a_1 = 3$, $a_2 = 5$, etc.

A.1.9 Partially commutative monoids. Suppose given an alphabet X and $I \subseteq X \times X$ a reflexive and symmetric relation called *independence*. The monoid with presentation

$$\langle \star \mid X \mid ab = ba \rangle_{(a,b) \in I}$$

is called a *trace monoid*, a *partially commutative monoid*, or a *heap monoid*. This family of monoids was introduced by Cartier and Foata [78] in order to study combinatorial problems of rearrangements and also studied in computer science [114, 123, 273] for the following reason. The elements of X can be interpreted as actions, and thus X^* as the set of possible sequences of actions. Sometimes, the order in which two actions a and b are performed does not matter, for instance when a and b consist in reading or writing at disjoint positions in the memory in a computer program: in the end, executing a then b, or b then a, will lead to the same state. This kind of situation is naturally modeled by having $(a, b) \in I$ and occurs when considering concurrent processes.

A.1.10 Baumslag–Solitar monoids. Given natural numbers m and n, the Baumslag-Solitar monoid $BS(m,n)$ [35] is the monoid presented by

$$\langle \star \mid a, b \mid ab^m = b^n a \rangle$$

which can be seen as a variant of the commutativity relation described above, $BS(1,1)$ being the free commutative monoid on two generators. The enveloping groups of those monoids provide examples of non-Hopfian groups, e.g., $BS(2,3)$.

A.1.11 Finite subsets. Given a set I, the set $\mathcal{P}_{\text{fin}}(I)$ of finite subsets of I is a monoid with union as multiplication and empty set as neutral element. It admits as presentation

$$\left\langle \star \,\middle|\, a_i \,\middle|\, a_i a_j = a_j a_i, a_i a_i = a_i \right\rangle_{i,j \in I}.$$

It is thus the free idempotent commutative monoid on the set I.

A.1.12 Coproduct and product. The monoids $\mathbb{N} \sqcup \mathbb{N}$ and $\mathbb{N} \times \mathbb{N}$ respectively admit the presentations

$$\langle \star \mid a, b \mid \ \rangle \qquad \text{and} \qquad \langle \star \mid a, b \mid ba = ab \rangle.$$

More generally, given 2-polygraphs P and Q, respectively presenting monoids M and N,

- their coproduct (or *free product*) $M + N$ is presented by

$$\langle \star \mid P_1, Q_1 \mid P_2, Q_2 \rangle,$$

- their product $M \times N$ is presented by

$$\langle \star \mid P_1, Q_1 \mid P_2, Q_2, R_2 \rangle$$

  where

$$R_2 = \{ba = ab \mid a \in P_1, b \in Q_1\}.$$

These constructions are detailed and generalized in Chapter 3.

A.1.13 Bicyclic monoid. The *bicyclic monoid* is the monoid presented by

$$\langle \star \mid a, b \mid ab = 1 \rangle.$$

If we read a as an "opening bracket" and b as a "closing bracket", then the words in the equivalence class of 1 are precisely the *well-bracketed words* (also called *Dyck words*). From this property follow applications in combinatorics and computer science (e.g., this monoid is the syntactic monoid of the language of Dyck words).

A.1.14 Integers. The additive group $\mathbb{Z}$ admits, as a monoid, the presentation

$$\langle \star \mid a, b \mid ab = 1, ba = 1 \rangle$$

where a and b can respectively be interpreted as 1 and -1.

A.1.15 Enveloping group. The forgetful functor **Grp** $\to$ **Mon**, from the category of groups to the category of monoids, admits a left adjoint, under which the image of a monoid is called the associated *enveloping group*. For instance, the additive group $\mathbb{Z}$ is the enveloping group of the additive monoid $\mathbb{N}$. As such, the previous example is an instance of a more general construction: given a polygraph P presenting a monoid M, its enveloping group G is presented as a monoid by

$$\left\langle \star \,\middle|\, P_1, P_1^- \,\middle|\, P_2, R_2 \right\rangle$$

where

$$P_1^- = \{a^- \mid a \in P_1\} \qquad \text{and} \qquad R_2 = \{aa^- = 1, a^-a = 1 \mid a \in P_1\}.$$

This construction is detailed and generalized in §3.2.1. By abuse of notation, given a word $u = a_1 \dots a_n$ in P_1^*, we sometimes write $u^- = a_n^- \dots a_1^-$.

Two distinct monoids can generate isomorphic enveloping groups. For instance, consider the monoid presented by

$$\langle \star \mid a, b \mid ab = bba, ba = aab \rangle.$$

This monoid is not the trivial one since for instance the classes of 1, a, and b contain only one element (there is no derivable relation involving those, the generating relations being between words of length 2 and 3). We can observe that the relation $ab = baab$ is derivable since $ab = bba = baab$, and similarly the relation $ba = abba$ is also derivable. In the enveloping group, we thus have $ba = baabb^-a^- = abb^-a^- = 1$, and similarly $ab = 1$. Therefore, $1 = ab = bba = b1 = b$ and similarly $1 = a$. The presented group is therefore the trivial one, which is also the enveloping group of the trivial monoid. Another example is given in §A.1.28.

A.1.16 Group presentation. As a variant of the previous construction, by a *group presentation*, one usually understands a pair

$$\langle P_1 \mid P_2 \rangle$$

where P_1 is a set and $P_2 \subseteq (P_1 \cup P_1^-)^*$, which implicitly means the group presented by the 2-polygraph

$$\left\langle \star \,\middle|\, P_1, P_1^- \,\middle|\, a^-a = 1, aa^- = 1, u = 1 \right\rangle_{a \in P_1, u \in P_2}.$$

Here, we only consider relations of the form $u = 1$, since having a relation $u = v$ is equivalent (by Tietze transformations) to having the relation $u^- v = 1$.

A.1.17 Positive rational numbers. We have seen a presentation of the multiplicative monoid $\mathbb{N} \setminus \{0\}$ in §A.1.8. The multiplicative group $\mathbb{Q}^{>0}$ of strictly positive rational numbers is its enveloping group and, from the construction of §A.1.15, we deduce that it admits the presentation

$$\langle \star \mid a_i, a_i^- \mid a_i a_j = a_j a_i, a_i^- a_i = 1, a_i a_i^- = 1 \rangle_{i,j \in \mathbb{N}}.$$

A.1.18 Non-negative rational numbers. Let us detail an example of a presentation coming from [205, Section 5.7]. The *additive* monoid $\mathbb{Q}^+$ of non-negative rational numbers can be presented by the 2-polygraph

$$P = \langle \star \mid a_i \mid (a_{i+1})^{i+1} = a_i \rangle_{i \in \mathbb{N} \setminus \{0\}}.$$

Here, any pair of generators commute, i.e., the relations $a_j a_i = a_i a_j$ are derivable since, supposing $j > i$, we have

$$a_j a_i = a_i^{j(j-1)\dots(i+1)} a_i = a_i a_i^{j(j-1)\dots(i+1)} = a_i a_j.$$

The interpretation of a generator a_i in $\mathbb{Q}^+$ is $f(x_i) = 1/i!$, which extends as a morphism of monoids $f\colon P^* \to \mathbb{Q}^+$. It is compatible with the relations since

$$f(a_{i+1}^{i+1}) = \frac{i+1}{(i+1)!} = \frac{1}{i!} = f(a_i)$$

and thus induces a morphism $f\colon \overline{P} \to \mathbb{Q}^+$, which we prove to be an isomorphism. The function f is surjective since for $(p, q) \in \mathbb{N} \times (\mathbb{N} \setminus \{0\})$, one has

$$\frac{p}{q} = \frac{p(q-1)!}{q!} = f\left(a_q^{p(q-1)!}\right).$$

Let us show that it is injective. Suppose given two words $u, v \in P_1^*$ such that $f(u) = f(v)$. Since the generators commute, the words u and v are, up to equivalence, of the form

$$u = a_1^{m_1} a_2^{m_2} \dots a_k^{m_k} \qquad \text{and} \qquad v = a_1^{n_1} a_2^{n_2} \dots a_k^{n_k}$$

with $k \in \mathbb{N}$ and $m_i, n_i \in \mathbb{N}$ for $0 \leqslant i \leqslant k$. We can also suppose that $m_k \neq 0$, up to exchanging the roles of u and v and lowering k. Suppose moreover that $k > 1$ (this hypothesis is not innocuous since it will lead to a contradiction). If $n_k \geqslant k$, we can use the relation $a_k^k = a_{k-1}$ to transform u into a word u' equivalent to u and satisfying $0 \leqslant m_k < k$. For this reason, we can also suppose, without loss of generality, that $0 < m_k < k$ and $0 \leqslant n_k < k$. Up to exchanging the roles of u and v, we can also suppose $m_k \geqslant n_k$. We have

$$0 = f(u) - f(u') = \frac{m_1 - n_1}{1!} + \frac{m_2 - n_2}{2!} + \ldots + \frac{m_k - n_k}{k!}$$

and therefore

$$\begin{aligned} 0 &= (k-1)!\,(f(u) - f(u')) \\ &= (k-1)!\left(\frac{m_1 - n_1}{1!} + \frac{m_2 - n_2}{2!} + \ldots + \frac{m_{k-1} - n_{k-1}}{(k-1)!}\right) + \frac{m_k - n_k}{k} \\ &= m + \frac{m_k - n_k}{k} \end{aligned}$$

with $m \in \mathbb{Z}$. Since $0 < m_k < k$, $0 \leqslant n_k < k$ and $m_k \geqslant n_k$, we have $0 < (m_k - n_k)/k < 1$, and thus a contradiction. Therefore, we have $k = 1$ and $m_k = f(u) = f(v) = n_k$, i.e., $u = v$. The morphism $f\colon \overline{P} \to \mathbb{Q}^+$ is thus injective. From the results of §A.1.15, we can deduce the following presentation of the additive monoid $\mathbb{Q}$, which is its enveloping group:

$$\left\langle \star \mid a_i, b_i \mid a_{i+1}^{i+1} = a_i, a_i b_i = 1, b_i a_i = 1 \right\rangle_{i \in \mathbb{N} \setminus \{0\}}.$$

A.1.19 Symmetric groups. The *symmetric group* S_{n+1} is the group of bijections (or *permutations*) from the set $\{0, \ldots, n\}$ to itself, with multiplication given by composition and unit by identity. More geometrically, it can also be defined as the group of symmetries of an n-simplex. Considered as a monoid, it admits a presentation where the set of generators is $P_1 = \{a_0, \ldots, a_{n-1}\}$ and the relations are

- $a_i a_i = 1$, for $0 \leqslant i < n$,
- $a_i a_{i+1} a_i = a_{i+1} a_i a_{i+1}$, for $0 \leqslant i < n - 1$,
- $a_i a_j = a_j a_i$, for $0 \leqslant i < j < n$ with $j > i + 1$,

as found out by Moore [288]. Here, the generator a_i should be interpreted as the transposition exchanging i with $i+1$, written $(i(i+1))$, whose graph can be pictured as

The presentation is studied in detail in §5.2.7.

The above presentation is based on the transpositions $(i(i+1))$ as generators a_i, but other choices of generators are possible and will give rise to other presentations [94, Section 6.2], for instance:

- the generators $a_i = (in)$, with $0 \leqslant i < n$, induce a presentation with relations

$$a_i^2 = 1 \qquad (a_i a_{i+1})^3 = 1 \qquad (a_i a_{i+1} a_i a_j)^2 = 1$$

for $1 \leqslant i, j \leqslant n-1$ with $j \neq i$ and $j \neq i+1$, where $a_n = a_0$ by convention,

- the generators $a_0 = (1n)$ and $a_i = (0in)$ for $1 \leqslant i < n$ induce a presentation with relations

$$a_0^2 = 1 \qquad\qquad a_j^3 = 1 \qquad\qquad (a_i a_j)^2 = 1$$

for $0 \leqslant i < j < n$,

- the two generators $a = (01)$ and $b = (012\ldots n)$ induce a presentation with relations

$$a^2 = 1 \quad b^{n+1} = 1 \quad (ba)^n = 1 \quad (ab^n ab)^3 = 1 \quad (ab^{n+1-j}ab^j) = 1$$

for $1 \leqslant j < n-1$.

A.1.20 Alternating groups. The *alternating group* A_{n+1} is the subgroup of S_{n+1} consisting of symmetries of even signature (recall that the *signature* of a symmetry is the parity of the number of transpositions used to express it). It admits a presentation with generators $P_1 = \{a_0, \ldots, a_{n-1}\}$ and relations

$$a_j^3 = 1 \qquad\qquad (a_i a_j)^2 = 1$$

for $0 \leqslant i < j < n-1$. Here, a generator a_i should be interpreted as the permutation $(i(n-1)n)$, see [94, Section 6.3] for details.

A.1.21 Braid groups and monoids. We introduce notations for the spaces $I = [0,1]$, $X = \mathbb{R}^2$ and $Y = I \times X$. Suppose given $n \in \mathbb{N}$ and continuous functions

$$b_i : I \to Y$$

with $0 \leqslant i < n$, which are mutually disjoint (i.e., for $t \in I$, $b_i(t) = b_j(t)$ implies $i = j$) and with fixed endpoints (say, $b_i(0) = (0, i, 0)$ and $b_i(1) = (1, i, 0)$). This induces a subspace β of $Y = I \times X$ defined as

$$\beta = \{(t, b_i(t)) \mid t \in I, 0 \leqslant i < n\}.$$

Such a subspace is called a (*geometric*) *braid* with n strands. Graphically, a braid with three strands can be pictured as follows, where the first coordinate is pictured vertically:

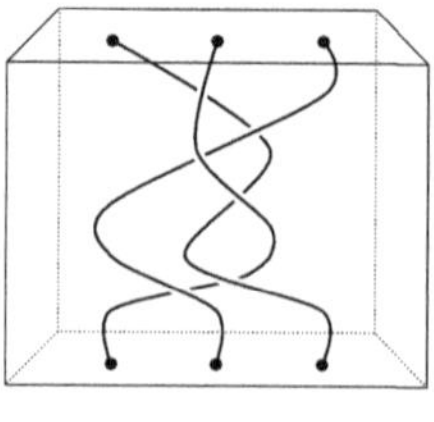

Braids are considered up to endpoint-preserving isotopy: we identify two braids β and β' for which there exists a continuous function

$$h\colon I \to Y^Y$$

such that $h(0) = 1_Y$, $h(1)(\beta) = \beta'$ and, for every $t \in I$, the function $h(t)$ is a homeomorphism, whose restriction to $(\{0\} \times X \sqcup \{1\} \times X) \subseteq Y$ is the identity, such that $h(t)(\beta)$ is a braid. The set B_n of braids with n strands (up to isotopy) forms a group, where the composition is given by concatenation of strands, and the identity is given by the braid induced by constant functions $b_i(t) = (i, 0)$.

An alternative description can be given as follows. Given a topological space X (typically, $X = \mathbb{R}^2$ in the following), the n-element *configuration space* is

$$C_n X = \left\{(x_1, \ldots, x_n) \,\middle|\, x_i = x_j \text{ implies } i = j\right\}.$$

The n-element *unlabeled configuration space* $D_n X = C_n X/\Sigma_n$ is the quotient of $C_n X$ under the action of the symmetric group permuting the coordinates. The n-strand *braid group* can equivalently be defined as the fundamental group of this space: $B_n = \pi_1(D_n\mathbb{R}^2)$.

Before presenting the braid group, we first present a submonoid of this group. The *(positive) braid monoid* B^+_{n+1} with $n+1$ strands, admits a presentation with $P_1 = \{a_0, \ldots, a_{n-1}\}$, where a_i can be pictured as

(with i strands on the left), and the relations are

- $a_i a_{i+1} a_i = a_{i+1} a_i a_{i+1}$, for $0 \leqslant i < n-1$,
- $a_i a_j = a_j a_i$, for $0 \leqslant i < i+1 < j < n$.

For instance, with $n+1 = 4$, the generators a_0, a_1, a_2 respectively correspond to the following braids with four strands:

$a_0 =$  $a_1 =$  $a_2 =$ 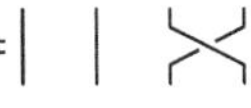

and the relations are

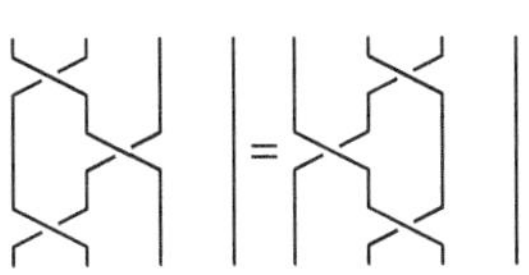

$a_0 a_1 a_0 = a_1 a_0 a_1$

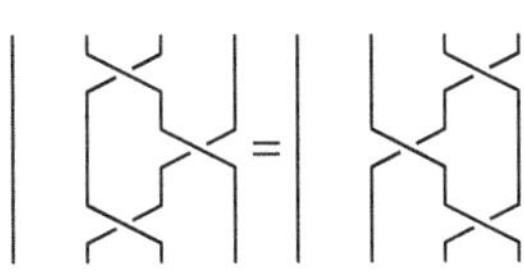

$a_1 a_2 a_1 = a_2 a_1 a_2$

and

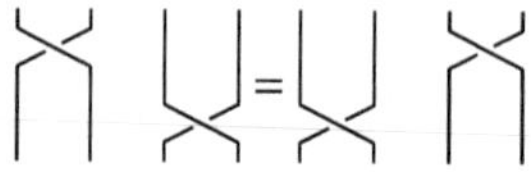

$$a_0a_2 = a_2a_0$$

We can expect that the full braid group can be recovered by moreover adding the other kind of crossings

$a_i^- =$ 

which are inverse to the crossings a_i. Indeed, the braid group B_{n+1} is the enveloping group of the monoid B^+_{n+1}, from which a presentation can be constructed, see §A.1.15:

- generators are a_i, a_i^- for $0 \leqslant i < n$,
- relations are those of the braid monoid plus $a_ia_i^- = 1$ and $a_i^-a_i = 1$ for $0 \leqslant i < n$.

This presentation is due to Artin [17].

From the above presentation, we see that the symmetric group can be obtained from the braid group by identifying a_i with a_i^-. There is thus a projection morphism $B_n \to S_n$ and we write P_n for its kernel, which is called the *pure braid group*: the group P_n can be seen as the submonoid of B_n consisting of braids which become identities if we interpret them as symmetries, and we have an exact sequence

$$1 \to P_n \to B_n \to S_n \to 1.$$

This group can also more directly be described as the fundamental group of a configuration space: $P_n = \pi_1(C_n\mathbb{R}^2)$. A presentation of P_n can be given with generators a_{ij} for $0 \leqslant i < j < n$, as well as their formal inverses a_{ij}^-. Considering P_n as a subgroup of B_n, the generators a_{ij} can be expressed in terms of the generators of B_n as

$$a_{ij} = a_ia_{i+1}\ldots a_{j-2}a_{j-1}^2a_{j-2}^-\ldots a_{i+1}^-a_i^-.$$

Graphically,

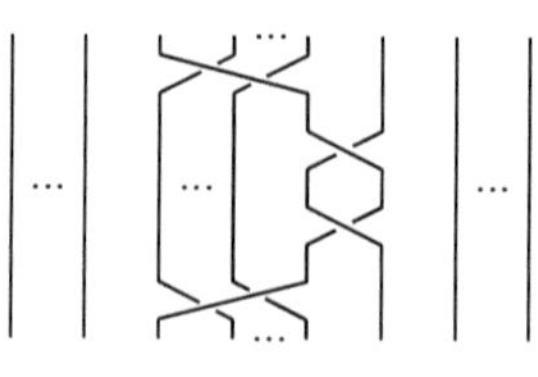

The relations satisfied by those are

$$a_{kl}a_{ij}a_{\overline{kl}} = \begin{cases} a_{ij} & \text{if } l < i \text{ or } j < k, \\ a_{\overline{il}}a_{ij}a_{il} & \text{if } i < j = k < l, \\ a_{\overline{ij}}a_{\overline{ik}}a_{ij}a_{ik}a_{ij} & \text{if } i < k < j = l, \\ a_{\overline{il}}a_{\overline{ik}}a_{il}a_{ik}a_{ij}a_{\overline{ik}}a_{\overline{il}}a_{ik}a_{il} & \text{if } i < k < j < l. \end{cases}$$

A.1.22 Hyperoctahedral group. The hyperoctahedral group B_n (or C_n) is the group of symmetries of a cube of dimension n. It admits a presentation with generators a_i, with $0 \leqslant i < n$ and relations

- $a_0a_1a_0a_1 = a_1a_0a_1a_0$
- $a_ia_{i+1}a_i = a_{i+1}a_ia_{i+1}$ for $0 < i < n-1$,
- $a_ia_j = a_ja_i$ for $0 \leqslant i < i+1 < j < n$.

This group can also be described as the group of signed symmetries, i.e., $n{\times}n$ orthogonal matrices with integer entries, i.e., $n{\times}n$ matrices with coefficients in $\{-1, 0, 1\}$ containing exactly one non-null coefficient on each row and each column, see §C.11.4 for details.

A.1.23 Progressive ribbon group. Given $n \in \mathbb{N}$, the monoid R^+_{n+1} of positive progressive ribbons with $n+1$ strands admits a presentation with generators a_i, with $0 \leqslant i < n$, and b_i, with $0 \leqslant i \leqslant n$, subject to the relations

- $a_ia_{i+1}a_i = a_{i+1}a_ia_{i+1}$ for $0 \leqslant i < n-1$,
- $a_ia_j = a_ja_i$ for $0 \leqslant i < i+1 < j < n$,
- $b_ia_i = a_ib_{i+1}$ for $0 \leqslant i < n$,
- $b_{i+1}a_i = a_ib_i$ for $0 \leqslant i < n$,
- $b_ia_j = a_jb_i$ for $0 \leqslant i < n$, $0 \leqslant j \leqslant n$, and $i \notin \{i, i+1\}$.

Graphically, an element of this monoid can be depicted as n ribbons in the space, the generators are respectively

$a_i =$ 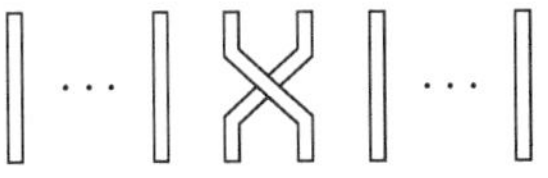 $b_i =$ 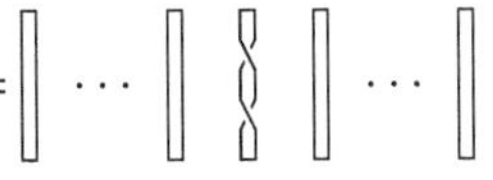

and the relations have natural graphical interpretations, which the reader is encouraged to draw. The group R_n of progressive ribbons is the enveloping group of R^+_n. Note that the relations satisfied by the generators a_i are precisely the relations of the braid group. In fact, there is an obvious action of the braid group B_n on the set $[n] = \{0, \ldots, n-1\}$ of braids and R_n can be obtained as the wreath product $R_n = \mathbb{Z} \wr B_n$ of the additive group of integers with the braid group. The groups R_n, as well as their presentation, are detailed in [208], as well as in §C.11.5.

A.1.24 Dihedral groups. The dihedral group D_n is the group of symmetries of a regular polygon with n vertices. It admits a presentation by the 2-polygraph with 1-generators $r_0, \ldots, r_{n-1}$ and $s_0, \ldots, s_{n-1}$, and relations

$$r_0 = 1 \qquad r_i r_j = r_{i+j} \qquad r_i s_j = s_{i+j} \qquad s_i r_j = s_{i-j} \qquad s_i s_j = r_{i-j}$$

for all indices $0 \leqslant i, j < n$, with addition and subtraction taken modulo n. Alternatively, it can be presented by the 2-polygraph

$$\langle \star \mid r, s \mid r^n = 1, s^2 = 1, rsrs = 1 \rangle$$

(the relationship with previous generators is given by $r = r_1$, $s = s_0$, $r_i = r^i$, and $s_i = r^i s$). In terms of symmetries of a polygon with n vertices, s corresponds to a symmetry and r a rotation of $2\pi/n$, and more generally s_i and r_i respectively correspond to a reflection along the i-th axis and a rotation of $2\pi i/n$:

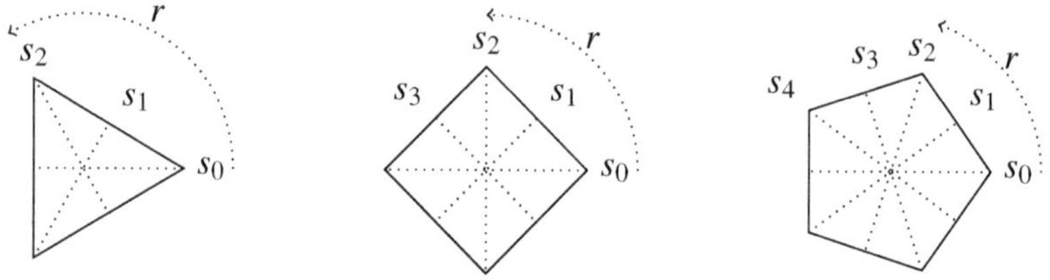

The case of D_3 is detailed in Example 5.1.4.

A.1.25 Artin monoids and Coxeter groups. Given a finite set S of generators, a *Coxeter matrix* M is a function which to every pair of elements $(s, t) \in S \times S$ associates $m_{st} \in \mathbb{N} \sqcup \{\infty\}$ such that $m_{st} = m_{ts}$, $m_{ss} = 1$, and $m_{st} > 1$ for every $s, t \in S$ with $s \neq t$. Such a matrix induces

- an *Artin monoid* with presentation

  $$\langle \star \mid s \mid \langle st \rangle^{m_{st}} = \langle ts \rangle^{m_{ts}} \rangle_{s,t \in S}$$

  where $\langle st \rangle^k$ denotes the alternating product of s and t of length k starting with k (e.g., $\langle st \rangle^5 = ststs$) and, by convention, there is no relation when $m_{st} = \infty$ or $m_{ts} = \infty$,
- an *Artin group*, which is the group freely generated by the above monoid, and
- a *Coxeter group* with presentation as a monoid

  $$\langle \star \mid s \mid \langle st \rangle^{m_{st}} = \langle ts \rangle^{m_{ts}}, ss = 1 \rangle_{s,t \in S}$$

or equivalently

$$\langle \star \mid s \mid (st)^{m_{st}} = 1 \rangle_{s,t \in S}.$$

Since the set S is finite, writing n for its cardinal, we can assume that it is of the form $S = \{i \in \mathbb{N} \mid 0 \leqslant i < n\}$, which is convenient in the following. A Coxeter matrix M is often pictured as a Dynkin diagram consisting of a labeled non-oriented simple graph with n vertices x_i, with $0 \leqslant i < n$, with an edge between x_i and y_i labeled by m_{ij}. By convention, the looping edges on vertices, as well as edges with label 2, are not drawn, and edges with label 3 are not labeled. Artin monoids are further detailed in Section B.1.

For instance, the Coxeter matrix such that $m_{i(i+1)} = 3$, and $m_{ij} = 2$ for $|i - j| > 2$, can be represented by the diagram

The Artin monoid is the braid monoid with $n + 1$ strands (see §A.1.21) and the Coxeter group is the symmetric group on $n + 1$ elements (see §A.1.19). Similarly, the hyperoctahedral group B_n is the Coxeter group associated to

(with n vertices) and the dihedral group D_n is the Coxeter group associated to

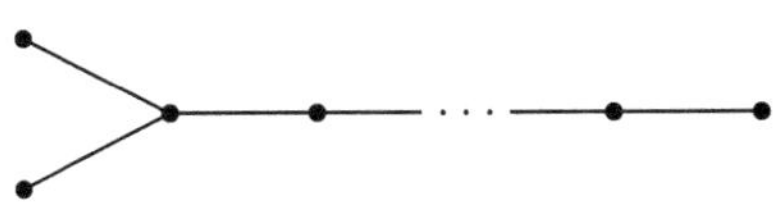

(with n vertices). A detailed presentation of these groups can be found in several places such as [146].

A.1.26 Plactic, Chinese, and Sylvester monoids. The *plactic monoid* P_n of rank n in type A was introduced by Knuth in [217] and further developed by Lascoux and Schützenberger [242]. It admits a presentation with $\{1, \ldots, n\}$ as generators, and relations

$$zxy = xzy, \text{ for } 1 \leqslant x \leqslant y < z \leqslant n, \qquad yzx = yxz, \text{ for } 1 \leqslant x < y \leqslant z \leqslant n.$$

Its elements are in bijection with semistandard Young tableaux [217, 254], from which stem applications in representation theory [141, 250]. Plactic monoids are further detailed in Section B.2.

Many variants of plactic monoids exist, such as the Chinese monoids. These monoids were introduced in [117] through the following presentation, called Chinese presentation. The *Chinese monoid of rank* $n > 0$, denoted by Ch_n, is the quotient of the free monoid $\{1, \dots, n\}^*$ by the congruence generated by the following relations:

$$zyx = zxy = yzx \quad \text{for all} \quad 1 \leqslant x \leqslant y \leqslant z \leqslant n. \tag{A.2}$$

The *Sylvester monoids* Syl_n [185] are also a variant, with the same generators and relations

$$zxuy = xzuy,$$

for $1 \leqslant x \leqslant y < z \leqslant n$ and $u \in \{1, \dots, n\}^*$.

A.1.27 Thompson group F. We write $I \subseteq \mathbb{R}$ for the interval $I = [0, 1]$. An element $x \in I$ is called *dyadic* when it is of the form $\frac{m}{2^n}$ for some $m, n \in \mathbb{N}$. The Thompson group F [74, 275] has as elements the homeomorphisms $f \colon I \to I$ which

- are piecewise linear with slopes 2^n for some $n \in \mathbb{Z}$,
- with a finite number of breakpoints,
- such that the coordinates of the breakpoints are dyadic.

The multiplication gf of two elements $f \colon I \to I$ and $g \colon I \to I$ is given by their composite $g \circ f$. For instance, we have the three functions whose graphs are depicted below are elements of F, respectively called x_0, x_1, and x_2:

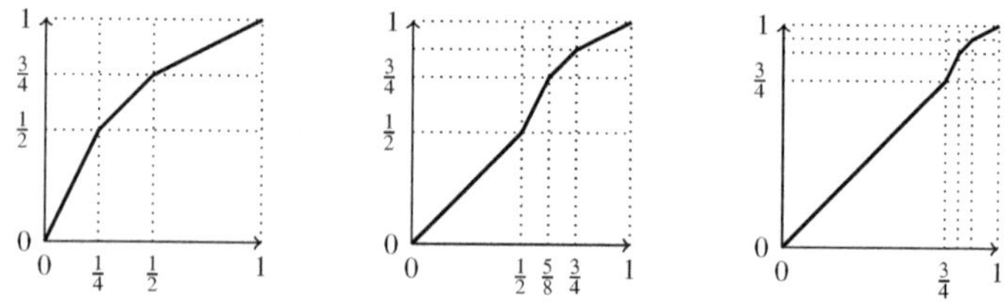

The explicit definition of x_0 is

$$x_0(t) = \begin{cases} 2t & \text{if } 0 \leqslant t \leqslant \frac{1}{4} \\ t + \frac{1}{4} & \text{if } \frac{1}{4} \leqslant t \leqslant \frac{1}{2} \\ \frac{t+1}{2} & \text{if } \frac{1}{2} \leqslant t \leqslant 1 \end{cases}$$

and more generally, given $i \in \mathbb{N}$, one can define a function $x_i \colon I \to I$ in F by

$$x_i(t) = \begin{cases} t & \text{if } 0 \leqslant t \leqslant t_i \\ x_0(2^i(t - t_i))/2^i + t_i & \text{if } t_i \leqslant t \leqslant 1 \end{cases}$$

where $t_i = 1 - \frac{1}{2^i}$ and $i \neq 0$. The group F is generated by the elements x_i and can be described as the enveloping group of the *Thompson monoid* F^+, which is presented by

$$F^+ = \langle \star \mid x_i \mid x_{j+1}x_i = x_ix_j \rangle_{i<j\in\mathbb{N}}.$$

More explicitly, this means that, starting from the above presentation, we should add a formal inverse x_i^- to each generator x_i, as explained in §A.1.15. Insights on this presentation, as well as alternative descriptions of F, are given in Section C.9. Note that the generator x_2 is superfluous, since the relation $x_2x_0 = x_0x_1$ can be replaced by $x_2 = x_0x_1x_0^-$. More generally, given $i \geqslant 2$, one has $x_i = x_0^{n-1}x_1(x_0^-)^{n-1}$, which shows that the group is generated by x_0 and x_1. Following this trail, the group F also admits the following finite presentation with only two generators and their inverses [74]:

$$\langle \star \mid a,b,a^-,b^- \mid [ab^-,a^-ba],[ab^-,a^-a^-baa],aa^-,a^-a,bb^-,b^-b \rangle$$

where a and b respectively correspond to x_0 and x_1 in the previous presentation, $[u,v]$ is a notation for the commutator uvu^-v^-, and we simply write u for a relation $u = 1$. The Thompson group F is FP_∞ [61].

Two other variants of the group F were introduced by Thompson [74]. The group T is a "cyclic variant" whose elements are piecewise linear functions $f\colon S^1 \to S^1$, where S^1 denotes the circle of unit perimeter. The group V is a "symmetric variant" consisting of functions $f\colon S^1 \to S^1$ which are right-continuous, bijective, and piecewise-linear satisfying similar requirements as before. There are natural inclusion morphisms $F \hookrightarrow T \hookrightarrow V$. Typical elements of T and V are

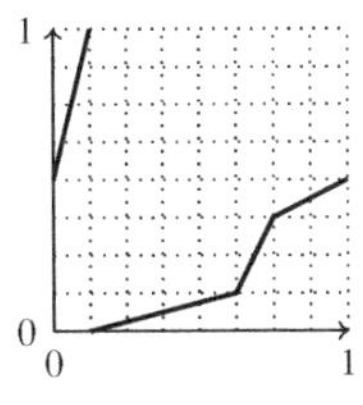

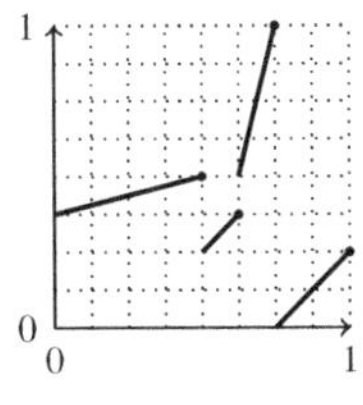

Both T and V also admit finite presentations. Generalizations of those groups were also introduced by using n-ary trees instead of binary ones [180], by using I^n instead of I for the domain of the endomorphisms [57], and by using braidings instead of symmetries [58].

A.1.28 Wirtinger monoids. A *knot* is an embedding $\kappa\colon S^1 \to \mathbb{R}^3$ of the circle into space. Suppose given a diagram representing such a knot on the plane, e.g., the trefoil knot:

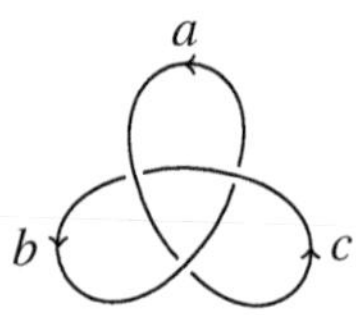

We label the arcs (hereby a, b, c) and orient them following a conventional orientation of S^1. This induces a presentation of a monoid, with arcs as generators, and a relation

$$ab = ca$$

for each crossing

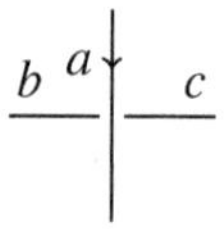

(the orientation of b or c is not relevant). For instance, the monoid associated to the trefoil knot is

$$\langle \star \mid a, b, c \mid ab = ca, bc = ab, ca = bc \rangle.$$

The enveloping group of the monoid is the fundamental group of the knot, i.e., the fundamental group of $\mathbb{R}^3 \setminus \kappa(S^1)$, and the presentation is called the *Wirtinger presentation*. This group only depends on the knot, up to isotopy (not on the diagram used to construct it or its orientation for instance). As a side note, this group is also the enveloping group of the monoid

$$\langle \star \mid a, b \mid bab = aba \rangle$$

which is not isomorphic to the above monoid.

Let us see an application of those presentations. Two knots are equivalent when their complements are isomorphic. Moreover, the fundamental group of the complement is an invariant of the equivalence classes. This shows that the trefoil is not isomorphic to the unknot: for instance, one has an abelian fundamental group whereas the other does not.

A.1.29 Temperley–Lieb monoids. Given $n \in \mathbb{N}$, the n-th *Temperley–Lieb monoid* is presented by the 2-polygraph with 1-generators d and u_i, indexed by $0 \leqslant i < n$, together with relations, for $0 \leqslant i, j < n$:

- $du_i = u_i d$,
- $u_i u_i = du_i$,

– $u_i u_j u_i = u_i$ whenever $|j - i| = 1$,
– $u_j u_i = u_i u_j$ whenever $|j - i| > 1$.

Given a ring R and a parameter $\delta \in R$, the *Temperley–Lieb algebra* $A_n(\delta)$ is the free R-algebra, generated by the u_i as above, satisfying the above relations with $d = \delta$. More details can be found in [1, 207, 215, 341], as well as §C.10.7.

A.1.30 Brauer monoids. Given $n \in \mathbb{N}$, the n-th *Brauer monoid* [55, 227] is presented by the 2-polygraph with 1-generators a_i and u_i, indexed by $0 \leqslant i < n$, subject to the 12 families of relations, indexed by $0 \leqslant i, j < n$,

– for $0 \leqslant i < n$,

$$a_i a_i = 1 \qquad u_i u_i = 1 \qquad a_i u_i = u_i \qquad u_i a_i = u_i$$

– for $|j - i| = 1$,

$$a_i a_j a_i = a_j a_i a_j \qquad u_i u_j u_i = u_i \qquad a_i u_j a_i = a_j u_i a_j \qquad u_i a_j u_i = u_i$$

– for $|j - i| > 1$,

$$a_j a_i = a_i a_j \qquad u_j u_i = u_i u_j \qquad u_j a_i = a_i u_j$$

– for $0 \leqslant i < n - 3$,

$$a_i a_{i+1} u_i u_{i+2} = a_{i+2} a_{i+1} u_i u_{i+2}$$

In particular, the relations satisfied by the a_i are precisely those of the symmetric group, see §A.1.19, and the relations satisfied by the u_i are those of Temperley–Lieb monoids with $d = 1$, see §A.1.29. More details about Brauer algebras, as well as a graphical illustration of the relations can be found in §C.10.8.

A.1.31 Graphs. Consider the monoid presented by

$$\langle \star \mid s, t \mid ss = st = s, ts = tt = t \rangle.$$

A right action of this monoid on a set X is precisely a directed graph: the vertices are elements $x \in X$ such that $x \cdot s = x \cdot t = x$, and other elements $y \in X$ are edges with source $y \cdot s$ and target $y \cdot t$.

A.1.32 Tseitin monoid. The *Tseitin monoid* is the one whose set of 1-generators is $\{a, c, b, d, e\}$, subject to the relations

$$ac = ca \qquad bc = cb \qquad eca = ce \qquad ccae = cca$$

$$ad = da \qquad bd = db \qquad edb = de$$

The reason why it is interesting is that it admits a small presentation, and yet it was shown to have undecidable word problem by Tseitin [346] (see Section 4.2), meaning that, writing P for the above 2-polygraph, there is no algorithm which, given any two 1-cells $u, v \in P_1^*$, answers whether $u \approx^P v$ holds or not. In fact, a 1-polygraph with only three relations exhibiting this property can be found [272].

A.2 Presentations of Categories

An example of a presentation of a category which is not a monoid was already given in Example 2.3.12: the walking isomorphism category. There are possible variations on this construction of the "walking something", as we show below, although most interesting categorical constructions (e.g., adjunctions) are 2-categorical and will require using 3-polygraphs.

A.2.1 The walking retract. A *retract* in a category C consists of a pair of morphisms $a\colon x \to y$ and $b\colon y \to x$ such that $b \circ a = 1_x$. In this case, a is called a *section* and b a *retraction*. The *walking retract* is the category presented by

$$\langle\, x, y \mid a\colon x \to y, b\colon y \to x \mid ab = 1_x \,\rangle.$$

It can be described as the category

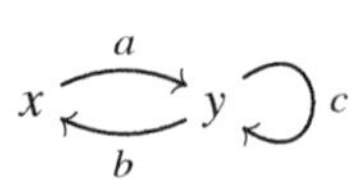

with two objects x, y and three non-trivial morphisms

$$a\colon x \to y \qquad\qquad b\colon y \to x \qquad\qquad c\colon y \to y$$

with non-trivial compositions

$$ab = 1_x \qquad\qquad ac = a \qquad\qquad ba = c \qquad\qquad cc = c$$

A.2.2 Interval objects. Given an object x in a category, a *cylinder* on x is an object y together with morphisms $s, t\colon x \to y$ and $p\colon y \to x$ such that $p \circ s = 1_x = p \circ t$. A typical example of cylinder object on a topological space X is the space $X \times I$, where $I = [0, 1]$ is the standard interval, with $s(x) = (x, 0)$, $t(x) = (x, 1)$ and $p(x, a) = x$, see [211] for details. The theory of an object together with a cylinder is the category presented by

$$\langle\, x, y \mid s\colon x \to y, t\colon x \to y, p\colon y \to x \mid sp = 1_x, tp = 1_x \,\rangle.$$

Note that it is a variation on previous case, since it consists of two morphisms with a common retraction. Depending on the applications, the cylinder can be equipped with various extra morphisms such as

- *reversion*: a morphism $r\colon y \to y$ satisfying

  $$r \circ s = t \qquad r \circ t = s \qquad r \circ r = 1_x \qquad s \circ r = s$$

  typically, the endomorphism of $X \times I$ defined by $r(x, a) = (x, 1 - a)$,
- *concatenation*: morphisms $c^-, c^+\colon y \to y$ satisfying the relations

  $$s \circ c^- = s \qquad t \circ c^+ = t \qquad p \circ c^- = p \qquad p \circ c^+ = p$$

  and moreover, in the presence of a reversion

  $$r \circ c^- = c^+ \circ r \qquad r \circ c^+ = c^- \circ r$$

  typically, on a topological cylinder $X \times I$, we define $c^-(x, a) = (x, a/2)$ and $c^+(x, a) = (x, (a + 1)/2)$.

A.2.3 The walking factorization. A *factorization* of a morphism $c\colon x \to z$ consists of a pair of morphisms $a\colon x \to y$ and $b\colon y \to z$ such that $c = b \circ a$. The *walking factorization* is the category presented by

$$\langle\, x, y, z \mid a\colon x \to y, b\colon y \to z, c\colon x \to z \mid ab = c \,\rangle.$$

It can be described as the category with $\{0, 1, 2\}$ as objects, one morphism $i \to j$ whenever $i \leqslant j$ and no morphism $i \to j$ whenever $i > j$.

A.2.4 The walking n-span. Given a natural number n, consider the poset S_n whose elements are

$$S_n = \{x_0^-, x_0^+, x_1^-, x_1^+, \ldots, x_{n-1}^-, x_{n-1}^+, x_n\}$$

ordered by $x_i^\alpha < x_j^\beta$ whenever $i < j$, for any $\alpha, \beta \in \{-, +\}$, and x_n is the maximal element. For $n = 0, 1, 2, 3$, the Hasse diagram of the poset is

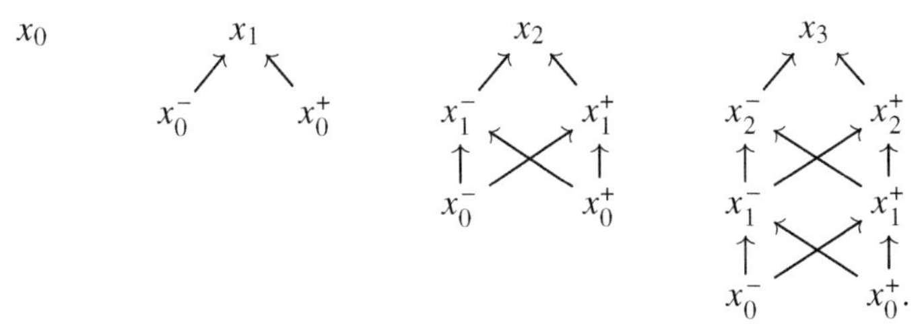

This poset can be seen as a category, still denoted S_n, with the above set of objects and an arrow $x \to y$ when $x \leqslant y$. Given a category C, a functor $S_n \to C$

(resp. $S_n^{\mathrm{op}} \to C$) is called an *n-cospan* (resp. *n-span*) [29]. In particular, a 1-span is a span in the usual sense. A typical n-cospan in the category **Top** is given by sending x_i^α to the i-disk D^i, the inclusion $x_{i-1}^- \to x_i^\alpha$ (resp. $x_{i-1}^+ \to x_i^\alpha$) corresponding to the inclusion of the $(i-1)$-disk into the lower (resp. upper) hemisphere of the sphere S^{i-1} bounding the disk D^i. The category S_n admits a presentation with

- 0-generators: x_i^-, x_i^+, x_n for $0 \leqslant i < n$,
- 1-generators: $x_{i-1}^\alpha \to x_i^\beta$ and $x_{n-1}^\alpha \to x_n$ for $0 < i < n$ and $\alpha, \beta \in \{-,+\}$,
- 2-generators:

$$
\begin{array}{ccc}
 & x_i^\gamma & \\
\nearrow & & \nwarrow \\
x_{i-1}^\beta & \Rightarrow & x_{i-1}^{\beta'} \\
\nwarrow & & \nearrow \\
 & x_{i-2}^\alpha &
\end{array}
\qquad\qquad
\begin{array}{ccc}
 & x_n & \\
\nearrow & & \nwarrow \\
x_{n-1}^\beta & \Rightarrow & x_{n-1}^{\beta'} \\
\nwarrow & & \nearrow \\
 & x_{n-2}^\alpha &
\end{array}
$$

for $1 < i < n$ and $\alpha, \beta, \beta', \gamma \in \{-,+\}$.

A.2.5 The globe category. Given an integer $n \geq 0$, we write $\mathbb{O}^{(n)}$ for the category with $\{0, \ldots, n\}$ as objects and two non-trivial morphisms $\sigma_j^i, \tau_j^i : j \to i$ for every $j < i$, with composition given by $f \circ \sigma_i^j = \sigma_i^k$ and $f \circ \tau_i^j = \tau_i^k$ for every morphism $f : j \to k$. Writing σ_i and τ_i for σ_i^{i+1} and τ_i^{i+1}, it can be pictured as

$$
0 \underset{\tau_0}{\overset{\sigma_0}{\rightrightarrows}} 1 \underset{\tau_1}{\overset{\sigma_1}{\rightrightarrows}} \cdots \underset{\tau_{n-1}}{\overset{\sigma_{n-1}}{\rightrightarrows}} n.
$$

This category is called the *category of globes of dimension* $\leq n$ and a presheaf on $\mathbb{O}^{(n)}$ an *n-globular set*, see Section 14.1. The category $\mathbb{O}^{(n)}$ admits the presentation

$$
\left\langle\, 0, \ldots, n \;\middle|\; \sigma_i, \tau_i : i \to i+1 \;\middle|\; \sigma_{i+1}\sigma_i = \tau_{i+1}\sigma_i, \sigma_{i+1}\tau_i = \tau_{i+1}\tau_i \,\right\rangle.
$$

A.2.6 The augmented simplicial category. The *augmented simplicial category* Δ_+ is the category where an object is a natural number $n \in \mathbb{N}$ and a morphism $f : m \to n$ is a weakly monotone function $f : [m] \to [n]$, where $[n]$ denotes the set $\{0, \ldots, n-1\}$. A presentation of this category is detailed in §4.5.6.

A.2.7 Presentations of monoidal categories. Many categories of interest are actually monoidal categories. It is generally much easier to present them as monoidal categories (as opposed to as a category) and then deduce a presentation of those as categories if one insists on having that, see §10.3.10. For instance, one can construct a presentation of the augmented simplicial category Δ_+, see §10.3.2, and recover the presentation evoked in previous section, see Example 10.3.11.

Appendix B
Examples of Coherent Presentations of Monoids

In this chapter, we present examples of coherent presentations of monoids in the sense developed in Chapter 7. In particular, we focus on families of monoids which occur in algebra and whose coherent presentations are computed using the rewriting method that extends Squier's and Knuth–Bendix's completion procedures into a homotopical completion-reduction procedure as presented in Chapter 7. Theorem B.1.2 shows how coherent presentations of monoids can be used to make explicit the actions of monoids on small categories. We apply this construction to the case of Artin monoids in Section B.1. In particular, we prove that the Tits-Zamolodchikov 3-generators extend the Artin presentation into a coherent presentation (Theorem B.1.9) and, as a byproduct, we give a constructive proof of a theorem of Deligne, Theorem [110, Theorem 1.5], on the actions of an Artin monoid on a category. We also give coherent presentations of plactic and Chinese monoids in Section B.2.

B.1 Artin Monoids

We provide here coherent presentations of Artin monoids, already introduced in §A.1.25. We also explain that these provide an explicit description of actions of those monoids on categories. The construction is done in two stages. Given an Artin monoid $B^+(W)$ on a Coxeter group W, first we consider the coherent presentation $\mathrm{Gar}_3(W)$ constructed on the Garside presentation of $B^+(W)$. Then we apply the homotopical completion-reduction on this coherent presentation by examining a family of generating triple confluences. We thus obtain a coherent presentation $\mathrm{Art}_3(W)$ of $B^+(W)$ that extends the Artin presentation.

B.1.1 Action of monoids on categories. Let M be a monoid seen as a 2-category with exactly one 0-cell $\star$, with the elements of M as 1-cells,

0-composition given by product in M and with identity 2-cells only, see Lemma A.1.2. An *action* T of the monoid M on a category is a pseudofunctor $T\colon M \to \mathbf{Cat}$. More explicitly, such an action is specified by

- a category $C = T(\star)$,
- an endofunctor $T(u)\colon C \to C$ for every element u of M,
- a natural isomorphism $T_{u,v}\colon T(u)T(v) \Rightarrow T(uv)$ for every pair (u, v) of elements of M and a natural isomorphism $T_\star\colon 1_C \Rightarrow T(1)$,

satisfying the following conditions:

- for every triple (u, v, w) of elements of M, the following diagram commutes:

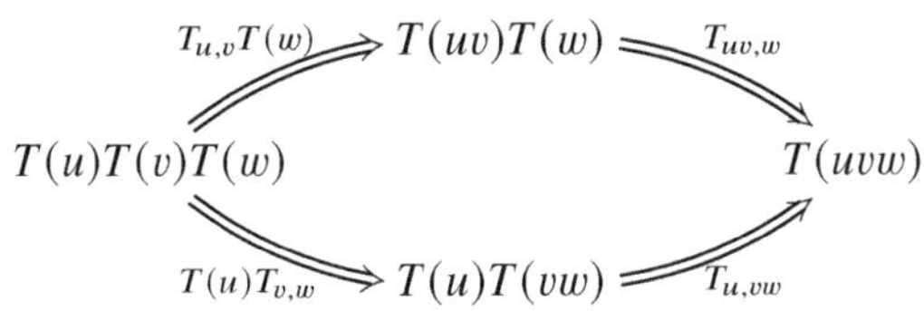

- for every element u of M, the following two diagrams commute:

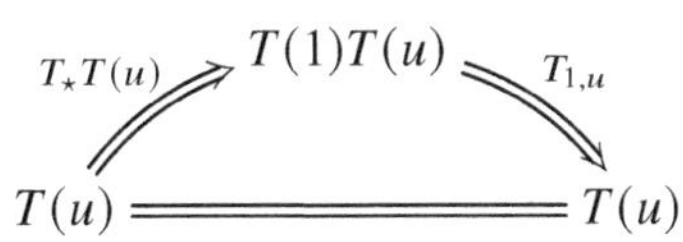

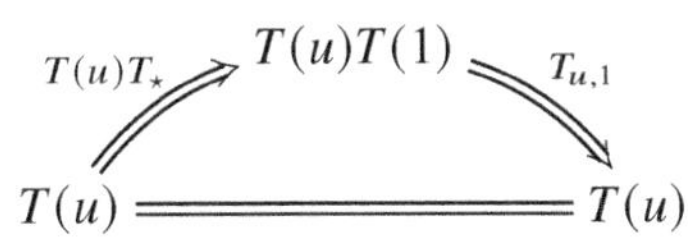

.

Such an action corresponds to a 2-*representation* of M in **Cat**, as defined by Elgueta in [121]. We denote by $\mathbf{Rep}_2(M, \mathbf{Cat})$ the category of actions of M on categories, equipped with suitable morphisms, as detailed in [145, Section 5.1]. The following result is proved in [145, Theorem 5.1.6]:

B.1.2 Theorem. *Suppose given a monoid M and a coherent presentation of M by a $(3, 1)$-polygraph P. We have an equivalence of categories*

$$\mathbf{Rep}_2(M, \mathbf{Cat}) \cong \mathbf{Cat}_2(\overline{P}, \mathbf{Cat})$$

between actions of M on categories and 2-functors from the $(2, 1)$-category presented by P to **Cat**.

Deligne already observed that this equivalence holds for Garside's presentation of spherical Artin monoids [110, Theorem 1.5]. In the rest of this section, we present an application of Theorem B.1.2 to give an equational description of an action of an Artin monoid on a category using Zamolodchikov relations.

B.1.3 Coxeter groups. Recall from §A.1.25 that a *Coxeter group* is a group W that admits a presentation with a finite set S of generators and with one relation

$$(st)^{m_{st}} = 1 \qquad \text{(B.1)}$$

with $m_{st} \in \mathbb{N} \sqcup \{\infty\}$, for every s and t in S, with the following requirements and conventions:

- $m_{st} = \infty$ means that there is, in fact, no relation between s and t,
- $m_{st} = 1$ if and only if $s = t$.

The last requirement implies that $s^2 = 1$ holds in W for every s in S. As a consequence, the group W can also be seen as the monoid with the same presentation. As in §A.1.25, we denote by $\langle st \rangle^n$ the element of length n in the free monoid S^*, obtained by multiplication of alternating copies of s and t:

$$\langle st \rangle^0 = 1, \qquad\qquad \langle st \rangle^{n+1} = s \langle ts \rangle^n.$$

When $s \neq t$ and $m_{st} < \infty$, we use this notation and the relations $s^2 = t^2 = 1$ to write (B.1) as a *braid relation*:

$$\langle st \rangle^{m_{st}} = \langle ts \rangle^{m_{st}}. \tag{B.2}$$

A *reduced expression* of an element u of W is a representative of minimal length of u in the free monoid S^*, and we write $l(u)$ for the length of any of the reduced expressions of u, called the *length* of u. The Coxeter group W is finite if and only if it admits an element of maximal length [56, Theorem 5.6]. In that case, this element is unique, called the *longest element of* W and is denoted by $w_0(S)$. For $I \subseteq S$, the subgroup of W spanned by the elements of I is denoted by W_I. It is a Coxeter group with generating set I. If W_I is finite, we denote by $w_0(I)$ its longest element.

We recall that the *Artin monoid* associated to W is the monoid denoted by $B^+(W)$, generated by S and subject to the braid relations (B.2). This presentation, seen as a 2-polygraph, is denoted by $\mathrm{Art}_2(W)$ and called *Artin's presentation*. This is the same as the one of W, except for the relations $s^2 = 1$.

B.1.4 Length notation. For every u and v in W, we have $l(uv) \leqslant l(u) + l(v)$ and we will use the following notations

$$\begin{array}{ccc} u \frown v & \Leftrightarrow & l(uv) = l(u) + l(v), \\ u \overset{\times}{\ } v & \Leftrightarrow & l(uv) < l(u) + l(v). \end{array}$$

We generalize the notation for a greater number of elements of W. For example, in the case of three elements u, v, and w of W, we write $u \frown v \frown w$ when both equalities $l(uv) = l(u) + l(v)$ and $l(vw) = l(v) + l(w)$ hold. This case splits in the following two mutually exclusive subcases:

$$\overbrace{u \frown v \frown w} \quad \Leftrightarrow \quad \begin{cases} u \frown v \frown w \\ l(uvw) = l(u) + l(v) + l(w), \end{cases}$$

$$u \frown v \overset{\times}{} \frown w \quad \Leftrightarrow \quad \begin{cases} u \frown v \frown w \\ l(uvw) < l(u) + l(v) + l(w). \end{cases}$$

B.1.5 Garside's coherent presentation. Let W be a Coxeter group. We call *Garside's presentation of* $B^+(W)$ the 2-polygraph $\mathrm{Gar}_2(W)$ whose 1-cells are the elements of $W \setminus \{1\}$ and with one 2-generator

$$\alpha_{u,v} \colon u|v \Rightarrow uv$$

whenever $l(uv) = l(u) + l(v)$ holds. Here, we write uv for the product in W and $u|v$ for the product in the free monoid over W. We denote by $\mathrm{Gar}_3(W)$ the extended presentation of $B^+(W)$ obtained from $\mathrm{Gar}_2(W)$ by adjunction of one 3-generator

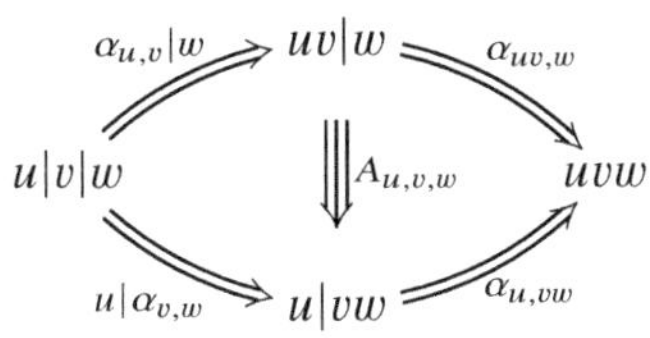

for all u, v, and w of $W \setminus \{1\}$ with $\overbrace{u \frown v \frown w}$.

B.1.6 Homotopical completion-reduction of Garside's presentation. The coherent presentation $\mathrm{Gar}_3(W)$ can be computed by coherent completion-reduction of the 2-polygraph $\mathrm{Gar}_2(W)$, as we now explain, see [145].

Let $<$ denote the strict order on the elements of the free monoid W^* that first compares their length as elements of W^*, then the length of their components, starting from the right. The order relation $\leqslant$ generated by $<$ by adding reflexivity is a termination order on $\mathrm{Gar}_2(W)$: for every 2-generator $\alpha_{u,v}$ of $\mathrm{Gar}_2(W)$, we have $u|v > uv$. Hence the 2-polygraph $\mathrm{Gar}_2(W)$ terminates, so that its coherent completion is defined (see Section 7.5). By applying the coherence completion-reduction procedure (see §7.5.6), one can obtain a coherent extension of the Garside presentation $\mathrm{Gar}_2(W)$, as detailed in [145, Proposition 3.2.1]. The resulting $(3,1)$-polygraph $\mathrm{KBS}(\mathrm{Gar}_2(W))$ has one 0-generator, one 1-generator for every element of $W \setminus \{1\}$, the 2-generators

$$\alpha_{u,v} \colon u|v \Rightarrow uv \qquad\qquad \beta_{u,v,w} \colon u|vw \Rightarrow uv|w$$

for every u and v of $W \setminus \{1\}$ with $u \frown v$ and every u, v, and w of $W \setminus \{1\}$ with $\overset{\times}{u \frown v \frown w}$, and the nine following families of 3-generators

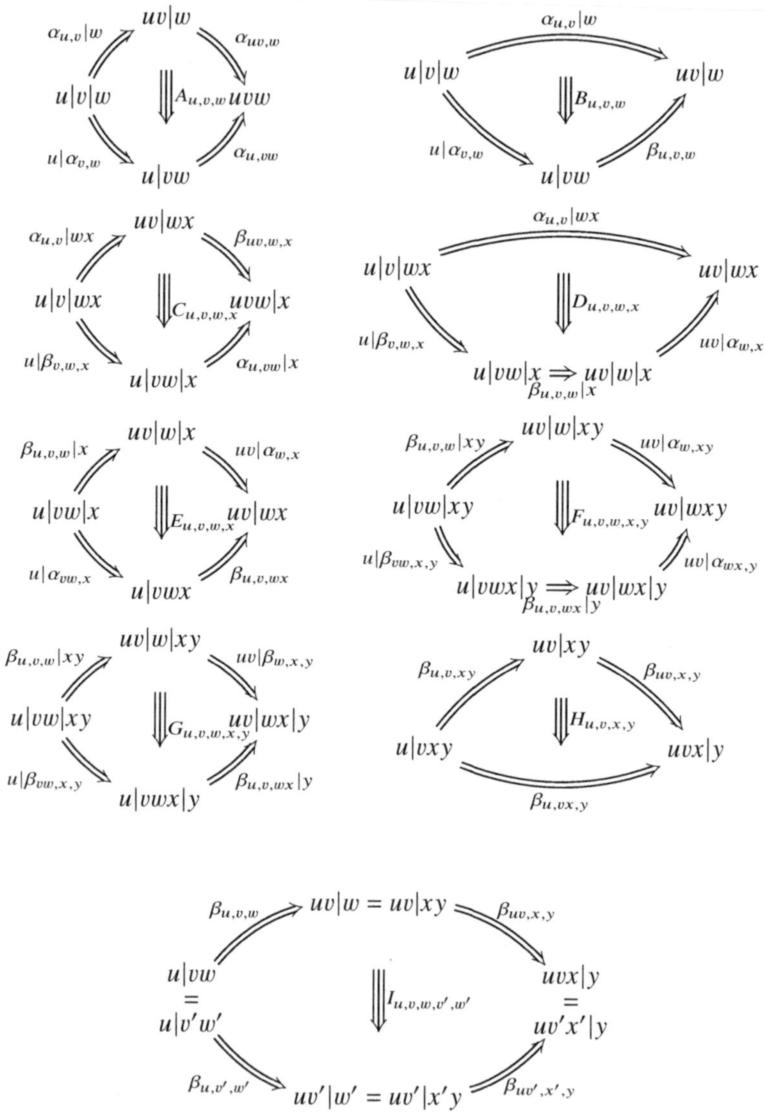

These 3-generators are families indexed by all the possible elements of $W \setminus \{1\}$, that can be deduced by the sources and targets of the 2-cells. For example, there is one 3-cell $A_{u,v,w}$ for all elements u, v, and w in $W \setminus \{1\}$ with $\overset{\frown}{u \frown v \frown w}$, and one 3-cell $F_{u,v,w,x,y}$ for all elements u, v, w, x, and y in $W \setminus \{1\}$ with

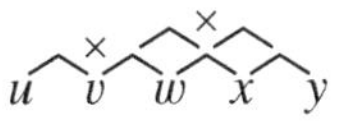 .

By considering a family of generating triple confluences associated to some of the triple critical branchings of $\mathrm{KBS}(\mathrm{Gar}_2(W))$, one can reduce this family of 3-cells and obtain the following result [145, Theorem 6.4.3]:

B.1.7 Theorem. *For every Coxeter group W, the Artin monoid $B^+(W)$ admits* $\mathrm{Gar}_3(W)$ *as a coherent presentation.*

The $(3, 1)$-polygraph $\mathrm{Gar}_3(W)$ is called *Garside's coherent presentation* of the Artin monoid $B^+(W)$.

B.1.8 Artin's coherent presentation of Artin monoids. Let W be a Coxeter group with a totally ordered set S of generators. The homotopical reduction method of §7.5.6 can be used on Garside's coherent presentation $\mathrm{Gar}_3(W)$ to contract it into a smaller coherent presentation associated to Artin's presentation [145].

We consider the presentation of the Artin monoid $B^+(W)$ by the 2-polygraph $\mathrm{Art}_2(W)$ with one 0-generator, the elements of S as 1-generator and one 2-generator

$$\gamma_{s,t} : \langle ts \rangle^{m_{st}} \Rightarrow \langle st \rangle^{m_{st}},$$

for all $t > s$ in S such that m_{st} is finite. The following result extends the 2-polygraph $\mathrm{Art}_2(W)$ into a coherent presentation of the Artin monoid $B^+(W)$, called the *Artin coherent presentation* of $B^+(W)$ [145, Theorem 4.1.1]. This coherent presentation is obtained using the homotopical reduction §7.5.4 on Garside's coherent presentation $\mathrm{Gar}_3(W)$.

B.1.9 Theorem. *For every Coxeter group W, the Artin monoid $B^+(W)$ admits the coherent presentation* $\mathrm{Art}_3(W)$ *made of Artin's presentation* $\mathrm{Art}_2(W)$ *and one* 3-*generator $Z_{r,s,t}$ for all elements $t > s > r$ of S such that the subgroup $W_{\{r,s,t\}}$ is finite.*

Artin's coherent presentation has exactly one k-cell, $0 \leqslant k \leqslant 3$, for every subset I of S of rank k such that the subgroup W_I is finite. The shape of the 3-generators $Z_{r,s,t}$ of the polygraph $\mathrm{Art}_3(W)$ are obtained by projection of the 3-generators of the polygraph $\mathrm{Gar}_3(W)$ given in §B.1.6 and depend on the type of the Coxeter type of the parabolic subgroup $W_{\{r,s,t\}}$. According to the classification of finite Coxeter groups [53, Chapter VI, §4, Theorem 1], there are five cases, shown below:

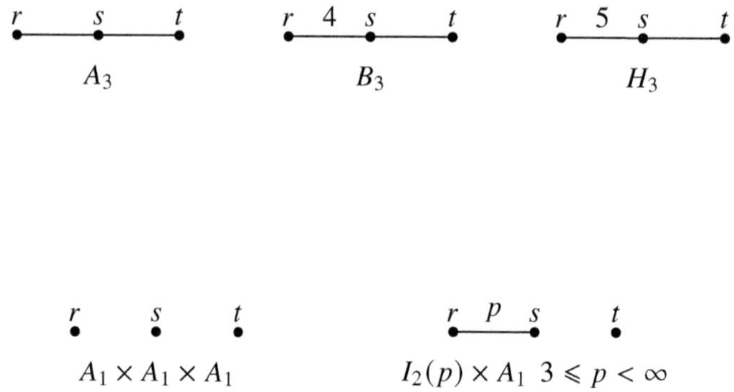

According to these types, the 3-generators $Z_{r,s,t}$ have the following shapes

– type A_3:

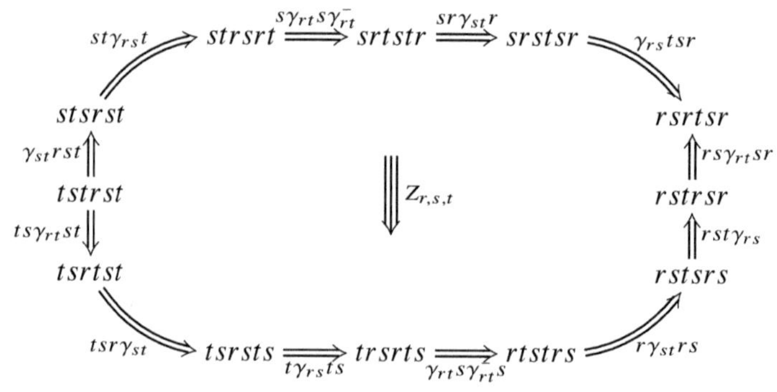

– type B_3:

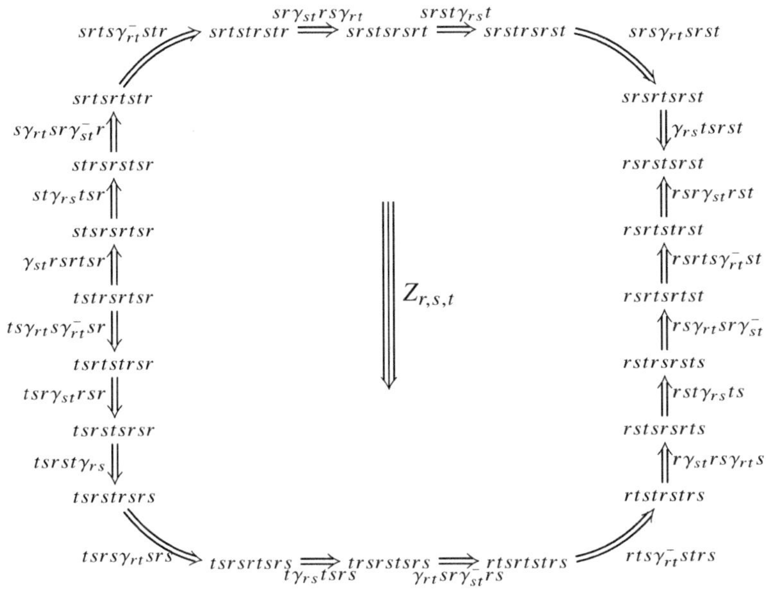

– type $A_1 \times A_1 \times A_1$:

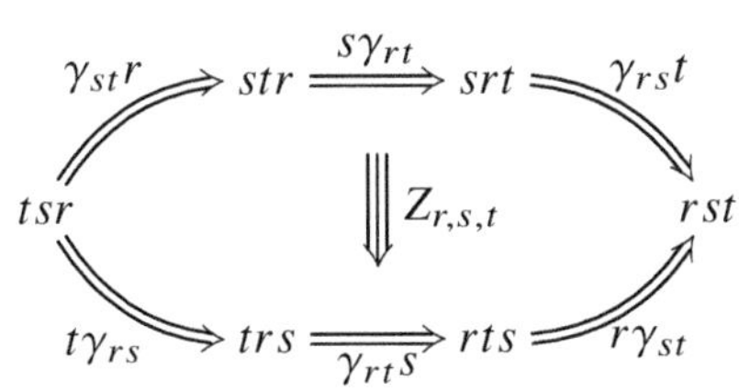

– type H_3:

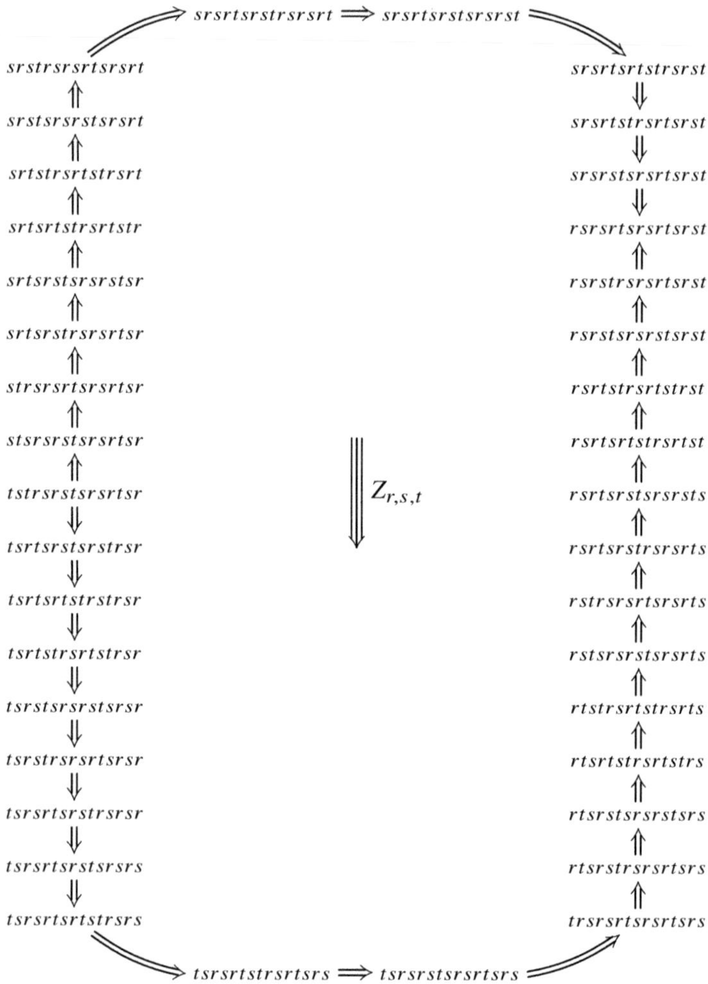

– type $I_2(p) \times A_1$, $p \geqslant 3$:

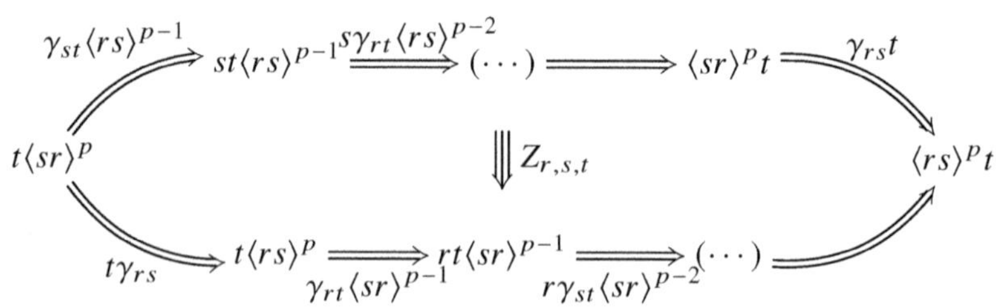

B.1.10 Action of braid monoids on categories. Following Theorem B.1.2 that, up to equivalence, the actions of a monoid M on categories are the same as the 2-functors from $\overline{P}$ to **Cat**, where P is any coherent presentation of M. As an application of Theorem B.1.9, we establish the relationship between coherent presentations of Artin monoids and Deligne's notion of an action on a category. In particular, Deligne's Theorem [110, Theorem 1.5] is equivalent to Theorem B.1.7.

The extended presentation $\mathrm{Gar}_3(W)$ is a coherent presentation of the Artin monoid $B^+(W)$. We thus get Deligne's Theorem [110, Theorem 1.5] for any Artin monoid as a consequence of Theorems B.1.2 and B.1.7. Moreover, Theorem B.1.9 gives a similar result in terms of Artin's coherent presentation $\mathrm{Art}_3(W)$, formalizing [110, Paragraph 1.3] on the actions of B_4^+ on a category.

B.2 Plactic and Chinese Monoids

We provide here coherent presentations of plactic and Chinese monoids whose presentations are recalled in §A.1.26. First, we recall a few points concerning the combinatorics of plactic monoids.

B.2.1 Rows, columns, and tableaux. A *row* is a non-decreasing word $x_1 \ldots x_k$ in the free monoid $\{1, \ldots, n\}^*$, i.e., with $x_1 \leqslant x_2 \leqslant \ldots \leqslant x_k$. A *column* is a decreasing word $x_p \ldots x_1$ in $\{1, \ldots, n\}^*$, i.e., with $x_p > x_{p-1} > \ldots > x_1$. We denote by $\mathrm{col}(n)$ the set of non-empty columns. A row $x_1 \ldots x_k$ *dominates* a row $y_1 \ldots y_l$, and we denote $x_1 \ldots x_k \rhd y_1 \ldots y_l$, if $k \leqslant l$ and $x_i > y_i$, for $1 \leqslant i \leqslant k$. Any word w in $\{1, \ldots, n\}^*$ has a unique decomposition as a product of rows of maximal length $u_1 \ldots u_k$, and it is called a *tableau* if $u_1 \rhd u_2 \rhd \ldots \rhd u_k$.

B.2.2 Schensted's algorithm. The *Schensted algorithm* computes for each w in $\{1, \ldots, n\}^*$ a tableau, denoted by $P(w)$, called the *Schensted tableau* of w, and constructed from the following steps [320]. Given u a tableau written as a product of rows of maximal length $u = u_1 \ldots u_k$ and $1 \leqslant y \leqslant n$, it computes the tableau $P(uy)$ as follows. If $u_k y$ is a row, the result is $u_1 \ldots u_k y$. If $u_k y$ is not a row, then suppose $u_k = x_1 \ldots x_l$ with $1 \leqslant x_i \leqslant n$ and let j minimal such that $x_j > y$, then the result is $P(u_1 \ldots u_{k-1} x_j) v_k$, where $v_k = x_1 \ldots x_{j-1} y x_{j+1} \ldots x_l$. The tableau $P(w)$ is computed from the empty tableau and iteratively applying the Schensted algorithm. In this way, $P(w)$ is the row reading of the planar representation of the tableau computed by the Schensted algorithm.

B.2.3 Knuth's 2-polygraph. Let us give an orientation to the Knuth relations set out in §A.1.26 with respect to the lexicographic order, thus forming a 2-polygraph, denoted by $\mathrm{Knuth}_2(n)$, whose 1-generators are $1, \ldots, n$ and the 2-generators are

$$\eta_{x,y,z}\colon zxy \Rightarrow xzy, \qquad \text{for } 1 \leqslant x \leqslant y < z \leqslant n,$$

$$\varepsilon_{x,y,z}\colon yzx \Rightarrow yxz, \qquad \text{for } 1 \leqslant x < y \leqslant z \leqslant n.$$

The congruence on the free monoid $\{1, \ldots, n\}^*$ generated by this polygraph is called the *plactic congruence of rank n* and the 2-polygraph $\mathrm{Knuth}_2(n)$ is a presentation of the plactic monoid P_n, [217, Theorem 6].

B.2.4 Pre-column presentation. One adds to the presentation $\mathrm{Knuth}_2(n)$ one superfluous generator c_u for any u in $\mathrm{col}(n)$. We denote by $\mathrm{Col}_1(n)$ the set of *column generators* c_u for any u in $\mathrm{col}(n)$ and by

$$\gamma_u\colon c_{x_p} \ldots c_{x_1} \Rightarrow c_u,$$

the defining relation for the column generators $u = x_p \ldots x_1$ in $\mathrm{col}(n)$ of length greater than 2. In the free monoid $\mathrm{Col}_1(n)^*$, the Knuth relations can be written in the following form

$$\eta^c_{x,y,z}\colon c_z c_x c_y \Rightarrow c_x c_z c_y, \qquad \text{for } 1 \leqslant x \leqslant y < z \leqslant n,$$

$$\varepsilon^c_{x,y,z}\colon c_y c_z c_x \Rightarrow c_y c_x c_z, \qquad \text{for } 1 \leqslant x < y \leqslant z \leqslant n.$$

The 2-polygraph $\mathrm{Knuth}^c_2(n)$ whose 1-generators are columns and 2-generators are the defining relations γ_u for columns generators and the Knuth relations $\eta^c_{x,y,z}$ and $\varepsilon^c_{x,y,z}$ is a presentation of the monoid P_n. We define the 2-polygraph $\mathrm{PreCol}_2(n)$ with column generators and the 2-generators are

$$\alpha'_{x,zy}\colon c_x c_{zy} \Rightarrow c_{zx} c_y, \qquad \text{for } 1 \leqslant x \leqslant y < z \leqslant n,$$

$$\alpha'_{y,zx}\colon c_y c_{zx} \Rightarrow c_{yx} c_z, \qquad \text{for } 1 \leqslant x < y \leqslant z \leqslant n,$$

$$\alpha'_{x,u}\colon c_x c_u \Rightarrow c_{xu}, \qquad \text{for } xu \in \mathrm{col}(n) \text{ and } 1 \leqslant x \leqslant n,$$

where $\alpha'_{x,zy}$ and $\alpha'_{y,zx}$ correspond respectively to the Knuth relations $\eta^c_{x,y,z}$ and $\varepsilon^c_{x,y,z}$. The 2-polygraph $\mathrm{PreCol}_2(n)$ is a presentation of the monoid P_n, then called the *pre-column presentation*.

B.2.5 Coherent column presentation. Given columns u and v, if the planar representation of the Schensted tableau $P(uv)$ is not the tableau obtained as the concatenation of the two columns u and v, we write $u \overset{\times}{} v$. In this case, the tableau $P(uv)$ contains at most two columns [73, Lemma 3.1]. We write $u^{\times 1} v$

(resp. $u \overset{\times 2}{} v$) if the tableau $P(uv)$ has one column (resp. two columns). When $u \overset{\times}{} v$, we define a 2-generator

$$\alpha_{u,v} \colon c_u c_v \Rightarrow c_w c_{w'}$$

where $w = uv$ and $c_{w'} = 1$, if $u \overset{\times 1}{} v$, and w and w' are respectively the left and right columns of the tableau $P(uv)$, if $u \overset{\times 2}{} v$.

The 2-polygraph $\mathrm{Col}_2(n)$ whose set of 1-generators is $\mathrm{Col}_1(n)$ and the 2-generators are the $\alpha_{u,v}$ is a finite convergent presentation of the monoid P_n, called the *column presentation* [73]. Using the coherent completion procedure defined in §7.5.2, this polygraph is extended into the *column coherent presentation* $\mathrm{Col}_3(n)$ of the monoid P_n [173, Theorem 3.2.2]. Its 3-generators, given by the confluence diagrams of the critical branchings of the 2-polygraph $\mathrm{Col}_2(n)$, have the following hexagonal form

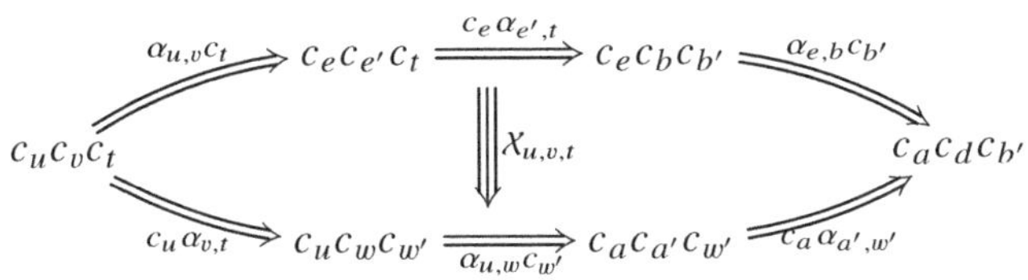

for any columns u, v, and t such that $u \overset{\times}{} v \overset{\times}{} t$.

B.2.6 Pre-column coherent presentation. Using the homotopical reduction procedure (§7.5.6), the coherent presentation $\mathrm{Col}_3(n)$ can be reduced into a smaller coherent presentation of P_n as follows. Firstly, we apply a homotopical reduction on the $(3,1)$-polygraph $\mathrm{Col}_3(n)$ with a collapsible part defined by some of the generating triple confluences of the 2-polygraph $\mathrm{Col}_2(n)$. In this way, we reduce the coherent presentation $\mathrm{Col}_3(n)$ of the monoid P_n into the coherent presentation $\overline{\mathrm{Col}_3}(n)$ of P_n, whose underlying 2-polygraph is $\mathrm{Col}_2(n)$ and the 3-generators $X_{u,v,t}$ are those of $\mathrm{Col}_3(n)$, but with u is of length 1. Then we reduce the coherent presentation $\overline{\mathrm{Col}_3}(n)$ into a coherent presentation $\mathrm{PreCol}_3(n)$ obtained from $\mathrm{PreCol}_2(n)$ by adjunction of the 3-generator $R_{\Gamma_3}(C'_{x,v,t})$ where

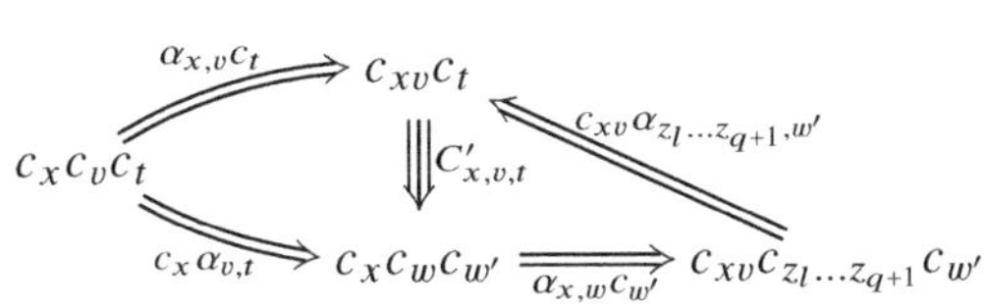

with $x^{\times 1} v^{\times 2} t$, and the 3-generator $R_{\Gamma_3}(D_{x,v,t})$ where

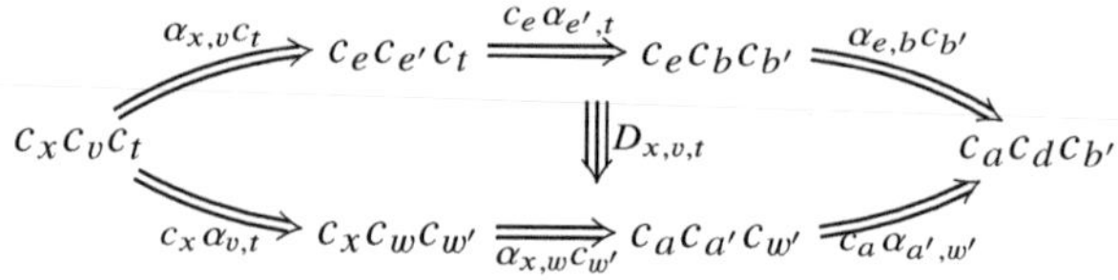

with $x^{\times 2} v^{\times 2} t$ and where the homotopical reduction R_{Γ_3} eliminates a collapsible part Γ_3 of $\overline{\mathrm{Col}_3}(n)$. In this way, we prove that the $(3,1)$-polygraph $\mathrm{PreCol}_3(n)$ is a coherent presentation of the monoid P_n [173, Theorem 4.3.4]. For instance, the coherent presentation $\mathrm{Col}_3(2)$ has only one 3-generator

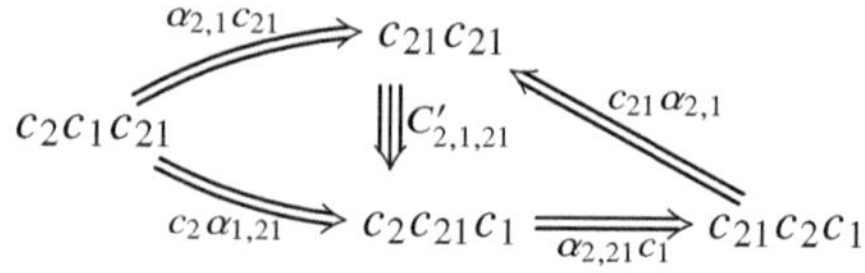

.

In this case, the $(3,1)$-polygraphs $\mathrm{PreCol}_3(2)$ and $\mathrm{Col}_3(2)$ coincide.

B.2.7 Knuth's coherent presentation. The coherent presentation $\mathrm{PreCol}_3(n)$ can be reduced into a coherent presentation of the monoid P_n whose underlying 2-polygraph is $\mathrm{Knuth}_2(n)$. We define an extended presentation $\mathrm{Knuth}_3(n)$ of the monoid P_n obtained from $\mathrm{Knuth}_2(n)$ by adjunction of the following set of 3-generators

$$\left\{ \mathcal{R}(C'_{x,v,t}) \;\middle|\; x^{\times 1} v^{\times 2} t \right\} \cup \left\{ \mathcal{R}(D_{x,v,t}) \;\middle|\; x^{\times 2} v^{\times 2} t \right\},$$

where $\mathcal{R}\colon \overline{\mathrm{Col}_3}(n)^\top \to \mathrm{Knuth}_3^c(n)^\top$ is a Tietze transformation, see [173, Section 4.4]. This gives a coherent presentation of the plactic monoid on the Knuth generators.

B.2.8 Theorem ([173, Theorem 4.4.7]). *For $n > 0$, the $(3,1)$-polygraph* $\mathrm{Knuth}_3(n)$ *is a coherent presentation of the monoid P_n.*

B.2.9 Example. For instance, the Knuth coherent presentation of the monoid P_2 has generators c_1 and c_2 subject to the Knuth relations

$$\eta^c_{1,1,2}\colon c_2c_1c_1 \Rightarrow c_1c_2c_1 \quad \text{and} \quad \varepsilon^c_{1,2,2}\colon c_2c_2c_1 \Rightarrow c_2c_1c_2$$

and the following 3-generator

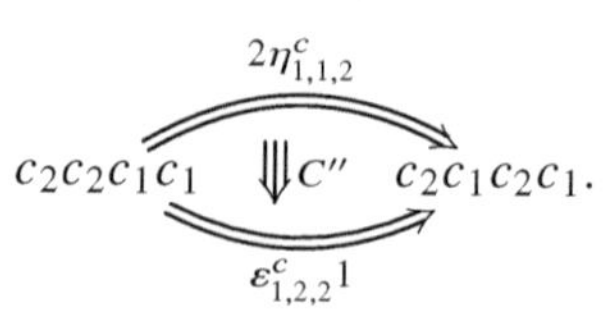

Note that the Knuth coherent presentation of the monoid P_2 corresponds to the coherent presentation that one can compute directly using the fact that the 2-polygraph $\mathrm{Knuth}_2(2)$ is convergent.

B.2.10 Chinese monoids. Using the completion-reduction method, as applied previously for the plactic monoid, we calculate a coherent presentation of the Chinese monoid Ch_n of rank $n > 0$ whose presentation is recalled in (A.2). We do not give here the details of this construction, developed in [174], but only present the method to obtain the form of the 3-generators of this coherent presentation.

Chinese relations (A.2) generate the Chinese congruence, denoted by $\sim_{C_n}$, and interpreted in [79] in terms of Chinese staircases. A *Chinese staircase* is a collection of boxes in right-justified rows, filled with non-negative integers, whose rows and columns are indexed with elements of $\{1, \dots, n\}$ from top to bottom and from right to left respectively, and where the i-th row contains i boxes, for $1 \leqslant i \leqslant n$. We denote by $R(t)$ the reading of a Chinese staircase t, row by row from right to left and from top to bottom. A Schensted-like insertion algorithm, denoted by $\leftsquigarrow_r$, is introduced in [79], and consists in inserting an element of $\{1, \dots, n\}$ into a Chinese staircase from the right. From a word $w = x_1 x_2 \dots x_k$, we associate a Chinese staircase $[\![w]\!]$ obtained by insertion of w in the empty staircase $\emptyset$ by application of $\leftsquigarrow_r$ step by step from left to right:

$$[\![w]\!] := (\dots((\emptyset \leftsquigarrow_r x_1) \leftsquigarrow_r x_2) \leftsquigarrow_r \dots) \leftsquigarrow_r x_k.$$

Chinese staircases satisfy the *cross-section property* for the congruence $\sim_{C_n}$, i.e., for all words w and w', $w \sim_{C_n} w'$ if and only if the insertion algorithm yields the same Chinese staircase: $[\![w]\!] = [\![w']\!]$, [79]. The elements of the monoid Ch_n can thus be identified with the Chinese staircases, which also form a monoid, whose product is defined by setting $t \star_r t' := (t \leftsquigarrow_r R(t'))$, for all Chinese staircases t and t'.

We construct a finite convergent presentation $\mathrm{Chin}_2(n)$ of the monoid Ch_n, whose 1-generators are *columns* on $\{1, \dots, n\}$ of length at most 2 and *square generators*, as defined in [174, Section 4.1], and whose 2-generators are

$$\gamma_{u,v} \colon c_u c_v \Rightarrow c_e c_{e'},$$

for all columns c_u and c_v such that $c_u c_v$ does not form a Chinese staircase and $c_u \star_r c_v$ is equal to the Chinese staircase composed by the columns c_e and $c_{e'}$. Note that the polygraph $\mathrm{Chin}_2(n)$ is obtained from the relations (A.2) by applying Tietze transformations, defined in Section 5.1, which consist in adding column generators and associated relations.

By definition of the 2-generators, the source of each critical branching of $\mathrm{Chin}_2(n)$ has the form $c_u c_v c_t$, for columns c_u, c_v, c_t such that $c_u c_v$ and $c_v c_t$ are not Chinese staircases. Their confluence diagrams are then obtained by applying the 2-generators γ, see [174, Theorem 5.6] for a detailed proof. Following Squier's homotopical theorem, Theorem 7.3.5, the 2-polygraph $\mathrm{Chin}_2(n)$ extends into a coherent presentation $\mathrm{Chin}_3(n)$ of the monoid Ch_n by adjunction of 3-generators with the following decagonal form

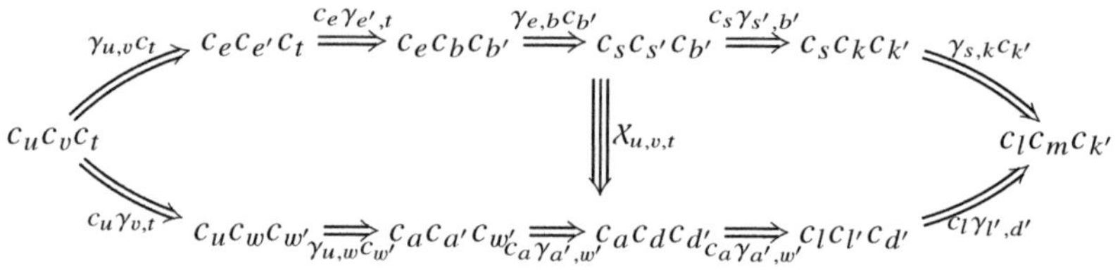

for all columns c_u, c_v, c_t such that $c_u c_v$ and $c_v c_t$ are not normal forms with respect to $\mathrm{Chin}_2(n)$, and where the γ denote either a 2-generator of $\mathrm{Chin}_2(n)$ or an identity. This proves the following result.

B.2.11 Theorem. *For $n > 0$, the $(3, 1)$-polygraph $Col_3(n)$ is a finite coherent convergent presentation of the Chinese monoid Ch_n.*

Appendix C
A Catalogue of 3-Polygraphs

In this chapter, we give some examples of presentations of 2-categories by 3-polygraphs. In many examples, the presented 2-categories are in fact monoidal categories and, actually, PROs. For those, we consider presentations by 3-polygraphs P with only one 0-generator $\star$ and one 1-generator a, so that we only need to provide the 2-generators (which we simply call *generators*) and 3-generators (which we call *rules* or *relations*). Moreover, we simply write n instead of a^n for a 1-cell in P_1^*.

C.1 Braids and Symmetries

C.1.1 Positive braids. The category $\mathbf{B}^+$ of *positive braids* contains all positive braid groups B_n^+, see §A.1.21. It has natural numbers as objects, every positive braid $b \in B_n^+$ induces a morphism $n \longrightarrow n$, for every $n \in \mathbb{N}$, and composition and identities are induced by multiplication and units of the monoids B_n^+. Otherwise said, if we consider the monoids B_n^+ as one-object categories (see §A.1.1), we have

$$\mathbf{B}^+ = \coprod_{n \in \mathbb{N}} B_n^+.$$

The expected monoidal structure makes a PRO out of it. As such, it admits a presentation by a 3-polygraph with one 2-generator $\gamma\colon 2 \longrightarrow 2$, called *braiding* and pictured as

one 3-generator corresponding to the Yang–Baxter relation

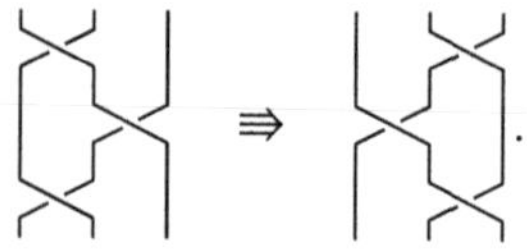

(C.1)

C.1.2 Braids. The category **B** of *braids* is defined similarly as

$$\mathbf{B} = \coprod_{n \in \mathbb{N}} B_n$$

where B_n is the n-th braid group, see §A.1.21, so that this category is a groupoid. It admits a presentation with two 2-generators $\gamma, \gamma^- : 2 \to 2$,

and three relations: the Yang–Baxter relation (C.1), as well as

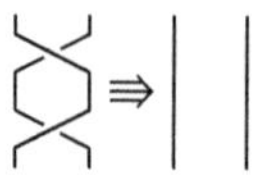

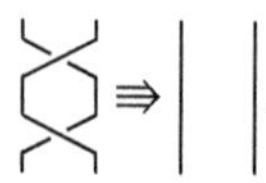

As an application, consider the ring $R = \mathbb{Z}[t, t^{-1}]$ of Laurent polynomials in one variable t. The category $\mathbf{Vect}_R$ of finitely generated free R-modules is monoidal with the usual tensor product. Writing P for the above presentation, we interpret a 1-cell n of P as R^n, the 2-generators γ and γ^- as the morphisms $R^2 \to R^2$ whose matrix representations are respectively

$$[\gamma] = \begin{pmatrix} 1-t & t \\ 1 & 0 \end{pmatrix} \qquad\qquad [\gamma^-] = \begin{pmatrix} 0 & 1 \\ t^{-1} & 1-t^{-1} \end{pmatrix}$$

This interpretation can be checked to be compatible with the relations of the presentation and thus induces a monoidal functor $f : \mathbf{B} \to \mathbf{Vect}_R^{\mathbf{Vect}_R}$, which is known as the *Burau representation* [69]. As a side note, this representation is not faithful [41] (i.e., there are distinct braids with the same image), but others are, such as the Lawrence-Krammer representation [42, 225, 244].

C.1.3 Permutations. The category **S** of *permutations* (or sometimes *symmetries*) is the monoidal category

$$\mathbf{S} = \coprod_{n \in \mathbb{N}} S_n$$

where S_n is the n-th symmetric group considered as a one-object category, see §A.1.19. Alternatively, the morphisms $m \to n$ can be described as the bijections $[m] \to [n]$ where $[n]$ is the set $\{0, \dots, n-1\}$. A presentation for this category can be obtained from the presentation of **B** by adding the relation

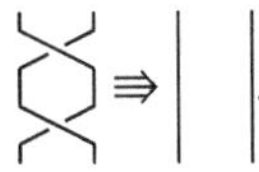

In this case, the relation 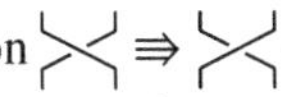 can be derived, which is why we generally remove one of the two generators, and note the remaining one as

 (C.2)

which is often called *transposition* in this context. The two relations of the presentation are thus

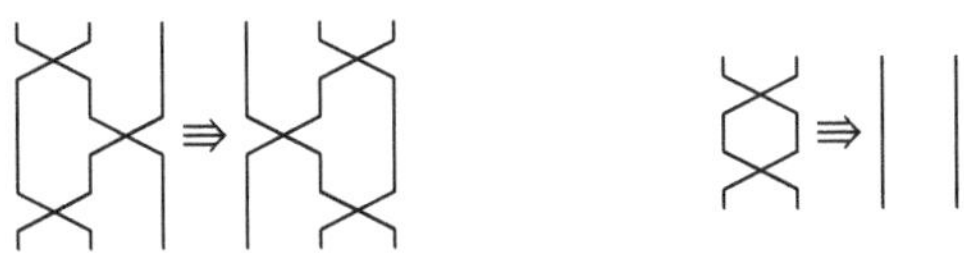 (C.3)

A notion of canonical form (which is in fact a normal form for the above rewriting system) was presented in §10.4.1: it consists of morphisms of the form

 or 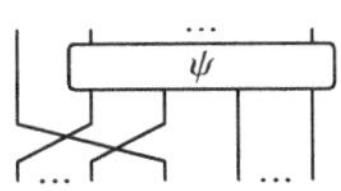

where the diagram on the left is the empty diagram, and ψ is a canonical form.

C.1.4 Free braided and symmetric monoidal categories. A *braided strict monoidal category* $(C, \otimes, i, \gamma)$ is a strict monoidal category $(C, \otimes, i)$ equipped with an invertible natural transformation of components

$$\gamma_{u,v} \colon v \otimes u \to u \otimes v$$

called *braiding*, which is compatible with the monoidal structure: for all objects $u, v, w \in C$,

$$\gamma_{u \otimes v, w} = (u \otimes \gamma_{v,w}) \circ (\gamma_{w,u} \otimes v), \qquad \gamma_{i,u} = 1_u,$$
$$\gamma_{u, v \otimes w} = (\gamma_{u,v} \otimes w) \circ (v \otimes \gamma_{u,w}), \qquad \gamma_{u,i} = 1_u.$$

A *symmetric monoidal category* is a braided monoidal category in which the braiding moreover satisfies $\gamma_{v,u} \circ \gamma_{u,v} = 1_{u \otimes v}$ for every object $u, v \in C$, in which case it is called a *symmetry*. A *braided monoidal functor* $f \colon C \to D$ between braided monoidal categories C and D, with γ^C and γ^D as respective braidings, is a monoidal functor which preserves the braiding, i.e., $f(\gamma^C_{u,v}) = \gamma^D_{fu,fv}$ for every objects $u, v \in C$. A *symmetric monoidal functor* is a braided monoidal functor between symmetric monoidal categories.

Given a monoidal category C, there always exists a free braided monoidal category $\tilde{C}$: it is a braided monoidal category equipped with a monoidal functor $C \to \tilde{C}$ such that, given a braided monoidal category D, a monoidal functor $C \to D$ extends uniquely as a braided monoidal functor $\tilde{C} \to D$. A similar statement holds for symmetric, instead of braided, monoidal categories. When C is presented by a 3-polygraph, $\tilde{C}$ admits the following presentation.

C.1.5 Theorem. *Suppose given a 3-polygraph P presenting a monoidal category $\overline{P}$, i.e., $P_0 = \{\star\}$. The free braided monoidal category on $\overline{P}$ is presented by the 3-polygraph Q such that*

$$
\begin{aligned}
Q_0 &= \{\star\}\\
Q_1 &= P_1\\
Q_2 &= P_2 \sqcup \left\{\gamma_{a,b} : ba \Rightarrow ab, \gamma^-_{a,b} : ab \Rightarrow ba \;\middle|\; a, b \in P_1\right\}\\
Q_3 &= P_3 \sqcup \left\{G_{a,b}, G'_{a,b}, L_{a,\alpha}, R_{a,\alpha} \;\middle|\; a, b \in P_1, \alpha \in Q_2\right\}
\end{aligned}
$$

with relations

$$G_{a,b} : \gamma^-_{a,b} * \gamma_{a,b} \Rrightarrow 1_{ab} \qquad\qquad G'_{a,b} : \gamma_{a,b} * \gamma^-_{a,b} \Rrightarrow 1_{ab}$$

for $a, b \in P_1$ and, for $\alpha : u' \to u$,

$$L_{a,\alpha} : \alpha a * \gamma_{a,u} \Rrightarrow \gamma_{u',a} * a\alpha \qquad\qquad R_{a,\alpha} : a\alpha * \gamma_{u,a} \Rrightarrow \gamma_{a,u'} * \alpha a$$

Above, $\gamma_{a,u} : ua \Rightarrow au$ is a notation for the morphism defined inductively on u by

$$\gamma_{a,i} = 1_a \qquad\qquad \gamma_{a,b} = \gamma_{a,b} \qquad\qquad \gamma_{a,bu} = b\gamma_{a,u} * \gamma_{a,b}u$$

and similarly, for $\gamma_{u,a} : au \Rightarrow ua$. Graphically, $\gamma_{a,b}$, $\gamma_{a,u}$, and $\gamma_{u,a}$ are respectively depicted as

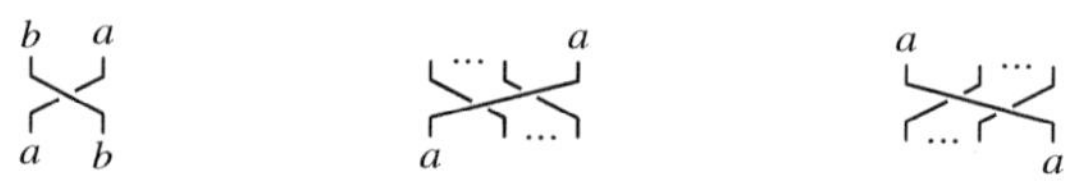

and the relations $L_{a,\alpha}$ and $R_{a,\alpha}$ are respectively

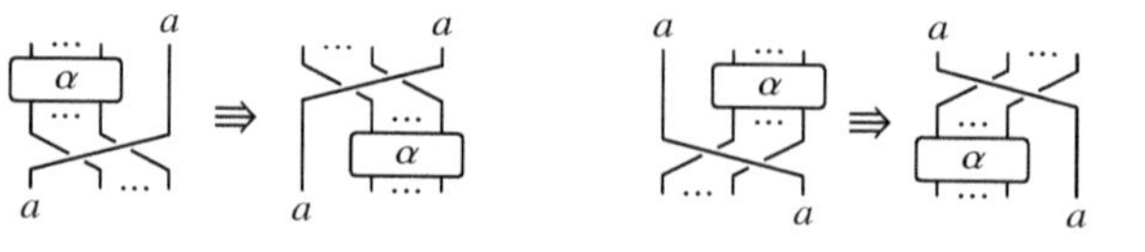

(C.4)

The braiding on $\overline{Q}$ is the one which is given on objects $a, b \in Q_1$ by $\gamma_{a,b}$. The free symmetric monoidal category is presented by the 3-polygraph obtained *from Q by removing the 2-generators $\gamma^-_{a,b}$ and the associated relations $G_{a,b}$ and $G'_{a,b}$, and adding the relations*

$$I_{a,b} \colon \gamma_{b,a} * \gamma_{a,b} \Rrightarrow 1_{ab}$$

indexed $a, b \in P_1$ (see also §12.5.5).

In particular, we see that **B** is the free braided monoidal category on the terminal category: there is only one 1-generator a, one invertible 2-generator $\gamma_{a,a}$, and the two relations $L_{a,\gamma_{a,a}}$ and $R_{a,\gamma_{a,a}}$ both correspond to the Yang–Baxter relation (C.1). Similarly, **S** is the free symmetric monoidal category on the terminal category. It can also be shown that **B** (resp. **S**) is the terminal braided (resp. symmetric) monoidal category.

In every presentation of a braided (resp. symmetric) monoidal category, the generators $\gamma_{a,b}$ are definable and the relations $L_{a,\alpha}$ and $R_{a,\alpha}$ (resp. and $I_{a,b}$) are derivable. Up to Tietze equivalence, we can thus suppose that every presentation of a braided (resp. symmetric) monoidal category is of the form given in the above theorem. For this reason, in a presentation P with only one 1-generator, we often say that a 2-generator $\gamma \colon 2 \to 2$ is a *symmetry* when it is involutive (i.e., satisfies the equation on the right of (C.3)) and satisfies relations (C.4) for every 2-generator α (in particular, for $\alpha = \gamma$, the Yang–Baxter relation has to be satisfied).

C.2 Monoids

Consider the category **F** where an object is a natural number and a morphism $f \colon m \to n$ is a function $f \colon [m] \to [n]$. Alternatively, this category can be described as the skeleton of the full subcategory of **Set** on finite sets. It can be equipped with a tensor product similar to the one of the simplicial category, see §10.3.2, thus making it a PRO. This category contains interesting subcategories, with the same objects, closed under tensor product, with the following morphisms:

- the category $\mathbf{F}_\eta$ of injective functions,
- the category $\mathbf{F}_\mu$ of surjective functions,
- the category **S** of bijections,
- the category Δ_+ of non-decreasing functions,
- the category Δ_η of injective non-decreasing functions,
- the category Δ_μ of surjective non-decreasing functions.

Details for this section can be found in [72, 230, 235].

C.2.1 Non-decreasing injections. The PRO Δ_η of injective non-decreasing functions admits a presentation with one generator $\eta\colon 0 \to 1$, called *unit* and pictured as

and no relation. The generator is interpreted as the terminal function $0 \to 1$ and, for instance, the interpretation of the diagram on the left is the injective non-decreasing function $f\colon [4] \to [6]$ whose graph is depicted on the right:

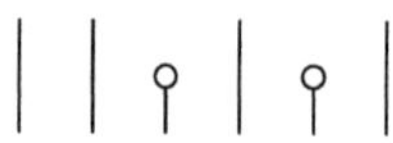

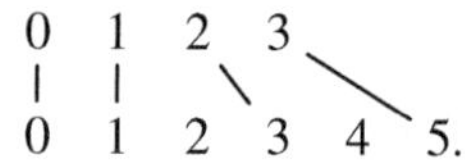

The opposite PRO $\Delta_\eta^{\mathrm{op}}$ can be described as the category whose morphisms are non-decreasing partial surjective functions.

C.2.2 Non-decreasing surjections. The PRO Δ_μ of non-decreasing surjective functions (already encountered in §23.3.10) admits a presentation with one generator $\mu : 2 \to 1$, called *multiplication* and pictured as

and one relation

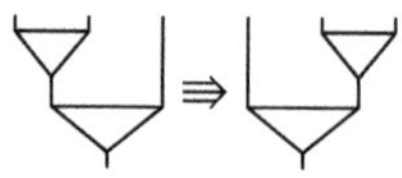

The generator can be interpreted as the terminal function $2 \to 1$, whose graph is

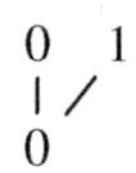

and, for instance, the interpretation of the diagram on the left is depicted on the right:

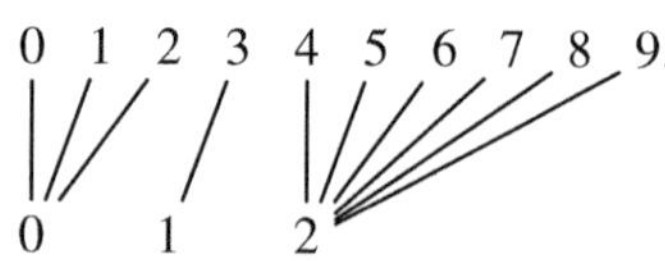

This category is thus the theory for *semigroups*: a monoidal functor from Δ_μ to **Set** (with the monoidal structure induced by cartesian product) consists of a set equipped with an associative binary operation. The rewriting system is convergent and normal forms can be described as the canonical form given by

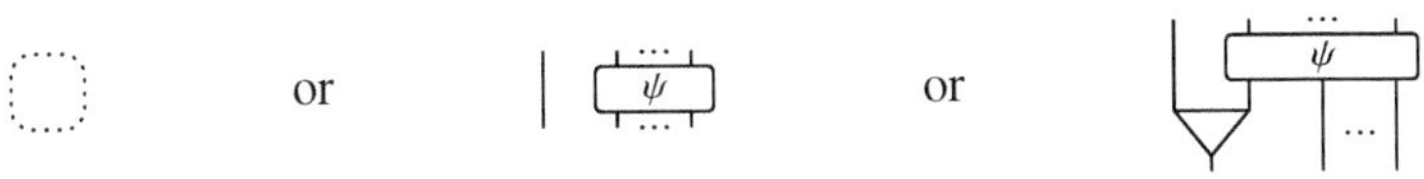

where ψ is a canonical form (the diagram on the left is the empty diagram).

C.2.3 Non-decreasing functions. The PRO Δ_+ (the augmented simplicial category) can be presented with two generators $\eta\colon 0 \to 1$ and $\mu\colon 2 \to 1$, pictured as above, together with the rules of §§C.2.1 and C.2.2:

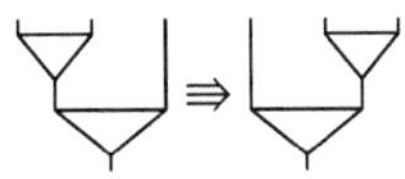

as well as the additional rules

These define a distributive law ℓ between the two previous categories so that

$$\Delta_+ = \Delta_\mu \otimes_\ell \Delta_\eta.$$

The rewriting system is convergent, see §10.3.2, and normal forms can be described as the following canonical forms:

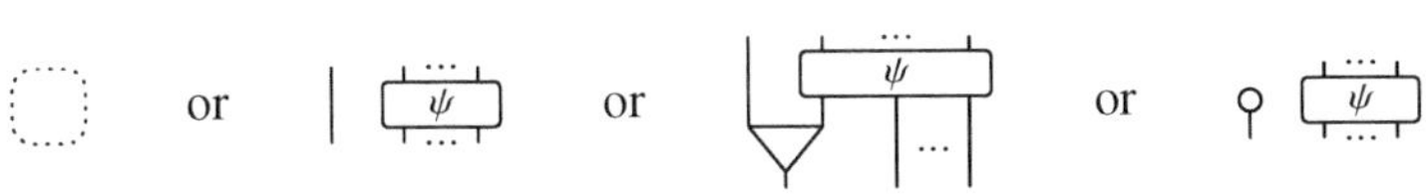

where ψ is a canonical form; moreover, in the third case, we suppose that ψ is not of the form given in the fourth case.

An alternative notion of canonical form (which is not a normal form for a rewriting system, for similar reasons as in Example 10.3.8) is

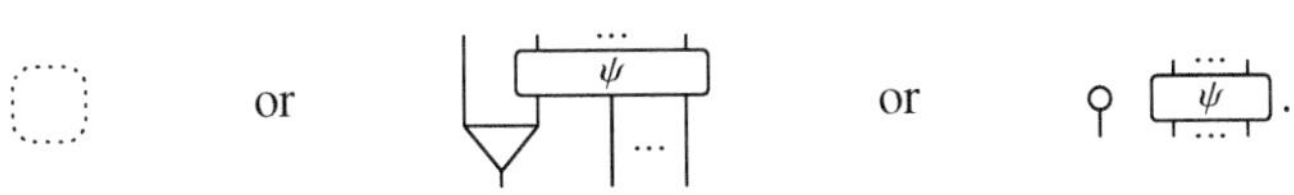

This means that morphisms in Δ_+ can be shown to be in bijection with 2-cells of the above form. Note that, here, the canonical form of an identity does not consist of wires only.

From this presentation, one sees that Δ_+ is the theory for *monoids*: the monoidal functors from Δ_+ to a monoidal category C correspond to monoids C (see Example 10.1.5). We can also deduce that it is the free cocartesian category on the terminal category, see Sections 13.4 and C.2.10.

Writing P for the above polygraph, the morphisms from n to 1 in P_2^*, for some $n \in \mathbb{N}$ can be ordered by $\phi \leqslant \psi$ whenever there exists a rewriting path $\phi \Rrightarrow \psi$. The resulting poset is a well-studied lattice called the n-th *Tamari lattice* [139].

C.2.4 Injections. The PRO $\mathbf{F}_\eta$ admits a presentation with two generators

satisfying the relations for symmetries (§C.1.3) and non-decreasing injections (§C.2.1)

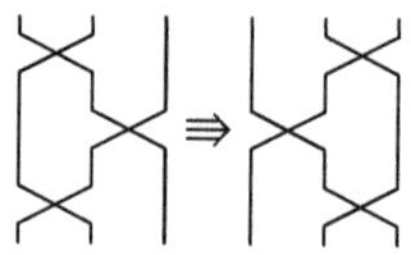

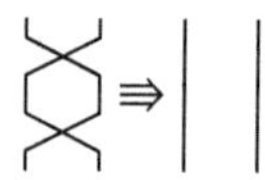

as well as the compatibility relations

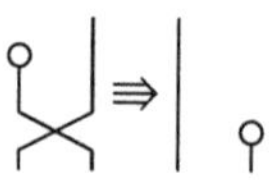

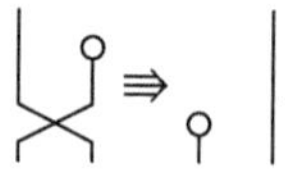

Those define a distributive law ℓ between $\mathbf{S}$ and Δ_η, so that

$$\mathbf{F}_\eta = \mathbf{S} \otimes_\ell \Delta_\eta.$$

The resulting presentation is convergent with normal forms being given by

or

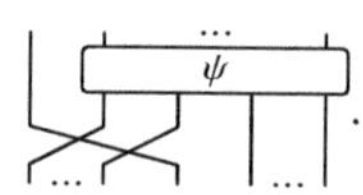

Note that the second (or the first) compatibility relation is redundant since we have

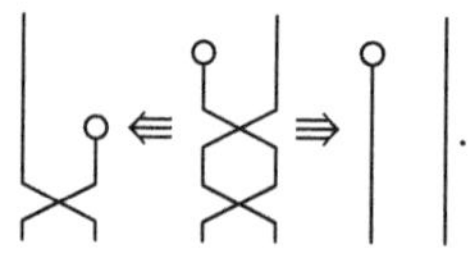

C.2.5 All functions. The PRO $\mathbf{F}$ corresponds to commutative monoids. It admits a presentation with three generators

and relations which consist of those for monoids (§C.2.3)

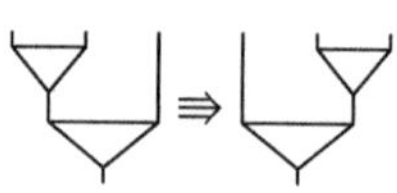

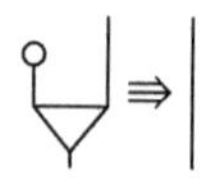

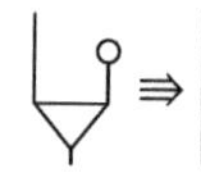

those for symmetries (§C.1.3)

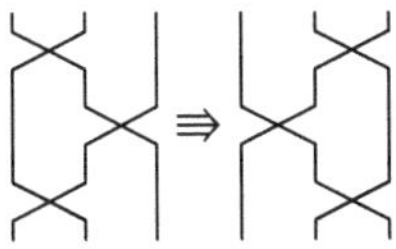

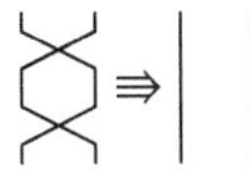

compatibility relations

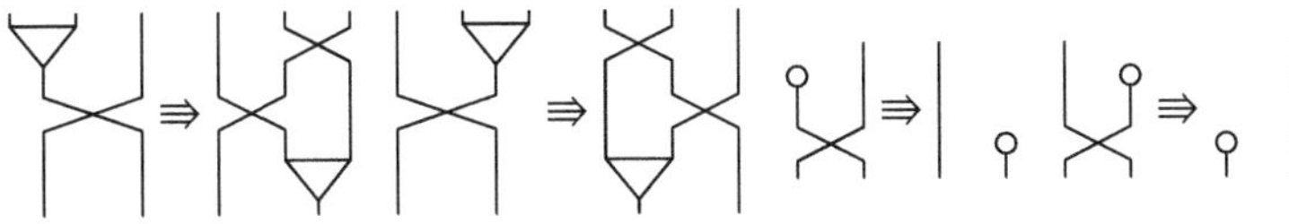 (C.5)

and the commutativity relation

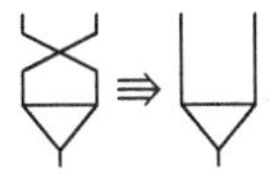 (C.6)

This rewriting system is convergent. A notion of canonical form (which does not exactly coincide with normal forms for similar reasons as in Example 10.3.8) can be given by

 or 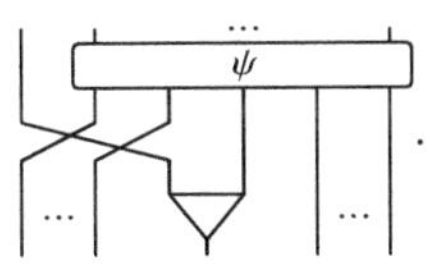

This monoidal category is the theory for *commutative monoids*: a symmetric monoidal functor to a symmetric monoidal category C corresponds to commutative monoid in C. It can be obtained as a composite PROP $\mathbf{F} = \mathbf{F}_\mu \otimes_\ell \mathbf{F}_\eta$ along the expected distributive law.

C.2.6 Partial functions. Consider the PRO $\mathbf{F}_\varepsilon$ where a morphism $f\colon m \to n$ is a partial function $f\colon [m] \to [n]$. It admits a presentation with four generators

satisfying the relations of §C.2.5 together with

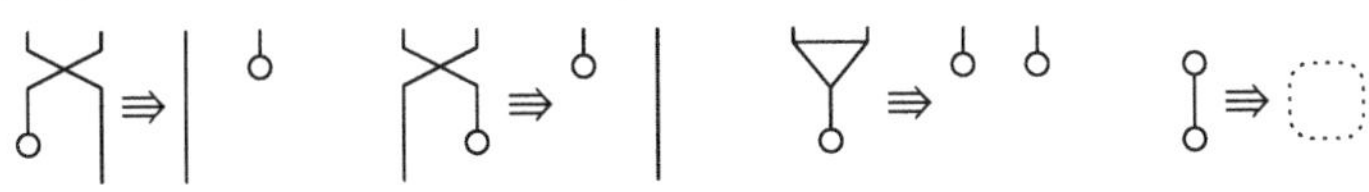

The resulting rewriting system is convergent and canonical forms can be given by

 or 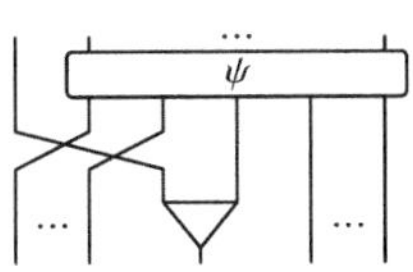 or 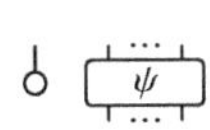.

The category can be described as a composite PROP $\mathbf{F}_\varepsilon = \mathbf{F}_\eta^{\mathrm{op}} \otimes_\ell \mathbf{F}$ along the expected distributive law, as well as the composite PRO $\mathbf{F} = \Delta_\eta^{\mathrm{op}} \otimes_\ell \mathbf{F}$ along the expected distributive law. In the last case, the associated factorization system given by Proposition 3.3.4 is the usual factorization of a partial function $f\colon X \to Y$ as a partial non-decreasing injection (given by the canonical partial function $X \to \mathrm{dom}(f)$ where $\mathrm{dom}(f) \subseteq X$ is the domain of f) followed by a total function (the restriction of f to its domain).

Similar presentations exists for other variants of the PRO (e.g., partial non-decreasing surjective functions). In particular, for the category of partial injective non-decreasing functions $\Delta_{\eta\varepsilon}$, we obtain a decomposition as

$$\Delta_{\eta\varepsilon} = \Delta_\eta^{\mathrm{op}} \otimes_\ell \Delta_\eta = \mathrm{Cospan}(\Delta_\eta).$$

C.2.7 Symmetric monoids. The theory of *symmetric monoids* is obtained from the theory of commutative monoids of §C.2.5 by removing the commutativity relation (C.6): by Theorem C.1.5, this is the free symmetric monoidal category on Δ_+, and therefore it is the PROP of monoids. The relations induce a distributive law ℓ so that this PRO can be obtained as $\mathbf{S} \otimes_\ell \Delta_+$, see [230]. A direct description of this category can be given as the PRO where a morphism $f\colon m \to n$ is a function $f\colon [m] \to [n]$ together with total order on each of the sets $f^{-1}(i)$ for $0 \leqslant i < n$, equipped with suitable composition [298].

C.2.8 Braided monoids. As a variant of the theory for commutative monoids, one can consider the one *braided monoids*. It admits a presentation with four generators

and the laws are similar to those of commutative monoids (§C.2.5) excepting that the transpositions satisfy the laws for braidings (§C.1.2) instead of symmetries (§C.1.3). As in §C.2.7, one could also considered the variant without the commutativity relation, i.e., the free braided monoidal category on Δ_+, but in practice braided monoids are always understood commutative.

C.2.9 Comonoids. Dual categories are also interesting. For instance, Δ_+^{op} is the theory representing comonoids and $\mathbf{F}^{\mathrm{op}}$ cocommutative comonoids.

C.2.10 Free cartesian categories. The category $\mathbf{F}^{\mathrm{op}}$ can be characterized as being the free cartesian category on the terminal one (and of course the case of $\mathbf{F}$ is dual, but the opposite is more commonly used from this perspective). We briefly describe here the situation and refer the reader to Section 13.4, where cartesian categories are considered in detail.

Any cartesian category has an underlying symmetric monoidal category, where the tensor product is induced by the cartesian product and unit is the terminal object. Not every symmetric monoidal category can be obtained in this way: those who can are characterized by the fact that every object is equipped with a structure of commutative comonoid, in a natural way, see Theorem 13.4.3. From this observation, one can derive the characterization of free cartesian categories given in Theorem 13.4.5. In particular, the category $\mathbf{F}^{\mathrm{op}}$ is the free category on the terminal category.

C.3 Distributive Laws

C.3.1 Distributive laws between monads. A monad is a particular case of a monoid, as already mentioned in Example 10.1.5. Namely, a 2-functor from the 2-category Δ_+ (which represents monoids, see §C.2.3) to the 2-category **Cat** (of categories, functors, and natural transformations) amounts to the data of

- a category C (the image of $\star$),
- a functor $T\colon C \to C$ (the image of a),
- natural transformations $\mu\colon T \circ T \Rightarrow T$ and $\eta\colon 1_C \Rightarrow T$ (the respective images of μ and η)

satisfying axioms so that (T, μ, η) is a monad on C (see §3.3.12).

As a variant, a theory corresponding to pairs of monads on a same category can easily be constructed. The category $\Delta_+ \sqcup \Delta_+$, the coproduct of Δ_+ with itself as monoidal categories admit a presentation with two 1-generators a, b and four 2-generators

$$\mu_a\colon aa \Rightarrow a \qquad \eta_a\colon 1 \Rightarrow a \qquad \mu_b\colon bb \Rightarrow b \qquad \eta_b\colon 1 \Rightarrow b$$

respectively pictured as

such that the pair (μ_a, η_a) satisfies the laws of monoids, see §C.2.3, as well as the pair (μ_b, η_b).

More interestingly, the previous theory can be modified in order to present a pair of monads on a same category, together with a distributive law between them. We recall that a *distributive law* between two monads T and U consists of a natural transformation $\ell\colon T \circ U \Rightarrow U \circ T$ satisfying four suitable axioms, see

§3.3.12 or [36]. This theory **DLaw** can be obtained from the one for $\Delta_+ \sqcup \Delta_+$ by adding a 2-generator $\lambda \colon ba \Rrightarrow ab$, pictured as

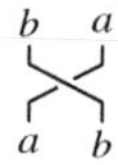

and the four relations

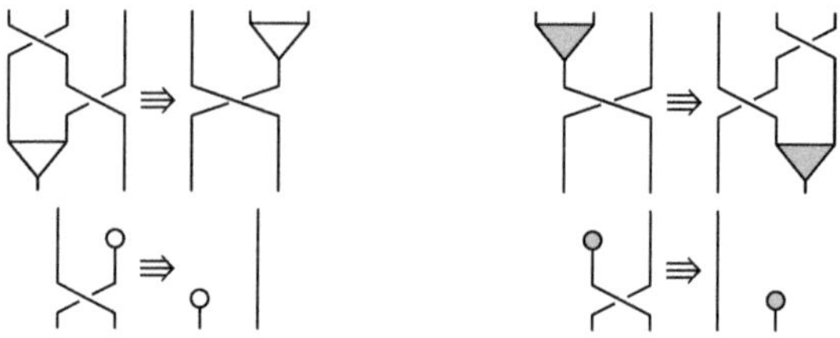

This presentation is convergent. It was shown by Beck [36] that a distributive law between monads T and U induces a structure of monad on the composite endofunctor $U \circ T$. This can be rephrased in the above setting as follows:

C.3.2 Theorem. *In* **DLaw***, the two following morphisms induce the structure of a monoid on ab:*

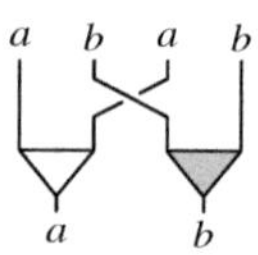

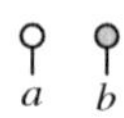

Proof. Associativity is shown by the following derivation

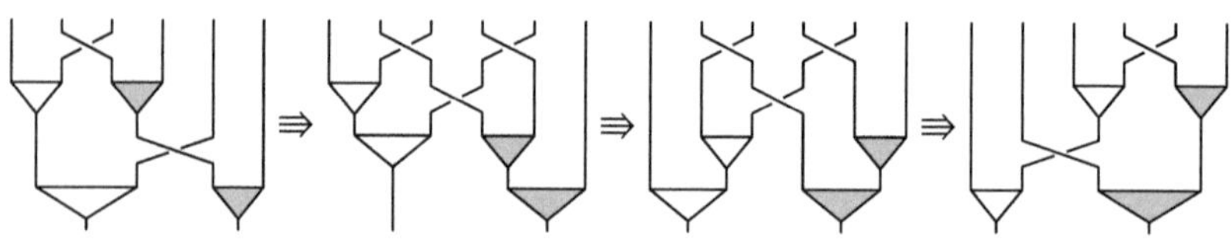

and left and right neutrality by

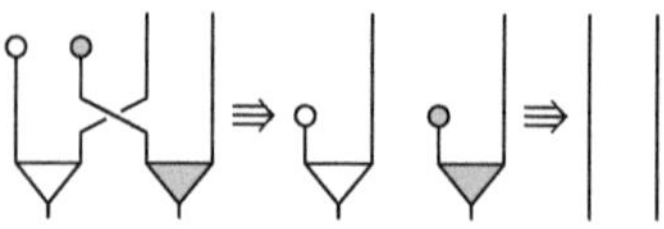 and 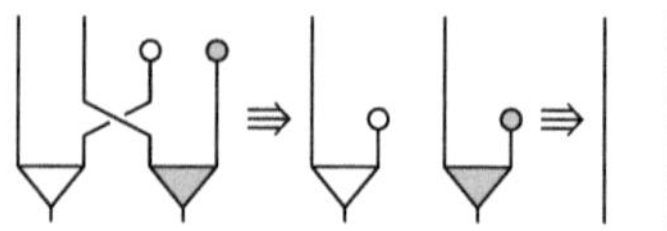

which concludes the proof. □

For instance, on **Set**, the monad of rings can be obtained as $U \otimes_\ell T$ where T is the monad of free monoids, U and the monad of free abelian groups and $\ell \colon TU \Rightarrow UT$ is the usual distributive law, which sends the formal expression of the form $(a+b)(c+d)$ to $ac+ad+bc+bd$.

C.3.3 Iterated distributive laws. It is useful to iterate this construction: given three monads S, T, and U on a category C and distributive laws between S and T, S and U, and T and U, we would like to have a monad structure on the composite $U \circ T \circ S$. This is the case where the distributive laws satisfy a suitable compatibility axiom [82]. The theory **IDLaw** for *iterated distributive laws* axiomatizes such a situation. It has three 1-generators a, b, and c, nine 2-generators

$$\mu_a : aa \Rightarrow a \qquad \mu_b : bb \Rightarrow b \qquad \mu_c : cc \Rightarrow c \qquad \ell_{ab} : ba \Rightarrow ab$$
$$\eta_a : 1 \Rightarrow a \qquad \eta_b : 1 \Rightarrow b \qquad \eta_c : 1 \Rightarrow c \qquad \ell_{ac} : ca \Rightarrow ac$$
$$\ell_{bc} : cb \Rightarrow bc$$

pictured as

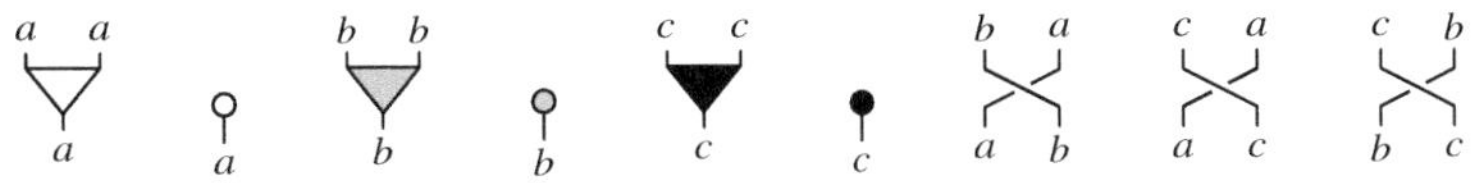

satisfying relations expressing that (a, μ_a, η_a), (b, μ_b, η_b), and (c, μ_c, η_c) are monoids, ℓ_{ab}, ℓ_{bc}, and ℓ_{ac} are distributive laws, and the additional axiom

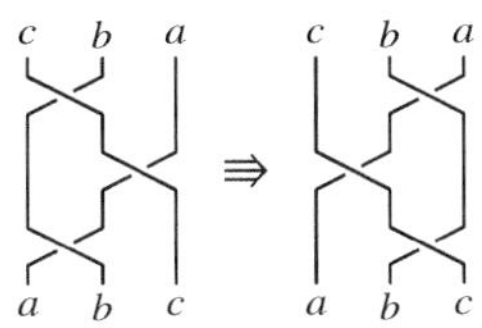

reminiscent of the Yang–Baxter relation. This is illustrated in §C.5.3.

In order to compose four or more monads, one might at first think that we should need new axioms, but it is in fact enough to assume the above axioms for every triple of monads in order to compose an arbitrary number of those, see [82] and §3.3.16.

C.3.4 Other distributive laws. Variants of the notion of distributive law have been studied in the literature. For instance, the expected notion of distributive law between a monad and a comonad is studied in [305]. A weaker notion of distributive law has also been studied by Street [340]: here the two relations involving units have been replaced by the unique relation

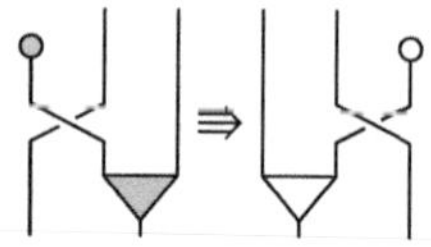

or, equivalently, the two relations

C.3.5 Linear non-linear terms. A functor $T\colon \mathbf{F} \to \mathbf{Set}$ is an abstract way to encode a collection of terms. Namely, for $n \in \mathbb{N}$, the set Tn can be seen as the collection of terms $t(x_0, \ldots, x_{n-1})$ with n free variables, and given a morphism $f\colon m \to n$, i.e., a function $f\colon [m] \to [n]$, the function $Tf\colon Tm \to Tn$ is the reindexing function, sending a term $t(x_0, \ldots, x_{n-1})$ to the term $t(x_{f(0)}, \ldots, x_{f(n-1)})$ obtained by replacing the variable x_i by $x_{f(i)}$, see for instance [131]. Similarly, a functor $T\colon \mathbf{S} \to \mathbf{Set}$ encodes a collection of "linear" terms: Tn consists of terms in which each x_i, for $0 \leqslant i < n$, occurs exactly once and, for this reason, we can only permute variables, and not merge two of them or forget one of them.

Now suppose that we are interested in a "mixed" situation where a term can have both linear and non-linear variables. A linear variable can always be considered as a non-linear one by forgetting about the fact that it should occur exactly once. Those are naturally modeled by the following theory [128, 194], with two 1-generators a and b, six 2-generators

$$\mu_a\colon aa \Rightarrow a \qquad \eta_a\colon 1 \Rightarrow a \qquad \sigma\colon b \Rightarrow a$$
$$\gamma_{aa}\colon aa \Rightarrow aa \qquad \gamma_{ab}\colon ba \Rightarrow ab \qquad \gamma_{bb}\colon bb \Rightarrow bb$$

pictured as

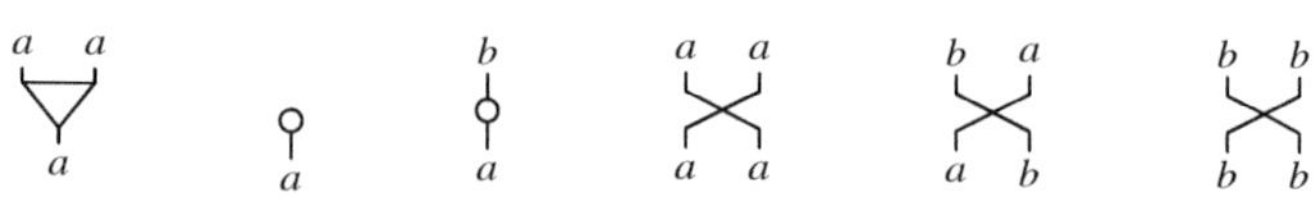

such that the axioms of symmetries hold for whichever typing of the wires, $(\mu_a, \eta_a, \gamma_{aa})$ satisfies the axioms of commutative monoids, and the axioms

hold for whichever typing of the wires. The object a thus corresponds to a non-linear variable (which can duplicated, erased, and exchanged with other variables), the object b to linear variable (which can only be exchanged with other variables), and the morphism σ to the fact that we can consider a linear variable as a non-linear one.

C.4 Bialgebras

C.4.1 Matrices. Given a semiring R, we write $\mathbf{M}_R$ for the category whose objects are natural numbers and a morphism $f\colon m \to n$ is an $n{\times}m$-matrix with coefficients in R, with usual composition and identities. This category is a PRO with the usual direct sum of matrices.

When R is a ring, the category $\mathbf{M}_R$ is equivalent to the category of finite-dimensional R-modules and R-linear maps. When $\Bbbk$ is a field, the category $\mathbf{M}_\Bbbk$ is equivalent to $\mathbf{Vect}_\Bbbk$, the category of vector spaces and $\Bbbk$-linear maps. Writing K for the semiring of small cardinals, with disjoint union as addition and cartesian product as product, the category $\mathbf{M}_K$ is equivalent to the full subcategory of Span($\mathbf{Set}$) (the category of isomorphism classes of spans, see §3.3.14) on finite sets.

C.4.2 Multirelations. Given two sets X and Y, a *multirelation* from X to Y is a function $X \times Y \to \mathbb{N}$ such that the set

$$\{(x, y) \in X \times Y \mid f(x, y) \neq 0\}$$

is finite. Any relation $R \subseteq X{\times}Y$ is canonically seen as the multirelation f such that $f(x, y) = 1$ if $(x, y) \in R$ and $f(x, y) = 0$ otherwise. Conversely, a multirelation can be thought of as a relation with multiplicities: given $(x, y) \in X \times Y$, $f(x, y)$ is called the *multiplicity* of the relation (x, y). Given two multirelations $f\colon X \to Y$ and $g\colon Y \to Z$, their composite is given by

$$(g \circ f)(x, z) = \sum_{y \in Y} f(x, y) \times g(y, z)$$

(note that the sum only involves a finite number of non-zero terms) and the identity on a set X is such that $1_X(x, x') = 0$ if $x \neq x'$ and $1_X(x, x) = 1$. We write **MRel** for the category of sets and multirelations. This is a monoidal category when equipped with the tensor product induced by disjoint union. A morphism $f\colon X \to Y$ in this category can be seen as a span

$$X \xleftarrow{s} R \xrightarrow{t} Y \tag{C.7}$$

where R is a finite set, such that the image under f of $(x, y) \in X \times Y$ is the cardinal of the following set:

$$f(x, y) = |\{r \in R \mid x = s(r) \text{ and } y = t(r)\}|$$

making **MRel** a subcategory of Span(**Set**), the category of sets and isomorphism classes of spans of functions.

The full subcategory of **MRel** on finite sets is equivalent to $\mathbf{M}_{\mathbb{N}}$, the category of matrices with coefficients in $\mathbb{N}$. Namely, a morphism $f\colon m \to n$ is an $n{\times}m$-matrix with coefficients in $\mathbb{N}$, which corresponds to the multirelation from $[m]$ to $[n]$ such that the multiplicity of $(i, j) \in [m] \times [n]$ is $f(i, j)$. Note that the category $\mathbf{M}_{\mathbb{N}}$ is isomorphic to Span(**F**). This category can also be described as the PRO where a morphism $f\colon m \to n$ is a morphism of free finitely generated monoids $f\colon \mathbb{N}^m \to \mathbb{N}^n$, see [298].

C.4.3 Bialgebras. The PRO $\mathbf{M}_{\mathbb{N}}$ admits a presentation with generators

$$\mu\colon 2 \Rightarrow 1 \qquad \eta\colon 0 \Rightarrow 1 \qquad \delta\colon 1 \Rightarrow 2 \qquad \varepsilon\colon 1 \Rightarrow 0 \qquad \gamma\colon 2 \Rightarrow 2$$

pictured as

and relations expressing that γ is a symmetry (§C.1.4), (μ, η, γ) is a commutative monoid (§C.2.5), $(\delta, \varepsilon, \gamma)$ is a cocommutative comonoid (§C.2.9) and the compatibility relations

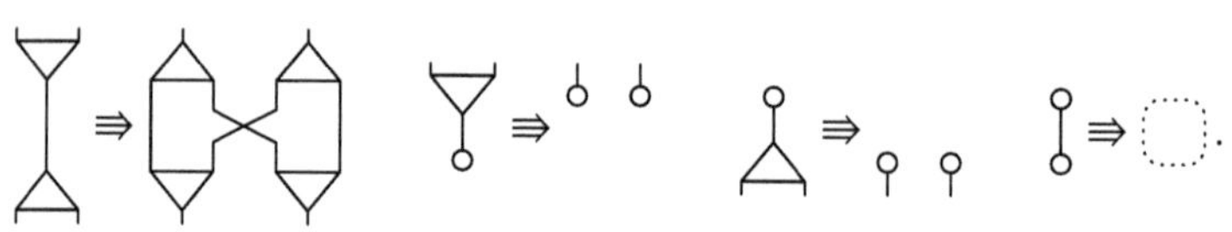

The interpretations of the generators are the following multirelations, represented as matrices

$$\mu = \begin{pmatrix} 1 & 1 \end{pmatrix} \qquad \delta = \begin{pmatrix} 1 \\ 1 \end{pmatrix} \qquad \gamma = \begin{pmatrix} 0 & 1 \\ 1 & 0 \end{pmatrix}$$

i.e., graphically,

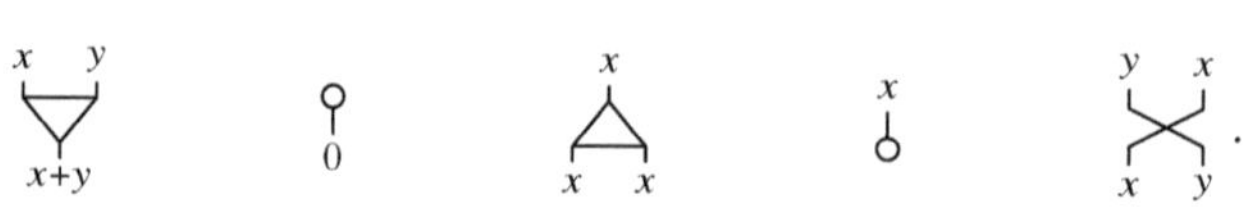

A notion of canonical form is given by

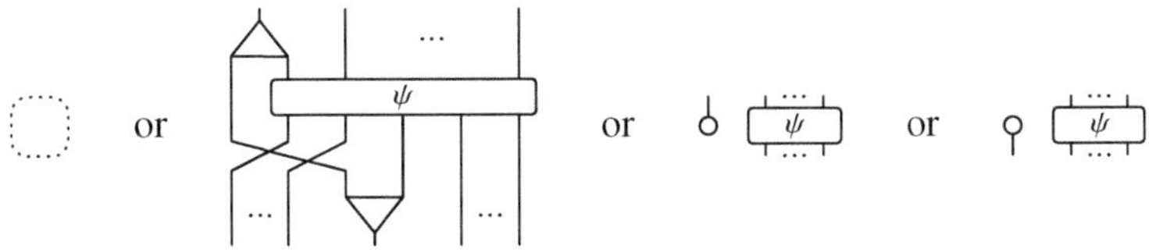

up to some relations, for which normal forms can be given, see [282]. For instance, the multirelation $f: 3 \to 4$ whose graph is shown on the left (we link two points as many times as their multiplicity in the multirelation) is represented by the canonical form on the right:

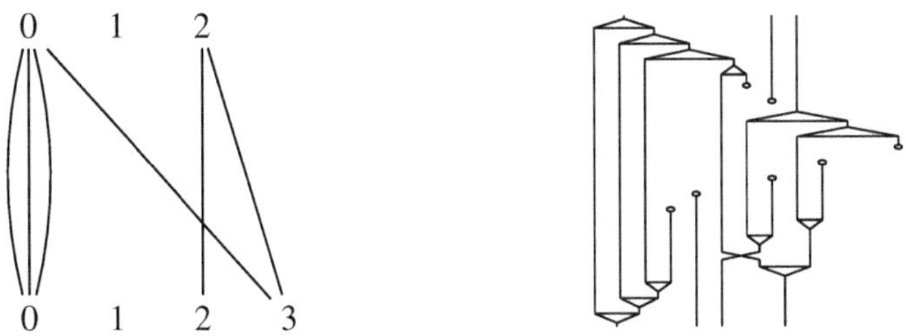

This example should make more clear the correspondence between the category and its presentation, and details can be found in various places [230, 235, 282, 298]. This presentation can be obtained from the one of $\mathbf{F}$ given in §C.2.5, using the methods of Section 10.5 for composing PROPs: we have $\mathrm{Span}(\mathbf{F}) = \mathbf{F}^{\mathrm{op}} \otimes_\ell \mathbf{F}$ where the distributive law $\ell: \mathbf{F} \otimes_{\mathbf{S}} \mathbf{F}^{\mathrm{op}} \to \mathbf{F} \otimes_{\mathbf{S}} \mathbf{F}^{\mathrm{op}}$ is given by pullback, see [230].

This is the theory for an algebraic structure called a *bicommutative bialgebra*, or *bimonoid*. More generally, if we drop the requirement that the monoid and the comonoid structures should be commutative, we obtain the theory of *bialgebras* (which are said to be commutative/cocommutative/bicommutative when the monoid/comonoid/both structures are commutative). Bialgebras are generally considered in the category $\mathbf{Vect}_{\Bbbk}$ of vector spaces over a fixed field $\Bbbk$. For instance, given a monoid $(M, \cdot, 1)$, the vector space $\Bbbk M$ generated by the set M is canonically a cocommutative bialgebra: writing e_a with $a \in M$ for a basis vector, the interpretations of the various morphisms are given by

$$
\begin{aligned}
\mu &: \Bbbk M \otimes \Bbbk M \to \Bbbk M & \qquad \eta &: 1 \to \Bbbk M \\
&\quad e_a \otimes e_b \mapsto e_{a \cdot b} & &\quad 1 \mapsto e_1
\end{aligned}
$$

$$
\begin{aligned}
\delta &: \Bbbk M \to \Bbbk M \otimes \Bbbk M & \quad \varepsilon &: \Bbbk M \to 1 & \quad \gamma &: \Bbbk M \otimes \Bbbk M \to \Bbbk M \otimes \Bbbk M \\
&\quad e_a \mapsto e_a \otimes e_a & &\quad e_1 \mapsto 1 & &\quad e_a \otimes e_b \mapsto e_b \otimes e_a
\end{aligned}
$$

C.4.4 Relations. Consider the category **Rel** whose objects are sets and where a morphism from X to Y is a relation from X to Y, i.e., a subset $R \subseteq X \times Y$. Explicitly, the composite of two relations $R \subseteq X \times Y$ and $S \subseteq Y \times Z$ is the relation $S \circ R \subseteq X \times Z$ defined by

$$S \circ R = \{(x,z) \in X \times Z \mid \exists y \in Y, (x,y) \in R \wedge (y,z) \in S\}.$$

This category is monoidal with tensor product given on objects by disjoint union of sets. It may be described as a variant of category of Span(**Set**) where morphisms are isomorphism classes of spans of the form (C.7), such that s and t are jointly monic, i.e., two distinct x and y give rise to distinct pairs of images $(s(x), t(x))$ and $(s(y), t(y))$.

The full subcategory whose objects are finite sets is equivalent to the category $\mathbf{M}_{\mathbb{B}}$ of matrices over the semiring of booleans (with $\vee$ as addition and $\wedge$ as multiplication). It admits a presentation obtained from the theory of bicommutative bialgebras by further adding the relation

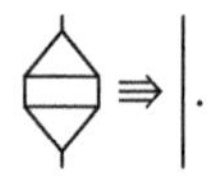

(C.8)

If we remember that the theory of bicommutative bialgebras is a presentation of the category $\mathbf{M}_{\mathbb{N}}$ of matrices with coefficients in $\mathbb{N}$, adding the relation (C.8) amounts to quotient coefficients in $\mathbb{N}$ in matrices by the relation $1+1 = 1$. Otherwise said, we describe $\mathbf{M}_{\mathbb{B}}$ as the quotient of $\mathbf{M}_{\mathbb{N}}$ under the relation identifying two morphisms $f, g \colon m \to n$ which have the same non-zero coefficients, i.e., when $f(i,j) = 0$ iff $g(i,j) = 0$ for every $(i,j) \in [m] \times [n]$. This is the theory for bialgebras which are called *special*, *relational* [193], or *qualitative* [282]. Note that the category of relations can be described as the following pushout in **MonCat** (the category of monoidal categories an strict monoidal functors):

$$\begin{array}{ccc} \mathbf{F}_\eta \sqcup \mathbf{F}_\eta^{\mathrm{op}} & \longrightarrow & \mathrm{Span}(\mathbf{F}_\eta) \\ \downarrow & & \vdots \\ \mathrm{Cospan}(\mathbf{F}_\eta) & \cdots\cdots\to & \mathbf{M}_{\mathbb{B}} \end{array}$$

where the arrows are the canonical inclusions, from which the presentation of $\mathbf{M}_{\mathbb{B}}$ can be deduced, see [132] for details and generalizations of this situation.

C.4.5 Matrices over $\mathbb{Z}/2\mathbb{Z}$. The category $\mathbf{M}_{\mathbb{Z}/2\mathbb{Z}}$ of natural numbers and matrices with coefficients over the ring $\mathbb{Z}/2\mathbb{Z}$ admits a presentation obtained from the theory of bialgebras by adding the relation

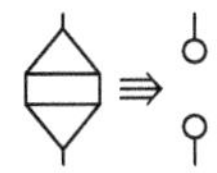

which amounts to quotient coefficients in $\mathbb{N}$ by $1 + 1 = 0$. This is the theory for bialgebras which are *anti-separable* [235]. More generally, matrices over $\mathbb{Z}/n\mathbb{Z}$ can be obtained by replacing the above relation by a relation of the form

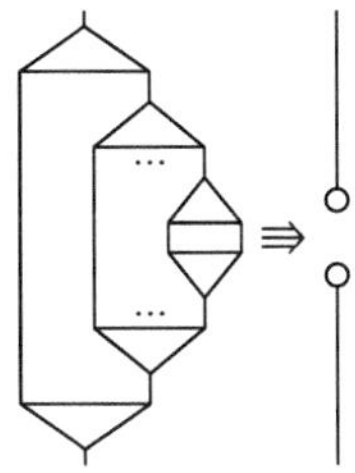

where the diagram on the left contains $n - 1$ morphisms  and morphisms . An alternative confluent presentation is given in [235, Section 3.2] and shown to be terminating in [158, Section 7].

C.4.6 Matrices over $\mathbb{Z}$, Hopf algebras. The category $\mathbf{M}_{\mathbb{Z}}$ of natural numbers and matrices over $\mathbb{Z}$ admits a presentation obtained from the theory of bialgebras by adding a generator $\sigma\colon 1 \to 1$, called *antipode*, pictured as

and interpreted as the 1×1-matrix (-1), together with the relations

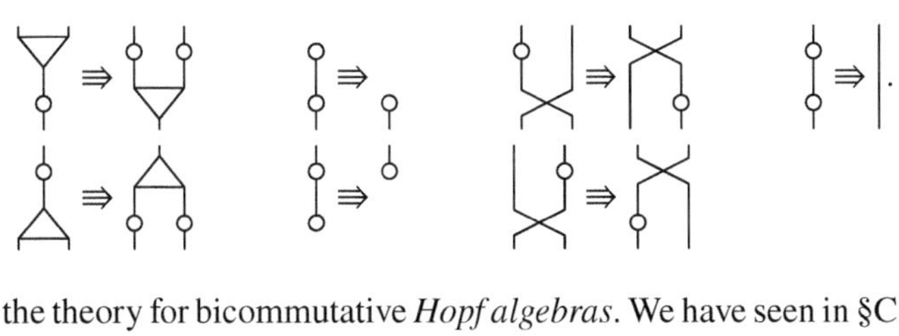

This is the theory for bicommutative *Hopf algebras*. We have seen in §C.4.3 that every monoid induces a bialgebra; when this monoid is a group G, bialgebra $\Bbbk G$ is canonically a Hopf algebra, the interpretation of the antipode being given by inverses:

$$\begin{aligned}\sigma\colon \Bbbk G &\to \Bbbk G\\ e_a &\mapsto e_{-a}.\end{aligned}$$

C.4.7 Matrices over an arbitrary semiring. More generally, given a semiring R, the PRO $\mathbf{M}_R$ admits a presentation containing the theory of matrices plus a generator

for every $a \in R$, interpreted as the $1{\times}1$-matrix (a), and relations

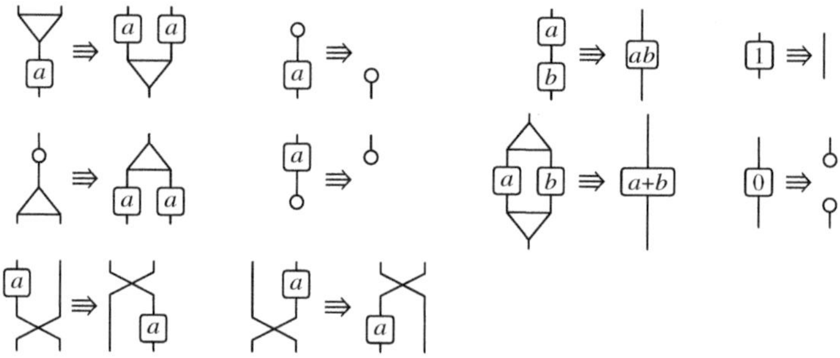

see [235]. This is the theory for *bicommutative R-linear bialgebras*. For instance the diagram on the left corresponds to the linear transformation $\mathbb{R}^3 \to \mathbb{R}^2$ associated to the matrix depicted on the right:

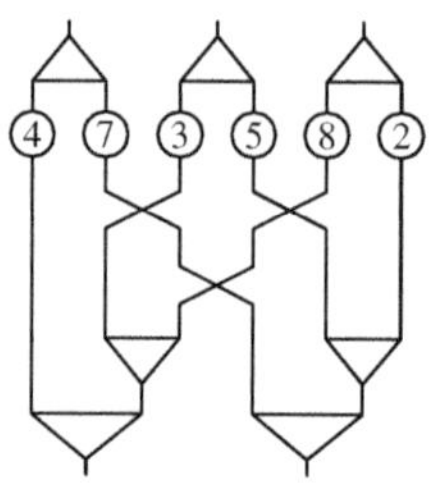

$$\begin{pmatrix} 4 & 3 & 8 \\ 7 & 5 & 2 \end{pmatrix}$$

Note that we recover the laws for Hopf algebras (§C.4.6) by setting

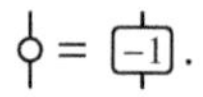

C.4.8 Variants. Other variants have been studied in the literature. We can mention presentation of invertible, orthogonal, special orthogonal, unitary, and special unitary matrices [235] and stochastic matrices [140].

C.5 Coefficients

C.5.1 The free monoidal category on a category. Given a category C, the *free monoidal category* $C^{\otimes}$ it generates is the monoidal category whose monoid of objects is $C_0^{\otimes}$, the free monoid over the objects of C, the monoid of morphisms is $C_1^{\otimes}$ the free monoid over the morphisms of C, with

$$f_1 \dots f_n : a_1 \dots a_n \to b_1 \dots b_n$$

whenever $f_i : a_i \to b_i$ for $1 \leqslant i \leqslant n$, composition is given pointwise, i.e.,

$$(g_1 \dots g_n) \circ (f_1 \dots f_n) = (g_1 \circ f_1) \dots (g_n \circ f_n)$$

with identities $1 \dots 1$ and tensor product is given by composition in the free monoid, i.e., concatenation.

There is an obvious functor $C \to C^{\otimes}$, which is such that for every functor $C \to D$, where D a monoidal category, there is a unique strict monoidal functor $C^{\otimes} \to D$ making the following diagram commute:

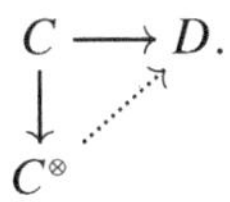

Given a presentation of a category C by a 2-polygraph P, the monoidal category $C^{\otimes}$ admits a presentation by the 3-polygraph Q with

$$Q_0 = \{\star\} \qquad Q_1 = P_0 \qquad Q_2 = P_1 \qquad Q_3 = P_2$$

sometimes referred to as the *suspension* of the 2-polygraph P.

C.5.2 Monoid actions. As a particular case, when M is a monoid considered as a category with one object, an algebra for the theory $M^{\otimes}$ in a monoidal category C, consists of an *action* of M, i.e., an object x of the monoidal category C together with a morphism of monoids $M \to C(x, x)$.

For instance, given a monoid $(M, \times, 1)$, we can consider its standard presentation, see §2.3.14:

$$\langle\, \star \,|\, a \,|\, \alpha_{u,v} : uv \Rightarrow (u \times v) \,\rangle$$

the resulting presentation of the free monoidal category generated by M has a presentation with a generator

for every element $u \in M$, with relations

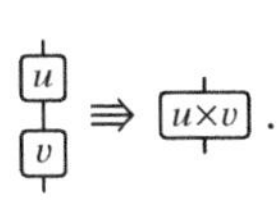

More generally, when a monoid M admits a presentation by a 2-polygraph P, the free monoidal category generated by M admits a presentation with 2-generators

indexed by $a \in P_1$, together with a relation

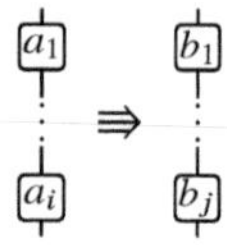

for every relation $a_1 \dots a_i \Rrightarrow b_1 \dots b_j$ in P_2.

C.5.3 Thee free linear category. Suppose given a monoid M and write R for the semiring it freely generates. As explained above, the monoid M freely generates a monoidal category, and we write here $M^{\otimes}$ for the free symmetric monoidal category generated by this monoidal category. We have seen in §C.4.3 that the PROP $\mathbf{M}_{\mathbb{N}}$ can be seen as a composite PROP $\mathbf{F}^{\mathrm{op}} \otimes_{\ell} \mathbf{F}$. More generally, the PROP $\mathbf{M}_R$ can be seen as a composite PROP:

$$\mathbf{M}_R = \mathbf{F}^{\mathrm{op}} \otimes_{\mathbf{S}} M^{\otimes} \otimes_{\mathbf{S}} \mathbf{F}.$$

The tensor product above is an iterated distributive law, in the sense of §C.3.3: it is induced by three distributive laws

$$\begin{aligned} \ell_1 &: M^{\otimes} \otimes_{\mathbf{S}} \mathbf{F}^{\mathrm{op}} \to \mathbf{F}^{\mathrm{op}} \otimes_{\mathbf{S}} M^{\otimes} \\ \ell_2 &: \mathbf{F} \otimes_{\mathbf{S}} \mathbf{F}^{\mathrm{op}} \to \mathbf{F}^{\mathrm{op}} \otimes_{\mathbf{S}} \mathbf{F} \\ \ell_3 &: \mathbf{F} \otimes_{\mathbf{S}} M^{\otimes} \to M^{\otimes} \otimes_{\mathbf{S}} \mathbf{F} \end{aligned}$$

such that the diagram

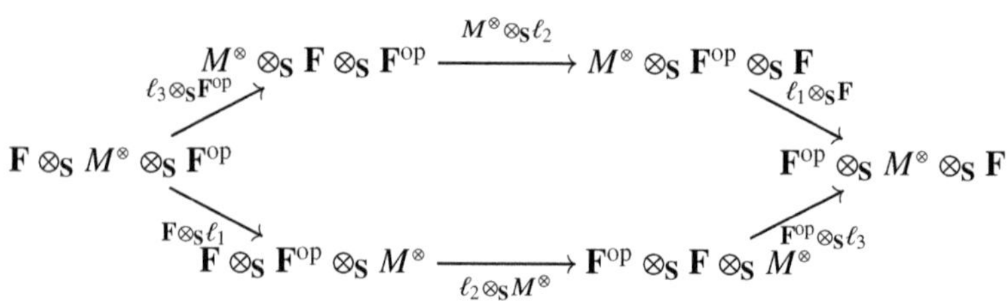

commutes. More generally, when the semiring R is not freely generated by a monoid, the PROP $\mathbf{M}_R$ admits a description of the form

$$\mathbf{M}_R = (\mathbf{F}^{\mathrm{op}} \otimes_{\mathbf{S}} M^{\otimes} \otimes_{\mathbf{S}} \mathbf{F})/\sim$$

meaning that it can be obtained from the above PROP by further quotienting by the congruence $\sim$ generated by the two relations

see [48] for details.

C.6 Frobenius Algebras

In this section, we study Frobenius algebras, which are "dual" to bialgebras in a sense that will be made precise. The axioms were first discovered by Lawvere [246] (and rediscovered in [77]) and a nice introduction to the subject can be found in Kock's book [224].

C.6.1 Presentation. The theory of *Frobenius algebras* is the PRO presented by four generators

$$\mu\colon 2 \to 1 \qquad \eta\colon 0 \to 1 \qquad \delta\colon 1 \to 2 \qquad \varepsilon\colon 1 \to 0$$

respectively pictured as

such that (μ, η) is a monoid, (δ, ε) is a comonoid, and the two following relations are satisfied

(C.9)

The theory of *commutative* Frobenius algebras is obtained by taking the free symmetric algebra as described in §C.1.4 (i.e., adding a generator $\gamma\colon 2 \to 2$ pictured as usual, see (C.2), together with the axioms for symmetries (C.3), the compatibility relations between symmetry and the monoid structure (C.5), as well as the comonoid structure), and adding the commutativity relations

The theory of *special* or *separable* Frobenius algebras (commutative or not) can be obtained by further adding the relation

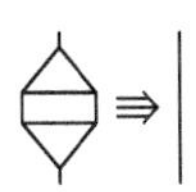

(C.10)

and the theory of *extraspecial* Frobenius algebras by further adding to the theory of special Frobenius algebras the relation

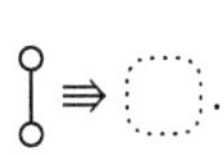

Other variants can be found in §C.11.9.

Note that these axiomatizations are not claimed to be minimal. For instance, associativity in presence of other axioms implies coassociativity:

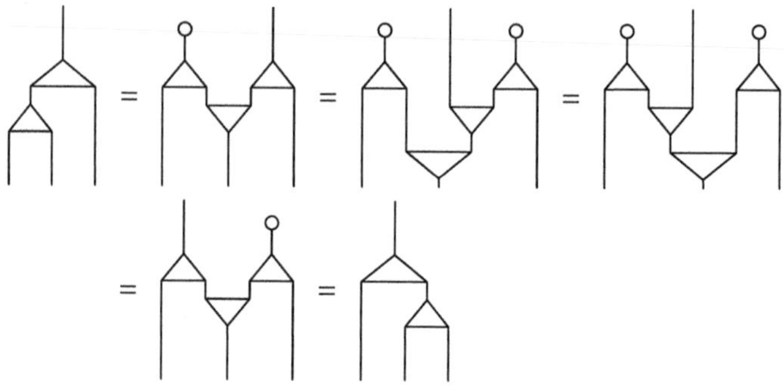

and the situation is similar for unitality and commutativity.

It is conjectured that the rewriting system for Frobenius algebras and its variants is terminating and can be completed into a finite convergent rewriting system [135]. Note that there are indexed critical pairs such as

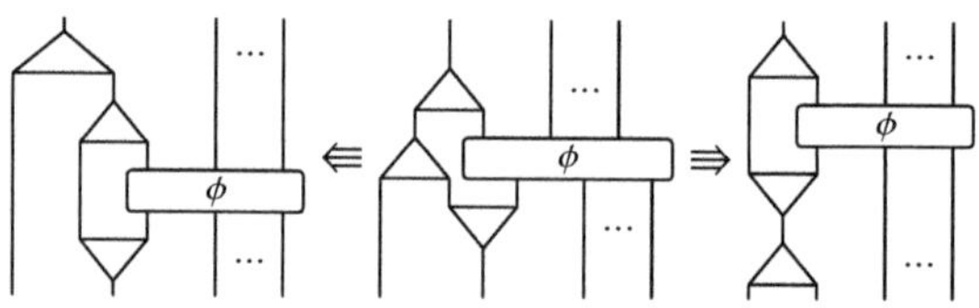

but we expect to be able to handle those using the techniques of Section 10.4.

Even in the absence of a notion of normal form, we can introduce the following notion of canonical form for morphisms. We define a family of morphisms $\phi_{m,g,n}$, indexed by $m, g, n \in \mathbb{N}$, defined by

$$\phi_{m,g,n} = \mu_m * (\delta * \mu) * (\delta * \mu) * \ldots * (\delta * \mu) * \delta_n$$

= 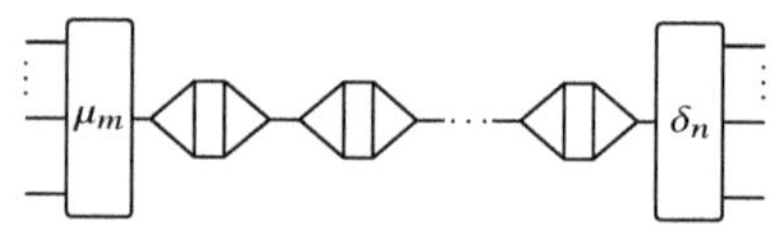 (C.11)

(the diagram is drawn horizontally to save space) where $\mu_m : m \to 1$ is a right comb of multiplications, defined inductively by

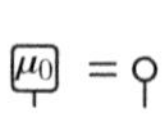

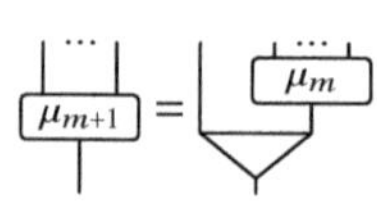

and dually $\delta_n : 1 \to n$ is a right comb of comultiplications, and there are g occurrences of $\delta * \mu$ in the middle: m and n are respectively the arity and coarity of the morphism and g is called its *genus* for reasons explained below. It can then be shown that, in the bicommutative case, any morphism rewrites to a tensor product of such morphisms (up to composing with a symmetry). In the case of special Frobenius algebras, the normal forms are tensor products of morphism of the form $\phi_{m,0,n}$, and in the case of extraspecial Frobenius algebras the normal forms are the same excepting that $\phi_{0,0,0}$ is not allowed to occur.

C.6.2 Frobenius algebras. A Frobenius algebra is typically considered in the category $\mathbf{Vect}_{\Bbbk}$: it consists of a vector space A which is both an algebra (i.e., equipped with associative and unital morphisms $\mu\colon A \otimes A \to A$ and $\eta\colon \Bbbk \to A$), a coalgebra (i.e., equipped with coassociative and counital morphisms $\delta\colon A \to A \otimes A$ and $\varepsilon\colon A \to \Bbbk$) satisfying the compatibility axioms (C.9).

There are many alternative characterizations of those [224, 339]. For instance, it can also be defined as an algebra (A, μ, η), equipped with a non-degenerate bilinear form $\sigma\colon A \otimes A \to A$ which is associative, in the sense that $\sigma \circ (\mu \otimes A) = \sigma \circ (A \otimes \mu)$, see §C.10.9 for details; or as an algebra (A, μ, η) equipped with a linear form $\varepsilon\colon A \to \Bbbk$ such that $\varepsilon(ab) = 0$ for every $a \in A$ implies $b = 0$. Namely, one can transform the first into the second definition, and vice-versa, by defining $\varepsilon(a) = \sigma(1, a)$ and $\sigma = \varepsilon \circ \mu$.

C.6.3 Frobenius as cospans. The category presented by the theory of special Frobenius algebras is the category $\mathrm{Cospan}(\Delta_+)$ of isomorphism classes of cospans of non-decreasing functions. Namely, it is composed of the presentation of Δ_+^{op} (comonoids), the presentation of Δ_+ (monoids), and compatibility axioms can be obtained as pushouts in Δ_+:

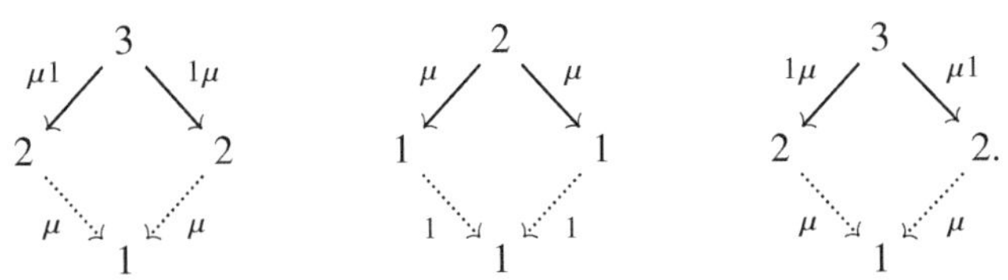

The diagram in the middle corresponds to the separability axiom (C.10), whereas the two other to the Frobenius axioms (C.9). The presentation can thus be deduced from the one of Δ_+, by using the fact that $\mathrm{Cospan}(\Delta_+) = \Delta_+ \otimes_\ell \Delta_+^{\mathrm{op}}$ where the distributive law $\ell\colon \Delta_+^{\mathrm{op}} \otimes \Delta_+ \to \Delta_+ \otimes \Delta_+^{\mathrm{op}}$ is given by pushout. Similarly, the theory of special commutative Frobenius algebras presents the category $\mathrm{Cospan}(\mathbf{F}) = \mathbf{F} \otimes_\ell \mathbf{F}^{\mathrm{op}}$ of isomorphism classes of cospans of functions. In this sense it is dual to the theory $\mathrm{Span}(\mathbf{F})$ of commutative bialgebras described in §C.4.3 (by analogy, one could be tempted to consider $\mathrm{Span}(\Delta_+)$ as a theory for non-commutative bialgebras, but this is not well-defined because Δ_+ does not have all pullbacks).

C.6.4 Corelations. The theory of extraspecial commutative Frobenius algebras corresponds to the variant of the category Cospan(**F**) whose morphisms are jointly surjective cospans (considered up to isomorphism), i.e., cospans of the form $X \xrightarrow{f} Y \xleftarrow{g} Z$ such that for every $y \in Y$ there exists $x \in X$ such that $f(x) = y$ or there exists $z \in Z$ such that $g(z) = y$, see [95]. Dually to the case of relations, see §C.4.4, these morphisms can alternatively be described as follows. Given two sets X and Y, a *corelation* $f \colon X \to Y$ is a partition of $X \sqcup Y$. Given two corelations $f \colon X \to Y$ and $g \colon Y \to Z$, their composite $g \circ f \colon X \to Z$ is defined as the restriction to $X \sqcup Z$ of the finest partition of $X \sqcup Y \sqcup Z$ which is coarser than f (resp. g) when restricted to $X \sqcup Y$ (resp. $Y \sqcup Z$). The identity corelation $1_X \colon X \to X$ is the diagonal corelation. We write **Corel** for the category of natural numbers and corelations: this category is isomorphic to the category of cospans described above. For instance, we have the following composition of corelations

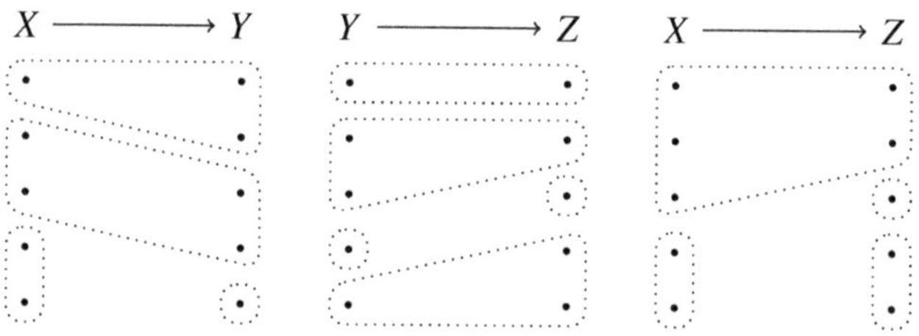

which corresponds to the following composition in the theory of extraspecial commutative Frobenius algebras

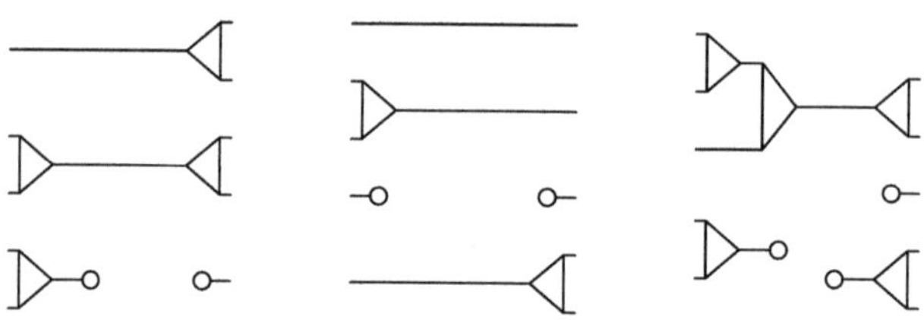

Any morphism $f \colon m \to n$ in Cospan(**F**), which consists of a cospan

$$m \xrightarrow{f_1} p \xleftarrow{f_2} n$$

up to isomorphism, can be uniquely be written as

$$f = f' *_0 \circ\!\!-\!\!\circ *_0 \dots *_0 \circ\!\!-\!\!\circ$$

where $f' \colon m \to n$ is a corelation and the number of instances of $\circ\!\!-\!\!\circ$ indicates the number of elements of $[p]$ which are neither in the image of f_1 nor in the image of f_2, i.e., measures the deficiency of surjectivity of f.

C.6.5 2-cobordisms. We briefly recall the well-known description of the category presented by the theory of commutative Frobenius algebras in geometrical terms. A clear and detailed account of the situation can be found in [224]. Fix a natural number $n \in \mathbb{N}$. Given a smooth oriented manifold with boundary Σ, its boundary decomposes as $\partial\Sigma = \partial^-\Sigma \sqcup \partial^+\Sigma$, where the two components are determined according to the orientation, and we say that Σ is an *n-cobordism* from $\partial^-\Sigma$ to $\partial^+\Sigma$. One can build a category $\mathbf{Cob}_n$ whose objects are oriented smooth $(n-1)$-manifolds, morphisms are n-cobordisms, considered up to diffeomorphism, and composition is given by gluing (i.e., taking pushouts) along common boundaries. This category is symmetric monoidal, with tensor product given on objects and morphisms by disjoint union, and people usually consider *topological quantum field theories*, which are symmetric monoidal functors $\mathbf{Cob}_n \to \mathbf{Vect}_{\Bbbk}$.

Here, we will be interested in $\mathbf{Cob}_2$: an object is a disjoint union of circles and a morphism consists of "trousers" between those, such as on the left below

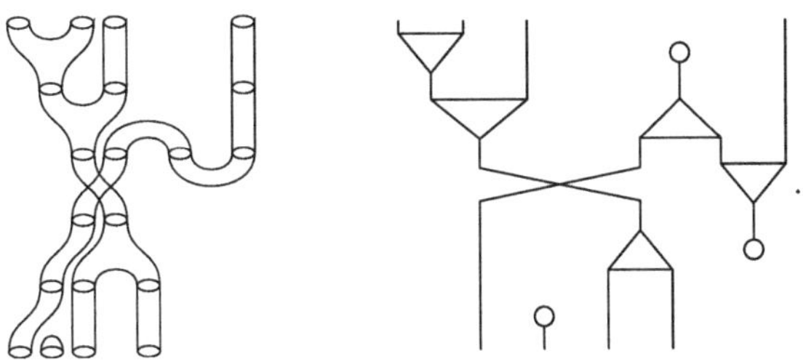

(C.12)

The category $\mathbf{Cob}_2$ admits the theory of commutative Frobenius algebras as presentation. Namely, the generators are interpreted as

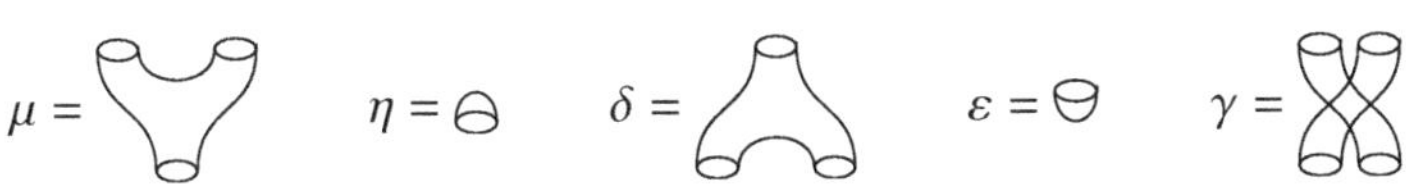

so that, for instance, the morphism on the left of (C.12) corresponds to the diagram on the right. Verifying that the relations are satisfied in $\mathbf{Cob}_2$ is direct. Conversely, in order to show that they are sufficient, one can use the classical result that the connected morphisms (i.e., diffeomorphism classes of connected compact oriented surfaces with boundaries) are characterized by their number of inputs, outputs, and genus (roughly, the number of holes): these are thus in bijection with normal forms (C.11) since the interpretation of $\phi_{m,g,n}$ is a cobordism of genus g with m inputs and n outputs.

Non-commutative Frobenius algebras admit a similar description by 2-dimensional thick tangles [243].

C.7 Linear Relations

Suppose fixed a field $\mathbb{k}$. The category $\mathbf{LinRel}_{\mathbb{k}}$ has vector spaces over $\mathbb{k}$ as objects, a morphism $f\colon X \to Y$ is a subspace of the vector space $X \oplus Y$, composition of $f\colon X \to Y$ and $g\colon Y \to Z$ is given by relational composition

$$g \circ f = \{x \oplus z \in X \oplus Z \mid \exists y \in Y, x \oplus y \in f \wedge y \oplus z \in g\}$$

and identity on X is the diagonal in $X \oplus X$. A morphism $f\colon X \to Y$ in this category is a relation between the underlying sets X and Y, which is closed under addition and multiplication by scalars and thus called a *linear relation*. This category is typically used to provide semantics to networks such as those found in electric circuits [23] or control [22, 47].

It is shown in [22, 48] that this category admits a presentation with generators

for $a \in \mathbb{k}$ such that

-  is a bicommutative linear bialgebra (§C.4.7),
- 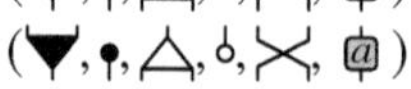 is a bicommutative linear bialgebra (§C.4.7),
-  is a bicommutative extraspecial Frobenius algebra (§C.6.1),
-  is a bicommutative extraspecial Frobenius algebra (§C.6.1),
- the following compatibility relations hold:

(C.13)

where  is a notation for . The generators should be interpreted as the following linear relations:

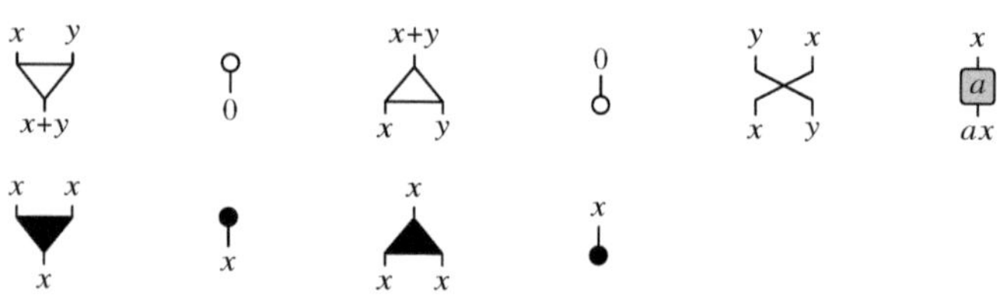

meaning that the interpretation of  is the linear relation

$$\{(x \otimes y) \oplus (x + y) \mid x, y \in \mathbb{k}\} \subseteq (\mathbb{k} \otimes \mathbb{k}) \oplus \mathbb{k}$$

and so on. For instance, the two compatibility relations (C.13) can be read as the fact that the space of pairs (x, y) such that $x + y = 0$ coincides with the space of pairs $(x, -x)$, with x and y ranging over $\mathbb{k}$. This theory is sometimes called the theory of *interacting Hopf algebras* over $\Bbbk$ and has applications to control theory.

In [48], it is shown that this category can be obtained as the following pushout in **MonCat**:

$$\begin{array}{ccc} \mathbf{Vect}_{\Bbbk} \sqcup \mathbf{Vect}_{\Bbbk}^{\mathrm{op}} & \longrightarrow & \mathrm{Span}(\mathbf{Vect}_{\Bbbk}) \\ \downarrow & & \vdots \\ \mathrm{Cospan}(\mathbf{Vect}_{\Bbbk}) & \cdots\cdots\rightarrow & \mathbf{LinRel}_{\Bbbk} \end{array}$$

from which the above presentation can be obtained. Another possible description is mentioned in §C.10.9.

C.8 Interchange

C.8.1 Interchange algebra. The theory of *interchange algebras* [234, 253] models algebraic structures with two multiplications satisfying the exchange law (which is sometimes also called the *interchange law*) holds. It can be presented by the 3-polygraph P with generators

$$P_0 = \{\star\} \qquad P_1 = \{a\} \qquad P_2 = \{\gamma \colon aa \Rightarrow aa, \mu \colon aa \Rightarrow a, \nu \colon aa \Rightarrow a\}$$

and relations expressing that γ is a symmetry (§C.1.4) together with the relation

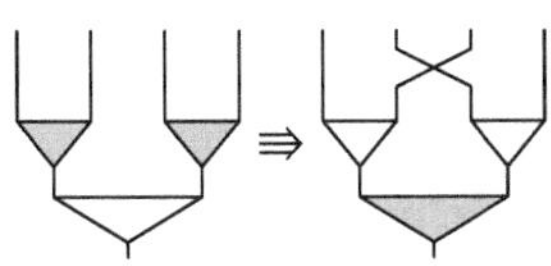

where the white (resp. gray) triangle corresponds to μ (resp. ν). An interchange algebra is *associative* when both μ and ν are. It is *unital* when equipped with two morphisms $1 \Rightarrow a$ acting as a left and right unit for μ and ν, respectively. For instance, an associative unital interchange algebra in the cartesian category **Set** is precisely a 2-category with only one 0-cell and one 1-cell. By the Eckmann-Hilton argument [119], this is the same as a commutative monoid.

C.8.2 Iterated monoidal categories. A variant of interchange algebras with n distinct monoid structures instead of two is the following one. An *n-fold monoid* is a set equipped with n-distinct products which are suitably compatible. More precisely, the corresponding theory can be presented by the 3-polygraph P with generators $P_0 = \{\star\}$, $P_1 = \{a\}$ and

$$P_2 = \{\gamma \colon aa \Rightarrow aa, \eta \colon 1 \Rightarrow a, \mu_0 \colon aa \Rightarrow a, \dots, \mu_{n-1} \colon aa \Rightarrow a\}$$

such that for every $0 \leqslant i < j < n$, η, μ_i, and μ_j satisfy the axioms of associative unital interchange algebras. Again, by the Eckmann-Hilton argument, this theory is equivalent to the one of commutative monoids. However, the theory of *n-fold pseudomonoids* (where equalities are replaced by coherent isomorphisms) is more interesting: its algebras in the cartesian 2-category **Cat** are *n-fold monoidal categories*. This structure is the one one obtains by considering monoids in the category of monoids in the category of monoids in ... in the category **Cat** with the monoidal structure induced by cartesian product, see [25].

C.8.3 Interchange bialgebra. An *interchange bialgebra* [253] is a PROP which is both an interchange algebra and an interchange coalgebra, with the following compatibility relations

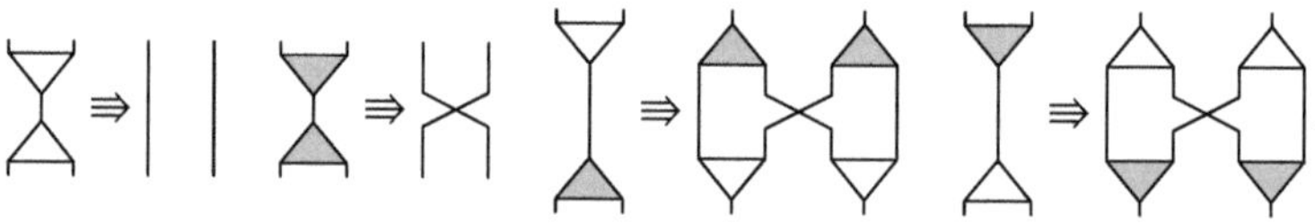

The most studied variant of this notion is the one of interaction nets, which are detailed in §C.12.5.

C.9 Idempotent Objects

An *idempotent object* in a monoidal category is an object x equipped with an isomorphism $x \otimes x \to x$. The theory for idempotent objects is the PRO $\mathbf{T}$, called the *Thompson category*, presented by the 3-polygraph P with $P_0 = \{\star\}$, $P_1 = \{a\}$, with 2-generators $\mu\colon aa \Rightarrow a$ and $\delta\colon a \Rightarrow aa$, depicted as

and subject to the relations

A morphism is a binary *tree* (resp. a *cotree*) when it has a as target (resp. source) and is a composite of generators μ (resp. δ) only. It is observed in [130] that the monoid of automorphisms $\mathbf{T}(a, a)$ is isomorphic to the Thomson group F, already presented in §A.1.27. Namely, we can recover the generators of the usual presentation as defined by induction on i by

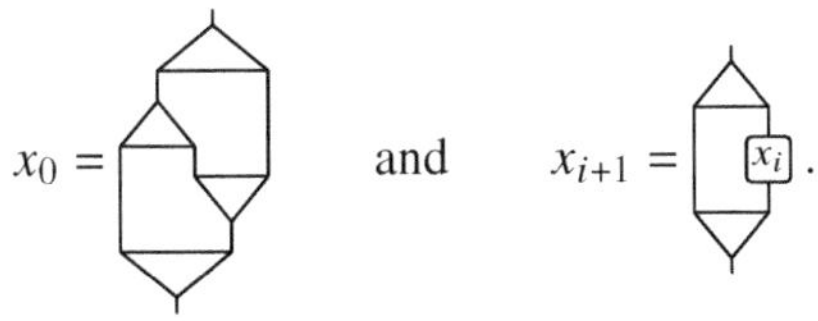

More generally, any morphism $a \to a$ of **T** decomposes as a cotree followed by a tree respectively encoding the dyadic partitions of the input and of the output of the corresponding morphism $I \to I$ in the Thompson group by specifying when the interval should be split in two halves. For instance, the morphism on the left corresponds to the function on the right:

By post-composition, the elements of F act on trees which are "large enough" (the action is partially defined). For instance, we have the following action of x_0:

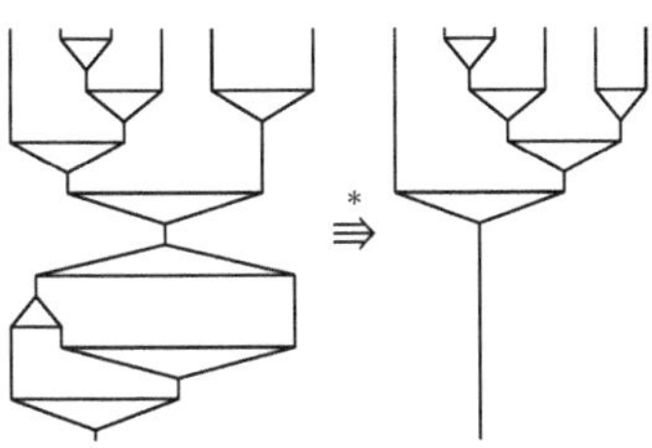

From this point of view, the generator x_0 can be pictured as

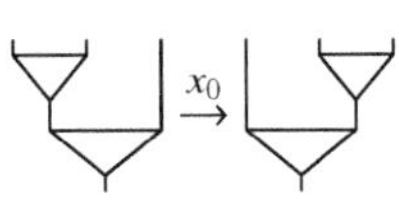

since it will "replace" a prefix of a tree as on the left with a prefix as on the right. Such a transformation is sometimes called an *associative law* and the group F can be described as the group of those with expected composition. For instance, Mac Lane's pentagon is

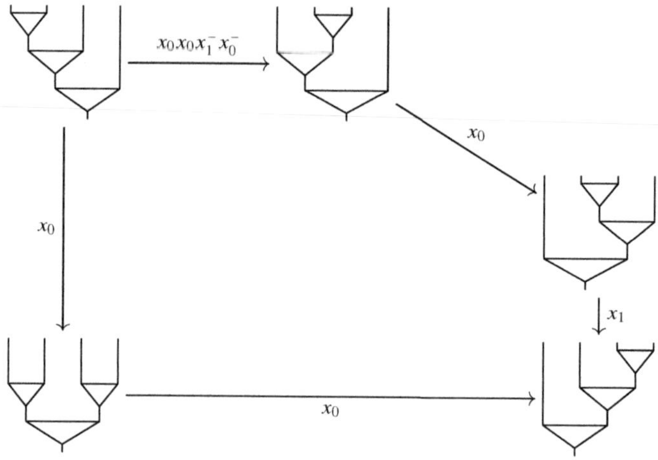

Similarly, the Thompson group V can be recovered as the group of automorphisms on a in the free symmetric monoidal category on $\mathbf{T}$. This entails that V can be described as the group of automorphisms of the free Cantor algebra on a singleton [59]: we recall that a *Cantor algebra* is a set A equipped with a bijection $\alpha \colon A \to A \times A$.

C.10 Dualities

C.10.1 Adjunctions. An *adjunction* consists of two functors $f \colon C \to D$ and $g \colon D \to C$ together with natural transformations $\eta \colon 1_C \Rightarrow fg$ and $\varepsilon \colon gf \Rightarrow 1_D$ such that $(\eta f) * (f\varepsilon) = 1_f$ and $(g\eta) * (\varepsilon g) = 1_g$. In such a situation f is called a *left adjoint* to g, and g a *right adjoint* to f.

The theory of *adjunctions* is the 2-category $\mathbf{Adj}$ presented by the 3-polygraph P with generators

$$P_0 = \{c, d\} \qquad\qquad P_1 = \{f \colon c \to d, g \colon d \to c\}$$

and

$$P_2 = \{\eta \colon 1_c \Rightarrow fg, \varepsilon \colon gf \Rightarrow 1_d\}$$

pictured as

$\eta =$ 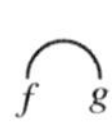

$\varepsilon =$ 

and respectively called *unit* and *counit*, together with the two relations

(C.14)

often called *zigzag* or *triangle identities*. This presentation is convergent, the normal forms being the horizontal composites of η, ε, and identities. A 2-functor $F\colon \mathbf{Adj} \to \mathbf{Cat}$ corresponds precisely to an adjunction in the usual sense.

The 2-category **Adj** is studied in [319]. In particular, if we consider the 2-category $\mathbf{Adj}(c, c)$, which is the full sub-2-category of **Adj** on the 0-cell c, we have an isomorphism of 2-categories (or of monoidal categories)

$$\mathbf{Adj}(c, c) \simeq \Delta_+$$

Namely, the monoid of 1-cells is freely generated by $fg\colon c \to c$ (thus isomorphic to $\mathbb{N}$), and one can define a structure of monoid on fg whose multiplication and unit are respectively $f\varepsilon g$ and η, from which the isomorphism can easily be deduced. For instance, the non-decreasing function pictured on the left corresponds to the 2-cell on the right in the theory of monoids

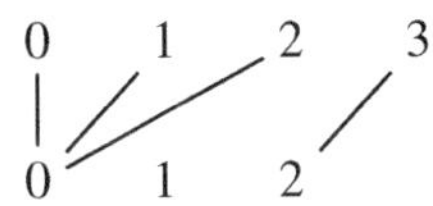

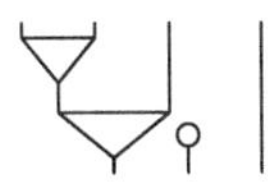

and to the following 2-cell in $\mathbf{Adj}(c, c)$:

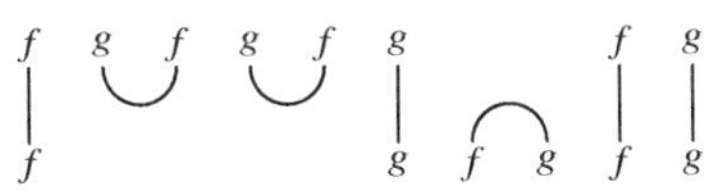

Similarly, we have $\mathbf{Adj}(d, d) \simeq \Delta_+^{\mathrm{op}}$, and $\mathbf{Adj}(c, d)$ and $\mathbf{Adj}(d, c)$ are the subcategories of Δ_+ whose objects are non-zero natural numbers and morphisms are the last-element (resp. first-element) preserving functions. Note that Δ_+^{op} is isomorphic to the subcategory of Δ_+ whose objects are non-zero natural numbers and morphisms are preserving both first and last element.

C.10.2 Duality. A *duality* in a monoidal category C is an adjunction in C, considered as a 2-category. It consists of two objects x and x^* together with morphisms

$$\eta_x\colon 1 \to xx^* \qquad\qquad \varepsilon_x\colon x^*x \to 1$$

satisfying the zigzag relations (C.14). In this case x is called a *left dual* of x^*, and x^* a *right dual* of x. Two left (resp. right) duals of a given object are necessarily isomorphic. An object x is *self-dual* when it admits a right dual x^* which is isomorphic to x.

A monoidal category is *right-autonomous* (resp. *left-autonomous*) when every object x admits a right dual x^* (resp. a left dual ${}^*x$). It is *strictly* so when duals are chosen so that $(xy)^* = y^*x^*$, $1^* = 1$, $\eta_1 = 1_1 = \varepsilon_1$, $\eta_{xy} = x\eta_y x^* \circ \eta_x$, and $\varepsilon_{xy} = \varepsilon_y \circ y^*\varepsilon_x y$; without loss of generality, we consider that this is always the case in the following. It is *autonomous* (or *rigid*) when it is both left- and right-autonomous. A *compact closed category* is a symmetric monoidal category which is autonomous. For instance, the category $\mathbf{Vect}_{\mathbb{k}}$ of $\mathbb{k}$-vector spaces is compact closed, the dual x^* of a vector space x being its linear dual. A coherence theorem for compact closed categories was shown by Kelly and Laplaza [216]. In particular, in a compact closed category, we have isomorphisms ${}^*x \simeq x^*$, $x^{**} \simeq x$, which we will be considered as equalities in the following. Given a morphism $f\colon x \to x$ of a compact closed category, the morphism

$$\mathrm{tr}(f) = \varepsilon_{x^*} \circ f x^* \circ \eta_x$$

is called its *trace*: there is a general axiomatization of trace in symmetric monoidal categories [210], which every compact closed category canonically possesses. In particular, the morphism $\mathrm{tr}(1_x)$ is often called the *dimension* of x. The category is *loop-free* when every object x is 1-dimensional, i.e., $\mathrm{tr}(x) = 1_1$.

A right-autonomous category is *pivotal* when equipped with a monoidal natural isomorphism $x \simeq x^{**}$, considered as an equality in the following. A nice survey of the flavors of categories with duals can be found in [324].

C.10.3 The free compact category. Suppose given a symmetric monoidal category C presented by a 3-polygraph P. The free compact category on C admits a presentation by the 3-polygraph Q with generators

$$\begin{aligned}
Q_0 &= P_0 = \{\star\}\\
Q_1 &= \{a^n \mid a \in P_1 \text{ and } n \in \mathbb{Z}\}\\
Q_2 &= \left\{f\colon u^0 \Rightarrow v^0 \mid f\colon u \Rightarrow v \in P_2\right\} \sqcup\\
&\qquad \left\{\eta_{a^n}\colon 1 \Rightarrow a^n a^{n+1}, \varepsilon_{a^n}\colon a^{n+1}a^n \Rightarrow 1 \mid a^n \in Q_1\right\}
\end{aligned}$$

where, for $u = a_1 \ldots a_k$, we write u^n for $a_1^n \ldots a_k^n$. The generators η_{a^n} and ε_{a^n} are respectively pictured

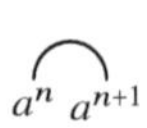

and

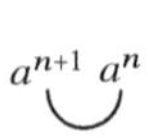

.

The relations are those in P_3 together with the zigzag relations (C.14) satisfied by η_{a^n} and ε_{a^n} for every $a \in P_1$ and integer n.

Here, the object a^0 corresponds to a in the original category and, for $n \in \mathbb{N}$, a^n (resp. a^{-n}) corresponds to $a^{**\cdots*}$ (resp. ${}^{*\cdots**}a$), with $*$ applied n times. Thanks to this presentation, one can for instance deduce that the canonical monoidal functor from a monoidal category into its free compact category is faithful [283].

C.10.4 The free compact closed category. Suppose given a symmetric monoidal category C presented by a 3-polygraph P. The free compact closed category on C admits a presentation by the 3-polygraph Q with generators

$$Q_0 = P_0 = \{\star\}$$
$$Q_1 = \{a, a^* \mid a \in P_1\}$$
$$Q_2 = \{f : u \Rightarrow v \mid f : u \Rightarrow v \in P_2\} \sqcup \{\eta_a : 1 \Rightarrow aa^*, \varepsilon_a : a^*a \Rightarrow 1 \mid a \in P_1\}$$

and relations being those in Q_3 together with the zigzag relations for each η_a and ε_a.

The theory for a pair of adjoint endofunctors can be obtained from the category **Adj** of §C.10.1 by identifying the two objects, and its presentation can be obtained from the one of **Adj** by identifying the 0-generators c and d. The above results show that this theory is the free compact closed category on the terminal category.

C.10.5 The free "self-dual" compact closed category. As a variant of the situation described in previous section, given a symmetric monoidal category C presented by a polygraph P, we call the *free self-dual compact closed category* on C, the category presented by the 3-polygraph Q with generators

$$Q_0 = P_0 = \{\star\}$$
$$Q_1 = P_1$$
$$Q_2 = \{f : u \Rightarrow v \mid f : u \Rightarrow v \in P_2\} \sqcup \{\eta_a : 1 \Rightarrow aa, \varepsilon_a : aa \Rightarrow 1 \mid a \in P_1\}$$

and relations being those in Q_3 together with the zigzag relations for each η_a and ε_a, and the relations

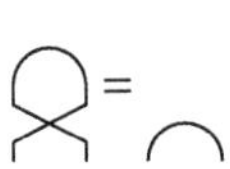

$\gamma_{a,a} \circ \eta_a = \eta_a$

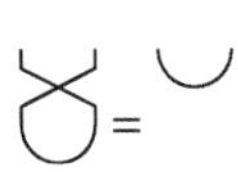

$\varepsilon_a \circ \gamma_{a,a} = \varepsilon_a$

for every 1-generator $a \in P_1$. In the resulting category, only the generators are self-dual, for instance the dual of the 1-cell ab is ba. The axiomatization of self-dual categories is quite subtle, see [323], and it is not clear that this construction is an instance of those. However, it is quite useful in the following, so we use it without claiming a universal property.

C.10.6 1-cobordisms. The category of n-cobordisms was presented in §C.6.5. In this section we study the case $n = 1$ and give a presentation of the monoidal category $\mathbf{Cob}_1$. Its objects are disjoint unions of oriented 0-dimensional manifolds, i.e., points together with an orientation – or +, and it admits a presentation by the polygraph P with $P_0 = \{\star\}$, $P_1 = \{-, +\}$, 2-generators being

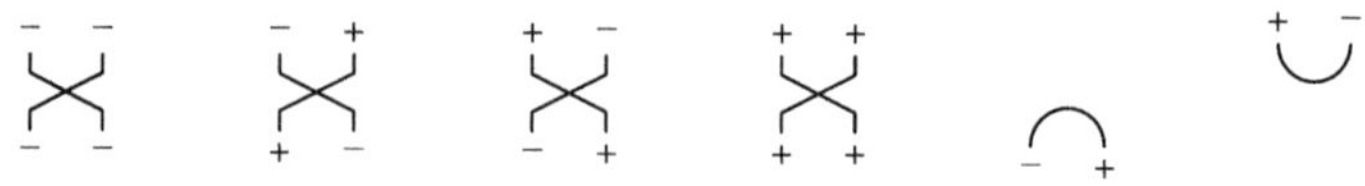

such that the four first generators equip the monoidal category with a symmetric structure, see §C.1.4, and the two last generators satisfy the axioms for dualities. In other words, $\mathbf{Cob}_1$ is the free compact closed symmetric monoidal category on one object.

C.10.7 The Temperley–Lieb category. The *Temperley–Lieb category* **TL**, introduced and studied in [1], is the PRO generated by

subject to the relations

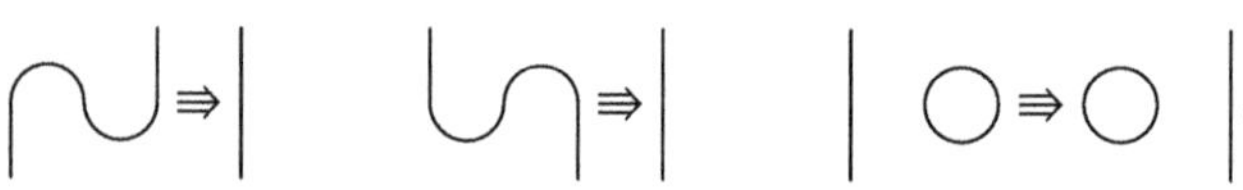

It is thus the free monoidal category containing a self-dual object, satisfying the last relation above. Note that this presentation is not terminating because of the loop

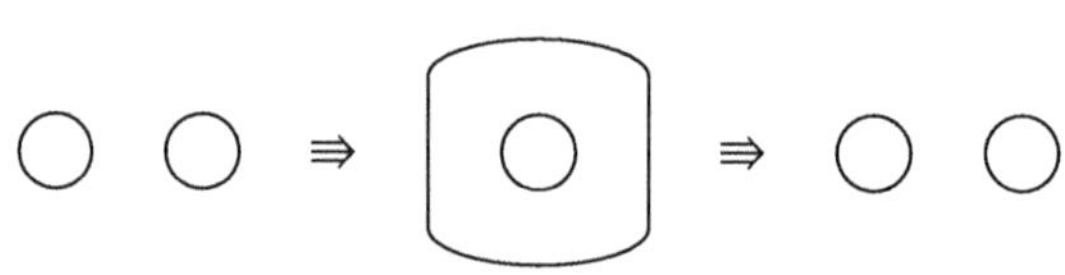

it can however be shown to be quasi-terminating [6], in the sense of §1.3.11. Given $n \in \mathbb{N}$, the monoid of endomorphisms $\mathbf{TL}(n, n)$ is generated by

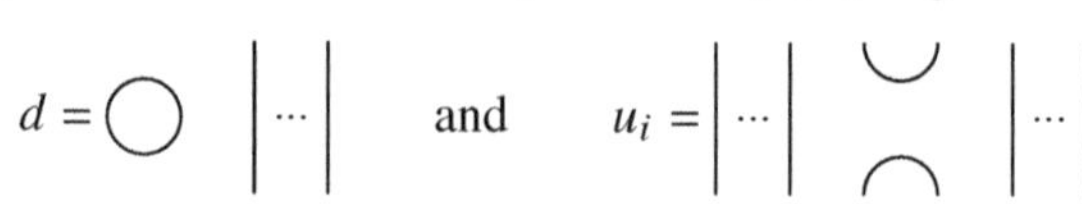

with $0 \leqslant i \leqslant n - 2$, where u_i is composed of the identity on i on the left and on $n - i - 2$ on the right, subject to relations of §A.1.29, making it the n-th

Temperley–Lieb monoid, see §A.1.29. For instance, with $n = 4$, we have the relations

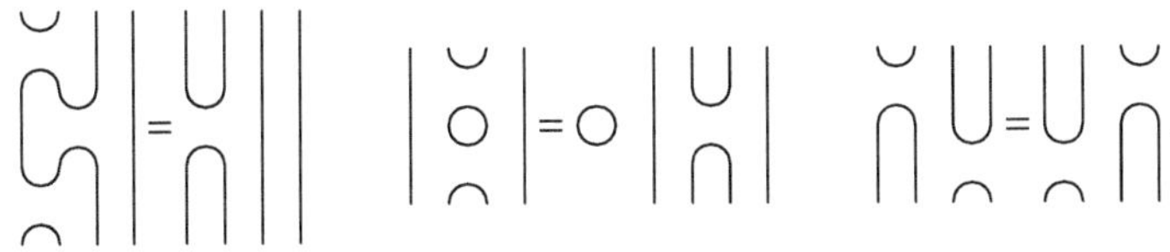

$u_0u_1u_0 = u_0$ $\qquad$ $u_1u_1 = du_1$ $\qquad$ $u_0u_2 = u_2u_0$

C.10.8 Chord diagrams. Consider the full subcategory C of **Corel** (see §C.6.4) whose morphisms are corelations $X \overset{f}{\rightarrow} Y \overset{g}{\leftarrow} Z$ which are *one to one*, meaning that for every $y \in Y$ the cardinal of $f^{-1}(y) \cup g^{-1}(y)$ is precisely 2. A morphism can be represented by drawing a line between two elements of the source X or the target Z which are sent to the same element of Y by f or g, as on the left below:

(C.15)

such diagrams are sometimes called *chord diagrams* or *Brauer linkings* [55]. The resulting PRO, which we call the *Brauer category*, has a presentation with generators

such that the last one is a symmetry (§C.1.4) and the relations

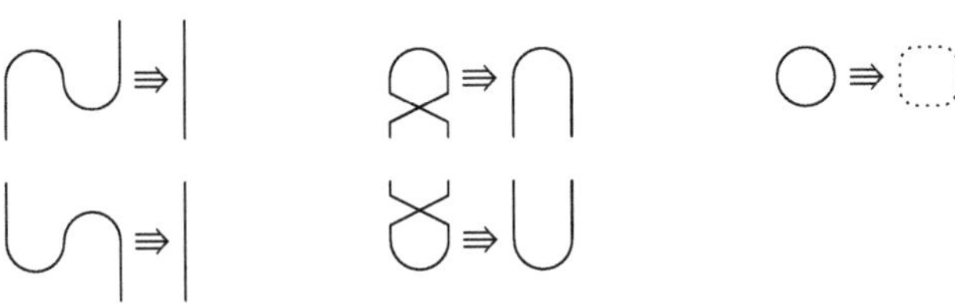

(C.16)

hold, see [192]. The morphism corresponding to the diagram on the left of (C.15) is shown on the right. This category is the free loop-free self-dual compact closed symmetric monoidal category on an object, in sense of §C.10.5. Given $n \in \mathbb{N}$, the monoid of endomorphisms on n is precisely the n-th Brauer monoid, as described in §A.1.30. In terms of the presentation of the monoid, the generators a_i and u_i are respectively interpreted as

$a_i =$  $\qquad$ $u_i =$ 

and for instance, we have the derivation of the following relations:

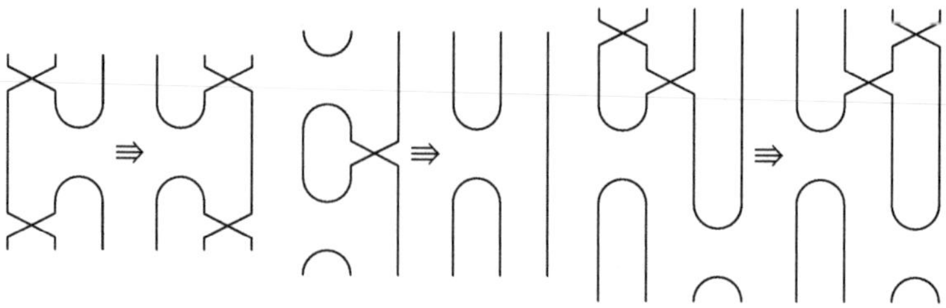

$a_0u_1a_0 = a_1u_0a_1$ $\qquad$ $u_0a_1u_0 = u_0$ $\qquad$ $a_0a_1u_0u_2 = a_2a_1u_0u_2$

A variant without loop-freeness can be obtained by dropping the last rule of (C.16). The resulting category corresponds to the subcategory of Cospan(Δ_+) whose morphisms satisfy a similar condition as above.

C.10.9 Frobenius algebras and dualities. A Frobenius algebra is always canonically equipped with a notion of duality and can even be characterized in terms of this structure, see [224] for details.

C.10.10 Proposition. *The following theories present the same category:*

1. *The theory of Frobenius algebras, generated by* ∇, ⫯, Δ, *and* ⫰ *such that*
   - (∇, ⫯) *is a monoid,*
   - (Δ, ⫰) *is a comonoid,*
   - *the Frobenius relations* (C.9) *hold between* Δ *and* ∇.
2. *The theory generated by* ∇, ⫯, ⫰, ∩ *such that*
   - (∇, ⫯) *is a monoid,*
   - *the following relations hold:*

3. *The theory generated by* ∇, ⫯, ∩, ∪ *such that*
   - (∇, ⫯) *is a monoid,*
   - (∩, ∪) *is a duality,*
   - *the following relation holds:*

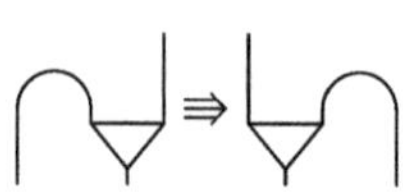

or equivalently the following relation holds:

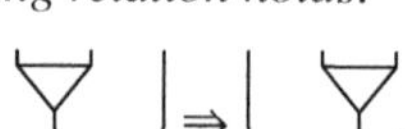

Proof. The equivalence between theories can be derived using Tietze transformations. We sketch the equivalence between the first and last one. In the theory (1), we can define

and conversely, in the theory (3), we can define

In both cases, the required relations are derivable. □

Various extensions of this result are possible in order to take into account variations on the notion of Frobenius algebra (commutative, special, etc.).

Following the same ideas as previously, in the theory of interacting Hopf algebras (Section C.7) a self-duality can be defined by

and the structure can be axiomatized as follows taking these as generators [22]:

C.10.11 Proposition. *The following theories present the same category:*

1. *The theory of interacting Hopf algebras.*
2. *The theory generated by* 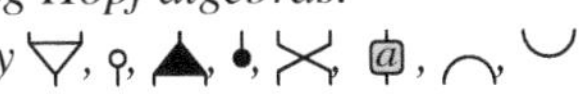 *such that*
   -  *is a bicommutative linear bialgebra,*
   - 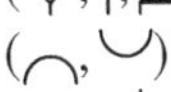 *is a duality,*
   - 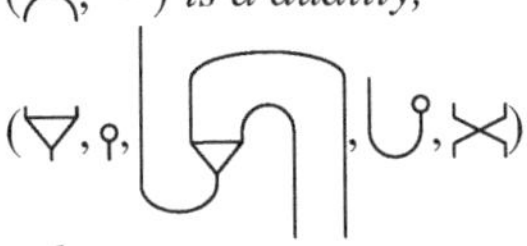 *is a bicommutative extraspecial Frobenius algebra,*
   - 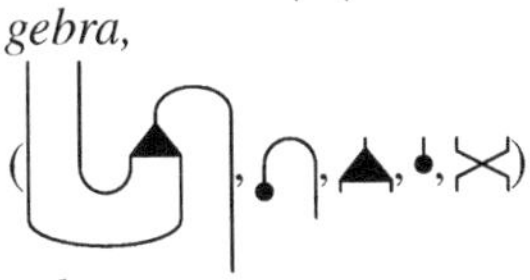 *is a bicommutative extraspecial Frobenius algebra.*

C.10.12 Tangles. A category of tangles can be defined as a variation of the category of braids, see §§A.1.21 and C.1.1, intuitively by allowing wires to loop. Given natural numbers $m, n \in \mathbb{N}$, a *tangle* from m to n is an embedding

$$t: T \to X$$

where T is a 1-manifold with boundary and $X = \mathbb{R}^2 \times [0, 1]$, such that the image of the boundary ∂T of T is of the form

$$t(\partial T) = \{(0, i, 0) \mid i \in \mathbb{N}, 0 \leqslant i < m\} \sqcup \{(0, i, 1) \mid i \in \mathbb{N}, 0 \leqslant i < n\}$$

where the natural number m (resp. n) is called the source (resp. target) of the tangle. Graphically, a tangle from 4 to 2 can be pictured as

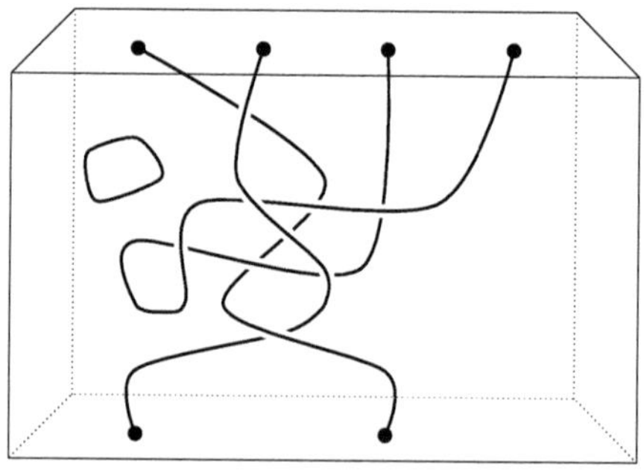

Tangles are considered up to endpoint-preserving isotopy: we identify two tangles $t\colon T \to X$ and $t'\colon T' \to X$ for which there exists a continuous map $h\colon [0, 1] \to X^X$ such that $h(0) = 1_X$, $h(1) \circ t = t'$ and, for every $t \in [0, 1]$, $h(t)$ is a homeomorphism such that $h(t)(\partial T) = h(0)(\partial T)$. We write **Tang** for the PRO with tangles as morphisms with expected composition (corresponding to linking wires) and tensor product (corresponding to juxtaposition of diagrams).

The category **Tang** is the free braided monoidal category on a self-dual object, which moreover satisfies the first Reidemeister move (C.17) below, see [137, 138]. Explicitly, this means that it admits a presentation with generators

subject to the relations

- first Reidemeister move:

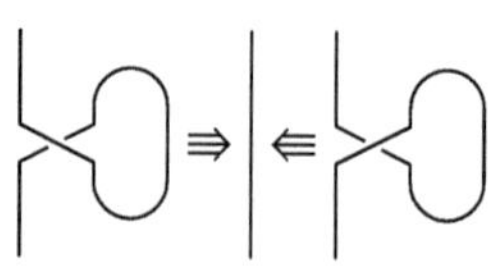

(C.17)

- second Reidemeister move:

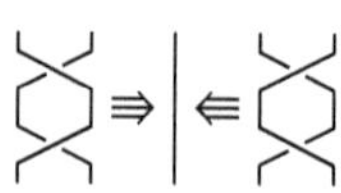

– third Reidemeister move (aka Yang–Baxter rule):

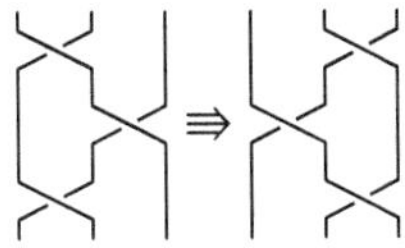

– zigzag relations:

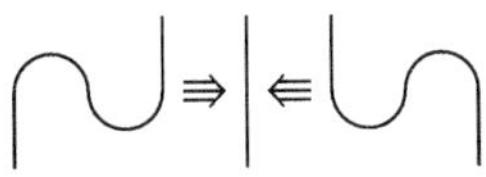

– the naturality relations:

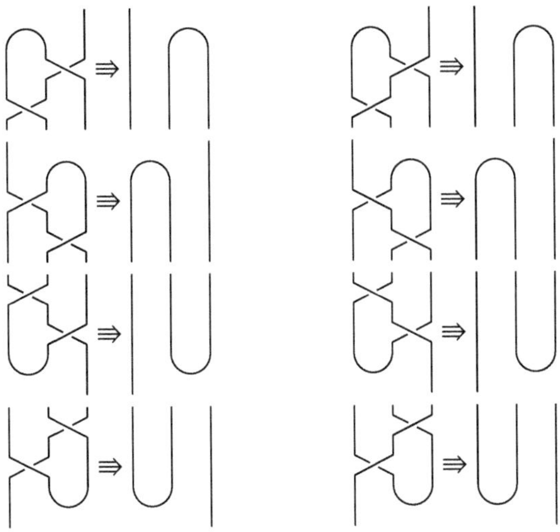

or equivalently, the sliding relations:

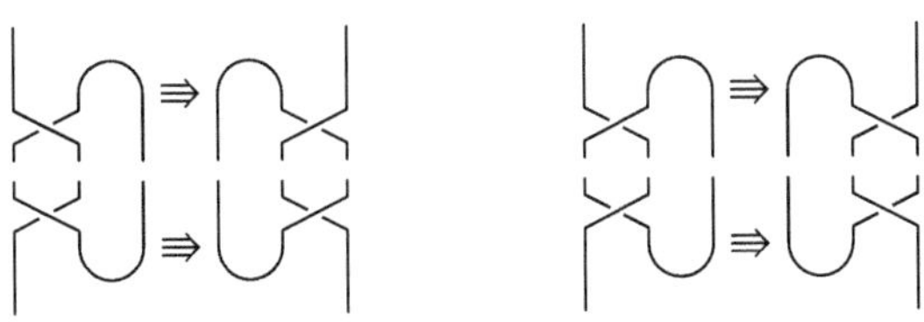

If we replace relation (C.17) by

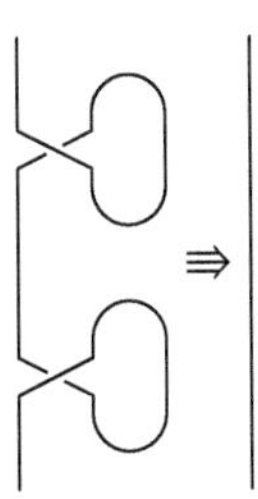

(C.18)

we present the category of *framed tangles* or *ribbons*: in this category, morphisms correspond to pieces of ribbon (instead of wire), i.e., embeddings

$$r\colon T \times [0,1] \to X$$

where T is a 1-manifold with boundary, e.g.,

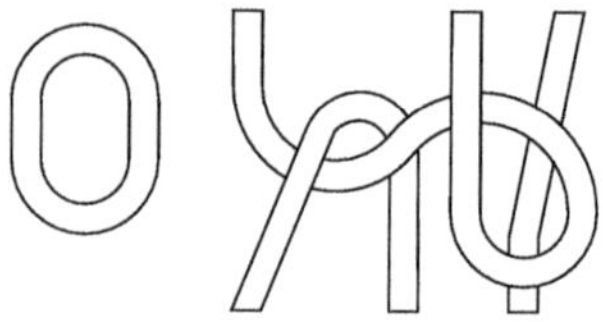

considered up to endpoint-preserving isotopy. Note that the relation (C.17) is not satisfied since trying to strengthen the loop introduces a "twist" on the rope

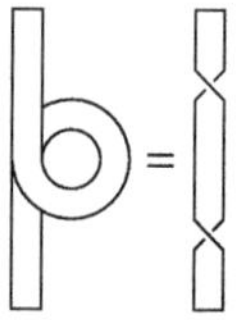

and is thus not the identity ribbon. If we remove the relation (C.17) (without adding (C.18)), we present the category of tangles up to regular isotopy, a variant of isotopy which forces ribbons to always be flat against the plane, which prevents the identity (C.18) from holding:

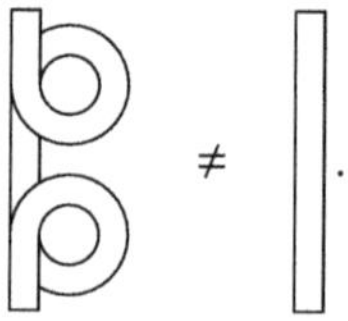

A tangle from 0 to 0 is called a *link*, and a *knot* is a link $t\colon T \to X$ such that T is the 1-sphere. The Reidemeister moves were originally introduced for those [309].

A variant of this category can be obtained by considering oriented tangles, see [137, 138, 214, 324]: the objects of this category are sequences of "−" and "+", and morphisms are oriented tangles up to endpoint-preserving planar isotopy, e.g.,

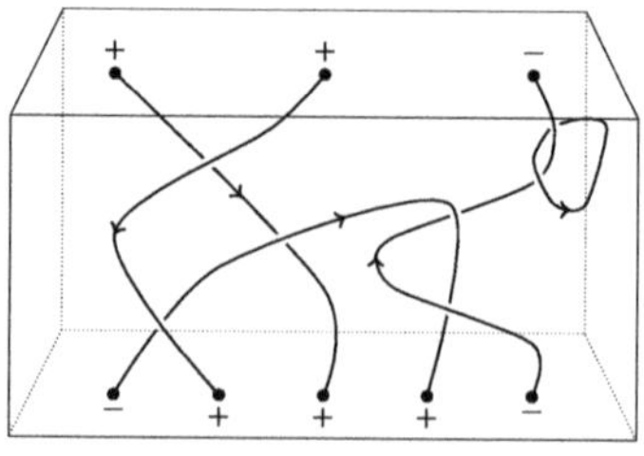

Of course, ribbon variants of this category can also be considered. The resulting categories are pivotal (instead of being self-dual).

C.10.13 First-order logic. The dependencies between quantifiers in proofs for first-order logic are characterized by the structure of free pivotal category on a bialgebra, see [282].

C.11 Endomorphisms

C.11.1 The theory of endomorphisms. Consider the PRO whose morphisms are endomorphisms $f: n \to n$ consisting of a list $(f_1, \ldots, f_n)$ of n natural numbers. The composite of two morphisms $f: n \to n$ and $g: n \to n$ is given by pointwise addition $(f_1 + g_1, \ldots, f_n + g_n)$, identities are lists $(0, \ldots, 0)$, and tensor product is given by concatenation of lists. This PRO admits a presentation with one generator

and no relation, the generator being interpreted as the list (1). An algebra for this theory in a monoidal category consists of an object x together with an *endomorphism* $f: x \to x$ on x. This is a particular case of a presentation of the free monoidal category generated by the monoid $\mathbb{N}$, see §C.5.2: since $\mathbb{N}$ is free on one generator, its action is entirely determined by an endomorphism (corresponding to the action of 1).

C.11.2 Actions of a set. Suppose fixed a set L. Previous situation can be generalized by considering the PRO where the morphisms are endomorphisms $f: n \to n$ consisting of a list $(f_1, \ldots, f_n)$ of elements of L^*, the free monoid over L: composition is given by pointwise concatenation in L^* and tensor by concatenation of lists. It admits a presentation with generators

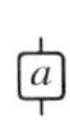

indexed by $a \in L$ and no relation. This is the theory for an *action of* L: it consists of an object x together with a family $f_a: x \to x$ of endomorphisms of x indexed by $a \in L$. Again, this is a particular case of a presentation of the free monoidal category generated by the monoid L^* (the free monoid on the set L). We recover the theory of §C.11.1 in the particular case where L is a singleton.

C.11.3 Involutions. The monoidal theory of involutions admits a presentation with one generator

and relation

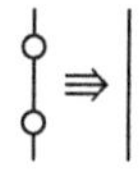

It is the free monoidal category on the monoid $\mathbb{Z}/2\mathbb{Z}$.

C.11.4 The hyperoctahedral category. The symmetric monoidal theory of involutions admits a presentation with two generators

and relations

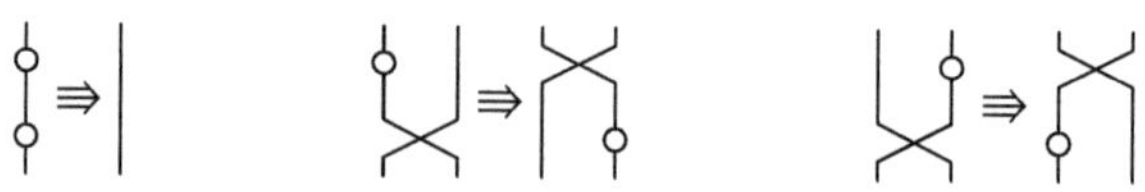

together with usual relations for symmetries, see §C.1.3.

Given $n \in \mathbb{N}$, the monoid of endomorphisms on n is in fact a group called the *hyperoctahedral group* or *signed symmetric group*, and noted C_n, see §A.1.22. It can be described as the group of *signed permutations* of a set with n elements, i.e., $(n{\times}n)$-matrices where each column and each row contains one non-null coefficient which is either 1 or -1, with usual multiplication and identities, the two generators being respectively interpreted as

$$\begin{pmatrix}-1\end{pmatrix} \qquad \text{and} \qquad \begin{pmatrix}0 & 1\\ 1 & 0\end{pmatrix}.$$

It can also be described as the wreath product

$$C_n = S_2 \wr S_n$$

of symmetric groups. The presentation of C_n given in §A.1.22 can be recovered as the associated presentation of category (in the sense of §10.3.10), where the generators a_0 and a_i (with $0 < i < n$) respectively correspond to

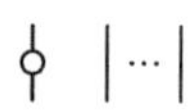

and

For instance, the relation $a_0 a_1 a_0 a_1 = a_1 a_0 a_1 a_0$ corresponds to

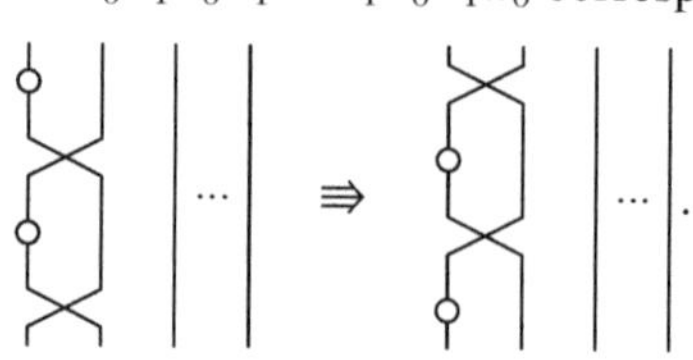

In this category, every morphism factors uniquely as a symmetry (i.e., a matrix containing only 0 and 1) followed by a diagonal matrix. This provides a factorization system inducing a description of the category as $\mathbf{S} \otimes_\ell \mathbb{N}^\otimes$ where $\mathbb{N}^\otimes$ is the free monoidal category over the monoid $\mathbb{N}$, considered as a category, and ℓ is the distributive law induced by the factorization system. Of course, this category being self-dual, we also have a decomposition as $\mathbb{N}^\otimes \otimes_\ell \mathbf{S}$.

C.11.5 Progressive ribbons. The category $\mathbf{R}$ is the subcategory of the category of ribbons up to endpoint-preserving isotopy described in §C.10.12, with the same objects, where

- we restrict to ribbons which are *progressive*, i.e., always "go down", so that the ribbon on the left is valid but not the one on the right:

- we restrict to ribbons which always show the "same face" at the boundary, so that the ribbon on the left is valid, but not the one on the right:

For instance, we have the following morphism $3 \to 3$:

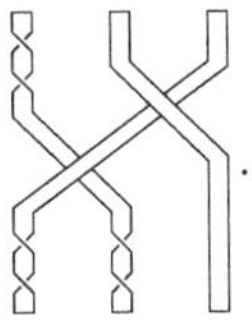

Note that the ribbons can be twisted. This category admits a presentation with generators

such that the two first generators are mutually inverse and the two last generators form a braiding (Theorem C.1.5). For instance the above morphism corresponds to the diagram

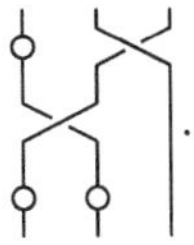

This category can be shown to be the free balanced category on an object [208]. We recall that a *balanced category* is a braided monoidal category equipped

with a natural family of isomorphisms $\theta_u : u \to u$, called *twists*, such that $\theta_i = i$ and the following diagram commutes for every pair of objects u and v:

$$\begin{array}{ccc} v \otimes u & \xrightarrow{\gamma_{u,v}} & u \otimes v \\ {\scriptstyle \theta_{v\otimes u}} \downarrow & & \downarrow {\scriptstyle \theta_u \otimes \theta_v} \\ v \otimes u & \xleftarrow[\gamma_{v,u}]{} & u \otimes v \end{array}.$$

Graphically,

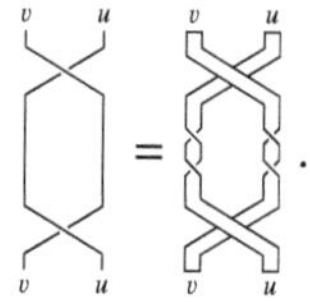

Given $n \in \mathbb{N}$, the group of endomorphisms of n is the group R_n of progressive ribbons with n strands described in §A.1.23:

$$\mathbf{R} = \coprod_{n \in \mathbb{N}} R_n$$

It can be obtained as the wreath product of the additive group of integers with the n-th braid group:

$$R_n = \mathbb{Z} \wr C_n.$$

C.11.6 The pearl necklace. The *pearl necklace* is the PRO, introduced in §12.2.8, presented by the 3-polygraph with three generators

subject to the four relations

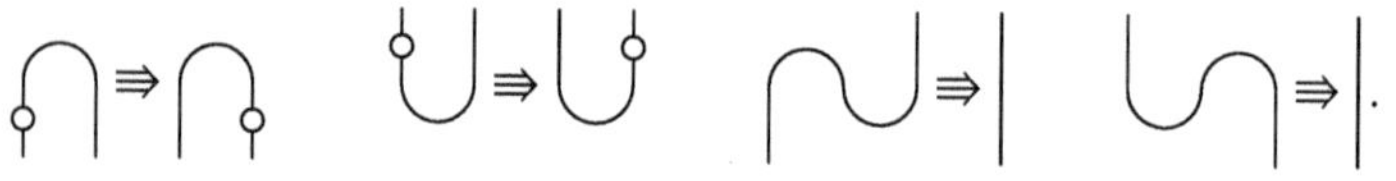

This is the free monoidal category on a self-dual object together with an endomorphism, see also §§C.10.5 and C.11.1. This presentation was introduced and studied in [161] as a first example of a convergent finite presentation which does not have a finite derivation type, see §12.2.8.

C.11.7 Directed acyclic graphs. A *directed acyclic graph*, or *DAG*, is a graph

$$G = G_0 \overset{s}{\underset{t}{\leftleftarrows}} G_1$$

which is acyclic, i.e., every path from a vertex to itself is empty. Any $n \in \mathbb{N}$ can canonically be seen as a DAG with the set $[n] = \{0, \ldots, n-1\}$ as set of vertices and no edge; we write $[n]$ for this graph. A vertex $x \in G_0$ of a DAG G is *minimal* (resp. *maximal*) when there is no edge with x as target (resp. as source). Given a set $V \subseteq G_0$, the *restriction* of G to V is the subgraph of G with V as vertices, whose edges are the edges G such that both their source and their target belong to V. Given a set $V \subseteq G_0$, the graph obtained by *hiding* V in G is the graph obtained from G by adding a new edge $x \to z$ for every vertex $y \in V$ and pair of edges $x \to y$ and $y \to z$ in G and then restricting the resulting graph to $G_0 \setminus V$.

We can build a PROP **DAG** of DAGs, where an object is a natural number and a morphism from m to n is a cospan

$$[m] \xrightarrow{f} G \xleftarrow{g} [n]$$

where G is a finite graph, f and g are injective morphisms of graphs such that for every $i \in [m]$ (resp. $i \in [n]$) the vertex $f(i)$ (resp. $g(i)$) is a minimal (resp. maximal) vertex in G, and the images of f and g are disjoint. A vertex of G is a *source* (resp. *target*, resp. *internal*) *vertex* when it is in the image of f (resp. in the image of g, resp. neither in the image of f nor the image of g). We will picture by

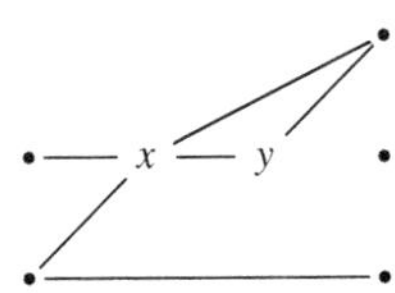

a morphism with x and y as internal vertices, two source vertices, and three target vertices. The composite of two morphisms $G: m \to n$ and $H: n \to o$ is given by computing the pushout

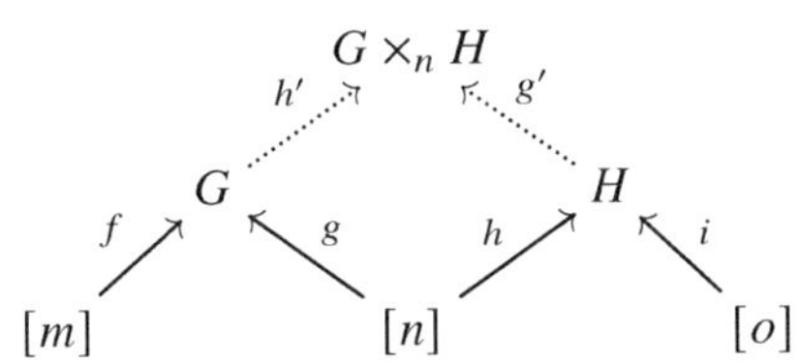

and then hiding $h'(g([n]))$ in the resulting graph, with the expected cospan morphisms induced from $h' \circ f$ and $g' \circ i$. For instance, we have the following composition of morphisms:

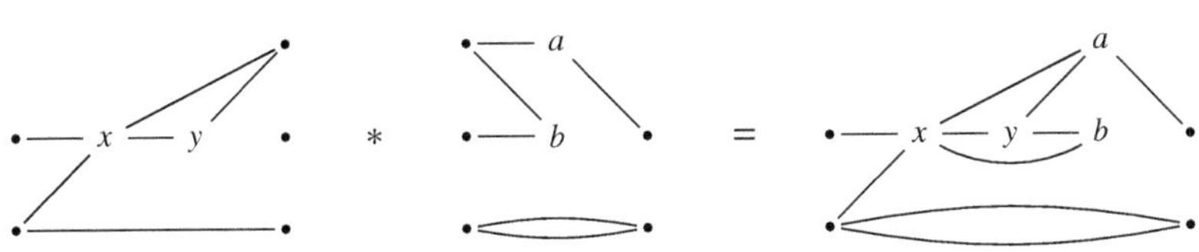

The identity on n is the graph obtained from $[n] \sqcup [n]$ by adding, for every $i \in [n]$ a vertex from the first copy of i to the second one, e.g., the identity on 3 is

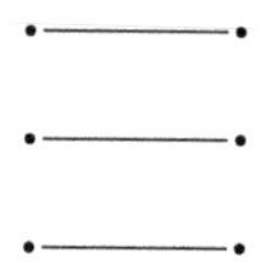

Tensor product is given on morphisms by disjoint union and symmetries are the expected ones. It is shown in [129] that this PROP admits a presentation with generators

such that the five first generators satisfy the axioms for bicommutative bialgebras (§C.4.3) and the penultimate one is a symmetry (§C.1.4). In this language, the composition of the two above morphisms can be written

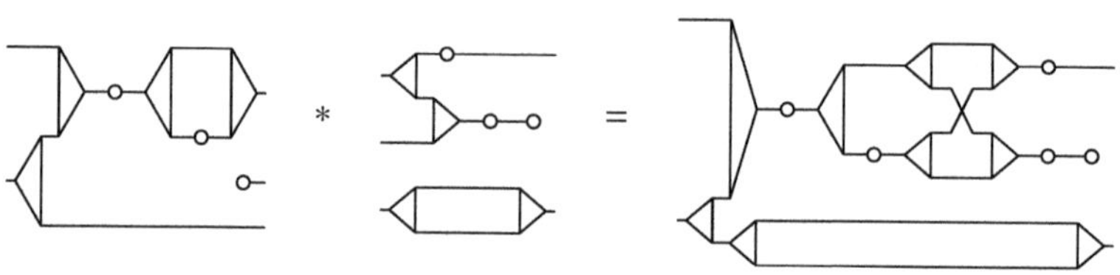

If we restrict the morphisms of the category **DAG** to simple graphs, i.e., forbidding multiple edges with same source and same target, we can obtain a presentation of the resulting category by further adding the relation (C.8) of special bialgebras.

C.11.8 Posets. As a variant of previous situation, consider the PROP whose objects are integers and morphisms $m \to n$ are cospans

$$[m] \xrightarrow{f} E \xleftarrow{g} [n]$$

of finite posets, where $[n]$ is the discrete poset with n elements, and the morphisms are injective non-decreasing functions with disjoint images, such that the images of f (resp. g) are minimal (resp. maximal) elements of the poset. Composition is obtained from the pushout by removing the elements which are identified in the interface, as in §C.11.7. This category admits the same presentation as the category of simple graphs, with the extra *transitivity* relation

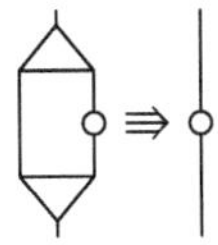

see [284] for details.

C.11.9 Frobenius. We now consider a variant of the categories introduced in the previous section, as well as a generalization of the theory of special commutative Frobenius algebras, which is the category Cospan(**F**) as explained §C.6.3.

A formalization of "circuits" (such as electric circuits made of electronic components) was introduced and studied in [315] by considering the category **Circ**, which is the full subcategory of the category Cospan(**FinGraph**) whose objects are integers $n \in \mathbb{N}$, seen as graphs with $[n]$ as vertices and no edge, morphisms being isomorphism classes of cospans of finite graphs. A morphism $G\colon m \to n$ in this category is a graph G (considered up to isomorphism) together with two functions $f\colon [m] \to G$ and $g\colon [n] \to G$, composition is given by pushout and identities are cospans of the form $[n] \longrightarrow [n] \longleftarrow [n]$ with both morphisms being identities. The resulting category is a PROP which admits a presentation with generators

such that the five first generators satisfy the axioms for special commutative Frobenius algebras (§C.6.1) and the penultimate one is a symmetry (§C.1.4).

Given a set L of *labels* (thought of as the possible components of our circuits), we can consider a variant of the previous category where the vertices of graphs are labeled in L. It is shown to have a similar presentation as above, with the generator  being replaced by a family of generators 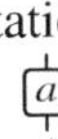, indexed by $a \in L$. From §C.11.2, we deduce that this is the theory for special Frobenius algebras equipped with an action of L [315, Proposition 3.2]. This category has found many applications to modeling networks [21, 23].

C.12 Nets

Some of the previous examples consist in categories whose morphisms are graphs of some sort (e.g., §§C.11.7 or C.11.9). The kind of constructions performed there can be adapted in order to build freely generated monoidal categories as follows by formalizing the networks occurring in string-diagrammatic representations of morphisms. Those networks will comprise nodes, which correspond to 2-generators, ports which represent the inputs and outputs of the generators and of the whole net, and wires which link ports together.

Throughout the section, we suppose fixed a *signature* 2-polygraph P with $P_0 = \{\star\}$. We recall that, given an element $u = a_0 \dots a_{n-1}$ of P_1^*, we write $|u| = n$ for its length.

C.12.1 Definition. Following [234], a *P-net* consists of

- a finite set N of *nodes*,
- a labeling function $\ell\colon N \to P_2$,
- finite totally ordered sets $X^- = \{x_0^-, \dots, x_m^-\}$ and $X^+ = \{x_0^+, \dots, x_n^+\}$ of *input* and *output ports*,
- a labeling function $\ell\colon X \to P_1$, where

  $$X = X^- \sqcup X^+ \sqcup X_\circ^- \sqcup X_\circ^+$$

  is the set of *ports* and

  $$X_\circ^- = \{v_i^- \mid v \in N, 0 \leqslant i < |s_1(\ell(v))|\}$$
  $$X_\circ^+ = \left\{v_j^+ \,\middle|\, v \in N, 0 \leqslant j < |t_1(\ell(v))|\right\}$$

  are the sets of *inner input* and *output ports* respectively,
- a finite set W of *wires*,
- a labeling function $\ell\colon W \to P_1$,
- a *boundary* $\partial w \subseteq X$ for every wire $w \in W$,

such that

- for every node $v \in N$ such that $s_1(v) = a_1 \dots a_p$ and $t_1(v) = b_1 \dots b_q$, we have $\ell(v_i^-) = a_i$ and $\ell(v_j^+) = b_j$ for $0 \leqslant i < p$ and $0 \leqslant j < q$,
- for every $w \in W$, ∂w contains 0 or 2 elements, and in the case $\partial w = \{x, y\}$ we have $\ell(w) = \ell(x) = \ell(y)$,
- the sets ∂w form a partition of X, i.e., $X = \bigcup_{w\in W} \partial w$ and $w \neq w'$ implies $\partial w \cap \partial w' = \emptyset$.

The *source* (resp. *target*) of such a net is $\ell(x_0^-) \dots \ell(x_m^-)$ (resp. $\ell(x_0^+) \dots \ell(x_n^+)$).

For instance, with $P_1 = \{a, b\}$ and $P_2 = \{\alpha\colon aaaa \Rightarrow a, \beta\colon ab \Rightarrow ab\}$, the diagram on the left below

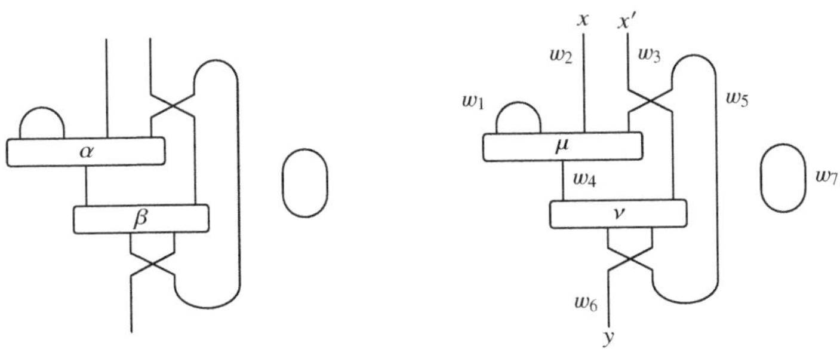

can be encoded as the net (figured on the right) with

$$N = \{\mu, \nu\} \qquad X^- = \{x, x'\} \qquad X^+ = \{y\} \qquad W = \{w_1, \dots, w_7\}$$

labels being

$$\ell(\mu) = \alpha \quad \ell(\nu) = \beta \quad \ell(x) = a \quad \ell(x') = b \quad \ell(\mu_i^-) = a \quad \ell(w_1) = a \quad \dots$$

and boundaries being

$$\partial w_1 = \{\mu_0^-, \mu_1^-\} \quad \partial w_2 = \{\mu_2^-, x\} \quad \partial w_3 = \{\nu_1^-, x'\} \quad \dots \quad \partial w_7 = \emptyset$$

As it can be observed for w_7 above, wires with empty boundary encode “loops”.

Nets are considered up to isomorphism, i.e., renaming of ports, wires, and nodes, preserving labels. Two nets can be composed by linking wires along inner boundary ports and removing those ports; two nets can also be tensored by juxtaposition. We write $\mathbf{Net}_P$ for the resulting monoidal category, with P_1^* as objects and nets as morphisms.

C.12.2 Proposition. *The monoidal category* $\mathbf{Net}_P$ *is the free self-dual compact closed category on P, in the sense of §C.10.5.*

Various other free constructions can be obtained by considering monoidal subcategories obtained by restricting the notion of net.

C.12.3 Traced categories. The free traced category on P, see [210], can be obtained by forbidding wires to link two input ports or two output ports. More formally, we restrict the category $\mathbf{Net}_P$ to nets such that for every wire $w \in W$, its boundary ∂w is either empty or of the form $\partial w = \{x, y\}$ with

$$(x, y) \quad \in \quad (X^- \times X_\circ^-) \cup (X_\circ^+ \times X_\circ^-) \cup (X_\circ^+ \times X^+)$$

see [175] for details.

C.12.4 Free symmetric monoidal categories. Given a net, we define a relation $<$ on nodes as the smallest transitive relation such that $\mu < \nu$ whenever there exists a wire w such that $\partial w = \left\{\mu_j^+, \nu_i^-\right\}$ for some indices i and j, i.e., some output port of μ is linked to some input port of ν. The free symmetric monoidal category on P can obtained by considering the subcategory of $\mathbf{Net}_P$ whose morphisms are nets such that the relation $<$ is acyclic, the condition of §C.12.3 is satisfied, and $\partial w \neq \emptyset$ for every wire $w \in W$.

C.12.5 Interaction nets. The morphisms in $\mathbf{Net}_P$ are called *interaction nets* whenever all the generators in P_2 are of the form $\alpha\colon a_1 \ldots a_n \Rightarrow b$, i.e., their target consists of only one generator. An *interaction rule* $A\colon \phi \Rrightarrow \psi$ consists of a pair of parallel nets $\phi\colon u \Rightarrow v$ and $\psi\colon u \Rightarrow v$ which are of the form

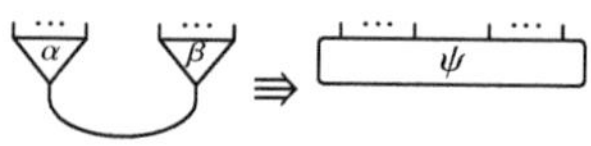

(in particular, v is always an identity). These were introduced by Lafont [232, 234] in order to provide a distributed model of computation, where there is no global synchronization. Because of the shape of the rules, the corresponding rewriting system is confluent (as a rewriting system on nets) since there is no critical branching. The signature P together with a set P_3 of interaction rules can be encoded as a 3-polygraph Q obtained from the free self-dual compact closed symmetric monoidal category on P, as described in §C.10.5, by adding the elements of P_3 as 3-generators.

As a particular case, *interaction combinators* [234] are the following interaction nets. The sets of 0- and 1-generators are respectively $P_0 = \{\star\}$ and $P_1 = \{a\}$. The 2-generators are

respectively called *constructor* (noted γ), *duplicator* (noted δ), and *eraser* (noted ε), and the interaction rules are

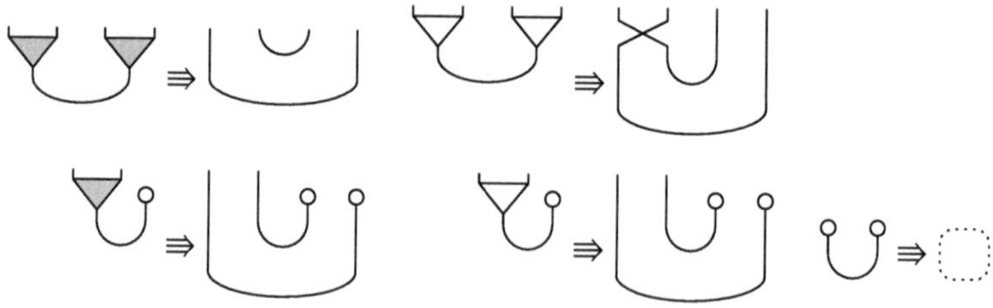

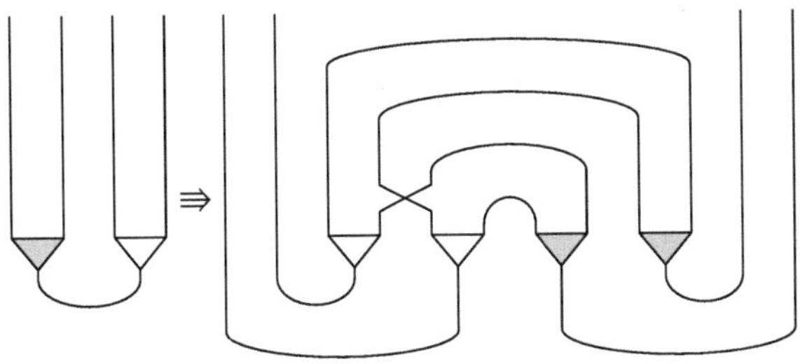

as well as symmetric ones. Lafont's original notation was

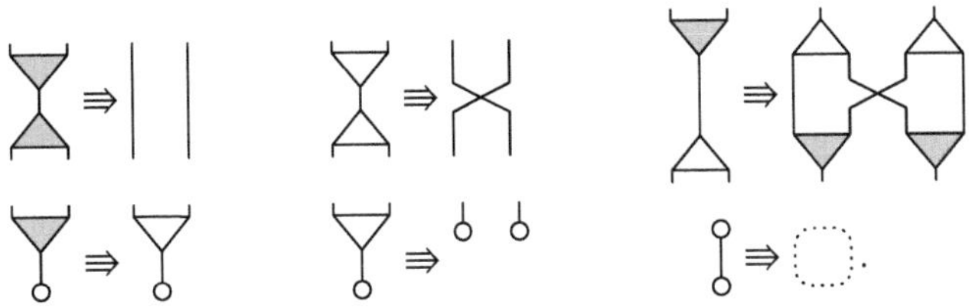

which suggests that they can also be formalized without resorting to the compact closure, at the cost of having to add two copies of each 2-generator (one going upward and one going downward). Interaction nets are "universal" in the sense that they can simulate any net [234, Theorem 1]. In particular, this implies that they are Turing-complete.

C.13 Simplicial and Cubical Categories

Given a category C, we write $\hat{C}$ for the category of *presheaves* on this category: the objects of $\hat{C}$ are functors $C^{\mathrm{op}} \to \mathbf{Set}$ and morphisms are natural transformations.

C.13.1 Simplicial categories. In §C.2, we have already described presentations of monoidal categories, whose associated presheaf categories are widely used in algebraic topology:

- for Δ_+ presented in §C.2.3, $\hat{\Delta}_+$ is the category of *augmented simplicial sets*,
- for Δ_η presented in §C.2.1, $\hat{\Delta}_\eta$ is the category of *augmented presimplicial* or *semisimplicial sets*,
- for $\mathbf{F}$ presented in §C.2.5, $\hat{\mathbf{F}}$ is the category of *augmented symmetric simplicial sets* [148].

The non-augmented variants can be obtained by taking presheaves on the same categories with the object 0 removed. For each of those, from their presentation as a monoidal category, we can deduce a presentation as a category by using the method of §10.3.10 and thus obtain an algebraic description of the associated presheaves.

For instance, from the presentation of the PRO Δ_η given in §C.2.1, we can deduce the following presentation of it as a category:

$$\left\langle n \,\middle|\, d_i^n : n \to n+1 \,\middle|\, d_j^n d_i^{n+1} \Rightarrow d_i^n d_{j+1}^{n+1} \right\rangle_{0 \leqslant i \leqslant j \leqslant n}$$

(see Example 10.3.11 for details). An *augmented presimplicial set* thus consists of a family of sets $(X_n)_{n \in \mathbb{N}}$ together with functions

$$\partial_i^n : X_{n+1} \to X_n$$

with $0 \leqslant i \leqslant n$, satisfying relations dual to those of the above presentation. The description of non-augmented presimplicial sets is similar, excepting that we constrain n to be strictly positive (there is no X_0 nor ∂_0^0). For those, an element of X_{n+1} can be interpreted geometrically as an n-simplex and the function ∂_i^n as the function associating to an n-simplex its i-th face, obtained by removing its i-th vertex; more generally, a simplicial set can be thought of as the result of gluing simplices. For instance, the space

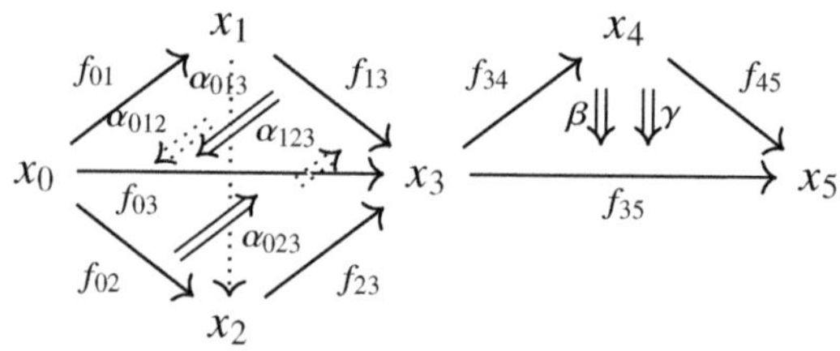

can be described by the simplicial set X with

$$X_1 = \{x_0, x_1, \ldots\} \quad X_2 = \{f_{01}, f_{02}, \ldots\} \quad X_3 = \{\alpha_{012}, \ldots, \beta, \gamma\} \quad X_4 = \{A\}$$

and $X_n = \emptyset$ for $n \geqslant 5$, face maps being

$$\partial_0^2(\beta) = f_{45} \qquad \partial_1^2(\beta) = f_{35} \qquad \partial_2^2(\beta) = f_{34}$$

and so on. A *simplicial set* X is moreover equipped with functions

$$\sigma_i^n : X_n \to X_{n+1}$$

with $0 \leqslant i < n$, sending an n-simplex to the corresponding n+1-simplex with degenerated i-th face. For instance, given a 1-simplex $x_0 \xrightarrow{f} x_1$, its images under σ_0^3 and σ_1^3 are respectively

Similarly, a *symmetric* augmented simplicial set X is equipped with functions

$$\gamma_f^n : X_n \to X_n$$

indexed by bijections $f : [n] \to [n]$ sending an n-simplex to the corresponding n-simplex with vertices renumbered according to f.

C.13.2 Cubical categories. Developments similar to previous section can be performed using cubes instead of simplices. The *precubical category* is the PRO $\square_\eta$ generated by

$$\eta^- : 0 \to 1 \qquad \text{and} \qquad \eta^+ : 1 \to 0$$

respectively depicted as

with no relation. As a category, $\square_\eta$ can be presented by

$$\left\langle n \;\middle|\; d_{n,i}^- : n \to n+1, d_{n,i}^+ : n \to n+1 \;\middle|\; d_{n,j}^{\epsilon} d_{n+1,i}^{\epsilon'} = d_{n,i}^{\epsilon'} d_{n+1,j+1}^{\epsilon} \right\rangle_{0 \leqslant i \leqslant j < n}$$

with $\epsilon, \epsilon' \in \{-,+\}$. A presheaf X on this category is called a *precubical set*: it consists of a family of sets $(X_n)_{n \in \mathbb{N}}$, whose elements can be interpreted as n-dimensional cubes, together with morphisms

$$\partial_i^\epsilon : X_{n+1} \to X_n$$

with $0 \leqslant i < n$ and $\epsilon \in \{-,+\}$, defined by

$$\partial_i^\epsilon = 1_{n-i} \otimes \eta^\epsilon \otimes 1_{n-i-1}$$

and associating to each $(n{+}1)$-cube its source ($\epsilon = -$) or target ($\epsilon = +$) face in the i-th direction. For instance, the space

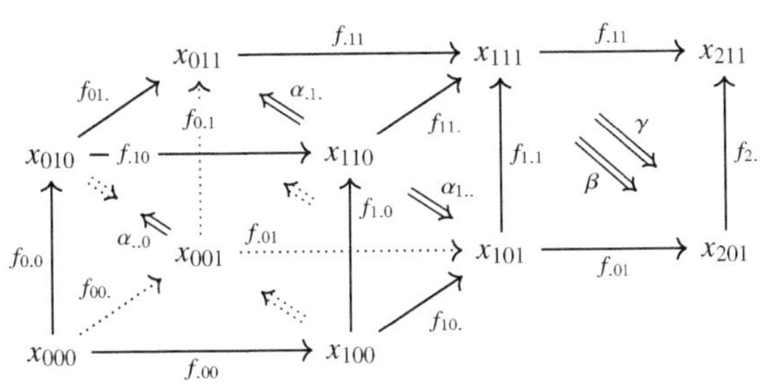

corresponds to the precubical set with

$$X_0 = \{x_{000}, x_{001}, \ldots\} \quad X_1 = \{f_{.00}, f_{.01}, \ldots\} \quad X_2 = \{\alpha_{..0}, \alpha_{..1}, \ldots\} \quad X_3 = \{A\}$$

and $X_n = \emptyset$ for $n \geqslant 4$. Face maps are given by

$$\partial_0^-(\beta) = f_{1.1} \qquad \partial_0^+(\beta) = f_{2.1} \qquad \partial_1^-(\beta) = f_{.01} \qquad \partial_1^+(\beta) = f_{.11}$$

and so on.

Many variations on this category (and thus on associated presheaves) are possible and considered in the literature, see [149] for a panorama which is briefly recalled here. The *cubical category* $\square$ is the PRO generated by

subject to the relations

(C.19)

A *cubical set* X is a presheaf on this category and comes equipped with morphisms

$$\sigma_i^n : X_n \to X_{n+1}$$

with $0 \leqslant i \leqslant n$, called *degeneracies*, sending an n-cube to the corresponding $(n+1)$-cube degenerated in dimension i. For instance, the images under σ_0 and σ_1 of the 1-cube $x \xrightarrow{f} y$ are respectively

The *symmetric cubical category* $\square_\gamma$ is the free symmetric monoidal category on the PRO $\square$, it is generated by

subject to the above relations (C.19) and those of symmetries (§C.1.3). A presheaf X on it is called a *symmetric cubical set* and is equipped with morphisms

$$\gamma_f^n : X_n \to X_n$$

indexed by bijections $f : [n] \to [n]$ sending an n-cube to the corresponding n-cube obtained by permuting the directions along f. For instance the image of the 2-cube on the left along the transposition $[2] \to [2]$ is the 2-cube on the right:

$$\begin{array}{ccc} x_{01} & \xrightarrow{f_{\cdot 1}} & x_{11} \\ {\scriptstyle f_{0\cdot}}\uparrow & \nwarrow & \uparrow{\scriptstyle f_{1\cdot}} \\ x_{00} & \xrightarrow[f_{\cdot 0}]{} & x_{10} \end{array} \qquad\qquad \begin{array}{ccc} x_{10} & \xrightarrow{f_{1\cdot}} & x_{11} \\ {\scriptstyle f_{\cdot 0}}\uparrow & \searrow & \uparrow{\scriptstyle f_{\cdot 1}} \\ x_{00} & \xrightarrow[f_{0\cdot}]{} & x_{01}\,. \end{array} \tag{C.20}$$

By definition, the *cartesian cubical category* $\square_\times$ is the free cartesian category on the PROP $\square_\gamma$: it is generated by

subject to the relations given in §C.2.10, which include (C.19). Note that this is the Lawvere theory for sets which are bipointed (i.e., equipped with two distinguished elements). A presheaf X on this category is a symmetric cubical set equipped with morphisms

$$\delta_i^n : X_{n+1} \to X_n$$

with $0 \leqslant i < n$, which to an $(n{+}1)$-cube associate a diagonal n-cube (in directions i and $i+1$). For instance, the image under δ_0^1 of the 2-cube on the left is a one cube as on the right

The *cubical category with connections* $\square_\kappa$ is the PRO generated by

subject to relations (C.19), as well as

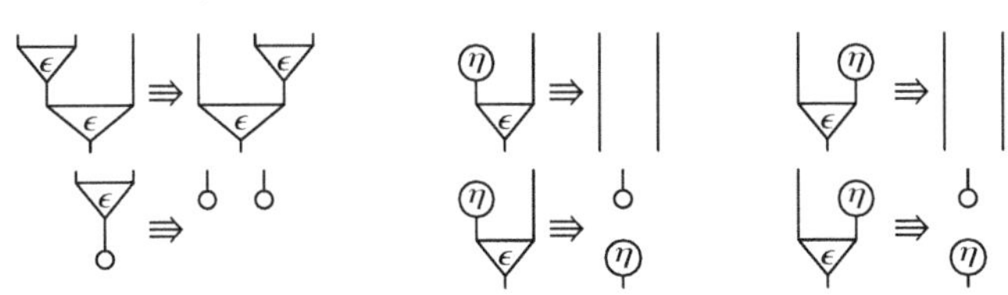

for $\eta, \epsilon \in \{-, +\}$ with $\eta \neq \epsilon$. A presheaf X on this category is a *cubical category with connections* and is equipped with morphisms

$$\kappa_i^\epsilon : X_n \to X_{n+1}$$

called *connections* which produce degenerated cubes, but in another way than degeneracies. For instance, the images under κ_0^- and κ_0^+ of the 1-cube $x \xrightarrow{f} y$

are shown on the left and those of the 2-cube (e.g., left of (C.20)) are shown on the right (omitting 2-cells for clarity):

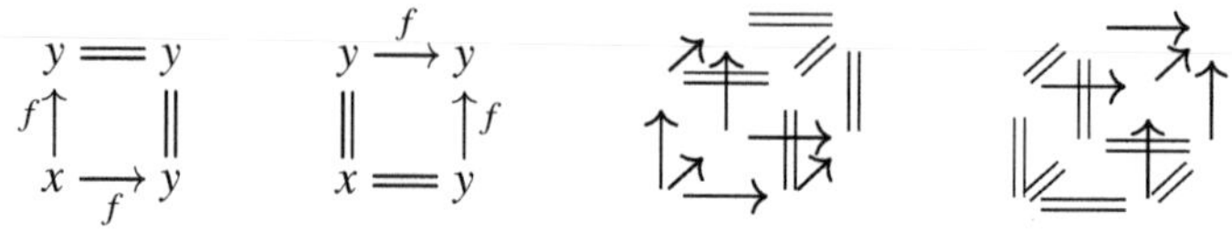

A symmetric variant can be obtained by taking the free symmetric monoidal category (in practice, one moreover asks both monoid structures to be commutative). The *cubical category with reversions* $\square_\rho$ is the PRO generated by

satisfying the relations of $\square_\kappa$ together with

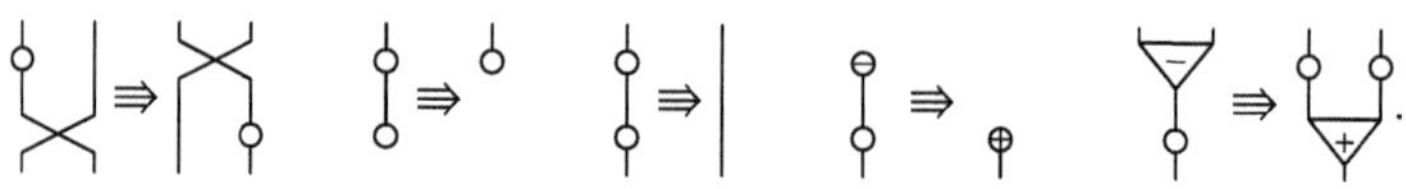

A presheaf X on this category is equipped with morphisms

$$\rho_i^n : X_n \to X_n$$

reversing cubes in the i-th direction. There is also a cartesian variant [90], where one usually requires axioms corresponding to the Lawvere theory of de Morgan algebras, i.e., bounded distributive lattices with an idempotent negation.

C.14 Quantum Processes

Let us briefly mention that presentations of PROPs are intensively used nowadays in the study of quantum processes, see [89] for an in-depth introduction. One of the most notable axiomatic approaches is the *ZX-calculus* [88], aiming at modeling operations on qubits, which is a presented PROP together with a canonical interpretation in the category **FdHilb** of finite-dimensional Hilbert spaces and linear maps, which is a PROP when equipped with the usual tensor product of vector spaces. The ZX-calculus is generated by

where α_n^m (resp. β_n^m) have m inputs and n outputs, such that various axioms are satisfied, among which the fact that the first generator induces a symmetry and the second and the third a self-duality. The generating object of this PROP is interpreted as $\mathbb{C}^2$. Using the traditional notation, we write $|0\rangle, |1\rangle$ for the standard basis (also called the Z-basis) of $\mathbb{C}^2$ and $|-\rangle, |+\rangle$ for the Bell basis (also called the X-basis) defined by $|-\rangle = \frac{1}{\sqrt{2}}(|0\rangle - |1\rangle)$ and $|+\rangle = \frac{1}{\sqrt{2}}(|0\rangle + |1\rangle)$. We also write $\langle 0|$ for the adjoint of $|0\rangle$, etc. The interpretation of the generators is then given by

 $= |0\rangle^{\otimes 2} + |1\rangle^{\otimes 2}$

 $= \langle 0|^{\otimes 2} + \langle 1|^{\otimes 2}$

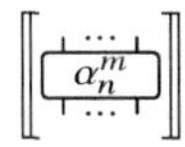 $= |0\rangle^{\otimes n}\langle 0|^{\otimes m} + e^{i\alpha}|1\rangle^{\otimes n}\langle 1|^{\otimes m}$

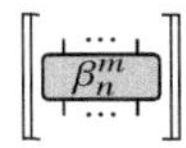 $= |+\rangle^{\otimes n}\langle +|^{\otimes m} + e^{i\alpha}|-\rangle^{\otimes n}\langle -|^{\otimes m}$

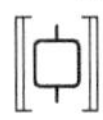 $= |0\rangle\langle +| + |1\rangle\langle -|$

The original axiomatization is not *complete* in general [103], meaning that the resulting functor to **FdHilb** is not faithful. However, the rules can be completed so that it is the case [291]: the proof is based on an alternative complete axiomatization called the *ZW-calculus* [167].

Appendix D
A Syntactic Description of Free n-Categories

In this chapter, we provide an explicit description of the free n-category P^* generated by an n-polygraph P. Variants of this construction can be found in [296, Deuxième partie], [264, Section 7], and [279, Section 4.1], but this section is mostly inspired by the work of Makkai [264], where the proofs of most assertions can be found, formulated in a slightly different language, see also [134, Section 2].

We first provide, in Section D.1, a formal definition of the syntax of n-categories, i.e., a description of the morphisms in an $(n+1)$-category freely generated by an n-polygraph, allowing reasoning by induction on its terms to prove results on free categories. It turns out that this syntax for n-categories, which corresponds to the one introduced in Chapter 14 and used throughout the book, is very "redundant", in the sense that there are many ways to express a composite of cells which will give rise to the same result, and is sometimes not very practical for this reason. In Section D.2, we provide an alternative syntax, which suffers less from these problems, by restricting compositions. Finally, in Section D.3, we briefly mention the word problem for free n-categories.

D.1 A Syntax for n-Categories

D.1.1 A syntax for free n-categories. Suppose fixed an n-polygraph P. We define two sets of terms as the smallest sets closed under certain operations.

- An *expression* is

$$x \qquad \text{or} \qquad 1_f \qquad \text{or} \qquad f *_{-i} g$$

  where x is an element of $\bigsqcup_{0 \leqslant i \leqslant n} P_n$, f and g are expressions and i is a natural number (which can be supposed to satisfy $0 < i \leqslant n$ without loss of generality).

– A *type* is either

$$\star \qquad \text{or} \qquad f \underset{T}{\to} g$$

for some type T and expressions f and g.

The expressions should be thought of as formal composites of generators; in particular, $f *_{-i} g$ corresponds to two formal n-cells f and g composed in dimension $n - i$ (thus the minus sign in the index) and types represent either the set of objects ($\star$) or a particular hom-set.

D.1.2 Type-theoretic syntax. A *judgment* is an expression of one of the following forms with the following meanings:

– well-formed type:

$$\vdash T$$

for some type T,

– well-typed term:

$$\vdash f : T$$

for some term f and type T,

– equivalent types:

$$\vdash T \simeq U$$

for some types T and U,

– equivalent terms:

$$\vdash f \simeq g : T$$

for some terms f and g and type T.

An *inference rule* is of the form

$$\frac{\vdash \Gamma_1 \qquad \dots \qquad \vdash \Gamma_n}{\vdash \Gamma}$$

where $\vdash \Gamma_i$ and $\vdash \Gamma$ are judgments, respectively called the *premises* and the *conclusion* of the inference rule. A judgment is *derivable* when it is the conclusion of an inference rule whose premises are all derivable. An expression f is derivable when the judgment $\vdash f : T$ is derivable for some type T, two terms f and g are equivalent if $\vdash f \simeq g : T$ is derivable for some type T, and two types T and U are equivalent when $\vdash T \simeq U$ is derivable.

We are going to describe a typing system CAT^P, i.e., a set of inference rules, on the above expressions and types. The notation CAT^P suggests that it depends on P: as we have seen, the expressions and types depend on P, but the rules will be uniform. This will allow us to define an n-category C^P, where C^P_i is the set of derivable expressions of dimension i (defined below), compositions are given by $*_{-j}$ and identities by 1. The main property this construction satisfies is the following one.

D.1.3 Theorem. *The n-category C^P is isomorphic to P^*.*

The result will be proved by induction on n: we thus suppose that the property is satisfied for strictly smaller values of n. In particular, given a cell $f \in P^*_i$ for $0 \leqslant i < n$, up to isomorphism, we can suppose that $P^*_i = C^{P_{\leqslant i}}$, and thus that f is the equivalence class of an expression in $\mathsf{CAT}^{P_{\leqslant i}}$. Clearly, a term in $\mathsf{CAT}^{P_{\leqslant i}}$ is a term in CAT^P, a valid derivation of $\mathsf{CAT}^{P_{\leqslant i}}$ is a valid derivation of CAT^P and finally the equivalence relation in $\mathsf{CAT}^{P_{\leqslant i}}$ coincides with the one of CAT^P (by Lemma D.1.10). In particular, given a cell $x \in P_{i+1}$, we abusively write $s_i x$ and $t_i x$ for an expression representing the source and target of x in P^*_i. This convention allows us to associate a type T_x to each generator $x \in P_i$ by

$$T_x = \begin{cases} \star & \text{if } i = 0, \\ s_{i-1}x \xrightarrow[T_{s_{i-1}x}]{} t_{i-1}x & \text{otherwise.} \end{cases}$$

D.1.4 Inference rules. Our typing system consists of the following inference rules:

– rules for types:

$$\frac{}{\vdash \star} \qquad\qquad \frac{\vdash f : T \qquad \vdash g : T}{\vdash f \xrightarrow[T]{} g}$$

– rules for terms:

$$\frac{\vdash f : T \qquad \vdash T \simeq T'}{\vdash f : T'}$$

$$\frac{}{\vdash x : T_x} \qquad \text{for } x \in \bigsqcup_i P_i$$

$$\frac{\vdash f : T}{\vdash 1_f : f \xrightarrow[T]{} f} \qquad\qquad \frac{\vdash f : g \xrightarrow[T]{} g' \qquad \vdash f' : g' \xrightarrow[T]{} g''}{\vdash f *_{-1} f' : g \xrightarrow[T]{} g''}$$

$$\frac{\vdash f : g \underset{T}{\longrightarrow} g' \qquad \vdash f' : h \underset{T'}{\longrightarrow} h' \qquad \vdash g *_{-i} h : U \qquad \vdash g' *_{-i} h' : U}{\vdash f *_{-(i+1)} f' : g *_{-i} h \underset{U}{\longrightarrow} g' *_{-i} h'}$$

– equivalence on types:

$$\frac{\vdash f \simeq f' : T \qquad \vdash g \simeq g' : T \qquad \vdash T \simeq T'}{\vdash (f \underset{T}{\longrightarrow} g) \simeq (f' \underset{T'}{\longrightarrow} g')}$$

– equivalence on terms:

$$\frac{\vdash f : T}{\vdash f \simeq f : T} \qquad \frac{\vdash f \simeq g : T}{\vdash g \simeq f : T} \qquad \frac{\vdash f \simeq g : T \qquad \vdash g \simeq h : T}{\vdash f \simeq h : T}$$

$$\frac{\vdash f \simeq f' : T}{\vdash 1_f \simeq 1_{f'} : f \underset{T}{\longrightarrow} f} \qquad \frac{\vdash f \simeq f' : T \qquad \vdash g \simeq g' : U \qquad \vdash f *_{-i} g : V}{\vdash (f *_{-i} g) \simeq (f' *_{-i} g') : V}$$

$$\frac{\vdash 1_g *_{-i} f : T}{\vdash (1_g *_{-i} f) \simeq f : T} \qquad \frac{\vdash f *_{-i} 1_g : T}{\vdash (f *_{-i} 1_g) \simeq f : T}$$

$$\frac{\vdash (f *_{-i} g) *_{-i} h : T}{\vdash (f *_{-i} g) *_{-i} h \simeq f *_{-i} (g *_{-i} h) : T}$$

$$\frac{\vdash 1_{f *_{-i} g} : T}{\vdash 1_{f *_{-i} g} \simeq 1_f *_{-i} 1_g : T}$$

$$\frac{\vdash (f *_{-j} f') *_{-i} (g *_{-j} g') : T}{\vdash (f *_{-j} f') *_{-i} (g *_{-j} g') \simeq (f *_{-i} g) *_{-j} (f' *_{-i} g') : T} \quad \text{for } i < j.$$

D.1.5 Remark. Note that, in the last rule, the side condition $i < j$ is important. For instance, consider the 2-polygraph corresponding to the following diagram:

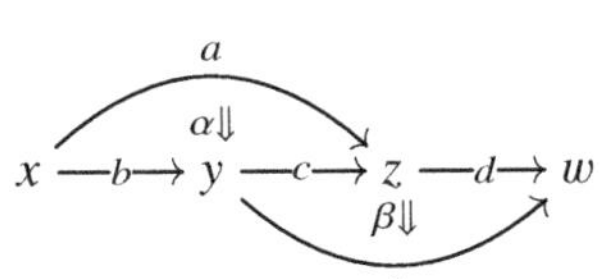

The expression on the left is derivable, but the one on the right is not:

$$(\alpha *_{-1} 1_d) *_{-2} (1_b *_{-1} \beta) \qquad\qquad (\alpha *_{-2} 1_b) *_{-1} (1_d *_{-2} \beta).$$

D.1.6 Admissible rules. A rule is *admissible* in the system when, whenever the premises are derivable, the conclusion is also derivable (with the above rules).

D.1.7 Lemma. *The following rules are admissible*

$$\frac{\vdash f \simeq g : T}{\vdash f : T} \qquad \frac{\vdash f : T \qquad \vdash f : U}{\vdash T \simeq U} \qquad \frac{\vdash f : T \qquad \vdash T \simeq U}{\vdash f : U}$$

In particular, the type of an expression is uniquely defined up to $\simeq$, and two equivalent expressions have equivalent types.

This ensures that we can meaningfully consider terms and types up to equivalence, and moreover, by definition of the equivalence, we have:

D.1.8 Lemma. *The relation $\simeq$ is a congruence on expressions.*

D.1.9 Dimension of cells. The *dimension* $\dim(T)$ of a type T is the natural number defined by

$$\dim(\star) = 0 \qquad\qquad \dim(f \underset{T}{\rightarrow} g) = \dim(T) + 1$$

and the dimension of a derivable expression f is the dimension of T for some derivable judgment $\vdash f : T$. It is easily shown that two equivalent types have the same dimension and, since by Lemma D.1.7 two equivalent expressions have equivalent types, two equivalent expressions have the same dimension, i.e., the dimension is well-defined on equivalence classes. A derivable expression of dimension k is sometimes called a *k-expression.*

D.1.10 Lemma. *Any derivable expression involving a generator $x \in P_i$ has dimension at least i.*

D.1.11 Construction of the free n-category. We define the category C^P as the category such that C_i^P, for $0 \leqslant i \leqslant n$, is the set of equivalence classes under $\simeq$ of derivable expressions of dimension i, composition of two i-cells f and g is defined by $f *_i g = f *_{-(n-i)} g$, and the identity on f is 1_f.

D.1.12 Lemma. *The n-category C^P is well-defined.*

D.1.13 Theorem. *The n-category C^P is isomorphic to P^*.*

D.2 Alternative Syntax for n-Categories

D.2.1 Composition in maximal codimension. In an n-category C, an i-cell x is *k-composable* with a j-cell y, with $0 \leqslant k < i \wedge j$, whenever $t_k x = s_k y$.

In this case, we can extend the composition operation to cells which do not necessarily have the same dimension, and define their *k-composite* as

$$x *_k y = 1_{i \vee j}(x) *_k 1_{i \vee j}(y)$$

which is a $(i \vee j)$-cell. It is moreover useful to adopt the following convention. We write $x * y$ to mean that we compose x and y in the maximal possible dimension:

$$x * y = x *_{(i \wedge j)-1} y.$$

We say that x and y are *composable*, when this composite is defined, i.e.,

$$t_{(i \wedge j)-1} x = s_{(i \wedge j)-1} y.$$

Perhaps surprisingly, all the compositions can be recovered from those compositions in maximal codimension:

D.2.2 Lemma. *Given cells x and y of respective dimensions i and j, and $0 \leqslant k < i \vee j - 1$, we have*

$$\begin{aligned} x *_k y &= (x * s_{k+1} y) *_{k+1} (t_{k+1} x * y) \\ &= (s_{k+1} x * y) *_{k+1} (x * t_{k+1} y) \end{aligned}$$

which allows to compute any composition by recurrence on k.

Proof. We have

$$\begin{aligned} x *_k y &= (x *_{k+1} 1_{t_{k+1} x}) *_k (1_{s_{k+1} y} *_{k+1} y) && \text{identity is neutral} \\ &= (x *_k s_{k+1} y) *_{k+1} (t_{k+1} x *_k y) && \text{exchange law} \\ &= (x * s_{k+1} y) *_{k+1} (t_{k+1} x * y) && \text{definition of } * \end{aligned}$$

and similarly for the second equality. □

In fact, the whole structure of n-category can be axiomatized using this operation [264, Section 8], as follows. Given an $(i+1)$-cell x, we write sx and tx instead of $s_i x$ and $t_i x$, respectively.

D.2.3 Proposition. *Given $n \in \mathbb{N} \cup \{\omega\}$, an n-category C consists of an n-globular set equipped with composition and identity partial operations as follows. Two cells $x \in C_i$ and $y \in C_j$, with $0 \leqslant i, j \leqslant n$ are said to be* composable *when*

$$t_{(i \wedge j)-1}(x) = s_{(i \wedge j)-1}(y).$$

The operations are

- compositions: *for all composable cells* $x \in C_i$ *and* $y \in C_j$, *there is an* $(i \vee j)$*-cell*

$$x * y$$

- identities: *for all* $x \in C_i$, *with* $0 \leqslant i < n$, *there is an* $(i{+}1)$*-cell*

$$1_x$$

and should satisfy

- sources and targets of compositions: *for all composable cells* $x \in C_i$ *and* $y \in C_j$, *with* $0 \leqslant i, j \leqslant n$,

$$s(x * y) = \begin{cases} sx * y & \text{if } i > j, \\ sx & \text{if } i = j, \\ x * sy & \text{if } i < j, \end{cases} \qquad t(x * y) = \begin{cases} tx * y & \text{if } i > j, \\ ty & \text{if } i = j, \\ x * ty & \text{if } i < j, \end{cases}$$

- sources and targets of identities: *for all* $x \in C_i$, *with* $0 \leqslant i < n$,

$$s(1_x) = x = t(1_x)$$

- associativity: *for all* $x \in C_i$, $y \in C_j$ *and* $z \in C_k$ *with* x *and* y *compatible,* y *and* z *compatible, and either* $i = j \leqslant k$ *or* $i = k \leqslant j$ *or* $j = k \leqslant i$,

$$(x * y) * z = x * (y * z)$$

- distributivity: *for all* $x \in C_i$, $y \in C_j$ *and* $z \in C_k$

$$x * (y * z) = (x * y) * (x * z)$$

if $i < j$ *and* $i < k$, *and* (y, z), (x, y), *and* (x, z) *are compatible, and*

$$(x * y) * z = (x * z) * (y * z)$$

if $i > k$ *and* $j > k$, *and* (x, y), (x, z), *and* (y, z) *are compatible.*
- unitality: *for all* $x \in C_i$ *and* $y \in C_j$ *with* $0 \leqslant i, j \leqslant n$,

$$1_x * y = \begin{cases} y & \text{if } i + 1 \leqslant j, \\ 1_{x*y} & \text{if } i + 1 > j, \end{cases} \qquad x * 1_y = \begin{cases} x & \text{if } i \geqslant j + 1, \\ 1_{x*y} & \text{if } i < j + 1, \end{cases}$$

whenever 1_x *and* y *(resp.* x *and* 1_y*) are composable and* $i < n$ *(resp.* $j < n$*),*
- commutativity: *for all* $x \in C_i$ *and* $y \in C_j$, *with* $t_{k-1}x = s_{k-1}y$, *where* $k = (i \wedge j) - 1$ *is supposed to satisfy* $k > 0$,

$$(x * s_k(y)) * (t_k(x) * y) = (s_k(x) * y) * (x * t_k(y)).$$

D.3 The Word Problem for Free n-Categories

D.3.1 The word problem. Given an n-polygraph P, the *word problem* for P consists in finding an algorithmic answer to the following decision problem:

Given two derivable expressions f and g, do we have $f \simeq g$?

It was shown by Makkai [264, Section 10] that this problem is decidable. We provide here the main arguments of his construction, after recalling the required notions and tools.

D.3.2 Multisupport. Given two multisets μ and ν with common domain (see §1.4.1), we write

- $\mu + \nu$ for their pointwise sum (noted $\mu \sqcup \nu$ in §1.4.1),
- $\mu - \nu$ for their pointwise difference (we only use this operation in situations where $\mu(x) \geqslant \nu(x)$ for every element x of the domain),
- $\mu \leqslant \nu$ whenever $\mu(x) \leqslant \nu(x)$ for every element x of the domain.

The *multisupport* $\mathrm{supp}^\sharp(f)$ of a derivable expression f is the multiset on $\bigsqcup_{i \leqslant n} P_i$ consisting of all occurrences of generators in f. Formally, $\mathrm{supp}^\sharp(f)$ is defined by induction on f as

$$\begin{cases} \{x\} & \text{if } f = x \text{ is a generator of type } \star, \\ \{x\} + \mathrm{supp}^\sharp(g) + \mathrm{supp}^\sharp(h) & \text{if } f = x \text{ is a generator of type } f \underset{T}{\to} g, \\ \mathrm{supp}^\sharp(g) & \text{if } f = 1_g, \\ \mathrm{supp}^\sharp(g) - \mathrm{supp}^\sharp(t_{-i}(g)) + \mathrm{supp}^\sharp(h) & \text{if } f = g *_{-i} h. \end{cases}$$

This definition is invariant under equivalence of expressions:

D.3.3 Lemma. *Given two derivable expressions f and g such that $f \simeq g$, we have* $\mathrm{supp}^\sharp(f) = \mathrm{supp}^\sharp(g)$.

D.3.4 Composition in maximal codimension. We recall the convention introduced in §D.2.1. Given two derivable expressions f and g, not necessarily of the same dimension, whose target and source coincide in dimension $k = (\dim(f) \wedge \dim(g)) - 1$, we write $f * g$ for their composite in dimension k, i.e., in their maximal codimension. Formally, we suppose that

$$t_{-(\dim(f)-k)}(f) = s_{-(\dim(g)-k)}(g)$$

and define

$$f * g = 1_f^{l-\dim(f)} *_{-(l-k)} 1_g^{l-\dim(g)}$$

which is a derivable expression of dimension $l = \dim(f) \vee \dim(g)$. By convention, $1^0_f = f$ and $1^{i+1}_f = 1_{1^i_f}$, and similarly for g.

D.3.5 Atoms and molecules. We mutually define notions of atom and molecule by recurrence on the dimension k as follows. Given $k \leqslant n$,

- an *atom* of dimension k is a derivable expression of the form

$$f_{k-1} * (f_{k-2} * (\dots (f_1 * x * g_1) \dots) * g_{k-2}) * g_{k-1} \tag{D.1}$$

  where x is a generator of dimension k (the *nucleus* of the atom) and each f_i and g_i is a molecule of dimension i,
- a *molecule* of dimension k is a derivable expression of the form

$$f_1 * \dots * f_p$$

  where each f_i is an atom of dimension k (in the case $p = 0$, it should be an expression of the form 1_f with f an expression of dimension $k - 1$).

In the following, we sometimes say *k-atom* (resp. *k-molecule*) for an atom (resp. molecule) of dimension k. Clearly, every k-atom is a particular k-molecule. Note that the notion of atom corresponds to the one of rewriting step (§16.7.1) and the notion of molecule to the one of rewriting path (§16.7.2).

Molecules provide canonical representatives of derivable expressions up to equivalence, in the sense that each equivalence class contains at least a molecule, see [264, Proposition 8.(12)] and Proposition 16.7.3:

D.3.6 Proposition. *Every k-expression is equivalent to a k-molecule.*

D.3.7 Example. In the polygraph P corresponding to the diagram

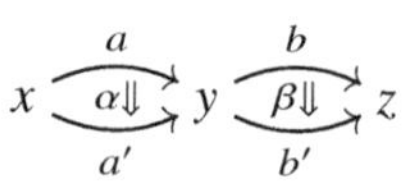

the composite $\alpha *_{-2} \beta$ is equivalent to the molecules

$$(1_x * \alpha * b) * (a' * \beta * 1_z) \qquad \text{and} \qquad (a * \beta * 1_z) * (1_x * \alpha * b').$$

D.3.8 Equivalence of molecules. Two molecules are equivalent when one can be obtained from the other by applying the exchange relation on adjacent atoms. Formally, we define an equivalence relation $\sim$ on k-molecules, as the smallest equivalence relation such that,

1. for all molecules f and h, and atoms g and g', we have

$$g \simeq g' \qquad \text{implies} \qquad f * g * h \sim f * g' * h$$

whenever all composites are defined,

2. for all k-atoms $f\colon f' \underset{T}{\to} f''$ and $g\colon g' \underset{U}{\to} g''$, for all k-atoms f_1, g_1, f_2, g_2 such that

$$f_1 = f * g' \qquad g_1 = f'' * g \qquad f_2 = f * g'' \qquad g_2 = f' * g \tag{D.2}$$

and for all k-molecules h and h', we have

$$h * f_1 * g_1 * h' \sim h * g_2 * f_2 * h'.$$

Graphically,

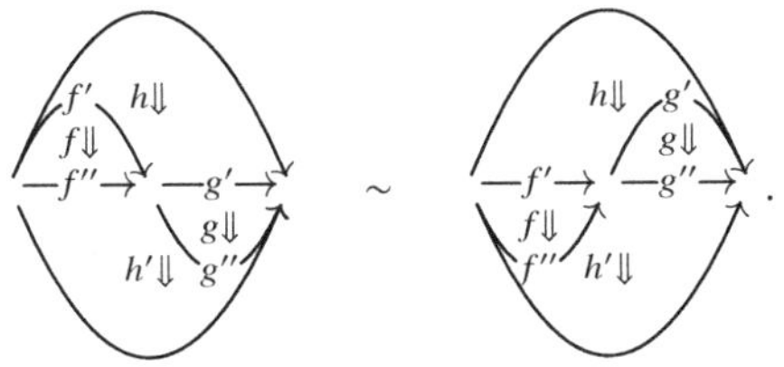

It is shown in [264, Section 9]:

D.3.9 Proposition. *Given two k-molecules f and g, we have $f \simeq g$ if and only if $f \sim g$.*

D.3.10 Contexts. Given $k > 0$ and two molecules h and h' with common type T, with $\dim(h) = \dim(h') = k - 1$, a *context* of type $h \underset{T}{\to} h'$ is a k-expression in which a formal variable of type $h \underset{T}{\to} h'$ occurs exactly once. Formally, we consider the polygraph P' obtained from P, by adding a new generator $\underline{x}$, i.e., $P'_k = P_k \sqcup \{\underline{x}\colon h \to h'\}$ and $P'_i = P_i$ for $i \neq k$. A context c is then a derivable expression in $\mathsf{CAT}^{P'}$ such that $\operatorname{supp}^\#(c)(\underline{x}) = 1$. The notion of context corresponds to the one already introduced in Section 16.5.

To every $(k{+}1)$-atom f of the form (D.1), with top dimensional generator $x\colon h \underset{T}{\to} h'$ of dimension $k + 1$, one can associate the k-context

$$f_{k-1} * (f_{k-2} * (\dots (f_1 * \underline{x} * g_1) \dots) * g_{k-2}) * g_{k-1}$$

where $\underline{x}$ is a formal variable of dimension k and of type T. We then say that f can be obtained by *substituting* the variable $\underline{x}$ by the generator x in c, what we write $f = c[x]$.

D.3.11 Example. Consider a 3-atom $f_2 * (f_1 * x * g_1) * g_2$:

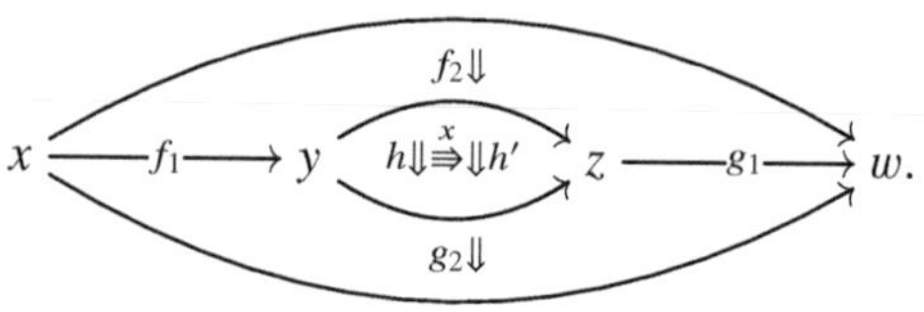

The associated context is $f_2 * (f_1 * \underline{x} * g_1) * g_2$:

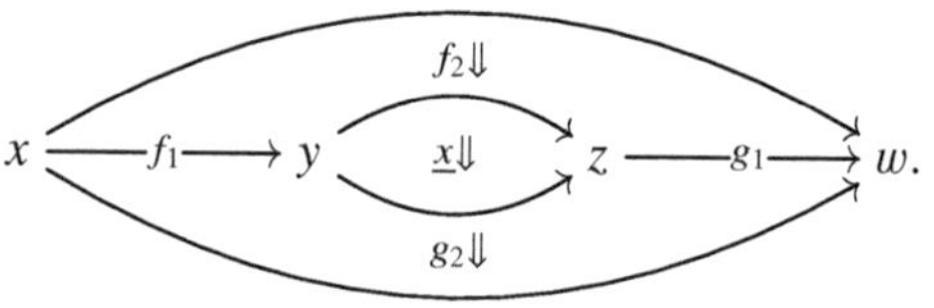

Equality on atoms can be characterized in the expected way, see [264, Section 10.(6)]:

D.3.12 Proposition. *Given two* $(k+1)$*-atoms* $f = c[x]$ *and* $g = d[y]$*, we have* $f \simeq g$ *if and only if* $x = y$ *and* $c \simeq d$.

D.3.13 Finiteness of equivalence classes. The main argument in order to show that the word problem is decidable is the observation that there is only a finite number of molecules with given generators of given multiplicities, see [264, Lemma 10.(10)]:

D.3.14 Proposition. *Suppose given a multiset* μ *on* $\bigsqcup_{i \leqslant n} P_i$*. There is a finite number of molecules* f *such that* $\mathrm{supp}^\sharp(f) \leqslant \mu$.

D.3.15 Decidability results. As expected, the proof of decidability of the word problem is performed by recurrence on the dimension k of the two cells f and g we aim at comparing. The base case is immediate, so that we focus on the inductive case. By Proposition D.3.6, whose proof is constructive, we can suppose that f and g are k-molecules. In Proposition D.3.9, we have reduced the equivalence of the molecules f to the existence of a sequence of molecules

$$f = f_1 \sim f_2 \sim \ldots \sim f_p = g \tag{D.3}$$

such that f_{i+1} can be obtained from f_i by exchanging adjacent atoms, which can be tested using equivalence of k-atoms for some decompositions. Moreover, in Proposition D.3.12, we have reduced the equivalence of two k-atoms to the equivalence of $(k-1)$-molecules (the associated contexts). In order to show our decidability result, it is therefore enough to show that

- the decompositions (D.2) in order to identify the possible exchange relations can be found in a finite search space,
- the sequence of molecules (D.3) can be found in a finite search space.

Both properties follow from the facts that the multisupport is preserved under equivalence (Lemma D.3.3) and there is only a finite number of molecules with given multisupport (Proposition D.3.14).

D.3.16 Theorem. *The word problem for polygraphs is decidable.*

As a consequence, it can be shown that

D.3.17 Theorem. *Whether an expression is derivable or not is decidable.*

An actual implementation was performed by Forest [133, 134].

Appendix E

Complexes and Homology

In this chapter, we recall basic notions on modules, abelian resolutions, and homology. We refer the reader to [241, 260, 317] for a deeper presentation and the proofs of the given results. Throughout the chapter, we fix a ring R, and we denote by 1_R the multiplicative identity in R.

E.1 Modules Over a Ring

E.1.1 Modules over a ring. A *left-R-module* consists of an abelian group $(V, +)$ together with a *left scalar multiplication*

$$\cdot\colon R \times V \to V$$

satisfying for all $r, s \in R$ and $x, y \in V$ the following four relations:

$$r \cdot (x + y) = r \cdot x + r \cdot y, \qquad (rs) \cdot x = r \cdot (s \cdot x),$$
$$(r + s) \cdot x = r \cdot x + s \cdot x, \qquad 1_R \cdot x = x.$$

In the following, we often write rx instead of $r \cdot x$. We will say *R-module*, or *module*, for a left R-module. The notion of *right R-module* can be defined similarly, based on a right action $\cdot\colon G \times R \to G$, satisfying dual axioms. All the notions presented in this appendix are defined in a similar fashion for right R-modules since every right R-module is a left R^{op}-module, where R^{op} is the opposite ring.

E.1.2 If V and W are two left-R-modules, a *morphism of R-modules* $f\colon V \to W$ is a morphism of abelian groups satisfying, for any $r \in R$ and $x \in V$,

$$f(r \cdot x) = r \cdot f(x).$$

The left-R-modules and their morphisms form a category denoted by $\mathbf{Mod}_R$. We denote by $\mathbf{Mod}_R(V, W)$ the abelian group of morphisms from V to W.

For a morphism $f\colon V \longrightarrow W$, we will denote by

$$\ker f := \{x \in V \mid f(x) = 0\},$$

the *kernel* of f and by

$$\operatorname{im} f := \{y \in W \mid y = f(x) \text{ for some } x \text{ in } V\},$$

the *image* of f. We will denote by $\operatorname{coker} f := W/\operatorname{im} f$ the *cokernel* of f.

E.1.3 Exact sequences. A pair of composable morphisms of modules

$$V' \xrightarrow{f} V \xrightarrow{g} V''$$

is *exact at* V if $\operatorname{im} f = \ker g$. A sequence of morphisms of modules

$$\cdots \longrightarrow V_{n+1} \xrightarrow{d_{n+1}} V_n \xrightarrow{d_n} V_{n-1} \longrightarrow \cdots$$

is *exact* if each adjacent pair (d_{i+1}, d_i) of morphisms is exact at V_i.

E.1.4 Example. The sequences

$$0 \longrightarrow V \xrightarrow{f} V', \qquad V \xrightarrow{f} V' \longrightarrow 0\,, \quad \text{and} \quad 0 \longrightarrow V \xrightarrow{f} V' \longrightarrow 0$$

are exact if and only if the morphism f is injective, surjective, and bijective, respectively. If the sequence $V' \xrightarrow{f} V \xrightarrow{g} V''$ is exact with f surjective and g injective, then V is the zero module.

E.1.5 Free modules. A R-module V is *free* if it is a direct sum of copies of R. If $V = \coprod_{i \in I} Rx_i$, with $R \simeq Rx_i$, the set $\{x_i \mid i \in I\}$ is called a *basis* of V. It follows that each element x in V has a unique decomposition

$$x = \sum_{i \in I} \lambda_i x_i,$$

where λ_i is in R and almost all λ_i are zero. Given a set X, there exists a free R-module having X as a basis, which is usually denoted $R[X]$.

E.1.6 Lemma. *Let $X = \{x_i \mid i \in I\}$ be a basis of a free module V. For every module W and every map $f\colon X \longrightarrow W$, there is a unique morphism of R-modules $\tilde{f}\colon V \longrightarrow W$ extending f, i.e., making the following triangle commute:*

$$\begin{array}{ccc} X & \longrightarrow & V \\ & {\scriptstyle f}\searrow & \downarrow{\scriptstyle \tilde{f}} \\ & & W \end{array}$$

where the morphism $X \longrightarrow V$ is the canonical inclusion.

One shows that every R-module is a quotient of a free R-module. As a consequence, any R-module V may be described by *generators* and *relations* in the following way. Given a free R-module F with basis X and a surjective morphism of R-modules $f\colon F \to V$, we say that X is a set of *generators of* V and the kernel $\ker f$ is called its submodule of *relations*.

E.1.7 Finitely generated modules. A R-module V is *finitely generated* if there is a finite subset $X = \{x_1, x_2, \ldots, x_n\}$ of V such that for every element x of V, there exist $r_1, r_2, \ldots, r_n$ in R with $x = r_1x_1 + r_2x_2 + \ldots + r_nx_n$. Then the set X is referred to as a *generating set* for V. The finite generators need not form a basis, since they need not be linearly independent over R. An R-module V is finitely generated if and only if there is a surjective morphism:

$$R^n \to V$$

for some n. That is, V is a quotient of a free module of finite rank.

E.1.8 Proposition. *Let F, V, and W be R-modules. If F is free, $e\colon V \to W$ is a surjective morphism and $f\colon F \to W$ is any morphism, then there exists a morphism $\tilde{f}\colon F \to V$ making the following triangle commute*

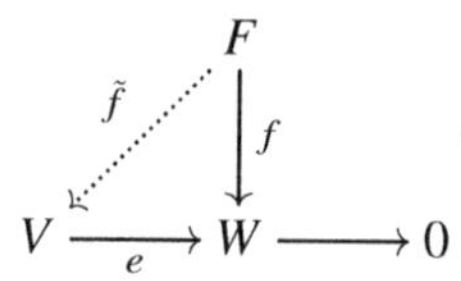

As a consequence, for any free R-module, the functor

$$\mathbf{Mod}_R(F, -)\colon \mathbf{Mod}_R \to \mathbf{Ab}$$

is *exact*, that is, for any exact sequence

$$0 \to V' \to V \to V'' \to 0$$

of R-modules, the induced sequence

$$0 \to \mathbf{Mod}_R(F, V') \to \mathbf{Mod}_R(F, V) \to \mathbf{Mod}_R(F, V'') \to 0$$

is exact.

E.1.9 Projective module. A projective module is a module which behaves as the free module F in Proposition E.1.8. More explicitly, a R-module P is *projective* if whenever $e\colon V \to W$ is a surjective morphism and $f\colon P \to W$

is any morphism, there exists a morphism $\tilde{f}\colon P \longrightarrow V$ making the following triangle commute:

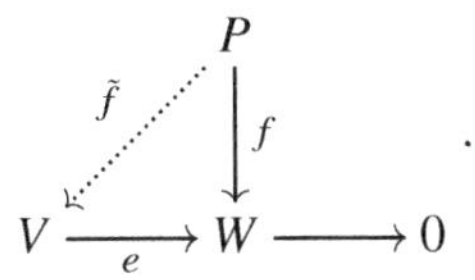

In particular, any free module is projective. The following result gives several ways to characterize projective modules.

E.1.10 Proposition. *The following conditions are equivalent for a R-module P:*

(i) *P is projective,*
(ii) *if $f\colon V \longrightarrow P$ is a surjective morphism, then there exists $h\colon P \longrightarrow V$ such that $fh = 1_P$,*
(iii) *if $f\colon V \longrightarrow P$ is a surjective morphism, then $V \simeq P \oplus \ker f$,*
(iv) *the functor $\mathbf{Mod}_R(P, -)\colon \mathbf{Mod}_R \longrightarrow \mathbf{Ab}$ is exact,*
(v) *P is a summand of a free module, that is, there exists a free R-module F such that $F \simeq P \oplus Q$, for some R-module Q.*

Note that the R-module Q in (v) is necessarily projective.

E.2 Chain Complexes

We recall in this section the notion of chain complex which is the fundamental object of study in homological algebra.

E.2.1 Chain complex. A *chain complex* in the category $\mathbf{Mod}_R$ is a sequence $(V_n)_{n\in\mathbb{Z}}$ of R-modules, together with a sequence $(d_n)_{n\in\mathbb{Z}}$ of morphisms

$$\cdots \longrightarrow V_{n+1} \xrightarrow{d_{n+1}} V_n \xrightarrow{d_n} V_{n-1} \longrightarrow \cdots$$

with the composite of adjacent morphisms being zero, that is $d_n d_{n+1} = 0$, for every n in $\mathbb{Z}$. Such a complex is denoted (V_*, d_*), or simply V_* or V. We say that V_n is the module of the complex V_* in *degree n*.

Given two chain complexes (V_*, d_*) and (V'_*, d'_*), a *chain map*

$$f\colon (V_*, d_*) \longrightarrow (V'_*, d'_*)$$

is a family of morphisms of R-modules $(f_n\colon V_n \longrightarrow V'_n)_{n\in\mathbb{Z}}$ making the following diagrams commute

$$\begin{array}{ccccccccc}
\cdots \longrightarrow & V_{n+1} & \xrightarrow{d_{n+1}} & V_n & \xrightarrow{d_n} & V_{n-1} & \longrightarrow \cdots \\
& \downarrow f_{n+1} & & \downarrow f_n & & \downarrow f_{n-1} & \\
\cdots \longrightarrow & V'_{n+1} & \xrightarrow{d'_{n+1}} & V'_n & \xrightarrow{d'_n} & V'_{n-1} & \longrightarrow \cdots .
\end{array}$$

Chain complexes of R-modules and chain maps form a category that we will denote by $\mathbf{Ch}_R$.

A chain complex V is *positive* if $V_n = 0$ for every $n < 0$. A positive complex looks like a sequence

$$\cdots \longrightarrow V_n \xrightarrow{d_n} V_{n-1} \longrightarrow \cdots \longrightarrow V_2 \xrightarrow{d_2} V_1 \xrightarrow{d_1} V_0.$$

Positive chain complexes of R-modules form a full subcategory of $\mathbf{Ch}_R$ denoted by $\mathbf{Ch}_{R,\geqslant 0}$.

E.2.2 Homology. Let V be a chain complex in $\mathbf{Mod}_R$, and $n \in \mathbb{Z}$. The morphisms d_n are called *boundary maps* of V, and the elements of the module V_n are called *n-chains*. We denote by $Z_n(V) = \ker d_n$ the module of *n-cycles* of V and by $B_n(V) = \operatorname{im} d_{n+1}$ the module of *n-boundaries* of V.

In the category $\mathbf{Mod}_R$, the equation $d_n d_{n+1} = 0$ is equivalent to the condition $B_n(V) \subseteq Z_n(V)$, that is any n-boundary of V is an n-cycle of V. The *n-th homology module of* V is the R-module defined as the quotient

$$H_n(V) = Z_n(V)/B_n(V).$$

An element of $H_n(V)$ is a coset $c + B_n(V)$, and called a *n-th homology class* of V.

Given a chain map $f: V \to V'$, and $n \in \mathbb{Z}$, we define a morphism of modules

$$H_n(f): H_n(V) \to H_n(V')$$

by setting $H_n(f)(c + B_n(V)) = f(c) + B_n(V')$. Then

$$H_n: \mathbf{Ch}_R \to \mathbf{Mod}_R$$

is a functor called *n-th homology functor*. We refer to [317, Proposition 6.8] for a detailed proof.

E.2.3 Acyclic complex. A complex (V, d) is an *exact sequence* if $H_n(V) = 0$ for every $n \in \mathbb{Z}$. This means that no n-cycles that are not n-boundaries. In this way, homology of the complex V measures the deviation of V from being an exact sequence. An exact sequence is also called an *acyclic complex*.

E.2.4 Chain homotopy. Let $f, g \colon V \to V'$ be two chain maps. We say that f and g are *(chain) homotopic*, denoted by $f \simeq g$, if there exists a family of morphisms $(s_n \colon V_n \to V'_{n+1})_{n\in\mathbb{Z}}$ such that the following relation holds for every $n \in \mathbb{Z}$

$$f_n - g_n = d'_{n+1}s_n + s_{n-1}d_n.$$

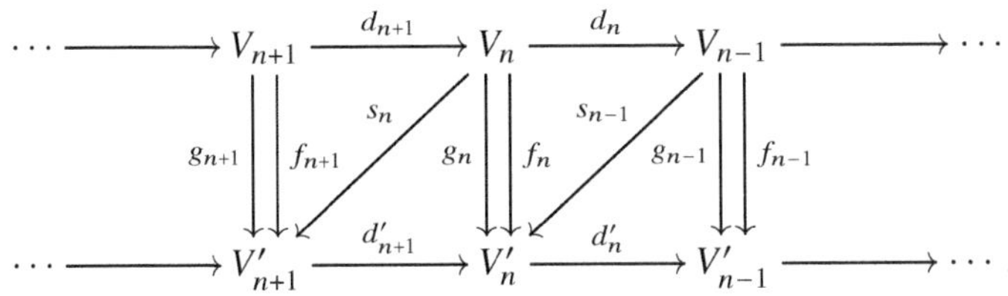

A chain map $f \colon V \to V'$ is *null-homotopic* if $f \simeq 0$, where 0 is the zero chain map.

It is easy to see that chain homotopy defines an equivalence relation between chain maps. Moreover, two chain homotopic maps induce isomorphisms in homology as states by the following result.

Two complexes V and V' are of the *same homotopy type*, or *homotopic*, if there exist chain maps

$$f \colon V \to V', \quad \text{and} \quad g \colon V' \to V,$$

such that $gf \simeq 1_V$ and $fg \simeq 1_{V'}$. In that case, the chain maps f and g are called *homotopy equivalences*.

E.2.5 Proposition. *Let $f, g \colon V \to V'$ be two chain maps such that $f \simeq g$, then for every $n \in \mathbb{Z}$,*

$$H_n(f) = H_n(g) : H_n(V) \to H_n(V').$$

Proof. Let $(s_n \colon V_n \to V'_{n+1})_{n\in\mathbb{Z}}$ be a homotopy between f and g. Given $x \in \ker d_n$, we have

$$(f_n - g_n)(x) = d'_{n+1}s_n(x) + s_{n-1}d_n(x) = d'_{n+1}s_n(x).$$

Hence, $f_n(x) - g_n(x) \in \operatorname{im} d'_{n+1}$ and $f_n(x) = g_n(x)$ holds in $H_n(V')$. □

E.2.6 Contracting homotopy. A chain complex (V, d) has a *contracting homotopy* if its identity $1_V = (1_{V_n} \colon V_n \to V_n)_{n\in\mathbb{Z}}$ is null-homotopic. That is, there is a family of morphisms $(i_n \colon V_n \to V_{n+1})_{n\in\mathbb{Z}}$ such that the following condition holds for every $n \in \mathbb{Z}$

$$1_{V_n} = d_{n+1}i_n + i_{n-1}d_n. \tag{E.1}$$

E.2.7 Proposition. *A complex V having a contracting homotopy is acyclic.*

Proof. Prove that $H_n(V) = 0$ for every $n \in \mathbb{Z}$. Which is the amount of proving the inclusion $Z_n(V) \subseteq B_n(V)$. This is a direct consequence of relation (E.1). Indeed, for any x in $Z_n(V)$, we have $x = d_{n+1}i_n(x)$. This proves the acyclicity of V. □

E.3 Resolutions

E.3.1 Resolutions. A *resolution* of an R-module V is an exact sequence of R-modules

$$\cdots \longrightarrow V_n \xrightarrow{d_n} V_{n-1} \longrightarrow \cdots \longrightarrow V_1 \xrightarrow{d_1} V_0 \xrightarrow{\varepsilon} V \longrightarrow 0. \qquad \text{(E.2)}$$

From the definition, the morphism ε is surjective and, for every $n \in \mathbb{N}$, we have

$$\operatorname{im} d_1 = \ker \varepsilon, \quad \text{and} \quad \operatorname{im} d_{n+1} = \ker d_n. \qquad \text{(E.3)}$$

Such a resolution is *projective* (resp. *free*) if all the modules V_n are projective (resp. free). Given a natural number n, a *partial resolution of length n of V* is defined in a similar way but with a bounded sequence $(V_k)_{0 \leqslant k \leqslant n}$ of R-modules:

$$V_n \xrightarrow{d_n} V_{n-1} \longrightarrow \cdots \longrightarrow V_1 \xrightarrow{d_1} V_0 \xrightarrow{\varepsilon} V \longrightarrow 0.$$

Note that by Proposition E.2.7 a way to prove that a complex (E.2) is a resolution of V is to construct a contracting homotopy

$$\cdots \longleftarrow V_{n+1} \xleftarrow{i_{n-1}} V_{n-1} \longleftarrow \cdots \longleftarrow V_1 \xleftarrow{i_0} V_0 \xleftarrow{i_{-1}} V$$

such that $d_0 i_{-1} = 1_V$.

E.3.2 Proposition. *Every R-module V has a free resolution.*

Proof. We take $V_0 = R[V]$ to be the free R-module generated by V and the morphism $\varepsilon \colon V_0 \to V$ to be the morphism extending the identity on V (whose existence is asserted in Lemma E.1.6). The morphism ϵ is clearly surjective, i.e., satisfies $\operatorname{im} \varepsilon = V$.

We then define $V_1 = R[\ker \varepsilon]$ to be the free module generated by $\ker \varepsilon$ and d_1 to be the morphism extending the canonical inclusion $\ker \varepsilon \to V_0$:

$$\begin{array}{ccccc} R[\ker \varepsilon] & \xrightarrow{d_1} & V_0 & \xrightarrow{\varepsilon} & V. \\ \uparrow & \nearrow & & & \\ \ker \varepsilon & & & & \end{array}$$

Similarly, by recurrence, if V_n and $d_n \colon V_n \to V_{n-1}$ are defined for $n \geqslant 1$, we set $V_{n+1} = R[\ker d_n]$ and $d_{n+1} \colon V_{n+1} \to V_n$ to be the extension of the inclusion $\ker d_n \to V_n$. The relations (E.3) are easily seen to be satisfied. □

E.3.3 Theorem. *Let*

$$\cdots \longrightarrow P_n \xrightarrow{d_n} P_{n-1} \longrightarrow \cdots \longrightarrow P_1 \xrightarrow{d_1} P_0$$

be a projective complex and

$$V_n \xrightarrow{d_n} V_{n-1} \longrightarrow \cdots \longrightarrow V_1 \xrightarrow{d_1} V_0$$

be an acyclic complex. Then, for every morphism $f \colon H_0(P_*) \to H_0(V_*)$, there exists a chain map $\tilde{f} \colon P_* \to V_*$ inducing f. Moreover, two chain maps inducing f are homotopic.

We refer to [182, Theorem 4.1] for a detailed proof of this result. In the same way as the proof of Proposition E.3.2, we prove that every R-module V has a projective resolution. Indeed, there exists a projective presentation of $V\colon\ 0 \to V_1 \to P_0 \to V \to 0$. Then there exists a projective presentation of $V_1\colon\ 0 \to V_2 \to P_1 \to V_1 \to 0$, and so on. This gives us the following projective and acyclic complex P_*:

$$\cdots \longrightarrow P_n \xrightarrow{d_n} P_{n-1} \longrightarrow \cdots \longrightarrow P_1 \xrightarrow{d_1} P_0$$

where the morphism d_n is defined as the composite $P_n \to V_n \to P_{n-1}$. As $H_0(P_*) = V$, this is a resolution a resolution of V.

From Theorem E.3.3 we deduce the following result, see [182, Theorem 4.3] for a detailed proof.

E.3.4 Proposition. *Two projective resolutions of a module V are of the same homotopy type.*

E.3.5 Quasi-isomorphisms and homotopy equivalences. Let V and V' be two chain complexes. A chain map $f \colon V \to V'$ is a *quasi-isomorphism* if it induces an isomorphism in homology, meaning that the induced map

$$H(f) \colon H(V) \to H(V')$$

is an isomorphism of graded modules.

A sufficient condition for a chain map $f \colon V \to V'$ to be a quasi-isomorphism is to be a *homotopy equivalence* in the sense that it admits an inverse up to homotopy, that is, a chain map $g \colon V' \to V$ such that gf and 1_V are homotopic, and fg and $1_{V'}$ are homotopic. The fact that a homotopy equivalence is a quasi-isomorphism follows from Proposition E.2.5.

E.3.6 Resolutions as quasi-isomorphisms. Note that a projective resolution (E.2) induces a chain map

$$\begin{array}{ccccccccc}
\cdots \longrightarrow & V_n & \xrightarrow{d_n} & V_{n-1} & \longrightarrow \cdots \longrightarrow & V_1 & \xrightarrow{d_1} & V_0 \\
 & \downarrow & & \downarrow & & \downarrow & & \downarrow \varepsilon \\
\cdots \longrightarrow & 0 & \longrightarrow & 0 & \longrightarrow \cdots \longrightarrow & 0 & \longrightarrow & V
\end{array}$$

between the chain complex (V_n, d_n) of projective R-modules and the chain complex consisting of V in degree 0 and the R-module 0 in other degrees.

E.3.7 Schanuel's lemma and finite homological type. Given two exact sequences of R-modules

$$0 \to K_1 \to P_1 \to V \to 0,$$
$$0 \to K_2 \to P_2 \to V \to 0,$$

where P_1 and P_2 are projective, Schanuel's lemma states that there is an isomorphism:

$$K_1 \oplus P_2 \quad \simeq \quad K_2 \oplus P_1.$$

We refer the reader to [317] for a detailed proof and applications of this result. The following proposition is an important generalization of this result which is useful for defining the finite homology type of modules in §F.3.1.

E.3.8 Proposition (Generalized Schanuel's Lemma). *Given two exact sequences of R-modules*

$$0 \to K \to P_k \to P_{k-1} \to \cdots \to P_1 \to P_0 \to V \to 0,$$
$$0 \to L \to Q_k \to Q_{k-1} \to \cdots \to Q_1 \to Q_0 \to V \to 0,$$

where all the P_i and Q_i are projective, we write

$$P_{\mathrm{od}} := \bigoplus_{i \text{ odd}} P_i, \quad P_{\mathrm{ev}} := \bigoplus_{i \text{ even}} P_i, \quad Q_{\mathrm{od}} := \bigoplus_{i \text{ odd}} Q_i, \quad \text{and} \quad Q_{\mathrm{ev}} := \bigoplus_{i \text{ even}} Q_i.$$

The following properties hold:

1. *If k is even, then $K \oplus Q_{\mathrm{ev}} \oplus P_{\mathrm{od}} \simeq L \oplus Q_{\mathrm{od}} \oplus P_{\mathrm{even}}$.*
2. *If k is odd then $K \oplus Q_{\mathrm{od}} \oplus P_{\mathrm{ev}} \simeq L \oplus Q_{\mathrm{ev}} \oplus P_{\mathrm{od}}$.*

Let us mention an important consequence of Proposition E.3.8 which will be used to define the finite homological type in Section F.3.

E.3.9 Corollary. *Given two exact sequences of R-modules*

$$0 \to K \to P_k \to P_{k-1} \to \cdots \to P_1 \to P_0 \to V \to 0,$$
$$0 \to L \to Q_k \to Q_{k-1} \to \cdots \to Q_1 \to Q_0 \to V \to 0,$$

where all the P_i and Q_i are finitely generated and projective, then the R-module K is finitely generated if and only if L is finitely generated.

E.4 Homology of Monoids

In this section we recall the homology of a monoid with integral coefficients. In particular, the homology of monoids is used in Chapter 9 to define homological finiteness conditions for the existence of finite convergent presentations.

E.4.1 Tensor product of modules. Given a right R-module C and a left R-module D, recall that their *tensor product* $C \otimes_R D$ is the group obtained as the quotient of the free abelian group over the set $C \times D$, whose elements are pairs noted $x \otimes y$ with $x \in C$ and $y \in D$, under the relations

$$(x + x') \otimes y = x \otimes y + x' \otimes y, \qquad (x \cdot r) \otimes y = x \otimes (r \cdot y),$$
$$x \otimes (y + y') = x \otimes y + x \otimes y',$$

for $x, x' \in C$, $y, y' \in D$, and $r \in R$. This construction is functorial: given a right R-module C and a morphism $f \colon D \to D'$ of left R-modules, there is an induced group morphism

$$C \otimes_R f \colon C \otimes_R D \to C \otimes_R D'$$

and this construction is compatible with composition and identities.

Given a monoid M, the trivial $\mathbb{Z}M$-module $\mathbb{Z}$ is canonically a right module with the right action given by $n \cdot u = n$ for $n \in \mathbb{Z}$ and $u \in M$. In the following, we will mostly be interested in computing the tensor product of the trivial module $\mathbb{Z}$ with a free $\mathbb{Z}M$-module $\mathbb{Z}M[X]$ on a set X, in which case the result is

$$\mathbb{Z} \otimes_{\mathbb{Z}M} \mathbb{Z}M[X] = \mathbb{Z}[X]$$

where $\mathbb{Z}[X]$ is the free abelian group over X (which coincides with the free $\mathbb{Z}$-module).

E.4.2 Homology of monoids with integral coefficients. Let M be a monoid. To a free resolution of the trivial $\mathbb{Z}M$-module $\mathbb{Z}$ by left $\mathbb{Z}M$-modules

$$\cdots \longrightarrow F_{n+1} \xrightarrow{d_{n+1}} F_n \longrightarrow \cdots \longrightarrow F_1 \xrightarrow{d_1} F_0 \xrightarrow{\varepsilon} \mathbb{Z},$$

by tensoring with the trivial right $\mathbb{Z}M$-module $\mathbb{Z}$, we associate the complex of $\mathbb{Z}$-modules:

$$\cdots \longrightarrow \mathbb{Z} \otimes_{\mathbb{Z}M} F_{n+1} \xrightarrow{\widetilde{d}_{n+1}} \mathbb{Z} \otimes_{\mathbb{Z}M} F_n \longrightarrow \cdots \longrightarrow \mathbb{Z} \otimes_{\mathbb{Z}M} F_1 \xrightarrow{\widetilde{d}_1} \mathbb{Z} \otimes_{\mathbb{Z}M} F_0,$$

where $\widetilde{d_k}$ denotes the map $1_{\mathbb{Z}} \otimes_{\mathbb{Z}M} d_k$, for all $k \geqslant 1$. Note that the $\mathbb{Z}$-module $\mathbb{Z} \otimes_{\mathbb{Z}M} F_n$ is obtained from F_n by trivializing the action of M, that is F_n quotiented by all relations $ux = x$, for u in M and x in F_n. In particular, if $F_n = \mathbb{Z}M[X]$, then $\mathbb{Z} \otimes_{\mathbb{Z}M} F_n = \mathbb{Z}[X]$ is the free $\mathbb{Z}$-module on X. We obtain a chain complex, because $d_n d_{n+1} = 0$ induces that $\widetilde{d}_n \widetilde{d}_{n+1} = 0$.

We define the *n-th homology group* of M with integral coefficient $\mathbb{Z}$ as the following $\mathbb{Z}$-module:

$$\mathrm{H}_n(M, \mathbb{Z}) = \ker \widetilde{d}_n / \operatorname{im} \widetilde{d}_{n+1},$$

with the convention that $\widetilde{d}_0 = 0$. By definition, for any monoid M, we have $\mathrm{H}_0(M, \mathbb{Z}) \simeq \mathbb{Z}$. Following results of the previous section, the homology does not depend on the choice of the resolution used to compute it.

E.4.3 Proposition. *The groups* $\mathrm{H}_n(M, \mathbb{Z})$ *do not depend on a particular choice of a free resolution, but only on the monoid* M.

Appendix F
Homology of Categories

F.1 Simplicial Homology and Nerve of a Category

F.1.1 Simplicial sets. We will denote by Δ the simplicial category, that is, the full subcategory of the category of posets whose objects are the

$$[n] = \{0 < 1 < \cdots < n\}$$

for $n \geq 0$. By definition, a *simplicial set* is a presheaf on Δ, i.e., a functor $\Delta^{\mathrm{op}} \to \mathbf{Set}$. We will denote by $\widehat{\Delta}$ the category of simplicial sets. If X is a simplicial set, the set $X_n = X([n])$ is called the set of *n-simplices* of X.

F.1.2 Homology of a simplicial set. Let $n \geq 1$. For i such that $0 \leq i \leq n$, denote by δ^i the unique injective order-preserving map

$$\delta^i_n \colon [n-1] \to [n]$$

not reaching i. If X is a simplicial set, we will denote by

$$d^n_i \colon X_n \to X_{n-1}$$

the map $X(\delta^i_n)$.

We define a functor

$$c \colon \widehat{\Delta} \to \mathbf{Ch}_{\mathbb{Z},\geqslant 0}$$

in the following way. If X is a simplicial set, $c(X_n)$ is the free abelian group on X_n and, if x is an n-simplex of X with $n > 0$, we set

$$d_n([x]) = \sum_{i=0}^{n} (-1)^n [d^i_n(x)].$$

A classical calculation shows that $c(X)$ is indeed a chain complex. If $f \colon X \to Y$ is a morphism of simplicial sets, the morphism $c(f) \colon c(X) \to c(Y)$ is defined by sending $[x]$, for x an n-simplex of X, to $[f_n(x)]$, where f_n denotes $f_{[n]}$.

The *homology of a simplicial set* X is the homology of the chain complex $c(X)$.

F.1.3 The nerve functor. As the category of posets embeds into the category **Cat** of small categories, we get a functor $i\colon \Delta \to \mathbf{Cat}$. This functor i induces the so-called *nerve functor*

$$N\colon \mathbf{Cat} \to \widehat{\Delta}$$

sending a category C to the simplicial set $\mathbf{Cat}(i(-), C)$. Explicitly, a 0-simplex of $N(C)$ is an object of C and an n-simplex, for $n > 0$, is a chain of n composable arrows in C.

F.1.4 Homology of categories. If C is a category, the *homology* of C is the homology of its nerve $N(C)$. In particular, if M is a monoid (or even more particularly a group), by considering M as a category with one object, we get a notion of homology of M.

F.1.5 Theorem. *If M is a monoid, then the homology of M, seen as a category with one object, coincides with the homology $\mathrm{H}_n(M, \mathbb{Z})$ of the monoid as defined in Section E.4.*

Proof. If M is a monoid, then the complex $c(N(M))$ is canonically isomorphic to the complex $\mathbb{Z}\otimes_{\mathbb{Z}M} B$, where B is the bar resolution of M, see [260, Chapter X, Section 5], whence the result. □

F.2 Homology of Categories With Coefficients

F.2.1 Modules of a category. A *(left) module* of a category C, or *(left) C-module*, is a functor from C to the category **Ab** of abelian groups [287]. The C-modules and the natural transformations between them form an abelian category with enough projectives denoted by $\mathbf{Mod}(C)$.

We denote by $\mathbb{Z}C$ the *free $\mathbb{Z}$-category* on C whose 0-cells are the ones of C and for all 0-cells p and q of C, the hom-set $\mathbb{Z}C(p, q)$ is the free $\mathbb{Z}$-module on the hom-set $C(p, q)$. It follows immediately from the definitions that the category $\mathbf{Mod}(C)$ is isomorphic to the category of additive functors from $\mathbb{Z}C$ to **Ab**.

Given a C-module M, if $x \in M(p)$ for some 0-cell p of C, then we say that x is an *element* of M. The *left action* on M is defined for a 1-cell $f\colon p \to q$ and x in $M(p)$, by $f \cdot x = M(f)(x)$. A family $X = (x_i)_{i\in I}$ of elements of M is called a *family of generators* for M if every element x of $M(p)$ can be written as

$$x = \sum_{i \in I} f_i \cdot x_i,$$

for 1-cells $f_i \colon p_i \to p$ in $\mathbb{Z}C$, where all but a finite number of f_i are zero. This amounts to say that the natural transformation

$$\phi_X \colon \bigoplus_{i \in I} \mathbb{Z}C(p_i, -) \to M,$$

with $x_i \in M(p_i)$ and which takes 1_{p_i} to x_i is an epimorphism in $\mathbf{Mod}(C)$. The family X is a *basis* for M if the natural transformation ϕ_X is an isomorphism. A C-module M is *free* if it has a basis. It is *finitely generated* if it has a finite set of generators. A finitely generated module M is thus the cokernel of a morphism of finitely generated free modules:

$$\bigoplus_{i \in I} \mathbb{Z}C(q_i, -) \longrightarrow \bigoplus_{j \in J} \mathbb{Z}C(p_j, -) \longrightarrow M \longrightarrow 0.$$

F.2.2 Natural systems. Given a small category C, the *category of factorizations* [34, 308] is the category, denoted by FC, whose 0-cells are the 1-cells of C and whose 1-cells from w to w' are pairs (u, v) of 1-cells of C such that the following diagram commutes in C:

$$\begin{array}{ccc} \cdot & \xrightarrow{w} & \cdot \\ {\scriptstyle u}\big\uparrow & & \big\downarrow{\scriptstyle v} \\ \cdot & \xrightarrow[w']{} & \cdot \end{array}$$

The triple (u, w, v) is called a *factorization of* w'. Composition in the category FC is defined by pasting, i.e., if $(u, v) \colon w \to w'$ and $(u', v') \colon w' \to w''$ are 1-cells of FC, then the composite $(u', v')(u, v)$ is defined by the pair $(u'u, vv')$:

$$\begin{array}{ccc} \cdot & \xrightarrow{w} & \cdot \\ {\scriptstyle u}\big\uparrow & & \big\downarrow{\scriptstyle v} \\ \cdot & \xrightarrow{w'} & \cdot \\ {\scriptstyle u'}\big\uparrow & & \big\downarrow{\scriptstyle v'} \\ \cdot & \xrightarrow[w'']{} & \cdot \end{array}$$

The identity of w is the pair $(1_{s(w)}, 1_{t(w)})$:

$$\begin{array}{ccc} \cdot & \xrightarrow{w} & \cdot \\ {\scriptstyle 1_{s(w)}}\big\uparrow & & \big\downarrow{\scriptstyle 1_{t(w)}} \\ \cdot & \xrightarrow[w']{} & \cdot \end{array}$$

An FC-module D is called a *natural system of abelian groups* on C. We will denote by D_w the abelian group that is the image of w by the functor D. If there is no confusion, we denote by uav the image of an element a of D_w through the homomorphism of groups $D(u,v)\colon D_w \to D_{w'}$. The category **Mod**(FC) is also denoted by $\mathbf{Nat}(C, \mathbf{Ab})$.

F.2.3 Homology of categories with coefficients. The cohomology of categories with values in natural systems was defined by Baues and Wirsching in [34] and by Wells in [354]. It generalizes Hochschild-Mitchell cohomology of categories with coefficients in bimodules [287], the cohomology with coefficient in left modules [308, 314], and the cohomology of the classifying space, see [34] for the correspondences.

Dually, we define the homology of a category C with values in a *contravariant natural system* D on C, that is an $(FC)^{\mathrm{op}}$-module, as follows. We consider the nerve $N(C)$ of C defined in §F.1.3, with boundary maps denoted by $d_i\colon N_n(C) \to N_{n-1}(C)$, for $0 \leqslant i \leqslant n$. For $s = (u_1, \ldots, u_n)$ in $N_n(C)$, we denote by $\overline{s}$ the composite 1-cell $u_1 \cdots u_n$ of C. For n in $\mathbb{N}$, the n-th chain group $C_n(C,D)$ is defined as the abelian group

$$C_n(C,D) = \bigoplus_{s \in N_n(C)} D_{\overline{s}}.$$

We denote by ι_s the embedding of $D_{\overline{s}} \hookrightarrow C_n(C,D)$. We define a boundary map $d\colon C_n(C,D) \to C_{n-1}(C,D)$ on the component $D_{\overline{s}}$ of $C_n(C,D)$ by setting

$$d\iota_s = \iota_{d_0(s)} u_{1*} + \sum_{i=1}^{n-1} (-1)^i \iota_{d_i(s)} + (-1)^n \iota_{d_n(s)} u_n^*$$

for $s = (u_1, \ldots, u_n)$ in $N_n(C)$ and where u_{1*} and u_n^* denote the morphisms $D(u_1, 1)$ and $D(1, u_n)$, respectively. The homology of the category C with coefficients in D is defined as the homology of the complex $(C_*(C,D), d_*)$:

$$\mathrm{H}_*(C,D) = \mathrm{H}_*(C_*(C,D), d_*).$$

Denoting $\mathrm{Tor}_*^{FC}(D, -)$ the left derived functor of the functor $D \otimes_{FC} -$, we can prove that there is an isomorphism

$$\mathrm{H}_*(C,D) \simeq \mathrm{Tor}_*^{FC}(D, \mathbb{Z}).$$

natural in D.

F.3 Categories of Finite Homological Type

F.3.1 Modules of finite homological type. For $n \in \mathbb{N} \cup \{\infty\}$, a C-module M is of *homological type* FP_n (where FP_n stands for finitely n-presented) if it admits a partial resolution of length n by finitely generated projective C-modules

$$\cdots \longrightarrow P_n \longrightarrow P_{n-1} \longrightarrow \cdots \longrightarrow P_0 \longrightarrow M \longrightarrow 0.$$

By general homological arguments, see §E.3.7, we have the following characterization of the property FP_n which is a consequence of Corollary E.3.9.

F.3.2 Proposition. *Let C be a category, M be a C-module, and n be a natural number. The following assertions are equivalent:*

1. *M is of homological type* FP_n.
2. *M admits a partial resolution of length n by finitely generated free C-modules:*

$$F_n \longrightarrow F_{n-1} \longrightarrow \ \cdots \ \longrightarrow F_0 \longrightarrow M.$$

3. *M is finitely generated and, for every natural number $k < n$ and every projective finitely generated partial resolution of M of length k*

$$P_k \xrightarrow{d_k} P_{k-1} \longrightarrow \cdots \longrightarrow P_0 \longrightarrow M \longrightarrow 0,$$

the C-module $\ker d_k$ *is finitely generated.*

F.3.3 Categories of finite homological type. The property for a category C to be of homological type FP_n is defined according to a category of modules over one of the categories in the following diagram

$$\begin{array}{ccccccc} & & & & C^{\mathrm{op}} & & \\ & & & \overset{p_1}{\nearrow\!\!\!\!\nearrow} & & \overset{q_1}{\searrow} & \\ FC & \xrightarrow{\ \partial\ } & C^{\mathrm{op}} \times C & & & & C^{\top} \\ & & & \underset{p_2}{\searrow\!\!\!\!\searrow} & & \underset{q_2}{\nearrow} & \\ & & & & C & & \end{array}$$

where ∂ is the boundary map, p_1 and p_2 are the projections of the cartesian product, $C^{\top}$ is the enveloping groupoid of C, and q_1 and q_2 are the canonical inclusion morphisms. A category C is of *homological type*

- FP_n if the constant natural system $\mathbb{Z}$ is of type FP_n,
- *bi*-FP_n if the $C^{\mathrm{op}} \times C$-module $\mathbb{Z}C$ is of type FP_n,
- *left*-FP_n if the constant C-module $\mathbb{Z}$ is of type FP_n,
- *right*-FP_n if the constant C^{op}-module $\mathbb{Z}$ is of type FP_n,
- *top*-FP_n if the constant $C^{\top}$-module $\mathbb{Z}$ is of type FP_n.

Using the fact that the property FP_n is preserved by left Kan extensions [163, Lemma 5.1.4], these finiteness homological properties of categories are related by the following implications [163, Proposition 5.2.4]:

$$\begin{array}{ccccc} & & \text{right-FP}_n & & \\ & \nearrow & & \searrow & \\ \mathrm{FP}_n \Longrightarrow \text{bi-FP}_n & & & & \text{top-FP}_n. \\ & \searrow & & \nearrow & \\ & & \text{left-FP}_n & & \end{array}$$

If C is a groupoid, all of these implications are equivalences [163, Proposition 5.2.6], but this is not the case in general. Indeed, Cohen constructed in [91] a right-FP_∞ monoid which is not left-FP_1: thus, top-FP_n, left-FP_n, and right-FP_n are not equivalent in general. Moreover, monoids with a finite convergent presentation are left-FP_∞ and right-FP_∞, see [8, 219, 326], but there exists a finitely presented monoid that is left-FP_∞ and right-FP_∞, but does not satisfy the homological finiteness condition FHT introduced by Pride and Wang [222]; since the properties FHT and bi-FP_3 are equivalent [223], it follows that left-FP_n and right-FP_n do not imply bi-FP_n in general.

Finally, we note the following consequence of the definition.

F.3.4 Proposition. *If a category C is of homological type* FP_n, *for a natural number n, then the abelian group $\mathrm{H}_k(C, \mathbb{Z})$ is finitely generated for every $0 \leqslant k \leqslant n$.*

F.3.5 Monoids of finite homological type. The notion of finite homological type for categories applies to monoids seen as categories with one object. In particular, the proofs to show that a monoid is of homological type left-FP_n in Chapter 9 are based on the following result, which is an immediate consequence of Proposition F.3.2.

F.3.6 Proposition. *Let n be a natural number. A monoid M is of homological type left-*FP_n *if and only if the trivial left $\mathbb{Z}M$-module $\mathbb{Z}$ is of homological type left-*FP_n.

Appendix G
Locally Presentable Categories

This appendix is a quick introduction to locally presentable categories. We refer the reader to the classical book [2] for a detailed presentation.

The notion of a locally presentable category is, in some sense, a formalization of what is an algebraic structure. When category theory is restricted to locally presentable categories, many things get simpler. In particular, there are characterizations of adjoint functors purely in terms of preservation of limits and colimits. Locally presentable categories also play an important role in the theory of model categories through the concept of combinatorial model categories.

There are many ways to define locally presentable categories. We start with the presentation in terms of sketches, which are somehow categories encoding the syntax of an algebraic structure. These sketches are used several times in the body of the book. We then give the intrinsic categorical characterization, defining on our way several notions that will be needed for the theory of model categories. Finally, we give the syntactic characterization.

G.1 Sketches

G.1.1 Cones. Let C be a category. By a *projective cone* in C, we will mean a triple (F, X, α), where $F\colon D \to C$ is a functor from some *small* category I, X is an object of C, and $\alpha\colon X \Rightarrow F$ is a natural transformation, where X is considered as a constant functor. The diagram F is then called the *base* of the cone and the object X the *tip* of the cone. We say that C is a *projective limit cone* if the morphism $X \to \varprojlim F$ induced by α is an isomorphism.

The notion of *inductive cone* is defined similarly by reversing the direction of the natural transformation α, that is, by considering a natural transformation

$\alpha\colon F \Rightarrow X$, and the notion of *inductive limit cone* by asking for the induced morphism $\varinjlim F \to X$ to be an isomorphism.

G.1.2 Sketches. A *sketch* is a triple (S, P, I), where S is a *small* category, P a *set* of projective cones in S and I a *set* of inductive cones in S. A sketch is *projective* (resp. *injective*) if $I = \emptyset$ (resp. $P = \emptyset$). By abuse of notation, we will often refer to the sketch (S, P, I) as S only, and we will talk of the projective cones of S (resp. of the inductive cones of S) for the elements of P (resp. of I).

A *model* of a sketch S in a category C consists of a functor $F\colon S \to C$ sending every projective cone of S to a projective limit cone in C and every inductive cone of S to an inductive limit cone in C. A *morphism of models of a sketch* S is just a natural transformation. We write $\mathrm{Mod}(S)$ for the category of models of a sketch S in the category of sets.

A *morphism of sketches* from a sketch S to a sketch S' is a functor $f\colon S \to S'$ sending every projective cone of S to a projective cone of S' and every inductive cone of S to an inductive cone of S'. If f is such a morphism, then precomposition by f induces a functor

$$f^*\colon \mathrm{Mod}(S') \to \mathrm{Mod}(S).$$

G.1.3 Sketchable categories. We say that a category C is *sketchable* (resp. *projectively sketchable*) if there exists a sketch (resp. a projective sketch) S such that C is equivalent to $\mathrm{Mod}(S)$.

In this section, we will focus on projectively sketchable categories.

G.1.4 Example. The category of graphs is projectively sketchable. Indeed, this category is nothing but the category of functors from the category

$$[0] \underset{t}{\overset{s}{\leftleftarrows}} [1]$$

to the category of sets, i.e., the category of models of a sketch without any projective or inductive cones. More generally, any diagram category is projectively sketchable.

G.1.5 Example. The category of monoids is projectively sketchable. Indeed, consider the category S with three objects

$$[0], \quad [1], \quad [2], \quad [3]$$

and generated by the morphisms

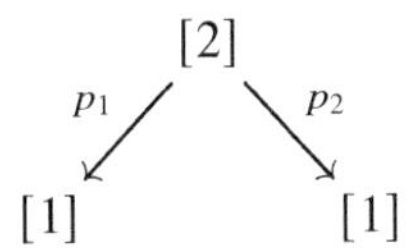

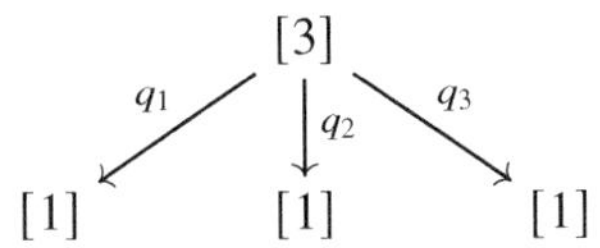

$$[3] \xrightarrow{q_{1,2}} [2] \qquad [3] \xrightarrow{q_{2,3}} [2]$$

$$[0] \xrightarrow{u} [1] \qquad [2] \xrightarrow{m} [1]$$

$$[1] \xrightarrow{\langle u,1\rangle} [2] \qquad [1] \xrightarrow{\langle 1,u\rangle} [2]$$

$$[3] \xrightarrow{\langle m,1\rangle} [2] \qquad [3] \xrightarrow{\langle 1,m\rangle} [2]$$

subject to the relations

$$p_1 q_{1,2} = q_1, \quad p_2 q_{1,2} = q_2, \quad p_2 q_{2,3} = q_2, \quad p_2 q_{2,3} = q_3,$$

$$p_1\langle u, 1\rangle = u, \quad p_2\langle u, 1\rangle = 1_{[1]}, \quad p_1\langle 1, u\rangle = 1_{[1]}, \quad p_2\langle u, 1\rangle = u,$$

$$p_1\langle m, 1\rangle = m q_{1,2} \quad p_2\langle m, 1\rangle = q_3 \quad p_1\langle 1, m\rangle = q_1, \quad p_2\langle u, m\rangle = m q_{2,3},$$

and

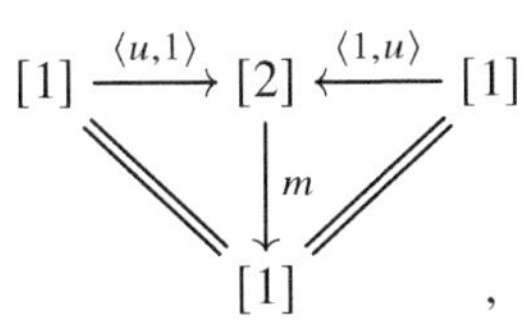

,

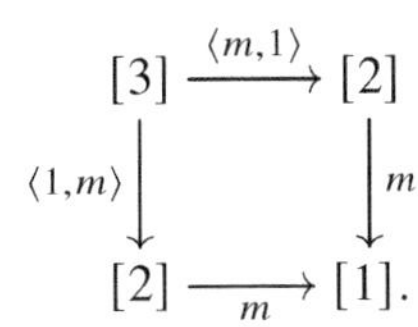

Endow S with the three projective cones

$$[0]$$

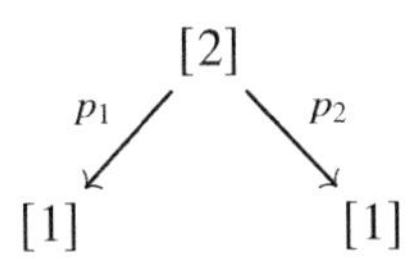

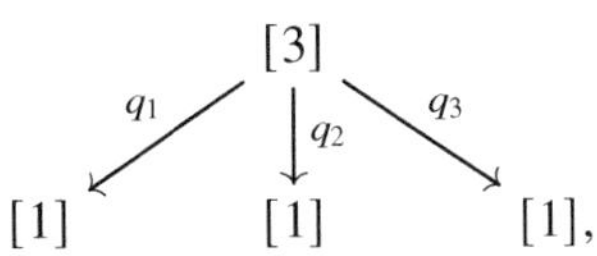

the first one being indexed by the empty category. Then we claim that the data of a model $F\colon S \to \mathbf{Set}$ of the sketch S in **Set** is equivalent to the data of a monoid of underlying set $M = F([1])$, multiplication $F(m)$ and unit given by $F(u)$ (the value of $F([i])$, for $0 \leq i \leq 3$, being forced to be sent to M^i).

More generally, any Lawvere theory defines a sketch whose models are the models of the starting Lawvere theory.

G.1.6 Example. The category **Cat** of small categories is projectively sketchable. Consider the full subcategory $S \subseteq \Delta^{\mathrm{op}}$ of the opposite category of the simplicial category (see §F.1.1) on the objects $[0]$, $[1]$, $[2]$, and $[3]$. If $[m] \hookrightarrow [n]$ is an inclusion of Δ whose image avoids exactly $i_0, \dots, i_k$, the corresponding morphism in Δ^{op} will be denoted by $d_{i_0,\dots,i_k} \colon [n] \to [m]$. With this notation, the commutative diagrams

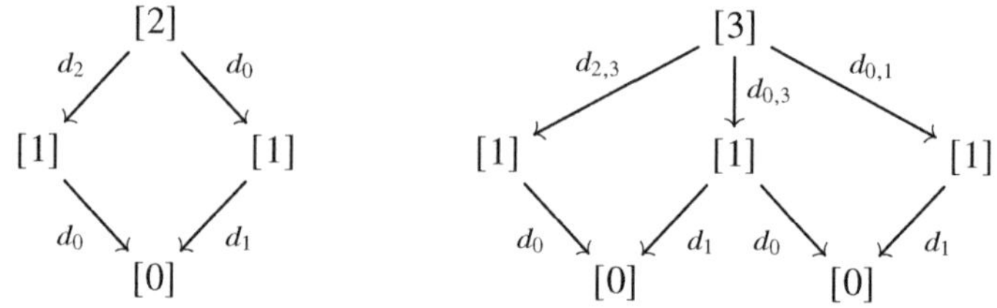

define two cones (which are actually limit cones) in S. One can show that the category of models of the sketch defined by S and these two cones is equivalent to **Cat**.

G.1.7 Example. The category $\mathbf{Cat}_\omega$ of small ω-categories is projectively sketchable (see Proposition 14.2.4).

G.1.8 Theorem. *If $f \colon S \to S'$ is a morphism of projective sketches, then the functor*

$$f^* \colon \mathrm{Mod}(S') \to \mathrm{Mod}(S).$$

admits a left adjoint

$$f_! \colon \mathrm{Mod}(S) \to \mathrm{Mod}(S').$$

Proof. See for instance [26, Section 4, Theorem 4.1]. □

G.1.9 Example. There is an obvious inclusion from the sketch of graphs defined in Example G.1.4 into the sketch of categories defined in Example G.1.6. This morphism of sketches induces the forgetful functor from small categories to graphs. The previous theorem shows the well-known fact that this forgetful functor admits a left adjoint, sending a graph to the free category on this graph.

G.1.10 Proposition. *A projectively sketchable category is complete and cocomplete.*

Proof. The completeness of such a category can be checked directly (limits are computed as in presheaves). Proving the cocompleteness is more involved, see [2, Example 3.11.8 and Theorem 1.38]. □

We end the section with a very powerful criterion due to Lair to prove that a functor is monadic in terms of sketches.

G.1.11 Theorem. *Let $f\colon S \to S'$ be a morphism of projective sketches satisfying the two following conditions:*

- *the base of any cone of S' factors though f,*
- *every object of S' not in the image of f is the tip of a cone of S'.*

Then the induced functor $f^*\colon \mathrm{Mod}(S') \to \mathrm{Mod}(S)$ is monadic.

Proof. By Theorem G.1.8, the functor f^* admits a right adjoint and the result is thus a particular case of [239, Corollary 1]. □

G.2 Locally Presentable Categories

Fix κ be a regular cardinal. Recall that the fact that κ is *regular* means that the category of sets of cardinal $< \kappa$ is closed under colimits of cardinality $< \kappa$.

G.2.1 κ-filtered diagrams. A small category I is said to be *κ-small* if the cardinal of its set of morphisms is strictly smaller than I. An $\aleph_0$-small category, $\aleph_0$ being the cardinal of countable sets, is called a *finite category*. A diagram $F\colon I \to C$ is said to be κ-small if I is κ-small. A *κ-filtered category* is a category I such that for every κ-small diagram $F\colon J \to I$, there exists an object i of I and cone $F \Rightarrow i$. An $\aleph_0$-filtered category is said to be *filtered*. A *κ-filtered diagram* is a diagram $F\colon I \to C$ such that I is κ-filtered.

G.2.2 κ-presentable objects. An object X of a category C is said to be *κ-presentable* if the functor

$$C(X,-)\colon C \to \mathbf{Set}$$

preserves κ-filtered colimits, that is, if for every κ-filtered diagram $F\colon I \to C$ the canonical map

$$\varinjlim_i C(X, Fi) \to C(X, \varinjlim_i Fi)$$

is a bijection. An $\aleph_0$-presentable object is called a *finitely presentable object*.

G.2.3 Locally presentable categories. A category C is said to be *locally κ-presentable* if it is cocomplete and if there exists a *set* S of κ-small objects of C such that every object of C can be obtained as a κ-filtered colimit of objects in S. A locally $\aleph_0$-presentable category is said to be *locally finitely* *presentable*. Essentially by definition, if a category C is locally κ-presentable, then it is locally κ'-presentable for every regular cardinal $\kappa' > \kappa$. A category C is *locally presentable* if it is locally κ-presentable for some regular cardinal κ.

G.2.4 Theorem. *A category is locally presentable if and only if it is equivalent to the category of models of some projective sketch. More precisely, if κ is a regular cardinal, a category is locally κ-presentable if and only it is the category of model of some projective sketch whose cones are over κ-small diagrams.*

Proof. See for instance [2, Corollary 1.52]. □

G.3 Essentially Algebraic Theories

G.3.1 Algebraic theory. An *algebraic theory* P consists of

- a set P_0 of *sorts*,
- a set P_1 of *operations* together with functions

$$s_0 \colon P_1 \to P_0^* \qquad\qquad t_0 \colon P_1 \to P_0$$

  associating to an operation its *arity* and *coarity*, where P_0^* denotes the free monoid on P_0,
- a set $P_2 \subseteq P_1^* \times P_1^*$ of *relations* consisting of pairs of terms with the same arity and coarity.

An element $(u, v) \in P_1^* \times P_1^*$ is often written $u = v$. Note that this notion corresponds to the one of term rewriting system, as introduced in §13.1.12.

G.3.2 Essentially algebraic theory. An *essentially algebraic theory* consists of an algebraic theory P together with a function

$$d \colon P_1 \to \mathcal{P}(P_1^* \times P_1^*)$$

which to every operation in $a \in P_1$ associates a subset $d(a)$ of $P_1^* \times P_1^*$, called the *domain* of a, specifying the relations under which the operation is defined. An operation $a \in P_1$ is *total* when $d(a) = \emptyset$, and a term is total when it is composed of total operations only. An essentially algebraic theory is required to satisfy the following condition: for every operation $a \in P_1$ and relation $(u, v) \in d(a)$, the terms u and v are total.

G.3.3 Model of an essentially algebraic theory. A *model* M of an essentially algebraic theory consists of

- a set M_s for every sort $s \in P_0$,
- a partial function

$$M_a \colon M_{s_1} \times \cdots \times M_{s_k} \to M_s$$

  for every operation

$$a \colon s_1 \ldots s_k \to s$$

  such that $M_a(m_1, \ldots, m_k)$ is defined if and only if

$$M_u(m_1, \ldots, m_k) = M_v(m_1, \ldots, m_k)$$

  for every $(u, v) \in d(a)$.

In the definition above, the interpretation M_u for a term u is defined by induction on u from the interpretation of operations in the expected way.

G.3.4 Example. The essentially algebraic theory P of categories has two sorts s_0 and s_1 in P_0, and four operations in P_1

$$s \colon s_1 \to s_0 \qquad t \colon s_1 \to s_0 \qquad c \colon s_1, s_1 \to s_1 \qquad e \colon s_0 \to s_1$$

with domains

$$d(s) = \emptyset \qquad d(t) = \emptyset \qquad d(c) = \{t(x_1) = s(x_2)\} \qquad d(e) = \emptyset$$

and relations

$$\begin{array}{lll} s \circ e(x_1) = x_1 & s \circ c(x_1, x_2) = s(x_1) & c(e(x_1), x_2) = x_2 \\ t \circ e(x_1) = x_1 & t \circ c(x_1, x_2) = t(x_2) & c(x_1, e(x_2)) = x_1 \end{array}$$

and

$$c(c(x_1, x_2), x_3) = c(x_1, c(x_2, x_3)).$$

Note that the terms $t(x_1)$ and $s(x_2)$ occurring in the domain of c are total as required. The models of this theory are precisely the small categories.

G.3.5 Theorem. *A category is locally finitely presentable if and only if it is equivalent to the category of models of an essentially algebraic theory.*

Proof. See [2, Theorem 3.36]. □

G.3.6 Remark. The above result can be generalized to locally κ-presentable categories by considering theories P with operations a with an arity of the form $s_0(a) \in P_0^{\kappa_a}$ for some cardinal $\kappa_a < \kappa$ (we recover the above case when all the κ_a are finite).

Appendix H
Model Categories

One of the goals of this book is to construct the so-called "folk" model category structure on the category of strict ω-categories. This is achieved in Chapter 21. The notion of model category, introduced by Quillen [307], constitutes a very general framework in which to study the homotopical properties of a category endowed with a class of weak equivalences.

Chapters 19 to 21 introduce along the way the main definitions of this theory. Nevertheless, for the convenience of the reader, we gather in this appendix these main definitions plus some complements. For more details, we refer the reader to the classical books [184, 187, 307] or to [313] for a recent panorama on the subject.

H.1 Definition

We start with some preliminary definitions.

H.1.1 2-out-of-3 property. A class of maps $\mathcal{W}$ in a category C is said to satisfy the *2-out-of-3 property* if for any commutative triangle

$$\begin{array}{ccc} X & \xrightarrow{\;f\;} & Y \\ & {\scriptstyle h}\searrow \quad \swarrow{\scriptstyle g} & \\ & Z & \end{array}$$

in C, if two morphisms among f, g, and h are in $\mathcal{W}$, then so is the third one. For instance, isomorphisms in a category satisfy the 2-out-of-3 property. More generally, any reasonable notion of "equivalence" in a category should satisfy this property.

H.1.2 Retracts. Let C be a category. We say that a morphism $f: X \to Y$ of C is a *retract* of a morphism $g: Z \to T$ of C if there exists a commutative diagram

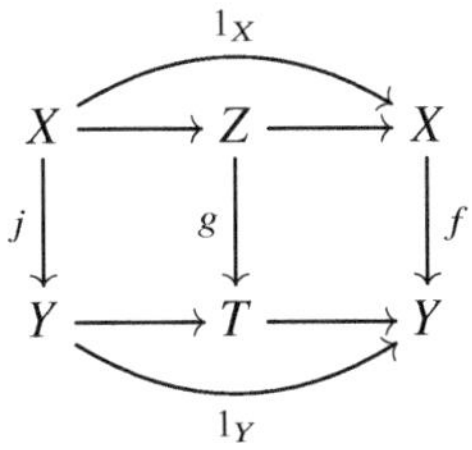

in C. We say that a class of morphisms of C is *closed under retracts* if any retract of an element of the class belongs to the class.

H.1.3 Lifting properties. Let $f: X \to Y$ and $g: Z \to T$ be two morphisms of C. One says that f *has the left lifting property* with respect to g or that g *has the right lifting property* with respect to f if for every commutative square

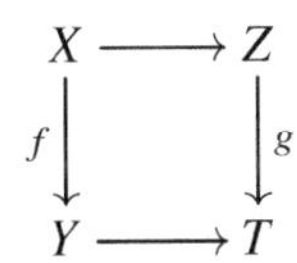

there exists a *lift*, that is, a morphism $h: Y \to Z$ making the two triangles

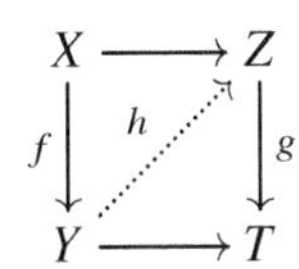

commute. More generally, one says that f has the left lifting property with respect to a class of maps $\mathcal{I}$ if it has the left lifting property with respect to every morphism in $\mathcal{I}$, and similarly for the right lifting property. We will denote by $l(\mathcal{I})$ and $r(\mathcal{I})$ the class of maps having the left or right lifting property with respect to a class $\mathcal{I}$.

We can now give the definition of a model category:

H.1.4 Model category. A *model category* is a category $\mathcal{M}$ endowed with three classes of maps: the *weak equivalences*, the *cofibrations*, and the *fibrations*; these data are required to satisfy the following axioms:

1. the category $\mathcal{M}$ is finitely complete and finitely cocomplete,
2. the class of weak equivalences satisfies the 2-out-of-3 property,

3. the class of weak equivalences, cofibrations, and fibrations are closed under retracts,
4. cofibrations have the left lifting property with respect to *trivial fibrations* (that is, maps that are both a fibration and a weak equivalence); trivial cofibrations (that is, maps that are both a cofibration and a weak equivalence) have the left lifting property with respect to fibrations, and
5. every map of $\mathcal{M}$ factors as a cofibration followed by a trivial fibration and as a trivial cofibration followed by a fibration.

H.1.5 Example. One of the motivating examples of Quillen is the following model category structure on the category **Top** of topological spaces:

- the weak equivalences are the *topological weak equivalences*, that is, the maps $f\colon X \to Y$ which induce a bijection $\pi_0(f)\colon \pi_0(X) \to \pi_0(Y)$ on path components and isomorphisms $\pi_n(f,x)\colon \pi_n(X,x) \to \pi_n(Y, f(x))$ on homotopy groups for every $n \geq 1$ and every base point x in X,
- the fibrations are the *Serre fibrations*, that is, the maps having the right lifting properties with respect to the inclusions

$$\begin{aligned} D^n &\hookrightarrow D^n \times I \\ x &\mapsto (x, 0) \end{aligned}$$

  of disks into cylinders, for $n \geq 1$,
- the cofibrations are the maps having the left lifting property with respect to maps that are both topological weak equivalences and Serre fibrations.

H.1.6 Example. Another motivating example of Quillen is the following model category structure on the category $\mathbf{Ch}_{\mathbb{Z},\geqslant 0}$ of chain complexes (see §E.2.1):

- the weak equivalences are the quasi-isomorphisms (see §E.3.5),
- the cofibrations are the monomorphisms f such that, for every $n \geq 0$, the cokernel of f_n is projective,
- the fibrations are the morphisms f such that, for every $n > 0$, f_n is an epimorphism.

H.1.7 Example. The category **Cat** of small categories can be endowed with the so-called "folk" model category structure:

- the weak equivalences are the equivalences of categories,
- the cofibrations are the functors injective on objects,
- the fibrations are the *iso-fibrations*, that is, the functors $f\colon C \to D$ such that for every object x of C and any isomorphism $v\colon f(x) \to y$ of D, there exists an isomorphism $u\colon x \to x'$ in C such that $f(u) = v$.

H.1.8 Example. Chapters 19 to 21 are devoted to the construction of the "folk" model category structure on the category $\mathbf{Cat}_\omega$ of strict ω-categories, see Theorem 21.1.2.

H.1.9 Cofibrant and fibrant replacements. Let $\mathcal{M}$ be a model category. An object X of $\mathcal{M}$ is said to be *cofibrant* if the unique morphism $\varnothing \to X$ from the initial object of $\mathcal{M}$ to X is a cofibration. Dually, one says that the object X is *fibrant* if the unique morphism $X \to *$ from X to the terminal object of $\mathcal{M}$ is a fibration.

If X is an object of $\mathcal{M}$, a *cofibrant replacement* of X is a cofibrant object QX of $\mathcal{M}$ endowed with a weak equivalence $QX \to X$. It follows immediately from the axioms of model categories that cofibrant replacements exist. Indeed, to produce one, it suffices to factor the morphism $\varnothing \to X$ as a cofibration followed by a trivial fibration. Dually, a fibrant replacement consists of a fibrant object RX together with a weak equivalence $X \to RX$.

H.1.10 Combinatorial model categories. A model category $\mathcal{M}$ is said to be *combinatorial* if

1. the underlying category of $\mathcal{M}$ is locally presentable (see §G.2.3),
2. there exist *sets* I and J of morphisms of $\mathcal{M}$ such that the class of cofibrations of $\mathcal{M}$ is $lr(I)$ and the class of trivial cofibrations of $\mathcal{M}$ is $lr(J)$.

Sets I and J, as in the definition, are called *generating cofibrations* and *generating trivial cofibrations*, respectively.

H.1.11 Remark. To build a model category, one often starts with a class of weak equivalences and sets I and J of candidates to be generators. This is what we did in Chapters 19 to 21 to build the folk model structure on $\mathbf{Cat}_\omega$.

H.1.12 Example. The model category structure on $\mathbf{Ch}_{\mathbb{Z},\geqslant 0}$ defined in Example H.1.6 can be proven to be combinatorial.

H.1.13 Example. The model category structure on $\mathbf{Cat}$ defined in Example H.1.7 is combinatorial. The set I can be taken to consist of the three functors

$$\varnothing \hookrightarrow \cdot, \qquad \{\cdot \quad \cdot\} \hookrightarrow \{\cdot \to \cdot\}, \quad \{\cdot \rightrightarrows \cdot\} \to \{\cdot \longrightarrow \cdot\},$$

and the set J to the functor

$$\{x\} \hookrightarrow \{x \xrightarrow{\sim} y\},$$

where the symbol $\sim$ denotes an isomorphism.

H.1.14 Example. The folk model structure on $\mathbf{Cat}_\omega$ is combinatorial. A set of generating cofibrations is given by the inclusions of spheres into disks (this is the set $\mathcal{I}$ of §19.2.1). It is harder to describe a set of generating trivial cofibrations, see the set $\mathcal{J}$ of §20.4.7.

H.1.15 Remark. The model category structure on **Top** described in Example H.1.5 is not combinatorial because **Top** is not locally presentable. Nevertheless, there exist sets I and J as in the second point of the definition. For instance, one can take as I the set of canonical inclusions of spheres into disks, and as J the set of inclusions of disks into cylinders as in the definition of Serre fibrations. These sets I and J do not satisfy the assumptions of the statement of the "small object argument" we gave (see Proposition 19.1.9). Nevertheless, one can check that the argument still applies. One says that this model category structure is *cofibrantly generated.*

H.2 The Homotopy Category

H.2.1 Localization. A *localizer* (also called *relative category*) is a category C endowed with a class $\mathcal{W}$ of morphisms called *weak equivalences*. The *homotopy category* of a localizer $(C, \mathcal{W})$ is the category $C[\mathcal{W}^{-1}]$ obtained from C by formally inverting arrows in $\mathcal{W}$. More precisely, the category $C[\mathcal{W}^{-1}]$ is the universal category endowed with a functor $p\colon C \to C[\mathcal{W}^{-1}]$ sending the elements of $\mathcal{W}$ to isomorphisms. Universal means that for any category $\mathcal{D}$ equipped with a functor $F\colon C \to \mathcal{D}$ sending the elements of $\mathcal{W}$ to isomorphisms, there is a unique functor $\widetilde{F}\colon C[\mathcal{W}^{-1}] \to \mathcal{D}$ making the triangle

$$\begin{array}{ccc} C & \xrightarrow{F} & \mathcal{D} \\ {\scriptstyle p}\downarrow & \nearrow_{\widetilde{F}} & \\ C[\mathcal{W}^{-1}] & & \end{array}$$

commute. We will often denote the category $C[\mathcal{W}^{-1}]$ by $\mathrm{Ho}(C)$, making implicit the class $\mathcal{W}$.

H.2.2 If $(C, \mathcal{W})$ is a localizer with C a small category, then the category $C[\mathcal{W}^{-1}]$ can be described in the following way: the objects of $C[\mathcal{W}^{-1}]$ are the object of C and its morphisms are sequences of zigzags of morphisms of C

$$\xrightarrow{f_1} \xleftarrow{f_2} \xleftarrow{f_3} \xrightarrow{f_4} \cdots \xleftarrow{f_n},$$

"This book is genius. As Neubauer points out, farmers markets just don't cut it anymore. Today's small farmer has to be more creative and shrewd. Neubauer is a smart farmer, and she shows a path forward: increasing value through pick-your-own agritourism. If you want to increase profits with less work, pick up a copy."

~ Ben Hartman

Megan Neubauer and her father, Jack, founded Pure Land Farm in 2012. After five years without a profit at farmers markets, they decided to convert Pure Land to a pick-your-own and quickly discovered the benefits of agritourism. By reallocating all the harvesting labor to visitors, and because the produce is sold upon harvest, Megan and Jack are able to spend more time on the work of growing, allowing them to produce much more food than they could pick, prep for market, and sell by themselves.

Pick-Your-Own Farming will help other farmers see the potential in agritourism not only for profitability, but to push the entire local food movement forward by educating the public about regenerative agriculture through the power of fun, immersive experiences on farms.

Megan Neubauer is a farmer in North Texas. She graduated with a BS in Biology from Boston University in 2004 and after several years working in biotech and scientific research, Megan and her father created Pure Land Farm in McKinney, Texas in 2012. Pure Land Farm uses regenerative growing methods on three acres of thornless blackberries and two acres of diversified fruits and vegetables, exclusively marketing their products using agritourism.

ACRES USA.
www.acresusa.com

$30.00
ISBN 978-1-60173-184-5
53000>
9 781601 731845

with all the backward morphisms in $\mathcal{W}$, modded out by the smallest congruence such that

$$\xrightarrow{f_1}\xrightarrow{f_2} \;=\; \xrightarrow{f_2\circ f_1} \qquad X \xrightarrow{1_X} X \;=\; 1_X \qquad X \xrightarrow{f} Y \xleftarrow{f} X \;=\; 1_X$$

$$\xleftarrow{f_1}\xleftarrow{f_2} \;=\; \xleftarrow{f_1\circ f_2} \qquad X \xleftarrow{1_X} X \;=\; 1_X \qquad Y \xleftarrow{f} X \xrightarrow{f} Y \;=\; 1_Y.$$

With appropriate set-theoretic foundations, one can adapt this construction to locally small categories, although the localization of a locally small category is not locally small in general.

H.2.3 Homotopy category. Any model category $\mathcal{M}$ has an underlying localizer $(\mathcal{M}, \mathcal{W})$ and thus a *homotopy category* $\mathrm{Ho}(\mathcal{M})$.

One of the goals of the theory of model categories is to get a good understanding of the homotopy category $\mathrm{Ho}(\mathcal{M})$. In particular, it admits a simpler description in terms of homotopies.

H.2.4 Homotopies. Let $\mathcal{M}$ be a model category. A *cylinder object* for an object A of $\mathcal{M}$ is a factorization

$$\begin{array}{ccc} A + A & \xrightarrow{(1_A, 1_A)} & A \\ & {\scriptstyle i}\searrow \quad \nearrow{\scriptstyle s} & \\ & IA & \end{array}$$

of the codiagonal map $A + A \to A$ as a cofibration i followed by a weak equivalence s. The components of the map i are denoted $i_0, i_1 \colon A \to IA$. Similarly, a *path object* for A is a factorization

$$\begin{array}{ccc} A & \xrightarrow{(1_A, 1_A)} & A \times A \\ & {\scriptstyle r}\searrow \quad \nearrow{\scriptstyle p} & \\ & A^I & \end{array}$$

of the diagonal map $A \to A \times A$ as a weak equivalence r followed by a fibration p. The components of the map p are denoted $p_0, p_1 \colon A^I \to A$.

Given morphisms $f, g \colon A \to B$ of $\mathcal{M}$, a *left homotopy* from f to g is a morphism $h \colon IA \to B$, for some cylinder object IA of A, such that $hi_0 = f$ and $hi_1 = g$. If such a homotopy exists, we say that f and g are *left homotopic*. Dually, a *right homotopy* from f to g is a morphism $k \colon A \to B^I$, for some path object B^I of B, such that $p_0 k = f$ and $p_1 k = g$. If such a homotopy exists, we say that f and g are *right homotopic*.

H.2.5 Proposition. *Let $\mathcal{M}$ be a model category. In the full subcategory $\mathcal{M}_{\mathrm{cf}}$ of $\mathcal{M}$ whose objects are both cofibrant and fibrant,*

1. *the relation "being left homotopic" coincides with the relation "being right homotopic",*
2. *the relation "being (left or right) homotopic" is a congruence.*

H.2.6 Proposition. *The homotopy category* $\mathrm{Ho}(\mathcal{M})$ *of a model category $\mathcal{M}$ is equivalent to the category $\mathcal{M}_{\mathrm{cf}}/\!\sim$ obtained from $\mathcal{M}_{\mathrm{cf}}$ by quotienting the morphisms by the relation "being (left or right) homotopic".*

In particular, the homotopy category of a model category is locally small.

H.3 Derived Functors

H.3.1 Homotopical functors. Let $(\mathcal{C}, \mathcal{W}_{\mathcal{C}})$ and $(\mathcal{D}, \mathcal{W}_{\mathcal{D}})$ be two localizers. A functor $F\colon \mathcal{C} \to \mathcal{D}$ is said to be *homotopical* if it preserves weak equivalences, that is, if it sends the weak equivalences of $\mathcal{C}$ to weak equivalences of $\mathcal{D}$. In this case, by the universal property of the localization, the functor F induces a functor

$$\overline{F}\colon \mathrm{Ho}(\mathcal{C}) \to \mathrm{Ho}(\mathcal{D})$$

making the square

$$\begin{array}{ccc} \mathcal{C} & \xrightarrow{F} & \mathcal{D} \\ {\scriptstyle p_{\mathcal{C}}}\downarrow & & \downarrow{\scriptstyle p_{\mathcal{D}}} \\ \mathrm{Ho}(\mathcal{C}) & \underset{\overline{F}}{\dashrightarrow} & \mathrm{Ho}(\mathcal{D}) \end{array}$$

commute.

If $F\colon \mathcal{C} \to \mathcal{D}$ is not homotopical, there is in general no functor $\overline{F}$ making the above square commute. Nevertheless, one can seek for "best approximations" to this situation:

H.3.2 Derived functors. Let $(\mathcal{C}, \mathcal{W}_{\mathcal{C}})$ and $(\mathcal{D}, \mathcal{W}_{\mathcal{D}})$ be two localizers and let $F\colon \mathcal{C} \to \mathcal{D}$ be a functor. The *(total) left derived functor* of F, if it exists, is the universal pair consisting of a functor

$$\mathbb{L}F\colon \mathrm{Ho}(\mathcal{C}) \to \mathrm{Ho}(\mathcal{D})$$

and a natural transformation

$$\begin{array}{ccc} C & \xrightarrow{F} & \mathcal{D} \\ {\scriptstyle p_C}\downarrow & \overset{\lambda}{\nearrow\!\!\!\nearrow} & \downarrow{\scriptstyle p_{\mathcal{D}}} \\ \mathrm{Ho}(C) & \underset{\mathbb{L}F}{\dashrightarrow} & \mathrm{Ho}(\mathcal{D}). \end{array}$$

This means that for every other functor $G\colon \mathrm{Ho}(C) \to \mathrm{Ho}(\mathcal{D})$ and natural transformation

$$\begin{array}{ccc} C & \xrightarrow{F} & \mathcal{D} \\ {\scriptstyle p_C}\downarrow & \overset{\alpha}{\nearrow\!\!\!\nearrow} & \downarrow{\scriptstyle p_{\mathcal{D}}} \\ \mathrm{Ho}(C) & \underset{G}{\dashrightarrow} & \mathrm{Ho}(\mathcal{D}), \end{array}$$

there exists a unique natural transformation $\gamma\colon G \Rightarrow \mathbb{L}F$ such that α factors as $\alpha = \lambda \circ (\gamma * p_C)$. By abuse of language, one often refers to $\mathbb{L}F$ as the left derived functor of F.

Similarly, the *(total) right derived functor* of F, if it exists, is the universal functor $\mathbb{R}F\colon \mathrm{Ho}(C) \to \mathrm{Ho}(\mathcal{D})$ endowed with a natural transformation

$$\begin{array}{ccc} C & \xrightarrow{F} & \mathcal{D} \\ {\scriptstyle p_C}\downarrow & \overset{\rho}{\swarrow\!\!\!\swarrow} & \downarrow{\scriptstyle p_{\mathcal{D}}} \\ \mathrm{Ho}(C) & \underset{\mathbb{R}F}{\dashrightarrow} & \mathrm{Ho}(\mathcal{D}). \end{array}$$

H.3.3 Remark. If F is homotopical, then $\overline{F}$ (endowed with the identity natural transformation) is both the left and the right derived functor of F.

One important use of model categories is to provide tools to prove the existence of derived functors and to compute them.

H.3.4 Theorem. *Let $F\colon \mathcal{M} \to \mathcal{N}$ be a functor between model categories. Suppose that F sends trivial cofibrations between cofibrant objects to weak equivalences. Then F admits a left derived functor $\mathbb{L}F\colon \mathrm{Ho}(\mathcal{M}) \to \mathrm{Ho}(\mathcal{N})$. Moreover, if X is an object of $\mathcal{M}$, then $\mathbb{L}F(p_{\mathcal{M}}(X))$ is canonically isomorphic to $p_{\mathcal{N}}(F(QX))$, where QX is a cofibrant replacement of X.*

Similarly, if F sends trivial fibrations between fibrant objects to weak equivalences, then F admits a right derived functor that can be computed using fibrant replacements.

H.3.5 Quillen pairs. Let $\mathcal{M}$ and $\mathcal{N}$ be two model categories and let

$$F : \mathcal{M} \rightleftarrows \mathcal{N} : G$$

be an adjunction. One says that (F, G) is a *Quillen pair* or a *Quillen adjunction* if F preserves cofibrations and trivial cofibrations. This is equivalent to asking that G preserves fibrations and trivial fibrations. In this case, one also says that F is a *left Quillen functor* and that G is a *right Quillen functor*. Using the previous proposition, one can show that left Quillen functors admit left derived functors and right Quillen functors admit right derived functors.

H.3.6 Theorem. *If*

$$F : \mathcal{M} \rightleftarrows \mathcal{N} : G$$

is a Quillen pair, then

$$\mathbb{L}F : \mathrm{Ho}(\mathcal{M}) \rightleftarrows \mathrm{Ho}(\mathcal{N}) : \mathbb{R}G$$

is an adjunction.

H.3.7 Quillen equivalences. Let

$$F : \mathcal{M} \rightleftarrows \mathcal{N} : G$$

be a Quillen pair. One says that (F, G) is a *Quillen equivalence* if the adjunction

$$\mathbb{L}F : \mathrm{Ho}(\mathcal{C}) \rightleftarrows \mathrm{Ho}(\mathcal{D}) : \mathbb{R}G$$

is an adjoint equivalence.

H.3.8 Proposition. *Let*

$$F : \mathcal{M} \rightleftarrows \mathcal{N} : G$$

be a Quillen pair. The following assertions are equivalent:

1. *(F, G) is a Quillen equivalence,*
2. *for every cofibrant object X of $\mathcal{M}$ and every fibrant object Y of $\mathcal{N}$, a morphism $FX \to Y$ in $\mathcal{N}$ is a weak equivalence if and only if the adjoint morphism $X \to GY$ in $\mathcal{M}$ is a weak equivalence.*

References

[1] Samson Abramsky. Temperley-Lieb algebra: from knot theory to logic and computation via quantum mechanics. In *Mathematics of quantum computation and quantum technology*, pages 533–576. Chapman & Hall/CRC, 2007.

[2] Jiří Adámek and Jiří Rosický. *Locally presentable and accessible categories*. Cambridge University Press, 1994.

[3] Jiří Adámek, Jiří Rosický, and Enrico Vitale. Algebraic theories. *Cambridge Tracts in Mathematics*, 184:1, 2011.

[4] Pierre Ageron. The logic of structures. *Journal of Pure and Applied Algebra*, 79:15–34, 1992.

[5] Fahd Ali Al-Agl and Richard Steiner. Nerves of multiple categories. *Proceedings of the London Mathematical Society*, 66(1):92–128, 1993.

[6] Clément Alleaume. Rewriting in higher dimensional linear categories and application to the affine oriented Brauer category. *Journal of Pure and Applied Algebra*, 222(3):636–673, 2018.

[7] David J. Anick. On monomial algebras of finite global dimension. *Transactions of the American Mathematical Society*, 291(1):291–310, 1985.

[8] David J. Anick. On the homology of associative algebras. *Transactions of the American Mathematical Society*, 296(2):641–659, 1986.

[9] David J. Anick and Edward L. Green. On the homology of quotients of path algebras. *Communications in Algebra*, 15(1–2):309–341, 1987.

[10] Dimitri Ara. A Quillen Theorem B for strict ∞-categories. *Journal of the London Mathematical Society*, 100(2):470–497, 2019.

[11] Dimitri Ara, Andrea Gagna, Viktoriya Ozornova, and Martina Rovelli. A categorical characterization of strong Steiner ω-categories. *Journal of Pure and Applied Algebra*, 227(7):24, 2023.

[12] Dimitri Ara and Maxime Lucas. The folk model category structure on strict ω-categories is monoidal. *Theory and Applications of Categories*, 35:745–808, 2020.

[13] Dimitri Ara and Georges Maltsiniotis. Vers une structure de catégorie de modèles à la Thomason sur la catégorie des n-catégories strictes. *Advances in Mathematics*, 259:557–664, 2014.

[14] Dimitri Ara and Georges Maltsiniotis. Le type d'homotopie de la ∞-catégorie associée à un complexe simplicial. Preprint, 2015.

[15] Dimitri Ara and Georges Maltsiniotis. Joint et tranches pour les ∞-catégories strictes. *Mémoires de la Société Mathématique de France*, 165:1–213, 2020.

[16] Dimitri Ara and François Métayer. The Brown-Golasiński model structure on strict ∞-groupoids revisited. *Homology Homotopy and Applications*, 13(1):121–142, 2011.

[17] Emil Artin. Theorie der Zöpfe. *Abhandlungen aus dem Mathematischen Seminar der Universität Hamburg*, 4(1):47–72, 1925.

[18] Juergen Avenhaus and Klaus Madlener. Subrekursive Komplexität bei Gruppen. *Acta Informatica*, 9(1):87–104, 1977.

[19] Jürgen Avenhaus and Klaus Madlener. Algorithmische Probleme bei Einrelatorgruppen und ihre Komplexität. *Archive for Mathematical Logic*, 19(1):3–12, 1978.

[20] Franz Baader and Tobias Nipkow. *Term rewriting and all that.* Cambridge University Press, 1998.

[21] John C. Baez, Brandon Coya, and Franciscus Rebro. PROPs in network theory. *Theory and Applications of Categories*, 33(25):727–783, 2018.

[22] John C. Baez and Jason Erbele. Categories in control. *Theory and Applications of Categories*, 30(24):836–881, 2015.

[23] John C. Baez and Brendan Fong. A compositional framework for passive linear networks. *Theory and Applications of Categories*, 33(38):1158–1222, 2018.

[24] John C. Baez and Aaron Lauda. A prehistory of n-categorical physics. In *Deep beauty: understanding the quantum world through mathematical innovation*, pages 13–128. Cambridge University Press, 2011.

[25] Cornel Balteanu, Zbigniew Fiedorowicz, Roland Schwänzl, and Rainer Vogt. Iterated monoidal categories. *Advances in Mathematics*, 176(2):277–349, 2003.

[26] Michael Barr and Charles Wells. *Toposes, triples and theories*, volume 278 of Grundlehren der mathematischen Wissenschaften. Springer-Verlag, 1985.

[27] Michael Barr and Charles Wells. *Category theory for computing science*, volume 1 in Prentice Hall International Series in Computer Science. Prentice Hall, 1990.

[28] Michael A. Batanin. Computads for finitary monads on globular sets. *Contemporary Mathematics*, 230:37–58, 1998.

[29] Michael A. Batanin. Monoidal globular categories as a natural environment for the theory of weak n-categories. *Advances in Mathematics*, 136(1):39–103, 1998.

[30] Michael A. Batanin. Computads and slices of operads. *arXiv preprint math/0209035*, 2002.

[31] Michael A. Batanin. On the Penon method of weakening algebraic structures. *Journal of Pure and Applied Algebra*, 172(1):1–23, 2002.

[32] Günther Bauer and Friedrich Otto. Finite complete rewriting systems and the complexity of the word problem. *Acta Informatica*, 21(5):521–540, 1984.

[33] Hans-Joachim Baues and Mamuka Jibladze. Classification of abelian track categories. *K-Theory*, 25(3):299–311, 2002.

[34] Hans-Joachim Baues and Günther Wirsching. Cohomology of small categories. *Journal of Pure and Applied Algebra*, 38(2–3):187–211, 1985.

[35] Gilbert Baumslag and Donald Solitar. Some two-generator one-relator non-Hopfian groups. *Bulletin of the American Mathematical Society*, 68(3):199–201, 1962.

[36] Jon Beck. Distributive laws. In *Seminar on triples and categorical homology theory*, pages 119–140. Springer, 1969.

[37] Marek A. Bednarczyk, Andrzej M. Borzyszkowski, and Wieslaw Pawlowski. Generalized congruences: epimorphisms in **Cat**. *Theory and Applications of Categories*, 5(11):266–280, 1999.

[38] Tibor Beke. Categorification, term rewriting and the Knuth–Bendix procedure. *Journal of Pure and Applied Algebra*, 215(5):728–740, 2011.

[39] George M. Bergman. The diamond lemma for ring theory. *Advances in Mathematics*, 29(2):178–218, 1978.

[40] Robert Bieri. *Homological dimension of discrete groups*, Queen Mary College Mathematics Notes. Mathematics Department, Queen Mary College, London, 1976.

[41] Stephen Bigelow. The Burau representation is not faithful for $n = 5$. *Geometry & Topology*, 3(1):397–404, 1999.

[42] Stephen Bigelow. Braid groups are linear. *Journal of the American Mathematical Society*, 14(2):471–486, 2001.

[43] Garrett Birkhoff. On the structure of abstract algebras. *Mathematical Proceedings of the Cambridge Philosophical Society*, 31(4):433–454, 1935.

[44] Leonid A. Bokut. Imbeddings into simple associative algebras. *Algebra i Logika*, 15(2):117–142, 245, 1976.

[45] Leonid A. Bokut and Yuqun Chen. Gröbner–Shirshov bases and their calculation. *Bulletin of Mathematical Sciences*, 4(3):325–395, 2014.

[46] Leonid A. Bokut, Yuqun Chen, Weiping Chen, and Jing Li. New approaches to plactic monoid via Gröbner–Shirshov bases. *Journal of Algebra*, 423:301–317, 2015.

[47] Filippo Bonchi, Paweł Sobociński, and Fabio Zanasi. The calculus of signal flow diagrams I: linear relations on streams. *Information and Computation*, 252:2–29, 2017.

[48] Filippo Bonchi, Paweł Sobociński, and Fabio Zanasi. Interacting Hopf algebras. *Journal of Pure and Applied Algebra*, 221(1):144–184, 2017.

[49] Ronald V. Book. Confluent and other types of Thue systems. *Journal of the ACM (JACM)*, 29(1):171–182, 1982.

[50] Ronald V. Book and Friedrich Otto. *String-rewriting systems*. Springer, 1993.

[51] William W. Boone. The word problem. *Proceedings of the National Academy of Sciences of the United States of America*, 44(10):1061, 1958.

[52] Francis Borceux. *Handbook of categorical algebra 1: basic category theory*. Number 50 in Encyclopedia of Mathematics and Its Applications. Cambridge University Press, 2008.

[53] Nicolas Bourbaki. *Éléments de mathématique. Fascicule XXXIV. Groupes et algèbres de Lie. Chapitre IV : Groupes de Coxeter et systèmes de Tits. Chapitre V : Groupes engendrés par des réflexions. Chapitre VI : systèmes de racines*. Number 1337 in Actualités Scientifiques et Industrielles. Hermann, 1968.

[54] Dominique Bourn. Another denormalization theorem for abelian chain complexes. *Journal of Pure and Applied Algebra*, 66(3):229–249, 1990.

[55] Richard Brauer. On algebras which are connected with the semisimple continuous groups. *Annals of Mathematics*, 38(4):857–872, 1937.

[56] Egbert Brieskorn and Kyoji Saito. Artin-Gruppen und Coxeter-Gruppen. *Inventiones mathematicae*, 17:245–271, 1972.

[57] Matthew G. Brin. Higher dimensional Thompson groups. *Geometriae Dedicata*, 108(1):163–192, 2004.

[58] Matthew G. Brin. The algebra of strand splitting. I. A braided version of Thompson's group *V*. *Journal of Group Theory*, 10(6):757–788, 2007.

[59] Kenneth S. Brown. Finiteness properties of groups. *Journal of Pure and Applied Algebra*, 44(1):45–75, 1987.

[60] Kenneth S. Brown. The geometry of rewriting systems: a proof of the Anick–Groves–Squier theorem. In *Algorithms and classification in combinatorial group theory*, pages 137–163. Springer, 1992.

[61] Kenneth S. Brown and Ross Geoghegan. An infinite-dimensional torsion-free FP_∞ group. *Inventiones mathematicae*, 77(2):367–381, 1984.

[62] Ronald Brown and Marek Golasiński. A model structure for the homotopy theory of crossed complexes. *Cahiers de Topologie et Géométrie Différentielle Catégoriques*, 30(1):61–82, 1989.

[63] Ronald Brown and Philip J. Higgins. The equivalence of ∞-groupoids and crossed complexes. *Cahiers de Topologie et Géométrie Différentielle Catégoriques*, 22(4):371–386, 1981.

[64] Ronald Brown and Johannes Huebschmann. Identities among relations. In *Low-dimensional topology (Bangor, 1979)*, volume 48 of London Mathematical Society Lecture Note Series, pages 153–202. Cambridge University Press, 1982.

[65] Bruno Buchberger. *Ein Algorithmus zum Auffinden der Basiselemente des Restklassenringes nach einem nulldimensionalen Polynomideal (An Algorithm for Finding the Basis Elements in the Residue Class Ring Modulo a Zero Dimensional Polynomial Ideal).* PhD thesis, Mathematical Institute, University of Innsbruck, Austria, 1965. English translation in *Journal of Symbolic Computation*, Special issue on Logic, mathematics, and computer science: interactions, 41(3–4):475–511, 2006.

[66] Bruno Buchberger. Ein algorithmisches Kriterium für die Lösbarkeit eines algebraischen Gleichungssystems. *Aequationes Mathematicae*, 4:374–383, 1970.

[67] Bruno Buchberger. History and basic features of the critical-pair/completion procedure. *Journal of Symbolic Computation*, 3(1–2):3–38, 1987.

[68] Bruno Buchberger. An algorithm for finding the basis elements of the residue class ring of a zero dimensional polynomial ideal. *Journal of Symbolic Computation*, 41(3–4):475–511, 2006. Translated from the 1965 German original by Michael P. Abramson.

[69] Werner Burau. Über Zopfgruppen und gleichsinnig verdrillte Verkettungen. *Abhandlungen aus dem Mathematischen Seminar der Universität Hamburg*, 11(1):179–186, 1935.

[70] Albert Burroni. Algèbres graphiques. *Cahiers de Topologie et Géométrie Différentielle Catégoriques*, 22(3):249–265, 1981.

[71] Albert Burroni. Higher dimensional word problem. In *Category theory and computer science*, pages 94–105. Springer, 1991.

[72] Albert Burroni. Higher-dimensional word problems with applications to equational logic. *Theoretical Computer Science*, 115(1):43–62, 1993.

[73] Alan J. Cain, Robert D. Gray, and António Malheiro. Finite Gröbner–Shirshov bases for plactic algebras and biautomatic structures for plactic monoids. *Journal of Algebra*, 423:37–53, 2015.

[74] James W. Cannon, William J. Floyd, and Walter R. Parry. Introductory notes on Richard Thompson's groups. *Enseignement Mathématique*, 42:215–256, 1996.

[75] Aurelio Carboni and Peter Johnstone. Connected limits, familial representability and Artin glueing. *Mathematical Structures in Computer Science*, 5(04):441–459, 1995.

[76] Aurelio Carboni and Peter Johnstone. Corrigenda for "Connected limits, familial representability and Artin glueing". *Mathematical Structures in Computer Science*, 14(01):185–187, 2004.

[77] Aurelio Carboni and Robert F. C. Walters. Cartesian bicategories I. *Journal of Pure and Applied Algebra*, 49(1–2):11–32, 1987.

[78] Pierre Cartier and Dominique Foata. *Problemes combinatoires de commutation et réarrangements*. Springer, 1969.

[79] Julien Cassaigne, Marc Espie, Daniel Krob, Jean-Christophe Novelli, and Florent Hivert. The Chinese monoid. *International Journal of Algebra and Computation*, 11(3):301–334, 2001.

[80] Bruce Chandler and Wilhelm Magnus. *The history of combinatorial group theory*, volume 9 of Studies in the History of Mathematics and Physical Sciences. Springer-Verlag, 1982.

[81] Cyrille Chenavier, Benjamin Dupont, and Philippe Malbos. Confluence of algebraic rewriting systems. *Mathematical Structures in Computer Science*, 32(7):870–897, 2022.

[82] Eugenia Cheng. Iterated distributive laws. *Mathematical Proceedings of the Cambridge Philosophical Society*, 150(03):459–487, 2011.

[83] Eugenia Cheng. A direct proof that the category of 3-computads is not cartesian closed. *Cahiers de Topologie et Géométrie Différentielle Catégoriques*, LIV(1):3–12, 2013.

[84] Eugenia Cheng. Distributive laws for Lawvere theories. *Compositionality*, 2(1):1–30, 2020.

[85] Eugenia Cheng and Michael Makkai. A note on the Penon definition of n-category. *Cahiers de Topologie et Géométrie Différentielle Catégoriques*, 50(2):83–101, 2009.

[86] Alonzo Church. An unsolvable problem of elementary number theory. *American Journal of Mathematics*, 58(2):345–363, 1936.

[87] Florence Clerc and Samuel Mimram. Presenting a category modulo a rewriting system. In *26th International Conference on Rewriting Techniques and Applications (RTA 2015)*, volume 36 of *Leibniz International Proceedings in Informatics (LIPIcs)*. Schloss Dagstuhl-Leibniz-Zentrum fuer Informatik, 2015.

[88] Bob Coecke and Ross Duncan. Interacting quantum observables. In *International colloquium on automata, languages, and programming*, pages 298–310. Springer, 2008.

[89] Bob Coecke and Aleks Kissinger. *Picturing quantum processes*. Cambridge University Press, 2017.

[90] Cyril Cohen, Thierry Coquand, Simon Huber, and Anders Mörtberg. Cubical type theory: a constructive interpretation of the univalence axiom. In *21st International Conference on Types for Proofs and Programs (TYPES 2015)*. Schloss Dagstuhl-Leibniz-Zentrum fuer Informatik, 2018.

[91] Daniel E. Cohen. A monoid which is right FP_∞ but not left FP_1. *Bulletin of the London Mathematical Society*, 24(4):340–342, 1992.

[92] Jonathan A. Cohen. *Coherence for rewriting 2-theories*. PhD thesis, Australian National University, June 2008.

[93] Paul M. Cohn. *Universal algebra*. Harper & Row Publishers, 1965.

[94] Harold S. M. Coxeter and William O. J. Moser. *Generators and relations for discrete groups*. Springer, 1972.

[95] Brandon Coya and Brendan Fong. Corelations are the PROP for extraspecial commutative Frobenius monoids. *Theory and Applications of Categories*, 32(11):380–395, 2017.

[96] Sjoerd Crans. *On combinatorial models for higher dimensional homotopies*. PhD thesis, University of Utrecht, 1995.

[97] Sjoerd Crans. Quillen closed model structures for sheaves. *Journal of Pure and Applied Algebra*, 101(1):35–57, 1995.

[98] Robert Cremanns and Friedrich Otto. Finite derivation type implies the homological finiteness condition FP_3. *Journal of Symbolic Computation*, 18(2):91–112, 1994.

[99] Robert Cremanns and Friedrich Otto. For groups the property of having finite derivation type is equivalent to the homological finiteness condition FP_3. *Journal of Symbolic Computation*, 22(2):155–177, 1996.

[100] Pierre-Louis Curien and Samuel Mimram. Coherent presentations of monoidal categories. *Logical Methods in Computer Science*, 13(3):1–39, 2017.

[101] Haskell B. Curry. Grundlagen der kombinatorischen Logik. *American Journal of Mathematics*, 52(4):789–834, 1930.

[102] Haskell B. Curry, J. Roger Hindley, and Jonathan P. Seldin. *Combinatory logic, Volume II*. North-Holland Co., 1972.

[103] Christian Schröder de Witt and Vladimir Zamdzhiev. The ZX-calculus is incomplete for quantum mechanics. *arXiv preprint arXiv:1404.3633*, 2014.

[104] Max Dehn. Über die Topologie des dreidimensionalen Raumes. *Mathematische Annalen*, 69(1):137–168, 1910.

[105] Max Dehn. Über unendliche diskontinuierliche Gruppen. *Mathematische Annalen*, 71(1):116–144, 1911.

[106] Max Dehn. Transformation der Kurven auf zweiseitigen Flächen. *Mathematische Annalen*, 72(3):413–421, 1912.

[107] Patrick Dehornoy. Groups with a complemented presentation. *Journal of Pure and Applied Algebra*, 116(1–3):115–137, 1997.

[108] Patrick Dehornoy. On completeness of word reversing. *Discrete Mathematics*, 225(1):93–119, 2000.

[109] Patrick Dehornoy. *Foundations of Garside theory*. European Mathematical Society, 2015.

[110] Pierre Deligne. Action du groupe des tresses sur une catégorie. *Inventiones mathematicae*, 128(1):159–175, 1997.

[111] Nachum Dershowitz. Termination of rewriting. *Journal of Symbolic Computation*, 3(1–2):69–115, 1987.

[112] Nachum Dershowitz and Zohar Manna. Proving termination with multiset orderings. *Communications of the ACM*, 22(8):465–476, 1979.

[113] Volker Diekert. Complete semi-Thue systems for abelian groups. *Theoretical Computer Science*, 44(2):199–208, 1986.

[114] Volker Diekert and Grzegorz Rozenberg. *The book of traces*, volume 15. World Scientific, 1995.

[115] Vladimir Dotsenko and Anton Khoroshkin. Gröbner bases for operads. *Duke Mathematical Journal*, 153(2):363–396, 2010.

[116] Eduardo J. Dubuc. Adjoint triangles. In *Reports of the Midwest Category Seminar* II, pages volume 61 of Lecture Notes in Mathematics (LNM) 69–91. Springer, 1968.

[117] Gérard Duchamp and Daniel Krob. Plactic-growth-like monoids. In *Words, Languages and Combinatorics, II (Kyoto, 1992)*, pages 124–142. World Scientific, 1994.

[118] Benjamin Dupont and Philippe Malbos. Coherent confluence modulo relations and double groupoids. *Journal of Pure and Applied Algebra*, 226(10), 2022.

[119] Beno Eckmann and Peter J. Hilton. Group-like structures in general categories. I. Multiplications and comultiplications. *Mathematische Annalen*, 145(3):227–255, 1962.

[120] Samuel Eilenberg and Ross Street. Rewrite systems, algebraic structures, and higher-order categories. Untitled handwritten manuscript, 1986.

[121] Josep Elgueta. Representation theory of 2-groups on Kapranov and Voevodsky's 2-vector spaces. *Advances in Mathematics*, 213(1):53–92, 2008.

[122] Jörg Endrullis, Jan Willem Klop, and Roy Overbeek. Decreasing diagrams: two labels suffice. In *5th International Workshop on Confluence*, pages 41–45. Obergurgl, Austria, September 8–9, 2016.

[123] Lisbeth Fajstrup, Eric Goubault, Emmanuel Haucourt, Samuel Mimram, and Martin Raussen. *Directed algebraic topology and concurrency*, volume 138 of Springerbriefs in Applied Sciences and Technology. Springer, 2016.

[124] Jean-Charles Faugère. A new efficient algorithm for computing Gröbner bases (F_4). *Journal of Pure and Applied Algebra*, 139(1–3):61–88, 1999.

[125] Jean-Charles Faugère. A new efficient algorithm for computing Gröbner bases without reduction to zero (F_5). In *Proceedings of the 2002 International Symposium on Symbolic and Algebraic Computation*, pages 75–83. Association for Computing Machinery, 2002.

[126] Bertram Felgenhauer and Vincent van Oostrom. Proof orders for decreasing diagrams. In *24th International Conference on Rewriting Techniques and Applications (RTA 2013)*. Schloss-Dagstuhl-Leibniz Zentrum für Informatik, 2013.

[127] Richard P. Feynman. Space-time approach to quantum electrodynamics. *Physical Review*, 76(6):769, 1949.

[128] Marcelo Fiore. Towards a mathematical theory of substitution. In *Invited talk for the Annual International Conference on Category Theory*. Carvoeiro, Algarve, 2007.

[129] Marcelo Fiore and Marco Devesas Campos. The algebra of directed acyclic graphs. In *Computation, logic, games, and quantum foundations. The many facets of Samson Abramsky*, pages 37–51. Springer, 2013.

[130] Marcelo Fiore and Tom Leinster. An abstract characterization of Thompson's group F. *Semigroup Forum*, 80(2):325–340, 2010.

[131] Marcelo Fiore, Gordon Plotkin, and Daniele Turi. Abstract syntax and variable binding. In *Proceedings of the 14th Annual IEEE Symposium on Logic in Computer Science*, pages 193–202. IEEE, 1999.

[132] Brendan Fong and Fabio Zanasi. A universal construction for (co)relations. In *7th Conference on Algebra and Coalgebra in Computer Science (CALCO 2017)*. Schloss Dagstuhl-Leibniz-Zentrum fuer Informatik, 2017.

[133] Simon Forest. Cateq, 2020. Software available at `https://github.com/SimonForest/cateq`.

[134] Simon Forest. *Computational descriptions of higher categories*. PhD thesis, Institut Polytechnique de Paris, 2021.

[135] Simon Forest and Samuel Mimram. Coherence of Gray categories via rewriting. In *3rd International Conference on Formal Structures for Computation and Deduction (FSCD 2018)*, volume 108 of Leibniz International Proceedings in Informatics (LIPIcs). Schloss Dagstuhl–Leibniz-Zentrum fuer Informatik, 2018.

[136] Thomas Fox. Coalgebras and cartesian categories. *Communications in Algebra*, 4(7):665–667, 1976.

[137] Peter J. Freyd and David N. Yetter. Braided compact closed categories with applications to low dimensional topology. *Advances in Mathematics*, 77(2):156–182, 1989.

[138] Peter J. Freyd and David N. Yetter. Coherence theorems via knot theory. *Journal of Pure and Applied Algebra*, 78(1):49–76, 1992.

[139] Haya Friedman and Dov Tamari. Problèmes d'associativité: Une structure de treillis finis induite par une loi demi-associative. *Journal of Combinatorial Theory*, 2(3):215–242, 1967.

[140] Tobias Fritz. A presentation of the category of stochastic matrices. *arXiv preprint arXiv:0902.2554*, 2009.

[141] William Fulton. *Young tableaux: With applications to representation theory and geometry*. Number 35 in London Mathematical Society Student Texts. Cambridge University Press, 1997.

[142] Peter Gabriel and Friedrich Ulmer. *Lokal präsentierbare Kategorien*. Springer, 1971.

[143] Peter Gabriel and Michael Zisman. *Calculus of fractions and homotopy theory*, volume 35 of Ergebnisse der Mathematik und ihrer Grenzgebiete. Springer-Verlag New York, Inc., 1967.

[144] Andrea Gagna. Strict n-categories and augmented directed complexes model homotopy types. *Advances in Mathematics*, 331:542–564, 2018.

[145] Stéphane Gaussent, Yves Guiraud, and Philippe Malbos. Coherent presentations of Artin monoids. *Compositio Mathematica*, 151(5):957–998, 2015.

[146] Meinolf Geck and Götz Pfeiffer. *Characters of finite Coxeter groups and Iwahori-Hecke algebras*. Number 21 in London Mathematical Society Monographs. Oxford University Press, 2000.

[147] Nick D. Gilbert. Monoid presentations and associated groupoids. *International Journal of Algebra and Computation*, 8(02):141–152, 1998.

[148] Marco Grandis. Finite sets and symmetric simplicial sets. *Theory and Applications of Categories*, 8(8):244–252, 2001.

[149] Marco Grandis and Luca Mauri. Cubical sets and their site. *Theory and Applications of Categories*, 11(8):185–211, 2003.

[150] Hans Grauert. Über die Deformation isolierter Singularitäten analytischer Mengen. *Inventiones mathematicae*, 15:171–198, 1972.

[151] John W. Gray. *Formal category theory: adjointness for 2-categories*, volume 391 of Lecture Notes in Mathematics. Springer-Verlag, 1974.

[152] Edward L. Green. Noncommutative Gröbner bases, and projective resolutions. In *Computational methods for representations of groups and algebras*, volume 173 of Progress in Mathematics, pages 29–60. Birkhäuser, 1999.

[153] Alexander Grothendieck. *À la poursuite des champs (volume 1)*, volume 20 of Documents Mathématiques. Société Mathématique de France, 2022.

[154] John R. J. Groves. Rewriting systems and homology of groups. In *Groups – Canberra 1989*, volume 1456 of Lecture Notes in Mathematics, pages 114–141. Springer, 1990.

[155] Léonard Guetta. Homology of categories via polygraphic resolutions. *Journal of Pure and Applied Algebra*, 225(10):106688, 33, 2021.

[156] Yves Guiraud. Catex. Software available at `https://webusers.imj-prg.fr/~yves.guiraud/programmes/catex/`.

[157] Yves Guiraud. *Présentations d'opérades et systèmes de réécriture*. PhD thesis, Université Montpellier 2, 2004.

[158] Yves Guiraud. Termination orders for 3-dimensional rewriting. *Journal of Pure and Applied Algebra*, 207(2):341–371, 2006.

[159] Yves Guiraud. *Rewriting methods in higher algebra*. Habilitation à diriger des recherches, Université Paris 7, 2019.

[160] Yves Guiraud, Eric Hoffbeck, and Philippe Malbos. Convergent presentations and polygraphic resolutions of associative algebras. *Mathematische Zeitschrift*, 293(1–2):113–179, 2019.

[161] Yves Guiraud and Philippe Malbos. Higher-dimensional categories with finite derivation type. *Theory and Applications of Categories*, 22(18):420–478, 2009.

[162] Yves Guiraud and Philippe Malbos. Coherence in monoidal track categories. *Mathematical Structures in Computer Science*, 22(6):931–969, 2012.

[163] Yves Guiraud and Philippe Malbos. Higher-dimensional normalisation strategies for acyclicity. *Advances in Mathematics*, 231(3–4):2294–2351, 2012.

[164] Yves Guiraud and Philippe Malbos. Identities among relations for higher-dimensional rewriting systems. In *Operads 2009*, volume 26 of Séminaires et congrès, pages 145–161. Société Mathématique de France, 2013.

[165] Yves Guiraud and Philippe Malbos. Polygraphs of finite derivation type. *Mathematical Structures in Computer Science*, 28(2):155–201, 2018.

[166] Yves Guiraud, Philippe Malbos, and Samuel Mimram. A homotopical completion procedure with applications to coherence of monoids. In *RTA-24th International Conference on Rewriting Techniques and Applications-2013*, volume 21 of Leibniz International Proceedings in Informatics (LIPIcs), pages 223–238. Schloss Dagstuhl–Leibniz-Zentrum fuer Informatik, 2013.

[167] Amar Hadzihasanovic. A diagrammatic axiomatisation for qubit entanglement. In *Proceedings of the 2015 30th Annual ACM/IEEE Symposium on Logic in Computer Science (LICS)*, pages 573–584. IEEE Computer Society, 2015.

[168] Amar Hadzihasanovic. *The algebra of entanglement and the geometry of composition*. PhD thesis, University of Oxford, 2017.

[169] Amar Hadzihasanovic. Representable diagrammatic sets as a model of weak higher categories. *arXiv preprint arXiv:1909.07639*, 2019.

[170] Amar Hadzihasanovic. Weak units, universal cells, and coherence via universality for bicategories. *Theory and Applications of Categories*, 34(29):883–960, 2019.

[171] Amar Hadzihasanovic. A combinatorial-topological shape category for polygraphs. *Applied Categorical Structures*, 28(3):419–476, 2020.

[172] Nohra Hage. Finite convergent presentation of plactic monoid for type C. *International Journal of Algebra and Computation*, 25(8):1239–1263, 2015.

[173] Nohra Hage and Philippe Malbos. Knuth's coherent presentations of plactic monoids of type A. *Algebras and Representation Theory*, 20(5):1259–1288, 2017.

[174] Nohra Hage and Philippe Malbos. Chinese syzygies by insertions. *Semigroup Forum*, 104(1):88–108, 2022.

[175] Masahito Hasegawa, Martin Hofmann, and Gordon Plotkin. Finite dimensional vector spaces are complete for traced symmetric monoidal categories. In *Pillars of computer science*, pages 367–385. Springer, 2008.

[176] Simon Henry. Regular polygraphs and the Simpson conjecture. *arXiv preprint arXiv:1807.02627*, 2018.

[177] Simon Henry. Non-unital polygraphs form a presheaf category. *Higher Structures*, 3(1):248–291, 2019.

[178] Simon Henry and Samuel Mimram. Tietze equivalences as weak equivalences. *Applied Categorical Structures*, 30(3):453–483, 2022.

[179] Anne Heyworth and Christopher D. Wensley. Logged rewriting and identities among relators. In *Groups St Andrews 2001 in Oxford*, pages 256–276, Cambridge University Press, 2003.

[180] Graham Higman. *Finitely presented infinite simple groups*, volume 8 of Notes on Pure Mathematics. Department of Pure Mathematics, Department of Mathematics, IAS, Australian National University, 1974.

[181] Peter J. Hilton. A brief, subjective history of homology and homotopy theory in this century. *Mathematics Magazine*, 61(5):282–291, 1988.

[182] Peter J. Hilton and Urs Stammbach. *A course in homological algebra*, volume 4 of Graduate Texts in Mathematics. Springer-Verlag, second edition, 1997.

[183] Heisuke Hironaka. Resolution of singularities of an algebraic variety over a field of characteristic zero: II. *Annals of Mathematics*, 79(2):205–326, 1964.

[184] Philip S. Hirschhorn. *Model categories and their localizations*, volume 99 of Mathematical Surveys and Monographs. American Mathematical Society, 2003.

[185] Florent Hivert, Jean-Christophe Novelli, and Jean-Yves Thibon. The algebra of binary search trees. *Theoretical Computer Science*, 339(1):129–165, 2005.

[186] Gerhard Hochschild. On the cohomology groups of an associative algebra. *Annals of Mathematics*, 46(1):58–67, 1945.

[187] Mark Hovey. *Model categories*. Number 63 in Mathematical Surveys and Monographs. American Mathematical Society, 2007.

[188] Gérard Huet. Confluent reductions: abstract properties and applications to term rewriting systems. *Journal of the ACM (JACM)*, 27(4):797–821, 1980.

[189] Gérard Huet. A complete proof of correctness of the Knuth-Bendix completion algorithm. *Journal of Computer and System Sciences*, 23(1):11–21, 1981.

[190] Gérard Huet. Initiation à la théorie des catégories. Notes de cours du DEA "Fonctionalité, Structures de Calcul et Programmation" donné à l'Université Paris VII en 1983-84 et 1984-85, 1987.

[191] Gérard Huet and Jean-Jacques Lévy. Computations in orthogonal rewriting systems, II. In *Computational logic-essays in honor of Alan Robinson*, pages 415–443. MIT Press, 1991.

[192] Dominic J. D. Hughes. Linking diagrams for free. *arXiv preprint arXiv:0805.1441*, 2008.

[193] Martin Hyland and John Power. Symmetric monoidal sketches. In *Proceedings of the 2nd ACM SIGPLAN International Conference on Principles and Practice of Declarative Programming*, pages 280–288. ACM, 2000.

[194] Martin Hyland and Christine Tasson. The linear-non-linear substitution 2-monad. In *Proceedings of the 3rd Annual International Applied Category Theory Conference 2020*, volume 333 of EPTCS, pages 215–229. EPTCS, 2020.

[195] Mirai Ikebuchi. A homological condition on equational unifiability. In *46th International Symposium on Mathematical Foundations of Computer Science*, volume 202 of LIPIcs. Schloss Dagstuhl - Leibniz-Zentrum für Informatik, 2021.

[196] Mirai Ikebuchi. A lower bound of the number of rewrite rules obtained by homological methods. *Logical Methods in Computer Science*, 18(3), 2022.

[197] Kenji Iohara and Philippe Malbos. From analytical mechanics problems to rewriting theory through M. Janet's work. In *Two algebraic byways from differential equations: Gröbner bases and quivers*, volume 28 of Algorithms and Computation in Mathematics, pages 3–74. Springer, 2020.

[198] Kenji Iohara and Philippe Malbos. Maurice Janet's algorithms on systems of linear partial differential equations. *Archive for History of Exact Sciences*, 75(1):43–81, 2021.

[199] Maurice Janet. Sur les systèmes d'équations aux dérivées partielles. *Journal de Mathématiques Pures et Appliquées*, 8(3):65–151, 1920.

[200] Maurice Janet. Sur les systèmes d'équations aux dérivées partielles. *Comptes Rendus Hebdomadaires des Séances de l'Académie des Sciences*, 170:1101–1103, 1920.

[201] Maurice Janet. Sur les systèmes d'équations aux dérivées partielles et les systèmes de formes algébriques. *Comptes Rendus Hebdomadaires des Séances de l'Académie des Sciences, Paris*, 170:1236–1239, 1920.

[202] Matthias Jantzen. Semi Thue systems and generalized Church-Rosser properties. Technical report, Institut für Informatik, Universität Hamburg, 1982.

[203] Matthias Jantzen. A note on a special one-rule semi-Thue system. *Information Processing Letters*, 21(3):135–140, 1985.

[204] Matthias Jantzen. *Confluent string rewriting*, volume 14 of EATCS Monographs on Theoretical Computer Science. Springer, 1988.

[205] David L. Johnson. *Presentations of groups*. Number 15 in London Mathematical Society Student Texts. Cambridge University Press, 1997.

[206] Michael Johnson. The combinatorics of n-categorical pasting. *Journal of Pure and Applied Algebra*, 62(3):211–225, 1989.

[207] Vaughan F. R. Jones. A polynomial invariant for knots via von Neumann algebras. *Bulletin (New Series) of the American Mathematical Society*, 12(1):103–111, 1985.

[208] André Joyal and Ross Street. The geometry of tensor calculus, I. *Advances in Mathematics*, 88(1):55–112, 1991.

[209] André Joyal and Ross Street. Braided tensor categories. *Advances in Mathematics*, 102(1):20–78, 1993.

[210] André Joyal, Ross Street, and Dominic Verity. Traced monoidal categories. *Mathematical Proceedings of the Cambridge Philosophical Society*, 119(3):447–468, 1996.

[211] Klaus H. Kamps and Timothy Porter. *Abstract homotopy and simple homotopy theory*. World Scientific, 1997.

[212] Mikhail M. Kapranov and Masahico Saito. Hidden Stasheff polytopes in algebraic K-theory and in the space of Morse functions. In *Higher Homotopy Structures in Topology and Mathematical Physics: Proceedings of an International Conference June 13–15, 1996 at Vassar College, Poughkeepsie, New York, to Honor the Sixtieth Birthday of Jim Stasheff*, volume 227 of Contemporary Mathematics, page 191. American Mathematical Society, 1999.

[213] Deepak Kapur and Paliath Narendran. A finite Thue system with decidable word problem and without equivalent finite canonical system. *Theoretical Computer Science*, 35:337–344, 1985.

[214] Christian Kassel. *Quantum groups*, volume 155 of Graduate Texts in Mathematics. Springer-Verlag New York, 1995.

[215] Louis H. Kauffman. An invariant of regular isotopy. *Transactions of the American Mathematical Society*, 318(2):417–471, 1990.

[216] Gregory M. Kelly and Miguel L. Laplaza. Coherence for compact closed categories. *Journal of Pure and Applied Algebra*, 19:193–213, 1980.

[217] Donald E. Knuth. Permutations, matrices, and generalized Young tableaux. *Pacific Journal of Mathematics*, 34(3):709–727, 1970.

[218] Donald E. Knuth and Peter B. Bendix. Simple word problems in universal algebras. In *Computational problems in abstract algebra*, pages 263–297. Pergamon, 1970.

[219] Yuji Kobayashi. Complete rewriting systems and homology of monoid algebras. *Journal of Pure and Applied Algebra*, 65(3):263–275, 1990.

[220] Yuji Kobayashi. Finite homotopy bases of one-relator monoids. *Journal of Algebra*, 229(2):547–569, 2000.

[221] Yuji Kobayashi. Gröbner bases of associative algebras and the Hochschild cohomology. *Transactions of the American Mathematical Society*, 357(3):1095–1124, 2005.

[222] Yuji Kobayashi and Friedrich Otto. On homotopical and homological finiteness conditions for finitely presented monoids. *International Journal of Algebra and Computation*, 11(3):391, 2001.

[223] Yuji Kobayashi and Friedrich Otto. For finitely presented monoids the homological finiteness conditions FHT and bi-FP_3 coincide. *Journal of Algebra*, 264(2):327–341, 2003.

[224] Joachim Kock. *Frobenius algebras and 2-d topological quantum field theories*, volume 59 of London Mathematical Society Student Texts. Cambridge University Press, 2004.

[225] Daan Krammer. Braid groups are linear. *Annals of Mathematics*, 155(1):131–156, 2002.

[226] Łukasz Kubat and Jan Okniński. Gröbner–Shirshov bases for plactic algebras. *Algebra Colloquium*, 21(04):591–596, 2014.

[227] Ganna Kudryavtseva and Volodymyr Mazorchuk. On presentations of Brauer-type monoids. *Open Mathematics*, 4(3):413–434, 2006.

[228] Stephen Lack. Icons. *Applied Categorical Structures*, 18:289–307, 2010.

[229] Steve Lack. A Quillen model structure for 2-categories. *K-Theory*, 26(2):171–205, 2002.

[230] Steve Lack. Composing PROPs. *Theory and Applications of Categories*, 13(9):147–163, 2004.

[231] Steve Lack. A Quillen model structure for bicategories. *K-Theory*, 33(3):185–197, 2004.

[232] Yves Lafont. Interaction nets. In *Proceedings of the 17th ACM SIGPLAN-SIGACT Symposium on principles of Programming Languages*, pages 95–108. ACM, 1989.

[233] Yves Lafont. A new finiteness condition for monoids presented by complete rewriting systems (after Craig C. Squier). *Journal of Pure and Applied Algebra*, 98(3):229–244, 1995.

[234] Yves Lafont. Interaction combinators. *Information and Computation*, 137(1):69–101, 1997.

[235] Yves Lafont. Towards an algebraic theory of boolean circuits. *Journal of Pure and Applied Algebra*, 184(2):257–310, 2003.

[236] Yves Lafont and François Métayer. Polygraphic resolutions and homology of monoids. *Journal of Pure and Applied Algebra*, 213(6):947–968, 2009.

[237] Yves Lafont, François Métayer, and Krzysztof Worytkiewicz. A folk model structure on omega-cat. *Advances in Mathematics*, 224(3):1183–1231, 2010.

[238] Yves Lafont and Alain Prouté. Church-Rosser property and homology of monoids. *Mathematical Structures in Computer Science*, 1(3):297–326, 1991.

[239] Christian Lair. Condition syntaxique de triplabilité d'un foncteur algébrique esquissé. *Diagrammes*, 1:1–16, 1979.

[240] Charles-Ange Laisant. Sur la numération factorielle, application aux permutations. *Bulletin de la Société Mathématique de France*, 16:176–183, 1888.

[241] Serge Lang. *Algebra*, volume 211 of Graduate Texts in Mathematics. Springer Science & Business Media, 2012.

[242] Alain Lascoux and Marcel-Paul Schützenberger. Le monoïde plaxique. *Quaderni de la Ricerca Scientifica*, 109:129–156, 1981.

[243] Aaron D. Lauda. Frobenius algebras and planar open string topological field theories. *arXiv preprint math/0508349*, 2005.

[244] Ruth J. Lawrence. Homological representations of the Hecke algebra. *Communications in Mathematical Physics*, 135(1):141–191, 1990.

[245] William Lawvere. Functorial semantics of algebraic theories. *Proceedings of the National Academy of Sciences*, 50(5):869–872, 1963.

[246] William Lawvere. Ordinal sums and equational doctrines. In *Seminar on Triples and Categorical Homology Theory*, pages 141–155. Springer, 1969.

[247] Philippe Le Chenadec. A catalogue of complete group presentations. *Journal of Symbolic Computation*, 2(4):363–381, 1986.

[248] Derrick H. Lehmer. Teaching combinatorial tricks to a computer. In *Combinatorial analysis*, volume 10 of *Proceedings of Symposia in Applied Mathematics*, pages 179–193. American Mathematical Society, 1960.

[249] Jean-Jacques Lévy. *Réductions correctes et optimales dans le lambda-calcul*. PhD thesis, Université Paris VII, 1978.

[250] Peter Littelmann. A plactic algebra for semisimple Lie algebras. *Advances in Mathematics*, 124(2):312–331, 1996.

[251] Muriel Livernet and Birgit Richter. An interpretation of E_n-homology as functor homology. *Mathematische Zeitschrift*, 269(1–2):193–219, 2011.

[252] Jean-Louis Loday. Homotopical syzygies. In *Une dégustation topologique [Topological morsels]: homotopy theory in the Swiss Alps (Arolla, 1999)*, volume 265 of Contemporary Mathematics, pages 99–127. American Mathematical Society, 2000.

[253] Jean-Louis Loday. *Generalized bialgebras and triples of operads*. Number 320 in Astérisque. Société mathématique de France, 2008.

[254] M. Lothaire. *Algebraic combinatorics on words*, volume 90 of Encyclopedia of Mathematics and Its Applications. Cambridge University Press, 2002.

[255] Maxime Lucas. A coherence theorem for pseudonatural transformations. *Journal of Pure and Applied Algebra*, 221(5):1146–1217, 2017.

[256] Maxime Lucas. *Cubical categories for homotopy and rewriting*. PhD thesis, University Paris Diderot – Paris 7, 2018.

[257] Roger C. Lyndon and Paul E. Schupp. *Combinatorial group theory*. Springer, 2015.

[258] Saunders Mac Lane. Homologie des anneaux et des modules. In *Colloque de topologie algébrique, Louvain*, pages 55–80. Masson & Cie, 1956.

[259] Saunders Mac Lane. Natural associativity and commutativity. *Rice University Studies*, 49(4):28–46, 1963.

[260] Saunders Mac Lane. *Homology*. Classics in Mathematics. Springer-Verlag, 1995. Reprint of the 1975 edition.

[261] Saunders Mac Lane. *Categories for the working mathematician*, volume 5 of Graduate Texts in Mathematics. Springer-Verlag, second edition, 1998.

[262] Saunders MacLane. Categorical algebra. *Bulletin of the American Mathematical Society*, 71(1):40–106, 1965.

[263] Wilhelm Magnus, Abraham Karrass, and Donald Solitar. *Combinatorial group theory*. Dover Publications, second edition, 2004.

[264] Michael Makkai. The Word Problem for Computads. Available at `www.math.mcgill.ca/makkai/WordProblem/WordProblemCombined.pdf`, 2005.

[265] Michael Makkai and Marek Zawadowski. 3-Computads do not form a presheaf category. *Personal letter to Michael Batanin*, 2001.

[266] Philippe Malbos. *Critères de finitude homologique pour la non convergence des systèmes de réécriture de termes*. Theses, Université Montpellier II – Sciences et Techniques du Languedoc, January 2004.

[267] Philippe Malbos and Samuel Mimram. Homological computations for term rewriting systems. In *1st International Conference on Formal Structures for Computation and Deduction (FSCD 2016)*, volume 52 of Leibniz International Proceedings in Informatics (LIPIcs), pages 27:1–27. Porto, Portugal, 2016.

[268] Philippe Malbos and Samuel Mimram. Higher-dimensional cartesian normalization strategies. In preparation, 2023.

[269] Philippe Malbos and Isaac Ren. Shuffle polygraphic resolutions for operads. *Journal of the London Mathematical Society*, 107(1):61–122, 2023.

[270] Georges Maltsiniotis. Grothendieck ∞-groupoids and still another definition of ∞-categories. *arXiv preprint arXiv:1009.2331*, 2010.

[271] Andreï A. Markov. On the impossibility of certain algorithms in the theory of associative systems. *Doklady Akademii Nauk SSSR*, 55:583–586, 1947.

[272] Yuri Matiyasevich and Géraud Sénizergues. Decision problems for semi-Thue systems with a few rules. *Theoretical Computer Science*, 330(1):145–169, 2005.

[273] Antoni Mazurkiewicz. Concurrent program schemes and their interpretations. *DAIMI Report Series*, 6(78):PB–78, 1977.

[274] Stuart McGlashan, Elton Pasku, and Stephen J. Pride. Finiteness conditions for rewriting systems. *International Journal of Algebra and Computation*, 15(1):175–205, 2005.

[275] Ralph McKenzie and Richard J. Thompson. An elementary construction of unsolvable word problems in group theory. *Studies in Logic and the Foundations of Mathematics*, 71:457–478, 1973.

[276] Paul-André Melliès. Axiomatic rewriting theory VI: residual theory revisited. In *International Conference on Rewriting Techniques and Applications*, pages 24–50. Springer, 2002.

[277] Paul-André Melliès. Categorical semantics of linear logic. *Panoramas et syntheses*, 27:15–215, 2009.

[278] François Métayer. Resolutions by polygraphs. *Theory and Applications of Categories*, 11(7):148–184, 2003.

[279] François Métayer. Cofibrant objects among higher-dimensional categories. *Homology, Homotopy and Applications*, 10(1):181–203, 2008.

[280] Yves Métivier. About the rewriting systems produced by the Knuth-Bendix completion algorithm. *Information Processing Letters*, 16(1):31–34, 1983.

[281] Charles F. Miller. *On group-theoretic decision problems and their classification*. Number 68 in Annals of Mathematics Studies. Princeton University Press, 1971.

[282] Samuel Mimram. The structure of first-order causality. *Mathematical Structures in Computer Science*, 21(01):65–110, 2011.

[283] Samuel Mimram. Towards 3-dimensional rewriting theory. *Logical Methods in Computer Science (LMCS)*, 10(2):1–47, 2014.

[284] Samuel Mimram. Presenting finite posets. In *Proceedings 8th International Workshop on Computing with Terms and Graphs*, volume 183 of Electronic Proceedings in Theoretical Computer Science, pages 1–17. Open Publishing Association, 2015.

[285] Samuel Mimram. Satex, 2021. Software available at `https://github.com/smimram/satex/`.

[286] Samuel Mimram. Categorical coherence from term rewriting systems. In *8th International Conference on Formal Structures for Computation and Deduction (FSCD 2023)*, volume 260 of Leibniz International Proceedings in Informatics (LIPIcs), pages 16:1–16:17, Dagstuhl, Germany, 2023. Schloss Dagstuhl – Leibniz-Zentrum für Informatik.

[287] Barry Mitchell. Rings with several objects. *Advances in Mathematics*, 8(1):1–161, 1972.

[288] Eliakim H. Moore. Concerning the abstract groups of order $k!$ and $\frac{1}{2}k!$ holohedrically isomorphic with the symmetric and the alternating substitution-groups on k letters. *Proceedings of the London Mathematical Society*, 1(1):357–367, 1896.

[289] Teo Mora. An introduction to commutative and noncommutative Gröbner bases. *Theoretical Computer Science*, 134(1):131–173, 1994. In *Second International Colloquium on Words, Languages and Combinatorics* (Kyoto, 1992).

[290] Maxwell H. A. Newman. On theories with a combinatorial definition of "equivalence". *Annals of Mathematics*, 43(2):223–243, 1942.

[291] Kang Feng Ng and Quanlong Wang. A universal completion of the ZX-calculus. *arXiv preprint arXiv:1706.09877*, 2017.

[292] Maurice Nivat. Congruences parfaites et quasi-parfaites. In *Séminaire P. Dubreil, 25e année (1971/72), Algèbre, Fasc. 1, Exp. No. 7*, page 9. Secrétariat Mathématique, Paris, 1973.

[293] Pyotr S. Novikov. On the algorithmic unsolvability of the word problem in group theory. *Proceedings of the Steklov Institute of Mathematics*, 44:1–143, 1955.

[294] Friedrich Otto, Masashi Katsura, and Yuji Kobayashi. Infinite convergent string-rewriting systems and cross-sections for finitely presented monoids. *Journal of Symbolic Computation*, 26(5):621–648, 1998.

[295] Renée Peiffer. Über Identitäten zwischen Relationen. *Mathematische Annalen*, 121:67–99, 1949.

[296] Jacques Penon. Approche polygraphique des ∞-catégories non strictes. *Cahiers de Topologie et Géométrie Différentielle Catégoriques*, 40(1):31–80, 1999.

[297] Roger Penrose. Applications of negative dimensional tensors. In *Combinatorial mathematics and its applications*, pages 221–244. Academic Press, 1971.

[298] Teimuraz Pirashvili. On the PROP corresponding to bialgebras. *Cahiers de Topologie et Géométrie Différentielle Catégoriques*, 43(3):221–239, 2002.

[299] Henri Poincaré. Analysis situs. *Journal de l'École Polytechnique*, 1(2):1–123, 1895.

[300] Alexander Polishchuk and Leonid Positselski. *Quadratic algebras*, volume 37 of University Lecture Series. American Mathematical Society, 2005.

[301] Emil L. Post. Recursive unsolvability of a problem of Thue. *The Journal of Symbolic Logic*, 12(01):1–11, 1947.

[302] James F. Power. Thue's 1914 paper: a translation. *arXiv preprint arXiv:1308.5858*, 2013.

[303] John Power. A general coherence result. *Journal of Pure and Applied Algebra*, 57(2):165–173, 1989.

[304] John Power. An n-categorical pasting theorem. In *Category theory*, pages 326–358. Springer, 1991.

[305] John Power and Hiroshi Watanabe. Combining a monad and a comonad. *Theoretical Computer Science*, 280(1–2):137–162, 2002.

[306] Stephen J. Pride. Low-dimensional homotopy theory for monoids. *International Journal of Algebra and Computation*, 5(06):631–649, 1995.

[307] Daniel Quillen. *Homotopical algebra*, volume 43 of Lecture Notes in Mathematics. Springer, 1967.

[308] Daniel Quillen. Higher algebraic K-theory. I. *Lecture Notes in Mathematics*, 341:85–147, 1973.

[309] Kurt Reidemeister. *Knotentheorie*. Springer-Verlag, 1932.

[310] Kurt Reidemeister. Über Identitäten von Relationen. *Abhandlungen aus dem Mathematischen Seminar der Universität Hamburg*, 16:114–118, 1949.

[311] Bodo Renschuch, Hartmut Roloff, Georgij G. Rasputin, and Michael Abramson. Contributions to constructive polynomial ideal theory XXIII: forgotten works of Leningrad mathematician NM Gjunter on polynomial ideal theory. *ACM SIGSAM Bulletin*, 37(2):35–48, 2003.

[312] Charles Rezk. A model category for categories. Unpublished note, 2000.

[313] Emily Riehl. *Categorical homotopy theory*. Number 24 in New Mathematical Monographs. Cambridge University Press, 2014.

[314] Jan-Erik Roos. Sur les foncteurs dérivés de $\varprojlim$. Applications. *Comptes Rendus Hebdomadaires des Séances de l'Académie des Sciences*, 252:3702–3704, 1961.

[315] Robert Rosebrugh, Nicoletta Sabadini, and Robert F. C. Walters. Generic commutative separable algebras and cospans of graphs. *Theory and Applications of Categories*, 15(6):164–177, 2005.

[316] Robert Rosebrugh and Richard J. Wood. Distributive laws and factorization. *Journal of Pure and Applied Algebra*, 175(1):327–353, 2002.

[317] Joseph J. Rotman. *An introduction to homological algebra*. Universitext. Springer, second edition, 2009.

[318] Mutsumi Saito, Bernd Sturmfels, and Nobuki Takayama. *Gröbner deformations of hypergeometric differential equations*, volume 6 of Algorithms and Computation in Mathematics. Springer-Verlag, 2000.

[319] Stephen Schanuel and Ross Street. The free adjunction. *Cahiers de Topologie et Géométrie Différentielle Catégoriques*, 27(1):81–83, 1986.

[320] Craige Schensted. Longest increasing and decreasing subsequences. *Canadian Journal of Mathematics*, 13:179–191, 1961.

[321] Moses Schönfinkel. Über die Bausteine der mathematischen Logik. *Mathematische annalen*, 92(3–4):305–316, 1924.

[322] Peter Selinger. The lambda calculus is algebraic. *Journal of Functional Programming*, 12(6):549–566, 2002.

[323] Peter Selinger. Autonomous categories in which $A \cong A^*$. In *Proceedings of 7th Workshop on Quantum Physics and Logic (QPL 2010)*. Citeseer, 2010.

[324] Peter Selinger. A survey of graphical languages for monoidal categories. In *New structures for physics*, pages 289–355. Springer, 2010.

[325] Anatoliĭ I. Shirshov. Some algorithmic problems for Lie algebras. *Sibirskii Matematicheskii Zhurnal*, 3(2):292–296, 1962.

[326] Craig C. Squier. Word problems and a homological finiteness condition for monoids. *Journal of Pure and Applied Algebra*, 49(1):201–217, 1987.

[327] Craig C. Squier and Friedrich Otto. The word problem for finitely presented monoids and finite canonical rewriting systems. In *International Conference on Rewriting Techniques and Applications*, pages 74–82. Springer, 1987.

[328] Craig C. Squier, Friedrich Otto, and Yuji Kobayashi. A finiteness condition for rewriting systems. *Theoretical Computer Science*, 131(2):271–294, 1994.

[329] John Stallings. A finitely presented group whose 3-dimensional integral homology is not finitely generated. *American Journal of Mathematics*, 85(4):541–543, 1963.

[330] Richard Steiner. Omega-categories and chain complexes. *Homology, Homotopy and Applications*, 6(1):175–200, 2004.

[331] Richard Steiner. Opetopes and chain complexes. *Theory and Applications of Categories*, 26(19):501–519, 2012.

[332] John G. Stell. Modelling term rewriting systems by sesqui-categories. *Proceedings of Catégories, Algèbres, Esquisses et Néo-Esquisses*, 1994.

[333] Ross Street. Limits indexed by category-valued 2-functors. *Journal of Pure and Applied Algebra*, 8(2):149–181, 1976.

[334] Ross Street. The algebra of oriented simplexes. *Journal of Pure and Applied Algebra*, 49(3):283–335, 1987.

[335] Ross Street. Parity complexes. *Cahiers de Topologie et Géométrie Différentielle Catégoriques*, 32(4):315–343, 1991.

[336] Ross Street. Parity complexes: corrigenda. *Cahiers de Topologie et Géométrie Fifférentielle Catégoriques*, 35(4):359–361, 1994.

[337] Ross Street. Higher categories, strings, cubes and simplex equations. *Applied Categorical Structures*, 3(1):29–77, 1995.

[338] Ross Street. Categorical structures. *Handbook of Algebra*, 1:529–577, 1996.

[339] Ross Street. Frobenius monads and pseudomonoids. *Journal of Mathematical Physics*, 45(10):3930–3948, 2004.

[340] Ross Street. Weak distributive laws. *Theory and Applications of Categories*, 22(12):313–320, 2009.

[341] Harold N. V. Temperley and Elliott H. Lieb. Relations between the "percolation" and "colouring" problem and other graph-theoretical problems associated with regular planar lattices: some exact results for the "percolation" problem. *Proceedings of the Royal Society London*, 322(1549):251–280, 1971.

[342] Terese. *Term rewriting systems*. Cambridge University Press, 2003.

[343] Joseph Miller Thomas. *Differential systems*, volume 21 of American Mathematical Society. Colloquium Publications, 1937.

[344] Axel Thue. Probleme über Veränderungen von Zeichenreihen nach gegebenen Regeln. Christiana Videnskabs-Selskabs Skrifter, I. Matematisk-naturvidenskabelig Klasse, 10, 1914.

[345] Heinrich Tietze. Über die topologischen Invarianten mehrdimensionaler Mannigfaltigkeiten. *Monatshefte für Mathematik und Physik*, 19(1):1–118, 1908.

[346] Grigorii S. Tseitin. An associative calculus with an insoluble problem of equivalence. *Trudy Matematicheskogo Instituta im. VA Steklova*, 52:172–189, 1958.

[347] Victor A. Ufnarovskij. Combinatorial and asymptotic methods in algebra. In *Algebra, VI*, volume 57 of Encyclopaedia of Mathematical Sciences, pages 1–196. Springer, 1995.

[348] Victor A. Ufnarovskij. Introduction to noncommutative Gröbner bases theory. In *Gröbner bases and applications (Linz, 1998)*, volume 251 of London Mathematical Society Lecture Note Series, pages 259–280. Cambridge University Press, 1998.

[349] Dirk van Dalen. *L.E.J. Brouwer – topologist, intuitionist, philosopher*. Springer-Verlag, 2013.

[350] Vincent Van Oostrom. Confluence by decreasing diagrams. *Theoretical Computer Science*, 126(2):259–280, 1994.

[351] Vincent van Oostrom. Residuation = skolemised confluence. In *12th International Workshop on Confluence*, page 20, Obergurgl, Austria, 2023.

[352] Leopold Vietoris. Über den höheren Zusammenhang kompakter Raüme und eine Klasse von zusammenhangstreuen Abbildungen. *Mathematische Annalen*, 97(1):454–472, 1927.

[353] Xiaofeng Wang and Stephen J. Pride. Second order Dehn functions of groups and monoids. *International Journal of Algebra and Computation*, 10(4):425–456, 2000.

[354] Charles Wells. Extension theories for categories. Preliminary report, 1979.

[355] Alfred N. Whitehead. *A treatise on universal algebra: with applications*, volume 1 of Principles of Algebraic Symbolism. The University Press, 1898.

[356] Henry Whitehead. Combinatorial homotopy I. *Bulletin of the American Mathematical Society*, 55(3):213–245, 1949.

[357] Henry Whitehead. Combinatorial homotopy II. *Bulletin of the American Mathematical Society*, 55(5):453–496, 1949.

[358] Noson S. Yanofsky. Coherence, homotopy and 2-theories. *K-Theory*, 23(3):203–235, 2001.

[359] Fabio Zanasi. *Interacting Hopf algebras – the theory of linear systems*. PhD thesis, École normale supérieure de Lyon, 2015.

Index of Symbols

Subject Index